Sample Preparation in LC-MS Bioanalysis

WILEY SERIES ON PHARMACEUTICAL SCIENCE AND BIOTECHNOLOGY: PRACTICES, APPLICATIONS, AND METHODS

Series Editor:

Mike S. Lee
Milestone Development Services

Mike S. Lee (ed.) • *Integrated Strategies for Drug Discovery Using Mass Spectrometry*

Birendra Pramanik, Mike S. Lee, and Guodong Chen (eds.) • *Characterization of Impurities and Degradants Using Mass Spectrometry*

Mike S. Lee and Mingshe Zhu (eds.) • *Mass Spectrometry in Drug Metabolism and Disposition: Basic Principles and Applications*

Mike S. Lee (ed.) • *Mass Spectrometry Handbook*

Wenkui Li and Mike S. Lee (eds.) • *Dried Blood Spots – Applications and Techniques*

Wenkui Li, Wenying Jian, and Yunlin Fu (eds.) • *Sample Preparation in LC-MS Bioanalysis*

Sample Preparation in LC-MS Bioanalysis

Edited by

Wenkui Li, Wenying Jian, and Yunlin Fu

WILEY

This edition first published 2019

Registered Office
John Wiley & Sons, Inc., 111 River Street, Hoboken, NJ 07030, USA

Editorial Office
111 River Street, Hoboken, NJ 07030, USA

For details of our global editorial offices, customer services, and more information about Wiley products visit us at www.wiley.com.

Wiley also publishes its books in a variety of electronic formats and by print-on-demand. Some content that appears in standard print versions of this book may not be available in other formats.

Library of Congress Cataloging-in-Publication Data

Names: Li, Wenkui, 1964– editor. | Jian, Wenying, editor. | Fu, Yunlin, 1980– editor.
Title: Sample preparation in LC-MS bioanalysis / edited by Wenkui Li, Wenying Jian, Yunlin Fu.
Other titles: Sample preparation in liquid chromatography-mass spectrometry bioanalysis
Description: Hoboken, NJ : Wiley, [2019] | Series: Wiley series on pharmaceutical science and biotechnology | Includes bibliographical references. |
Identifiers: LCCN 2018046969 (print) | LCCN 2018055539 (ebook) | ISBN 9781119274308 (Adobe PDF) | ISBN 9781119274322 (ePub) | ISBN 9781119274292 (hardcover)
Subjects: LCSH: Liquid chromatography. | Mass spectrometry. | Biotechnology.
Classification: LCC QD79.C454 (ebook) | LCC QD79.C454 S275 2019 (print) | DDC 543/.84–dc23
LC record available at https://lccn.loc.gov/2018046969

Cover design: Wiley
Cover image: © motorolka/Shutterstock

Set in 10/12pt Warnock by SPi Global, Pondicherry, India

10 9 8 7 6 5 4 3 2 1

Contents

List of Contributors

Gilberto Alves, PhD
CICS-UBI – Health Sciences Research Centre
University of Beira Interior
Covilhã
Portugal

Miguel Ángel Bello-López, PhD
Department of Analytical Chemistry
Universidad de Sevilla
Sevilla
Spain

Matthew Barfield, PhD
Research and Development
GlaxoSmithKline Pharmaceuticals
Ware
UK

Michael G. Bartlett, PhD
Department of Pharmaceutical and Biomedical Sciences
University of Georgia
Athens, GA
USA

Babak Basiri, PhD
Department of Pharmaceutical and Biomedical Sciences
University of Georgia
Athens, GA
USA

Ian A. Blair, PhD
Department of Systems Pharmacology and
Translational Therapeutics
Perelman School of Medicine
University of Pennsylvania
Philadelphia, PA
USA

Chester L. Bowen, MS
Research and Development
GlaxoSmithKline Pharmaceuticals
Collegeville, PA
USA

Stacy Brown, PhD
Department of Pharmaceutical Sciences
Gatton College of Pharmacy at East Tennessee State
University
Johnson City, TN
USA

Pilar Campíns-Falcó, PhD
Química Analítica
Universitat de València
Burjassot
Spain

Jennifer Carmical, PharmD
Department of Pharmaceutical Sciences
Gatton College of Pharmacy at East Tennessee
State University
Johnson City, TN
USA

Zhongzhe Cheng, PhD
School of Pharmacy
Weifang Medical University
Weifang, Shandong
China

Theo de Boer, PhD
LC-MS Bioanalysis
Ardena Bioanalytical Laboratory (ABL)
Assen
The Netherlands

Myriam Díaz-Álvarez, MSc
Department of Environment
Instituto Nacional de Investigación y Tecnología
Agraria y Alimentaria (INIA)
Madrid
Spain

Fuyou Du, PhD
Department of Applied Chemistry
Guilin University of Technology
Guilin, Guangxi
China

Amílcar Falcão, PhD
Laboratory of Pharmacology
Faculty of Pharmacy
University of Coimbra
Coimbra
Portugal

Rut Fernández-Torres, PhD
Department of Analytical Chemistry
Universidad de Sevilla
Sevilla
Spain

Ana Fortuna, PhD
Laboratory of Pharmacology
Faculty of Pharmacy
University of Coimbra
Coimbra
Portugal

Yunlin Fu, MS
Pharmacokinetic Sciences
Novartis Institutes for BioMedical Research
East Hanover, NJ
USA

Hong Gao, PhD
Drug Metabolism & Pharmacokinetics
Vertex Pharmaceuticals
Boston, MA
USA

Rodrigo A. González-Fuenzalida, PhD
Química Analítica
Universitat de València
Burjassot
Spain

Rosa Herráez-Hernández, PhD
Química Analítica
Universitat de València
Burjassot
Spain

Bruce J. Hidy, BSc
R&D, PPD
Richmond, VA
USA

Samuel Hofbauer, BS
Department of Systems Pharmacology and Translational Therapeutics
University of Pennsylvania
Philadelphia, PA
USA

Mike (Qingtao) Huang, PhD
Clinical Pharmacology
Akros Pharma Inc.
Princeton, NJ
USA

Rand G. Jenkins, BSc (retired)
PPD
Mechanicsville, VA
USA

Allena J. Ji, PhD, NRCC, DABCC
Biomarkers & Clinical Bioanalyses-Boston, Sanofi
Framingham, MA
USA

Wenying Jian, PhD
Janssen Research & Development, LLC
Spring House, PA
USA

Hongliang Jiang, PhD
Tongji School of Pharmacy
Huazhong University of Science and Technology
Wuhan, Hubei
China

Neus Jornet-Martinez, PhD
Química Analítica
Universitat de València
Burjassot
Spain

Maria Kechagia, MSc
Chemistry Department
Aristotle University of Thessaloniki
Thessaloniki
Greece

Jaeah Kim, PhD
Department of Pharmaceutical and Biomedical Sciences
University of Georgia
Athens, GA
USA

Maria Kissoudi, MSc
Chemistry Department
Aristotle University of Thessaloniki
Thessaloniki
Greece

Fumin Li, PhD
R&D, PPD
Middleton, WI
USA

Ning Li, PhD
Department of Pharmaceutical Analysis
School of Pharmacy
Shenyang Pharmaceutical University
Shenyang, Liaoning
China

Wenkui Li, PhD
Pharmacokinetic Sciences
Novartis Institutes for BioMedical Research
East Hanover, NJ
USA

Ang Liu, PhD
Bioanalytical Sciences
Translational Medicine
Bristol-Myers Squibb
Princeton, NJ
USA

Rao N.V.S. Mamidi, PhD, DABT
Janssen Research & Development, LLC.
Raritan, NJ
USA

Yan Mao, PhD
Drug Metabolism & Pharmacokinetics
Boehringer Ingelheim Pharmaceuticals, Inc.
Ridgefield, CT
USA

Antonio Martín-Esteban, PhD
Department of Environment
Instituto Nacional de Investigación y Tecnología
Agraria y Alimentaria (INIA)
Madrid
Spain

Henri Meijering, MSc
LC-MS Bioanalysis
Ardena Bioanalytical Laboratory (ABL)
Assen
The Netherlands

Clementina Mesaros, PhD
Department of Systems Pharmacology and
Translational Therapeutics
University of Pennsylvania
Philadelphia, PA
USA

Akira Namera, PhD
Department of Forensic Medicine
Graduate School of Biomedical and Health Sciences
Hiroshima University
Hiroshima
Japan

Ragu Ramanathan, PhD
Medicine Design – ADME Sciences, Pfizer, Inc
Groton, CT
USA

María Ramos-Payán, PhD
Department of Analytical Chemistry
Universidad de Sevilla
Sevilla
Spain

Márcio Rodrigues, PhD
CICS-UBI – Health Sciences Research Centre
University of Beira Interior
Covilhã
Portugal

Guihua Ruan, PhD
Department of Applied Chemistry
Guilin University of Technology
Guilin, Guangxi
China

Takeshi Saito, PhD
Department of Emergency and Critical Care Medicine
Tokai University School of Medicine
Isehara
Japan

Ashkan Salamatipour, BS
Department of Systems Pharmacology and
Translational Therapeutics
University of Pennsylvania
Philadelphia, PA
USA

Victoria F. Samanidou, PhD
Chemistry Department
Aristotle University of Thessaloniki
Thessaloniki
Greece

Nico van de Merbel, PhD
PRA Health Sciences
Assen
The Netherlands

Cong Wei, PhD
Drug Metabolism & Pharmacokinetics, Vertex
Pharmaceuticals
Boston, MA
USA

Zongyu Wei, MS
Department of Applied Chemistry
Guilin University of Technology
Guilin, Guangxi
China

Naidong Weng, PhD
Janssen Research & Development, LLC.
Spring House, PA
USA

John Williams, PhD
Drug Metabolism & Pharmacokinetics
Vertex Pharmaceuticals
Boston, MA
USA

Xin Xiong, MS
Department of Pharmacy
Peking University Third Hospital
Beijing
China

Long Yuan, PhD
Bioanalytical Sciences
Bristol-Myers Squibb
Princeton, NJ
USA

Qiulian Zeng, MS
Department of Applied Chemistry
Guilin University of Technology
Guilin, Guangxi
China

Jun Zhang, PhD
Dynamega LLC
Lake Forest, IL
USA

Dafang Zhong, PhD
Shanghai Institute of Materia Medica
Chinese Academy of Sciences
Shanghai
China

Yunting Zhu, PhD
Shanghai Institute of Materia Medica
Chinese Academy of Sciences
Shanghai
China

Preface

Sample preparation is a pivotal part of the integral LC-MS bioanalysis, which has been heavily employed in the determination of drugs, drug metabolites, biomarkers, and other molecules of interest in various biological matrices (e.g. fluids or tissues) for decades. It has been playing an important role in a variety of human healthcare studies, ranging from drug discovery and development, therapeutic drug monitoring, to biomarker analysis. While highly sophisticated LC-MS systems with better sensitivity and higher bioanalytical throughput have been continuously introduced, challenges that remain unchanged are the sample preparation prior to LC-MS quantitation, for which data quality has direct impact on study conclusion.

The purpose of sample preparation is not only to make the analyte(s) of interest available in sample extracts at an appropriate concentration for MS detection but also to remove interfering matrix elements (e.g. phospholipids and salts) that, if not addressed properly, can alter MS response (e.g. signal suppression). In quantitative LC-MS bioanalysis, clean sample extracts means: (i) better chromatography, (ii) lower limit of quantification, (iii) decreased assay variability (due to reduced matrix effects), (iv) less chance of false-positive/negative results, (v) longer column lifetime, (vi) less instrument downtime, and (vii) minimized costs in manpower and equipment maintenance, etc. In practice, the best sample preparation strategies should always be considered, evaluated, and implemented whenever possible in developing a robust quantitative LC-MS bioanalytical method.

As a companion for the previously published *Handbook of LC-MS Bioanalysis: Best Practice, Experimental Protocols and Regulations* (Li, Zhang, and Tse, 2013, Wiley), the current book is to provide a timely and comprehensive update along with representative experimental protocols on all important sample preparation techniques for quantitative LC-MS bioanalysis of small and large molecules. The 26 chapters of the book are divided into three parts. The first part of the book is focused on not only the basic but also the contemporary sample preparation techniques in LC-MS bioanalysis. These include Protein Precipitation, Liquid–Liquid Extraction, and Solid-Phase Extraction (Chapter 1), Online Extraction and Column Switching (Chapter 2), Equilibrium Dialysis, Ultracentrifugation, and Ultrafiltration (Chapter 3), Phospholipid Depletion (Chapter 4), Salting-out Assisted Liquid–Liquid Extraction (SALLE) (Chapter 5), Supported Liquid Extraction (SLE) (Chapter 6), Immunocapture (Chapter 7), Microextraction (Chapter 8), Microsampling (Chapter 9), Extraction via Nanomaterials (Chapter 10), Extraction via Molecularly Imprinted Polymers (MIP) (Chapter 11), Stir-bar Sorptive Extraction (Chapter 12), Monolithic Spin Column Extraction (Chapter 13), Aptamer-based Sample Preparation (Chapter 14), and Sample Extraction via Electromembranes (Chapter 15).

In Part II, the current sample preparation techniques for LC-MS bioanalysis of biological sample matrices other than common whole blood, plasma, or serum are discussed in detail along with experimental protocols. These matrices include but are not limited to Tissues, Hair, Nail, Skins, and Bones (Chapter 16), Peripheral Blood Mononuclear Cells (Chapter 17), Urine, Cerebrospinal Fluid, Synovial Fluid, Sweat, Tears, and Aqueous Humor (Chapter 18), and Liposomal Samples (Chapter 19).

Part III of the book is focused on sample preparation for LC-MS bioanalysis of challenging molecules. This part starts with some Key Pre-analytical Considerations in Quantitative LC-MS Bioanalysis (Chapter 20), which is followed by Derivatization strategies for enhancing assay sensitivities in quantitative LC-MS bioanalysis of molecules with poor ionization efficiency (Chapter 21). Sample preparation for quantitative LC-MS bioanalysis of Lipids is captured in Chapter 22. In Chapter 23, detailed instructions and associated stepwise protocols are provided for LC-MS bioanalysis of peptides. Expanding from peptides, detailed instructions of sample preparation for LC-MS bioanalysis of Proteins, Oligonucleotides, and Antibody–drug Conjugates (ADCs) are captured in Chapters 24, 25, and 26, respectively.

Our purpose in committing to this project was to provide scientists in industry, academia, and regulatory agencies with all "practical tricks" in extracting various analyte(s) of interest from biological samples for LC-MS quantification according to the current health authority regulations and industry practices. In this book we are confident that we have accomplished our goal. The book represents a major undertaking which would not have been possible without the contributions of all the authors and the support of their families. We also wish to thank the terrific editorial staff at John Wiley & Sons and give a special acknowledgment to Michael Leventhal, Managing Editor; Vishnu Narayanan, Project Editor; Beryl Mesiadhas, Project Manager; S. Grace Paulin Jeeva, Production Editor; and Robert Esposito, Associate Publisher, at John Wiley & Sons, for their premier support of this project.

Wenkui Li, PhD
Wenying Jian, PhD
Yunlin Fu, MS

List of Abbreviations

2D	two-dimensional
3NPH	3-nitrophenylhydrazine
5-FU	5-fluorouracil
5-HETE	5-hydroxyeicosatetraenoic acid
AA	acrylamide
AA	alendronic acid
AAC	α1-antichymotrypsin
ACE	angiotensin I converting enzyme
ACE	automatic cartridge exchange
ACN	acetonitrile
ADA	anti-drug antibody
ADC	antibody–drug conjugate
ADME	absorption, distribution, metabolism, and excretion
ADP	adenosine diphosphate
ADS	alkyl-diol-silica
AFA	adaptive focused acoustics
AFMC	aptamer-functionalized monolithic column
AFMPC	aptamer-functionalized material-packed column
AFM	aptamer-functionalized material
AFOTCC	aptamer-functionalized open tubular capillary column
AFSC	aptamer-functionalized spin column
AG	2-arachidonoylglycerol
AGP	acid glycoprotein
AIBN	azo(bis) isobutyronitrile
AML	acute myeloid leukemia
AMP	adenosine monophosphate
APA	anti-peptide antibody
APCI	atmospheric pressure chemical ionization
Apt-AC	aptamer-based affinity column
Apt-AuNR	aptamer-functionalized gold nanorod
Apt-MM	aptamer-functionalized magnetic material
Apt-MNP	aptamer-functionalized magnetic nanoparticle
Apt-PANCMA	aptamer-functionalized poly(acrylonitrile-co-maleic acid)
Apt-PP-fiber	aptamer-based-polypropylene fiber
Apt-SA-SPE	aptamer-based surface affinity solid-phase extraction
Apt-SBSE	aptamer-functionalized stir-bar sorptive extraction
Apt-SPE	aptamer-based solid-phase extraction
Apt-SPME	aptamer-based solid-phase microextraction
ATP	adenosine triphosphate
AUC	area under the curve
AuNP	gold nanoparticle

BAL	bronchoalveolar lavage
BEAD	bead extraction and acid dissociation
BEH	bridged ethylene hybrid
BLQ	below limit of quantification
BNP	B-type natriuretic peptide
BP	bisphosphonate
BP-3	benzophenone-3
BPA	bisphenol A
BSA	bovine serum albumin
BSL-2	biosafety level-2
BSTFA	*N,O*-bis(trimethylsilyl)trifluoroacetamide
CAD	collision-activated dissociation
Cape	capecitabine
CCSHLLE	counter current salting-out homogenous liquid–liquid extraction
CDA	cytidine deaminase
CDI	carbonyl diimidazole
CDR	complementarity-determining region
CE	capillary electrophoresis
CE	cholestryl oleate
CHAPS	3-([3-cholamidopropyl]dimethylammonio)-1-propanesulfonate
CID	collision-induced dissociation
CIP	chiral imprinted polymer
CNBF	4-chloro-3,5-dinitrobenzotrifloride
CNS	central nervous system
CNT-PDMS	carbon nanotube–poly(dimethylsiloxane)
CNT	carbon nanotube
COXs	cyclooxygenases
CPT	cell preparation tube
CSF	cerebrospinal fluid
CV	coefficient of variation
CZE-C^4D	capillary zone electrophoresis with capacitively coupled contactless conductivity detection
D	distribution ratio
D2EHPA	di-(2-ethylhexyl)phosphoric acid
DAD	diode array detection
DADPA	diaminodipropylamine
DAG	diacylglycerol (1,3-dilinoleoyl-rac-glycerol)
DAR	drug-to-antibody ratio
DBS	dried blood spot
DCM	dichloromethane
DEHP	di-(2-ethylhexyl) phosphate
DEME	dynamic electromembrane extraction
DEX	dextromethorphan
DI	direct immersion
DIC	diclofenac
DIEA	diisopropylethylamine
DI-SDME	direct immersion single-drop microextraction
DLLE	dispersive liquid–liquid extraction
DLLME	dispersive liquid–liquid microextraction
DMBA	dimethylbutylamine
DMF	*N,N*-dimethylformamide
DMSO	dimethyl sulfoxide
DNS-Cl	dansyl chloride
DOPA	dihydroxyphenylalanine

DOR	dextrorphan
DP IV	dipeptidyl peptidase IV
DPBS	Dulbecco's phosphate-buffered saline
DPX	disposable pipette extraction
DSEA	dansyl sulfonamide ethyl amine
D-SPE	dispersive solid-phase extraction
DTT	dithiothreitol
DVB	divinylbencene
DXR	doxorubicin
EA	ethyl acetate
EBF	European Bioanalytical Forum
ECAPCI	electro capture atmospheric pressure chemical ionization
ED	equilibrium dialysis
EDC·HCl	1-ethyl-3-(3-dimethylaminopropyl) carbodiimide hydrochloride
EDC/NHS	*N*-(3-dimethylamnopropyl)-*N*-ethylcarbodiimide hydrochloride/*N*-hydroxysuccinimide
EDTA	ethylenediaminetetraacetic acid
EG	ethylene glycol
EGDMA	ethylene glycol dimethacrylate
EHS	ethylhexyl salicylate
ELISA	enzyme-linked immunosorbent assay
EME	electromembrane extraction
EME-DLLME	electromembrane extraction dispersive liquid–liquid microextraction
EME-LDS-USAEME	electromembrane extraction low-density solvent-based ultrasound-assisted emulsification electromembrane microextraction
EM-SPME	electromembrane-surrounded solid-phase microextraction
ENB	1-ethyl-2-nitrobenezene
EPR	enhanced permeation and retention
ESI	electrospray ionization
EtOH	ethanol
FA	fatty acid
FA	formic acid
FBAL	α-fluoro-β-alanine
FBS	fetal bovine serum
Fc	fragment crystallizable region/constant region
FcRn	human neonatal Fc receptor
FDA	Food and Drug Administration
FLD	fluorescence detection
FLM	free liquid membrane
Fmoc-Cl	9-florenylmethoxycarbonyl chloride
FNME	fiber-packed needle microextraction
GAC	green analytical chemistry
GC	gas chromatography
GC-FID	gas chromatography–flame ionization detection
GC-MS	gas chromatography–mass spectrometry
GIP	glucose-dependent insulintropic peptide
GLP-1	glucagon-like peptide-1
GMA	glycidylmethacrylate
GnRH	gonadotropin-releasing hormone
GPChos	glycerophosphatidylcholines
GPCs	glycerophosphatidylcholines
GPE	gum-phase extraction
GPI	glycosylphosphatidylinositol
HETP	height equivalent to a theoretical plate
HFIP	1,1,1,3,3,3-hexafluoro-isopropanol

HF-LPME	hollow fiber liquid-phase microextraction
HILIC	hydrophilic interaction liquid chromatography
HIV	human immunodeficiency virus
HMS	homosalate
HND-G	high nitrogen-doped graphene
HNE	human neutrophil elastase
HPIM	homemade polymer inclusion membrane
HPLC	high-performance liquid chromatography
HRMS	high-resolution mass spectrometry
HS	headspace
HSSBSE	headspace stir-bar sorptive extraction
HS-SDME	headspace single-drop microextraction
HTLC	high-turbulence liquid chromatography
IACUC	Institutional Animal Care and Use Committee
IAE	immunoaffinity extraction
IAM	iodoacetamide
IA-SPE	immunoaffinity solid-phase extraction
IC	immunocapture
iCAT	isotope-coded affinity tag
ICP-MS	inductively coupled plasma mass spectrometry
ID	internal diameter
IGF	insulin-like growth factor
IgG	immunoglobulin G
IL-21	interleukin-21
IMAC	immobilized metal ion affinity chromatography
IPA	isopropanol
IS	internal standard
ISET	integrated selective enrichment target
IS-MRM	in-source multiple reaction monitoring
ISR	incurred sample reanalysis
ISTD	internal standard
ITMS	ion trap mass spectrometry
iTRAQ	isobaric tags for relative and absolute quantification
IT-SPME	in-tube solid-phase microextraction
IUPAC	International Union of Pure and Applied Chemistry
IV	intravenous
IVT	*in vitro* transcription
IX-SPE	ion exchange-solid-phase extraction
Kb/p	blood to plasma ratio
Ke/p	red blood cell partition coefficient
LBA	ligand-binding assay
LC	liquid chromatography
LC-MS	liquid chromatography–mass spectrometry
LC-MS/MS	liquid chromatography–tandem mass spectrometry
LC-UV/FL	liquid chromatography with ultraviolet/fluorescence detection
LD	liquid desorption
LGPChos	lysoglycerophosphocholines
LLE	liquid–liquid extraction
LLOQ	lower limit of quantification
LOXs	lipoxygenases
LPCs	lyso-phosphatidylcholines
LPME	liquid-phase microextraction
LSC	liquid scintillation counting
MA	methyacrylate, methyl acrylate

MA	minodronic acid
MAA	methacrylamide
MAA	methacrylic acid
mAb	monoclonal antibody
MADB	poly(methacrylic acid-3-sulfopropyl ester potassium salt-co-divinylbenzene)
MAG	monoacylglycerol (1-stearyl-rac-l glycerol)
MALDI	matrix-assisted laser desorption ionization
MAX	mixed-mode anion exchange
MCV	mean cell volume
MCX	mixed-mode cation exchange
MDA-LDL	malondialdehyde-modified low-density lipoprotein
MDMA	3,4-methylenedioxy-*N*-methylamphetamine
MDS	myelodysplastic syndromes
MeOH	methanol
MEPS	microextraction by packed sorbent
MF	matrix factor
MI-MSPE	molecularly imprinted micro-solid-phase extraction
MIPs	molecularly imprinted polymers
MISPE	molecularly imprinted solid-phase extraction
MISPE-DPE	molecularly imprinted solid-phase extraction with differential pulsed elution
MISPE-PE	molecularly imprinted solid-phase extraction with pulsed elution
MIST	metabolites in safety testing
MIT	molecular imprinting technology
MLLE	micro-liquid–liquid extraction
MMA	methylmalonic acid
MMAE	monomethyl auristatin E
MMST	monolithic molecularly imprinted polymer sol–gel packed tip
MNP	magnetic nanoparticle
MPB	2-bromo-3′-methoxyacetophenone
mPGES-1	microsomal prostaglandin E synthase-1
MPS	3-methacryloyloxypropyltrimethoxysilane
MRM	multiple reaction monitoring
mRNA	messenger RNA
MS	mass spectrometry
MS/MS	tandem mass spectrometry
MSP	magnetic supraparticle
MSPD	matrix solid-phase dispersion
MSPE	magnetic solid-phase extraction
MTBE	methyl *tert*-butyl ether
MTBSTFA	*N*-(tert-butyldimethylsilyl)-*N*-methyl trifluoroacetamide
MW	molecular weights
MWCNT	multiwall carbon nanotube
MWCO	molecular weight cutoff
NAaPs	nucleic acid associated proteins
NAb	neutralizing antibody
NA	nucleic acid
NCEs	new chemical entities
NEM	*N*-ethylmaleimide
NHS	*N*-hydroxysuccinimide
NK	natural killer
NPOE	2-nitrophenyloctyl ether
NPPE	2-nitrophenyl pentyl ether
NPs	nanoparticles
NSB	nonspecific binding

NSE	neuron-specific enolase
NTproBNP	N-terminal pro-B-natriuretic peptide
OC	octocrylene
OD-PABA	ethylhexyl dimethyl *p*-aminobenzoate
ODS	octadecyl
OH-PAH	monohydroxylated polycyclic aromatic hydrocarbon
OH-PDMS	hydroxyl polydimethylsiloxane
OTT	open tubular trapping
OxLDL	oxidized low-density lipoprotein
P	partition ratio
PA	phosphatidic acid
PA	polyacrylate
PA-EG	poly(methyl methacrylate/ethyleneglycoldimethacrylate)
Pa-EME	parallel electromembrane extraction
PALME	parallel artificial liquid membrane extraction
PANCMA	poly(acrylonitrile-co-maleic acid)
PAR	peak area ratio
PBD	pyrrolobenzodiazepine
PBMC	peripheral blood mononuclear cell
PBS	phosphate-buffered saline
PBST	phosphate-buffered saline with Tween-20
PCA	perchloric acid
PCB	polychlorinated biphenyl
PCI	protein C inhibitor
PCs	phosphatidylcholines
PD	pharmacodynamics
PD	phospholipid depletion
PDMS	polydimethylsiloxane
PE	phosphoethanolamine
PEG	polyethylene glycol
PEME	pulsed electromembrane extraction
PEO	polyethylene oxide
PE	phosphatidylethanolamine
PFB	pentafluorobenzyl
PG	phosphatidylglycerol
PGs	prostaglandins
PHMB	4-(hydroxymercuri)benzoate
PI	phosphatidylinositol
PK	pharmacokinetics
PK/PD	pharmacokinetic/pharmacodynamic
PK/TK	pharmacokinetic/toxicokinetic
PKU	phenylketonuria
PLs	phospholipids
PMMA	pentamethylated minodronic acid
PMMA	poly(methyl methacrylate)
PMSF	phenylmethylsulfonyl fluoride
poly(GMA-co-EDMA)	poly(glycidyl methacrylate-coethylene dimethacrylate)
PP	polypropylene
PPB	plasma protein binding
PPESK	poly(phthalazine ether sulfone ketone)
PP-fiber	porous polymer-coated fiber
PPT	protein precipitation
PPY	polypyrrole
ProGRP	pro-gastrin releasing peptide

PS	phosphatidylserine
PTFE	polytetrafluorethylene
PTV	programmable temperature vaporize
PU	polyurethane foams
PUFA	polyunsaturated fatty acids
QC	quality control
QTOF	quadropole time-of-flight
QuEChERS	quick, easy, cheap, effective, rugged, safe extraction method
RA	risedronic acid
RAM	restricted access material
RBC	red blood cell
REC	extraction recovery
RED	rapid equilibrium dialysis
rhTRAIL	recombinant human tumor necrosis factor-related apoptosis-inducing ligand
RISC	RNA-induced silencing complex
ROS	reactive oxygen species
RP	reversed phase
RP-SPE	reversed-phase solid-phase extraction
RPV	rilpivirine
SA-EME	surfactant-assisted electromembrane extraction
SALLE	salting-out assisted liquid–liquid extraction
SAX	strong anion exchange
SBSE	stir-bar sorptive extraction
SCAP	sample card and prep
SCIT	(+)-(*S*)-citalopram
SCX	strong cation exchange
SDCIT	(+)-(*S*)-desmethylcitalopram
SDDCIT	(+)-(*S*)-didesmethylcitalopram
SDF	stromal cell-derived factor
SDME	single-drop microextraction
SDS-PAGE	sodium dodecyl sulphate–polyacrylamide gel electrophoresis
SDU	solvent delivery unit
sEGFR	soluble epidermal growth factor receptor
SELEX	systematic evolution of ligands by exponential enrichment
SF	synovial fluid
SHBG	sex hormone-binding globulin
SIL	stable isotope labeled
SIL-IS	stable isotopically labeled internal standard
SiNWA	silicon nanowire array
SISCAPA	stable isotope standards and capture by anti-peptide antibodies
SLE	supported liquid extraction
SLM	supported liquid membrane
SM	sphingomyelin
SPDE	solid-phase dynamic extraction
SPE	solid-phase extraction
SPME	solid-phase microextraction
SRM	selected reaction monitoring
SRM	single reaction monitoring
SSH	steroid sex hormone
SWCNT	single-wall carbon nanotube
TAG	triacylglycerol (1,3-dipalmitoyl,2oleoyl-glycerol)
TAHS	*p-N,N,N*-trimethylammonioanilyl *N′*-hydroxysuccinimidyl carbamate iodide
TBS	tris-buffered saline
TCA	trichloroacetic acid
TCAFMF	thermally controlled aptamer-functionalized microfluid

TCEP	tris(2-carboxyethyl)phosphine
TD	thermal desorption
TD	toxicodynamic
TDU	thermal desorption unit
TEA	triethylamine
TEHP	tris(2-ethylhexyl)phosphate
TEPA	tetraethylenepentamine
TFA	trifluoroacetic acid
TFC	turbulent flow chromatography
TFME	thin-film microextraction
Tg	thyroglobulin
THCA	11-nor-9-carboxy-Δ^9-tetrahydrocannabinol
THF	tetrahydrofuran
THU	tetrahydrouridine
Ti	titanium
TK	toxicokinetic
TK/TD	toxicokinetic/toxicodynamic
TLC	thin layer chromatography
TMMA	tetramethyl minodronic acid
TMS-DAM	trimethylsilyldiazomethane
TNFα	tumor necrosis factor alpha
TRAIL	tumor necrosis factor-related apoptosis-inducing ligand
Tris	tri(hydroxymethyl)aminomethane
TSV-DEME	two-step voltage dual electromembrane extraction
TXB_2	thromboxane B_2
Tyr	tyrosine
UC	ultracentrifugation
UF	ultrafiltration
UHPLC	ultra-high-performance liquid chromatography
ULOQ	upper limit of quantitation
UPLC	ultra performance liquid chromatography
UV	ultraviolet
VAMS	volumetric absorptive microsampling
VIDB	vinylimidazole–divinylbenzene
VPy	vinylpyridine
WAX	weak anion exchange
WBC	white blood cell
WCX	weak cation exchange
Zr	zirconium
β-NGF	beta-nerve growth factor
γ-MPTS	γ-mercaptopropyltrimethoxysilane
μ-EME	micro-electromembrane extraction
μ-SPE	micro-SPE

Part I

Current Sample Preparation Techniques in LC-MS Bioanalysis

1

Basic Sample Preparation Techniques in LC-MS Bioanalysis

Protein Precipitation, Liquid–Liquid Extraction, and Solid-Phase Extraction

Wenkui Li, Wenying Jian, and Yunlin Fu

1.1 Introduction

Bioanalysis is a subdiscipline of analytical chemistry for the determination of xenobiotics (chemically synthesized or naturally extracted drug candidates and genetically produced biological molecules and/or their metabolites or post-translationally modified products) and biotics (macromolecules such as proteins and DNA, small-molecule endogenous metabolites) in biological systems. The focus of bioanalysis in the human health industry is to provide a quantitative measurement of active drug and/or its metabolite(s) and/or biomarkers for the accurate assessment of pharmacokinetics, toxicokinetics, bioequivalence, bioavailability and/or exposure–response (e.g., pharmacokinetics/pharmacodynamics, toxicokinetics/toxicodynamics) relationships in support of drug discovery and development, and post-approval therapeutic drug monitoring (Unger et al. 2013).

Many techniques have been employed in bioanalysis, including liquid chromatography with ultraviolet/fluorescence detection (LC-UV/FL), liquid chromatography–mass spectrometry (LC-MS), and gas chromatography–mass spectrometry (GC-MS). Among these techniques, liquid chromatography–tandem mass spectrometry (LC-MS/MS) has been the most widely used and most reliable tool due to its high sensitivity and specificity. In LC-MS bioanalysis, a high-performance liquid chromatography (HPLC) or ultra-high-performance liquid chromatography (UHPLC) system is employed to separate the analyte(s) of interest from other unwanted matrix components in sample extracts based on the specific interactions between the analyte(s) of interest and the analytical LC column. The LC eluent is then introduced to a mass spectrometer for molecule-based separation via multiple reaction monitoring (MRM), also named as selected reaction monitoring (SRM). In MRM or SRM, the specific precursor ion(s) of analyte of interest is selected for collision-induced dissociation (CID) or collision-activated dissociation (CAD) to generate fragment ions that are also specific to the analyte(s) of interest, and usually the most abundant fragment ion is selected for MS detection. With the specificity and sensitivity provided by both the LC separation and the MRM of the MS system, LC-MS/MS has become one of the most suitable tools for quantitative bioanalysis (Unger et al. 2013).

The common matrix that is subjected to LC-MS bioanalysis includes various body fluids (e.g. plasma, serum, whole blood, saliva, tears, and urine) and organ tissues (e.g. kidneys, liver, lung, skin, and brain tissue). In general, these biological samples contain abundant various endogenous components like salts, small molecules, proteins, and lipids, or exogenous components such as formulation ingredients. In contrast, the analyte(s) of interest is often at very low concentration levels, typically in low $ng\,ml^{-1}$ concentration range and even at $pg\,ml^{-1}$ level for highly potent molecules. The presence of abundant (typically at $\mu g\,ml^{-1}$ to $mg\,ml^{-1}$ range) endogenous or exogenous components in the biological matrix, compounded by the very low concentration of the analyte(s) of interest, is definitely a challenge for bioanalytical scientists in the field in developing and validating a robust LC-MS bioanalytical assay (Unger et al. 2013).

In order to ensure adequate sensitivity, selectivity, and reproducibility of the LC-MS assay method for measuring analyte(s) of interest in biological samples, sample preparation, also known as sample pretreatment or sample cleanup, is a must step. Sample preparation in LC-MS bioanalysis is considered a pre-analytical separation process that involves selective isolation of analyte(s) of interest from the matrix, minimization or elimination of matrix components in the extracted samples, and, if necessary, enrichment of analyte(s) to ensure achievable assay sensitivity. An ideal sample preparation method should be able to reduce matrix effect to a minimal level

Sample Preparation in LC-MS Bioanalysis, First Edition. Edited by Wenkui Li, Wenying Jian, and Yunlin Fu.

while maintaining a reasonable and consistent extraction recovery (REC) (e.g. 80%). However, due to the many factors affecting matrix removal and analyte recovery, developing an optimal sample preparation procedure can be difficult, tedious, and labor-intensive, which makes it one of the most significant parts in the development of a robust LC-MS bioanalytical method. In this chapter, we describe the commonly used sample preparation techniques in LC-MS bioanalysis, namely protein precipitation (PPT), liquid–liquid extraction (LLE), and solid-phase extraction (SPE), with focus on the importance of understanding the physicochemical properties of the analyte(s) of interest that determines its extractability and the need of balancing REC and assay selectivity. There are a variety of techniques that are derived from these three basic sample preparation workflows as well as emerging methodologies that utilize advanced techniques for more selective and/or efficient extraction of analyte(s) of interest. They will be covered in detail in subsequent chapters of the book and therefore will not be discussed here. In addition, although dilution-and-shoot (DAS) of biological samples has been used in LC-MS bioanalysis, it has been primarily used in the area of drug discovery for the analysis of samples generated from *in vitro* system rather than *in vivo* studies, and therefore is not covered in this chapter.

1.2 Physicochemical Properties of Drugs and Their Metabolites

The first and foremost important aspect of sample preparation in LC-MS bioanalysis is to understand and utilize the relevant physicochemical properties of the analyte(s) of interest. These include but are not limited to hydrophilicity, lipophilicity, and protolytic properties ($\log P$, $\log D$, and $\mathrm{p}K_\mathrm{a}$, etc.). Understanding these properties can help select suitable sample preparation techniques and the associated experimental conditions for the analyte(s) of interest.

1.2.1 Hydrophilicity vs. Lipophilicity of Analyte(s)

Hydrophilicity refers to the ability of a molecule to dissolve in water and other hydrophilic solvents. The interaction of a hydrophilic or polar molecule with water and other polar molecules is more thermodynamically favorable than its interactions with oil or other hydrophobic/lipophilic molecules. A hydrophilic molecule is typically charge-polarized and capable of hydrogen bonding. In contrast, lipophilicity refers to the ability of a molecule to dissolve in fats, oils, lipids, and nonpolar solvents such as hexane or toluene. Such nonpolar solvents are themselves lipophilic. In the field of bioanalysis, lipophilicity, hydrophobicity, polarity and nonpolarity may have been used to describe the same tendency of a molecule and these terms are often used interchangeably. There are two key parameters to help understand the hydrophilicity vs. lipophilicity of the analyte(s) of interest:

The partition coefficient, P, is the concentration ratio of a molecule, specifically for unionized molecule, in two immiscible solvents at equilibrium. When one of the solvents is water and the other is a nonpolar one (e.g. *n*-octanol), the $\log P$ value is the logarithm of the concentration ratio of the unionized form of the molecule between the two immiscible phases. $\log P$ is considered a measure of lipophilicity or hydrophobicity of a given molecule, for which the higher the $\log P$ value, the more lipophilic or hydrophobic the molecule is.

$$\log P = \log_{10}\left(\frac{C_{\text{organic}}}{C_{\text{aqueous}}}\right),$$

where C_{organic} is the concentration of the neutral form of the molecule in the water-immiscible solvent; C_{aqueous} is the concentration of the neutral form of the molecule in the aqueous phase.

The distribution coefficient, D, is the ratio of the sum of the concentrations of both ionized and unionized forms of the molecule between the two immiscible phases, i.e. aqueous and organic phases. Since ionization of a molecule in aqueous phase is pH dependent, the $\log D$ value is also pH dependent. The aqueous phase is adjusted to certain pH for $\log D$ measurement.

$$\log D = \log_{10}\left(\frac{C_{\text{organic}}}{C_{\text{aqueous(ionized)}} + C_{\text{aqueous(neutral)}}}\right),$$

where C_{organic} is the concentration of the molecule (ionized and neutral) in the water-immiscible solvent, $C_{\text{aqueous (ionized)}}$ is the concentration of the ionized form of the molecule in the aqueous phase, and $C_{\text{aqueous (neutral)}}$ is the concentration of the neutral form of the molecule in the aqueous phase. Accordingly, in LLE and to some extent SPE, the REC of a given molecule under given conditions (pH, organic solvent, volume ratio between aqueous and organic) can be predicted (Liu and Aubry 2013).

1.2.2 Protolytic Properties of Analyte(s)

Protolytic refers to a reaction or process in solution that involves transfer of a proton from one molecule to another (one of the molecules usually belonging to the solvent). Many molecules show protolytic properties and are often present in both ionic/ionized (acidic [H^+ donor] or basic [H^+ acceptor]) and neutral forms in aqueous solution.

In chemistry, pH is the negative logarithm of the concentration of the hydrogen ion (pH = $-\log_{10}(H^+)$) and is used to specify the acidity or basicity of an aqueous solution. Solutions with a pH <7 are considered acidic while solutions with a pH >7 are considered basic. Pure water is neutral at pH 7 (25 °C). pK_a is the negative base-10 logarithm of the acid dissociation constant (K_a) of a solution (pK_a = $-\log_{10} K_a$) and is also defined as the pH where a molecule exists as 50% ionized and 50% unionized. pK_a is a property of a given molecule that tells us how acidic or basic it is. The lower the pK_a a molecule is, the stronger the acid it is. For example, the pK_a value of acetic acid is 4.8, while the pK_a value of a stronger acid, lactic acid is 3.8. The pK_a value of a given molecule can be related to its charge state in solution under a given pH. In this regard, the pH of an aqueous solvent has a great impact on degree of ionization of a given analyte and the choice of the sample preparation method to be employed in LC-MS bioanalysis. An acid analyte in an acid solution will generally not ionize. In contrast, it will ionize in a basic solution. Similarly, a basic analyte will generally not ionize in basic solution but will ionize in acidic solution. When pH is equal to pK_a, 50% of the analytes are in ionized form and 50% in unionized (neutral) form. However, when pH is <<pK_a, almost 100% of the acidic analytes are unionized but basic analytes are almost 100% ionized; when pH is >>pK_a, almost 100% of the acidic analytes are ionized but basic analytes are almost 100% unionized.

Accordingly, in LLE and SPE with reverse-phased (RP) stationary phase, the best REC can be obtained at a pH at which most of the analytes are not charged, i.e. neutral form. In contrast, in SPE with ion-exchange stationary phase, the best REC can be obtained when the analyte molecules are all ionized (charged) to interact with the charged stationary phase (Liu and Aubry 2013).

1.3 Pre-analytical Variables of Analyte(s) of Interest in Biological Matrix

LC-MS bioanalysis is dealing with a variety of biological matrices such as plasma, serum, whole blood, urine, CSF, tissue homogenates, etc. The composition and complexity (e.g. pH, nature, and concentration of proteins, lipids, and salts) of these matrices is significantly different from one to the other. Even for the same matrix, it can be largely different from one subject to the other, depending on the age, gender, disease stage, medication, and other factors of the subject. Understanding the above can generally be helpful in developing an overall sample preparation strategy. However, each analyte is considered a unique entity in LC-MS bioanalysis. In addition to understanding the physicochemical properties and others discussed in Section 1.2 and complexity of the matrix discussed above, understanding some analyte-specific variables, including stability, possible nonspecific binding, protein binding and/or blood-to-plasma ratio, and red blood cell partition can be pivotal in defining and/or optimizing a specific sample preparation method for LC-MS bioanalysis of the analyte(s) of interest.

1.3.1 Stability

Stability is an important pre-analytical variable for quantitative LC-MS bioanalysis of drug molecules and/or their metabolites in biological matrices. Instability of an analyte in any stage of the bioanalysis process, including sample collection, processing, storage, extraction, and LC-MS analysis, can result in under- or over-estimation of analyte exposure if an adequate preventive procedure is not in place (Li et al. 2011). Therefore, one should carefully examine the structural characteristics and the *in vivo* and/or *in vitro* biotransformation of any analyte of interest prior to developing a specific LC-MS bioanalysis method. Previous internal and/or external (i.e. literature) experiences with structurally similar compounds can be very useful for a general understanding of the potential instability alert. The possible instability of some analyte(s) that contains biologically or chemically labile moieties, e.g. thiol, ester, or catechol, etc., can be readily predicted and estimated for effectiveness of specific precautions throughout the entire method development, validation, and sample analysis process. However, instability may not be readily predictable for many analyte(s) unless necessary stability assessment has been conducted using QC and/or incurred samples against freshly prepared calibration standards. For example, an unstable conjugated metabolite could convert to parent drug leading to overestimation of parent drug concentration, a phenomenon that can only be estimated using incurred samples when the putative conjugated metabolite is not available. Instability of an unstable analyte may occur at one or several points of the bioanalytical process, i.e. sample collection, processing, storage, extraction, and LC-MS/MS analysis, etc. Some general strategies should be formulated with incorporation of one or more specific guidance (e.g. addition of enzyme inhibitors, pH modifiers, or antioxidants, etc.) in developing a robust LC-MS/MS quantitative bioanalytical method for the unstable molecules. For more details on stability strategies one can refer to Strategies in quantitative LC-MS/MS analysis of unstable small molecules in biological matrices (Li et al. 2011, 2013a).

1.3.2 Nonspecific Binding

Some matrix, e.g. urine or CSF, does not normally contain proteins and lipids that are present at ~8% in whole blood, plasma, or serum (Lentner 1981). The lack of protein and lipids in these samples can be associated with issues of nonspecific binding or container surface adsorption of drug molecules, especially those lipophilic/hydrophobic and those with high affinity to proteins (high protein binding), in LC-MS bioanalysis (Li et al. 2010). The nonspecific binding or container surface adsorption is often evidenced by the unusually low REC of the analyte(s) of interest and/or nonlinearity of the calibration curves or highly variable QC sample results. Unfortunately, the issue is often overlooked during the early stage of bioanalytical method development, especially when assay sensitivity is not an issue and/or both the calibration standards and QC samples are prepared in the same fashion, i.e. freshly spiked, daily prepared, or pre-pooled. In the latter cases, the unexpected low recovery of analyte would often be interpreted due to matrix effect or signal suppression because the problem, if any, may be masked by a similar degree of LC-MS signal loss of the analyte for both the calibration standards and QC samples that are prepared in the same fashion. In some cases, the issue might not be realized until after many failed feasibility runs. Failure to promptly assess and adequately address this issue would result in underestimated urine/CSF drug concentrations. Details on how to assess and address the issues due to nonspecific binding or container surface adsorption can be found in some good research and review articles (Fu et al. 2011; Ji 2013; Ji et al. 2010; Li et al. 2010). Some detailed strategies can be found in Chapter 18 of this book.

1.3.3 Protein Binding

Depending on the affinity of drug molecules to plasma protein, a portion of the drug molecules may become bound to plasma proteins, with the remainder being unbound. Therefore, a given drug molecule generally exists in two forms in blood, plasma or serum: bound and unbound. If the protein binding is reversible, then a chemical equilibrium will exist between the bound and unbound states of the drug molecules, such that: Protein + drug $\rightleftharpoons$ Protein–drug complex. Since albumin, a plasma protein, is alkalotic, acidic and neutral drug molecules will primarily bind to albumin. If albumin becomes saturated, then these molecules will bind to lipoprotein. In contrast, basic drug molecules will primarily bind to the acidic alpha-1 acid glycoprotein in whole blood, plasma, or serum.

In general, a LC-MS bioanalytical method is employed to measure the total analyte concentrations in the intended study sample matrix unless otherwise specified. Therefore, release of the analyte of interest from plasma proteins by interrupting the protein binding is considered a key step for good and consistent REC, regardless of choice of the sample preparation method and protein binding rate of the analyte. In this regard, the biological samples are generally treated with acid (e.g. acetic acid or formic acid), base (e.g. ammonium hydroxide), buffer, or organic to free the analyte from the plasma proteins prior to further processing (de Boer and Wieling 2013).

1.3.4 Blood-to-plasma Ratio and Red Blood Cell Partition

The blood to plasma ratio (Kb/p) of a drug is the ratio of its concentration in whole blood (containing both red blood cells, RBCs, and plasma) to the corresponding value in plasma, namely C_B/C_P. In contrast, the red blood cell partition coefficient (Ke/p) is the ratio of the drug concentration in the RBCs (i.e. not including plasma) to its concentration in plasma, namely C_{RBC}/C_P. In practice, when working with whole blood samples, the intended method of sample preparation should be capable of releasing the analyte from the RBCs (Brockman et al. 2007).

Temperature-dependent re-equilibration of drug molecules between RBCs and plasma is a known phenomenon (Dell 2004). Ideally, if plasma is the matrix of choice for bioanalysis, blood samples should be centrifuged to separate the plasma as soon as they are collected. Extended stay of the collected blood samples at 4 °C may be an issue if temperature-dependent re-equilibration occurs to the analyte(s) of interest. As a result of temperature-dependent re-equilibration, the analyte concentration in plasma obtained from blood that have been left on wet ice (~4 °C) for some time may no longer be the "original" one. For example, dehydronorketamine, a ketamine major metabolite, is stable in plasma under all normal conditions. Due to re-equilibration of compound into blood cells at 4 °C with time, a significant decrease in its plasma concentration was observed over time. In contrast, no change was seen in the concentrations of the plasma samples prepared from the blood that has been stored at ambient temperature. This phenomenon can be overcome by centrifuging the blood immediately after collection, so that the actual *in vivo* plasma analyte concentration can be maintained and determined (Hijazi et al. 2001). In the case where the collected blood samples have somehow stayed on wet ice for a period of time, the blood samples can be incubated at 37 °C for ~30 minutes to resume the equilibrium between RBCs and plasma for the analyte of interest (to mimic *in vivo* condition) prior to centrifugation of the blood for plasma (unpublished data).

1.4 Most Commonly Used Sample Preparation Methods in LC-MS Bioanalysis

The most commonly employed sample preparation methods in LC-MS bioanalysis generally include PPT, LLE, and SPE. Many other contemporary sample preparation methods, such as online extraction and column switching, phospholipids depletion, salting-out assisted LLE, supported liquid extraction (SLE), immunoextraction, microextraction, sample preparation via nanomaterials, molecularly imprinted polymers (MIP), aptamers or electromembranes, and stir-bar sorptive extraction, etc. are captured in other chapters of this book.

1.4.1 Protein Precipitation (PPT)

The common biological matrix, such as blood, plasma, or serum contains ~8% (w/w) of proteins as mentioned earlier. Direct injection of these samples onto the LC-MS system is generally not a good option for LC-MS bioanalysis. This is because the proteins in these samples could readily precipitate as a result of contact with organic solvents and/or buffers in the mobile phase. As a result of PPT in the LC system, the performance of the LC column will rapidly deteriorate unless the majority of the protein in the samples is removed.

Proteins are large biological molecules composed of amino acids that are linked to each other by peptide bonds. Under normal physiological conditions, a soluble protein has one or more peptide chains in a folded conformation. Within this conformation, the majority of hydrophobic amino acid residues are toward inside while the charged or hydrophilic amino acid residues are toward outside. In the inside of the conformation, the peptide chains are folded together mainly through hydrophobic interactions between hydrophobic amino acid residues; other interactions such as hydrogen bonds, salt bridges, or disulfide bonds also contribute to the folded structure of proteins. On the outside of the conformation, the charged or polar surface residues interact with the biological environment, where water forms a solvation layer that surrounds the protein. The formation of the solvation layer weakens the ionic interactions between proteins and decreases the likelihood of aggregation (Li and Bartlett 2014).

PPT is a simple, quick, and convenient sample preparation technique in LC-MS bioanalysis. In this process, a small volume of blood, plasma, serum, tissue homogenate, or other aqueous matrices is mixed with a certain volume of protein precipitant. When proteins in sample matrix/solution come in contact with precipitant, the conformation of the proteins is altered due to the interaction. This results in aggregation and precipitation of the proteins. As a result of conformation changes of the proteins, the analyte(s) of interest that are bound to the proteins are released into and stay in the solution. Upon centrifugation and/or filtration, the precipitated proteins are separated from the analyte(s) containing supernatant.

The equipment required for PPT is relatively simple. It includes low-cost and disposable centrifuge tubes or 96-well plates and a centrifuge. A bioanalytical scientist can find many published methods ready for use as the starting point. If method development and/or optimization are necessary, it is relatively simple and straightforward due to the simple nature of the technique. Briefly, optimization of a PPT method includes selection of precipitant and the amount of precipitant along with centrifugation and/or filtration.

A significant advantage of PPT is its high recovery as compared to other techniques, e.g. LLE and SPE. Since only proteins are hypothetically removed from the sample matrix by the method, small-molecule analyte(s) should remain in the solution and this yields a theoretical recovery of 100%. Such an advantage of PPT has made it very popular in the bioanalytical community.

Common protein precipitants include: (i) water-miscible organic solvents, (ii) acids, (iii) metal ions, or (iv) salts. Among these precipitants, the water-miscible organic solvents and acids are the most common ones.

1.4.1.1 Water-miscible Organic Solvents

Common water-miscible organic solvents include acetonitrile, acetone, ethanol, and methanol and they are 100% miscible in aqueous solution. When these solvents are added to blood, plasma, serum, or tissue homogenate, they quickly displace water molecules of the solvation layer of the proteins in the sample matrix. When the solvation layer becomes thinner and thinner, proteins get closer and closer to each other via attractive electrostatic or dipole interactions, leading to the aggregation. The protein aggregates grow by diffusive additions of other protein molecules and eventually reach a critical size for precipitation, forming protein sediments or floccules in the mixture of samples and precipitant (Li and Bartlett 2014).

Acetonitrile, acetone, ethanol, and methanol are all good precipitant with PPT efficiency in the order of acetonitrile > acetone > ethanol > methanol. Among these organic solvents, the most commonly used one is acetonitrile. It is worth noting that PPT efficiency via an organic solvent also depends on the volume of the organic solvent added. With acetonitrile as an example, when 0.2 ml of acetonitrile is added to 1.0 ml of plasma, only 13.4% of proteins are removed. With the volume of acetonitrile increased to 0.4, 0.6, 0.8, 1.0, 1.5, 2.0, 3.0, and

4.0 ml, the percentage (%) of protein removed is increased, respectively, to 14.8, 45.8, 88.1, 97.2, 99.4, 99.7, 99.8, and 99.8% (Blanchard 1981). It is apparent that by adding 2.0 ml of acetonitrile to 1.0 ml of plasma for PPT, the resulting supernatant is basically protein-free.

Although pure organic solvents can perfectly fit the purpose of PPT, it is a common practice to add a small volume of acids (e.g. formic acid, acetic acid) (Li et al. 2014) or bases (e.g. ammonium hydroxide) (Li et al. 2007b) to the sample matrix, followed by proper mixing to interrupt protein binding and/or change the charge states of the analyte by changing the pH of the sample matrix. The acids or bases can also be added to the organic solvent to prepare a PPT solution. The addition of a small volume of acids or bases can help improve REC. On the other hand, internal standards (IS) can be added to the organic protein precipitants. This allows for the addition of the IS at the same time as for protein precipitant. This practice not only improves the throughput by combining two steps together but also enhances the precision of the assay by pipetting a relatively large volume of IS working solution instead of spiking a relatively small volume. In addition, for certain analytes with high protein binding, it is recommended to add IS in aqueous solution to plasma sample first. This allows for additional time of equilibrium before addition of an organic precipitant. By doing so, the IS may better mimic the extraction of the analyte and therefore improve the assay accuracy and reproducibility (unpublished data).

As the strongest precipitant, use of acetonitrile can lead to the formation of very solid protein precipitates of large particle sizes. Being relatively weaker protein precipitants, the use of ethanol or methanol can lead to the formation of looser flocculent precipitates. However, there is a concern that using 100% acetonitrile as the protein precipitant may not give the highest recovery. Due to the very quick PPT using acetonitrile, some analyte(s) with high plasma protein binding might coprecipitate with plasma proteins if the analyte(s) are not disassociated with the proteins during the process of protein precipitation. In a study with no acids or base added to the sample matrix prior to PPT using acetonitrile, an increasing percentage of methanol (10, 20, 30, 40, 50, 60, and 70%) was added to acetonitrile for PPT prior to direct analysis via hydrophilic chromatography (HILIC)-MS/MS for atenolol in human plasma. The analysis of the samples showed that use of acetonitrile containing 10% methanol yielded a sample extract with LC-MS signal higher than that obtained using acetonitrile alone (Li et al. 2005).

1.4.1.2 Acids

The commonly used acids for PPT are trichloroacetic acid (TCA) (5–15%, TCA) and perchloric acid (PCA) (6%). Both reagents are highly efficient in precipitating proteins in the sample matrix (Hee et al. 2017). It is understood that protein denaturing is a key in PPT by TCA and PCA. Upon addition of the acids, the pH of the solution/sample matrix is greatly lowered and the protein conformation is drastically altered, resulting in the aggregation of proteins. Addition of 0.2 ml of TCA (10%) to 1 ml of the matrix is capable of removing >99% of the proteins in the plasma samples. One advantage of using acids for PPT is that the resulting supernatant (of filtrate) after centrifugation is still highly aqueous and the supernatant can be directly injected onto the LC-MS/MS system. The drawback of this approach is that the resulting supernatant is strongly acidic with pH < 2.0, which might be a concern of analyte instability in the sample extract (Li and Bartlett 2014; Pedersen-Bjergaard et al. 2015).

Other than the composition and pH of the organic protein precipitants discussed above, temperature is another factor that should be considered for efficient PPT. Ice-cold organic solvents are generally more efficient in precipitating proteins in the sample matrix. This approach is particularly useful when dealing with unstable analyte(s) (Li et al. 2011).

In the PPT process, upon addition of protein precipitant (discussed above), IS (in working solution or prepared in protein precipitant solution), and pH modifier (can be added to the protein precipitant), adequate vortex-mixing has to be applied to the sample mixture. Most proteins in the sample mixture are expected to precipitate. The next step is a physical process to separate precipitated proteins from the supernatant. Usually 5–10 minutes of centrifugation at ~1500 × g is sufficient to settle the protein precipitates. The higher centrifugation force and the longer centrifugation time are applied, the more solid the protein pellet will be formed in the bottom of the sample preparation device (individual tubes or 96-well plate). This makes it easier to transfer the supernatant without disturbing the pellet.

Since at least two volumes of organic solvents are commonly employed in PPT and the resulting supernatant generally has a high organic composition, direct LC-MS/MS analysis of the supernatant using reversed-phase columns may not be feasible. If assay sensitivity is permitted, dilution of the supernatant with aqueous solution (e.g. water) should be applied to lower the concentrations of organic solvents in the final sample extract prior to LC-MS/MS analysis. Otherwise, the resulting supernatant should be subjected to evaporation step, at which the organic solvents in the supernatant are evaporated under a stream of nitrogen and resulting sample residues are reconstituted in aqueous solutions (with proper pH, organic and/or acid concentrations) or starting mobile phase. The volume of the reconstitution solution used should be adjusted to ensure suitable assay sensitivity on the intended LC-MS/MS platform. In contrast, depending on the polarity and other physicochemical properties of the analyte(s) of interest, the resulting supernatant obtained from PPT may be directly subjected to analysis on HILIC-MS/MS that employs mobile phases with high organic composition. This high organic composition is, in general, compatible with sample extract (supernatant) (Jian et al. 2010b; Li et al. 2005).

A representative protocol of PPT with an organic solvent is shown below.

A representative example of PPT protocol (for the analysis of a steroid 11β-hydroxylase inhibitor, in human plasma by Li et al. 2014).

- A 0.100 ml aliquot of blank plasma (for matrix blanks, zero samples [blank matrix mixed with the IS only] or fresh calibration standards), QC samples, or unknown samples was added to a 2-ml, 96-well plate.
- To the standard samples, 0.100 ml of the respective standard working solution (in 50% aqueous methanol, v/v) was added.
- To the matrix blanks, zero samples, QC samples, and unknown samples, a 0.100 ml aliquot of 50% methanol in water (v/v) was added.
- This was followed by brief vortex-mixing.
- A 25 μl aliquot of the IS working solution in 50% methanol in water (v/v) was added to each well except for the matrix blank, to which a 25 μl aliquot of 50% methanol in water (v/v) was added.
- After adding 25 μl of 40% formic acid in water (v/v) to each well, the plate was vortex-mixed briefly.
- To each well, 0.500 ml of acetonitrile/ethanol/acetic acid (90/10/0.1, v/v/v) was added and the plate was covered and vortex-mixed for ~15 minutes on a pulse vortex-mixer with a motor speed setting of about 50 units.
- The sample plate was centrifuged at ~3500 rpm (~2000 × *g*) for ~10 minutes at ~10 °C.
- The supernatants (0.500 ml) were transferred via a TomTec Quadra 96 system (Hamden, CT, USA) to the corresponding wells in a 1-ml 96-well plate, evaporated to dryness under a stream of nitrogen at approximately 45 °C and the sample residues reconstituted with 100 μl of reconstitution solution (30% methanol in water, v/v).
- After brief vortex-mixing and centrifugation at ~3500 rpm (~2000 × *g*) for approximately five minutes remove potential un-dissolved contents in the sample, a 10 μl aliquot was injected onto the LC-MS/MS system.

Conventional PPT involves multisteps of manipulation, including centrifugation and transfer of the resulting supernatant. The overall throughput is low. Therefore, it is not ideal in support of analysis of a large number of study samples. One of the recent approaches to overcoming such a disadvantage is the use of membrane-based PPT filter plates, such as Unifilter® from Whatman, Strata® Impact™ from Phenomenex, Isolute® PPT+ from Biotage, and Sirocco™ from Waters. By using PPT plates, separation of the supernatant can be carried out by filtration without the need of supernatant transfer.

In practice, the 96-well PPT plate is placed on top of a 96-well collection plate. The PPT plate contains filters in the bottom of each well with a typical pore size of 0.2 μm. Upon centrifugation or application of vacuum, the supernatant from PPT flows through the filter into the collection plate, while the pellets are retained on the filter. By this approach, significant reduction in the sample preparation time (four-fold, from 4–5 hours to 1–1.5 hours) was reported as compared with the manual PPT method (96 samples each) (Kocan et al. 2006). Furthermore, the use of PPT plates yields cleaner sample extracts with a higher solvent recovery.

Issues related to solvent leaking with PPT plates have been reported. Among the PPT solvents tested, acetonitrile showed higher leaking than methanol. It has been also suggested that the precipitating solvent, filter material, pore size, centrifugation speed or vacuum strength, nonspecific binding of analyte to the plate, and matrix effects (ion suppression) should be concurrently considered when choosing a PPT plate. Table 1.1 summarizes commercially available PPT plates with respective technical details (Kole et al. 2011).

Matrix effect is considered a major issue related to PPT. This is because, upon PPT, all other matrix components, including phospholipids, are present in the sample extract. Phospholipids are an important class of biological compounds containing one or more phosphate groups. Their molecular structures have two major functional group regions, i.e. (i) a polar head group which includes an ionizable organic phosphate moiety as well as other polar groups of various types and (ii) one or two long-chain fatty acid ester groups which are hydrophobic. The highly ionic nature of phospholipids makes them responsible for influencing the ionization of the analyte(s) of interest and desolvation of the LC effluent droplets in the electrospray MS source, causing significant matrix effects.

Noticeably, some of the recently developed PPT plates are packed with materials that specifically retain phospholipids and, therefore, have been increasingly employed to remove the abundant phospholipids along with proteins in LC-MS bioanalysis. These plates include but are not limited to Captiva™ ND Lipids from Agilent, Ostro™ from Waters, Phree™ from Phenomenex, and

Table 1.1 Commercially available protein precipitation plate/tubes.

Company	Trade name	Format	Volume handling range (μl)	Membrane/depth filter	Phospholipids removal	Advantages offered
3M	Empore™	96-well	50–200	Polypropylene filter	—	98% removal of particles >10 μm
Biotage	Isolute PPT+	96-well	15–400	Depth filter (optimized porosity distribution)	—	Solvent First Technology
Glygen	Glysci	96-well	—	3 mm filter	Yes	Minimum sample loss, no leaking; the plate packed with $TiO_2 + ZrO_2$ can be used for the removal of phospholipids at the same time as the protein is precipitated
Millipore	MultiScreen® Deep well and multiscreen filter plates	96-well	50–200	Membrane; Durapore® (variety of membranes)	—	Low binding, low extractables, and high recoveries
Orochem Technologies	OC21PPT20	96-well	400	Variety of membranes	—	Solvent First Technology
Pall Corporation	AcroPrep™	96-well	350	PTFE membrane (variety of membranes)	—	Available in a variety of housing colors
Phenomenex	Phree	96-well	25–400	Membrane filter	Yes	The Phree sorbent selectively removes phospholipids from precipitated plasma samples
Phenomenex Inc.	Strata Impact	96-well	25–300	Membrane filter	—	Solvent Shielding Technology™
Sigma	HybridSPE	96-well	20–300	Upper 5 μM PFFE frit and Lower 0.2 μM hydrophobic filter, packed bed assembly acts as depth filter	Yes	The Zr atom coated on the silica surface acts as a Lewis acid which interacts with phosphate moiety of phospholipids, which is a strong Lewis base
ThermoFisher	HyperSep™	96-well	400	Dual frit	—	No wetting out, no leakage; nonspecific binding of compound to filter/plate
ThermoFisher	Pierce™	96-well	15–400	0.2 μM membrane	—	Leakfree, no dripping
Varian Inc.	Captiva ND^{Lipids}	96-well and single filter cartridges	50–200	Non-drip membrane	Yes	Proteins and surfactants removed
Waters	Ostro	96-well	5–200	Valve and cap mat	Yes	Effective for protein and phospholipids removal
Waters	Sirocco	96-well	200	Valve and cap mat	—	Sealing cap mat and patented valve technology
Whatman	Unifilter	96-well	200	Non-drip dual membrane (variety of membranes)	—	Available in two types: fast flow and standard

Source: Modified from Kole et al. (2011). Reproduced with permission of John Wiley & Sons.

HybridSPE® from Sigma. By using these plates, the resulting sample extracts are not only free of proteins, but also free of or with significantly reduced phospholipids. The sample extracts are readily available for the subsequent LC-MS analysis. In the case of using a HybridSPE PPT plate, once PPT occurs after the addition of acetonitrile (containing 1% formic acid), the sample extract passes through the HybridSPE packed bed. The packed bed consists of proprietary zirconia-coated silica particles. The zirconia sites exhibit Lewis acid (electron acceptor) properties that interact strongly with Lewis bases (electron donor). As mentioned above, phospholipids consist of zwitterionic phosphonate moieties (the polar head group) and a large hydrophobic tail (two fatty acyl groups that are hydrophobic). The phosphate group(s) on the phospholipids acts as a very strong Lewis base that interact strongly with zirconia atoms functionalized on the particle surface. Formic acid or other acids is a critical modifier in PPT via phospholipid removal PPT plate approach to improve the recovery of analytes of interest, particularly those acidic. The acid plays a critical role in preventing analyte retention without affecting phospholipids' retention on the packed bed (Ahmad et al. 2012).

Each of the commercially available PPT plates comes with standardized procedures recommended by the providers. The procedures include direct PPT on the plates (Pucci et al. 2009) or conventional PPT followed by loading the resulting supernatant onto the PPT plates (Asimakopoulos and Thomaidis 2015; Jiang et al. 2011; Zeng et al. 2010).

A representative example of PPT protocol using Sirocco PPT plate.

- A 200 μl aliquot of acetonitrile was added to each well of a Sirocco PPT plate with a 1-ml 96-well collection plate underneath.
- A 25 μl aliquot of IS working solution in acetonitrile : water (50/50, v/v) was added to each well except for the blanks, to which a 25 μl aliquot of acetonitrile : water (50/50, v/v) was added.
- A 25 μl aliquot of blank, zero, standard, or QC was added to the assigned well. The plate was covered and vortex-mixed for one minute on a pulse-vortex-mixer and then centrifuged for five minutes at ambient temperature.
- The filtrate in the collection plate was evaporated to dryness at a temperature of 45 °C under a flow of nitrogen. A 200 μl aliquot of reconstitution solution consisting of acetonitrile in water (10/90, v/v) with 0.1% formic acid was added to each well and the plate was vortex-mixed for five minutes.
- A 10 μl aliquot of the reconstituted extract was injected onto the LC-MS/MS system.

1.4.2 Liquid–Liquid Extraction (LLE)

LLE is another common sample preparation technique that has been widely used in LC-MS bioanalysis. The method involves the extraction of the analyte(s) of interest or unwanted interference components from one liquid phase (e.g. biological samples) to another immiscible liquid phase (e.g. organic solvent), resulting in sample clean-up.

In LLE, biological samples (plasma, serum, whole blood, and urine or tissue homogenate) are commonly mixed with additives (buffer, acids, or bases) to ensure efficient extraction of the target molecules. This is followed by addition of IS working solution and an organic solvent (extraction solvent), which is immiscible with water. Then the two-immiscible phase mixture in tubes/wells is shaken or vortex-mixed for a certain period of time to mix the sample and the organic solvent, during which the target molecules are transferred from the aqueous phase to the organic phase or vice versa. This is followed by centrifugation for phase separation. After centrifugation, the phase containing the target molecules can be collected for further processing and analysis (Liu and Aubry 2013).

1.4.2.1 Mechanism of LLE and Extraction Recovery

The mechanism of LLE can be explained using a simple phrase "like dissolves like" (Li and Bartlett 2014). A solute can be dissolved best in a solvent that has a similar polarity to itself. Nonpolar compounds have higher solubility in organic solvents than in water. In contrast, ionic or polar compounds have higher solubility in aqueous solutions than in organic solvents. When two immiscible solvents are present in one system, the solute currently being dissolved in the solvent with less solubility will diffuse across the liquid–liquid interphase of the two immiscible solvents to enter the one in which the solute solubility is higher. When analyte molecules are extracted from the aqueous phase into the immiscible organic phase or vice versa in LLE, interactions take place between the analyte molecules and the solvent molecules. The predominant interactions in LLE are the following (Li and Bartlett 2014; Pedersen-Bjergaard et al. 2015):

- **Hydrophobic interactions:** The interactions between the nonpolar analyte(s) and nonpolar organic solvent or solvent mixture in a polar (aqueous, usually water) solvent. Hydrophobic interactions are nonpolar attractive interactions between hydrocarbon moieties and related nonpolar molecular elements.
- **Dispersion interactions:** The interactions between a relatively nonpolar electron-rich molecule and a polar (or charged) molecule. It is generally a weak attractive interaction. The polar molecules in the system cause

electron density change in the nonpolar molecules by making them slightly polar. In other words, the polar molecules induce dipole moment in the nonpolar but electron-rich molecules.

- **Dipole interactions:** The interactions between two molecules with permanent dipole moment due to attractive electrostatic forces, by which the positive end of one molecule is attracted by the negative end of another molecule. Both components have high permanent dipole moments.
- **Hydrogen bonding interactions:** Interactions between the hydrogen atom that is part of a polar bond (hydrogen-bonding donor) and an electronegative atom with a lone pair of electrons such as O and N (hydrogen-bonding acceptor). Hydrogen bonding interaction is another type of dipole interaction.

As discussed in Section 1.2.1, **partition ratio** (*P*) of a given analyte in two immiscible phases can be expressed as follows: $P = [C_{\text{Organic}}]/[C_{\text{Aqueous}}]$, where $[C_{\text{Organic}}]$ is the concentration of the un-ionized analyte in the organic phase (e.g., *n*-octanol) while $[C_{\text{Aqueous}}]$ is the concentration of the un-ionized analyte in the aqueous phase (e.g., water) at equilibrium.

The *P* value is constant for a given analyte under given conditions of the two-phase (aqueous vs. immiscible organic solvent) system. In LLE, equilibrium of the analyte(s) of interest between the two immiscible phases is normally established at the end of the extraction. The higher the *P* value of the analyte(s) of interest, the more efficiently it is extracted from the aqueous samples (e.g. plasma) into the organic solvent.

For neutral analyte(s), the REC can be maximized by optimizing the type of extraction solvent. The pH value of the sample matrix does not affect the extraction efficiency. However, for acidic or basic analyte(s), the situation is different. Both acidic and basic molecules tend to dissociate in aqueous solution. The degree of dissociation (or ionization) is largely dependent on the pH of the solution.

For acidic analyte(s) ($HA \leftrightarrow H^+ + A^-$), as only the uncharged fraction of the molecules can be extracted from the aqueous phase into the organic phase whereas the charged species will remain in the aqueous phase during LLE, the **distribution ratio** (*D*) can be re-defined according to the following equation:

$$D = \frac{P \times [H^+]}{[[H^+] + K_a]}$$

where K_a is the dissociation constant ($K_a = [H^+] \times [A^-]/[HA]$) and *P* is the partition ratio as discussed above. Apparently, the ionization of an acidic analyte increases with decreasing $[H^+]$ and increasing pH in the solution.

Similarly, the following equation can be applied for extraction of a basic analyte ($A + H^+ \leftrightarrow AH^+$), which partly dissociates (ionizes) in the aqueous sample:

$$D = \frac{P \times K_a}{[[H^+] + K_a]}$$

where K_a is the dissociation constant for the corresponding acid AH^+. Apparently, the ionization of basic analyte(s) increases with increasing $[H^+]$ and decreasing pH in the aqueous solution.

In summary, acidic analyte(s) are charged in a basic solution. Their charged form is generally soluble in water (aqueous samples) and, therefore, they will mainly remain in the basified aqueous samples. These molecules are less charged or become neutral in an acidic solution and, therefore, they can be easily extracted from acidified aqueous samples into organic solvent. In contrast, basic analyte(s) are charged in an acidic solution. Their charged form is generally soluble in water (aqueous samples) and, therefore, they will mainly remain in the acidified aqueous samples. These molecules are less charged or become neutral in basic solution and, therefore, they can be easily extracted from basified samples into organic solvent. Taking the above into consideration, biological samples have to be acidified or basified for good REC in LLE for acidic or basic analyte(s).

REC is an important indicator in demonstrating the efficiency of a LLE method. As discussed above, under optimal pH conditions (i.e. low pH for acids and high pH for bases), the partition ratio of the analyte(s) of interest can be used to estimate its REC. For a given molecule to be extracted under optimal pH conditions (at which point it is in nearly 100% neutral form) from a fixed volume of biological sample (V_{Aqueous}) and into a fixed volume of organic solvent (V_{Organic}), the REC can be expressed by the following equation:

$$REC = 100 \times \frac{[P \times V_{\text{Organic}}]}{[P \times V_{\text{Organic}} + V_{\text{Aqueous}}]}.$$

Clearly, the REC in LLE can be increased by increasing the volume of extraction solvent and by selecting proper extraction solvent with a higher partition ratio (*P*) for the analyte(s) of interest. In practice, both the volume of extraction solvent and solvent selection have to be considered concurrently when developing a LLE method for sample extraction.

The partition ratio varies from a compound to the other in a given mixture of two immiscible solvents (i.e. aqueous vs. organic). In addition, it is also dependent on the experimental conditions, including temperature and matrix components. Although such information may not be readily available for the analyte(s) of interest in the intended organic solvent, the log *P*-value of the analyte(s) of interest should be considered as a reference for the possible extraction efficiency assessment in LLE. Theoretically, a compound with a higher log *P*-value can be more easily extracted by LLE than a compound with a lower log *P*-value because of the high partition ratio of the former. As a general rule, compounds with log *P*-values below 0 are very polar and difficult to extract with any organic solvent in LLE. Compounds with log *P*-values between 0 and 1 are relatively polar and not suitable for LLE. Compounds with log *P*-values above 1 are more hydrophobic, and they can be easily extracted with organic solvent in LLE (Li and Bartlett 2014; Liu and Aubry 2013; Pedersen-Bjergaard et al. 2015).

1.4.2.2 Solvent in LLE

The considerations of an organic solvent for LLE should include but are not limited to polarity (solubility/distribution for the analyte(s) of interest), density, viscosity, and water solubility. Listed in Table 1.2 are the physicochemical parameters of some organic solvents commonly used in LLE.

Generally, for a solvent selected for extraction, its solubility in water should be low to avoid substantial amount of the solvent being dissolved into the aqueous sample. In other words, the organic solvent selected for LLE has to be immiscible with the aqueous samples, so that a sharp interface can be formed between the organic phase and the aqueous phase at the equilibrium at the end of LLE. Only when the two phases are physically separated can one of them be removed to achieve separation of the analyte(s) of interest from the unwanted matrix components. As can be seen from the table, all the solvents have positive partition coefficient (log *P*) values, meaning that they are all nonpolar and immiscible with water (or have very low solubility in water).

The volatility and density of the solvent are also important factors to consider in LLE. Solvent with a lower boiling point is preferred because this makes evaporation of solvent after extraction much faster than the one with a higher boiling point. Solvents with a lower density than water float on top of the aqueous phase in LLE. This makes it easier to transfer the organic layer by either pipetting or via freezing the aqueous phase followed by pouring into new tubes/wells. Examples of solvents with density less than water are *n*-hexane, methyl *tert*-butyl ether, *n*-butanol, butyl chloride, and ethyl acetate (Table 1.2). In contrast, when solvent with density higher than water is used, the aqueous phase floats on top of the organic phase in LLE. This makes it difficult in sample manipulation. Examples of solvents with density higher than water are chloroform and dichloromethane (Table 1.2).

In addition to the above general rules, an important factor in solvent selection in LLE is the target analyte itself. Listed below are some general rules to consider when choosing solvent for LLE for the analyte(s) of interest (Li and Bartlett 2014; Liu and Aubry 2013; Pedersen-Bjergaard et al. 2015):

- **Analyte(s) with no or very little functional groups (highly hydrophobic):** The primary interaction between the analyte(s) and solvent appears to be hydrophobic interactions. Therefore, the best solvents should be very nonpolar solvents, such as *n*-hexane.
- **Analyte(s) with low to medium polarity (hydrophobic to slightly hydrophobic):** The primary interaction between the analyte(s) and solvent appears to be dispersion interactions and/or dipole interactions mixed with hydrophobic interactions and/or hydrogen bonding interactions. A slightly polar solvent should

Table 1.2 Commonly used organic solvents in LLE and their physicochemical parameters.

Solvent	Log *P*	Polarity index	Water solubility (w/w, %)	Boiling point (°C)	Density (g ml^{-1})	Viscosity (cp 20 °C)	Dipole moment
n-Hexane	3.9	0.1	0.014	68.7	0.655	0.31	0.08
Chloroform	2.3	4.1	0.795	61.2	1.483	0.58	1.15
Dichloromethane	1.5	3.1	0.132	39.8	1.325	0.45	1.14
Methyl *tert*-butyl ether	0.9	2.5	5	53–56	0.735	0.27	1.66
Diethyl ether	0.98	2.8	7.5	34.5	0.713	0.24	1.15
Heptane	4.274	0.1	0.01	98.4	0.684	0.42	0
Ethyl acetate	0.7	4.4	8.7	77.1	0.897	0.46	1.88
n-Butanol	0.839	4.0	6.3	117.7	0.810	2.95	1.66
n-Butyl chloride	2.39	4.0	0.05	78.5	0.88	0.0045	1.90

be considered, but the solvent should not be too polar. For slight to medium polar analyte(s), solvents with high dipole moments, such as dichloromethane, methyl-*tert* butyl ether, and ethyl acetate, are efficient in extraction of these molecules from aqueous phase. For extraction of basic analyte(s) containing nitrogen, solvents with hydrogen-bonding donor properties, e.g. chloroform, may be employed. However, the higher density of the solvent than water makes it less favorable. Other solvents like butyl chloride or methyl-*tert* butyl ether can be used as alternatives. For acidic analyte(s) with high hydrogen-bonding donor properties, hydrogen-bonding acceptor solvents, such as methyl-*tert* butyl ether, should preferably be considered.

- **Analyte(s) with medium to high polarity (slightly hydrophilic to hydrophilic):** In general, LLE is not recommended unless the analyte(s) of interest can be extracted after derivatization (Zeng and Cao 2018) or under modified extraction conditions. For example, tiotropium is a quaternary amine, which is extremely polar and positively self-charged. However, tiotropium is in the form of halogenated ion-pair salt, i.e. tiotropium bromide. Dichloromethane was found to be the solvent for selective extraction of tiotropium bromide from biological samples. It was also found that adding potassium iodide to the sample mixture until its saturation can facilitate LLE. Furthermore, the added potassium iodine increases the density of aqueous layer, causing the dichloromethane layer to stay in the top, which is convenient for solvent transfer (Chi et al. 2016).

Although LLE can be achieved by using a single solvent, the use of solvent mixtures has been often. By mixing solvents, the polarity of the extraction solvent can be modified for an improved REC with minimized matrix effect. Typical examples are the addition of alcohols (such as 1-propanol or 1-butanol) or acetonitrile to relatively nonpolar solvents like methyl-*tert* butyl ether (Xue et al. 2004), addition of isopropanol to *n*-hexane (Jian et al. 2010a), a mixture of diethyl ether and dichloromethane (Yin et al. 2015; Zhang et al. 2009a), a mixture of ethyl acetate and *n*-hexane (Jian et al. 2013), and a mixture of ethyl acetate, heptane, and dichloromethane (Oiestad et al. 2009).

1.4.2.3 General Procedures in LLE

Similar to PPT, LLE of biological samples involves three major steps: (i) addition of additives, organic solvents, and/or IS, (ii) mixing of the two immiscible phases for extraction, and (iii) separation of the two phases followed by transfer, evaporation, and reconstitution (Jian et al. 2010a).

1.4.2.3.1 Addition of Solvents, Additive(S), and/or Internal Standard

A general rule in LLE is to use a minimum two volumes of a preferred/selected organic solvent for extraction. In some cases, a mixture of two solvents is employed for better REC (Xue et al. 2006a) as discussed previously. Prior to this step, the aqueous biological sample needs to be treated with a certain amount of an acid (e.g. acetic acid, formic acid) (Cheruvu et al. 2018), a base (e.g. ammonium hydroxide) (Li et al. 2007a, 2007c, 2013a), or a buffer to adjust the pH of the samples (Li et al. 2006). As mentioned above, pH adjustment of the aqueous sample prior to LLE is important for acidic and basic analytes. A general recommendation is to adjust the pH value in the sample at least two units from the pK_a-value of the analyte(s) of interest. Thus, acidic analyte(s) should be preferably extracted from samples that are acidified to a pH value of two units below the pK_a-value of the analyte(s). For basic compounds, the pH in the sample should be adjusted two units higher than the pK_a-value of the analyte(s). However, under strongly acidic or alkaline conditions, certain analyte(s) might be subjected to degradation. In addition, precipitation of matrix components may occur in biological fluids under extreme pH conditions. Therefore, care should be taken when using extreme pH values in LLE.

1.4.2.3.2 Mixing

After addition of additives, organic solvents, and/or IS, the sample mixture needs to be mixed thoroughly to ensure the highest and consistent extraction efficiency. LLE is a thermokinetic process that involves diffusion of the analyte(s) of interest from the aqueous phase to the organic phase, which may take a certain period of time. Therefore, under a fixed LLE condition such as solvents, additives, pH, and volumes, the highest extraction efficiency can only be achieved after the distribution equilibrium of the analyte(s) of interest between the two immiscible phases has been reached, i.e. when the amount of analyte crossing the interphase does not increase any more. To facilitate the diffusion of the analyte(s) of interest to reach its equilibrium between the two phases, vortex-mixing is usually applied. Since the aqueous biological samples are immiscible with the organic solvents, turbulence due to high-speed vortex-mixing creates very small droplets of both phases. The formation of the droplets significantly increases the interface area between the two phases, thus facilitating the diffusion of the analyte(s) of interest across from one phase to the other, thus shortening the time that is needed to reach equilibrium. In general, a 5- to 10-minute of vortex-mixing is typically needed for efficient LLE of biological samples by organic solvents.

1.4.2.3.3 Phase Separation

After the distribution of the analyte(s) of interest has reached equilibrium between the two phases, separation of the organic and aqueous phases has to be carried out in order to isolate the analyte(s) of interest from those unwanted in the sample matrix or remove the impurities of interest from those in the remaining phase for further processing. Centrifugation is the most common technique for phase separation in LLE. In general, a 5- to 10-minute of centrifugation at ~2000 × *g* is sufficient to achieve sharp phase separation. In some cases, an extended centrifugation at a higher speed and/or longer centrifugation time might be necessary. The relative position of organic vs. aqueous phase is dictated by the density of both phases. Most commonly used organic solvents, such as *n*-hexane, methyl-*tert* butyl ether, *n*-butanol, butyl chloride, and ethyl acetate, have densities lower than that of water. They will become supernatant after phase separation. In contrast, chloroform and dichloromethane have densities higher than that of water and they stay in the bottom layer after centrifugation. However, in some cases, by adding salt(s), such as potassium iodide to the sample mixture until its saturation can significantly increase the density of aqueous layer, causing the dichloromethane layer to stay in the top, leading to a convenient solvent transfer (Chi et al. 2016).

Upon centrifugation, phase separation can be easily fulfilled by transferring the wanted phase or removing the unwanted phase. However, care has to be taken not to disturb the interphase between the two phases. In the case where the top layer is organic, the tubes/wells can be frozen in acetone containing dry ice. When the bottom aqueous layer becomes frozen, the top organic layer can be easily poured into new tubes or transferred to new wells. However, when the analyte(s) of interest stay in the bottom layer containing chloroform or dichloromethane, phase separation might be a challenge unless the density of the aqueous layer can be manipulated by adding salts but without interfering the outcome of the LLE (Chi et al. 2016).

1.4.2.3.4 Direct Analysis or Evaporation of Sample Extract Followed by Reconstitution Prior to LC-MS Analysis

With the organic and aqueous phases separated from each other, the extraction of the analyte(s) of interest from the aqueous samples can be considered finished. If the analyte(s) of interest are present in the top organic phase after LLE, upon transfer, direct LC-MS analysis using a HILIC column may be feasible as the solvent used for LLE is generally compatible with acetonitrile which is a common starting mobile phase in HILIC-MS bioanalysis (Jian et al. 2010b). However, assay sensitivity may be a challenge if the evaporation and reconstitution is not applied (Song and Naidong 2006; Xue et al. 2006a).

On the other hand, if the analyte(s) of interest stay in the top layer of the aqueous phase when an organic solvent such as chloroform or dichloromethane is employed in LLE to remove unwanted matrix components, the analyte-containing aqueous phase can be directly subjected to LC-MS analysis using RP column. Again, assay sensitivity may be a challenge in direct LC-MS bioanalysis. If needed, evaporation and reconstitution should be applied to the aqueous sample extracts to achieve the needed assay sensitivity (Onorato et al. 2010).

In most applications, direct RPLC-MS bioanalysis of the organic extract obtained from LLE is not feasible. This is because organic solvents are strong eluents in reversed-phase liquid chromatography and their presence in the sample extract to be injected will dramatically weaken the analyte retention on the column. Furthermore, analyte concentrations in the final extract are generally lower than those in the original samples as a large volume of organic solvents, e.g. more than twofold of the aqueous sample size, are usually used in LLE. Without proper concentration of the analyte(s) of interest, the needed assay sensitivity may not be achievable. Therefore, evaporation of organic solvent of the resulting supernatant and reconstitution of the resulting sample residues are commonly employed in LC-MS bioanalysis following LLE. In this step, the organic solvent in the sample extract is evaporated for complete dryness under a steam of nitrogen or by vacuumed centrifugation. The resulting sample residues are reconstituted in the reconstitution solution, which can be the starting mobile phase or a solution with polarity and pH similar to the starting mobile phase. The volume of the reconstitution solution added should be adjusted based on the needed assay sensitivity and the available instrument platform. Analyte enrichment may be achieved by decreasing the volume of the reconstitution solution in this step. Upon vortex-mixing, followed by centrifugation to push the possible insoluble residues to the bottom of the assay vial/well, the reconstituted sample extracts can be subjected to LC-MS analysis.

1.4.2.4 Application of LLE in LC-MS Bioanalysis

In general, there are three types of application of LLE for sample preparation in LC-MS bioanalysis, i.e. (i) forward extraction, (ii) backward extraction (or back extraction), and (iii) double extraction (or two-stage forward extraction). The forward extraction and backward extraction may be alternatively staged in a complex LLE procedure for higher assay selectivity or specificity for the analyte(s) of interest in the intended sample matrix. Different from the alternatively staged forward and backward extraction, double extraction is to carry out LLE for the same aqueous sample matrix initially with highly nonpolar solvent (e.g. hexane) and then with moderately nonpolar solvent (e.g. methyl-*tert* butyl ether or ethyl acetate) for an improved assay selectivity and/or minimized matrix effect. With the recent developments in robotic equipment and extraction plates, most of the LLE procedures can be automated.

1.4.2.4.1 Forward Extraction

The **forward extraction** generally refers to a LLE procedure by which the analyte(s) of interest stay in the organic layer after extraction. After discarding the aqueous phase, the resulting organic phase is subjected to direct HILIC-MS analysis or further processing (evaporation and reconstitution) as discussed above prior to LC-MS analysis. This approach has been widely used in many applications for nonpolar analyte(s).

A representative LLE protocol (for the analysis of NVP-1 by Li et al. 2013b)

- An aliquot of 100 μl of blank plasma (for matrix blanks and zero samples), calibration standard, QC sample, and study sample was added to the appropriate well of a 96-well assay plate.
- A 25 μl aliquot of the IS working solution in 50% aqueous methanol (v/v) was added to each well except to the matrix blanks, where a 25 μl aliquot of 50% aqueous methanol (v/v) was added.
- A 100 μl aliquot of a sodium carbonate (100 mM) solution in water was added to each individual well, and the plate was vortex-mixed for about 0.5 minute on a pulse-vortex-mixer with a motor speed setting of ~65 units.
- A 500 μl aliquot of methyl *tert*-butyl ether was added to each well, and the plate was covered and vortex-mixed for ~10 minutes using the same vortex-mixing setting as above.
- The sample plate was centrifuged at ~4000 × *g* for ~10 minutes at 4 °C. The supernatant (400 μl) from each well was transferred via a TomTec Quadra 96 system (Hamden, CT, USA) to the corresponding well in a 1 ml, 96-well plate. This was followed by evaporation of the supernatants to dryness under a stream of nitrogen at 45 °C. The sample residues were reconstituted with a 100 μl volume of reconstitution solution (50% aqueous methanol, v/v).
- After a brief vortex-mixing and centrifugation, a 10 μl volume of the reconstituted sample extracts was injected onto the LC-MS/MS system.

1.4.2.4.2 Backward Extraction

In situations that require higher assay selectivity or specificity for the analyte(s) of interest in the intended sample matrix, a single forward extraction may not be sufficient as interfering matrix compounds can also be extracted into the organic solvent. In such cases, it may be necessary to perform a backward extraction, by which the organic extract from the first extraction (done by forward extraction) is extracted with a new aqueous solvent with suitable pH. Although the use of backward extraction is less common as compared to forward extraction, it can be very useful and practical in leaving the unwanted hydrophobic interference molecules in organic layer while extracting the analyte(s) of interest into the aqueous phase for LC-MS analysis (Bolden et al. 2002; Constanzer et al. 1997) or extracting the interfering polar compounds into the aqueous phase while leaving the analyte(s) of interest in the organic phase for further processing prior to LC-MS analysis (Buhrman et al. 1996).

In a study for the analysis of dextromethorphan (DEX) and dextrorphan (DOR) using LC-MS, Bolden et al. (2002) reported a semiautomated backward extraction method. The procedure started by extracting DEX and DOR using diethyl ether in a 96-well rack of plastic tubes via the Microlab liquid handling system. Following extraction, the tubes in the 96-well rack were placed in a dry ice–acetone bath to freeze the plasma layer. After the plasma layer was frozen, a portion of the ether layer (400 μl out of 600 μl) was transferred (using the Microlab) to a 96-well rack of clean polypropylene tubes that already contained 200 μl of 1% formic acid. The DEX and DOR in the samples were back-extracted into the acidified water. Finally, the Microlab was used to transfer a portion (150 μl) of the acidified extract to clean autosampler vials prior to LC-MS analysis (Bolden et al. 2002).

In a separate study published by Buhrman et al. (1996), a combination of multiple steps of forward and backward extraction was developed and validated for the analysis of SR27417, a platelet-activating factor receptor antagonist, in human plasma. By this procedure, 1.5 ml of 0.1 M sodium carbonate and 7 ml of ethyl acetate were added to a 1 ml plasma sample, which was followed by rotary extraction and centrifugation. The resulting organic layer was mixed with 1 ml of 0.025 M sulfuric acid, followed by rotary extraction and centrifugation. The organic layer was removed and discarded. To the aqueous layer, 1.5 ml of 0.1 M sodium carbonate and 5 ml of ethyl acetate were added, followed by rotary extraction and centrifugation. The resulting organic layer was collected and evaporated. The resulting residues were reconstituted in 200 μl of 50/50 acetonitrile/water prior to LC-MS analysis. The final ethyl acetate extract, which contained the fewest co-eluting components, had the highest ion intensity for the analyte, with curve range of 5–2000 pg ml^{-1} and accuracy ranging from −11.6 to 2.61% of the nominal concentrations. In contrast, the liquid–liquid method using hexane alone had poor accuracy and precision below 20 pg ml^{-1} (Buhrman et al. 1996).

A representative backward LLE protocol (for the analysis of dorzolamide and its de-ethylated metabolite by Constanzer et al. 1997).

- To 1.0 ml of samples, a 0.5 ml of 10% TCA (for PPT) was added, followed by vortex-mixing for one minute.
- The samples were then buffered to pH 8.0 with 5 ml of 0.2 M phosphate buffer followed by LLE with 10 ml of ethyl acetate : toluene : isopropanol at 50/40/10 (v/v/v).
- The tubes were capped, rotate-mixed for 20 minutes, and centrifuged for 5 minutes.
- The organic layer was transferred to a 15-ml centrifuge tube containing 0.2 ml of phosphoric acid, followed by rotate-mixing for 15 minutes and centrifugation for 5 minutes.
- The organic layer was aspirated and discarded.
- The acidic layer containing the analyte(s) was additionally washed with 2 ml of hexane, followed by rotate-mixing for 10 minutes and centrifugation for 5 minutes. The organic layer was aspirated and discarded.
- 50 μl of the acidic layer was injected onto the LC-MS/MS system.

1.4.2.4.3 Double Extraction

This approach is often referred to as "two-stage forward extraction." In the first stage of extraction, a very nonpolar solvent, e.g. hexane, is employed to extract hydrophobic interfering compounds from the sample matrix and the extract is discarded. This is followed by extraction using a moderately hydrophobic solvent, e.g. acetyl acetate, to extract the analyte(s) of interest. In a study published by Hou et al. (2012), a double LLE method was employed to prepare brain tissue homogenate samples for the quantitative determination of a novel anti-Parkinson's disease candidate drug FLZ. Endogenous hydrophobic interferences were initially extracted by *n*-hexane, while the analyte remained in the sample matrix. With the hexane layer removed and discarded, the aqueous phase was further extracted by using ethyl acetate, yielding a clean sample extract with high purity and low matrix effects (Hou et al. 2012).

A typical example of double LLE protocol (for the analysis of an anti-Parkinson's disease drug candidate drug FLZ by Hou et al. 2012).

- To 0.5 ml of the brain homogenate (tissue : methanol 1 : 5) were added 20 μl of the IS working solution.
- The mixture was vortex-mixed for 20 seconds and evaporated to dryness with a gentle stream of nitrogen gas at 36 °C and the sample was reconstituted with 100 μl of 80% methanol.
- Then, 400 μl of water was added to the reconstitution (to facilitate LLE).
- After adding 1.5 ml of *n*-hexane, the mixture was vortex-mixed for 90 seconds and centrifuged at 1721 × *g* for 10 minutes, then the *n*-hexane layer was removed and discarded.
- The remainder was extracted with 1.5 ml ethyl acetate by vortex-mixing for 90 seconds. After centrifuging at 1721 × *g* for 10 minutes, the organic layer was transferred into another test tube and evaporated to dryness again.
- The residue was reconstituted in 50 μl 80% methanol by the ultrasonic method for 180 seconds and centrifuged at 16 654 × *g* for 10 minutes.
- Finally, 5 μl supernatant was injected onto the LC-MS/MS system.

As a sample preparation method with moderate selectivity, LLE is ideal for nonpolar to moderately polar analyte(s) with favorable distribution in water-immiscible organic solvent. However, when developing a bioanalytical assay for more than one analyte, particularly a drug plus its metabolites, a fine-tuning of the LLE conditions may be difficult to achieve acceptable recoveries for both molecules as the metabolite tends to be more polar than the parent molecule. In this case, other sample preparation procedures such as PPT or SPE should be considered from the very beginning of the assay development.

Depending on the assay needs, efforts should be made to balance REC and selectivity (matrix background removal) in LLE. When matrix interference is a concern, REC may be sacrificed for selectivity. In this case, use of a stable isotope labeled IS should be preferably used to compensate for the variation in sample preparation under nonoptimal REC. Nevertheless, efforts should be made to ensure a consistent REC across the low, middle, and high concentration levels.

1.4.2.5 Other Formats of LLE

There are two major alternatives of LLE, i.e. salting-out assisted LLE (SALLE) and supported LE (SLE) that have been commonly employed in LC-MS bioanalysis (Tang and Weng 2013; Valente and Rodrigues 2015).

SALLE employs water-miscible organic solvent (such as acetonitrile) and high concentration of salts, such as magnesium sulfate, potassium carbonate, sodium chloride, or ammonium acetate, to induce phase separation of the water-miscible organic phase (Zhang et al. 2009b). Compared to conventional LLE, SALLE has a broader application, covering molecules from low to high lipophilicity with a better analyte recovery. One apparent drawback of SALLE is the higher matrix effect as compared to conventional LLE. This is because SALLE extracts tend to contain more endogenous compounds. More details of SALLE can be found in Chapter 5.

SLE is an alternative LLE that is carried out on the high-surface interface of an inert solid support in a 96-well format. Briefly, biological samples are mixed with aqueous buffer and then loaded onto the solid support. Subsequently, water-immiscible solvent is added, passing through the solid support to extract the analyte(s) of interest. The extraction mechanism involved is mainly a partition of the analyte(s) between the organic solvent and the absorbed aqueous phase on the solid support. The advantages of SLE include no emulsion formation, easy automation, and high extraction efficiency (Wu et al. 2010). More details of SLE can be found in Chapter 6.

1.4.3 Solid-phase Extraction (SPE)

SPE is a powerful sample preparation technique that has been used for decades for the selective extraction and enrichment of trace analyte(s) of interest in various biological samples. The mechanism of SPE is similar to that of liquid chromatography, which is based on the affinity or interaction between solutes (analyte of interest) dissolved in a liquid (mobile phase) and sorbent materials (stationary phase). Due to the difference in physicochemical properties, different components in the liquid sample have different affinities or interactions with the sorbents in the stationary phase of the SPE device. By SPE, an liquid sample, after a proper treatment (dilution, pH adjustment, and/or addition of the IS), is loaded onto preconditioned/equilibrated column/cartridge/plate packed with appropriate sorbent materials; the analyte(s) of interest is retained by interacting with the sorbent materials (stationary phase) through different interaction mechanisms; the interfering matrix components either directly pass through the column/cartridge/plate during the loading step or are washed away during the wash steps with a proper solvent; the analyte(s) of interest is subsequently eluted from the column/cartridge/plate with a suitable elution solvent. The collected eluate is either directly subjected to LC-MS analysis or subjected to drying process to evaporate organic solvent. This is followed by reconstitution or starting mobile phase with a proper solvent prior to LC-MS analysis (Li and Bartlett 2014; Liu and Aubry 2013; Pedersen-Bjergaard et al. 2015).

With SPE, some issues associated with PPT and/or LLE can be prevented or minimized. These issues include but are not limited to (i) matrix effect in PPT and (ii) poor recovery due to incomplete phase separations in LLE. SPE products are commercially available from a variety of vendors in a wide variety of chemistries, sorbents, sizes, and formats to accommodate different sample sizes and bioanalytical applications.

1.4.3.1 SPE Stationary Phases (Sorbents)

SPE sorbent materials are commonly irregular-shaped rigid particles, with nominal sizes ranging from 8 to 70 μm (most commonly 40–60 μm), which allows for reasonable flow rates of the samples or solutions through the sorbent bed during loading, washing, and elution. Smaller particles of the SPE sorbent materials require higher pressure for processing. The exception is for the sorbent materials in plate formats. Because of the very short bed paths of such formats, the total resistance due to the small particle sizes is less than those of the typical sorbent bed formatted into column or cartridge.

Most SPE materials are fully porous in nature. The smaller porosity of the sorbent materials is associated with a higher total surface area and, consequently, a higher capacity per mass of the sorbent materials for active adsorption. Typically, the capacity of a given sorbent material in SPE is about a few percent (%) retained compounds per mass of sorbent (for example, 5 mg of compounds are retained by 100 mg of sorbents).

SPE sorbents can be broadly divided into two major classes, i.e. silica-based sorbents and polymer-based sorbents.

- **Silica-based sorbents:** By definition, silica-based sorbents were made by bonding different functional groups to silica, although base silica also plays an active role in some SPE workflows. Many commercially available silica-based sorbents are named using acronyms. Those acronyms describe the primary character of the respective functional group on the silica, including C18, C8, C6, C4, C2, phenyl, cyclohexyl, cyanopropyl, aminopropyl, diethyl amine, diol, propylsulfonic acid, phenylsulfonic acid, propyl carboxylic acid, and propyltrimethyl amine. C18-based SPE columns/cartridges/plates are very popular and most hydrophobic. Because of their strong hydrophobicity, they give strong retention to many compounds. However, it may not give the best selectivity. In order to obtain a better selectivity in SPE, one can use SPE column/cartridge/plate with less hydrophobic phases, such as C8-columns, which are also very popular.

Regardless of which bonding chemistry is used in making silica-based SPE sorbents, the number of unreacted (or residual) silanol species that remain on the sorbent surface is still high (>50% for a typical C18-bonded phase) due to the steric factors involved in binding the functional groups. These silanol species are capable of interacting with analyte molecules, often leading to unexpected retention effects and poor extraction efficiency. It is also well known that at pH ~4 of a solution, about 50% of the silanol species on the silica surface are ionized. Therefore, the solution pH can be adjusted to ensure ion suppression ($pH \leq 2$) or full ionization ($pH \geq 6$) of the silanol species. It should be noted that silica sorbents may be subjected to hydrolysis when exposed to extreme pH for an extended period of time. At $pH \leq 2$, the bonded surface species are susceptible to hydrolysis and efficiency of the sorbent can be drastically reduced. At $pH \geq 8$, the silica itself is susceptible to hydrolysis and the sorbent will rapidly deteriorate upon extended exposure (Li and Bartlett 2014; Liu and Aubry 2013; Pedersen-Bjergaard et al. 2015).

- **Polymer-based sorbents:** Various polymer-based sorbents are commercially available, covering a wide range of polarities and chemistries. The most common polymer sorbents are based on styrene–divinyl benzene copolymers. Further modifications via amination or sulfonation help create ion-exchange polymer sorbents. Incorporation of some polar functional groups makes the sorbents water-wettable, which offers additional possibilities for retention mechanisms. An important advantage of the polymer-based stationary phases is that their chemical properties do not change even if they dry out during the SPE procedure, and the initial conditioning step used for silica-based SPE column/cartridge/plate may not be required. Furthermore, compared to silica-based sorbents, most polymeric sorbents are stable over the entire pH range (Li and Bartlett 2014; Liu and Aubry 2013; Pedersen-Bjergaard et al. 2015).

1.4.3.2 Common SPE Platforms in LC-MS Bioanalysis

According to respective retention mechanisms, SPE commonly used in LC-MS bioanalysis can be classified into three categories, i.e. reversed-phase SPE, ion-exchange SPE, and mixed-mode SPE. Normal-phase SPE is designed to extract analyte(s) from organic extracts, very nonpolar solvents, and fatty oils, etc. It is not included in the current discussion as it is rarely applied in sample extraction for biological matrix such as plasma and urine.

- **Reversed-phase SPE:** It employs nonpolar stationary phases of the silica- or polymer-based sorbents, which retain most molecules with any hydrophobic character. The common functional groups of silica-based nonpolar sorbents include but are not limited to C18, C8, C6, C4, C2, C1, phenyl, cyclohexyl, and cyanopropyl. Among these groups, the short-chain one (e.g. C1, C2) is less retentive, but the long-chain one (e.g. C18) is highly retentive. Reversed-phase SPE is considered the least selective retention mechanism when compared to other type SPEs. Because of this, it is very useful for extracting analytes that are very diverse in structure within the same sample.
- **Ion-exchange SPE:** It utilizes ionic functional groups (strong or weak organic acids and bases bonded to the supporting base) of the sorbents and can be further classified as strong/weak cation/anion ion-exchange SPE:
  - **Strong cation exchange (SCX) SPE:** The sorbents typically contain sulfonic acids as the ionic groups for ion-exchange. The sulfonic acids have very low pK_a values and are negatively charged almost in the entire pH range.
  - **Weak cation exchange (WCX) SPE:** The sorbents often contain carboxylic acids as the ionic groups. The carboxylic acids are weak acids with pK_a values around 4.5–5.0 and are ionized only in parts of the pH range. If pH in the SPE column/cartridge/plate is adjusted to above pH 3, the ion exchanger is gradually turned "on." In contrast, if pH is reduced well below pH 3, the ion exchanger is turned "off."
  - **Strong anion exchange (SAX) SPE:** The sorbents typically contain quaternary amine moieties as the functional group. The quaternary amines are positively charged in the entire pH range.
  - **Weak anion exchange (WAX) SPE:** The sorbents often contain secondary or tertiary amines as the functional groups for ion-exchange. The secondary or tertiary amines are weak bases with pK_a values of ~10. Therefore, by adjusting $pH > 10$, the ion-exchangers are gradually turned "off." In contrast, if pH is adjusted to well below pH 10, i.e. pH 8 or lower, the ion-exchangers is turned "on."
- **Mixed-mode SPE**: Mixed-mode SPE is an extraction approach involving sorbents that exhibit two or more primary interactions for retaining the analyte(s) of interest. Commercially available mixed-mode sorbents can be silica- or polymer-based, and are typically produced by either bonding the sorbents concurrently with two different functional groups (e.g. C2, C8 vs. sulfate) or by blending discrete sorbent chemistries in appropriate ratios to create the combination of retention properties. The most commonly used mixed-mode sorbents have a hydrophobic functional group in combination with an ion-exchange functional group. The hydrophobic groups can be short chain (e.g. C2) which is less retentive or long chain (e.g. C18)

which is highly retentive. Ion-exchangers can be either cation- or anion-oriented. Mixed-mode SPE allows for compound retention by both the ionic and hydrophobic interactions.

1.4.3.3 General SPE Workflows

Technically, SPE is a form of low-pressure chromatography that can be carried out either off-line or online with the LC-MS system. When developing a SPE-based sample preparation method for LC-MS bioanalysis, in addition to physicochemical properties of the analyte(s) of interest, parameters that need to be considered include (i) the type and amount of sorbent, (ii) the sample volume (considering the required volume for initial analysis, repeat analysis, and incurred sample reanalysis), and (iii) loading, washing, and elution conditions (time, volume, and composition).

Similar to LC columns, a variety of SPE products are commercially available in various formats. These include but are not limited to single-use cartridge, multiple-well plate (96-well, 384-well), and online SPE column from some major vendors, including Waters (www.waters.com), ThermoFisher Scientific (www.thermofisher.com), Agilent (www.agilent.com), Biotage (www.biotage.com), Phenomenex (www.phenomenex.com), Sigma-Aldrich (https://www.sigma-aldrich.com). In general, each specific SPE product is supplied with a recommended procedure or method protocol from the vendor. These procedures or protocols should be considered a starting point to method development of the intended SPE method for the analyte(s) of interest. The overview, detailed mechanisms, and associated workflows of various SPE in sample preparation can also be found from CHROMacademy website (www.chromacademy.com).

1.4.3.3.1 Reversed-phase SPE

Reversed-phase SPE is a common approach for extracting relative nonpolar analyte(s) from aqueous sample solutions.

- **Mechanism of extraction:** Hydrophobic interactions between the analyte and the stationary phase. Retention of the analyte(s) of interest increases with increasing hydrophobicity of the analyte via van der Waals forces, which is primarily between hydrogen–carbon bonds in the stationary phase and hydrogen–carbon bonds in the analyte molecules. This interaction is promoted in the aqueous phase but suppressed or disrupted in the organic phase.
- **Applicable analyte(s):** Compounds with $\log P$ values >1.0, most commercially available silica-based or polymer-based reversed-phase SPE can be considered. However, for compounds with $\log P$ values ranging from −1.0 to 1.0, modified polymer-based reversed-phase SPE should be preferably considered as the polar modifications on polymer-based SPE sorbent materials allow for increased polar selectivity as compared to the silica-based C18 or C8 materials; for compounds with $\log P$ values <−0.1, their retention on reversed-phase SPE is expected to be poor, reversed-phase SPE is generally not recommended.
- **Sample pretreatment:** Samples need to be diluted with a proper buffer or water (for neutral compound). For ionizable compound, pH adjustment is important as the ionization state of the analyte(s) of interest can drastically change its retention and elution characteristics on a given RP SPE sorbent. A general rule is to adjust the sample pH to 2 pH units above (for basic compound) or below (for acidic compound) the pK_a value of the compound to ensure the compound is neutralized prior to being loaded onto SPE column/cartridge/plate. In addition to commonly used buffers, some common reagents, such as 2% formic, 2% TCA, 2% acetic acid, TFA, phosphoric acid, or ammonium hydroxide, $NaHCO_3$, Na_2CO_3, can be employed for sample treatment.
- **Sorbent conditioning and equilibration:** Sorbent conditioning is to "activate" or "wet" the SPE sorbent to ensure consistent interaction between the analyte(s) of interest and the functional groups of the SPE sorbents. The most common conditioning method involves application of one to two volumes of water-miscible organic solvent (typically methanol or acetonitrile) to reversed-phase sorbents. This is followed by equilibration of the SPE sorbent by introducing a solution similar to the sample loading in terms of solvent strength and pH in order to maximize retention of the analyte(s) of interest to the SPE sorbents. One to two volume of buffer (the same as used in sample pretreatment) or water (for neutral compound) is a common choice for equilibration of the reversed-phase sorbents.
- **Sample loading:** The key element of sample loading is the linear velocity when the sample passes through the SPE sorbents to ensure optimal retention. For off-line application, the samples are often allowed to flow through by natural gravity to ensure sample contact time for the analyte(s) of interest with the stationary phase. The aqueous sample can be loaded directly onto the equilibrated reversed-phase SPE column/cartridge/plate. The nonpolar nature of the analyte results in strong retention by hydrophobic interactions with the stationary phase of the SPE sorbents.
- **Wash:** Interference compounds in the biological sample matrix are often co-retained on reversed-phase SPE column/cartridge/plate with analyte(s) of interest after sample loading. The goal of washing is to selectively remove unwanted matrix components, while leaving the analyte(s) of interest retained on the SPE sorbents. During the washing step, pure water or buffer is generally used first to remove water-soluble interferences (e.g. polar compounds or salts) that have

no affinity for the SPE sorbents. This is followed by the use of a solvent, typically aqueous methanol with higher elution strength than water or buffer to clean the sample as much as possible but without washing away the analyte(s) of interest. This will require some level of evaluation and/or optimization, during which the strength of wash solvent is increased stepwise by increasing the percentage of organic solvent (e.g. methanol) in the washing solution and the analyte(s) of interest, if any, in the washings is analyzed to identify and/or confirm the optimal strength of the wash solvent. In general, the higher $\log P$ value of the analyte(s) of interest, the stronger its interaction with the reversed-phase SPE sorbent materials and the stronger wash solvent should be employed. A good estimate is that every unit increase in $\log P$ value of the analyte(s) of interest will loosely correspond to an increase in 10% in methanol (in water) in the wash solution. For example, testosterone has a $\log P$ value of 3.37, use of ~35% of methanol in wash solvent can lead to an optimal recovery of the analyte with minimum background interference. For an efficient optimization of wash solution, a solvent with its strength around the estimated percentage (e.g. 35% methanol) can be a good starting point by bracketing it with ±10% (i.e. 25, 35, and 45%) of methanol in water (Rummel 2012).

- **Elution:** Elution is to disrupt all of the hydrophobic interactions between the analyte(s) of interest and the functional groups of the reversed-phase SPE sorbents with an organic solvent (e.g. methanol or acetonitrile) or solvent combination. Otherwise, partial elution may occur. Partial elution is often associated with irreproducible LC-MS results and poor and inconsistent REC of the analyte(s) of interest. In general, a higher $\log P$ value compound requires a stronger solvent or solvent combination in elution in order to obtain the highest REC with the lowest elution volume. Therefore, in reversed-phase SPE, the solvent with lower polarity is a stronger elution solvent. In other words, choosing a more nonpolar solvent leads to better results. If the analyte(s) are basic or acidic compounds, pH of the elution solvent should be adjusted to charge the analyte(s), which further reduces the hydrophobic interactions. In this case, use of weaker elution solvents and/or reduced elution volumes may be feasible. In practice, a consistent and low flow rate during elution is commonly associated with an improved recovery and reproducibility. The elution step is considered an additional opportunity for sample cleanup. Use of a very strong elution solvent often results in elution of interferences along with the analyte(s) of interest. Thus, elution step should be performed with the weakest solvent possible that provides complete elution of the analyte(s) of interest but with no or minimum interfering compound in the final eluate. In this regard, optimization of the eluent solution should be carried out by decreasing the strength of the elution solvent stepwise from high (e.g. 100% methanol) to low and monitoring both recovery and matrix effect in the eluate (Rummel 2012).
- **Direct LC-MS analysis of eluate or evaporation of the eluate followed by reconstitution of the resulting residues prior to LC-MS analysis:** In reversed-phase SPE, the resulting eluate may be subjected to direct LC-MS analysis if the eluate has less organic than the starting mobile phase in reversed-phase LC-MS or has more organic than the starting mobile phase in HILIC-MS analysis. However, in many applications, post-treatment of the eluate is necessary to evaporate the organic solvent and reconstitute the resulting residues using the starting mobile phase prior to reversed-phase LC-MS analysis.**

A typical sample preparation protocol using a RP SPE plate (for the analysis of goserelin, an agonists that stimulate the pituitary gland, using LC-MS by Kim et al. (2010)).

- A 500 μl volume of rabbit plasma sample was placed in a 2-ml Eppendorf tube, and 50 μl of IS working solution and 50 μl 3 M hydrochloric acid were added, and then the mixture was vortexed for 30 seconds.
- The SPE cartridge was pretreated with 1 ml methanol, followed by 1 ml deionized water.
- The sample mixture was gently loaded onto the SPE cartridge and left for 30 seconds. The SPE cartridge was washed with 1 ml deionized water and then dried for 60 seconds before elution with 1 ml methanol.
- The eluate was collected in a clean microtube and was evaporated to dryness at 40 °C under a stream of nitrogen.
- The dry residue was reconstituted with 100 μl mobile phase A (0.05% acetic acid in deionized water/acetonitrile 85/15, v/v) and centrifuged at 13 000 rpm for 10 minutes.
- The clear supernatant was transferred to an autosampler vial, and 20 μl of the supernatant was injected onto the LC-MS/MS system.

1.4.3.3.2 Ion-exchange SPE

Ion-exchange SPE is a common approach for extracting ionizable analyte(s) from aqueous sample solutions.

- **Mechanism:** Electrostatic (ionic) interactions of charged functional groups of the analyte(s) to oppositely charged functional groups on the SPE sorbents. Compared to hydrophobic interactions employed in the reversed-phase SPE, ion exchange via electrostatic (ionic) interactions is a much stronger retention mechanism that allows for the use of more aggressive wash steps for sample cleanup. It is especially ideal for removing hydrophobic interfering compounds with a strong organic wash. If the analyte(s) of interest are retained by cation or anion exchange, 100% methanol, acetonitrile, or other solvent can be used as the organic wash without eluting the analyte(s) of interest. **Note:** Because the kinetic exchange processes between the functional groups of the analyte(s) of interest and the functional groups of the SPE sorbents in ion-exchange SPE are considerably slower than that for reversed-phase SPE, flow rates should be generally low.

- **Applicable analyte(s):** Compounds that can be positively or negatively charged. Strong cation exchangers (e.g. aliphatic sulphonic acids) are permanently negatively charged in aqueous solution and can be used to extract bases with a wide range of pK_a values. In contrast, strong anion exchangers (e.g. aliphatic quaternary amines) are permanently positively charged in aqueous solution and can be used to extract acids with a wide range of pK_a values. Weak cation exchangers (e.g. aliphatic carboxylic acids, pK_a typically 4.5–5) can be used for the extraction of bases (pK_a 9 and above) as long as there are 4 pH unit difference between bonded phase and analyte pK_a values. Similarly, weak anion exchangers (e.g. secondary or tertiary amines) can be used for extraction of acids. The 4 pH rule is an important factor that is often overlooked when developing an ion-exchange SPE method. As discussed earlier, for an analyte with an ionic group, in order to ensure it is either totally neutral or charged, the pH of the sample matrix/solution must be adjusted to a minimum of 2 pH units away from the pK_a value of the ionic group. This might be an issue with ion-exchange SPE. In the case where an anionic sorbent (cation exchange SPE) with a pK_a value of 5.0 is used to extract an analyte with a pK_a value of 7.0. Based on the 2 pH unit rule, in order to ensure that the analyte is fully charged, the pH of the sample matrix must be adjusted to 5.0 or lower. However, in order to ensure the SPE sorbents are also fully charged, the pH of the sorbent environment must be adjusted to pH 7 or higher. Apparently, the two requirements cannot be met at the same time unless the difference in the pK_a values between anion (sorbents or analyte) and cation (analyte or sorbents) counterparts are 4 or greater in ion-exchange SPE. Furthermore, the cation (sorbent or analyte) must have pK_a values higher than the anion (analyte or sorbents). Taking the above into account, understanding the pK_a of both the analyte(s) of interest and the sorbents will definitely help to decide on what kind of ion-exchange SPE sorbent should be chosen for the intended analyte(s).
- **Sample pretreatment:** Based on the pK_a values of the analyte(s) of interest, sample can be diluted with a buffer of appropriate pH to ensure the functional groups of the analyte(s) are positively or negatively charged so that the analyte(s) can be retained by cation or anion exchange. For basic compounds, the pH of the sample needs to be adjusted to 2 pH units below the pK_a. This turns the positive charge of the analyte "on" so that they can be retained by cation exchange. In contrast, for acidic compounds, the pH of the samples needs to be adjusted to 2 units above the pK_a to negatively charge the molecules. This turns the negatively charge of the analyte "on" so that they can be retained by anion exchange. It is recommended to adjust the pH of the sample matrix with a strong acid or base or strong buffer. By doing so, the possible increase in the ionic strength of the sample matrix can be minimum. For example, it is most likely to add a much smaller quantity of hydrochloric acid than acetic acid to drop the pH of a sample when an amine analyte is to be extracted using cation ion-exchange SPE. Of course, using a strong acid or base can be a concern of the stability of the analyte(s) of interest during sample processing. Therefore, necessary evaluation of the analyte stability under such an environment should be performed prior to implementation. In addition, some samples (e.g. urine) contain high concentration of salts, a large-scale dilution may be necessary to reduce the ionic strength prior to application of an ion-exchange SPE. Alternatively, a reversed-phase SPE should be applied first to remove the salts prior to the ion-exchange SPE procedure.
- **Sorbent conditioning and equilibration:** Condition the ion exchange SPE cartridge/plate/plate with one to two sorbent bed volumes of methanol or acetonitrile. This is followed by a few bed volumes of pure water prior to equilibration, which is carried out with a buffer or a salt solution containing a counter ion with affinity for the SPE sorbents lower than the analyte(s) of interest. For pH adjustment of the sorbents, it is necessary to raise the pH to a minimum of 2 pH units above the pK_a value for the cation sorbents or lower the pH to a minimum of 2 pH units below the pK_a of the anion sorbents. It is also important that the chemistry environment of the sorbents (solvent/ionic strength) should be similar to that of the sample as much as possible to ensure the extraction environment does not alter during sample loading.
- **Sample loading:** It is important that both the ion exchanger and the analyte(s) remain charged in the loading process to promote ionic interactions and retention. In the loading step, the pretreated samples can be loaded to the pretreated column/cartridge/plate slowly to ensure optimal retention without breakthrough. A

longer residence time is necessary to achieve a reasonable analyte orientation relative to the sorbent surface.

- **Wash:** Ion exchange sorbent can be washed using solvents with certain pH, ionic strength, or an alternative sorbent counter ion to remove interfering species. The initial wash can be the same solution used for sample pretreatment. This can be followed by 100% organic solvent to remove more hydrophobic interferences. The pH of the wash solvent can be adjusted to remove more interfering components by neutralizing the charge of these molecules. However, adequate control of pH and ionic strength is necessary to ensure the analyte(s) of interest and the sorbent surface remain charged to prevent premature elution. Therefore, washing step should be optimized by washing with several solvents at various pH values and analyzing the washings for evidence of premature elution.
- **Elution:** Elution is carried out through interrupting the ionic interactions between the analyte(s) of interest and the sorbents by (i) neutralizing either the analyte(s) or sorbents, (ii) using high ionic strength, or (iii) using a strong counter ion. The most common elution strategy is by pH manipulation, i.e. eluting with basic organic solvent at least pH 2 units above the pK_a of basic analyte or eluting with acidic organic solvent at least pH 2 units below the pK_a of acidic analyte(s). This turns the charge of the molecules "off," i.e. neutralization, and disrupts the ion exchange retention, releasing the compounds from the SPE sorbents. It is worth noting that neutralization of the analyte(s) of interest is more desirable than neutralization of the sorbents. Neutralization of SPE sorbents might release other unwanted compounds that are retained during the ion-exchange SPE process and these compounds might interfere with LC-MS. In addition, using high ionic strength or a strong counter ion for SPE elution can be a concern of interference with LC-MS detection in some cases.
- **Eluate:** Post-treatment of the eluate is often necessary to evaporate the organic solvent and reconstitute the resulting residues in the mobile phase prior to LC-MS analysis.

A typical sample preparation protocol using strong cation-exchange (SCX) SPE plate (for the determination of dopamine in human plasma by Zhang et al. 2016).

- To each sample, 10 µl of IS working solution was spiked before extraction.
- 200 µl of plasma sample was diluted with 400 µl of 0.1 N HCl and then loaded onto a previously conditioned SCX cartridge.
- The cartridge was washed with 0.4 ml of 0.1 N HCl and 1 ml of acetonitrile.
- The analyte was then eluted using 1 ml of acetonitrile–propionic anhydride (90/10, v/v) with 2% (v/v) pyridine.
- The eluent was incubated in an oven at 60 °C for 20 minutes, evaporated to dryness under nitrogen at 60 °C, and reconstituted with 100 µl of acetonitrile–water (10/90, v/v).
- Each sample was centrifuged at 18 000 × g for 15 minutes before injection.

A typical sample preparation protocol using strong anion-exchange (SAX) SPE plate (for the determination of methylmalonic acid [MMA] in human plasma by Schmedes and Brandslund 2006).

- SPE was performed automatically on a Gilson Aspec XL4 liquid handler.
- 750 µl of serum (calibrator, control, or sample) was mixed with 375 µl of IS working solution in water.
- The SAX columns were conditioned with 1 ml of methanol and 1 ml of water before 750 µl of the mixed sample (corresponding to 500 µl of serum) was loaded.
- The columns were washed sequentially with 1 ml each of water, methanol, and butanol before elution of the MMA with 300 µl of a 10 : 90 mixture of concentrated (37%) HCl and butanol.
- Conical vials were placed directly in the collection racks of the Gilson Aspec. After sample collection, 80 µl of formic acid (1 g l^{-1}) was added to each vial.

1.4.3.3.3 Mixed-mode SPE

The mixed-mode SPE has been designed to utilize two or more primary interaction mechanisms for the extraction of the analyte(s) of interest from biological samples with improved selectivity and recovery. Compared to reversed-phase or ion-exchange SPE alone, mixed-mode SPE offers some unique advantages in terms of applicability as most of the analyte(s) of interest may exhibit both ion-exchange and reversed-phase retention for SPE.

- **Mechanism:** In mixed-mode SPE, both hydrophobic interactions and ionic interactions are used for analyte retention.
- **Applicable analyte(s):** The use of mixed-mode sorbent is to extract an analyte that exhibits functional groups capable of interacting with both of the hydrophobic and ionic groups of the SPE sorbents. For example, basic analyte with certain degree of hydrophobicity may work well with a mixed-mode sorbent containing a cation exchanger and a hydrophobic

chemistry. When the analyte is loaded, the cation exchanger interacts with the basic functional group of the analyte, while the hydrophobic sorbent group interacts with the hydrophobic group(s) of the analyte. Mixed-mode SPE can be used to simultaneously extract acidic and basic compounds, for which two separate extractions must be carried out for the same sample by conventional SPE. With mixed-mode cation-exchange sorbents, the acidic and neutral molecules can be retained via the reversed-phase mechanism and basic compounds can be retained by the mixed-mode mechanism at low pH.

- **Sample pretreatment:** Depending on which mechanism is to be employed as the primary retention mechanism in mixed-mode SPE, sample pretreatment can be different. In other words, the samples must be treated according to the primary mechanism of the mixed-mode SPE to be employed in sample processing prior to LC-MS. If ion-exchange is considered the primary retention mechanism, similar approach as for ion-exchange mode SPE as above should be used for sample pretreatment. Based on the pK_a values of the analyte(s) of interest, dilute the sample with a buffer or a solvent of appropriate pH to ensure ionic functional groups of the analyte(s) are positively or negatively charged.
- **Sorbent conditioning and equilibration:** Typically the surface of the sorbents is first conditioned with methanol or acetonitrile. This is followed by equilibration of the SPE sorbents with buffer similar/identical in pH to buffer used in the sample pretreatment. This buffer may also be employed in ionizing the ion-exchange groups if the ion-exchange mechanism is the primary one for compound retention.
- **Sample loading:** Same as reversed-phase or ion-exchange SPE, the pretreated sample should be loaded slowly to ensure optimal retention. In the sample loading step, hydrophobic interactions are normally used as the primary source of retention. This is because hydrophobic interactions are less affected by variations in the sample matrix than ionic interactions.
- **Wash:** As the target analyte(s) are retained by both retention mechanisms, interferences that are retained only by a single mechanism can be easily washed off by suppression of such mechanism. Therefore, in a mixed-mode SPE, two separate wash steps are commonly employed. For basic and hydrophobic analyte(s) that are retained on a mixed-mode SPE sorbents via both cation-exchange and hydrophobic interaction mechanism, a ionic buffer is commonly used in the first wash step, by which ionic matrix components (e.g. salts) are generally removed. Such a wash step should have no influence on the retention of the analyte(s) via a hydrophobic mechanism. In the second wash, a pure organic solvent is used, which will remove matrix components that have only hydrophobic character but without affecting the ionic retention of the analyte(s). By using this double-wash approach, many interfering components can be very effectively removed from the mixed-mode sorbents without affecting the analyte(s). The same as for other retention mechanisms, maximum extract purity can be achieved by washing with the strongest solvent possible but without eluting the analyte(s) of interest.
- **Elution:** Elution can be carried out through suppressing both hydrophobic and ionic retention mechanisms simultaneously by using an elution solution that contains organic solvent (such as methanol, acetonitrile) and a strong acid or base. The organic solvent in the elution solution is employed to suppress the hydrophobic interactions while the bases (e.g. ammonium hydroxide) or acids (e.g. hydrochloric acid) are used to suppress the ionic interactions. Attention should be made to check the stability of the analyte of interest under such a strong acidic or basic condition. Different from the washing step above, the weakest solvent possible should be employed in elution that gives the highest REC for the analyte(s) of interest but without eluting the interfering compounds.
- **Eluate:** Post-treatment of the eluate is often necessary to evaporate the organic solvent and reconstitute the resulting residues in the mobile phase prior to LC-MS analysis.

A typical sample preparation protocol using a mixed-mode cation-exchange (MCX) SPE plate (for quantification of 5-azacytidine (5Ac) in plasma Zhao et al. 2004).

- Plasma samples (200 μl) were initially deproteinized with 1 ml of acetonitrile containing 5 ng ml^{-1} IS in a borosilicate glass tube.
- The mixture was then centrifuged at 2000 × *g* for 10 minutes at ambient temperature. The supernatant was dried with a gentle nitrogen stream at 30 °C. The residue was dissolved in 0.5 ml of 0.1 N HCl and transferred to an activated Oasis MCX 1 cc (30 mg) ion exchange SPE cartridge (the SPE was conditioned and equilibrated by washing with 1 ml of methanol and then 1 ml of Milli-Q water).
- The samples were cleaned using two wash steps: first by 1 ml twice of 0.1 N HCl and then with 1 ml of methanol.
- The analyte was eluted with 2 ml total (1 ml twice) of ammonium hydroxide in acetonitrile (5 : 95, v/v).
- The sample extracts were evaporated to dryness under nitrogen at 30 °C. The residue was reconstituted in 100 μl of methanol : water (1 : 9, v/v) by vortex-mixing (30 seconds). The reconstituted sample extract was

transferred to a 250 μl polypropylene autosampler vial, and a volume of 20 μl was injected onto LC-MS/MS (**Note:** Due to the instability of 5AC in plasma, all processing and handling of 5AC samples were performed on ice until the samples were dried and reconstituted).

A typical sample preparation protocol using a mixed-mode anion-exchange (MAX) SPE plate (for the simultaneous analysis of glucocorticosteroid fluticasone propionate and its metabolite fluticasone propionate 17β-carboxylic acid in human plasma by Nair et al. 2017).

- To an aliquot of 500 μl of sample, 25 μl of IS working solution was added. After vortex-mixing for 30 seconds, the samples were centrifuged at 13 148 × *g* for 5 minutes at 10 °C.
- Plasma samples were then loaded on Oasis MAX (30 mg, 1 ml) cartridges, which were preconditioned with 1.0 ml of methanol and 1.0 ml of water.
- The samples were washed sequentially with 1.0 ml of *n*-hexane, 1.0 ml of water, and 1 ml of 5% methanol in water.
- The cartridges were dried for 1.0 minute by applying nitrogen (1.72 × 105 Pa) at 2.4 l min^{-1} flow rate.
- Elution was performed using 1.0 ml of 1.0 mM TFA in methanol.
- The eluent was evaporated to dryness in pre-labeled vials under a gentle stream of nitrogen and then reconstituted with 200 μl of the mobile phase.
- The resulting solution was briefly vortexed and 10 μl was used for injection.

A typical sample preparation protocol using Waters Oasis WCX plate, a mixed-mode weak cation-exchange SPE plate (for the analysis of pasireotide in human plasma by Fu et al. 2016).

- A volume of 0.1 ml of blank plasma or freshly prepared calibration standards, QC samples, or unknown samples was added to a 1 ml 96-well plate.
- A 25 μl aliquot of the IS working solution in 50% acetonitrile in water (v/v) was added to all wells except for the matrix blanks, to which a 25 μl aliquot of 50% acetonitrile in water (v/v) was added.
- After adding 100 μl of 2% formic acid in water (v/v) to each well, the plate was vortex-mixed briefly and centrifuged.
- The sample mixtures in the above plate were loaded onto the corresponding wells of a Waters Oasis WCX μElution plate (96-well format, 30 mg, Milford, MA, USA), which were pre-equilibrated with 200 μl of methanol followed by 200 μl of water.
- The plate was washed with 200 μl of 5% ammonium hydroxide in water (v/v) followed by 200 μl of methanol. Finally, the μElution plate was eluted with 100 μl of 5% formic acid in methanol (v/v).
- The eluent was diluted with 50 μl of water. After brief vortex-mixing and centrifugation at ~3000 × *g* for approximately five minutes, a volume of 15 μl of the reconstituted sample extract was injected onto the LC-MS/MS system for analysis.

A typical sample preparation protocol using Waters Oasis WAX plate, a mixed-mode weak anion-exchange SPE plate (for the analysis of carbazochrome sodium sulfonate in human plasma by Hu et al. 2014).

- 20 μl of IS working solution and 100 μl of water (supplementary volume) were added to 1 ml plasma sample and vortex-mixed for 30 seconds.
- The Oasis WAX (30 mg, 1 ml) cartridges were conditioned with 1 ml of methanol and 1 ml of water.
- The plasma samples were applied to the cartridges. Following sample application, the cartridges were washed with 1 ml of 2% formic acid water solution, and subsequently washed with 1 ml of methanol.
- After drying, the analyte(s) were eluted with 1 ml of methanol containing 5% ammonium hydroxide.
- The eluate was evaporated to dryness at 40 °C using a gentle stream of nitrogen.
- The residue was reconstituted in 100 μl of the mobile phase, then vortex-mixed and centrifuged at 10 500 rpm for 10 minutes.
- A 5 μl aliquot of the resulting solution was injected onto the LC-MS/MS system for analysis.

Some general considerations when developing a SPE method (Li and Bartlett 2014; Liu and Aubry 2013; Pedersen-Bjergaard et al. 2015):

- **Capacity:** The capacity of silica-based SPE phases is around 5–10% of the sorbent mass per analyte and around 20–25% for polymer-based SPE phases. This information can be useful in judging the sorbent mass needed for the intended bioanalytical assay.
- **Speed:** Hydrogen bonding and electrostatic (ionic) interactions are "point to point" mechanism. A low speed in loading, washing, and/or elution is necessary to ensure good analyte retention and recovery.
- **Soaking:** Under certain circumstances, it is necessary to include soaking steps in a SPE procedure. During a soaking step, the solvent that is applied to the SPE column/cartridge/plate is allowed to remain

stationary (~30 seconds to 2 minutes for a typical 100 mg sorbent bed mass) without application of vacuum or pressure. This allows for establishment of chemical and physicochemical equilibrium between the solvent and sorbent materials during conditioning and sorbent surface activation steps. During elution step, soaking allows the analyte(s) of interest diffuse into the elution solvent and reach equilibrium prior to application of vacuum or pressure to facilitate elution. In general, soaking can help improve retention and elution efficiency and reduce the volume of elution solvent.

- **Drying step:** Often a drying step is included after the wash step. This can be particularly important when water-immiscible solvent is employed in the elution step. Any water that remains on the surface of the sorbents may exclude the immiscible elution solvent and cause low recovery of the analyte(s) of interest.

1.4.3.4 Other Formats of SPE

Other formats of SPE are available with various packing materials. They include but are not limited to: immunoaffinity SPE, MIP SPE, and restricted access material (RAM) SPE, dispersive SPE, disposable pipette extraction (DPX), solid-phase microextraction (SPME), microextraction by packed sorbent (MEPS), and stir-bar sorptive extraction (SBSE). The sample extraction mechanism of some of these SPE techniques is captured in separate chapters of this book (Liu and Aubry 2013; Pedersen-Bjergaard et al. 2015).

1.4.4 Combination of PPT, LLE, and/or SPE in LC-MS Bioanalysis

PPT, LLE, and SPE each discussed above has advantages and disadvantages over each other. None of them can fit all requirements of modern LC-MS bioanalysis with optimal REC and minimal matrix effect. Indeed, some special cases require more sophisticated approaches to achieve the needed high sensitivity or remove persistent interferences. The real troubling matrix interferences most likely have similar physicochemical properties to those of the analyte(s) of interest. Therefore, they are very likely to be co-extracted with the analyte(s) of interest using the single sample extraction methods. To address these issues, orthogonal or hybrid sample preparation techniques, such as PPT/LLE, PPT/SPE, and LLE/SPE, should be considered and implemented.

1.4.4.1 Combination of PPT and LLE

While PPT is considered a plug-and-play method in LC-MS bioanalysis, sometimes persistent endogenous interferences may be encountered in the retention time window for the analyte(s) of interest. A simple LLE step added after PPT may be helpful in mitigating such interference as reported by Wang et al. (2015), who identified a large interfering peak eluted close to the retention time of octreotide in LC-MS. After screening extraction solvent among diethyl ether, ethyl acetate, trichloromethane, dichloromethane, etc. 2.5-fold of dichloromethane was found to be able to completely eliminate the impurity peak with no apparent sacrifice in recovery for the analyte. Mixing 200 μl of supernatant from PPT with 200 μl of water for LLE with 500 μl dichloromethane yielded the best extraction efficiency and lowest background noise for octreotide in LC-MS/MS analysis (Wang et al. 2015).

1.4.4.2 Combination of PPT and SPE

There have been many applications where biological samples were subjected to PPT followed by the SPE process. Depending on the physicochemical properties and assay sensitivity needs, the supernatants from PPT may be treated with proper solvents prior to being loaded onto SPE with suitable sorbents, such as phospholipid removal SPE plates (Nilsson et al. 2014), reversed-phase sorbents (Zhang et al. 2014), or mixed-mode sorbents (Guo et al. 2015) for further processing prior to LC-MS analysis.

PPT with acetonitrile and methanol along with formic acid or ammonium hydroxide can effectively remove plasma proteins and other endogenous components, keeping the basic or acidic compounds protonated or deprotonated even in a high percentage of organic solvents. The supernatants from the PPT can be directly loaded onto an unconditioned 96-well mixed-mode cation or anion-exchange SPE plate without analyte breakthrough and cartridge blockage during the loading step. The loading step is considered a rinsing step to remove lipid-soluble components; therefore, wash steps may not be necessary. This simplified PPT/SPE procedure was reported to effectively eliminate time-consuming sorbent conditioning and wash steps required by conventional SPE (Xue et al. 2006b).

1.4.4.3 Combination of LLE and SPE

A single-step SPE may be sufficient for the extraction of the analyte(s) of interest with a variety of physicochemical properties. However, blockage of the SPE cartridges/discs often occur during the sample loading step because of viscosity of study samples, particularly urine samples. Depending on the polarity of the analyte(s) of interest, a simple treatment with LLE prior to SPE can lead to a significantly reduced baseline in the followed-up LC-MS bioanalysis (Shin et al. 2014).

A typical sample preparation protocol using combined PPT and SPE procedure (for the analysis of peptide glucagon in human plasma by Howard et al. 2017).

- 400 µl of plasma sample and 20 µl of IS working solution were added to the wells of 2 ml 96-well plate. The plate was vortex-mixed.
- PPT was conducted by adding 1.1 ml of acetonitrile : water : ammonium hydroxide (75 : 25 : 0.1, v/v/v). The plate was vortex-mixed, sonicated for 5 minutes, and then centrifuged for 10 minutes at 2300 × *g*.
- A total of 1.2 ml of supernatant was transferred using an automated liquid handler system (Quadra Tower, Tomtec, CT, USA) to a 2 ml plate and evaporated to dryness at 40 °C under nitrogen (~90 minutes).
- Samples were reconstituted in 800 µl of 2% ammonium hydroxide in water and then vortex-mixed before being extracted using SPE.
- A Bond Elut Plexa SPE plate (96-roundwell, 30 mg; Agilent Technologies, CA, USA) was conditioned using 1 ml of methanol and 1 ml of water.
- The samples were loaded, washed with 1 ml of 5% methanol in water, eluted with 2 × 225 µl of acetonitrile : water : formic acid (75 : 25 : 0.1, v/v/v) into a 1 ml Lo-bind plate and then evaporated under nitrogen at 40 °C, before being reconstituted in 200 µl of 0.2% formic acid in water.
- The plate was centrifuged for 10 minutes at 2300 × *g*, and 40 µl of sample was injected onto the LC-MS/MS system for analysis.

1.4.5 Summary

Physicochemical properties of the analyte(s), assay dynamic range, and available MS instrumentation platform should be jointly considered when developing a sample preparation procedure for the intended LC-MS bioanalysis. Although not discussed extensively, pre-analytical parameters, such as stability in the intended sample matrix, protein binding, possible nonspecific adsorption, and blood-to-plasma distribution of the analyte(s) of interest should be assessed, as part of assay development.

Three major platforms of sample preparation, i.e. PPT, LLE, and SPE have been discussed. Each technique has its unique advantage and also disadvantages over the other.

There is no doubt that among the three, PPT is the simplest, quickest, and most convenient technique and has been widely employed in LC-MS bioanalysis. The method is not only associated with less cost as compared to LLE and SPE but also associated with the speed and throughput, which is needed in routine analysis of a large number of study samples. In authors' view, with the advancement of MS instruments, less sample volume, as in the current trend of microsampling, will be needed for the intended assay sensitivity. Less sample volume is associated with less matrix effect, which, if any, should be compensated for by the stable labeled IS that is being increasingly used in nowadays' bioanalysis practice. Therefore, PPT will continue to be the primary choice of sample preparation in LC-MS bioanalysis. In comparison to PPT and SPE, LLE provides the most-effective sample cleanup by selectively extracting analyte(s) of interest and removing matrix effects-causing interferences. There is a clear trend of automated LLE platform which can significantly improve the throughout and efficiency of LLE sample preparation. Obviously, SPE technology will continue evolving. New sorbents will continue to be available to meet the needs of extracting different types of molecules from biological matrices. There is also a trend of development of SPE plate that fits for processing of small volume of biological samples. Finally, the combination of different platforms, i.e. PPT/LLE, PPT/SPE, or LLE/SPE, has been demonstrated to be very useful in dealing with challenging assay development.

References

Ahmad, S., Kalra, H., Gupta, A. et al. (2012). HybridSPE: a novel technique to reduce phospholipid-based matrix effect in LC-ESI-MS bioanalysis. *J. Pharm. Bioallied Sci.* 4 (4): 267–275.

Asimakopoulos, A.G. and Thomaidis, N.S. (2015). Bisphenol A, 4-t-octylphenol, and 4-nonylphenol determination in serum by Hybrid Solid Phase Extraction-Precipitation Technology technique tailored to liquid chromatography-tandem mass spectrometry. *J. Chromatogr. B Analyt. Technol. Biomed. Life Sci.* 986–987: 85–93.

Blanchard, J. (1981). Evaluation of the relative efficacy of various techniques for deproteinizing plasma samples prior to high-performance liquid chromatographic analysis. *J. Chromatogr.* 226 (2): 455–460.

de Boer, T. and Wieling, J. (2013). LC-MS bioanalysis of unbound drugs in plasma and serum. In: *Handbook of*

LC-MS Bioanalysis, Best Practices, Experimental Protocols, and Regulations (ed. W. Li, J. Zhang and F.L.S. Tse), 317–326. Hoboken, NJ: Wiley.

Bolden, R.D., Hoke, S.H. 2nd, Eichhold, T.H. et al. (2002). Semi-automated liquid-liquid back-extraction in a 96-well format to decrease sample preparation time for the determination of dextromethorphan and dextrorphan in human plasma. *J. Chromatogr. B Analyt. Technol. Biomed. Life Sci.* 772 (1): 1–10.

Brockman, A.H., Hatsis, P., Paton, M., and Wu, J.T. (2007). Impact of differential recovery in bioanalysis: the example of bortezomib in whole blood. *Anal. Chem.* 79: 1599–1603.

Buhrman, D.L., Price, P.I., and Rudewiczcor, P.J. (1996). Quantitation of SR 27417 in human plasma using electrospray liquid chromatography-tandem mass spectrometry: a study of ion suppression. *J. Am. Soc. Mass Spectrom.* 7 (11): 1099–1105.

Cheruvu, H.S., Yadav, N.K., Valicherla, G.R. et al. (2018). LC-MS/MS method for the simultaneous quantification of luteolin, wedelolactone and apigenin in mice plasma using Hansen solubility parameters for liquid-liquid extraction: application to pharmacokinetics of *Eclipta alba* chloroform fraction. *J. Chromatogr. B Analyt. Technol. Biomed. Life Sci.* 1081–1082: 76–86.

Chi, J., Li, F., and Jenkins, R. (2016). Ultrasensitive sub-pg/ml determination of tiotropium bromide in human plasma by 2D-UHPLC-MS/MS: challenges and solutions. *Bioanalysis* 8 (5): 385–395.

Constanzer, M.L., Chavez, C.M., and Matuszewski, B.K. (1997). Low level determination of dorzolamide and its de-ethylated metabolite in human plasma by liquid chromatography with atmospheric pressure chemical ionization tandem mass spectrometry. *J. Pharm. Biomed. Anal.* 15 (7): 1001–1008.

Dell, D. (2004). Labile metabolites. *Chromatographia* 59 (Suppl. 2): S139–S148.

Fu, Y., Li, W., Smith, H.T., and Tse, F.L. (2011). An investigation of incurred human urine sample reanalysis failure. *Bioanalysis* 3 (9): 967–972.

Fu, Y., Xia, Y.Q., Flarakos, J. et al. (2016). Differential mobility spectrometry coupled with multiple ion monitoring in regulated LC-MS/MS bioanalysis of a therapeutic cyclic peptide in human plasma. *Anal. Chem.* 88 (7): 3655–3661.

Guo, W., Li, G., Yang, Y. et al. (2015). LC-MS/MS analysis of pramipexole in mouse plasma and tissues: elimination of lipid matrix effects using weak cation exchange mode based solid-phase extraction. *J. Chromatogr. B Analyt. Technol. Biomed. Life Sci.* 988: 157–165.

Hee, K.H., Leaw, Y.K.J., Ong, J.L., and Lee, L.S. (2017). Development and validation of liquid chromatography tandem mass spectrometry method quantitative determination of polymyxin B1, polymyxin B2, polymyxin B3 and isoleucine-polymyxin B1 in human plasma and its application in clinical studies. *J. Pharm. Biomed. Anal.* 140: 91–97.

Hijazi, Y., Bolon, M., and Boulieu, R. (2001). Stability of ketamine and its metabolites norketamine and dehydronorketamine in human biological samples. *Clin. Chem.* 47 (9): 1713–1715.

Hou, J., Qu, F., Wu, C. et al. (2012). Quantitative determination and pharmacokinetic study of the novel anti-Parkinson's disease candidate drug FLZ in rat brain by high performance liquid chromatography-tandem mass spectrometry. *J. Pharm. Biomed. Anal.* 66: 232–239.

Howard, J.W., Kay, R.G., Jones, B. et al. (2017). Development of a UHPLC-MS/MS (SRM) method for the quantitation of endogenous glucagon and dosed GLP-1 from human plasma. *Bioanalysis* 9 (9): 733–751.

Hu, X., Zheng, Y., Wu, G. et al. (2014). Optimization of solid phase extraction clean up and validation of quantitative determination of carbazochrome sodium sulfonate in human plasma by liquid chromatography-electrospray ionization tandem mass spectrometry. *J. Chromatogr. B Analyt. Technol. Biomed. Life Sci.* 951–952: 1–6.

Ji, A.J. (2013). LC-MS bioanalysis of drugs in urine. In: *Handbook of LC-MS Bioanalysis, Best Practices, Experimental Protocols, and Regulations* (ed. W. Li, J. Zhang and F.L.S. Tse), 307–316. Hoboken, NJ: Wiley.

Ji, A.J., Jiang, Z., Livson, Y. et al. (2010). Challenges in urine bioanalytical assays: overcoming nonspecific binding. *Bioanalysis* 2 (9): 1573–1586.

Jian, W., Edom, R., Weng, N. et al. (2010a). Validation and application of an LC-MS/MS method for quantitation of three fatty acid ethanolamides as biomarkers for fatty acid hydrolase inhibition in human plasma. *J. Chromatogr. B Analyt. Technol. Biomed. Life Sci.* 878 (20): 1687–1699.

Jian, W., Edom, R.W., Xu, Y., and Weng, N. (2010b). Recent advances in application of hydrophilic interaction chromatography for quantitative bioanalysis. *J. Sep. Sci.* 33 (6–7): 681–697.

Jian, W., Edom, R.W., Xue, X. et al. (2013). Quantitation of leukotriene B(4) in human sputum as a biomarker using UPLC-MS/MS. *J. Chromatogr. B Analyt. Technol. Biomed. Life Sci.* 932: 59–65.

Jiang, H., Zhang, Y., Ida, M. et al. (2011). Determination of carboplatin in human plasma using HybridSPE-precipitation along with liquid chromatography-tandem mass spectrometry. *J. Chromatogr. B Analyt. Technol. Biomed. Life Sci.* 879 (22): 2162–2170.

Kim, M.K., Lee, T.H., Suh, J.H. et al. (2010). Development and validation of a liquid chromatography-tandem mass spectrometry method for the determination of goserelin in rabbit plasma. *J. Chromatogr. B Analyt. Technol. Biomed. Life Sci.* 878 (24): 2235–2242.

Kocan, G., Quang, C., and Tang, D. (2006). Evaluation of protein precipitation filter plates for high-throughput LC-MS/MS bioanalytical sample preparation. *Am. Drug Discov.* 1 (3): 21–24.

Kole, P.L., Venkatesh, G., Kotecha, J., and Sheshala, R. (2011). Recent advances in sample preparation techniques for effective bioanalytical methods. *Biomed. Chromatogr.* 25 (1–2): 199–217.

Lentner, C. (ed.) (1981). *Geigy Scientific Tables*, 8th, revised and enlarged ed., p. 53 (vol. 1) and p. 65 (vol. 3). West Caldwell, NJ: CIBA-GEIGY.

Li, P. and Bartlett, M. (2014). A review of sample preparation methods for quantitation of small-molecule analytes in brain tissue by liquid chromatography tandem mass spectrometry (LC-MS/MS). *Anal. Methods* 6: 6183–6207.

Li, W., Fu, Y., Flarakos, J., and Zhang, D. (2013a). LC-MS bioanalysis of ester prodrugs and other esterase labile molecules. In: *Handbook of LC-MS Bioanalysis, Best Practices, Experimental Protocols, and Regulations* (ed. W. Li, J. Zhang and F.L.S. Tse), 431–446. Hoboken, NJ: Wiley.

Li, W., Li, Y., Francisco, D.T., and Naidong, W. (2005). Hydrophilic interaction liquid chromatographic tandem mass spectrometric determination of atenolol in human plasma. *Biomed. Chromatogr.* 19 (5): 385–393.

Li, W., Lin, H., Flarakos, J., and Tse, F.L. (2013b). Practical approaches to incurred sample LC-MS/MS reanalysis: confirming unexpected results. *Anal. Chem.* 85 (4): 2405–2413.

Li, W., Luo, S., Hayes, M. et al. (2007a). Determination of *N*-methyl-4-isoleucine-cyclosporin (NIM811) in human whole blood by high performance liquid chromatography-tandem mass spectrometry. *Biomed. Chromatogr.* 21 (3): 249–256.

Li, W., Luo, S., Li, S. et al. (2007b). Simultaneous determination of ribavirin and ribavirin base in monkey plasma by high performance liquid chromatography with tandem mass spectrometry. *J. Chromatogr. B Analyt. Technol. Biomed. Life Sci.* 846 (1–2): 57–68.

Li, W., Luo, S., Rebello, S. et al. (2014). A semi-automated LC-MS/MS method for the determination of LCI699, a steroid 11β-hydroxylase inhibitor, in human plasma. *J. Chromatogr. B Analyt. Technol. Biomed. Life Sci.* 960: 182–193.

Li, W., Luo, S., Smith, H.T., and Tse, F.L. (2007c). Simultaneous determination of midazolam and 1′-hydroxymidazolam in human plasma by liquid chromatography with tandem mass spectrometry. *Biomed. Chromatogr.* 21 (8): 841–851.

Li, W., Luo, S., Smith, H.T., and Tse, F.L. (2010). Quantitative determination of BAF312, a S1P-R modulator, in human urine by LC-MS/MS: prevention and recovery of lost analyte due to container surface adsorption. *J. Chromatogr. B Analyt. Technol. Biomed. Life Sci.* 878 (5–6): 583–589.

Li, W., Rettig, J., Jiang, X. et al. (2006). Liquid chromatographic-electrospray tandem mass spectrometric determination of clarithromycin in human plasma. *Biomed. Chromatogr.* 20 (11): 1242–1251.

Li, W., Zhang, J., and Tse, F.L. (2011). Strategies in quantitative LC-MS/MS analysis of unstable small molecules in biological matrices. *Biomed. Chromatogr.* 25 (1–2): 258–277.

Liu, G. and Aubry, A.F. (2013). Best practices in biological sample preparation for LC-MS bioanalysis. In: *Handbook of LC-MS Bioanalysis, Best Practices, Experimental Protocols, and Regulations* (ed. W. Li, J. Zhang and F.L.S. Tse), 139–164. Hoboken, NJ: Wiley.

Nair, S.G., Patel, D.P., Sanyal, M. et al. (2017). Simultaneous analysis of glucocorticosteroid fluticasone propionate and its metabolite fluticasone propionate 17β-carboxylic acid in human plasma by UPLC-MS/MS at sub pg/mL level. *J. Pharm. Biomed. Anal.* 135: 1–7.

Nilsson, K., Andersson, M., and Beck, O. (2014). Phospholipid removal combined with a semi-automated 96-well SPE application for determination of budesonide in human plasma with LC-MS/MS. *J. Chromatogr. B Analyt. Technol. Biomed. Life Sci.* 970: 31–35.

Oiestad, E.L., Johansen, U., Stokke Opdal, M. et al. (2009). Determination of digoxin and digitoxin in whole blood. *J. Anal. Toxicol.* 33 (7): 372–378.

Onorato, J.M., Chen, L., Shipkova, P. et al. (2010). Liquid-liquid extraction coupled with LC/MS/MS for monitoring of malonyl-CoA in rat brain tissue. *Anal. Bioanal. Chem.* 397 (7): 3137–3142.

Pedersen-Bjergaard, S., Gjelstad, A., and Halvorsen, T.G. (2015). Sample preparation. In: *Bioanalysis of Pharmaceuticals-Sample Preparation, Separation Techniques and Mass Spectrometry* (ed. S.H. Hansen and S. Pedersen-Bjergaard), 73–121. Chichester, UK: Wiley.

Pucci, V., Di Palma, S., Alfieri, A. et al. (2009). A novel strategy for reducing phospholipids-based matrix effect in LC-ESI-MS bioanalysis by means of HybridSPE. *J. Pharm. Biomed. Anal.* 50 (5): 867–871.

Rummel, M. (2012). The partition coefficient (log P), distribution coefficient (log D) and acid dissociation constant (pK_a): knowing your compound for solid-phase extraction method development in reversed-phase mode. *The Column* 8 (2): 2–7.

Schmedes, A. and Brandslund, I. (2006). Analysis of methylmalonic acid in plasma by liquid chromatography-tandem mass spectrometry. *Clin. Chem.* 52 (4): 754–757.

Shin, J.H., Na, Y.C., Chung, J.H. et al. (2014). Quantitative analysis of heterocyclic amines in urine by liquid chromatography coupled with tandem mass spectrometry. *Anal. Biochem.* 447: 169–176.

Song, Q. and Naidong, W. (2006). Analysis of omeprazole and 5-OH omeprazole in human plasma using hydrophilic interaction chromatography with tandem mass spectrometry (HILIC-MS/MS) – eliminating evaporation and reconstitution steps in 96-well liquid/liquid extraction. *J. Chromatogr. B Analyt. Technol. Biomed. Life Sci.* 830 (1): 135–142.

Tang, Y.Q. and Weng, N. (2013). Salting-out assisted liquid-liquid extraction for bioanalysis. *Bioanalysis* 5 (12): 1583–1598.

Unger, S., Li, W., Flarakos, J., and Tse, F. (2013). Role of LC-MS bioanalysis in drug discovery, development and therapeutic drug monitoring. In: *Handbook of LC-MS Bioanalysis, Best Practices, Experimental Protocols, and Regulations* (ed. W. Li, J. Zhang and F.L.S. Tse), 3–14. Hoboken, NJ: Wiley.

Valente, I.M. and Rodrigues, J.A. (2015). Recent advances in salt-assisted LLE for analyzing biological samples. *Bioanalysis* 7 (17): 2187–2193.

Wang, Q., Liang, Y., Rao, T. et al. (2015). PK study of octreotide based on LC-MS/MS combining protein precipitation and impurity extraction technique. *Bioanalysis* 7 (7): 885–894.

Wu, S., Li, W., Majumdar, T. et al. (2010). Supported liquid extraction in combination with LC-MS/MS for high-throughput quantitative analysis of hydrocortisone in mouse serum. *Biomed. Chromatogr.* 24 (6): 632–638.

Xue, Y.J., Akinsanya, J.B., Liu, J., and Unger, S.E. (2006b). A simplified protein precipitation/mixed-mode cation-exchange solid-phase extraction, followed by high-speed liquid chromatography/mass spectrometry, for the determination of a basic drug in human plasma. *Rapid Commun. Mass Spectrom.* 20 (18): 2660–2668.

Xue, Y.J., Liu, J., and Unger, S. (2006a). A 96-well single-pot liquid-liquid extraction, hydrophilic interaction liquid chromatography-mass spectrometry method for the determination of muraglitazar in human plasma. *J. Pharm. Biomed. Anal.* 41 (3): 979–988.

Xue, Y.J., Pursley, J., and Arnold, M.E. (2004). A simple 96-well liquid-liquid extraction with a mixture of acetonitrile and methyl t-butyl ether for the determination of a drug in human plasma by high-performance liquid chromatography with tandem mass spectrometry. *J. Pharm. Biomed. Anal.* 34 (2): 369–378.

Yin, L., Meng, X., Zhou, X. et al. (2015). Simultaneous determination of carboprost methylate and its active metabolite carboprost in dog plasma by liquid chromatography-tandem mass spectrometry with positive/negative ion-switching electrospray ionization and its application to a pharmacokinetic study. *J. Chromatogr. B Analyt. Technol. Biomed. Life Sci.* 998–999: 8–14.

Zeng, M. and Cao, H. (2018). Fast quantification of short chain fatty acids and ketone bodies by liquid chromatography-tandem mass spectrometry after facile derivatization coupled with liquid-liquid extraction. *J. Chromatogr. B Analyt. Technol. Biomed. Life Sci.* 1083: 137–145.

Zeng, W., Xu, Y., Constanzer, M., and Woolf, E.J. (2010). Determination of sitagliptinin human plasma using protein precipitation and tandem mass spectrometry. *J. Chromatogr. B Analyt. Technol. Biomed. Life Sci.* 878 (21): 1817–1823.

Zhang, S., Han, J., Leng, G. et al. (2014). An LC-MS/MS method for the simultaneous determination of goserelin and testosterone in rat plasma for pharmacokinetic and pharmacodynamic studies. *J. Chromatogr. B Analyt. Technol. Biomed. Life Sci.* 965: 183–189.

Zhang, D., Wu, L., Chow, D.S. et al. (2016). Quantitative determination of dopamine in human plasma by a highly sensitive LC-MS/MS assay: application in preterm neonates. *J. Pharm. Biomed. Anal.* 117: 227–231.

Zhang, J., Wu, H., Kim, E., and El-Shourbagy, T.A. (2009a). Salting-out assisted liquid/liquid extraction with acetonitrile: a new high throughput sample preparation technique for good laboratory practice bioanalysis using liquid chromatography-mass spectrometry. *Biomed. Chromatogr.* 23 (4): 419–425.

Zhang, H., Zhong, D., Zhang, Z. et al. (2009b). Liquid chromatography/tandem mass spectrometry method for the quantification of deserpidine in human plasma: application to a pharmacokinetic study. *J. Chromatogr. B Analyt. Technol. Biomed. Life Sci.* 877 (27): 3221–3225.

Zhao, M., Rudek, M.A., He, P. et al. (2004). Quantification of 5-azacytidine in plasma by electrospray tandem mass spectrometry coupled with high-performance liquid chromatography. *J. Chromatogr. B Analyt. Technol. Biomed. Life Sci.* 813 (1–2): 81–88.

2

Online Extraction and Column Switching Techniques in LC-MS Bioanalysis

Yan Mao and Mike (Qingtao) Huang

2.1 Introduction

Bioanalysis plays an important role in drug discovery and development. The information generated by bioanalysis is widely used in the entire drug discovery and development process, which includes drug candidate selection and optimization, preclinical and clinical studies, and post-marketing therapeutic drug monitoring (Unger et al. 2013). Many scientific decisions in drug development are dependent upon the accurate quantification of drug candidates and/or their metabolites in biological samples from various studies. Since the commercialization of the atmospheric pressure ionization interface in the mid-1990s, liquid chromatography-tandem mass spectrometry (LC-MS/MS) has been the method of choice for performing bioanalysis of small-molecule drug candidates. LC-MS/MS provides high selectivity and sensitivity for the analysis of the analytes of interest. However, it is also subject to potential interferences from other matrix components that are co-eluted with the analytes. Therefore, sample preparation is necessary to remove proteins and other endogenous components from the biological samples for a robust and reliable analysis by LC-MS/MS (Jemal and Xia 2006). Protein precipitation, liquid–liquid extraction (LLE), and solid-phase extraction (SPE) are the traditional techniques used to extract analytes from biological matrices. Manual operations of these sample extraction procedures are labor intensive and time consuming, which, in many cases, has become the bottleneck and speed-limiting step in method development and sample analysis. To overcome this, robotic liquid handlers and 96-well plates have been employed to automate or semiautomate sample extraction procedures, which include sample aliquoting, dilution, extraction, solvent evaporation, and reconstitution. These automated or semiautomated sample preparation procedures are carried out independently from LC-MS/MS, referred to as off-line extraction.

As an integral approach, online extraction combines sample cleanup, analyte isolation and enrichment, and LC-MS/MS analysis into one platform (Jemal 2000). For a typical online extraction–LC-MS/MS method, samples in biological fluids with minimal pretreatment or without pretreatment are injected directly onto a sample extraction column. The analytes of interest are retained via various mechanisms depending on the type of the extraction column, while other unwanted matrix components are washed away from the column and diverted to waste. After column switching, the analytes of interest are eluted from the extraction to an analytical column for LC-MS/MS analysis. A 6- or 10-port column switching valve is commonly used to direct the eluent from the extraction column to waste at the extraction step and to the analytical column at the elution step. Therefore, the whole process from extraction to mass spectrometric analysis is fully automated without any manual interventions. This integrated process provides a significant improvement in overall sample throughput by saving both time and labor (Xu et al. 2009). In the past 20 years, online extraction methods have been widely used and become an alternative to automated off-line techniques (PPT, LLE, or SPE, etc.) in quantitative LC-MS bioanalysis (Wei et al. 2015).

In this chapter, in addition to a review of the various hardware configurations of online extraction systems, the most commonly used online extraction techniques are discussed in detail. These techniques include turbulent/high flow chromatography, restricted access media (RAM), monolithic chromatography, hydrophilic interaction liquid chromatography (HILIC), immunoaffinity sorbents, online SPE, online SPE-MS, and online dried blood spot (DBS) extraction. Furthermore, general strategies for online extraction method development are also discussed.

Sample Preparation in LC-MS Bioanalysis, First Edition. Edited by Wenkui Li, Wenying Jian, and Yunlin Fu.

2.2 System Configuration

2.2.1 Single-column System

As published by Ayrton et al. (1997), the single-column system has no analytical column included. The extraction column serves two functions – sample extraction and analytical separation. The single-column configuration provides the simplest and fastest online extraction. However, it was also found that this system gave no or little chromatographic separation (Jemal et al. 1998) and the co-eluted unwanted components from the matrix may cause ion suppression or enhancement (Jemal 2000). Due to these drawbacks, the single-column approach has not been widely used.

2.2.2 Dual-column System

The hardware configuration of a simple homemade online extraction system is depicted in Figure 2.1. The system consists of an autosampler, two LC pumps, an extraction column, a six-port switching valve, an analytical column, and a mass spectrometer detector (Huang et al. 2006). The autosampler, analytical column, and mass spectrometer used in this system are the same as those in a regular LC-MS/MS system. A program-controlled electrically actuated high-pressure six-port switching valve is configured into the system to control the two-column switching modes, i.e. straight-flushing column switching (Figure 2.1, Huang et al. 2006) and back-flushing column switching (Figure 2.2, Mao et al. 2007). The back-flushing is used to minimize peak broadening of analytes and maximize the washing of strongly retained matrix components. Two binary high-pressure mixing HPLC pumps are configured to the system. Pump 1 is used to deliver a high flow of the mobile phase through the extraction column to load and wash the sample and to equilibrate the extraction column. Pump 2 is used to deliver a gradient flow to elute the analytes from the extraction column and to perform the separation on the analytical column.

2.2.3 Staggered Parallel Online Extraction Systems

In LC-MS/MS bioanalysis, an assay consists of a series of individual injections, LC separation, and MS/MS detection of sample extracts in a sequential manner. The majority of the MS/MS instrument time is spent on either waiting for the analytes to be eluted from the analytical column or waiting for the column to be re-equilibrated to the initial condition prior to the next injection cycle. The mass spectrometer is actually idle during the waiting period. For online extraction, the extraction column often adds more time for re-equilibration between injections. In order to increase efficiency and sample throughput, some laboratories developed systems to analyze more than one sample simultaneously during the period of a single chromatographic run time (Wu 2001). These systems,

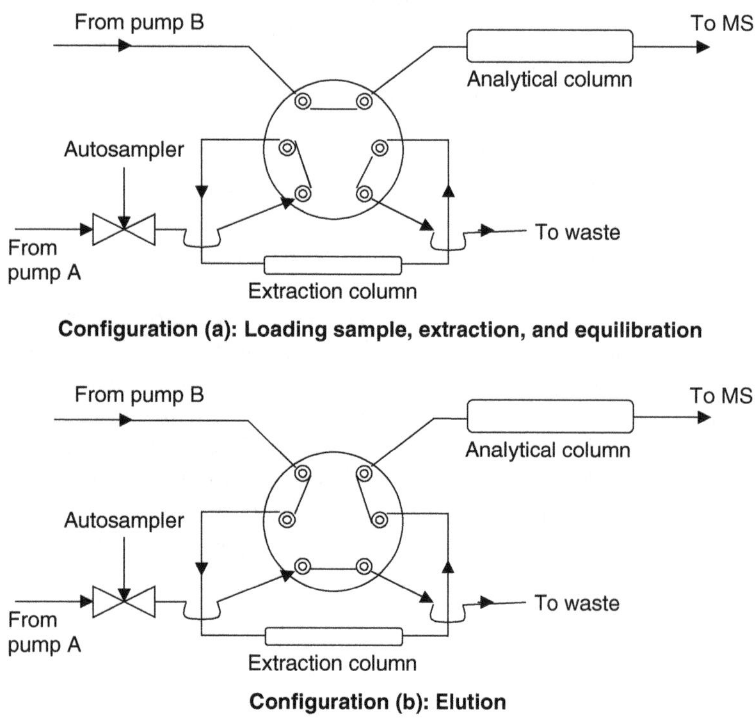

Figure 2.1 A schematic representation of online extraction LC-MS/MS system with straight-flushing column switching.

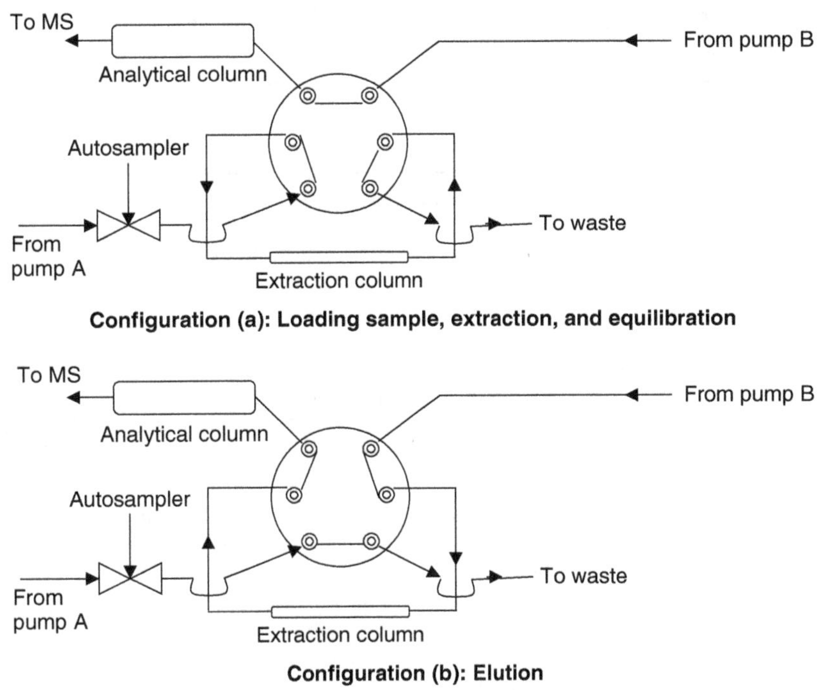

Figure 2.2 A schematic representation of online extraction LC-MS/MS system with back-flushing column switching.

often referred to as staggered parallel analysis, consisted of multiple extraction columns and analytical columns in parallel with staggered injections at specified intervals within the chromatographic run time. The mass spectrometer data collection was scheduled for each injection. For instance, a two-extraction column and two-analytical column configuration allowed the use of one extraction column for analyte extraction while the other was being equilibrated in preparation for the next injection cycle. Likewise, while separation was being carried out in one analytical column, the other one was undergoing equilibration. Consequently, in the total injection-to-injection cycle time window, there was no need for the extra time for re-equilibration of both the extraction and analytical columns. This resulted in a shorter run time and higher sample throughput of the intended assay sequence (Xia et al. 2001). An even more aggressive configuration was also reported. The system consisted of four extraction columns, four analytical columns, and four-sprayer electrospray interface in one mass spectrometer (Deng et al. 2001). These systems were typically constructed by the users themselves with commercially available HPLC systems, multichannel switching valves, online extraction columns, and analytical columns. In this regard, the users had to deal with the plumbing and electrical connections and make sure the switching valves work properly. However, due to the complicated controlling programs, flow paths, and valve-switching procedures, the staggered parallel multiplexing systems tended to be less robust than the simple one-extraction-column and one-analytical-column system, as discussed in Section 2.2.2.

2.3 Commonly Used Online Extraction Techniques

The core component of any online extraction system is the extraction column. Generally, the properties of the sorbents packed inside the extraction column are associated with the performance of the method. In the past two decades, many different types of extraction sorbents have been developed for online extraction. Table 2.1 summarizes the main mechanisms and representative applications for each type of the sorbents.

2.3.1 Turbulent/High Flow Chromatography

Turbulent flow chromatography is one of the most commonly used online extraction techniques. Its principles and numerous applications have been well reviewed (Rudewicz 2011). The technique employs specialized online extraction columns with small internal diameter (ID, typically 0.5 or 1 mm), which is packed with large particles of a stationary phase material (typically 30–50 μm) and is run at a high flow rate of a mobile phase (typically 1.2 ml min^{-1} for 0.5 mm ID columns and 4 ml min^{-1} for 1 mm ID columns). Under these conditions, it is generally believed that a turbulent flow is formed, which results in a rapid removal of proteins and

Table 2.1 Summary of extraction mechanisms and representative applications of commonly used online extraction techniques in quantitative LC-MS/MS bioanalysis.

Online extraction technique	Extraction mechanism	Examples of application	References
Turbulent/high flow chromatography	Columns with small internal diameter packed with large particles at a very high flow rate of mobile phase, resulting in rapid removal of large molecules in the matrix and retention of the small-molecule analytes of interest.	A drug candidate and its epimer metabolite Nefazodone and its three metabolites MK-0974	Huang et al. (2006) Mao et al. (2007) Xu et al. (2009)
Restricted access media (RAM)	RAM have small pores with sizes ranging from 6 to 25 nm and use size exclusion and hydrophilic shielding to restrict large molecules from accessing the inner adsorption surface and allow the small molecule analytes to access the pores and adsorb to the inner surface.	Estrogens	Beinhauer et al. (2015)
Monolithic chromatography	Monolithic columns consist of one piece of silica rod or organic polymer with flow-through pores. The smaller pores provide large surface areas for sufficient separation capacity and the larger pores reduce flow resistance.	Amprenavir and Atazanavir	Xu et al. (2006)
Hydrophilic interaction liquid chromatography (HILIC)	HILIC uses hydrophilic stationary phase (e.g. bare silica and cyclodextrin-based packings) and is similar to normal-phase chromatography. Analytes are eluted with a relatively hydrophobic binary mobile phase. The retention of a polar analyte on a HILIC column is proportional to its polarity.	Taspoglutide	Heinig and Wirz (2009)
Immunoaffinity extraction (IAE)	IAE uses antibody–antigen interactions to isolate the analyte from the matrix. The antibodies are typically immobilized by either covalent binding or noncovalent adsorption onto a pressure-resistant solid support that can be packed into an extraction column. The immobilized antibodies can remove a specific analyte from other matrix components with high selectivity and sensitivity.	Type II collagen neoepitope Pepsin and pepsinogen	Berna et al. (2006) Neubert et al. (2010)
Disposable extraction cartridges – online SPE	In an online SPE system, the extraction cartridge is for a single use or at most for several. The extraction mechanism is the same as for off-line SPE depending on the type of the sorbents used (e.g. C18, C8, C4, and phenyl).	Propranolol and diclofenac	Alnouti et al. (2005)
Online dried blood spot (DBS) extraction	The compounds are first eluted by flow-through desorption from the DBS card online onto an SPE cartridge and then eluted from the SPE cartridge online onto an analytical column followed by MS analysis. The fully automated DBS extraction methods extract compounds from filter paper cards without DBS punching and sample pretreatment.	Mephedrone, phentermine, amphetamine, and other stimulates	Verplaetse and Henion (2016)
Online SPE-MS	This online SPE-MS system couples SPE cartridge directly to mass spectrometer without liquid chromatographic separation to achieve fast SPE-MS analysis. The extraction mechanism is the same as for online SPE.	Structurally diverse compounds	Jian et al. (2011a)

other large molecules in the matrix, and the retention of the small-molecule analytes of interest. Arguably, it has also been reported that turbulent flow did not occur under these conditions (Ayrton et al. 1998). Therefore, in addition to turbulent flow, other names (e.g. ultra-high flow and high flow) have been used in the literature for this type of online extraction methods (Ayrton et al. 1998; Jemal et al. 1998; Wu et al. 2001).

A typical experimental procedure for turbulent flow chromatography using the dual-column system (Figures 2.1 and 2.2) includes valve switching periodically between analyte extraction and elution for LC separation and MS/MS detection. Once a plasma/serum sample with internal standard(s) is injected, it is loaded onto the extraction column with a 100% aqueous mobile phase delivered by pump 1 at a high flow rate of 3.0–5.0 ml min^{-1}. The proteins and other matrix components in the sample are washed away and the analytes are retained on the extraction column. The eluent is directed to waste to protect the mass spectrometer from matrix contamination. This loading and washing stage lasts for <0.5 minutes (Figures 2.1 and 2.2, configuration (a)). The switching valve is then switched to the other position, at which the extraction column is in line with the analytical column, mass spectrometer, and the flow path of pump 2 (Figures 2.1 and 2.2, configuration (b)). Analytes are eluted from the extraction column to the analytical column with a gradient organic/aqueous mobile phase delivered by pump 2 at a flow rate of 0.5 ml min^{-1}. This elution stage completes in ~1.0 minute before the switching valve is switched back to the original position (Figures 2.1 and 2.2, configuration (a)). The extraction column is then reconditioned under the same condition as the extraction stage. This is the equilibration stage. At the same time, the analytes are being separated on the analytical column and subsequently analyzed by the mass spectrometer. A total run time is ~2.0 minutes.

2.3.2 Restricted Access Media

The fundamental characteristics and applications of RAM as the sorbent materials of extraction columns have been well described in a recent review article (Yang et al. 2013). The particles packed within RAM columns have small pores with sizes ranging from 6 to 25 nm (Yang et al. 2013). The outer surface of the particles functions as both size exclusion and hydrophilic shielding to restrict large molecules such as proteins from accessing the inner adsorption surfaces. The small molecule analytes of interest and other small molecules in the matrix are able to access the pores and be adsorbed to the inner surfaces by diverse mechanisms including hydrophobic interaction. Therefore, when the sample is uploaded, proteins and large molecules quickly pass through the column and are washed out to waste, while the small-molecule analytes are retained and eluted for following analysis.

The most commonly used RAM support is alkyl-diol-silica (ADS) sorbent (LiChrospher™ ADS), which features conventional C4-, C8-, and C18-bonded phases in the inner pores. Other RAM columns have also been introduced for online extraction of biological samples (Cassiano et al. 2012). For example, attaching ion-exchange functional groups to the inner surface of RAM allows for retention of charged or polar molecules through electrostatic interactions. In a recent report by Beinhauer et al. (2015), an online weak cation exchange (WCX) RAM-based extraction column (Shi-Pack MAYI-WCX(G) HP, 10 mm × 4.6 mm) was used to capture trace estrogens from human and fetal bovine serum. After derivatization, the extracted samples were injected onto the WCX RAM extraction column. Proteins and other un-retainable components in the matrix were washed out to waste with water/methanol (98/2, v/v) at a flow rate of 2 ml min^{-1}. To increase the sensitivity of the method, a multiple injection-loading scheme was utilized to inject the derivatized samples. Ten consecutive injections each with 30 µl sample loading and 1 minute washing per injection were made to inject a 300 µl sample for accumulated trapping and continuous washing. After multiple injections and washes were completed, the trapped estrogens were reverse-eluted onto a C18 analytical column for separation and LC-MS/MS analysis. The analytical column was set at 50 °C and the separation was achieved by a gradient method with water/methanol mobile phases containing 2.5% formic acid (v/v) and 20 mM ammonium formate at a flow rate of 0.5 ml min^{-1}. The total cycle time was 15 minutes per sample. The lower limit of quantitation was 5 pg ml^{-1}.

To enhance extraction selectivity, molecularly imprinted polymers (MIP) have also been applied in RAM particles. MIPs possess the same three-dimensional shape and functional group positions of the imprinted molecules. Because the recognition site is complementary to the structure and spatial orientation of the imprinted molecule, the RAM MIP particles are ideal for chiral analysis. By using an imprinted chiral extraction center of MIP, the desired enantiomer can be directly isolated from biological matrices. For example, an RAM MIP particle imprinted for propranolol utilized glycerol monomethacrylate and glycol dimethacrylate as a hydrophilic surface modification to analyze various β-blockers in biological fluids by direct injection online extraction (Sanbe and Haginaka 2003).

2.3.3 Monolithic Materials

Monolithic columns, sometimes also called continuous bed, have been used as an alternative to particulate columns for fast chromatographic separation due to their high permeability and low backpressure (Chen et al. 2009). They consist of a one-piece silica rod or organic polymer with flow-through pores and have a unique biporous structure. The smaller pores provide large surface areas for sufficient separation capacity, while the larger pores considerably reduce flow resistance (Minakuchi et al. 1996). These two structural characteristics together allow for efficient separation at a high flow rate and make monoliths suitable for online extraction of biological fluids. In online extraction systems, monolithic columns can be used in either single-column mode, where the column performs both the extraction and separation roles, or two-column mode, where the column serves as the analytical column for separation in conjunction with a different extraction column. Huang et al. (2006) reported the use of a monolithic column as the analytical column coupled with high-flow direct-injection LC-MS/MS to quantify a drug and its epimer metabolite in rat plasma. An Oasis column was used as the extraction column. The plasma samples (5 µl each) were injected onto the extraction column run with a mobile phase of aqueous 0.05% formic acid at a flow rate of 4.0 ml min^{-1}. Chromatographic separation on baseline was achieved on the monolithic column for the two analytes with a mobile phase of acetonitrile/aqueous 0.05% formic acid at a flow rate of 3.2 ml min^{-1}. The results were compared with those obtained using a conventional C18 column. Although both systems had the same sample extraction, LC and MS conditions, the run time was much shorter with the monolithic column than that with the conventional C18 column. As a result, the method with monolithic column increased the sample throughput by a factor of 2, with equivalent assay accuracy, precision, sensitivity, and chromatographic separation. Xu et al. (2006) described an online extraction procedure using monolithic sorbent as the extraction material to determine Amprenavir and Atazanavir in human plasma by LC-MS/MS. They used a short monolithic C18 4.6 mm × 10 mm cartridge for high flow extraction at 4 ml min^{-1}. The plasma samples were extracted via protein precipitation with acetonitrile, and then the supernatant was diluted and loaded onto the monolithic cartridge. A conventional C18 column was used as the analytical column for separation with a mobile phase of acetonitrile/2 mM ammonium acetate in 0.1% formic acid (70/30, v/v) at a flow rate of 0.3 ml min^{-1}. The run time was four minutes for each sample. The backpressure of the monolithic extraction cartridge stayed the same after 450 samples being extracted. The monolithic sorbent online extraction method showed similar accuracy and precision, as compared with an automated 96-well LLE assay for the same compounds. Besides hydrophobic interactions for the reverse-phase mode, monolithic materials can be modified for other functionalities. For example, a weak cation exchange monolithic column was used for the simultaneous determination of alpha-adrenergic receptor antagonists (terazosin, alfuzosin, prazosin, and doxzosin) in human plasma (Wei et al. 2007). In addition, the monolithic materials are also well suited for miniaturized online extraction systems, as they are made of a single piece of separation medium and more easily prepared in narrow columns by attaching to the inner wall of the column. This is ideal for the analysis of analyte of interest with very low sample volume (Luo et al. 2007).

2.3.4 Hydrophilic Interaction Liquid Chromatography

Most of the reported online extraction systems are featured with a reversed-phase extraction column coupled with a reversed-phase analytical column. The combination is compatible with various mobile phase compositions for different classes of compounds. However, quantitative analysis of highly polar compounds by reversed-phase LC-MS/MS has often been a challenge due to their poor chromatographic retention. In order to achieve reasonable retention for those polar compounds, a highly aqueous mobile phase is required for reversed-phase chromatography, which might cause column phase collapse (dewetting of the stationary phase) and sensitivity decrease in MS/MS detection. To improve LC-MS/MS performance for the analysis of polar compounds, HILIC has been introduced as a valuable technique (Jian et al. 2011b; Naidong 2003). In principle, HILIC is similar to normal-phase chromatography. In HILIC, analyte interacts with a hydrophilic stationary phase such as bare (non-bonded, underivatized) silica and cyclodextrin-based packings, and are eluted with a relatively hydrophobic binary mobile phase. The retention of polar analytes on a HILIC column is proportional to their polarities and inversely proportional to the polarity of the mobile phase. Water-miscible polar organic solvents such as acetonitrile and methanol were typically used for HILIC to achieve good retention. The organic content in the mobile phase is typically greater than 70%. This highly volatile organic mobile phase is usually ideal for analyte desolvation and ionization, which leads to an increased sensitivity for MS/MS detection. Because water is the strongest elution solvent in HILIC, samples extracted by online SPE can be eluted with high organic mobile phase directly onto a HILIC column. Deng et al. (2005) coupled an online reversed-phase high-flow extraction system with a silica column operated under HILIC conditions for quantitation of polar compounds in biological fluids. The online extraction system employed a Cohesive column (50 mm× 1 mm) as the extraction column and a Hypersil silica column as the analytical column. The plasma samples were loaded on to the extraction column with an aqueous 0.1% formic acid solvent at a flow rate of 4 ml min^{-1}. After one minute, the valve that connects the extraction and the analytical columns in tandem in the flow path of the separation pump was switched to the elution position. The polar compounds retained on the extraction column were eluted onto the analytical column with a highly aqueous mobile phase. It was observed that the relatively large volume of water transferred from the extraction column to the analytical column with an isocratic elution resulted in broad and tailing chromatographic peaks. To overcome this issue, a gradient elution was applied to the separation pump, starting from a high organic content (90% acetonitrile in water with 0.2% formic acid) and ending at a low organic content (10% acetonitrile in water with 0.2% formic acid). The incorporation of a gradient elution resulted in much better peak shapes of the analyte of interest. The orthogonal separation effect from this online reversed-phase SPE–HILIC configuration minimized the matrix effects and significantly improved sensitivity for highly polar compounds. A similar approach was employed for the quantitation of morphine, morphine-3-glucuronide, and morphine-6-glucuronide in microdialysis and plasma, for which a combination of HyPurity C18 guard column (10 mm × 3 mm) and ZIC-HILIC analytical column was utilized (Bengtsson et al. 2005). Another case was for the determination of fentanyl in dog plasma with the use of a Waters Oasis hydrophilic–lypophilic balanced extraction column and a Waters Atlantis HILIC Silica analytical column (Chang et al. 2007). HILIC column can be used for the extraction of hydrophilic compounds; however, the samples must be loaded with an aqueous solvent containing high organic percentage. In an online HILIC-reserved phase LC configuration, a HILIC column serves as an extraction column to extract the analyte for chromatography on a reversed-phase analytical column. For optimal separation on the reversed-phase analytical column, the analytes need to be eluted from the HILIC column with low organic solvent. However, this low organic solvent is not compatible with HILIC column elution conditions, which normally require an aqueous solvent containing 45–95% acetonitrile. This limits the use of HILIC column for online extraction. Because of this reason, there have been only a few applications using this approach in combination with reversed-phase chromatography for the analysis of analytes of interest, such as cerebral neuropeptides in rat brain tissue (Mihailova et al. 2008) and synthetic peptide drug taspoglutide in human plasma (Heinig and Wirz 2009).

2.3.5 Immunoaffinity Extraction

Immunoaffinity extraction (IAE) technique, also referred to as immunoaffinity solid-phase extraction (IA-SPE), utilizes antibody–antigen interactions to isolate the analytes of interest from the matrix. Typically, antibodies are immobilized by either covalent binding or noncovalent adsorption onto a pressure-resistant solid support that can be packed into an extraction column (Moser and Hage 2010). The immobilized antibodies can specifically remove an analyte from the matrix, which provides high selectivity and associated high sensitivity for the intended assay.

Berna et al. (2006) developed and validated an online IAE-LC-MS/MS assay with column switching for the determination of a type II collagen peptide biomarker, NET2C, in rat urine. The online IAE was demonstrated as the most suitable form of sample preparation for the analysis of low levels of NET2C in urine matrix. The IA column was packed with the Poros column sorbent coupled with the 9A4 antibody, which is specific for the collagenase-generated C-terminal neoepitope. The Poros IA column had the capacity of handling a relatively large volume of rat urine (0.5 ml injection^{-1}) and more than 2000 injections before replacement.

Neubert et al. (2010) reported an online immunoaffinity LC-MS/MS system that combined high-flow peptide immunoaffinity enrichment and nanoflow LC-MS/MS for the measurement of total salivary pepsin/pepsinogen in human saliva. The online IAE columns were built in-house using an antipeptide antibody against the pepsin C-terminal sequence, DRANNQVGLAPVA (target peptide). Saliva was heat inactivated to quench residual enzymatic activity followed by digestion with endoproteinase AspN. The equilibration of IAE column was carried out by running 25 μM ml^{-1} ammonium formate (pH 7.0) at a flow rate of 0.3 ml min^{-1}. A 230 μl of sample mixture was loaded in three minutes using the same buffer and flow rate. The target peptide and its stable isotope-labeled internal standard (IS) were retained on the IAE column, while other matrix components were washed to waste. The extracted target peptide and IS were first eluted from the IAE column to a PepMap100 C18 cartridge using 0.2% formic acid at 0.3 ml min^{-1}, and then eluted from the C18 cartridge onto a PepMap100 C18 analytical nanocolumn at a flow rate of 400 nl min^{-1}. A Sciex 4000 Qtrap mass spectrometer with a Nanospray II interface was used for MS/MS analysis. The assay was validated for a range of 4.08–2980 pmol l^{-1} for human pepsin/pepsinogen. This online IA enrichment combined with nanoflow LC-MS/MS approach provided a sensitive platform for protein/peptide bioanalysis at low picomolar levels.

In an ideal instrumentation setup (Dufield and Radabaugh 2012), the IAE-LC-MS/MS system is composed of two pumps (one binary HPLC and one quaternary), one autosampler capable of large volume injections (1–2 ml typically), two 6- or 10-port two-position switching valves, and one quadruple or Q-Trap mass spectrometer. For IAE with trapping/enrichment function, there are one IA column for extraction, one SPE (trapping) column for enrichment, and one analytical column for analysis. The sample is loaded at high volume through autosampler onto IAE column, which allows the analytes of interest to be extracted while the flow goes to waste. The loading mobile phase is typically 25–50 mM ammonium acetate at pH 7.0. After about three minutes, the extracted analytes on the IA column are eluted to the SPE column by lowering the pH using 0.1–1% formic acid. The analytes are concentrated and desalted on the SPE column for three minutes and then eluted onto the analytical column for MS/MS analysis. It is important to adjust the concentrations of acids and organics of the mobile phase to maintain the integrity of the IAE column with antibody bound to the immunosorbents. The authors pointed out that different SPE columns showed significantly different impacts on assay sensitivity. The selection of trapping column, therefore, plays an important role in the method development as well.

2.3.6 Disposable Extraction Cartridges – Online SPE

In general, online extraction columns or cartridges can be divided into two categories: reusable and disposable. The online extraction columns described in Sections 2.3.1–2.3.5 are reusable ones, which can be reused for several hundred samples. A reusable extraction column is “fixed” in the system configuration for continuous extraction of study samples during an analytical run. In contrast, a disposable extraction cartridge is replaced by a new one after extracting a single sample or several samples via programmed procedures during an analytical run. Due to the sophisticated mechanical design and software engineering requirement, the systems with disposable cartridge settings are usually made by vendor. An example of the online extraction systems using disposable extraction cartridges is the Prospekt-2/Symbiosis system (Spark Holland, The Netherlands), which consists of an autosampler, a cartridge exchange unit, and a pumping system. In this system, extraction cartridges are loaded into 96-well trays. Cartridge exchange and valve switching are performed by the Automated Cartridge Exchange module. A robot is used to move an unused extraction cartridge online from the cartridge tray and sends it back into the tray after extraction.

Alnouti et al. (2005) investigated the online SPE Symbiosis System for fast bioanalytical method development, method validation, and sample analysis. The online SPE LC-MS/MS method was developed using the automated mode for the determination of model compounds, propranolol and diclofenac, in rat plasma. For SPE cartridge selection and extraction condition optimization, a method development tray with 12 different types of cartridges, which represented a wide spectrum of packing material polarities, was automatically screened with a generic extraction method. The screening process was initially to identify one type of cartridge with satisfactory recovery, retention, and peak shape. Then, optimization was carried out for washing and elution conditions by varying solvent composition and flow rate to obtain desired retention and elution. Finally, a C18 HD (2 mm × 10 mm) cartridge was selected based on the highest recovery, retention, and satisfactory peak shape for both propranolol and diclofenac. The automated method development only required four hours. For the plasma sample extraction process, a new cartridge was first conditioned with organic solvent followed by aqueous solvents. After that, the sample was transferred from the autosampler loop injector onto the conditioned cartridge by transfer solvent. The loaded cartridge was flushed with a highly aqueous solvent to wash out salts and large molecules such as proteins from the plasma sample. Then, the cartridge was moved by a robotic arm to a different position in the system for elution. During this period, another new cartridge was placed in the conditioning position for conditioning, loading, and washing. Reusing the same cartridge up to 20 times showed no significant changes in the analyte recovery and the cartridge backpressure.

2.3.7 Online Extraction of Dried Blood Spot (DBS) Samples

In the past 10 years, the DBS technique has gained increased interest in the bioanalytical community. Many laboratories across the pharmaceutical industry have investigated the DBS sampling as an alternative to conventional plasma/serum/blood sampling for preclinical and clinical studies (Henion et al. 2013; Li and Tse 2010). By using this technique, a 5–30 µl of whole blood is collected onto a paper/cellulose card. Upon complete dryness, the DBS samples can be transferred to bioanalytical laboratories for analysis. For preclinical studies using rodent animals, by using DBS sampling, blood volumes and animal numbers can be significantly reduced for pharmacokinetic and toxicokinetic assessment. This is perfectly in line with the 3R principles (reduction, refinement, and replacement). Clinically, this approach of blood collection offers a number of advantages over conventional whole blood, plasma, and serum sample collection techniques (Li and Tse 2010). These include but are not limited to: (i) collecting clinical blood (e.g. PK) samples from patients in remote or developing regions in resource-limited settings and (ii) permitting flexible or frequent blood sample collection in a point-of-care setting.

However, from bioanalytical perspectives, it is quite challenging to develop an automated LC-MS assay for quantitative analysis of DBS samples. To analyze DBS samples, additional steps must be taken in the sample preparation process to extract the analytes of interest from the collection cards into a liquid. These steps typically include punching a disc from the center of the DBS, transferring it into an extraction tube or a 96-well plate well, and extracting the analytes with solvent containing an IS. The punching step is very tedious, labor intensive, and time consuming, especially when a large number of samples are prepared. Although DBS punching robots are commercially available, manual transfer steps still remain in the sample preparation workflow, as the DBS punching and sample extraction procedures are carried out on different instruments. To overcome these barriers in DBS analysis, online DBS extraction methodologies have been explored to eliminate DBS punching and sample transfer steps.

Fully automated online DBS extraction methods have been reported by Ganz et al. (2012), Oliveira et al. (2014), and Verplaetse and Henion (2016). The Sample Card And Prep (SCAP) DBS System (Prolab Instruments GmbH, Reinach, Switzerland) was used by both Ganz et al. and Oliveira et al. A robotic gripper executed the transfer of DBS cards from the sample rack to the pneumatically controlled clam module. DBS cards were integrated into the LC flow path for online extraction by the clamp module. Ganz et al. used three columns, i.e. two extraction columns for successive online DBS sample cleanup and enrichment and one analytical column for chromatographic separation. Oliveira et al. employed one extraction column and one analytical column coupled with a high-resolution quadrupole time-of-flight mass spectrometer. In Verplaetse and Henion's work, hardware instrumentation was from Spark Holland (Emmen, The Netherlands). The DBS cards were manually placed into the card racks of DBS card autosampler. The card racks were inserted into the DBS card autosampler by robotic arm and the card was clamped. The compounds were eluted by flow-through desorption from the DBS card online onto the SPE cartridge. The targeted analytes were then eluted from the SPE cartridge onto the LC column followed by MS/MS analysis. These fully automated DBS extraction methods enabled the online extraction of compounds from filter paper cards without the need of DBS punching, sample pretreatment, and sample transfer steps.

2.3.8 SPE-MS

Besides the online extraction-LC-MS techniques discussed above, some bioanalytical laboratories have investigated another approach that couples SPE cartridge directly with mass spectrometry detection without liquid chromatographic separation to achieve higher analysis speed. In a typical online SPE-MS method, a SPE cartridge can be used for several hundreds or thousands of injections. One of the commercially available online SPE-MS systems is the Agilent RapidFire™ High-throughput Mass Spectrometry System, which uses a microscale and reusable SPE, and is fully automated. For a typical run cycle, the sample is injected by a sample aspirator and loaded onto the online SPE cartridge with an aqueous mobile phase, which also serves as A washer to wash away the non-retaining components such as salts, buffer, and proteins to waste. Then, the SPE cartridge is backflushed with a relatively high organic content mobile phase to directly elute the analyte to the mass spectrometer for analysis. Typically, the entire run cycle that consists of injection, load/wash, elution, and re-equilibrium takes only 5–10 seconds per sample, which is significantly faster than other LC-associated online extraction systems. The online SPE cartridges packed with different sorbents, e.g. C18, C8, C4, and phenyl, are commercially available for analytes with a variety of physicochemical properties. The online SPE system can be programmed to switch automatically between six cartridges during a run without interrupting the injection sequence. This online SPE-MS system has mostly been applied in various *in vitro* high-throughput screening assays including cell culture and microsomal preparations for therapeutic targets and new drug candidates in early drug discovery. To assess the feasibility of utilizing this online SPE-MS system for *in vivo* bioanalysis, Jian et al. (2011a) analyzed a set of samples from preclinical and clinical studies using some generic SPE operating conditions. A generic and nonselective protein precipitation method was used as sample pretreatment. Critical bioanalytical parameters such as matrix effects and carry-over were evaluated for structurally diverse compounds. The evaluation of the matrix effect showed significant ionization suppression for most compounds tested, indicating insufficient removal of the matrix interference by the online SPE system. Nevertheless, the matrix effect was effectively compensated by the isotope-labeled ISs, suggesting that the selection of reliable ISs is critical to the success of the online SPE-MS assays. Carry-over was found to be variable among the tested compounds and could be reduced by inserting matrix blanks without sacrificing the assay efficiency due to the ultrafast analysis speed. The results showed that the presence of a stable metabolite of a test compound had no impact on the quantitation of the parent compound. However, due to the lack of chromatographic separation, in-source back-conversion of conjugated metabolites could be an issue for the fast SPE-MS format, resulting in overestimation of the parent concentration. The experiments demonstrated that this back-conversion could be reduced by optimizing the ion source conditions to minimize in-source breakdown. Overall, the results from the online SPE-MS for the test compounds correlated well with those from the conventional LC-MS analysis. Comparable accuracy, precision, linearity, and sensitivity were achieved with analysis speed 20–30 times higher. The authors concluded that the online SPE-MS system exhibited high potential for high-throughput bioanalysis, especially for the early stages of drug discovery where a large number of drug candidates need to be dosed and analyzed for their pharmacokinetic properties, or for the late-phase clinical studies, for which the analyte(s) and the metabolite(s) are normally well characterized and the stable isotope-labeled IS(s) is readily available.

2.4 Considerations for Utilizing Online Extraction Techniques

2.4.1 Advantages and Limitations

As an alternative to off-line SPE methods, online extraction techniques have some noticeable advantages. By performing column conditioning, washing, and elution in an integrated system automatically, online extraction significantly reduces sample preparation time and hands-on labor work with a higher throughput as compared to conventional off-line approach. On the other hand, online extraction techniques can help minimize potential analyte loss due to nonspecific adsorption of analyte of interest onto the surface of sample tube or vial. Nonspecific adsorption often takes place after the analyte has been extracted from the biological matrix, e.g. in the evaporation step, where the analyte has lost its binding to the matrix components such as proteins. This can result in poor assay performance and loss of sensitivity. Using online extraction, the analyte in biological matrix was directly injected onto the online extraction system, which reduces the risk of analyte loss due to nonspecific adsorption. In contrast to off-line extraction method by which only a small aliquot of the sample extract is injected onto the analytical columns for LC-MS analysis, by using online extraction, the entire sample extract from the extraction column is transferred to the analytical column. In addition, online extraction is carried out in a closed system that limits the exposure of analysts to hazardous samples or solvents. Online SPE extraction is preferable for the analysis of volatile compounds in the study samples. This was demonstrated by the measurement of 18 perfluorinated organic acids and amides at trace levels in human serum using the Symbiosis system (Kuklenyik et al. 2005), which was equipped with a HySphere HD C18 cartridge (1 mm × 10 mm) and a C8 analytical column. An off-line SPE method was developed prior to setting up the online SPE method. However, due to the compound loss during evaporation and preconcentration processes after the SPE, only 13 out of 18 perfluorinated organic acids and amides were detected. The online SPE LC-MS/MS helped to eliminate these steps, which resulted in increased recovery, sensitivity, and precision for all 18 analytes.

While online extraction systems offer many advantages as discussed above, limitations do exist. First, it requires more time for method development, which includes setting up, tuning, and optimization of various parts of the system. Second, some additional effort is often needed to reduce carryover that is particularly associated with reusable extraction columns. Third, it is often not straightforward to transfer an online extraction-based method to a different laboratory, in particular if the method is based on homemade online extraction systems.

2.4.2 Strategies for Online Extraction Method Development

Although online extraction techniques have been widely used in bioanalysis, method development can be challenging. This is because that many user-adjustable parameters can affect method performance and need to be optimized during method development. Xu et al. (2008) described strategies for efficient method development of online extraction assays. They developed and validated two turbulent flow chromatography assays for determination of MK-0974 in human plasma and urine using a Cohesive Aria™ 2300 system (Cohesive Technologies Inc., Franklin, MA, USA). This online extraction system included two quaternary Flux pumps, a valve module, and a CTC HTS autosampler. Several issues, such as inconsistent recovery, carryover, and nonspecific binding of the analyte to urine container, were successfully addressed. Based on this work, the authors proposed a four-step strategy for efficient online extraction method development. The first step, the same as other LC-MS/MS assay development, was to evaluate the LC-MS/MS conditions under a single-column configuration. Several different analytical columns were screened and FluophasePFP (50 mm × 3 mm, 5 μm) from Thermo-Hypersil keystone was selected. Desirable peak shape and retention were obtained for MK-0974 when 60% acetonitrile in 0.1% formic acid aqueous solution was used as the mobile phase. The second step was a quick start with online extraction by selecting the basic elements of Cohesive high-turbulence liquid chromatography (HTLC) system under Cohesive quick-elution mode. The extraction columns, HTLC C18 and Cyclone, with different available column dimensions (0.5 mm × 50 mm and 1.0 mm × 50 mm) were compared. HTLC C18 (0.5 mm × 50 mm) was chosen because a similar recovery with less carryover and less consumption of solvent was featured using this column as compared to Cyclone. At the same time, mobile-phase composition and gradients were evaluated to retain the analytes on the extraction column after sample loading and washing steps and to separate them on the analytical column with a minimal carryover. At the end of this step, the online extraction HTLC/LC-MS/MS assay was established. The third step was to optimize the recovery and reduce matrix effects by selecting appropriate sample pretreatment, mobile-phase composition, and timing in loading and eluting steps. Prior to being injected onto the system, the plasma samples were treated with acid to disturb the protein binding for consistent recovery and with <15% of acetonitrile to ensure no protein precipitation. To evaluate the loading and eluting mobile phases, the extraction column was connected to the mass spectrometer directly. Neat solution was injected into the extraction column and eluted with mobile phases of different strengths to check on analyte breakthrough. The loading and eluting mobile phases were optimized to retain the analyte on the extraction column to the maximum and transfer the analyte from the extraction column to the analytical column completely. The loading and eluting time were also optimized to keep the run time short, but long enough to wash away the unwanted matrix components and transfer the analyte to the analytical column. The last step, also the most challenging one, was to minimize the carryover by optimizing the washing solutions and procedures for the autosampler, extraction column, analytical column, and switching valves. The injection needle position on the autosampler had to be carefully adjusted to minimize the dead volume at the injection valve. The needle wash solvents and washing program were optimized to completely clean up the protein from the syringe and injection valve. The wash mobile phases and washing procedures for the extraction column, analytical column, and switching valve were programmed into the HTLC method. With these washing steps, the carryover was reduced to <20% of the lower limit of quantification for both plasma and urine assays with 1000-fold dynamic range. For the urine assay, 37–80% analyte loss was observed in urine QCs during method development, which indicated adsorption of analyte to the container surface. Different amounts of bovine serum albumin (BSA) addition were tested and 1% BSA (urine volume) was selected to bring urine QC accuracy back to around 100% of nominal.

Another consideration is for LC-MS bioanalysis of analytes that form labile metabolites, such as acyl-glucuronides, N-oxides, and sulfate conjugates. These metabolites may back-convert to the parent drug by in-source fragmentation (Jemal et al. 2010). If the drug is not chromatographically separated from its conjugated metabolites or N-oxide metabolite, the concentration of the parent drug would be overestimated. This interference of metabolite to the drug analysis may not be predictable at the very early phase of drug development because the metabolite profile of the drug has not been fully understood at this stage. It is almost impossible to know the chromatographic conditions and elution time required to chromatographically separate the drug from its metabolites without authentic reference standards. Jemal et al. (2010) proposed a strategy using incurred samples to check the possible presence of metabolite interference with drug analysis in LC-MS during method development. A pooled incurred sample can be prepared from high-dose samples of different animals or human subjects, including different post-dosing time points. A portion of the pooled-incurred sample is extracted using the protein precipitation method. The sample extract is then analyzed using the chromatographic conditions of the method under development with the selected reaction monitoring (SRM) transitions of known and potential metabolites, in addition to the SRM transition of the drug. This procedure is then repeated using different chromatographic conditions and different columns to ensure chromatographic separation of the drug from its known or potential metabolites. If an additional peak appears in the drug SRM channel, it would indicate the presence of a metabolite which is not separated from the drug under the initially used chromatographic conditions.

2.5 Summary

The primary goal of sample preparation is to clean up, isolate, concentrate the target analytes from the sample matrix, and to constitute or reconstitute them in a form amenable to the subsequent LC-MS analysis. With integration of sample extraction, separation, and analysis into one system, the online extraction and column switching techniques discussed in this chapter have been widely used for high throughput in quantitative LC-MS/MS bioanalysis in support of drug discovery and development and therapeutic drug monitoring. A variety of online extraction columns and analytical columns are commercially available and these columns can be used to create various formats of online extraction and separation methods to fit the purpose of the intended assay. In most of the applications, an analytical column is necessary to achieve efficient chromatographic separation of analytes from other unwanted matrix components, and more importantly, to reduce the likelihood of matrix effects and to eliminate the interference of metabolites, particularly those that can be back-converted to the parent drug molecules.

References

Alnouti, Y., Srinivasan, K., Waddell, D. et al. (2005). Development and application of a new on-line SPE system combined with LC-MS/MS detection for high throughput direct analysis of pharmaceutical compounds in plasma. *J Chromatogr A.* 1080 (2): 99–106.

Ayrton, J., Dear, G.J., Leavens, W.J. et al. (1997). The use of turbulent flow chromatography/mass spectrometry for the rapid, direct analysis of a novel pharmaceutical compound in plasma. *Rapid Commun. Mass Spectrom.* 11 (18): 1953–1958.

Ayrton, J., Dear, G.J., Leavens, W.J. et al. (1998). Optimisation and routine use of generic ultra-high flow-rate liquid chromatography with mass spectrometric detection for the direct on-line analysis of pharmaceuticals in plasma. *J. Chromatogr. A* 828 (1–2): 199–207.

Beinhauer, J., Bian, L., Fan, H. et al. (2015). Bulk derivatization and cation exchange restricted access media-based trap-and-elute liquid chromatography-mass spectrometry method for determinaiton of trace estrogens in serum. *Anal. Chim. Acta* 858: 74–81.

Bengtsson, J., Jansson, B., and Hammarlund-Udenaes, M. (2005). On-line desalting and determination of morphine, morphine-3-glucuronide and morphine-6-glucuronide in microdialysis and plasma samples using column switching and liquid chromatography/tandem mass spectrometry. *Rapid Commun. Mass Spectrom.* 19 (15): 2116–2122.

Berna, M., Schmalz, C., Duffin, K. et al. (2006). Online immunoaffinity liquid chromatography/tandem mass spectrometry determination of a type II collagen peptide biomarker in rat urine: investigation of the impact of collision-induced dissociation fluctuation on peptide quantitation. *Anal. Biochem.* 356 (2): 235–243.

Cassiano, N.M., Barreiro, J.C., Oliveira, R.V., and Cass, Q.B. (2012). Direct bioanalytical sample injection with 2D LC-MS. *Bioanalysis* 4 (22): 2737–2756.

Chang, Y.W., Yao, H.T., Chao, Y.S., and Yeh, T.K. (2007). Rapid and sensitive determination of fentanyl in dog plasma by on-line solid-phase extraction integrated with a hydrophilic column coupled to tandem mass spectrometry. *J. Chromatogr. B* 857 (2): 195–201.

Chen, L., Wang, H., Zeng, Q. et al. (2009). On-line coupling of solid-phase extraction to liquid chromatography – a review. *J. Chromatogr. Sci.* 47 (8): 614–623.

Deng, Y., Zeng, H., Unger, S.E., and Wu, J.T. (2001). Multiple-sprayer tandem mass spectrometry with parallel high flow extractiono and parallel separation for high-throughput quantitation in biological fluids. *Rapid Commun. Mass Spectrom.* 15 (17): 1634–1640.

Deng, Y., Zhang, H., Wu, J.T., and Olah, T.V. (2005). Tandem mass spectrometry with online high-flow reversed-phase extraction and normal-phase chromatography on silica columns with aqueous-organic mobile phase for quantitation of polar compounds in biological fluids. *Rapid Commun. Mass Spectrom.* 19 (20): 2929–2934.

Dufield, D. and Radabaugh, M. (2012). Online immunoaffinity LC/MS/MS. Ageneral method to increase sensitivity and specificity: how do you do it and what do you need? *Methods* 56 (2): 236–245.

Ganz, N., Singrasa, M., Nicolas, L. et al. (2012). Development and validation of a fully automated online human dried blood spot analysis of bosentan and its metabolites using the sample card and prep DBS system. *J. Chromatogr. B* 885-886: 50–60.

Heinig, K. and Wirz, T. (2009). Determination of taspoglutide in human and animal plasma using liquid chromatography-tandem mass spectrometry with orthogonal column-switching. *Anal. Chem.* 81 (10): 3705–3713.

Henion, J., Oliveira, R.V., and Chace, D.H. (2013). Microsample analysis via DBS: challenges and opportunities. *Bioanalysis* 5 (20): 2547–2565.

Huang, M.-Q., Mao, Y., Jemal, M., and Arnold, M. (2006). Increased productivity in quantitative bioanalysis using a monolithic column coupled with high-flow direct-injection liquid chromtography/tandem mass spectrometry. *Rapid Commun. Mass Spectrom.* 20 (11): 1709–1714.

Jemal, M. (2000). High-throughput quantitative bioanalysis by LC/MS/MS. *Biomed. Chromatogr.* 14 (6): 422–429.

Jemal, M., Ouyang, Z., and Xia, Y.Q. (2010). Systematic LC-MS/MS bioanalytical method development that incorporates plasma phospholipids risk avoidance, usage of incurred sample and well thought-out chromatography. *Biomed. Chromatogr.* 24 (1): 2–19.

Jemal, M. and Xia, Y.Q. (2006). LC-MS development strategies for quantitative bioanalysis. *Curr. Drug Metab.* 7 (5): 491–502.

Jemal, M., Xia, Y.Q., and Whigan, D.B. (1998). The use of high-flow high performance liquid chromatography coupled with positive and negative ion electrospray tandem mass spectrometry for quantitative bioanalysis via direct injection of the plasma/serum samples. *Rapid Commun. Mass Spectrom.* 12 (19): 1389–1399.

Jian, W., Romm, M.V., Edom, R.W. et al. (2011a). Evaluation of a high-throughput online solid phase extraction – tandem mass spectrometry system for in vivo bioanalytical studies. *Anal. Chem.* 83 (21): 8259–8266.

Jian, W., Xu, Y., Edom, R.W., and Weng, N. (2011b). Analysis of polar metabolites by hydrophilica interaction chromatography-MS/MS. *Bioanalysis* 3 (8): 899–912.

Kuklenyik, Z., Needham, L.L., and Calafat, A. (2005). Measurement of 18 perfluorinated organic acids and amides in human serum using on-line solid-phase extraction. *Anal. Chem.* 77 (18): 6085–6091.

Li, W. and Tse, F.L.S. (2010). Dried blood spot sampling in combination with LC/MS/MS for quantitative analysis of small molecules. *Biomed. Chromatogr.* 24 (1): 49–65.

Luo, Q., Page, J.S., Tang, K., and Smith, R.D. (2007). MicroSPE-nanoLC-ESI-MS/MS using 10 μm-i.d. silica-based monolithic columns for proteomics. *Anal. Chem.* 79 (2): 540–545.

Mao, Y., Huang, M.-Q., Xia, Y.Q., and Jemal, M. (2007). High-throughput quantitation of nefazodone and its metabolites in human plasma by high flow direct-injection LC-MS/MS. *J. Pharm. Biomed. Anal.* 43 (5): 1808–1819.

Mihailova, A., Malerod, H., Wilson, S.R. et al. (2008). Improving the resolution of neuropeptides in rat brain with on-line HILIC-RP compared to on-line SCX-RP. *J. Sep. Sci.* 31 (3): 459–467.

Minakuchi, H., Nakanishi, K., Soga, N. et al. (1996). Octadecylsilylated porous silica rods as separation media for reversed-phase liquid chromatography. *Anal. Chem.* 68 (19): 3498–3501.

Moser, A.C. and Hage, D.S. (2010). Immunoaffility chromatography: an introduction to applications and recent developments. *Bioanalysis* 2 (4): 769–790.

Naidong, W. (2003). Bioanalytical liquid chromatography tandem mass spectrometry methods on underivatized silica columns with aqueous/organic mobile phases. *J. Chromatogr. B* 796 (2): 209–224.

Neubert, H., Gale, J., and Muirhead, D. (2010). Online high-flow peptide immunoaffinity enrichment and nanoflow LC-MS/MS: assay development for total salivary pepsin/pepsinogen. *Clin. Chem.* 56 (9): 1413–1423.

Oliveira, R.V., Henion, J., and Wickremsinhe, E. (2014). Fully-automated approach for online dried blood spot extraction and bioanalysis by two-dimensional-liquid chromatography coupled with high-resolution quadrupole time-of-flight mass spectrometry. *Anal. Chem.* 86 (2): 1246–1253.

Rudewicz, P.J. (2011). Turbulent flow bioanalysis in drug metabolism and pharmacokinetics. *Bioanalysis* 3 (14): 1663–1671.

Sanbe, H. and Haginaka, J. (2003). Restricted-access media-molecularly imprinted polymer for propranolol and its application to direct injection analysis of β-blockers in biological fluids. *Analyst* 128 (6): 593–597.

Unger, S., Li, W., Flarakos, J., and Tse, F.L.S. (2013). Roles of LC-MS bioanalysis in drug discovery, development, and therapeutic drug monitoring. In: *Handbook of LC-MS Bioanalysis, Best Practices, Experimental Protocols, and Regulations* (ed. W. Li, J. Zhang and F. Tse), 3–13. Hoboken, NJ: Wiley.

Verplaetse, R. and Henion, J. (2016). Hematocrit-independent quantitation of stimulants in dried blood spots: pipet versus microfluidic-based volumetric sampling coupled with automated flow-through desorption and online solid phase extraction-LC-MS/MS. *Anal. Chem.* 88 (13): 6789–6796.

Wei, D., Li, M., King, K.W., and Yang, L. (2015). Online and automated sample extraction. *Bioanalysis* 7 (17): 2227–2233.

Wei, X., Yin, J., Yang, G. et al. (2007). On-line solid-phase extraction with a monolithic weak cation-exchange column and simultaneuous screening of aladrenergic receptor antagonists in human plasma. *J. Sep. Sci.* 30 (17): 2851–2857.

Wu, J.T. (2001). The development of a staggered parallel separation liquid chromatography/tandem mass spectrometry system with on-line extraction for high-throughput screening of drug candidates in biological fluids. *Rapid Commun. Mass Spectrom.* 15 (2): 73–81.

Wu, J.T., Zeng, H., Deng, Y., and Unger, S.E. (2001). High-speed liquid chromatography/tandem mass spectrometry using a monolithic column for high-throughput bioanalysis. *Rapid Commun. Mass Spectrom.* 15 (13): 1113–1119.

Xia, Y.Q., Hop, C.E., Liu, D.Q. et al. (2001). Parallel extraction columns and parallel analytical columns coupled with liquid chromatography/tandem mass spectrometry for on-line simultaneous quantification of a drug candidate and its six metabolites in dog plasma. *Rapid Commun. Mass Spectrom.* 15 (22): 2135–2144.

Xu, R.N., Fan, L., Kim, G.E., and El-Shourbagy, T.A. (2006). A monolithic-phase based on-line extraction approach for determination of pharmaceutical components in human plasma by HPLC-MS/MS and a comrison with liquid-liquid extraction. *J. Pharm. Biomed. Anal.* 40 (3): 728–736.

Xu, Y., Willson, K.J., Anderson, M.D.G. et al. (2008). Elimination of diasteromer interference to determine Telcagepant (MK-0974) in humna plasma usning on-line turbulent-flow technology and off-line solid-phase extraction coupled with liquid chromatography/

tandem mass spectrometry. *J. Chromatogr. B* 877 (16-17): 1634–1642.

Xu, Y., Willson, K.J., and Musson, D.G. (2009). Strategies on efficient method development of on-line extraction assays for determination of MK-0974 in human plasma and urine using turbulent-flow chromatography and tandem mass spectrometry. *J. Chromatogr. B* 863 (1): 63–73.

Yang, S.H., Fan, H., Classon, R.J., and Schug, K.A. (2013). Restricted access media as a streamlined approach toward on-line sample preparation: recent advancements and applications. *J. Sep. Sci.* 36 (17): 2922–2938.

3

Equilibrium Dialysis, Ultracentrifugation, and Ultrafiltration in LC-MS Bioanalysis

Theo de Boer and Henri Meijering

3.1 Introduction

The binding of a drug to plasma proteins, e.g. albumin, α-1-acid glycoprotein (AGP), and sex hormone-binding globulin (SHBG), influences its distribution, elimination, and pharmacological activity and the free drug concentration correlates better to safety and efficacy than the total plasma concentration when the free fraction of the drug is subjected to change as for instance in special clinical populations (Shackleton 2011). The latter is acknowledged by regulatory guidelines requesting the assessment of (*in vitro*) protein binding prior to the start of Phase I clinical trials (ICH 2009; EMA 2012; FDA 2017). To analyze the free concentration of a drug rather than the total concentration in biological matrix (serum, plasma, blood, or tissue), specific techniques have to be used to separate the free drug fraction from the bound fraction. Several techniques have been described in the literature (Wells 2003; Cohen 2005; De Boer and Wieling 2013). These include equilibrium dialysis (ED, traditionally using radio-labeled compounds), ultrafiltration (UF), ultracentrifugation (UC), solid-phase microextraction (SPME), turbulent flow chromatography, capillary electrophoresis, surface plasmon resonance technology (biosensors, e.g. Biacore®), and high-performance liquid chromatography (HPLC) using analytical columns containing immobilized plasma proteins. The Biacore system monitors molecular interactions in real time using a label technology that responds to changes in the concentration of molecules at a sensor surface as molecules bind or dissociate from the surface. Although the Biacore system determines both the percent binding and the dissociation constants, it requires expensive equipment and specific operating skills. In addition to the aforementioned techniques for the determination of the unbound drug concentration in plasma, the following approaches can be used as an alternative:

1) Analysis of the drug in saliva: Because protein-bound drugs cannot enter saliva, the measured concentration of the drug in saliva (collected by passive drooling) may represent the free drug concentration in plasma, although the binding of the drug to specific saliva proteins can substantially affect the free concentration as postulated by Fiers et al. (2014).
2) Analysis of the drug in cerebrospinal fluid (CSF): It is suggested that the free brain concentration of drugs affecting the Central Nervous System (CNS) may be estimated from the drug concentration in the CSF (Liu et al. 2008).
3) Steroids: mathematical algorithms have been developed to calculate the unbound concentration of testosterone based on the total concentration of the testosterone and the binding affinities with SHBG and albumin (Vermeulen et al. 1999). It should be noted that the calculations are highly dependent on the accuracy of the SHBG and albumin quantification and, moreover, on the association constants applied (Hackbarth et al. 2011).

Most recently a white paper was published by the European Bioanalytical Forum (EBF) describing the views and recommendations of the organization regarding the conduct of plasma protein binding (PPB) studies in drug discovery and development (Buscher et al. 2014). The white paper recommends:

- For *in vitro* PPB studies in drug discovery, ED is preferred as it can be used for nearly all small-molecule compounds. For compounds with limited stability,

Sample Preparation in LC-MS Bioanalysis, First Edition. Edited by Wenkui Li, Wenying Jian, and Yunlin Fu.

UF might be the method of choice. For these initial experiments, a limited in-study qualification of the analytical method should be sufficient.

- For *in vitro* PPB studies in drug development, the decision on which technique to use should be carefully evaluated based on physical and chemical characteristics of the compound and the experience gained in drug discovery. In general, ED can be used for highly hydrophilic drugs and drugs with high PPB (85–100%), UF can be used for labile, highly hydrophilic drugs with medium to low PPB (<85%), and UC can be used for sticky (hydrophobic) compounds. Both radiolabeled and non-labeled compounds can be used and the analytical method should be adequately qualified, although no need to claim GLP (no formal safety study).
- For *in vivo* PPB in human plasma samples originating from healthy subjects or special populations, the EBF suggested to use the same technique that was used for the *in vitro* PPB during drug development. However, it was recommended to adequately validate the analytical method especially with regard to matrix effects and stability of metabolites (e.g. back conversion to parent drug).

In this Chapter, the three most commonly used techniques for the separation of the free and protein-bound fraction of a drug (in plasma) in quantitative LC-MS bioanalysis are described: ED, UF, and UC. The techniques are described in separate sections and each section covers the principles of the technique and provides an overview of the apparent advantages and disadvantages. Furthermore, for each technique a generic experimental procedure for the application of the described technique is given.

3.2 Challenges and Considerations

Today's drug discovery and drug development in the pharmaceutical industry demands rapid determination of drug candidates and/or their metabolites in the intended matrix. To comply with this demand, high-throughput versions (e.g. 96-well format) of the traditional techniques like UF and ED have been developed and implemented. However, nonspecific binding (NSB) of drugs, particularly highly lipophilic drugs, to the plastic container or membrane hinders the success of those automated and high-throughput applications. To address these issues, it is recommended to use filters and membranes that are made of Teflon® as NSB is significantly reduced by using this material. In addition to NSB, other factors need to be considered as well. Special consideration should be given to changes in environmental conditions (temperature and pH) during the entire sample handling process that encompasses the time from sample collection and long-term storage (Howard et al. 2010; Curran et al. 2011). Moreover, during long-term storage of the study samples, lipolysis can occur, leading to elevated levels of free fatty acids that may cause elevated free drug concentrations of small molecules (Howard et al. 2010). Instability of the compound in plasma can be minimized by adding enzyme inhibitors (e.g. sodium fluoride) directly after blood withdrawal. The recommended anticoagulant is ethylenediaminetetraacetic acid (EDTA) as concerns have been raised that the anticoagulant heparin may interfere with protein binding (Cohen 2005). Finally, it should be realized that because no real plasma sample with a known *in vivo* concentration of the free drug is available, the trueness of the quantified free drug concentration is somehow questionable (Li et al. 2011).

3.3 Experimental Procedures

This section describes the application and execution of three widely used techniques for the separation of the free and protein-bound fraction of a drug in plasma in quantitative LC-MS bioanalysis:

- ED (including rapid equilibrium dialysis, RED)
- UF (including 96-well format)
- UC

Independent from the applied technique, it is recommended to verify the sample processing method by assessing the unbound fraction of a positive control substance, e.g. levamisole (Sahagun et al. 1997). PPB experiments can be conducted using radiolabeled or non-radiolabeled compounds. While non-radiolabeled compounds require scientifically qualified or fully validated LC-MS methods to measure the low drug concentration levels, radiolabeled compounds can be quantified using Liquid Scintillation Counting (LSC), which is very straightforward. However, when using radiolabeled compounds it should be noted that the purity of the radiolabeled compounds should be high (≥99%), especially when the PPB is high (≥99%). During development of an LC-MS method for the analysis of unbound drug concentrations in plasma, special attention should be given to the selection of the devices (brand, type), the stability of the drug during incubation (dialysis, filtration, and centrifugation), reproducibility of the applied technique, NSB of the drug molecules to the device and the membrane, incubation time, pH, freeze–thaw stability, and long-term stability of the drug in the obtained dialysate or filtrate (Li et al. 2011; Xu et al. 2014).

3.3.1 Equilibrium Dialysis

ED is a concentration-driven process that separates free from protein-bound drugs in small volumes of plasma by diffusion through the pores of a selective and permeable membrane with varying molecular-weight cutoffs (MWCO), e.g. 8K or 12K MWCO (Figure 3.1). ED is considered to be the "gold standard" as it studies the true nature of the protein interaction because no external forces (e.g. pressure) are applied. The method is therefore less susceptible to experimental artifacts and is widely used for registration purposes. To control the environmental conditions, ED is conducted at 37 °C to mimic *in vivo* conditions and is preferably conducted in a carbon dioxide (usually 5%) controlled incubator to control pH.

Traditional ED requires large volumes of plasma and long equilibrium time. To overcome these drawbacks, RED devices (e.g. Thermo Scientific Pierce) that have been introduced are suitable for high-throughput analysis. Due to the high surface-to-volume ratio of the membrane compartment (Figure 3.2) which allows rapid dialysis, potential complications like volume shifts and protein leakage can be minimized by reducing the equilibration time considerably (Van Liempd et al. 2011; Xu et al. 2014). Moreover, the RED devices are constructed of low binding material (e.g. Polytetrafluoroethylene) that ensures minimal NSB to the apparatus. Furthermore, it has been postulated that NSB has a negligible impact on the unbound fraction of the drug in ED due to the nature of the dialysis experiment. NSB occurs in both the plasma and dialysate chamber to a similar extent after which a new equilibrium will be established, although at a lower concentration than intended (Nilsson 2013). The advantages and disadvantages of the ED method are summarized in Table 3.1.

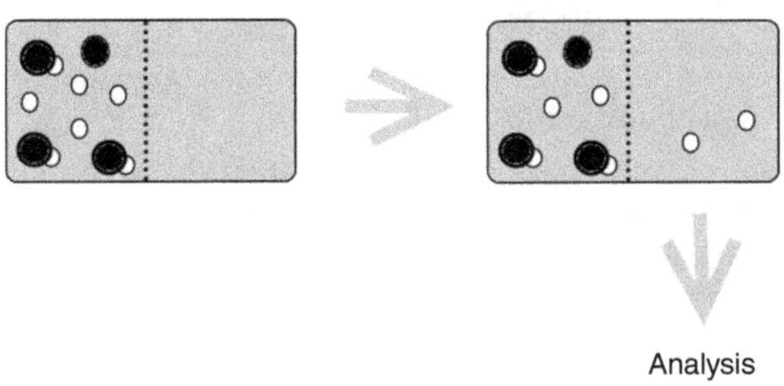

Figure 3.1 Schematic representation of the equilibrium dialysis methodology. *Source:* Reproduced from De Boer and Wieling (2013) with permission from John Wiley and Sons.

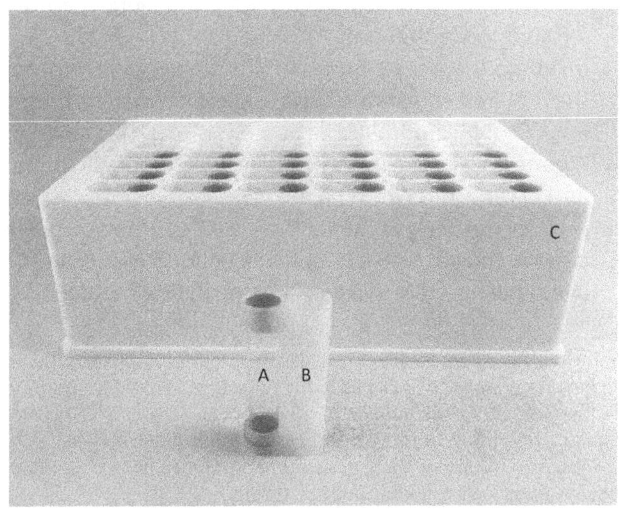

Figure 3.2 Rapid equilibrium dialysis (RED) device. Each insert is comprised of two side-by-side chambers: (A) O-ring-sealed vertical cylinder of dialysis membrane with varying molecular-weight cutoffs (MWCO) to which the plasma is added; (B) Polytetrafluoroethylene (PTFE or Teflon) chamber to which the dialysate buffer is added; and (C) 96-well base plate designed for 48 inserts.

Table 3.1 Advantages and disadvantages of (rapid) equilibrium dialysis (ED) for sample preparation.

Advantages	Disadvantages
Studies the true nature of the protein interaction. Method is less susceptible to experimental artifacts	Time consuming: 3–48 h for conventional equilibrium dialysis (limited number of samples), 2–6 h for RED (96-well format)
Low nonspecific binding, but is believed not to affect the percentage of free drug	Not suitable for unstable compounds
Temperature and pH can be controlled by conducting the experiments in a CO_2 incubator	Volume shift and protein leakage due to the osmotic effect of the macromolecules, although significantly reduced by using RED
Highly accurate	Possible drug insolubility in dialysate
Available in 96-well format (RED)	Cannot be used for highly hydrophobic compounds
Low sample volume (RED)	Large sample volume (conventional ED)
Inexpensive and easy to perform	

Suggested Experimental Procedure for Rapid Equilibrium Dialysis (RED):

- Place the base plate in 20% methanol for 10 minutes and rinse with deionized water.
- Soak RED inserts (8K MWCO) in deionized water for 10 minutes (repeat twice). Do not allow membranes to become dry.
- The pH of the plasma samples should be 7.4±0.1. If needed, the pH can be adjusted with acid or by bubbling CO_2 gas through the plasma.
- Pipet 50–500 µl plasma into chamber 1 and an equivalent volume of dialysis buffer[1] to the other chamber (Vendor-specific requirements should be followed).
- For assessment of NSB, pipette spiked plasma sample into chamber 1 and spiked buffer sample into chamber 2 (same concentration).
- Incubate the device on an orbital shaker (500 rpm) for the required equilibrium time (typically two to six hours) in a CO_2 (5%) controlled incubator set at 37 °C.
- Pipette an aliquot (same volumes) from dialysis buffer and plasma into two separate vials.
- Add to the dialysis buffer an equal amount of blank plasma and vice versa to obtain a uniform matrix and to avoid the need for two validated analytical methods.[2]

[1] The dialysis buffer usually consists of an isotonic phosphate buffer to which NaCl is added to diminish the difference in ionic strength on either side of the membrane (Gibbs- Donnan Effect), e.g. 100 mM sodium phosphate and 150 mM sodium chloride (PBS; pH 7.4).
[2] Endogenous compounds (e.g. testosterone) require two independent assays as no analyte-free plasma is available.

- Simultaneously incubate spiked plasma and spiked dialysate buffer under equivalent conditions for the duration of the equilibrium experiment to assess stability.
- Prepare calibration standards and QC samples for an appropriate dynamic range in dialysis buffer/plasma (1 : 1).
- Add internal standard and analyze the samples, calibrators, and QC samples using an appropriate method (e.g. LC-MS).
- The percentage NSB in plasma and buffer compartment are calculated using the following equations:

$$\%\text{NSB}(p) = 100 \times \left(1 - \frac{C_p}{C_{\text{plasma}}}\right)$$

$$\%\text{NSB}(b) = 100 \times \left(1 - \frac{C_b}{C_{\text{buffer}}}\right)$$

where C_b and C_p are the concentration measured in the buffer compartment and plasma compartment, respectively, and C_{buffer} and C_{plasma} are the initially spiked concentrations in buffer and plasma, respectively.
- The percentage unbound and the percentage bound are calculated using the following equations:

$$\%\text{Free} = 100 \times \left(\frac{C_b}{C_p}\right)$$

$$\%\text{Bound} = 100\% - \%\text{Free}$$

where C_b and C_p are the concentration measured in the buffer compartment and plasma compartment, respectively.

3.3.2 Ultrafiltration

UF separates free from protein-bound drugs according to molecular weight and size using low-adsorptive hydrophilic membranes with a specified MWCO value (Figure 3.3) and a centrifugal force. Compared to ED, UF is much more suitable for high-throughput analysis. The advantages and disadvantages of the UF technique are summarized in Table 3.2. NSB due to the filter membranes can be minimized by prewashing the UF devices with a cationic surfactant for basic compounds, e.g. 5% benzalkonium chloride (Larsen et al. 2012) or with 0.5–5% Tween-80 for neutral and acidic compounds (Lee et al. 2003). To minimize NSB due to the ultrafiltrate unit, Taylor and Harker (2006) added control plasma retentate to the obtained ultrafiltrate. When using automated (96-well) UF devices, special attention should be given to the reproducibility of the ultrafiltrate volume across the plate ("edge effect") as it may affect the analysis results (Zhang and Musson 2006). Moreover, evaluation of multiple UF devices, such as Ultrafree®, Microcon®, Centrifree®, EMD MultiScreen®, Ultracel®, devices from Milipore (Bedford, MA, USA), can be helpful to overcome specific restrictions like protein leakage.

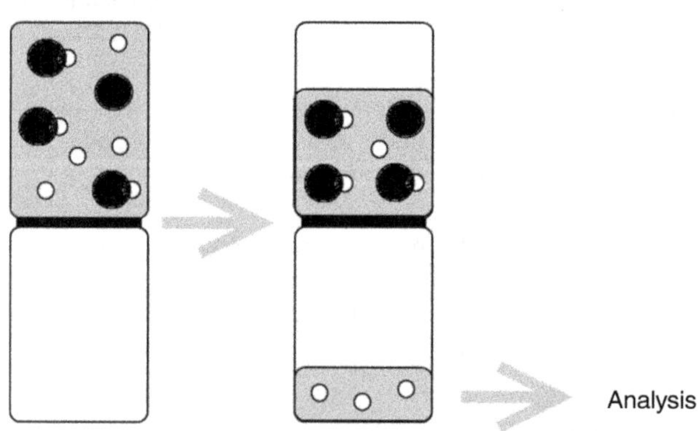

Figure 3.3 Schematic representation of the ultrafiltration methodology. *Source:* Reproduced from De Boer and Wieling (2013) with permission from John Wiley and Sons.

Table 3.2 Advantages and disadvantages of ultrafiltration for sample preparation.

Advantages	Disadvantages
Highly accurate	Uses external forces (pressure). Susceptible to experimental artifacts, e.g. protein leakage.
High-throughput potential (96-well format)	Temperature and pH cannot be easily controlled
Relatively fast and easily implemented	Nonspecific binding affects the unbound fraction
Can be used for unstable compounds	Susceptible to "Edge Effect" when using 96-well plates
Inexpensive and easy to perform	Possible drug insolubility in ultrafiltrate
	Cannot be used for highly hydrophobic compounds

Suggested Experimental Procedure for ultrafiltration (UF):

- Soak filter devices in water for 10 minutes and decant the water (repeat twice).
- Pipette 50–500 µl plasma into the UF device (e.g. Millipore Amicon® Ultra, 10K MWCO).
- For assessment of NSB, fortify blank ultrafiltrate with the analyte and repeat the UF process.
- Spin the UF devices at 37 °C in a fixed-angle rotation centrifuge for 20 minutes. Use the maximum allowable speed (e.g. 1500 × *g*) for the centrifuge to obtain sufficient volume of ultrafiltrate in a reasonable time frame.
- Simultaneously incubate spiked plasma and spiked ultrafiltrate under the same conditions for the duration of the UF experiment to assess stability.
- Prepare calibration standards and QC samples for an appropriate dynamic range.
- Add internal standard and analyze the samples, calibrators, and QC samples using an appropriate method (e.g. LC-MS).
- The percentage NSB is calculated using the following equation:

$$\%\text{NSB} = 100 \times \left(\frac{C_{\text{BD}} - C_{\text{BF}}}{C_{\text{BD}}} \right)$$

where C_{BD} is the total drug concentration in ultrafiltrate before centrifugation and C_{BF} is the concentration measured in the ultrafiltrate after centrifugation.

- The percentage unbound and the percentage bound are calculated using the following equations:

$$\%\text{Free} = 100 \times \left(\frac{C_{\text{SF}}}{C_{\text{SD}}(1 - \text{NSB})} \right)$$

$$\%\text{Bound} = 100\% - \%\text{Free}$$

where C_{SF} is the drug concentration in the plasma filtrate after centrifuge, C_{SD} is the nominal plasma concentration, and NSB is the nonspecific binding factor.

3.3.3 Ultracentrifugation

UC separates "large" molecules from "small" molecules by gravitational forces (200 000–300 000 × *g* for 6 h at 37 °C, e.g. Beckman L8-70 M Floor Model Ultra Centrifuge). This may be an alternative method for compounds with very high NSB (hydrophobic drugs) or for compounds unable to diffuse through the dialysis membrane (Figure 3.4). Via UC, the plasma is separated into three distinct layers: (i) the upper layer: contains very low-density lipoproteins; (ii) the middle layer (just below the surface): protein-free layer; and (iii) the bottom layer: contains the plasma proteins. The sample for analysis should be very carefully taken (by an automated process) from the middle layer to ensure that the layers are not disturbed. It is recommended to determine the protein content in the collected sample, where a protein content of <1% of the initial plasma concentration seems acceptable. The advantages and disadvantages of the method are summarized in Table 3.3.

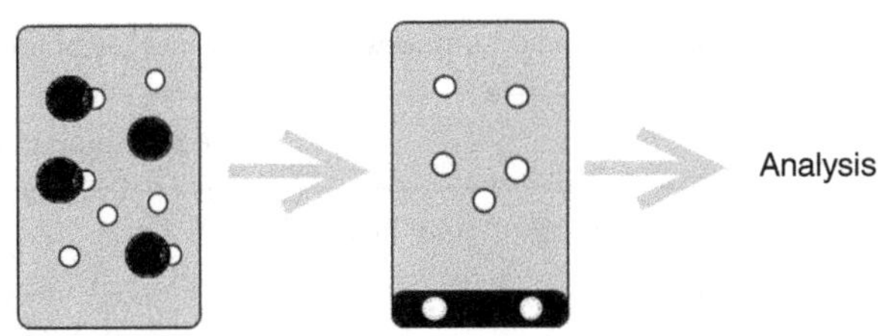

Figure 3.4 Schematic representation of the ultracentrifugation methodology. *Source:* Reproduced from De Boer and Wieling (2013) with permission from John Wiley and Sons.

3.4 Summary

A comparison among the three commonly used sample preparation techniques for the measurement of unbound drug in plasma is given in Table 3.4. From the table it can be concluded that when there is no plasma instability of the drug, ED would be the method of choice as it studies the true nature of the protein interactions

Table 3.3 Advantages and disadvantages of ultracentrifugation for sample preparation.

Advantages	Disadvantages
Suitable for highly hydrophobic compounds	Uses external forces (pressure). Susceptible to experimental artifacts: co-sedimentation of free and bound drugs (MW >300 Da), possible back diffusion
Highly accurate	Temperature can but pH cannot be easily controlled
No NSB to membranes or plastic device	No high-throughput potential
No volume shift and protein leakage	Time consuming and limited number of samples
	Cannot be used for unstable compounds
	Expensive equipment and specific operator skills needed
	Possible drug insolubility in supernatant
	Large sample volume

Suggested Experimental Procedure for ultracentrifugation (UC):

- Pipette 1000–1500 μl plasma into vendor specific tubes (e.g. polyallomer, nitrocellulose tubes).
- Spin the tubes in the ultracentrifuge (300 000 × *g* at 37 °C for six hours), acceleration of the centrifuge must be set to slow and the break must be turned off.
- Carefully take the sample from the middle layer.
- Determine the total protein content in the collected sample (e.g. BCA Protein Assay Kit, Thermo Scientific™).
- Determine NSB by adding spiked buffer solution to the tubes.
- Simultaneously incubate spiked plasma at 37 °C for the duration of the UC experiment to assess plasma stability.
- Prepare calibration standards and QC samples for an appropriate dynamic range.
- Add internal standard and analyze the samples, calibrators, and QC samples using an appropriate method (e.g. LC-MS).
- The percentage NSB is calculated using the following equation:

$$\%NSB = 100 \times \left(1 - \frac{C_b}{C_{buffer}}\right)$$

where C_b is the concentration measured in the buffer solution after incubation in the appropriate tube and C_{buffer} is the initially spiked concentrations in buffer.
- The percentage unbound and the percentage bound are calculated using the following equations:

$$\%\text{Free} = 100 \times \left(\frac{C_{SF}}{C_{SD}(1 - NSB)}\right).$$

$$\%\text{Bound} = 100\% - \%\text{Free}$$

where C_{SF} is the drug concentration in the supernatant after centrifuge, C_{SD} is the nominal plasma concentration, and NSB is the nonspecific binding factor.

Table 3.4 Comparison of the advantages and disadvantages of three methods used for the assessment of protein binding of drugs.

Topic	Equilibrium dialysis	Ultrafiltration	Ultracentrifugation
Studies the true nature of the protein interaction	Yes	No	No
Risk for nonspecific binding (NSB)	Minimal, but is believed not to affect the percentage of free drug	Yes	Minimal
Susceptible to experimental artifacts	Volume shift (minimized with RED) Protein leakage	"Edge Effect" 96-well plates	co-sedimentation, possible back diffusion.
Time consuming	Conventional: Yes (3–48 h) RED: Yes (2–6 h)	No	Yes (≥6 h)
High-throughput potential (96-well format)	Conventional: No RED: Yes	Yes	No
Temperature and pH can be controlled by conducting the experiments in a CO_2 incubator	Conventional: Not easily RED: Yes	No	Only temperature
Inexpensive and easy to perform	Yes	Yes	No
Low sample volume	Conventional: No RED: Yes	Yes	No
Can be used for unstable compounds	No	Yes	No
Can be used for highly hydrophobic compounds	No	No	Yes
Risk for insolubility of the drug in dialysate, ultrafiltrate, or supernatant	Yes	Yes	Yes

despite a major disadvantage of a long processing time that is necessary to reach equilibrium (ca. 3–48 h). The use of RED reduces the processing time significantly (two to six hours) and is considered to be an excellent alternative. If there is instability of the drug that cannot be controlled by addition of stabilizers, high-throughput UF (96-well format) would be the method of choice. In cases where NSB is the limiting factor, UC could be employed, although it requires expensive equipment and specific operating skills.

References

Buscher, B., Laakso, S., Mascher, H. et al. (2014). Bioanalysis for plasma protein binding studies in drug discovery and drug development: view and recommendations of the European bioanalysis forum. *Bioanalysis* 6 (5): 673–682.

Cohen, L.H. (2005). Plasma protein binding methods in drug discovery. In: *Methods in Pharmacology and Toxicology Optimization in Drug Discovery: In Vitro Methods* (ed. Z. Yan and G.W. Caldwell), 111–122. Totowa, NJ: Caldwell Humana Press Inc.

Curran, R.E., Claxton, C.R.J., Hutchinson, L. et al. (2011). Control and measurement of plasma pH in equilibrium dialysis: influence on drug plasma protein binding. *Drug Metab. Dispos.* 39 (3): 551–557.

De Boer, T. and Wieling, J. (2013). LC-MS bioanalysis of unbound drugs in plasma and serum. In: *Handbook of LC-MS Bioanalysis, Best Practices, Experimental Protocols, and Regulations* (ed. W. Li, J. Zhang and F. Tse), 317–325. Hoboken, NJ: Wiley.

European Medicine Agency (EMA) (2012). Committee for Human Medicinal Products (CHMP). *Guideline on the Investigation of Drug Interactions* (Adopted by CHMP: 21 June 2012).

Fiers, T., Delanghe, J., T'Sjoen, G. et al. (2014). A critical evaluation of salivary testosterone as a method for the assessment of serum testosterone. *Steroids* 86: 5–9.

Food and Drug Administration (FDA) (2017). Center for Drug Evaluation and Research (CDER). *In Vitro Metabolism- and Transporter-Mediated Drug-Drug Interaction Studies.* Draft Guidance, February 2017.

Hackbarth, J.S., Hoyne, J.B., Grebe, S.K., and Singh, R.J. (2011). Accuracy of calculated free testosterone differs between equations and depends on gender and SHBG concentration. *Steroids* 76: 48–55.

Howard, M.L., Hill, J.J., Galluppi, G.R., and McLean, M.A. (2010). Plasma protein binding in drug discovery and development. *Comb. Chem. High Throughput Screen.* 13 (2): 170–187.

International Conference of Harmonization (ICH) (2009). *Guidance on Nonclinical Safety Studies for the Conduct of Human Clinical Trials and Marketing Authorization for Pharmaceuticals.* ICH M3 (R2).

Larsen, H.S., Chin, P.K.L., Begg, E.J., and Jensen, B.P. (2012). Quantification of total and unbound concentrations of lorazepam, oxazepam and temazepam in human plasma by ultrafiltration and LC-MS/MS. *Bioanalysis* 3 (8): 843–852.

Lee, K.J., Mower, R., Hollenbeck, T. et al. (2003). Modulation of nonspecific binding in ultrafiltration protein binding studies. *Pharm. Res.* 20 (7): 1015–1021.

Li, W., Lin, H., Smith, H.T., and Tse, F.L.S. (2011). Developing a robust ultrafiltration-LC-MS/MS method for quantitative analysis of unbound vadimezan (ASA404) in human plasma. *J. Chromatogr. B Analyt. Technol. Biomed. Life Sci.* 879: 1927–1933.

Liu, X., Chen, C., and Smith, B.J. (2008). Progress in brain penetration evaluation in drug discovery and development. *J. Pharmacol. Exp. Ther.* 325: 349–356.

Nilsson, L.B. (2013). The bioanalytical challenge of determining unbound concentration and protein binding for drugs. *Bioanalysis* 5 (24): 3033–3050.

Sahagun, A., Fernandez, N., Teran, M.T. et al. (1997). Plasma protein binding of levamisole in several species. *Methods Find. Exp. Clin. Pharmacol.* 19 (3): 185–187.

Shackleton, G. (2011). Special populations: protein binding aspects. In: *Drug Discovery and Evaluation: Methods in Clinical Pharmacology*, 1e (ed. H.G. Vogel, J. Maas and A. Gebauer), 67–71. Berlin: Springer-Verlag.

Taylor, S. and Harker, A. (2006). Modification of the ultrafiltration technique to overcome solubility and non-specific binding challenges associated with the measurement of plasma protein binding of corticosteroids. *J. Pharm. Biomed. Anal.* 41 (1): 299–303.

Van Liempd, S., Morrison, D., Sysmans, L. et al. (2011). Development and validation of a higher-throughput equilibrium dialysis assay for plasma protein binding. *J. Lab. Autom.* 16: 56–67.

Vermeulen, A., Verdonck, L., and Kaufman, J.M. (1999). A critical evaluation of simple methods for the estimation of free testosterone in serum. *J. Clin. Endocrinol. Metab.* 84: 3666–3672.

Wells, D.A. (2003). *High Throughput Bioanalytical Sample Preparation. Methods and Automation Strategies*, 1e. Amsterdam, The Netherlands: Elsevier.

Xu, X., Rose, A., Demers, R. et al. (2014). Quantitative determination of free/bound atazanavir via high-throughput equilibrium dialysis and LC-MS/MS, and the application in ex vivo samples. *Bioanalysis* 6 (23): 3169–3182.

Zhang, J. and Musson, D.G. (2006). Investigation of high–throughput ultrafiltration for the determination of an unbound compound in human plasma using liquid chromatography and tandem mass spectrometry with electrospray ionization. *J. Chromatogr. B Analyt. Technol. Biomed. Life. Sci.* 843: 47–56.

4

Phospholipid Depletion Techniques in LC-MS Bioanalysis

Stacy Brown and Jennifer Carmical

4.1 Introduction

Phospholipids (PLs) are a primary component of cell membranes, and are found in many types of biological samples, including plasma, urine, tears, cerebrospinal fluid, synovial fluid, and tissues (Teo et al. 2015). Analysis of biological samples for the purpose of determination of drugs, metabolites, and other xenobiotics is often hindered by the presence of these PLs, especially when using high-performance liquid chromatography (HPLC) coupled with mass spectrometry (MS) (Furey et al. 2013; Pike and Kushon 2012; Rossmann et al. 2015). Plasma PLs are also becoming a field of inquiry for their potential use as biomarkers for disease states, among other reasons (Azzouz et al. 2005; Calderon-Santiago et al. 2014; Ham et al. 2006; Petkovic et al. 2001; Postle et al. 2007). Therefore, improvements in isolating, removing, quantifying, and qualifying these molecules have become increasingly important.

PLs are most commonly composed of fatty acids attached to glycerol and phosphate molecules, but a number of other substituents can be added to create the array of PLs needed for different biological functions. The phosphate "head group" lends a negative charge, while other components, such as quaternary amines confer a positive charge, as shown in Figure 4.1 (Shen et al. 2005). The coupling of polar and nonpolar molecular characteristics contributes to interference in a variety of quantitative and qualitative LC-MS assays. This affects the quality of analytical data, and has resulted in recommendations by the United States Department of Health and Human Services Food and Drug Administration (FDA) Center for Drug Evaluation and Research, which enjoins researchers to preemptively consider the impact of PLs in data analysis (Services 2001). As drug development trends toward higher specificity, the fields of pharmacokinetics and therapeutic drug monitoring demand analytical methods free from the interference of PLs, with maximum efficiency and accuracy (Aubry 2011; De Baere et al. 2015; Rossmann et al. 2015). Removing PLs or compensating for the variations due to the presence of PLs in biological samples has consequently become essential for optimizing LC-MS bioanalysis techniques.

4.2 Impact of Phospholipids on Bioanalytical Methods

Srinivas suggests that PLs may be the largest single-source of matrix effects in bioanalytical assays (Srinivas 2009). Several methods have been developed to avoid or minimize the detrimental effects due to PLs on different types of sample analysis. Ion suppression or enhancement of the mass spectrometric signal is the most common type of interference caused by PLs in the sample extracts that are co-eluted with the target analyte (Chambers et al. 2007). These effects are commonly referred to as matrix effects, and can vary widely depending on the type of sample being analyzed, the method of sample preparation, and the instrument used (Shen et al. 2005; Xia and Jemal 2009). Matrix effects can be attributed to the impact of the matrix on the analyte signal, via enhancement or suppression. This phenomenon, termed *absolute matrix effects*, is defined as a difference in LC-MS response between a neat solution and a spiked post-extraction sample. Additionally, matrix effects can be associated with difference in response among various lots of spiked post-extraction samples, called *relative matrix effects*. In cases of absolute and relative matrix effects, the accuracy and precision of the method can be compromised, especially at low analyte concentrations, in instances of co-elution of matrix and analyte, and cases where the internal standard does not track the analyte well (Ahmad et al. 2012; Ismaiel et al. 2007). An example of the impact of absolute matrix effects is seen in Figure 4.2 (Ismaiel et al. 2007). These investigators noted an ionization suppression of hydrocodone and pseudoephedrine, and their

Sample Preparation in LC-MS Bioanalysis, First Edition. Edited by Wenkui Li, Wenying Jian, and Yunlin Fu.

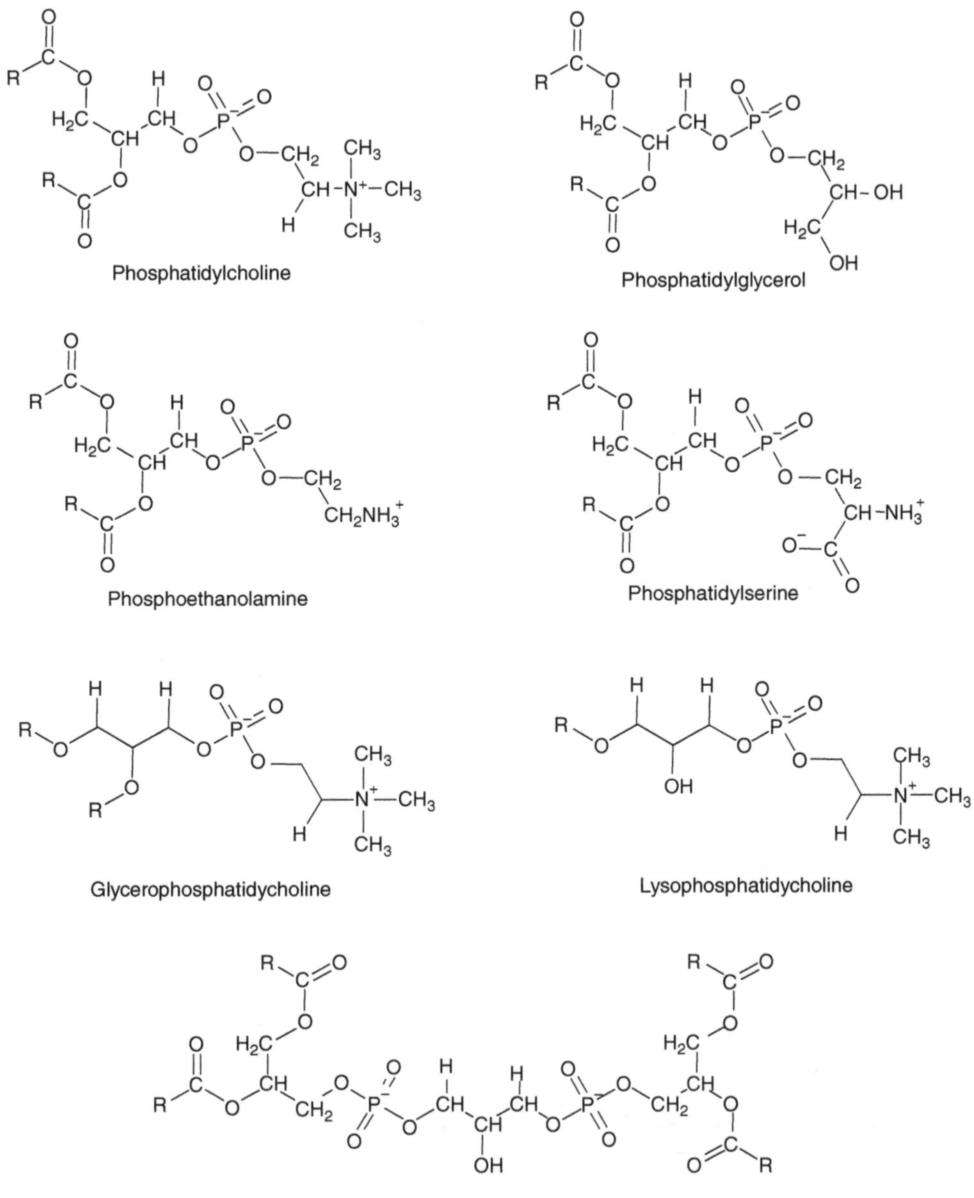

Figure 4.1 Frameworks for common phospholipids found in membranes and plasma. *Source:* From Berg et al. (2002).

corresponding deuterium-labeled internal standards, which chromatographically coincided with elution of phosphatidylcholine (PC). This co-elution was confirmed using a comparison of an infused PC solution (Figure 4.2) with extracted blank plasma (Figure 4.2b), while monitoring *m/z* 184, a characteristic fragment ion of PC (Ismaiel et al. 2007). The authors also noted the potential for relative matrix effects, by confirming retention times of PL peaks ranging from 3.7 to 4.2 minutes in 15 different sources of human plasma (Ismaiel et al. 2007).

Plasma and serum samples are the most affected by PLs in LC-MS bioanalysis, due to the possibility of ~17 000 different types of lipids and fatty acids being present in the matrix (Teo et al. 2015). The number and type of PLs present in these matrices can vary widely, depending on the species, metabolism, and diet of the animal from which the sample was obtained (Gao et al. 2006; Shen et al. 2005). The most common PL species are phosphatidylcholines (PCs), glycerophosphatidylcholines (GPCs or GPChos), lyso-phosphatidylcholines (LPCs), phosphatidylethanolamines (PEs), phosphatidylinositol (PI), phosphatidylserine (PS), and sphingomyelin (SM). GPChos and PCs are the most abundant species present in plasma (Chambers et al. 2007); however, one study found that more interference was caused by lyso-PLs due to their earlier elution than the more abundant GPCho

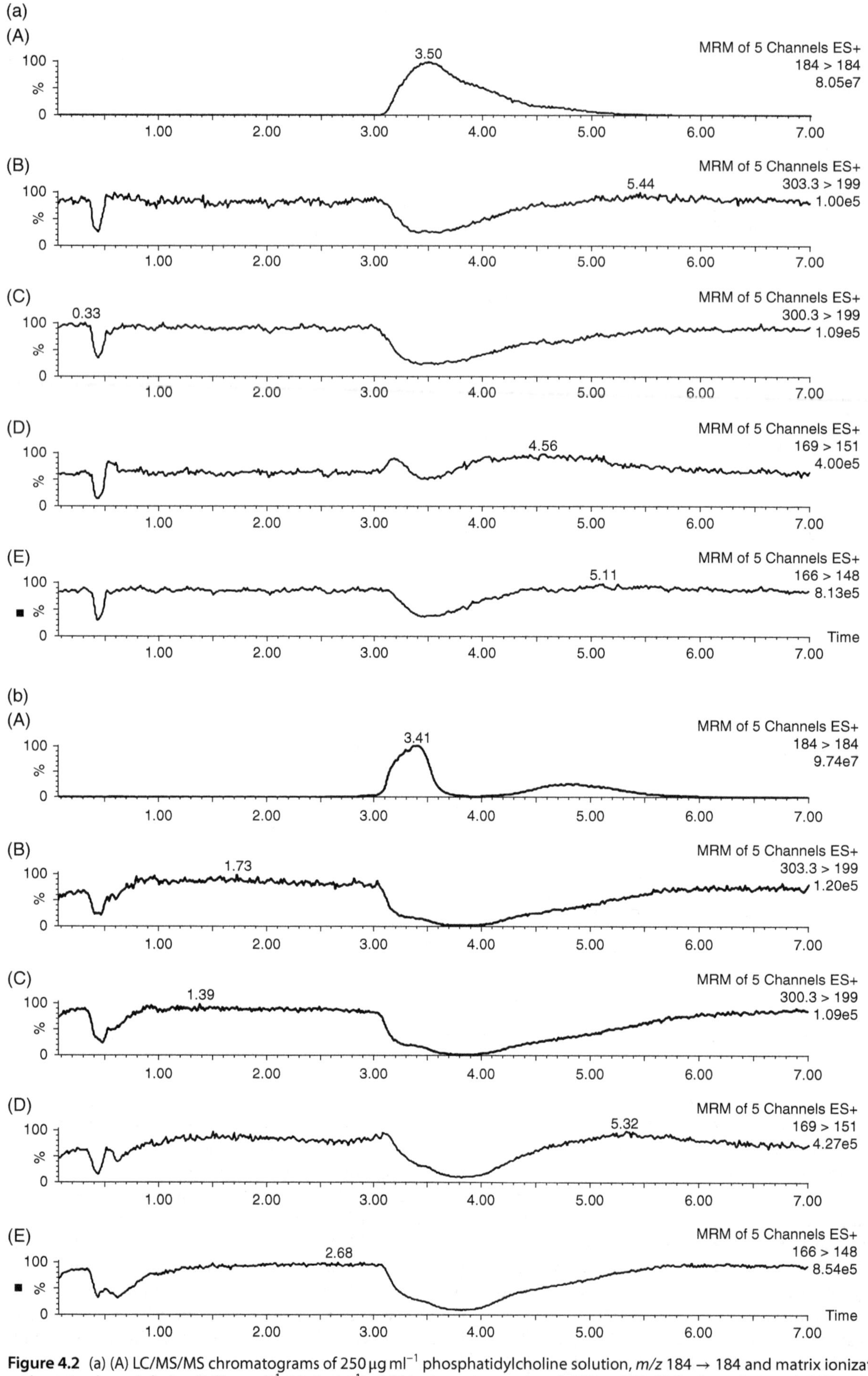

Figure 4.2 (a) (A) LC/MS/MS chromatograms of 250 μg ml^{-1} phosphatidylcholine solution, *m/z* 184 → 184 and matrix ionization effects with post-column infusion (100 ng ml^{-1}, 10 μl min^{-1}) of (B) hydrocodone – d_3, *m/z* 303 → 199; (C) hydrocodone, *m/z* 300 → 199; (D) pseudoephedrine – d_3, *m/z* 169 → 155; and (E) pseudoephedrine, *m/z* 166 → 148. (b) (A) LC/MS/MS analysis of extracted plasma, *m/z* 184 → 184 and matrix ionization effects with post-column infusion (100 ng ml^{-1}, 10 μl min^{-1}) of (B) hydrocodone – d_3, *m/z* 303 → 199; (C) hydrocodone, *m/z* 300 → 199; (D) pseudoephedrine– d_3, *m/z* 169 → 155; and (E) pseudoephedrine, *m/z* 166 → 148. *Source:* Reprinted with permission from Ismaiel et al. (2007). Reproduced with permission of Elsevier.

and PC PLs (Xia and Jemal 2009). This strengthens the notion that each analysis needs to be tailored to the individual characteristics of the analyte and matrix involved.

4.3 Investigating Matrix Effects Associated with Phospholipids

A seminal paper on investigation of matrix effects proposes the use of post-column infusion of target analyte while chromatographically running a matrix blank through the analytical method (Bonfiglio et al. 1999). While this paper does not specifically mention PLs, the technique of post-column infusion to identify matrix effects is widely accepted as an essential part of bioanalytical method development prior to validation. Post-column infusion involves infusing a standardized solution of the analytes during MS analysis to identify regions of ionization suppression or enhancement caused by the matrix ions. This gives the chromatographer the opportunity to resolve the analytes from the regions of ion suppression/enhancement to avoid errors in quantification due to matrix ions (Ahmad et al. 2012; Dams et al. 2003). Only the components, whether they are endogenous metabolites, PLs, proteins, or other compounds, that co-elute with the analytes(s) will affect their signal intensity produced in the MS detection (Bonfiglio et al. 1999). Recently, Matuszewski (2006) proposed that the calibration curve slope precision (matrix versus standard) could be an alternative tool for quantifying matrix effects, with a benchmark of <3–4%.

A popular method of determining the presence of matrix effects is in-source multiple reaction monitoring (IS-MRM), which is an approach utilized by Little et al. (2006) to specifically monitor ion transitions of GPChos during analysis LC-MS or LC coupled with tandem mass spectrometry (LC-MS/MS). This involves using the positive ESI mode to monitor *m/z* 184 → 184 mass transitions for GPChos, as well as the *m/z* 104 → 104 transition for 2-lyso GPChos, as shown in Figure 4.3 (Little et al. 2006). The authors also learned that SMs exhibit the *m/z* 184 → 184 ion fragmentation signature. This method

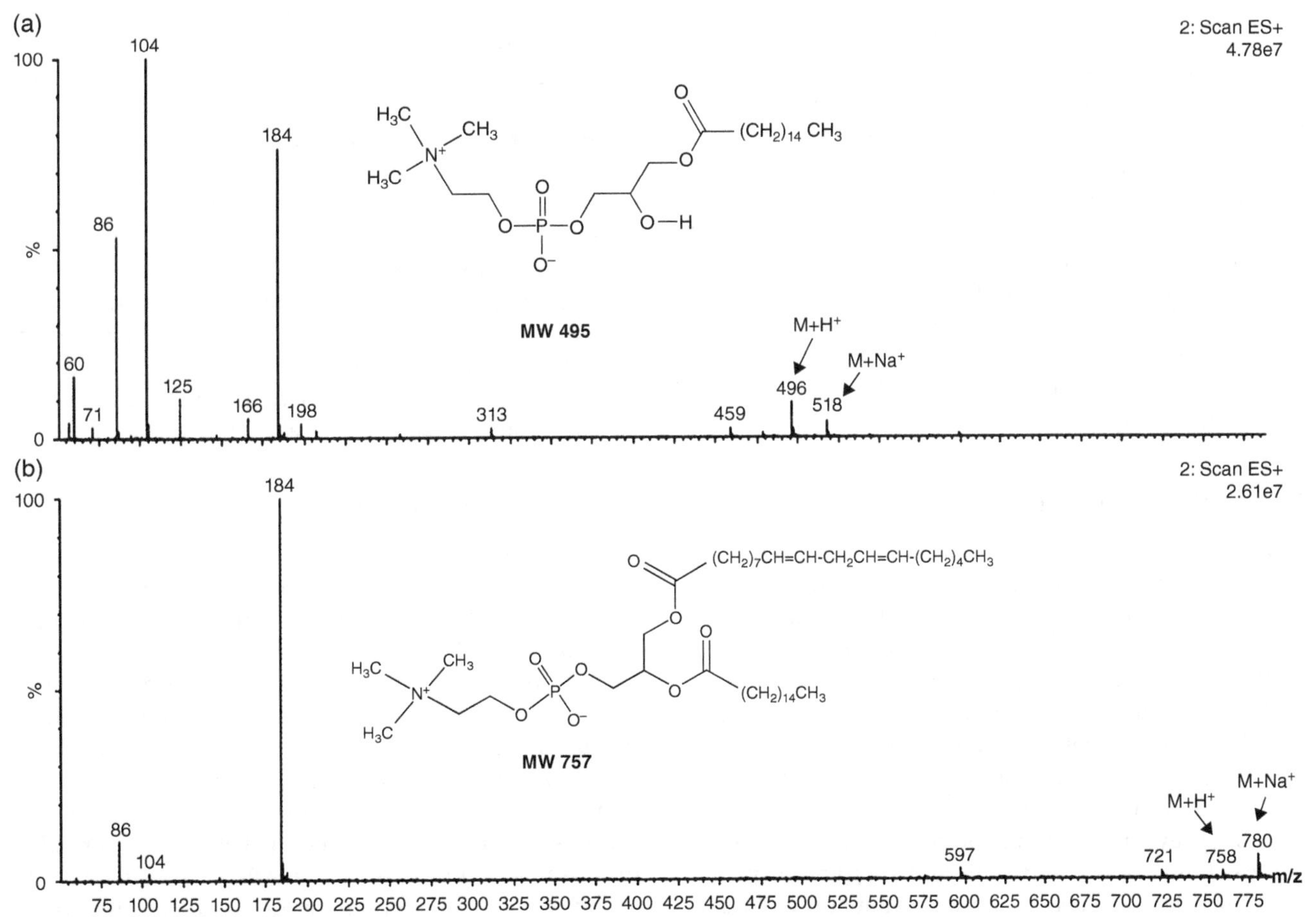

Figure 4.3 In-source CID spectra for GPChos at 90V cone voltage on Quattro Micro mass spectrometer: (a) 2-lysoGPCho and (b) diacyl GPCho. *Source:* Reprinted with permission from Little et al. (2006). Reproduced with permission of Elsevier.

helps determine at what point the PLs will elute during the cycle, so that the LC gradient can be modified to have the analyte elute at a separate retention time for better quantification and qualification. The ion monitoring method has been successfully used in many studies as a tool in modifying the chromatography to avoid matrix effects (Jemal et al. 2010). Class-specific MRM transitions have also been identified for PLs other than GPChos, SM, or lyso-GPChos (Chambers et al. 2007; Ismaiel et al. 2010).

4.4 Minimizing Matrix Effects Associated with Phospholipids

4.4.1 Sample Dilution

Several techniques have been developed to reduce matrix effects during sample analysis. The simplest technique is sample dilution, in which an organic solvent is added in order to allow for less surface tension on individual droplets and to decrease viscosity of the samples (Rayleigh effects) during the electrospray ionization (ESI) step of MS analysis (Furey et al. 2013). This allows for smaller droplets, which provide greater surface area for ionization (Ferrer et al. 2011). Each sample/analyte may require different levels of dilution, as there is currently no consensus on the optimal dilution factor needed to avoid matrix effects. Some studies have used up to 1000-fold dilution. However, this may not always be practical, and may change the limits of quantification to an unusable level (Stahnke et al. 2012). A 15-fold dilution was sufficient to reduce matrix effects in samples of tomato, orange, and leek juice for the quantification of 53 different pesticides; however, these findings may not be translatable to biological samples (Ferrer et al. 2011).

4.4.2 Column Manipulations

Column-switching and pre-column back-flushing are effective methods of removing PLs from the samples during analysis by determining exactly when the PLs will appear and switching to waste while eluting analyte on another column during that time period (Ismaiel et al. 2011). PLs tend to adhere to analytical columns, especially hydrophobic columns, and can then elute later in the analysis (Kushon and Pike 2013; Silvester and Smith 2012). Therefore, back-flushing is used to clear the column and prevent contamination of subsequent samples. This can both extend column life and improve the quality of the analysis (Ismaiel et al. 2011). Many different solvents have been tested to determine the best choice for clearing PLs that have adhered to the column. These include acetonitrile (ACN), methanol (MeOH), dichloromethane (DCM), isopropanol (IPA), chloroform, *n*-hexane, and various mixtures of these, some including water. One study claims that none of these work well (Ismaiel et al. 2007) alone, but a combination of water/ACN/MeOH in a 10 : 45 : 45 ratio effectively removed PLs associated with five major mass transitions in human plasma samples (Ismaiel et al. 2011). Finally, a high-percentage ACN mobile phase cleared several PLs from hydrophilic interaction chromatography (HILIC) columns following injection of PPT-treated plasma (Ekdahl et al. 2013).

4.4.3 Internal Standards

Use of stable isotopically labeled internal standards (SIL-IS) of the analytes of interest is a common technique for compensating for matrix effects, as well as for tracking extraction recovery (Antignac et al. 2005; Postle et al. 2007). Zhou et al. (2005) suggested that this ability to compensate for matrix effects is mediated by an analyte's level of protein binding. In this respect, highly protein-bound analytes need to be extracted by giving sufficient equilibrium for protein binding or using strong conditions to break protein binding so that SIL-IS will behave like their unlabeled counterparts. The use of SIL-IS is ideal for quantitative LC-MS bioanalysis because it has the same retention time as the analyte and it will experience the same impact from matrix effects as its non-labeled counterpart. However, cost of these isotopically-labeled analytes can be limiting for some labs, especially in cases where custom-synthesis is required. Internal standards that are not stable isotopically labeled, though structurally similar to the analyte, should be considered independent of the target analytes for ion suppression or enhancement effects as they may behave differently in the LC-MS process (Ahmad et al. 2012; Furey et al. 2013).

4.4.4 Ionization Choice

ESI is the most common ionization mode in mass spectrometric methods, but atmospheric pressure chemical ionization (APCI) can be used as an alternative (Trufelli et al. 2011). APCI tends to be less affected by the presence of matrix ions compared to ESI, since the ionization occurs after the analytes are in the gas phase (Brouwers 2011; Ismaiel et al. 2008). Additionally, Liang et al. (2003) showed that APCI tends to enhance ionization potential for a group of test compounds while ESI suppressed ionization. Souverain et al. (2004) found that the less matrix effects seen with APCI were independent of the type of sample preparation used. For an assay of chlorpheniramine in human plasma, the mean matrix effect associated with APCI was 75% less than that when samples were run in the ESI mode (Ismaiel et al. 2008). These authors attributed the improvement in matrix effects to an apparent decrease in ionization efficiency of GPChos and 2-lyso GPChos using APCI (Ismaiel et al. 2008). However, APCI can be less sensitive for quantification of analytes than ESI, and the high temperatures inherent to APCI can degrade the samples, including the analyte molecules. In instances where APCI is not a viable choice to help reduce matrix effects, polarity switching between positive and negative ESI can sometimes be useful for reducing the impact of PLs on ion signal (Li et al. 2013a). As many PLs dissociate into a common fragment ions of *m/z* 184 or 104 in positive ESI mode, analytes compatible with negative ESI can still be quantified in the presence of PLs in the sample, while impact of PLs being simultaneously monitored in positive ESI (Li et al. 2013a).

4.4.5 Other Experimental Modifications

Several creative method modifications have been evaluated to help avoid PL-related matrix effects. Li et al. (2013b) utilized echinus-like Fe_3O_4@TiO_2 core–shell-structured microspheres to selectively remove interfering PCs and LPCs in a MALDI-TOF-MS analysis. In an extensive study of reverse-phase chromatography based LC-MS methods, Silvester and Smith (2012) discovered that the use of MeOH as the organic solvent significantly reduced matrix effects due to PLs versus DCM, a trend repeated on eight different column types. Neville et al. (2012) demonstrated that in cases where a highly lipophilic chromatographic column is not necessary, utilization of a BEH phenyl-phase column would greatly reduce PL retention, thus minimizing their effect on target analytes. Alternatively, HILIC columns have the potential to alter selectivity of analytes versus PLs, with a wide variety of retentive characteristics across different HILIC scaffoldings (Ekdahl et al. 2013). A study utilizing MALDI-TOF MS discovered significant interference in the negative mode caused by PLs, and strongly recommended PL separation prior to analysis (Schiller et al. 2002). However, when using MALDI-TOF MS in positive mode, different classes of PLs were easily identified (Petkovic et al. 2001); both of these studies found 2,5-dihydroxybenzoic acid (DHB) to be the preferred matrix for MALDI-TOF MS. Lytle et al. (2000) evaluated piperidine as a post-column modifier and they found that the compound promoted deprotonation of the target PLs, thus making it easier to detect PLs in negative-mode ESI. Finally, many investigators chose to monitor the MS/MS ion transitions of specific PLs to account for their impact on analyte signal (Bylda et al. 2014; Ismaiel et al. 2010; Little et al. 2006; Schiller et al. 2002; Trufelli et al. 2011).

4.5 Removing Phospholipids Prior to LC-MS Analysis

While changing analytical columns, mobile phases and other chromatographic conditions can help separate the PLs from the analyte(s), avoiding co-elution does not eliminate the tendency for these compounds to accumulate on the analytical column (Chambers et al. 2007; Silvester and Smith 2012). The most successful approach of minimizing matrix effects due to the presence of PLs is to remove them from the samples prior to analysis (Ask et al. 2016; Trufelli et al. 2011). Several approaches exist for PL removal as a component of sample preparation, each with its own unique advantages and disadvantages, depending on the analyte and matrix.

4.5.1 Protein Precipitations

Protein precipitation (PPT) through the use of acids, bases, or organic solvents followed by centrifugation is one of the simplest methods of sample preparation. However, this method only removes proteins without extensively removing PLs (Ahmad et al. 2012; Chambers et al. 2007). Some studies indicated that the choice of PPT solvent can play a role in the extent of PL removal, with ACN precipitates containing fewer PLs compared to MeOH extracts (Chambers et al. 2007). Other studies indicated that the volume of PPT solvent could manipulate the level of resulting matrix effects (Liu et al. 2007). In general, the degree of ionization suppression found in PPT-only sample extracts far exceeds those obtained via liquid–liquid extraction (LLE, Bonfiglio et al. 1999). When combined with another sample preparation strategy, the hybrid sample preparation method can be very efficient in PL removal. In a study of PPT using acidified ACN versus a hybrid SPE-PPT removal plate, matrix effects due to PLs in human plasma sample extracts were decreased from −34.8 to −5.1% (Pucci et al. 2009).

4.5.2 Liquid–Liquid Extraction and Supported Liquid Membranes

LLE can provide significant PL removal from samples, but it has drawbacks of difficult automation, high organic solvent consumption, and loss of analyte molecules that are electrostatically bound to PLs. The most common solvents used in LLE include methyl *tert*-butyl ether (MTBE), diethyl ether, chloroform, MeOH, DCM, toluene, and water (in differing ratios) (Reis et al. 2013). Modifications of LLE solvents include using additives such as acetic acid, hydrochloric acid, formic acid, citric acid, ethyl acetate (EA), and *n*-butyl chloride, as well as replacing chloroform with MTBE or butanol (Bylda et al. 2014; Jemal et al. 2010; Teo et al. 2015). Most of these modifications are intended to enhance analyte recovery, or are for combatting matrix effects due to a particular species of PL (Michopoulos et al. 2011). For example, for highly hydrophilic compounds that are not easily recovered using solvents alone, adding a high concentration of salt to the sample increases the polarity of the aqueous phase. This leads to superior phase separation when using a water-miscible solvent for LLE, a process often referred to as salt-assisted LLE or SALLE (Bylda et al. 2014). In a 2008 review, Srinivas (2009) cautioned against discarding a possible LLE solvent based solely on its reputation for enhancing matrix effects. To support this argument, the author cites contrasting studies regarding the utility of EA versus diethyl ether. Ding et al. (2007) cites the high polarity of EA for contributing to extraction of endogenous compounds that co-elute with metabolites of suplatast tosilate. However, in an assay of mitiglinide, EA was preferred over other LLE solvents for extraction of the drug from plasma (Zhang et al. 2008).

A novel approach to LLE was introduced in 2013, utilizing principles of hollow fiber liquid phase microextraction (HF-LPME) to extract analytes across a supported liquid membrane (SLM) (Gjelstad et al. 2013). This technique, coined parallel artificial liquid membrane extraction (PALME), enables an LLE-like extraction that is easily automated using 96-well plates (Ask et al. 2016; Gjelstad et al. 2013). The 96-well format has a tandem sequence of acceptor and donor plates. The acceptor plate contains organic solvent-impregnated membranes that create the SLMs, and the samples are loaded into the donor plates. When the acceptor plates are filled with an acceptor solution and stacked onto the donor plates, a pH gradient between the donor (sample) liquid and the acceptor solution drives extraction through the SLM (Gjelstad et al. 2013). In a 2016 study by Ask et al. (2016), PALME was found to be comparable to the HybridSPE® technology and superior to PPT for PL removal from human plasma samples.

4.5.3 Supported Liquid Extraction

Supported liquid extraction (SLE), a process that is mechanistically similar to LLE but requires less solvent use, involves using a porous solid material (usually diatomaceous earth) on which aqueous samples will adsorb, facilitating removal of PLs from the final sample during sample extraction. This process has been shown to be superior to PPT in removal of PLs from plasma (Michopoulos et al. 2011), and others have suggested that SLE performs as well as HybridSPE for removal of GPChos and SMs from plasma (Ismaiel et al. 2010). PLs have the capacity to bind to the adsorption material through polar interactions with the head group, as well as hydrophobic interactions with the fatty acid side chains (Ismaiel et al. 2010), thus the SLE solvent choice is important to ensure analyte recovery while leaving PLs adsorbed on the porous solid materials. For the purpose of extracting PLs, DCM and MTBE have been shown to be effective for this process, reportedly removing ~99% of PLs (Bylda et al. 2014), with IPA and EA also tested. Ismaiel et al. (2010) reported that SLE with DCM did extract GPChos and SMs into the final sample, indicating that other solvents (or combinations), such as MTBE, ether, EA, and hexane, may be better suited to remove PLs. Mixtures of these and other organic solvents have shown variable results, depending on the type of PL isolated, the pH of the solvent mixture, and the specific matrix. For example, acidified solvents (MTBE, DCM, ether, and EA/hexane with FA) or those with no pH adjustment resulted in more PLs present in the final sample extract as compared to their alkaline counterparts (Ismaiel et al. 2010).

4.5.4 Electrostatic Removal

An unusual approach to PL removal lies in the utilization of colloidal silica combined with cationic solutions to electrostatically interact with these interfering components of biological matrices. The seminal work in the area was conducted by Schoener et al. (2006), who demonstrated that a combination of Ludox AS-40® with a $MnCl_2$ solution removes endogenous PC and lyso-PC from plasma. Other investigators explored alternative cationic solutions, including metals in the lanthanide series. Ultimately, $LaCl_3$ stood out as an effective cation to interact with the phosphate group of plasma PLs, thus effectively removing them without compromising the recovery of target analytes (Michopoulos et al. 2011; Wu et al. 2008).

4.5.5 Solid-Phase Extraction

Solid-phase extraction (SPE) is an efficient mode of sample preparation for LC-MS bioanalysis, providing a high level of PL removal, as shown in Figure 4.4 (Chambers et al. 2007). Compared to LLE, SPE is generally more selective and easily automated, and requires small volumes of organic solvents. Many of the SPE columns targeted for phospholipid depletion (PD) contain either Zr or Ti. Zirconium (Zr) and titanium (Ti) preferentially bind to PLs in non-biological matrices and have also been shown useful in PLs removal in biological matrix (Li et al., 2013; Lin et al. 2012). Products designed for PD include HybridSPE (Supelco), Phree® (Phenomenex), and Ostro® (Waters). In addition to providing SPE specific to scavenging PLs, these products aid in PPT as part of the sample preparation workflow, thus resulting in their designation as SPE-PPT technologies (Ahmad et al. 2012; Omnia and Karnes 2013). The packing in these PD-SPE

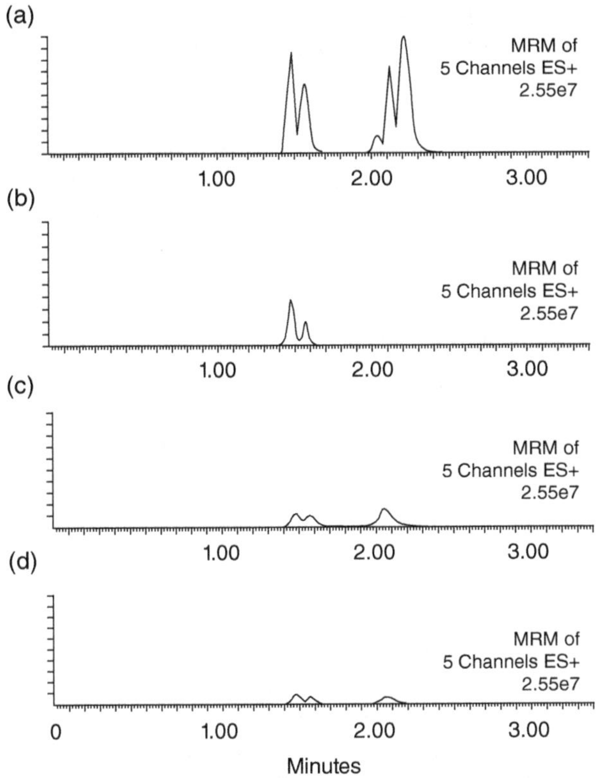

Figure 4.4 TICs of MRM transitions for five phospholipids remaining in final extracts after sample preparation by (a) ACN PPT, (b) reversed-phase polymeric SPE, (c) silica-based pure cation exchange, and (d) mixed-mode cation exchange SPE; HPLC-MS/MS with pH 10 mobile phase. *Source:* Reprinted with permission from Chambers et al. (2007). Reproduced with permission of Elsevier.

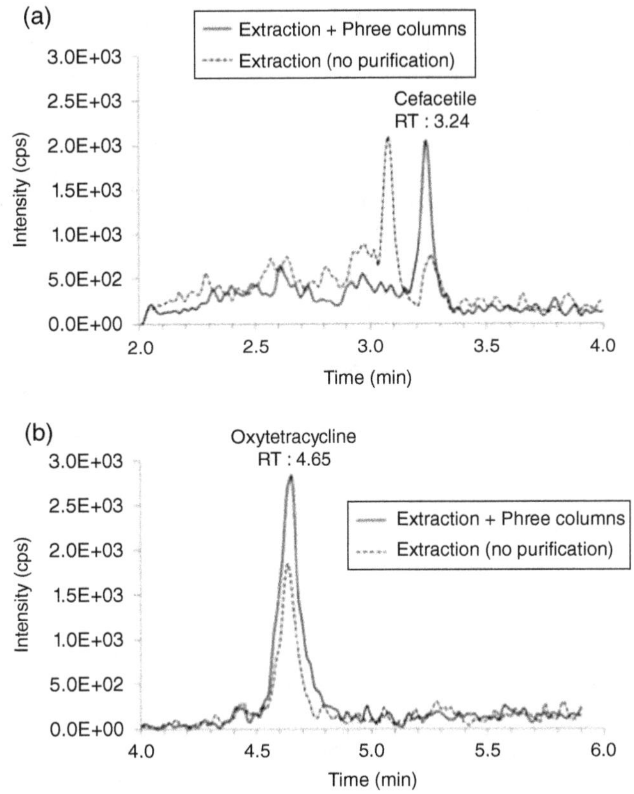

Figure 4.5 The UHPLC-MS/MS chromatograms for matrix (sturgeon) samples spiked with 5 μg kg^{-1} standards of cefacetrile (a) and oxytetracycline (b), prepared by conventional ACN extraction (PPT) and ACN extraction (PPT) followed by a cleanup stage on PL removal columns. *Source:* Reprinted with permission from Reinholds et al. (2016)). Reproduced with permission of Elsevier.

products works as a Lewis acid causing the PL phosphate moiety, which acts as a Lewis base, to adsorb to the column (Calderon-Santiago et al. 2014; Ferreiro-Vera et al. 2012; Pucci et al. 2009). Several research groups have indicated that this favorable Lewis acid–base mechanism is effective in selectively retaining PLs without leading to analyte loss (Ahmad et al. 2012; Cepurnieks et al. 2015; Omnia and Karnes 2013). In addition, utilization of nanoparticles for this application has potential to improve PL retention versus the ZrO_2 powder-based products (Li et al. 2013; Lin et al. 2012; Reinholds et al. 2016) due to a greater surface area of the former. In an extensive study of PPT, LLE, and SPE for PLs profiling in human serum, Ferreiro-Vera et al. (2012) concluded that the HybridSPE cartridges using MeOH elution resulted in the highest selectivity and sensitivity of the analyte(s) while avoiding interference from PLs. Likewise, Ismaiel et al. (2010) identified the HybridSPE column as an effective means to remove most PLs, as well as cholesterol and triacylglycerols in human plasma, as compared to several other sample preparation approaches. Finally, Reinholds et al. (2016) demonstrated the utility of another zirconium oxide SPE product, Phree from Phenomenex®, in aquaculture sample cleanup. Extraction with and without Phree pretreatment is shown in Figure 4.5 (Reinholds et al. 2016). The sensitivity facilitated by the PL removal enabled quantification of a wide range of antibiotics in these samples, relevant to assessment of consumer risk (Reinholds et al. 2016).

SPE pretreatment methods that effectively remove PLs can positively impact chromatographic workflow for large sample numbers by allowing simpler separation conditions, higher throughput, and longer column lifetimes (Danaceau et al. 2014). SPE can remove other plasma components, such as proteins, in addition to PLs, additionally diminishing matrix effects (Nilsson et al. 2014). In a comparison of PPT, LLE, and SPE techniques, Neville et al. (2012) demonstrated ~100% combined PL removal using PD-specific products, Ostro and HybridSPE technologies. HybridSPE was used to improve detection of perfluoroalkyl acids in environmental samples compared to turbulent flow chromatography (TFC) cleanup (Mazzoni et al. 2016). The effect of PL accumulation on reverse-phase columns was mitigated through the use of Waters Ostro cartridges for the analysis of 10 drugs and metabolites (Danaceau et al. 2014). In a head-to-head comparison of Waters Ostro versus Phenomenex Phree technologies as a pretreatment for untargeted metabolomics experiments, both PD products performed equally at minimizing matrix effects versus PPT alone (Tulipani et al. 2015). Alternatively, Van Damme et al. (2014) reported the use of HILIC-SPE with acidified acetone pretreatment as a comparable approach to PD-specific SPE phases. In this study, the investigators spiked human plasma with several drugs, including antivirals and antipsychotics, and compared extractions using HILIC-SPE with acetone pretreatment to HybridSPE, noting only marginal difference in analyte recoveries with comparable PL removal.

4.6 Example Methods that Demonstrate Successful Phospholipid Removal

Table 4.1 provides a summary of strategies used for monitoring PLs, both as analytes and as matrix components. Some of the entries in this table utilized focused PL removal through new sample preparation technologies. The implementation of novel PD technologies resulted in an improved recovery or decreased ionization suppression for the analyte(s) of interest. In a metabolic profiling study of bile acids in blood, Han et al. (2015) noted superior recovery and reproducibility for seven of nine analytes using HybridSPE sample preparation versus PPT and conventional SPE. For a metabolomics investigation in rainbow trout, David et al. (2014) successfully used the Phenomenex Phree PL removal pretreatment

Table 4.1 Summary of sample preparation strategies involved in monitoring phospholipids in biological samples and utilization of sample preparation to minimize matrix effects associated with phospholipids.

Study referenced	Sample preparation	Analysis	Matrix	Application notes
Ask et al. (2016)	PALME compared to HybridSPE and PPT	UHPLC-MS/MS (+ESI)	Human plasma	PALME showed equivalent PPL removal compared to HybridSPE to reduce matrix effects in analysis of model analytes (fluoxetine, fluvoxamine, and quetiapine)
Mazzoni et al. (2016)	HybridSPE Phospholipid Ultra cartridges (off-line) and online TFC	UHPLC-MS/MS (−ESI)	Environmental solid samples (bivalve tissue, yolk, sediment, fish tissue)	Novel utilization of PL removal coupled with online TFC for analysis of perfluoroalkyl acid in environmental solid extracts
Reinholds et al. (2016)	Phree extraction with ACN	UHPLC-MS/MS (+ESI)	Fish tissues (sturgeon, trout)	Multiple classes of antibiotics; addition of Phree® sample preparation enabled detection sufficient for consumer risk assessment
De Baere et al. (2015)	PPT, LLE (Diethyl ether), SPE: Ostro 96-well, HybridSPE	LC-MS/MS (+ESI)	Animal plasma (chicken, turkey, pig, calf), lung tissue, pulmonary epithelial lining fluid (PELF) of turkey poults	Gamithromycin; recovery was maximized and matrix effects minimized using Ostro and HybridSPE
Han et al. (2015)	HybridSPE compared to reverse-phase SPE (Waters HLB)	UHPLC-MS/MS (−ESI)	Human, mouse serum	30 unconjugated and 20 conjugated (glycine/taurine) bile acids; reproducibility and recovery optimized with HybridSPE®
Kyle et al. (2015)	PPT and SPE (Strata X-Drug B) versus Phree and SPE	UHPLC-MS/MS (+ESI)	Human umbilical cord plasma	Buprenorphine and metabolites, methadone; PL removal improved recovery of methadone and norbuprenorphine
Rossmann et al. (2015)	SPE using Oasis® HLB	LC-MS/MS (+ESI)	Urine, plasma, wastewater	33 pharmaceutical substances; monitored general matrix effects using post-column infusion

Table 4.1 (Continued)

Study referenced	Sample preparation	Analysis	Matrix	Application notes
Tulipani et al. (2015)	PPT followed by Waters Ostro versus Phenomenex Phree	LC-MS/MS (±ESI)	Aqueous standard metabolite mixture, reference plasma, human plasma	Both PL removal products performed equally and were superior to PPT alone; useful component in workflow for untargeted metabolomics
Calderon-Santiago et al. (2014)	PPT, HybridSPE	LC-QTOF MS/MS (±ESI)	Human serum	Monitoring PC, LPC, and SM toward biomarker panel development in atherosclerosis
Danaceau et al. (2014)	PPT, Ostro 96-well plates	UPLC-MS/MS (+ESI)	Human plasma	10 drugs, 10 metabolites of those drugs (selected for varying chemical properties); significant PL accumulation on the column noted with PPT removal
David et al. (2014)	Phree® PL removal, PPT, SPE: Strata-X, Strata-X-C, Strata-X-AW	nUHPLC-nESI-TOFMS (+) and (−) modes	Fish plasma (rainbow trout)	63 compounds of varying chemistries; PL removal+SPE versus PPT had the same range of metabolome coverage but SPE allowed for direct injection of more concentrated sample into nano-system.
Nilsson et al. (2014)	PPT and PL removal with Waters Ostro filter plate and SPE (HLB)	UPLC-MS/MS (+ESI)	Human plasma	Budesonide; PL removal necessary because of chromatographic elution near lipids
Van Damme et al. (2014)	HILIC-SPE with acidified acetone as pretreatment	LC-MS with Turbo Ion Spray source	Human plasma	Sulfamethazine, zidovudine, risperidone, sulfadimethoxine, propranolol, carbamazepine, amitriptyline, terfenadine, and trifluoroperazine indomethacin; HILIC-SPE proved to be viable alternative to HybridSPE
Ekdahl et al. (2013)	PPT	Eight different HILIC columns, LC-MS/MS (+ESI)	Human plasma	Urea, creatinine, PC, LPC, SM, sodium ion, chloride ion, choline, and proline betaine; pair-wise and PCA used to determine optimum column
Ismaiel et al. (2013)	LLE (MeOH, ACN, MTBE, IPA, ethyl ether, DCM, and *n*-hexane)	LC-MS/MS (+ESI)	Dried blood spots (DBS) cards	Monitoring of GPChos, TAG, cholesterols for determination of matrix effects during DBS analysis
Li et al. (2013b)	SPE: HybridSPE, Adsorption: Echinus-like Fe_3O_4@TiO_2 core-shell structured microspheres	MALDI-TOF/MS	Human plasma	Monitoring PCs, LPCs for PL removal toward a metabolomics study (steroids)
Ahmad et al. (2012)	HybridSPE compared to PPT and SPE; 96-well plate format	LC-MS/MS (+ESI)	Rat plasma	Clenbuterol enantiomers (analytes); monitored for GPChos, LPCs
Ferreiro-Vera et al. (2012)	PPT, LLE (MeOH/chloroform or EtOH/chloroform), SPE: HybridSPE®	HPLC-TOF/MS (±ESI)	Human serum	Confirmed 50 PLs in (−) ESI, 159 PLs in (+) ESI for PL profiling; matching to Human Metabolome Database
Kushon and Pike (2013)	PPT versus Phree	HPLC or UHPLC LC-MS/MS	Plasma	Amoxapine analyte; monitoring PC, LPC for matrix effects between two sample preparation approaches
Neville et al. (2012)	PPT, LLE (DCM), ultrafiltration, multiple SPE technologies compared, including H-PPT, Waters Ostro, Captiva®, HLB, MCX, Strata™-X, Evolute® ABN	UPLC-MS/MS (+ESI)	Rat/dog/human plasma	H-PPT, Ostro, Captiva, and ultrafiltration all removed 100% of PLs, PLs least retained on BEH phenyl-phase columns

(Continued)

Table 4.1 (Continued)

Study referenced	Sample preparation	Analysis	Matrix	Application notes
Silvester and Smith (2012)	PPT, sample dilution	LC-MS/MS (±ESI)	Rat/dog/human plasma	PL profiling in reverse-phase chromatography using four model analytes (warfarin, rosuvastatin, caffeine, and propranolol)
Aubry (2011)	LLE (MTBE), PPT	LC-MS/MS (+ESI)	Plasma	BSM-790052
Ismaiel et al. (2011)	SPE: Oasis WCX; column-switching and pre-column back-flushing for PL removal	UHPLC-MS (+ESI)	Human plasma	Octreotide analyte, monitoring GPChos, LPCs to assess effectiveness of online PL removal using guard column and switching valve
Michopoulos et al. (2011)	SPE: HyperSep™ Retain PEP, Retain CX, Phenyl SPE cartridges, SLE: Thermo Scientific SLE cartridges, PPT: ACN	HPLC-MS (+ESI)	Rat Plasma	Five different phosphocholines
Ismaiel et al. (2010)	PPT (4 solvents); LLE (12 solvents); SLE (12 conditions); SPE (6 phases, including MCX, WAX, MAX, WCX, HLB, and H-PPT)	LC-MS (+ESI)	Human plasma, human erythrocytes	Chlorpheniramine, loratadine, monitored PC, LPC, PE, PG, PA (phosphatidic acid), PI, SM, PS, CE, MAG, DAG, TAG for the establishment of a matrix-effect library; H-PPT effectively removed most PLs
Pucci et al. (2009)	PPT and HybridSPE	LC-MS (+ESI)	Rat/human plasma	Zirconia-coated particles in HybridSPE plates have high affinity for PLs; recovery of three model analytes improved with SPE-PPT combination
Xia and Jemal (2009)	PPT	LC-MS/MS (±ESI)	Human plasma	Diclofenac, simvastatin, pravastatin, buspirone, metformin (analytes); monitored for 10 different PL species in (+) ESI (4) and (−) ESI (6) to compare MS/MS techniques to monitor PLs
Ismaiel et al. (2008)	LLE (MTBE)	LC-MS/MS (+ESI/+APCI)	Human plasma	Chlorpheniramine analyte, diphenhydramine internal standard; monitored GPChos, 2-lyso-GPChos to compare matrix effects of APCI versus ESI
Wu et al. (2008)	Colloidal silica (Ludox AS-40) combined with cation solutions; Ultimate combination of lanthanum chloride ($LaCl_3$) with Ludox AS-40®	LC-MS/MS (+ESI)	Rat/dog plasma	"Model analytes" representing a wide range of chemical features; monitoring endogenous phospholipids PC, PE, and SM
Chambers et al. (2007)	PPT, LLE (MTBE), SPE: Oasis MCX, Oasis MAX, Oasis WCX, Oasis WAX, Oasis HLB, and an experimental silica-based pure cation exchange SPE product	HPLC or UPLC, MS/MS with (+ESI)	Rat plasma	Compared recovery of eight small molecule drugs and metabolites; monitored PCs, GPChos for matrix effect contribution
Ismaiel et al. (2007)	LLE (MTBE)	LC-MS/MS (+ESI)	Human plasma	Hydrocodone, pseudoephedrine, monitored m/z 184 to examine matrix effects
Gao et al. (2006)	LLE (Chloroform/MeOH)	LC-MS (−ESI)	Sprague–Dawley rat peritoneal lining	PE, PI, PS, PC, SM, LPC, and >90 PL constituents toward rapid screening/profiling assay development

Table 4.1 (Continued)

Study referenced	Sample preparation	Analysis	Matrix	Application notes
Ham et al. (2006)	LLE (Chloroform)	MALDI-TOF MS, IMAC	Rabbit tears	Monitoring PC, SM, LPC, PS toward comparison of normal versus dry-eye animals
Little et al. (2006)	PPT	UPLC-MS/MS (+ESI)	Animal plasma (mouse, rat, rabbit, dog, monkey, human)	Tamoxifen, itraconazole, diphenhydramine HCl, 4-hydroxytamoxifen, n-desmethyltamoxifen, hydroxyitraconazole, GPChos, GPEtns (glycero-PEs), GPSers (glycero-PSs)
Shen et al. (2005)	SPE, including Oasis HLB MCX, Varian SPEC® SCX and MP1	LC-MS (+ESI)	Human plasma	Desloratadine and metabolites following dosing with Clarinex™, SPEC SCX showed 30× greater PL removal versus MP1 and 7× greater than Oasis HLB
Dams et al. (2003)	PPT, SPE with Clean-Screen® DAU columns	LC-MS/MS (+APCI or + ESI)	Urine, oral fluid, plasma	Morphine
Schiller et al. (2002)	Combine samples with MALDI matrix (0.5 M 2,5-dihydroxybenzoic acid (DHB) in MeOH with 0.1% TFA as matrix)	MALDI-TOF MS	Egg yolk, human plasma	Exploring detectability of PLs in the presence of DHB in (+) and (−) modes

technology, that, when combined with additional SPE, allowed for injections of high concentration of sample extracts into the nanoflow system. In quantification of buprenorphine, methadone, and metabolites from umbilical cord blood, Kyle et al. (2015) reported that removal of proteins and PLs using Phree columns prior to anion-exchange SPE improved recovery of the most lipophilic analyte, methadone, as well as the hydroxylated metabolite of buprenorphine, but did not improve the recovery of the highly polar glucuronide metabolites. For quantification of gamithromycin in agriculturally relevant matrices, the use of Ostro sample pretreatment or HybridSPE provided superior analyte recovery and minimized matrix effects versus PPT or LLE with diethyl ether (De Baere et al. 2015). In a previously published budesonide assay, the presence of PLs contributed significant matrix effects, which were ameliorated with the integration of Ostro pretreatment (Nilsson et al. 2014). Finally, in an LC-MS assay of amoxapine, the Phree pretreatment for plasma samples showed reduced ionization suppression compared to ACN precipitation (Pike and Kushon 2012).

4.7 Conclusions

PLs of varying species are ubiquitous in plasma and serum samples as well as other biological fluids, and removal of them from biological samples is necessary to maximize assay sensitivity, specificity, and reliability. Identification and monitoring of the molecule/product ions of the specific PLs in question allows for adjustment of elution times in the chromatography process. Use of a SIL-IS can assist in compensating for matrix effects, but these approaches do not prevent PL accumulation on the analytical column. Utilization of alternative column chemistries, such as BEH Phenyl or HILIC, may play a role in minimizing PL retention, thus making it easier to chromatographically resolve analytes from these interfering compounds (Ekdahl et al. 2013; Neville et al. 2012). PLs are retained on hydrophobic reverse-phase columns, thus likely to affect drug and metabolite analytes suitable for such stationary phases, as shown in Figure 4.6, where late-eluting analytes and metabolites may be impacted by the PLs (Little et al. 2006). Many studies have shown the utility of SPE for PL removal. Specifically, technologies like HybridSPE, sometimes referred to as PD-SPE (Han et al. 2015) rise to the top in terms of maximizing analyte recovery while simultaneously reducing the impact caused by PLs on the mass spectrometric signal (Ferreiro-Vera et al. 2012; Ismaiel et al. 2010). Alternatively, PALME was shown to be as effective as HybridSPE for elimination of matrix effects in the extraction of analytes such as fluoxetine, fluvoxamine, and quetiapine from human plasms (Ask et al. 2016). Overall, sample preparation tools with high specificity for PL removal provide excellent cleanup for biological samples with some reports indicating 100% of PL removal (Neville et al. 2012).

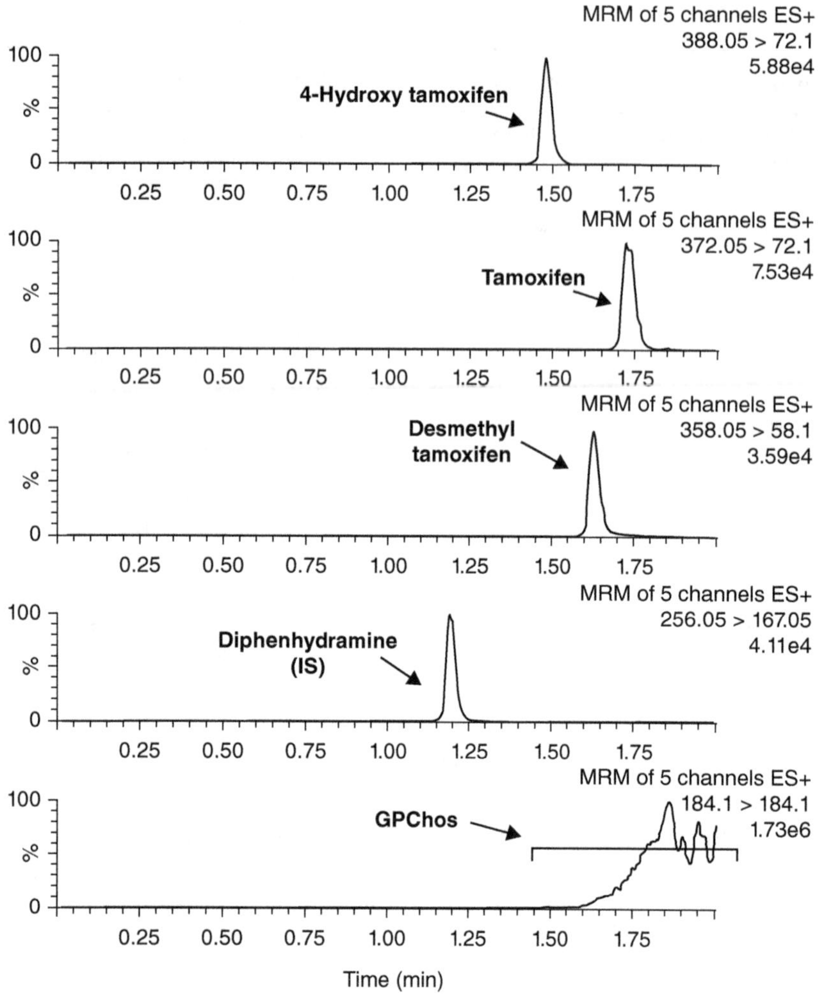

Figure 4.6 LC-MS/MS separation of tamoxifen and its metabolites using steady-state gradient method, 10 ng ml^{-1} calibrator, 2.8 minute analysis cycle, separation performed on a Varian MonoChrom C18 column, 3 μm, 50 mm × 2.0 mm. *Source:* Reprinted with permission from Little et al. (2006). Reproduced with permission of Elsevier.

Acknowledgement

The authors would like to acknowledge the Bill Gatton College of Pharmacy for their ongoing support of student research.

References

Ahmad, S., Kalra, H., Gupta, A. et al. (2012). HybridSPE®: a novel technique to reduce phospholipid-based matrix effect in LC-ESI-MS bioanalysis. *J. Pharm. Bioallied Sci.* 4 (4): 267–275.

Antignac, J.-P., de Wasch, K., Monteau, F. et al. (2005). The ion suppression phenomenon in liquid chromatography–mass spectrometry and its consequences in the field of residue analysis. *Anal. Chim. Acta* 529 (1–2): 129–136.

Ask, K.S., Bardakci, T., Parmer, M.P. et al. (2016). Parallel artificial liquid membrane extraction as an efficient tool for removal of phospholipids from human plasma. *J. Pharm. Biomed. Anal.* 129: 229–236.

Aubry, A.F. (2011). LC-MS/MS bioanalytical challenge: ultra-high sensitivity assays. *Bioanalysis* 3 (16): 1819–1825.

Azzouz, N., Shams-Eldin, H., and Schwarz, R.T. (2005). Removal of phospholipid contaminants through precipitation of glycosylphosphatidylinositols. *Anal. Biochem.* 343 (1): 152–158.

Berg, J.M., Tymoczko, J.L., and Stryer, L. (2002). *Biochemistry*, 5e. New York, NY: WH Freeman.

Bonfiglio, R., King, R.C., Olah, T.V., and Merkle, K. (1999). The effects of sample preparation methods on the variability of the electrospray ionization response for model drug compounds. *Rapid Commun. Mass Spectrom.* 13 (12): 1175–1185.

Brouwers, J.F. (2011). Liquid chromatographic-mass spectrometric analysis of phospholipids. Chromatography, ionization and quantification. *Biochim. Biophys. Acta* 1811 (11): 763–775.

Bylda, C., Thiele, R., Kobold, U., and Volmer, D.A. (2014). Recent advances in sample preparation techniques to overcome difficulties encountered during quantitative analysis of small molecules from biofluids using LC-MS/MS. *Analyst* 139 (10): 2265–2276.

Calderon-Santiago, M., Priego-Capote, F., Galache-Osuna, J.G., and Luque de Castro, M.D. (2014). Analysis of serum phospholipid profiles by liquid chromatography-tandem mass spectrometry in high resolution mode for evaluation of atherosclerotic patients. *J. Chromatogr. A* 1371: 154–162.

Cepurnieks, C., Rjabova, J., Zacs, D., and Bartkevics, V. (2015). The development and validation of a rapid method for the determination of antimicrobial agent residues in milk and meat using ultra performance liquid chromatography coupled to aqudrupole-orbitrap mass spectrometry. *J. Pharm. Biomed. Anal.* 102: 184–192.

Chambers, E., Wagrowski-Diehl, D.M., Lu, Z., and Mazzeo, J.R. (2007). Systematic and comprehensive strategy for reducing matrix effects in LC/MS/MS analyses. *J. Chromatogr. B Analyt. Technol. Biomed. Life Sci.* 852 (1–2): 22–34.

Dams, R., Huestis, M.A., Lambert, W.E., and Murphy, C.M. (2003). Matrix effect in bio-analysis of illicit drugs with LC-MS/MS: influence of ionization type, sample preparation, and biofluid. *J. Am. Soc. Mass Spectrom.* 14 (11): 1290–1294.

Danaceau, J., Yu, H., Chambers, E., and Fountain, K.J. (2014). Matrix effects in metabolite quantification for MIST assessment: the impact of phospholipid removal and HPLC column particle size. *Bioanalysis* 6 (6): 761–771.

David, A., Abdul-Sada, A., Lange, A. et al. (2014). A new approach for plasma (xeno)metabolomics based on solid-phase extraction and nanoflow liquid chromatography-nanoelectrospray ionisation mass spectrometry. *J. Chromatogr. A* 1365: 72–85.

De Baere, S., Devreese, M., Watteyn, A. et al. (2015). Development and validation of a liquid chromatography-tandem mass spectrometry method for the quantitative determination of gamithromycin in animal plasma, lung tissue and pulmonary epithelial lining fluid. *J. Chromatogr. A* 1398: 73–82.

Ding, L., Ding, L., Zhou, X. et al. (2007). Liquid chromatography/electrospray ionization mass spectrometry method for the determination of the active metabolite M-1 of suplast tosilate in human plasma. *Biomed. Chromatogr.* 21: 1297–1302.

Ekdahl, A., Johansson, M.C., and Ahnoff, M. (2013). Tracing and separating plasma components causing matrix effects in hydrophilic interaction chromatography-electrospray ionization mass spectrometry. *J. Chromatogr. B Analyt. Technol. Biomed. Life Sci.* 923-924: 83–91.

Ferreiro-Vera, C., Priego-Capote, F., and Luque de Castro, M.D. (2012). Comparison of sample preparation approaches for phospholipids profiling in human serum by liquid chromatography-tandem mass spectrometry. *J. Chromatogr. A* 1240: 21–28.

Ferrer, C., Lozano, A., Aguera, A. et al. (2011). Overcoming matrix effects using the dilution approach in multiresidue methods for fruits and vegetables. *J. Chromatogr. A* 1218 (42): 7634–7639.

Furey, A., Moriarty, M., Bane, V. et al. (2013). Ion suppression; a critical review on causes, evaluation, prevention and applications. *Talanta* 115: 104–122.

Gao, F., Tian, X., Wen, D. et al. (2006). Analysis of phospholipid species in rat peritoneal surface layer by liquid chromatography/electrospray ionization ion-trap mass spectrometry. *Biochim. Biophys. Acta* 1761 (7): 667–676.

Gjelstad, A., Rasmussen, K., Parmer, M., and Bjergaard, S.P. (2013). Parallel artificial liquid membrane extraction: micro-scale liquid-liquid extraction in the 96-well format. *Bioanalysis* 5: 1377–1385.

Ham, B.M., Cole, R.B., and Jacob, J.T. (2006). Identification and comparison of the polar phospholipids in Normal and dry eye rabbit tears by MALDI-TOF mass spectrometry. *Invest. Ophthalmol. Vis. Sci.* 47 (8): 3330–3338.

Han, J., Lui, Y., Wang, R. et al. (2015). Metabolic profiling of bile acids in human and mouse blood by MS/MS in combination with phospholipid-depletion solid-phase extraction. *Anal. Chem.* 87: 1127–1136.

Ismaiel, O.A., Halquist, M.S., Elmamly, M.Y. et al. (2007). Monitoring phospholipids for assessment of matrix effects in a liquid chromatography-tandem mass spectrometry method for hydrocodone and pseudoephedrine in human plasma. *J. Chromatogr. B Analyt. Technol. Biomed. Life Sci.* 859 (1): 84–93.

Ismaiel, O.A., Halquist, M.S., Elmamly, M.Y. et al. (2008). Monitoring phospholipids for assessment of ion enhancement and ion suppression in ESI and APCI LC/MS/MS for chlorpheniramine in human plasma and the importance of multiple source matrix effect evaluations. *J. Chromatogr. B Analyt. Technol. Biomed. Life Sci.* 875 (2): 333–343.

Ismaiel, O.A., Jenkins, R.G., and Karnes, H.T. (2013). Investigation of endogenous blood lipids components that contribute to matrix effects in dried blood spot samples by liquid chromatography-tandem mass spectrometry. *Drug Test. Anal.* 5 (8): 710–715.

Ismaiel, O.A., Zhang, T., Jenkins, R.G., and Karnes, H.T. (2010). Investigation of endogenous blood plasma phospholipids, cholesterol and glycerides that contribute to matrix effects in bioanalysis by liquid chromatography/

mass spectrometry. *J. Chromatogr. B Analyt. Technol. Biomed. Life Sci.* 878 (31): 3303–3316.

Ismaiel, O.A., Zhang, T., Jenkins, R., and Karnes, H.T. (2011). Determination of octreotide and assessment of matrix effects in human plasma using ultra high performance liquid chromatography-tandem mass spectrometry. *J. Chromatogr. B Analyt. Technol. Biomed. Life Sci.* 879 (22): 2081–2088.

Jemal, M., Ouyang, Z., and Xia, Y.Q. (2010). Systematic LC-MS/MS bioanalytical method development that incorporates plasma phospholipids risk avoidance, usage of incurred sample and well thought-out chromatography. *Biomed. Chromatogr.* 24 (1): 2–19.

Kushon, S. and Pike, E. (2013). *Phenomenex. Rapid Improvements for LCMSMS Analysis without Additional Method Development Using Phree® Phospholipid Removal Plates.* www.phenomenex.com (accessed September 2016).

Kyle, A.R., Carmical, J., Shah, D. et al. (2015). UHPLC-MS/MS quantification of buprenorphine, norbuprenorphine, methadone, and glucuronide conjugates in umbilical cord plasma. *Biomed. Chromatogr.* 29 (10): 1567–1574.

Li, F., Ewles, M., Pelzer, M. et al. (2013a). Case studies: the impact of nonanalyte components on LC-MS/MS-based bioanalysis: strategies for identifying and overcoming matrix effects. *Bioanalysis* 5 (19): 2409–2441.

Li, H., Shi, X., Qiao, L. et al. (2013b). Synthesis of a new type of echinus-like Fe3O4@TiO2 core-shell-structured microspheres and their applications in selectively enriching phosphopeptides and removing phospholipids. *J. Chromatogr. A* 1275: 9–16.

Liang, H.R., Foltz, R.L., Meng, M., and Bennett, P. (2003). Ionization enhancement in atmospheric pressure chemical ionization between target drugs and stable-isotope-labeled internal standards in quantitative liquid chromatographic/tandem mass spectrometry. *Rapid Commun. Mass Spectrom.* 17: 2815–2821.

Lin, Y.F., Chen, J.H., Hsu, S.H. et al. (2012). The synthesis of Lewis acid ZrO2 nanoparticles and their applications in phospholipid adsorption from Jatropha oil used for biofuel. *J. Colloid Interface Sci.* 368 (1): 660–662.

Little, J.L., Wempe, M.F., and Buchanan, C.M. (2006). Liquid chromatography-mass spectrometry/mass spectrometry method development for drug metabolism studies: examining lipid matrix ionization effects in plasma. *J. Chromatogr. B Analyt. Technol. Biomed. Life Sci.* 833 (2): 219–230.

Liu, F., Xu, Y., Huang, J. et al. (2007). Sensitive liquid chromatography/mass spectrometry assay for the quantification of azithromycin in human plasma. *Biomed. Chromatogr.* 21: 1272–1278.

Lytle, C.A., Gan, Y.D., and White, D.C. (2000). Electrospray ionization/mass spectrometry compatible reversed-phase separation of phospholipids: piperidine as a post column modifier for negative ion. *J. Microbiol. Methods* 41: 227–234.

Matuszewski, B.K. (2006). Standard line slopes as a measure of a relative matrix effect in quantitative HPLC-MS. *Bioanalysis* 830: 293–300.

Mazzoni, M., Polesello, S., Rusconi, M., and Valsecchi, S. (2016). Liquid chromatography mass spectrometry determination of perfluoroalkyl acids in environmental solid extracts after phospholipid removal and on-line turbulent flow chromatography. *J. Chromatogr. B* 1453: 62–70.

Michopoulos, F., Edge, A.M., Hui, Y.T. et al. (2011). Extraction methods for the removal of phospholipids and other endogenous material from a biological fluid. *Bioanalysis* 3 (24): 2747–2455.

Neville, D., Houghton, R., and Garrett, S. (2012). Efficacy of plasma phospholipid removal during sample preparation and subsequent retention under typical UHPLC conditions. *Bioanalysis* 4 (7): 795–807.

Nilsson, K., Andersson, M., and Beck, O. (2014). Phospholipid removal combined with a semi-automated 96-well SPE application for determination of budesonide in human plasma with LC-MS/MS. *J. Chromatogr. B Analyt. Technol. Biomed. Life Sci.* 970: 31–35.

Omnia, I.A. and Karnes, T.H. (2013). Hybrid stationary phases: the answer to all of your matrix effects problems? *Bioanalysis* 5: 2735–2737.

Petkovic, M., Schiller, J., Muller, M. et al. (2001). Detection of individual phospholipids in lipid mixtures by matrix-assisted laser desorption/ionization time-of-flight mass spectrometry: phosphatidylcholine prevents the detection of further species. *Anal. Biochem.* 289 (2): 202–216.

Pike, E. and Kushon, S. (2012). An improved method for eliminating ion suppression and other phospholipid-induced interferences in bioanalytical samples. *LCGC* 10 (4): 28–31.

Postle, A.D., Wilton, D.C., Hunt, A.N., and Attard, G.S. (2007). Probing phospholipid dynamics by electrospray ionisation mass spectrometry. *Prog. Lipid Res.* 46 (3–4): 200–224.

Pucci, V., Di Palma, S., Alfieri, A. et al. (2009). A novel strategy for reducing phospholipids-based matrix effect in LC-ESI-MS bioanalysis by means of HybridSPE. *J. Pharm. Biomed. Anal.* 50 (5): 867–871.

Reinholds, I., Pugajeva, I., Perkons, I., and Bartkevics, V. (2016). The application of phospholipid removal columns and ultra-high performance liquid chromatography-tandem quadrupole mass spectrometry for quantification of multi-class antibiotics in aquaculture samples. *J. Pharm. Biomed. Anal.* 128: 126–131.

Reis, A., Rudnitskaya, A., Blackburn, G.J. et al. (2013). A comparison of five lipid extraction solvent systems for lipidomic studies of human LDL. *J. Lipid Res.* 54 (7): 1812–1824.

Rossmann, J., Gurke, R., Renner, L.D. et al. (2015). Evaluation of the matrix effect of different sample matrices for 33 pharmaceuticals by post-column infusion. *J. Chromatogr. B Analyt. Technol. Biomed. Life Sci.* 1000: 84–94.

Schiller, J., Suss, R., Petkovic, M. et al. (2002). Negative-ion matrix-assisted laser desorption and ionization time-of-flight mass spectra of complex phospholipid mixtures in the presence of phosphatidylcholine: a cautionary note on peak assignment. *Anal. Biochem.* 309 (2): 311–314.

Schoener D, Noren M, Lehman J, Lin P (2006). Reduction of phospholipid matrix effects in electrospray ionization by adding colloidal silica and manganese chloride. *54thASMS Conference*, Seattle (2006).

Services UDoHaH. (2001). *FDA 2001 Guidance for Industry Bioanalytical Method Validation*, Administration FaD (ed.).

Shen, J.X., Motyka, R.J., Roach, J.P., and Hayes, R.N. (2005). Minimization of ion suppression in LC-MS/MS analysis through the application of strong cation exchange solid-phase extraction (SCX-SPE). *J. Pharm. Biomed. Anal.* 37 (2): 359–367.

Silvester, S. and Smith, L. (2012). Profiling phospholipid elution in reversed-phase LC-MS/MS bioanalytical methods in order to avoid matrix effects. *Bioanalysis* 4 (8): 879–895.

Souverain, S., Rudaz, S., and Veuthey, J.L. (2004). Matrix effect in LC-ESI-MS and LC-APCI-MS with off-line and on-line extraction procedures. *J. Chromatogr. A* 1058 (1–2): 61–66.

Srinivas, N.R. (2009). Dodging matrix effects in liquid chromatography tandem mass spectrometric assays-compilation of key learnings and perspectives. *Biomed. Chromatogr.* 23 (5): 451–454.

Stahnke, H., Kittlaus, S., Kempe, G., and Alder, L. (2012). Reduction of matrix effects in liquid chromatography-electrospray ionization-mass spectrometry by dilution of the sample extracts: how much dilution is needed? *Anal. Chem.* 84 (3): 1474–1482.

Teo, C.C., Chong, W.P.K., Tan, E. et al. (2015). Advances in sample preparation and analytical techniques for lipidomics study of clinical samples. *TrAC Trends Anal. Chem.* 66: 1–18.

Trufelli, H., Palma, P., Famiglini, G., and Cappiello, A. (2011). An overview of matrix effects in liquid chromatography-mass spectrometry. *Mass Spectrom. Rev.* 30 (3): 491–509.

Tulipani, S., Mora-Cubillos, X., Jáuregui, O. et al. (2015). New and vintage solutions to enhance the plasma metabolome coverage by LC-ESI-MS untargeted metabolomics: the not-so-simple process of method performance evaluation. *Anal. Chem.* 87 (5): 2639–2647.

Van Damme, T., Lachova, M., Lynen, F. et al. (2014). Solid-phase extraction based on hydrophobil interactions liquid chromatography with acetone as eluent for eliminating matrix effects in the analysis of biological fluids by LC-MS. *Anal. Bioanal. Chem.* 406: 401–407.

Wu, S.T., Schoener, D., and Jemal, M. (2008). Plasma phospholipids implicated in the matrix effect observed in liquid chromatography/tandem mass spectrometry bioanalysis: evaluation of the use of colloidal silica in combination with divalent or trivalent cations for the selective removal of phospholipids from plasma. *Rapid Commun. Mass Spectrom.* 22 (18): 2873–2881.

Xia, Y.Q. and Jemal, M. (2009). Phospholipids in liquid chromatography/mass spectrometry bioanalysis: comparison of three tandem mass spectrometric techniques for monitoring plasma phospholipids, the effect of mobile phase composition on phospholipids elution and the association of phospholipids with matrix effects. *Rapid Commun. Mass Spectrom.* 23 (14): 2125–2138.

Zhang, Y., Ding, L., Tian, Y. et al. (2008). Liquid chromatography/electrospray ionization tandem mass spectrometry for the quantification of mitiglinide in human plasma: validation and its application to pharmacokinetic studies. *Biomed. Chromatogr.* 22: 873–878.

Zhou, S., Song, Q., Tang, Y., and Naidong, W. (2005). Critical review of development, validation, and transfer for high throughput bioanalytical LC-MS/MS methods. *Curr. Pharm. Anal.* 1 (1): 3–14.

5

Salting-out Assisted Liquid–Liquid Extraction (SALLE) in LC-MS Bioanalysis

Jun Zhang and Xin Xiong

5.1 Introduction

Protein precipitation (PPT), liquid–liquid extraction (LLE), and solid-phase extraction (SPE) are the three most widely used sample preparation techniques in bioanalysis (Chang et al. 2007). Among the three common procedures, PPT is the simplest one. It is applicable to both hydrophilic and hydrophobic compounds. However, the resulting supernatant contains a significant amount of soluble components, which might interfere with the chromatography or suppress the ionization in mass spectrometry for the target analyte. LLE and SPE can generate cleaner extracts than PPT. However, each technique will involve more extraction steps than PPT. In the conventional LLE procedure, a water-immiscible organic solvent needs to be vigorously mixed with biological samples, and after centrifugation, the organic phase needs to be transferred, evaporated, and reconstituted. It must be noted that hydrophilic compounds are difficult to be extracted. SPE is suitable for most of the compounds. But its procedure is relatively more complicated than LLE, these steps include sorbent conditioning, sample buffering and loading, multiple washing, and analyte eluting. The evaporation and reconstitution of elute are often needed. In addition, the operation cost of SPE is high. Recently, salting-out assisted liquid–liquid extraction (SALLE) as a homogeneous LLE technique has shown some distinct advantages over conventional PPT or LLE in bioanalysis. By SALLE, extraction of analytes of interest and cleaning up of the biological sample background can be carried out simultaneously. A wide polarity range of the compounds can be extracted from biological matrices with SALLE. The extracts can be directly injected onto several analytical systems, such as reversed-phase liquid chromatography (LC) or LC-hyphenated techniques without a need of evaporation and reconstitution.

A typical SALLE procedure is shown in Figure 5.1. A water-miscible organic solvent, such as acetonitrile, is added to biological samples. This is followed by the addition of a concentrated salt solution. After vortex-mixing and centrifugation, the organic solvent is separated from sample solution and the target analytes are extracted into the organic phase. After collection of the organic layer, the extracts can be injected directly or after dilution with water or buffer onto reversed-phase LC.

5.2 Considerations in Developing a SALLE Method

The type and amount of salting-out reagent, water-miscible organic solvent, and the pH value of the SALLE system are the main factors that affect the efficiency of SALLE. Those items need to be studied systematically by evaluating extraction recovery and matrix effect for the analytes of interest under various compositions of the above items. A key criterion is that the extraction recovery and matrix effect should be consistent at different concentration levels over the dynamic range. In Table 5.1 a compilation of the most relevant applications of SALLE for bioanalysis since 2006 is presented.

5.2.1 Salts

Three factors should be considered when investigating the salting-out effect of salts (Matkovich and Christian 1973) in extraction solvent. Firstly, the solubility of the salt in organic extraction solvent should be low. Secondly, the solubility of the salt in water should be high to reach maximum interaction with the water molecules. Lastly, the salt must have high ionic strength in the solution. Based on their ability to precipitate hydrophilic substances, the empirical order of various ions is as follows: $Mg^{2+} > Ca^{2+} > Sr^{2+} > Ba^{2+} > Li^{+} > Na^{+} > K^{+} > Rb^{+} > Cs^{+}$ for cations and citrate > tartrate > $SO_4^{2-} > CH_3COO^{-} > Cl^{-} > NO_3^{-} > Br^{-} > I^{-} > CNS^{-}$ for anions (Glasstone 1946). Considering the order above, $MgSO_4$ is the first choice,

Sample Preparation in LC-MS Bioanalysis, First Edition. Edited by Wenkui Li, Wenying Jian, and Yunlin Fu.

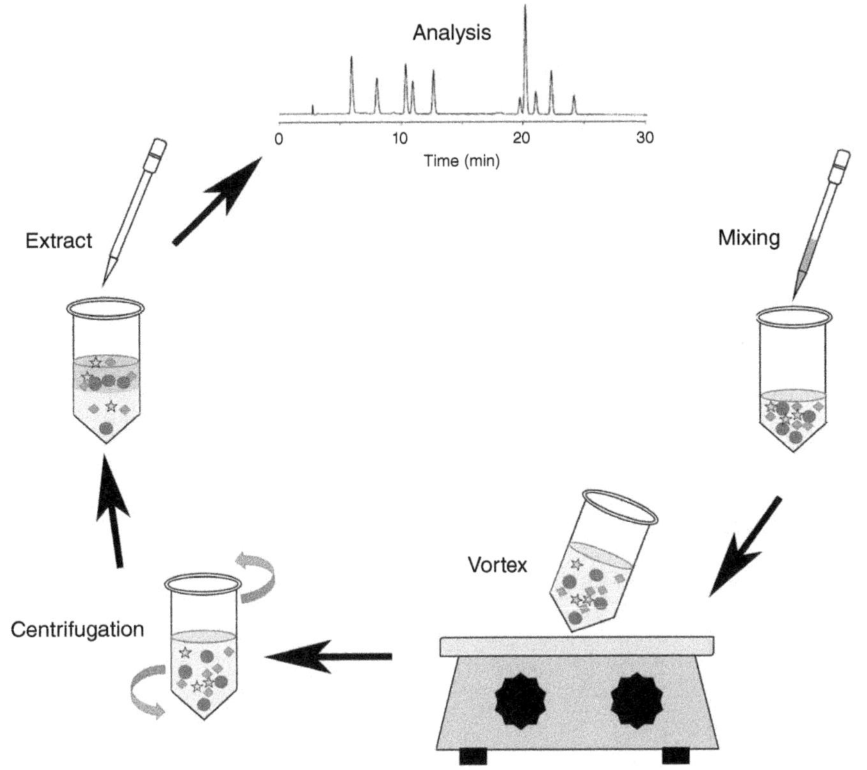

Figure 5.1 Schematic diagram of the SALLE procedure.

and has been reported in many published applications (Zhang et al. 2009b; Zhao et al. 2012; Song et al. 2013). K_2CO_3, NaCl, and $(NH_4)_2SO_4$ are also the salts commonly used as salting-out agents. However, the selection of a suitable salt is also affected by other factors, which includes analyte chemical property, type of sample, type of organic solvent, and the analytical instrument. For example, the impurities present in inorganic salts will introduce interferences to organic metal drug analysis. In this case, a hydrophobic salt should be used as a phase-separation agent for sample preparation (Nagosa et al. 1997).

Ammonium salts have been well studied since the salts are compatible with both LC and MS detection. Nanita and Padivitage (2013) investigated five ammonium salts used as salting-out agents for the extraction of pesticides from food and biological matrices. Two particular advantages of NH_4Cl over other ammonium salts were that saturated NH_4Cl solution has an acidic pH (~4) and the mixing process of NH_4Cl/acetonitrile/water is endothermic which can improve recoveries and analyte stabilization. Ammonium acetate, a MS-friendly salt, had been reported as the favorite choice for biological sample analysis by LC-MS (Wu et al. 2008; Zhang et al. 2009a, 2010; Yanes and Lovett 2012; Xiong et al. 2015; Xiong and Yang 2015). Although a small portion of the salts may dissolve into the organic phase, the impact of the salts on subsequent LC-MS analysis will be insignificant. In addition, the presence of ammonium in the sample extract helps minimize the formation of sodium adduct and enhance the formation of $[M+NH_4]^+$ in the electrospray process. Wu et al. (2008) found that clean and clear phase separation can be achieved at a 1 M concentration of salt in biological samples. Addition of a salt at significantly higher concentration (10 M) may result in unsatisfying precision in LC-MS analysis.

Wang et al. (2008) made an interesting observation that a high concentration of sugar can also induce a phase separation between water-miscible organic solvent and aqueous solution. This phenomenon is called "sugaring-out." Sugaring-out may provide an entirely new platform to the extraction of organic compounds and other products from the aqueous phase. The sugaring-out extraction method has been successfully applied to various matrices (Tsai et al. 2010; Zhang et al. 2013). Glucose, fructose, xylose, arabinose, sucrose, maltose, or polysaccharides could be used as appropriate sugaring-out reagents. A glucose concentration of $15\,g\,l^{-1}$ or higher can induce phase separation of a mixture of acetonitrile and water (1,1, v/v). Compared with SALLE, sugar does not react with analyte nor change the extraction environment (e.g. pH). In addition, it does not cause damage to the MS. Sugaring-out is still new and can be used as a SALLE substitution in bioanalysis. More evaluation should be conducted in order to expand the application scope.

Table 5.1 Salting-out assisted liquid–liquid extraction procedures for bioanalysis.

Compounds	Matrix	Salt	Solvent in SALLE	pH	Other extraction	References
Lopinavir, Ritonavir	Human, dog, and rabbit plasma	$MgSO_4$ (2 M)	ACN	—	—	Zhang et al. (2009b)
Multi-mycotoxin biomarkers	Pig urine	$MgSO_4$ (2 M)	ACN/FA(99 : 1)	—	Previous LLE	Song et al. (2013)
Entecavir	Human plasma	$MgSO_4$ (37.5%)	ACN	—	Posterior LLE	Zhao et al. (2012)
Pesticides	Human plasma and urine	NH_4Cl (saturated)	ACN	~4.0	—	Nanita and Padivitage (2013)
Hydrophobic drug candidate ABT-869 and its hydrophilic metabolite A-849529	Human plasma	NH_4AC (2 M)	ACN	—	—	Wu et al. (2008)
Unknown investigation new drug	Human plasma	NH_4AC (6 M)	ACN	—	—	Zhang et al. (2009a)
Simvastatin and simvastatin acid	Human plasma	NH_4AC (5 M)	ACN	4.5	—	Zhang et al. (2010)
Cannabinoid metabolites	Human urine	NH_4AC (10 M)	ACN	—	–	Yanes and Lovett (2012)
Anandamide	Human plasma	NH_4AC (5 M)	ACN	—	—	Xiong et al. (2015)
Trimetazidine	Rat plasma	NH_4AC (5 M)	ACN	—	—	Xiong and Yang (2015)
Lopinavir, Ritonavir (plus six proprietary compounds)	Human plasma	Glucose (2 M)	ACN	—	—	Zhang et al. (2013)
Lopinavir, Ritonavir	Human plasma	$ZnSO_4$ (3 M)	ACN	—	—	Myasein et al. (2009)
Atorvastatin, ortho-hydroxyatorvastatin, and para-hydroxyatorvastatin	Human plasma	NH_4AC (6 M)	ACN	6.0	—	Yong et al. (2015)
Warfarin	Human urine	NaCl (5 M)	ACN	4.0	—	Wang et al. (2011)
Hydrophobic porphyrins	Human urine	NaCl (5 M)	ACN	3.5	—	Li and Huie (2006)
Amphetamine, methamphetamine	Human urine	NaCl	ACN	~10	Extract cleanup by DLLME	Akramipour et al. (2016)
Sulfonamides	Honey, river water, and human urine	NaCl (300 mg ml^{-1})	ACN (a 1-ml syringe adapted with a tip was employed as the phase separation device)	6.5	—	Liu et al. (2010a)
Dorzolomide, timol	Aqueous humor	Phosphate (100 mmol l^{-1})	ACN (a 1-ml syringe adapted with a capillary tube was employed as the phase separation device)	11.9	—	Mohamed et al. (2014)
Sulfanilamide	Milk, honey, plasma, blood, and urine	NaCl (250 mg ml^{-1})	ACN (a coupled 1-ml syringe system was employed as the phase separation device)	7	—	Sereshti et al. (2014)

5.2.2 Water-Miscible Solvent

Water-miscible solvent is another important factor that needs to be evaluated for the best outcome of the SALLE method. In a study conducted by Tabata et al. (1994), fourteen water-miscible polar solvents were investigated for their separation from aqueous solutions in the presence of sodium chloride at a concentration of 4 M as the salting-out agent. Among the fourteen solvents, acetone, acetonitrile, 1,4-dioxanem, tetrahydrofuran, 1-propanol, and 2-propanol showed phase separation. However, 1-propanol and 2-propanol have larger volumes of organic phases after salting-out than their initial volumes. This is because that propanol interacts strongly with water via hydrogen-bonding, the separated propanol phase after salting-out contained a large percentage of water (2-propanol: 24.2% w/w). Furthermore, the phase-separated solvents contain a lot of chloride ions and the concentration of chloride ions increases with increasing water concentration in the organic phase. Among the above solvents evaluated, the least water residue (6.0% w/w) and salt concentration (0.16×10^{-2} M) were found in the organic phase after salting-out using tetrahydrofuran as the solvent. This was followed by acetonitrile, which was associated with a relatively high volume of organic phase with a moderate amount of water (10.3% w/w) and salt (2.5×10^{-2} M).

Isopropanol, acetonitrile, and acetone are the most popular solvents for SALLE. Although isopropanol paired with potassium carbonate (Horning et al. 1972) or ammonium carbonate (Horning et al. 1974) showed high extraction recovery for multiple basic drugs, many other unwanted matrix components were extracted simultaneously. Acetone has the same drawback as isopropanol. Acetonitrile extract is comparatively cleaner than isopropanol and acetone, which is the most commonly used extracting solvent in SALLE. There are several reasons for this choice. Firstly, during SALLE for sample preparation, acetonitrile readily precipitates proteins in biological samples and extracts both hydrophilic and hydrophobic compounds from aqueous solution. Secondly, with phase separation, acetonitrile led to a cleaner layer than the acetonitrile supernatant from PPT (Myasein et al. 2009). Finally, acetonitrile is compatible with most of the reversed-phase LC systems, so that the extracts can be directly analyzed by LC and LC-hyphenated techniques without evaporation and reconstitution. So far, in most published SALLE methods, acetonitrile has been chosen as the first choice of extracting solvent.

5.2.3 pH

The pH of the aqueous phase is an important factor that needs to be considered for the optimal extraction recovery of the analyte of interest. It is well known that ionized compounds have the lowest partition in organic solvents. In order to achieve maximum extraction recovery, the pH of aqueous phase of the extraction system should be adjusted to ensure the analytes of interest remain un-charged. Ideally, the pH in the aqueous matrix should be adjusted to at least two units away from the pK_a values of the ionizable group (Wieling et al. 1993). This is known as the "$pK_a \pm 2$ rule." In a study conducted by Wang et al. (2011) for the measurement of racemic warfarin ($pK_a = 5$ in aqueous solutions) in human urine, the influence of pH in the SALLE system was investigated. The pH (1–10) of the salting-out solution (5 M NaCl) was adjusted by mixing it with 100 mM phosphoric acid and 100 mM sodium phosphate. The result showed that the extraction efficiency for racemic warfarin increased gradually with the pH of the sample solution increased from 1.0 to 4.0. The maximum recovery (92.7%) was achieved at pH 4.0, after which the extraction efficiency decreased rapidly. At pH between 7.0 and 10, the extraction recovery fell below 60%. In a separate case study, Li and Huie (2006) investigated the effect of pH on the salting-out extraction efficiency of hydrophobic and hydrophilic porphyrins in urine. The best extraction efficiency of three hydrophobic porphyrins was achieved at pH 3.5 (with extraction efficiency between 75 and 90%), while the extraction efficiency for the most hydrophilic porphyrins (with carboxylic acid groups) remained relatively low (<20%) throughout the studied pH range (1.5–4.5).

It is known that stability of certain unstable compounds can be improved by the adjustment of pH, which can be pivotal when dealing with pH-sensitive molecules, such as acidic metabolites (to ester), certain N-oxides, glucuronide metabolites, and lactones. Zhang et al. (2010) reported a SALLE method for the measurement of simvastatin and its metabolite simvastatin acid in human plasma. Temperature (2–4 °C) and acidity (pH 4.5) were well controlled throughout the sample preparation process to minimize the conversion of simvastatin and its metabolite simvastatin acid. In a separate case study, Yong et al. (2015) reported a SALLE procedure to extract atorvastatin and its metabolites from human plasma. Acetonitrile and 6 M ammonium acetate were chilled prior to extraction was carried out. The addition of ammonium acetate (pH 6.0) not only helped induce separation of an obvious acetonitrile layer but also control the inter-conversion between the lactone and hydroxyl acid forms.

It is worth noting that pH adjustment should be carried out with caution because the biological matrices of interest may contain instable metabolites that are prone to break down with pH change. This may lead to inaccurate determination of analytes.

5.3 Combination of SALLE with Other Extraction Techniques

Although sample extracts generated using SALLE can be analyzed directly, it should be noted that the enrichment factor using SALLE may often be too low for the ultra-trace residue analysis. Recently, Farajzadeh et al. (2015) introduced a new method of SALLE named Counter Current Salting-out Homogenous Liquid–Liquid Extraction (CCSHLLE) in combination with Dispersive Liquid–Liquid Micro-Extraction (DLLME) for the extraction and preconcentration of pesticides from aqueous samples (2015). Aqueous solution of the analyte containing acetonitrile and 1,2-dibromoethane was transferred into a narrow bore tube which is filled partially with NaCl. During passing the solution through the tube, fine droplets of the organic phase are produced at the interface of solution and salt, and form a separated layer on the aqueous phase. The collected organic phase was removed and injected into de-ionized water for more enrichment of the analytes. In this method, acetonitrile was used as a coextraction solvent in the first step (CCSHLLE) and as a disperser solvent in the second step (DLLME). 1,2-Dibromoethane was selected as a suitable extraction solvent which extracted the selected analyte during the CCSHLLE step and preconcentrated them in the DLLME step. The results demonstrated that the proposed method achieved higher enrichment factor (3480–3800) compared with other preconcentration methods. Not only the combination of CCSHLLE and DLLME result in a high enrichment factor but also it can be used in complex matrices. Akramipour et al. (2016) modified this method for the analysis of amphetamines in human urine. In this method, NaCl was initially selected as a separation reagent to fill into a small column. Urine containing acetonitrile was transferred onto the column. Fine droplets of acetonitrile were produced due to the salting-out effect when it passed through the column. The droplets formed a separated organic layer on the aqueous phase. The organic phase was removed and mixed with 30 μl of 1-undecanol (extraction solvent). Next, 5 ml of K_2CO_3 solution (2% w/v) was rapidly injected into the above mixture for further enrichment of the analytes. A high enrichment factor (157–168) in complex matrices was successfully achieved without any pretreatment or dilution using this combined method. This system could be further applied to determine drugs from complex biological and pharmaceutical matrices under different analytical platforms.

SALLE has also been reported to combine with a conventional LLE (a water-immiscible organic solvent) to improve the recovery of polar compounds (Song et al. 2013). After extracted by a conventional LLE, the remaining aqueous phase was further extracted by SALLE. The result showed that more polar compounds were recovered, which could not be recovered using the conventional LLE (e.g. the recovery for fumonisin B1 changed from 0% with ethyl acetate extraction to ~100% when the complementary SALLE were performed).

5.4 Matrix Effect in SALLE

Matrix effect is a major issue in LC-MS bioanalysis, for which matrix suppression/enhancement is often investigated. In SALLE, non-volatile salts can cause suppression or enhancement of analyte signal on MS and damage the mass spectrometer, and high amounts of solid salts can cause deterioration of LC columns and mass spectrometers. To address this issue, column switch technique was incorporated into the method to remove salt before injection to MS (Zhao et al. 2012). Matrix effect on SALLE was compared with that on SPE, and the result showed that ion suppression patterns were similar between the two methods.

Phospholipids are identified as the major source of matric effect at present. In general, the organic layer from SALLE has similar or higher content of phospholipid than LLE. This is because the acetonitrile layer from SALLE contains trace salt and water which is associated with a higher extraction rate for phospholipids as compared to conventional LLE (Garofolo et al. 2010). Adding a small amount of hydrophobic solvent (chloroform) to acetonitrile to induce phase separation had been reported, and the matrix effects were compared with LLE and PPT (Liu et al. 2010b). The organic layer from the modified SALLE was cleaner with least matrix effect than others because less polar matrix components such as phospholipids were extracted to the modified acetonitrile phase than the typical salt-out acetonitrile phase.

To avoid the potential matrix suppression, good chromatography should be employed to separate the phospholipid interference from the analyte of interest. In this regard, hydrophilic interaction liquid chromatography (HILIC) should be considered a valued tool (Tan et al. 2012). By modifying the composition of the mobile phase, the retention time for the analytes can be adjusted to ensure a good separation from the phospholipid interferences. Since SALLE can efficiently extract polar compounds into the salted-out acetonitrile layer, and HILIC is powerful in retaining and separating polar compounds with acetonitrile as an ideal injection solvent, we can predict that SALLE with HILIC-MS detection will be developed and applied to the determination of polar compounds. However, at present only one literature had reported HILIC-UV method with SALLE as the sample preparation procedure for the determination of β-blockers and their metabolites in human urine (Magiera et al. 2016).

5.5 Miniaturization and Automatization

Miniaturization is a new trend in sample preparation for less consumption of toxic reagents, simplified experimental operation, and possibly reduced time of analysis. According to the published applications (Liu et al. 2010a; Mohamed et al. 2014; Sereshti et al. 2014), microsyringes have been commonly used as the organic solvent holder in miniaturized LLE. By the similar approach, Liu et al. (2010a) used a 1-ml syringe as the device for the extraction of sulfonamides from different matrices using SALLE. After phase separation, the upper layer was narrowed into the needle tip by pushing the plunger. Then the upper layer solvent could be easily collected. The consumption of organic solvent was minimized by this miniaturization. In a separate report, Mohamed et al. (2014) developed a 1 ml syringe adapted with a capillary tube as the phase separation device and the method was applied to the extraction of two antiglaucoma drugs in aqueous humor. In the meantime, Sereshti et al. (2014) employed a coupled 1-ml syringe system (two syringes connected via their tips) to perform SALLE of sulfanilamide from complex matrix samples. These literatures are summarized in Table 5.1. Automated liquid handler and 96-well format have been reported to promote the automatization and miniaturization of SALLE extraction procedure. Varied biological matrices such as urine, blood, and plasma have been well demonstrated by Zhang et al. (2009a, 2010).

5.6 Summary

Sample preparation is a key component in any bioanalytical procedure. As one of the many tools in sample preparation, SALLE can be employed for the simultaneous extraction of a wide range of compounds with different polarity but with less inferencing components as compared to the conventional PPT method. Care has to be taken in managing the interference due to phospholipids in the sample extract. Combination of SALLE with other techniques, such as LC and LC-hyphenated techniques, can help alleviate the possible impact due to interference compounds in the sample extract. Moreover, because of the high organic percentage in the sample extracts using SALLE, incorporation of HILIC-MS into the workflow is considered as an alternative in managing matric interference while maintaining the high throughput.

References

Akramipour, R., Fattahi, N., Pirsaheb, M., and Gheini, S. (2016). Combination of counter current salting-out homogenous liquid-liquid extraction and dispersive liquid-liquid microextraction as a novel microextraction of drugs in urine samples. *J. Chromatogr. B* 1012–1013: 162–168.

Chang, M.S., Ji, Q.C., Zhang, J., and El-Shourbagy, T.A. (2007). Historical review of sample preparation for chromatographic bioanalysis: pros and cons. *Drug Dev. Res.* 68 (3): 107–133.

Farajzadeh, M.A., Feriduni, B., and Mogaddam, M.R.A. (2015). Development of counter current salting-out homogemous liquid-liquid extraction for isolation and preconcentration of some pesticides from aqueous samples. *Anal. Chim. Acta* 885: 122–131.

Garofolo F, Bergeron M, Milton F, Mathieu L (2010). Selection of buffer additives for liquid-liquid extraction (LLE) procedure to remove matrix effect. Presented at: 2010 AAPS Annual Meeting and Exposition, 14–18 November, New Orlenas, LA, USA.

Glasstone, S. (1946). *Textbook of Physical Chemical*, 2e. New York, USA: D. Van Nostrand company, Inc.

Horning, M.G., Boucher, E.A., Stafford, M., and Horning, E.C. (1972). A rapid procedure for the isolation of drugs and metabolites from plasma. *Clin. Chim. Acta* 37: 381–386.

Horning, M.G., Gregory, P., Nowling, J. et al. (1974). Isolation of drugs and drug metabolites from biological fluids by use of salt-solvent pairs. *Clin. Chem.* 20 (2): 282–287.

Li, Q. and Huie, C.W. (2006). Coupling of acetonitrile deproteinization and salting-out extraction with acetonitrile stacking for biological sample clean-up and the enrichment of hydrophobic compounds (porphyrins) in capillary electrophoresis. *Electrophoresis* 27 (21): 4219–4229.

Liu, J., Jiang, M., Li, G. et al. (2010a). Miniaturized salting-out liquid-liquid extraction of sulfonamides from different matrices. *Anal. Chim. Acta* 679 (1–2): 74–80.

Liu, G., Zhou, N., Zhang, M. et al. (2010b). Hydrophobic solvent induced phase transition extraction to extract drugs from plasma for high performance liquid chromatography-mass spectrometric analysis. *J. Chromatogr. A* 1217: 243–249.

Magiera, S., Kolanowska, A., and Baranowski, J. (2016). Salting-out assisted extraction method coupled with hydrophilic interaction liquid chromatography for determination of selected (–)blockers and their metabolites in human urine. *J. Chromatogr. B* 1022: 93–101.

Matkovich, C.E. and Christian, G.D. (1973). Salting-out of acetone from water-basis of a new solvent extraction system. *Anal. Chem.* 45: 1915–1921.

Mohamed, A.I., Abdel-Wadood, H.M., and Mousa, H.S. (2014). Simultaneous determination of dorzolomide and timolol in aqueous humor: a novel salting out liquid-liquid microextraction combined with HPLC. *Talanta* 130: 495–505.

Myasein, F., Kim, E., Zhang, J. et al. (2009). Rapid, simultaneous determination of lopinavir and ritonavir in human plasma by stacking protein precipitations and salting-out assisted liquid/liquid extraction, and ultrafast LC-MS/MS. *Anal. Chim. Acta* 651 (1): 112–116.

Nagosa, Y., Taniguchi, M., and Miura, J. (1997). Hydrophobic salt-induced phase-separation and extraction for sample preparation in chromatographic determination of heavy metal ions as dibenzyldithiocarbamates. *Anal Chem Acta.* 356 (s2–3): 135–140.

Nanita, S.C. and Padivitage, N.L.T. (2013). Ammonium chloride salting-out extraction/cleanup for trace-level quantitative analysis in food and biological matrices by flow injection tandem mass spectrometry. *Anal Chmi Acta.* 768: 1–11.

Sereshti, H., Khosraviani, M., and Amini-Fazl, M.S. (2014). Miniaturized salting-out liquid-liquid extraction in a coupled-syringe system combined with HPLC-UV for extraction and determination of sulfanilamide. *Talanta* 121: 199–204.

Song, S., Ediage, E.N., Wu, A., and Saeger, S.D. (2013). Development and application of salting-out assisted liquid/liquid extraction for multi-mycotoxin biomarkers analysis in pig urine with high performance liquid chromatography/tandem mass spectrimetry. *J. Chromatogr. A* 1292: 111–120.

Tabata, M., Kumamoto, M., and Nishimoto, J. (1994). Chemical properties of water-miscible solvents separated by salting-out and their application to solvent extraction. *Anal. Sci.* 10 (3): 383–388.

Tan, B., Negahban, A., McDonald, T. et al. (2012). Utilization of hydrophilic-interaction LC to minimize matrix effects caused by phospholipids. *Bioanalysis* 4 (16): 2049–2058.

Tsai, W., Chuang, H., Chen, H. et al. (2010). Application of sugaring-out extraction for the determination of sulfonamides in honey by high-performace liquid chromatography with fluorescence detection. *J. Chromatogr. A* 1217 (49): 7812–7815.

Wang, M., Cai, Z., and Xu, L. (2011). Coupling of acetonitrile deproteinization and salting-out extraction with acetonitrile stacking in chiral capillary electrophoresis for the determination of warfarin enantiomers. *J Chromatogra A.* 1218 (26): 4045–4051.

Wang, B., Ezejias, T., Feng, H., and Blaschek, H. (2008). Sugaring-out: a novel phase separation and extraction system. *Chem. Eng. Sci.* 63 (9): 2595–2600.

Wieling, J., Dijkstra, H., Mensink, C.K. et al. (1993). Chemometrics in bioanalytical sample preparation: a fractionated combined mixture and factorial design for the modelling of the recovery of five tricyclic amines from plasma after liquid-liquid extraction prior to high-performance liquid chromatography. *J. Chromatogr.* 629 (2): 181–199.

Wu, H., Zhang, J., Norem, K., and El-Shourbagy, T.A. (2008). Simultaneous determination of a hydrophobic drug candidate and its methabolite in human plasma with salting-out assisted liquid/liquid extraction using a mass spectrometry friendly salt. *J. Pharm. Biomed. Anal.* 48 (4): 1243–1248.

Xiong, X. and Yang, L. (2015). Salting-out-assisted liquid-liquid extraction with acetonitrile for the determination of trimetazidine in rat plasma using liquid chromatography-mass spectrometry. *Biomed. Chromatogr.* 29 (2): 268–274.

Xiong, X., Zhang, L., Cheng, L., and Mao, W. (2015). High-throughput salting-out assisted liquid-liquid extraction with acetonitrile for the determination of anandamide in plasma of hemodialysis patients with liquid chromatography tandem mass spectrometry. *Biomed. Chromatogr.* 29 (9): 1317–1324.

Yanes, E.G. and Lovett, D.P. (2012). High-throughput bioanalytical method for analysis of synthetic cannabinoid metabolites in urine using salting-out sample preparation and LC-MS/MS. *J. Chromatogr. B Analyt. Technol. Biomed. Life Sci.* 909 (1): 42–50.

Yong, Y., Xu, Q.F., Zhou, L. et al. (2015). High-throughput salting-out-assisted liquid-liquid extraction for the simultaneous determination of atorvastatin, ortho-hydroxyatorvastatin, and Para-hydroxyatorvastatin in human plasma using ultrafast liquid chromatography with tandem mass spectrometry. *J. Sep. Sci.* 38 (6): 1026–1034.

Zhang, J., Myasein, F., Wu, H., and El-Shourbagy, T.A. (2009a). High-throughput salting-out assisted liquid-liquid extraction and ultrafast LC for the same day delivery of bioanalytical data in a first-in-human study. *Bioanalysis* 1 (4): 715–719.

Zhang, J., Myasein, F., Wu, H., and El-Shourbagy, T.A. (2013). Sugaring-out assisted liquid/liquid extraction with acetonitrile for bioanalysis using liquid chromatography-mass spectrometry. *Microchem. J.* 108: 198–202.

Zhang, J., Ramona, R., Gage, E. et al. (2010). High-throughput salting-out assisted liquid/liquid extraction with acetonitirle for the simultaneous determination of simvastatin and simvastatin acid in human plasma with liquid chromatography. *Anal. Chim. Acta* 661 (2): 167–172.

Zhang, J., Wu, H., Kim, E., and El-Shourbagy, T.A. (2009b). Salting-out assisted liquid/liquid extraction with acetonitrile: a new high throughput sample preparation technique for good laboratory practice bioanalysis using liquid chromatography-mass spectroscopy. *Biomed. Chromatogr.* 23 (4): 419–425.

Zhao, F.J., Tang, H., Zhang, Q.H. et al. (2012). Salting-out homogeneous liquid-liquid extraction approach applied in sample pre-processing for the quantitative determination of entecavir in human plasma by LC-MS. *J. Chromatogr. B* 881–882 (2): 119–125.

6

Supported Liquid Extraction (SLE) in LC-MS Bioanalysis

Zhongzhe Cheng and Hongliang Jiang

6.1 Introduction

Sample preparation is a critical procedure that requires much attention in bioanalysis, which is usually associated with liquid chromatography mass spectrometry-based determination (Thevis et al. 2013). The wide heterogeneity of the analytes (in terms of polarity, pK_a, or size) and the complexity of the matrices often results in matrix effect and interference issues in bioanalysis, which makes appropriate sample cleanup critical (Dominguez-Romero et al. 2014). In addition, it is often necessary to balance between good recoveries for the targeted analytes and cleanliness of the extract (Dominguez-Romero et al. 2014).

Supported liquid extraction (SLE) serves as a new variant to conventional liquid–liquid extraction (LLE). SLE employs special porous sorbents to absorb the aqueous sample before the sorbents are eluted with water-immiscible solvent. Both SLE and LLE are effective ways to extract compounds from biological matrices with minimal matrix effect and interference, but SLE is relatively easier to be automated with minimized issues related to liquid handling when compared to LLE (Jiang et al. 2008a).

SLE was first reported in the late 1976 by Joachim and co-workers who described a liquid–solid elution method on a chromatography column packed with dry supporting material (Breiter et al. 1976). In the past decades, much progress has been made in both manufacturing the SLE devices and evaluation and implementation of the technique in quantitative bioanalysis. SLE is commercially available in plate and cartridge formats (Singleton 2012). To date, SLE has been employed to extract analytes from various aqueous matrices, such as urine, serum, plasma, blood, and oral fluids (Breiter et al. 1976; Dominguez-Romero et al. 2014; Rositano et al. 2016; Valen et al. 2016). In addition, SLE can be used for extraction of solid samples after appropriate homogenization (Nguyen et al. 2010). Application of SLE has covered areas of pharmacokinetics (Rao et al. 2013; Pernice et al. 2016), toxicokinetics (Tweed et al. 2012), clinical study support (Meunier et al. 2015; De Meulder et al. 2016), endogenous components analysis (Keevil 2013; Jenkinson et al. 2016), doping control (Dominguez-Romero et al. 2014), drugs of abuse testing (Valen et al. 2016), and forensic toxicological analysis (Gao et al. 2016). Several comprehensive reviews on the principles and advancements of sample preparation techniques for bioanalysis related to SLE have been published (Franke and de Zeeuw 1998; Singleton 2012; Bylda et al. 2014). This chapter highlights the principle and application of SLE in sample preparation in bioanalysis.

6.2 Principle of SLE

SLE utilizes specially cleaned and sized porous sorbents, such as a modified form of diatomaceous earth (De Korompay et al. 2008). The diatomaceous earth is mainly composed of SiO_2 and chemically inert with high porosity and surface area, but does not interact chemically with the aqueous sample. It is capable of being loaded to the limit of their water-absorption capacity without releasing the absorbed water upon elution with water-immiscible solvents (Breiter et al. 1976; Nave et al. 2007). The aqueous mixture containing analytes spreads over the surface of the sorbents and is adsorbed. Then, a water-immiscible organic solvent is applied to the plate and/or cartridge to extract and elute analytes (De Korompay et al. 2008). In this process, an efficient LLE occurs between the very thin aqueous stationary film formed on the sorbents after sample loading and a water-immiscible organic solvent (Wu et al. 2010). The large surface area of the extraction interface and flow through nature of the technique leads to an efficient extraction process. Subsequently, the analytes come into contact with fresh solvent as the organic phase travels through the packing bed, an efficient and continuous "LLE" takes place (Breiter et al. 1976). SLE is ideal for non-polar to moderately polar analytes. Typical extraction

Sample Preparation in LC-MS Bioanalysis, First Edition. Edited by Wenkui Li, Wenying Jian, and Yunlin Fu.

solvents used in SLE are methyl *tert*-butyl ether (MTBE) (Dominguez-Romero et al. 2014; Meunier et al. 2015; Rositano et al. 2016), ethyl acetate (EtOAc) (Li et al. 2011; Tweed et al. 2012; Owen and Keevil 2013; Scheidweiler and Huestis 2014), dichloromethane (DCM) (Rao et al. 2013; De Meulder et al. 2016; Gao et al. 2016), or mixed solvents (O'Maille et al. 2008; Edel et al. 2013; Jenkinson et al. 2016; Valen et al. 2016). Other water-immiscible extraction solvents have also been used, such as 1-chlorobutane (Gong et al. 2012), hexane (Liu et al. 2009), and chloroform (De Korompay et al. 2008).

6.3 Advantages and Limitation of SLE in Quantitative LC-MS Bioanalysis

6.3.1 Advantages

- **Efficient extraction process**
  Appropriate recovery of the analytes of interest without or with a minimum of interferences is critical for sample preparation to ensure the needed sensitivity and selectivity for an intended assay. Table 6.1 summarizes a list of analytes that have been successfully analyzed in biological samples by using SLE. Some of these investigations suggested that LLE can be replaced by SLE with similar recoveries for the intended analyses (Singleton 2012). In general, SLE gives at least comparable or higher extraction recovery than LLE (Wu et al. 2010; Gong et al. 2012), which thanks to the efficient and continuous extraction that takes place on the large surface area of the small sorbent particles of diatomaceous earth in SLE (Logan and Stafford 1989; Jiang et al. 2009). As a result of the enhanced extraction recovery, sample volume can be reduced without sacrificing assay sensitivity.
- **No emulsion formation**
  The absence of emulsion is another advantage of SLE (Zhang et al. 2013). Instead of mixing the two immiscible phases by shaking or vertexing, the aqueous phase is immobilized on the inert diatomaceous earth-based sorbents and the water-immiscible organic phase flows through the sorbents under gravity (Wu et al. 2010; Zhang et al. 2013). As SLE prevents any potential formation of emulsion, it minimizes the loss of analytes recovery and improves assay precision and accuracy (Zhang et al. 2013).
- **Easy for automation**
  Traditional LLE is often labor-intensive and time-consuming, especially when individual glass tubes are used. Even with the use 96-well format, LLE still requires manual intervention at several stages in the

Table 6.1 Current applications involving SLE related to bioanalysis.

Compound	Technique	Species	Matrix	Curve range (ng ml^{-1})	Recovery (%)	References
Rosiglitazone *N*-desmethyl rosiglitazone	LC-MS/MS	Human	Plasma	1.00–500 (both analytes)	~80 71	O'Maille et al. (2008)
Alprenolol	Chiral LC-MS/MS	Human	Plasma	0.500–500 (both enantiomers)	84	Jiang et al. (2008a)
Molindone	Chiral LC-MS/MS	Human	Plasma	0.100–100	75	Jiang et al. (2008b)
PD 0332991	LC-MS/MS	Mouse	Tumor	0.100–100	93–95	Nguyen et al. (2010)
Fluoxetine Norfluoxetine	LC-MS/MS	Human	Plasma	0.05–20 (both analytes)	70.8 82.3	Li et al. (2011)
Indolocarbazoles	LC-MS/MS (negative mode)	Human	Plasma	0.05–200	62–66	Wang et al. (2002)
Hydrocortisone	LC-MS/MS	Mouse	Serum	2.00–2000	90–110	Wu et al. (2010)
Acebutolol Diacetolol	Chiral LC-MS/MS	Human	Plasma	0.0500–50.0 (both analytes)	95 95	Jiang et al. (2009)
Erlotinib	HILIC LC-MS/MS	Human	Plasma	2.00–2000	100	Pan et al. (2010)
Amitraz	UHPLC-Q-TOF	Human	Blood	0.3–1000	79–92	Gao et al. (2016)
JWH-250 N-5-carboxypentyl (β-glucuronidasehydrolysis)	LC-MS/MS	Human	Urine	0.2–50	87	Scheidweiler and Huestis (2014)
Methylamphetamine Methylenedioxymethylamphetamine Delta-9-tetrahydrocannabinol	LC-MS/MS	Human	Oral fluid	5–625 5–625 0.9–56	81 85 67	Rositano et al. (2016)

sample preparation process, such as capping/sealing, uncapping/unsealing, and centrifugation (Wu et al. 2010). Compared with LLE, sample loading and analytes elution with SLE in 96-well format can be conducted by a robotic liquid-handling system in an automated manner without off-line shaking or mixing, and centrifuging steps required (Jiang et al. 2009; Pan et al. 2010). Therefore, the entire sample preparation process for SLE can be significantly simplified and automated by eliminating the off-line steps and manual interventions. Consequently, potential issues including cross-contamination and operation errors due to manual manipulations in LLE can be minimized, thus increasing the ruggedness and reproducibility of assays (Zhang et al. 2013). In addition, the time for sample extraction in 96-well format can be shortened from over 60 minutes in LLE to approximately 35 minutes in SLE, leading to significant improvement in sample throughput (Jiang et al. 2009; Wu et al. 2010).

Although solid-phase extraction (SPE) can generally provide clean extracts through selection of an appropriate sorbent materials and optimization of the corresponding washing and elution conditions, the complexity in multi-step operation and time-consuming method development make it less favorable in high-throughput bioanalytical sample analysis (Zhang et al. 2013). Unlike SPE, SLE does not require solvent preconditioning, washing, and waste collecting steps in sample preparation. Aqueous sample can be directly loaded on the SLE column or 96-well plate where the aqueous solution containing analytes is absorbed on the packing materials, and then analytes are eluted using water-immiscible organic solvents (Edel et al. 2013).

- **Reduced matrix effect**
  Phospholipids are abundant in plasma, blood or serum and were identified as a major cause of matrix effect in ESI-based LC-MS/MS analysis (Little et al. 2006; Silvester and Smith 2012). However, chromatographic separation of phospholipids from the analyte(s) of interest may be difficult due to the wide ranges of polarity of the phospholipids in sample matrix (Little et al. 2006; Chambers et al. 2007). Therefore, the effective removal of phospholipids via sample preparation becomes essential to reduce/eliminate the matrix effect. A comprehensive study on phospholipids removal through protein precipitation (PPT), LLE, SLE, and SPE revealed that SLE was more effective than LLE in eliminating most phospholipids under all eluting solvents or solvent mixtures evaluated (Ismaiel et al. 2010). The final extract resulting from SLE with different eluting solvents was demonstrated with less than 0.04% of the glycerophosphocholines (GPChos) and sphingomyelins (SMs) levels and less than 0.006% of the lysoglycerophosphocholines (LGPChos) levels (except for DCM and ether with PBS) as compared to corresponding extract obtained using PPT with acetonitrile as the solvent (Ismaiel et al. 2010). In addition, LLE methods using different organic solvents showed 0.2–17% (GPChos and SMs) and 0.01–0.96% (LGPChos) levels when compared to the extract prepared using PPT with acetonitrile. The SPE procedure yields more than 0.1% of the GPChos and SMs levels and 0.07% of the LGPChos level as compared to the PPT with acetonitrile (Ismaiel et al. 2010). In a separate study, Jiang et al. (2012) indicated that the matrix effect largely correlates with the elution of phospholipids in chromatogram and can be minimized or eliminated via SLE by choosing appropriate elution solvent or solvent mixtures. In addition, SLE has been reported to be similar to LLE with regard to eliminating matrix effect (Nguyen et al. 2010; Wu et al. 2010). Moreover, cleaner chromatographic baseline of sample extract from SLE was also reported when compared with that from LLE, which helped improve assays' selectivity with reduced interferences (Wu et al. 2010). In summary, SLE is more effective than LLE in phospholipid removal and is at least comparable with LLE in reduction of matrix effect and other interferences.
- **Solvent compatibility**
  SLE is fully compatible with any water-immiscible organic solvents or solvent mixtures regardless of density (Breitenbucher et al. 2001; Ismaiel et al. 2010). This property allows for more options in selection of eluting solvent or solvent mixtures for improving recovery and minimizing the matrix effect in SLE extraction. Moreover, there is no need for dry-down and reconstitution steps, when the hydrophilic interaction liquid chromatography (HILIC) HPLC-MS/MS system is used, affording the needed capacity for high-throughput bioanalysis (Pan et al. 2010; De Meulder et al. 2016).
- **Transferable methods**
  The water-immiscible extraction solvents used in LLE can also be used for SLE procedures. If a LLE method exists, similar extraction conditions (solvent and pH conditions) are easily transferable to SLE.

6.3.2 Limitation

As described previously, the principle of SLE is similar to LLE and based on the extraction of the targeted analytes from an aqueous phase into a water-immiscible organic solvent. Hence, SLE is not suitable for compounds with high polarity. Since drug metabolism often leads to formation of metabolite(s) with increased hydrophilicity, especially for phase II reactions such as glucuronidation (Wen and Zhu 2015), bioanalytical method development using SLE can be challenging for simultaneous determination of both the parent drug and its conjugated metabolites with acceptable extraction recovery (Scheidweiler and Huestis 2014; Jagerdeo and Schaff 2016). In addition, SLE plates or cartridges introduce additional cost to sample analysis, which might be a concern for some users with limited budget and resources (Zhang et al. 2013).

6.4 Key Consideration in Developing Robust SLE-LC-MS Bioanalytical Method

The physicochemical properties of compounds, such as pK_a and $\log P$ values, are key factors that affect acidic or basic compounds' partition into the organic phase in SLE (Jiang et al. 2012). Thus, pH control of the aqueous sample is important when dealing with acidic or basic analytes that are poorly soluble in water-immiscible organic solvents when charged. On the basis of pK_a and $\log P$ values of analytes, manipulation of the pH values of aqueous samples can suppress the charge on any acidic or basic functional groups of the molecule(s) to improve the extraction efficiency of those ionizable acidic and basic analytes. For acidic analytes, it is often favorable to acidify the sample with an appropriate buffer prior to sample loading to ensure that the analyte is present in neutral form and therefore more soluble in the organic phase. In contrast, for basic analytes, recoveries may be improved by adding a small amount of basic buffer to increase the sample pH value prior to sample loading. In addition, for amphoteric analyte, the loading buffer at a pH value corresponding to the isoelectric point of the analyte should be used. Jiang et al. (2012) conducted a systematic evaluation of SLE in improving extraction

Table 6.2 Extraction recovery of 10 representative compounds under various extraction conditions using PPT and SLE.

	Recovery (%)[a,b]									
Extraction/elution solvent	**Dic**	**Pro**	**Amb**	**Car**	**Ate**	**Met**	**Dau**	**Pac**	**Cel**	**Fex**
Protein precipitation										
Methanol	120	113	119	120	110	84.2	107	98.7	114	112
Acetonitrile	116	82.8	103	108	111	68.8	78.2	95.5	138	101
SLE-EtOAc										
Dilution with 100 μl 1% FA in H_2O	69.1	77.6	92.2	84.0	0.6	2.0	4.3	64.4	49.5	86.7
Dilution with 100 μl 0.1% FA in H_2O	78.5	101.1	3.7	84.0	4.3	58.7	60.7	75.8	65.9	76.6
Dilution with 100 μl H_2O	34.4	93.0	2.7	83.9	7.3	89.8	76.1	72.4	64.3	63.9
Dilution with 100 μl 0.5 M NH_4OH	14.2	84.2	2.0	81.7	71.7	80.7	63.1	58.9	77.3	14.5
SLE-MTBE										
Dilution with 100 μl 1% FA in H_2O	74.9	73.2	88.0	80.5	0.2	0.2	0.1	71.8	60.6	6.1
Dilution with 100 μl 0.1% FA in H_2O	64.1	76.3	0.5	79.9	0.2	29.1	0.8	67.8	47.3	21.3
Dilution with 100 μl H_2O	23.7	78.9	0.4	74.3	0.2	79.1	25.3	68.5	61.4	19.3
Dilution with 100 μl 0.5 M NH_4OH	2.9	68.3	0.1	72.0	8.3	80.7	36.4	54.8	61.2	1.8
SLE-DCM										
Dilution with 100 μl 1% FA in H_2O	73.0	57.9	86.5	86.6	0.5	2.0	0.4	88.2	73.3	12.3
Dilution with 100 μl 0.1% FA in H_2O	0.7	58.7	1.9	87.7	2.9	82.0	60.0	59.5	71.6	12.0
Dilution with 100 μl H_2O	0.5	63.5	1.4	88.0	4.6	79.7	60.8	63.9	66.6	8.1
Dilution with 100 μl 0.5 M NH_4OH	0.9	66.1	0.4	90.1	53.1	83.7	48.8	64.4	70.0	0.5
SLC-DCM/IPA (95 : 5,v/v)										
Dilution with 100 μl 1% FA in H_2O	79.2	56.8	94.0	90.0	0.7	10.2	4.0	74.2	80.5	83.6
Dilution with 100 μl 0.1% FA in H_2O	18.8	65.6	20.3	88.2	6.7	89.5	81.4	67.5	84.4	92.2
Dilution with 100 μl H_2O	6.0	63.8	6.6	92.5	11.1	94.6	72.7	74.6	81.1	84.1
Dilution with 100 μl 0.5 M NH_4OH	0.8	64.5	1.7	87.5	99.4	92.3	53.8	44.8	87.4	8.9

Diclofenac sodium (Dic); Progesterone (Pro); Ambrisentan (Amb); Carbamazepine (Car); Atenolol (Ate); Metoprolol tartrate (Met); Daunorubicin (Dau); Paclitaxel (Pac); Celecoxib (Cel); Fexofenadine (Fex).

[a] Calculated as the percentage of mean peak area of extracted sample over post-extraction-spiked blank extract at the concentration with 100% recovery.

[b] All values are means of three independent extractions and analysis. RSD% is <15%.

efficiency/recovery under various extraction conditions with 10 model compounds in LC-MS/MS bioanalysis (Table 6.2). Model compounds with diverse physicochemical properties where log P ranges from 0.1 to 6.24 and pK_a ranges from 4.0 to 11.1 were evaluated. The pH value of the aqueous sample was adjusted with various loading buffers, i.e. 1% formic acid (FA) in H_2O, H_2O or 0.5 M NH_4OH, and analyte was eluted with MTBE, EtOAc, DCM, or DCM/IPA (95 : 5, v/v). For neutral compounds with appropriate log P values including progesterone (3.58), carbamazepine (2.10), and paclitaxel (3.96), no significant difference in recovery was observed when different loading buffers were assessed (Table 6.2). As a weak base with a pK_a of 11.1, celecoxib was present as a charged state at physiological pH, as well as under buffered conditions. Similar recoveries were observed with different loading buffers. The recoveries of acidic analytes, such as diclofenac (pK_a 4.15), ambrisentan (pK_a 4.0), and fexofenadine (pK_a 4.25, 9.53, zwitterionic compound with two pK_a values), were dramatically improved under acidic extraction conditions (Table 6.2). Compared with neutral conditions, acidification increases recoveries by approximately 44.1–72.5%, 85.1–89.5% and 2.0–22.8% for diclofenac, ambrisentan, and fexofenadine, respectively, with different eluting solvents (Table 6.2). In contrast, basic analytes, such as atenolol (pK_a 9.6), metoprolol (pK_a 9.7), and daunorubicin (pK_a 8.3), a barely higher recovery was noticed when extracted under basic extraction conditions (Table 6.2).

As an alternative to traditional LLE, SLE is ideal for nonpolar to moderately polar analytes, which have favorable distribution in water-immiscible organic solvent. For some polar analytes, typical extraction solvents may not be suitable due to their low solubility in the typical water-immiscible solvents. Therefore, a challenge for SLE sample preparation in LC-MS/MS bioanalysis is how to extend the range of extractable analytes. For polar molecules that do not elute well with typical extraction solvents, certain percentage of polar (water-miscible) modifiers, such as isopropanol (IPA) in a less polar-eluting solvent such as DCM and hexane, can be used to improve extraction efficiency and analyte recovery. Fox examples, a mixture of eluting solvents of DCM and IPA at a ratio of 95 : 5 affords much higher extraction recovery for fexofenadine (84.1%) than DCM (8.1%), MTBE (19.3%), or EtOAc alone (63.9%) (Jiang et al. 2012). This mixture of eluting solvents was also used to extract 19 benzodiazepines, 12 opiates, cocaine, and three metabolites in both blood and urine specimens (Jagerdeo and Schaff 2016). However, the use of higher percentage of polar (water miscible) modifiers in the mixture of eluting solvents can lead to co-extraction of some endogenous matrix components, such as phospholipids, which have been identified as a major source of matrix effect in LC-MS/MS analysis (Little et al. 2006; Silvester and Smith 2012).

It is worth noting that, when developing methods for LC-MS/MS bioanalysis, volatile buffers (mass spectrometry friendly) such as ammonium hydroxide, acetate, or formate are recommended.

6.5 Representative Protocols

6.5.1 Material, Supplies, and Equipment

SLE is commercially available in plate and cartridge format. 96-well formats are suitable for processing low-volume biological samples while cartridge format is for handling a relatively large sample volume (Zhang et al. 2013). Because plate and cartridge processing is performed by gravity-guided flow-through, a pulse of vacuum or positive pressure can be used to load the sample and maximize solvent recovery. In addition, both plates and cartridges are well suited for use in manual or automated.

6.5.2 Protocols for SLE Cartridge and Plate Processing

SLE is a simple sample preparation technique. The major steps include sample pretreatment, sample loading, soaking, and analytes elution (Figure 6.1).

1) Pretreat sample with internal standard and loading buffer.
2) Load sample onto the SLE sorbent.
3) Wait for about five minutes for sample to soak the bedding materials and completely absorbed.
4) Add water-immiscible organic solvent and allow solvent flow for two to five minutes under gravity.
5) Apply vacuum or pressure for 10–30 seconds to complete elution.
6) Dry-down eluate and reconstitute as needed.

- **Extraction of acebutolol and its active metabolite-diacetolol enantiomers in human plasma by 96-well format SLE**

Step 1: Transfer 100 μl of each calibration standard, quality control (QC) sample, zero sample (plasma blank with internal standard), and blank matrix control sample in a 96-well plate.
Step 2: 25.0 μl of the intermediate internal standard solution of acebutolol-d_5 racemate at 5.00 ng ml^{-1} were added to individual wells containing samples with the exception of the blank control samples, to which 25.0 μl of acetonitrile/water (1/1, v/v) were added.
Step 3: 100 μl of 0.5% NH_4OH in water with pH 10.5 were then added to each sample.

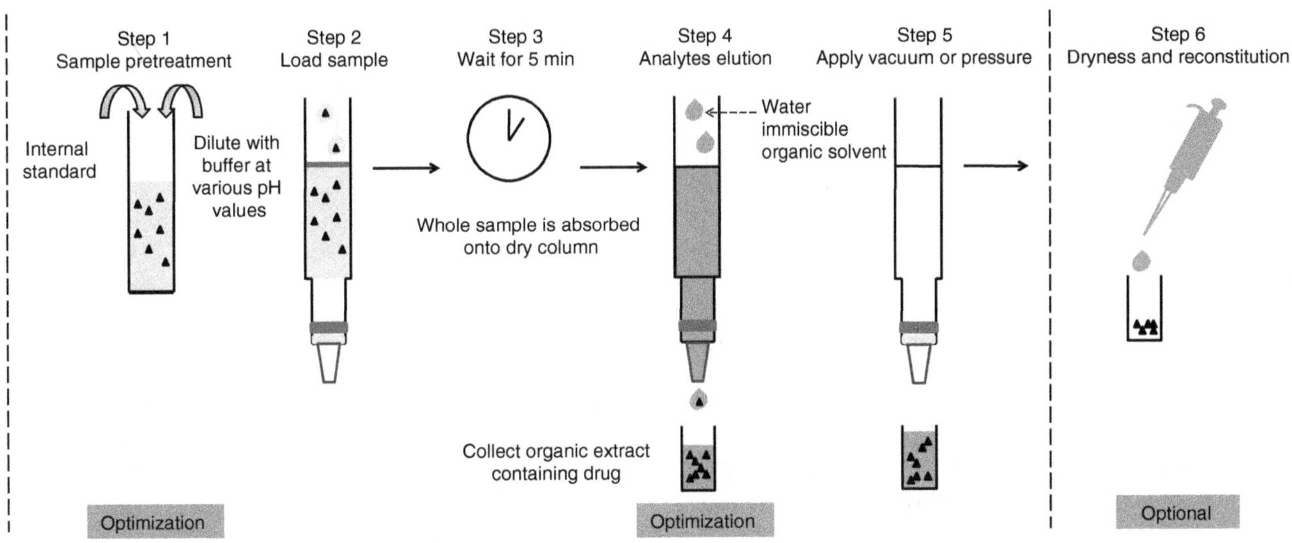

Figure 6.1 Procedure for supported liquid extraction. *Source:* Image courtesy of Scott Klepfer © 2012 Biotage AB. Singleton (2012).

Step 4: 200 μl of each sample in the above 96-well plate were transferred to individual wells of a 2 ml 96-well SLE plate using an automated SPE system.

Step 5: Minimum positive pressure was applied to facilitate the sample absorption into the cartridge within 10 seconds using a positive pressure processor. Prior to adding water-immiscible organic solvent for elution, two to five minutes were needed for complete analyte absorption to the SLE sorbents.

Step 6: 800 μl of ethyl acetate were added to each well of the SLE plate to elute the analytes. A minimum positive pressure can be used to speed-up the elution. The complete elution of 800 μl of ethyl acetate takes approximately five minutes.

Step 7: The eluate was evaporated to dryness using a 96-well sample concentrator (SPE DRY-96) set at 50 °C for about 5–10 minutes.

Step 8: Then the samples were reconstituted in 200 μl of acetonitrile/water (15/85, v/v) for LC-MS/MS analysis (Jiang et al. 2009).

- **Extraction of hydrocortisone in mouse serum by 96-well format SLE**

Step 1: 100 μl of each calibration standard sample, QC sample, zero sample (serum blank with internal standard), and blank control sample was aliquoted into individual wells of a 2 ml 96-well plate.

Step 2: 25 μl of the internal standard working solution (500 ng ml^{-1} in 50% aqueous methanol) was added to all wells except for the blank sample wells, to which 25 μl of 50% aqueous methanol was added.

Step 3: After brief vortex-mixing, a 100 μl aliquot of deionized water was added to each well of the plate via a TomTec system (Hamden, CT, USA), followed by a brief mixing using the TomTec.

Step 4: Thereafter, a 200 μl aliquot of each sample in the above 96-well plate was transferred to the corresponding wells of a 2 ml 96-well SLE plate using TomTec.

Step 5: The plate was allowed to sit for at least five minutes, with no pressure or vacuum applied. The contents of the wells were allowed to pass through the SLE adsorbent via gravity.

Step 6: Then 600 μl (2 × 300 μl) of MTBE was added to each well of the SLE plate. After two minutes, a vacuum of ~10 Hg was applied to the plate for a complete elution of the analyte from the SLE plate.

Step 7: The eluate was evaporated to dryness using a 96-well sample concentrator (SPE Dry-96, Biotage) set at 45 °C for about 10 minutes.

Step 8: The resulting residues were reconstituted in 75 μl of 50% aqueous methanol for LC-MS/MS analysis (Wu et al. 2010).

6.6 Summary

SLE is closely related to conventional LLE, but is a flow-through technique (Breitenbucher et al. 2001). The technique is based on using a porous sorbent (diatomaceous earth) as the support material, where an efficient LLE takes place between the very thin aqueous "stationary film" formed after sample loading and a water-immiscible organic elution solvent (Ismaiel et al. 2010). SLE has been shown to effectively remove the majority of phospholipids when the extraction conditions are carefully optimized (Jiang et al. 2012). SLE provides comparable or better data quality as compared to LLE for LC-MS/MS-based assays (Zhang et al. 2013). In addition, SLE can be conducted in a more convenient, faster, and automated manner to achieve higher throughput while minimizing potential contaminations and errors (Zhang et al. 2013). Solvent compatibility extends options in the selection of eluting solvents to improve recovery and minimize matrix effect. Finally, it is easy to directly transfer LLE method to SLE which makes a minimal method development. Therefore, the application of SLE in LC-MS/MS-based bioanalysis, especially in high-throughput format, will come into prominence and gain increasing acceptance in routine sample preparation in the future.

References

Breitenbucher, J.G., Arienti, K.L., and McClure, K.J. (2001). Scope and limitations of solid-supported liquid-liquid extraction for the high-throughput purification of compound libraries. *J. Comb. Chem.* 3 (6): 528–533.

Breiter, J., Helger, R., and Lang, H. (1976). Evaluation of column extraction: a new procedure for the analysis of drugs in body fluids. *Forensic Sci.* 7 (2): 131–140.

Bylda, C., Thiele, R., Kobold, U., and Volmer, D.A. (2014). Recent advances in sample preparation techniques to overcome difficulties encountered during quantitative analysis of small molecules from biofluids using LC-MS/MS. *Analyst* 139 (10): 2265–2276.

Chambers, E., Wagrowski-Diehl, D.M., Lu, Z., and Mazzeo, J.R. (2007). Systematic and comprehensive strategy for reducing matrix effects in LC/MS/MS analyses. *J. Chromatogr. B Analyt. Technol. Biomed. Life Sci.* 852 (1–2): 22–34.

De Korompay, A., Hill, J.C., Carter, J.F. et al. (2008). Supported liquid-liquid extraction of the active ingredient (3,4-methylenedioxymethylamphetamine) from ecstasy tablets for isotopic analysis. *J. Chromatogr. A* 1178 (1–2): 1–8.

De Meulder, M., Waldron, M.P., Li, L. et al. (2016). Development and validation of HILIC-ESI/MS/MS methods for simultaneous quantitation of several antipsychotics in human plasma and blood. *Bioanalysis* 8 (8): 765–794.

Dominguez-Romero, J.C., Garcia-Reyes, J.F., and Molina-Diaz, A. (2014). Comparative evaluation of seven different sample treatment approaches for large-scale multiclass sport drug testing in urine by liquid chromatography-mass spectrometry. *J. Chromatogr. A* 1361: 34–42.

Edel, A.L., Aliani, M., and Pierce, G.N. (2013). Supported liquid extraction in the quantitation of plasma enterolignans using isotope dilution GC/MS with application to flaxseed consumption in healthy adults. *J. Chromatogr. B Analyt. Technol. Biomed. Life Sci.* 912: 24–32.

Franke, J.P. and de Zeeuw, R.A. (1998). Solid-phase extraction procedures in systematic toxicological analysis. *J Chromatogr B Biomed Sci Appl.* 713 (1): 51–59.

Gao, X., Guo, H., Zhang, W.W., and Gu, C. (2016). Simultaneous determination of Amitraz and its metabolites in blood by support liquid extraction using UPHLC-QTOF. *J. Anal. Toxicol.* 40 (6): 437–444.

Gong, Z., Chandler, K., Webster, S. et al. (2012). Simple and rapid determination of norethindrone in human plasma by supported liquid extraction and ultra performance liquid chromatography with tandem mass spectrometry. *Talanta* 91: 77–82.

Ismaiel, O.A., Zhang, T., Jenkins, R.G., and Karnes, H.T. (2010). Investigation of endogenous blood plasma phospholipids, cholesterol and glycerides that contribute to matrix effects in bioanalysis by liquid chromatography/mass spectrometry. *J. Chromatogr. B Analyt. Technol. Biomed. Life Sci.* 878 (31): 3303–3316.

Jagerdeo, E. and Schaff, J.E. (2016). Rapid screening for drugs of abuse in biological fluids by ultra high performance liquid chromatography/Orbitrap mass spectrometry. *J. Chromatogr. B Analyt. Technol. Biomed. Life Sci.* 1027: 11–18.

Jenkinson, C., Taylor, A.E., Hassan-Smith, Z.K. et al. (2016). High throughput LC-MS/MS method for the simultaneous analysis of multiple vitamin D analytes in serum. *J. Chromatogr. B Analyt. Technol. Biomed. Life Sci.* 1014: 56–63.

Jiang, H., Cao, H., Zhang, Y., and Fast, D.M. (2012). Systematic evaluation of supported liquid extraction in reducing matrix effect and improving extraction efficiency in LC-MS/MS based bioanalysis for 10 model pharmaceutical compounds. *J. Chromatogr. B Analyt. Technol. Biomed. Life Sci.* 891–892: 71–80.

Jiang, H., Jiang, X., and Ji, Q.C. (2008a). Enantioselective determination of alprenolol in human plasma by liquid chromatography with tandem mass spectrometry using cellobiohydrolase chiral stationary phases. *J. Chromatogr. B Analyt. Technol. Biomed. Life Sci.* 872 (1–2): 121–127.

Jiang, H., Li, Y., Pelzer, M. et al. (2008b). Determination of molindone enantiomers in human plasma by high-performance liquid chromatography-tandem mass spectrometry using macrocyclic antibiotic chiral stationary phases. *J. Chromatogr. A* 1192 (2): 230–238.

Jiang, H., Randlett, C., Junga, H. et al. (2009). Using supported liquid extraction together with cellobiohydrolase chiral stationary phases-based liquid chromatography with tandem mass spectrometry for enantioselective determination of acebutolol and its active metabolite diacetolol in spiked human plasma. *J. Chromatogr. B Analyt. Technol. Biomed. Life Sci.* 877 (3): 173–180.

Keevil, B.G. (2013). Novel liquid chromatography tandem mass spectrometry (LC-MS/MS) methods for measuring steroids. *Best Pract. Res. Clin. Endocrinol. Metab.* 27 (5): 663–674.

Li, Y., Emm, T., and Yeleswaram, S. (2011). Simultaneous determination of fluoxetine and its major active metabolite norfluoxetine in human plasma by LC-MS/MS using supported liquid extraction. *Biomed. Chromatogr.* 25 (11): 1245–1251.

Little, J.L., Wempe, M.F., and Buchanan, C.M. (2006). Liquid chromatography-mass spectrometry/mass spectrometry method development for drug metabolism studies: examining lipid matrix ionization effects in plasma. *J. Chromatogr. B Analyt. Technol. Biomed. Life Sci.* 833 (2): 219–230.

Liu, G., Liu, J., Hara, K. et al. (2009). Rapid determination of cyanide in human plasma and urine by gas chromatography-mass spectrometry with two-step derivatization. *J. Chromatogr. B Analyt. Technol. Biomed. Life Sci.* 877 (27): 3054–3058.

Logan, B.K. and Stafford, D.T. (1989). Liquid/solid extraction on diatomaceous earth for drug analysis in postmortem blood. *J. Forensic Sci.* 34 (3): 553–564.

Meunier, C., Blondelle, D., Faure, P. et al. (2015). Development and validation of a method using supported liquid extraction for aldosterone determination in human plasma by LC-MS/MS. *Clin. Chim. Acta* 447: 8–15.

Nave, F., Cabrita, M.J., and da Costa, C.T. (2007). Use of solid-supported liquid-liquid extraction in the analysis of polyphenols in wine. *J. Chromatogr. A* 1169 (1–2): 23–30.

Nguyen, L., Zhong, W.Z., Painter, C.L. et al. (2010). Quantitative analysis of PD 0332991 in xenograft mouse tumor tissue by a 96-well supported liquid extraction format and liquid chromatography/mass spectrometry. *J Pharm. Biomed. Anal.* 53 (3): 228–234.

O'Maille G, Pai S M, Tao X, Douglas G T, Jr., Jenkins R G. An improved LC-ESI-MS-MS method for simultaneous quantitation of rosiglitazone and N-desmethyl rosiglitazone in human plasma. *J. Pharm. Biomed. Anal.* 2008; 48(3):934–939.

Owen, L.J. and Keevil, B.G. (2013). Supported liquid extraction as an alternative to solid phase extraction for LC-MS/MS aldosterone analysis? *Ann. Clin. Biochem.* 50 (5): 489–491.

Pan, J., Jiang, X., and Chen, Y.L. (2010). Automatic supported liquid extraction (SLE) coupled with HILIC-MS/MS: an application to method development and validation of Erlotinib in human plasma. *Pharmaceutics* 2 (2): 105–118.

Pernice, T., Bishop, A.G., Guillen, M.J. et al. (2016). Development of a liquid chromatography/tandem mass spectrometry assay for the quantification of PM01183 (lurbinectedin), a novel antineoplastic agent, in mouse, rat, dog, Cynomolgus monkey and mini-pig plasma. *J. Pharm. Biomed. Anal.* 123: 37–41.

Rao, R.N., Prasad, K.G., Kumar, K.V.S., and Ramesh, B. (2013). Diatomaceous earth supported liquid extraction and LC-MS/MS determination of elvitegravir and ritonavir in rat plasma: application to a pharmacokinetic study. *Anal. Methods* 5 (23): 6693–6699.

Rositano, J., Harpas, P., Kostakis, C., and Scott, T. (2016). Supported liquid extraction (SLE) for the analysis of methylamphetamine, methylenedioxymethylamphetamine and delta-9-tetrahydrocannabinol in oral fluid and blood of drivers. *Forensic Sci. Int.* 265: 125–130.

Scheidweiler, K.B. and Huestis, M.A. (2014). Simultaneous quantification of 20 synthetic cannabinoids and 21 metabolites, and semi-quantification of 12 alkyl hydroxy metabolites in human urine by liquid chromatography-tandem mass spectrometry. *J. Chromatogr. A* 1327: 105–117.

Silvester, S. and Smith, L. (2012). Profiling phospholipid elution in reversed-phase LC-MS/MS bioanalytical methods in order to avoid matrix effects. *Bioanalysis* 4 (8): 879–895.

Singleton, C. (2012). Recent advances in bioanalytical sample preparation for LC-MS analysis. *Bioanalysis* 4 (9): 1123–1140.

Thevis, M., Thomas, A., Pop, V., and Schanzer, W. (2013). Ultrahigh pressure liquid chromatography-(tandem) mass spectrometry in human sports drug testing: possibilities and limitations. *J. Chromatogr. A* 1292: 38–50.

Tweed, J.A., Walton, J., and Gu, Z. (2012). Automated supported liquid extraction using 2D barcode processing for routine toxicokinetic portfolio support. *Bioanalysis* 4 (3): 249–262.

Valen, A., Leere Oiestad, A.M., Strand, D.H. et al. (2017). Determination of 21 drugs in oral fluid using fully automated supported liquid extraction and UHPLC-MS/MS. *Drug Test. Anal.* **9** (5):808–823.

Wang, A.Q., Zeng, W., Musson, D.G. et al. (2002). A rapid and sensitive liquid chromatography/negative ion tandem mass spectrometry method for the determination of an indolocarbazole in human plasma using 96-well diatomaceous earth plates for solid-liquid extraction [correction of using internal standard (IS) 96-well diatomaceous earth plates for solid-liquid extraction]. *Rapid Commun. Mass Spectrom.* 16 (10): 975–981.

Wen, B. and Zhu, M. (2015). Applications of mass spectrometry in drug metabolism: 50 years of progress. *Drug Metab. Rev.* 47 (1): 71–87.

Wu, S., Li, W., Mujamdar, T. et al. (2010). Supported liquid extraction in combination with LC-MS/MS for high-throughput quantitative analysis of hydrocortisone in mouse serum. *Biomed. Chromatogr.* 24 (6): 632–638.

Zhang, Y., Cao, H., and Jiang, H. (2013). Supported liquid extraction versus liquid-liquid extraction for sample preparation in LC-MS/MS-based bioanalysis. *Bioanalysis* 5 (3): 285–288.

7

Immunocapture in LC-MS Bioanalysis

Ang Liu

7.1 Introduction

Currently, immunocapture (or immunoextraction, immunoaffinity capture) techniques are widely used for sample cleanup and enrichment of protein-based biotherapeutics and biomarkers. The combination of immunocapture (IC) with subsequent LC-MS/MS analysis provides a powerful tool in the development of quantitative pharmacokinetics (PK) and pharmacodynamics (PD) assays. The principle of IC is based on the highly specific interaction of a ligand (analyte of interest) with its antibody (capture reagent) to form a unique immune complex. Therefore, an IC LC-MS/MS assay is also known as a hybrid approach which involves ligand binding assay (LBA) and LC-MS/MS quantitation. Here, the experimental workflow, the selection of capture reagents and platforms, the performance evaluation, and the recent applications of IC are discussed.

7.2 Experimental Workflow and Optimization

The power of IC lies in the ability of capture reagent to separate the analyte of interest from a complex mixture such as a biological sample with a high degree of specificity. The antibody as an IC reagent is usually immobilized onto a solid support in a manner that its binding capacity still retains. The biological sample is incubated with or passed through the support and the unbound components are washed away. The retained analyte of interest is then eluted from the support by weakening the antibody–antigen interaction, for example, via changing pH or ionic strength. A typical experimental workflow of IC is shown in Figure 7.1. Generally, it involves the following steps:

1) Immobilize the IC reagent onto the solid support by the formation of covalent chemical bonds or strong non-covalent binding between the particular functional group on the reagent to the reactive group on the support.
2) Incubate the biological sample (e.g. plasma or serum) with the IC support to allow the target analyte in the sample to bind to the immobilized capture reagent.
3) Remove any unbound sample component from the support using appropriate buffers that maintain the binding interaction between the target analyte and the capture reagent.
4) Dissociate the target analyte from the immobilized capture reagent by altering the buffer conditions so that the binding interaction no longer occurs.

The IC reagents can be immobilized onto supports by using a variety of chemistry and techniques that range from direct covalent binding to indirect methods. A description of the immobilization of capture reagent to generate an immunoaffinity support is beyond the scope of this chapter, but detailed techniques can be found in some previous work (Kim and Hage 2006; Moser and Hage 2010). An ideal situation is to attach the reagent to the support in a way that it does not affect the activity of the binding sites and the accessibility of the sites to the target analyte. In addition, good chemical stability must be maintained to resist the harsh elution condition.

The avidin–biotin or streptavidin–biotin interaction is the strongest non-covalent interaction ($K_d = 10^{-15}$ M) that is ever known. The bond formation between biotin and streptavidin is rapid and specific, and once formed, the bonding is unaffected by extremes of various conditions, such as pH, temperature, organic solvents, and other denaturing agents. Most of the antibodies and ligands can be easily labeled with biotin in a laboratory setting using commercially available reagents. Therefore, the biotin–streptavidin interaction is a commonly used strategy for the immobilization of capture antibodies. The most popular biotinylation technique is to incubate the

Sample Preparation in LC-MS Bioanalysis, First Edition. Edited by Wenkui Li, Wenying Jian, and Yunlin Fu.

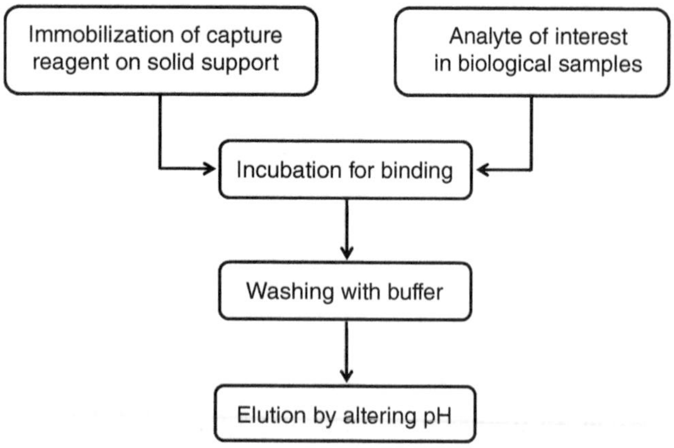

Figure 7.1 Summary of a typical immunocapture procedure.

capture antibody with *N*-hydroxysuccinimide-*D*-biotin (NHS-activated biotin) at pH 7–9 to form stable amide bonds with its primary amines. A 10- to 20-fold molar excess of biotin reagent is typically used to label 1–10 mg antibody and the excess biotin can be removed using desalting column. Approximately, biotin incorporation of 4–6 biotins per antibody molecule can be obtained. The biotinylated antibodies can usually be stored for long term before use. For sample extraction, the biotinylated capture reagents then can be immobilized onto streptavidin-coated support by incubation at room temperature. The popular solid supports used for IC are magnetic beads, agarose beads, and other column packing materials.

In general, the conditions for binding and elution of IC (antibody–antigen affinity interactions) are predictable. As the binding affinity in *in vivo* systems occurs at physiological pH and ionic strength, aqueous buffer such as phosphate-buffered saline (PBS) is commonly used as a binding buffer to treat biological samples. In some cases, pretreatment of samples prior to binding may be required, such as dilution, pH adjustment, buffer exchange, or ultrafiltration. Selection of sample volume depends on the capacity of the capture reagents, the total amount of target analyte, and the complexity of the sample where other matrix components may interact with the capture reagents. Usually the total amount of potential binding components in the sample should not exceed the capacity of the capture reagents. Following binding, the unbound components are removed using washing buffers. The wash buffers are often PBS containing moderate concentration of salt which reduces nonspecific binding without causing the elution of the analyte of interest. Washing may involve multiple steps if nonspecific interaction is suspected. Low concentration of gentle surfactants, such as 0.05% Tween 20, is usually added in washing buffer to minimize the nonspecific binding.

Table 7.1 Binding, washing, and elution buffers commonly used for immunocapture.

Buffer	Examples
Binding	PBS, pH 7.4
	0.05% Tween in PBS, pH 7.4
	0.01% BSA in PBS, pH 7.4
Washing	0.05% Tween in PBS, pH 7.4
	0.1% BSA and 2 mM EDTA in PBS, pH 7.4
	0.1% BSA and 500 mM sodium chloride in PBS, pH 7.4
Elution	0.1 M glycine•HCl, pH 2.5
	12 mM hydrochloric acid with 0.5% BSA, pH 2.5
	2.5% formic acid in 10% acetonitrile, pH 2.5
	0.1 mM citric acid, pH 2.5

Finally, elution buffer is added to release and recover the captured target molecules from the capture reagents. Elution can be accomplished by raising or lowering the pH or altering the ionic strength to disrupt the binding interaction. Generally, pH and ionic strength conditions are gentle on antibody and protein function, while denaturants and organic are harsh. The most widely used elution buffer is 0.1 M glycine•HCl, pH 2.5, which is usually a good condition to start with. This buffer effectively dissociates most antibody–antigen binding and protein–protein interactions without permanently affecting protein structure and function. The IC could be reused; however, a gentle but complete elution is required. In practice, the reproducibility of the reused capture needs to be evaluated to ensure accurate quantitation. In addition, blocking agents, for examples, surfactants and bovine serum albumin, can be added to the elution buffer to help reduce/eliminate the potential protein adsorption of eluted analyte onto the collection plate. The eluted sample is usually neutralized to a condition which is favorable for the following step such as enzyme digestion. Examples of commonly used buffers for IC are listed in Table 7.1.

7.3 Considerations on the Selection of Capture Reagents and the Limitations

The selection of the reagent for IC is critical. The ideal IC reagent specifically binds to the target analyte with high affinity but with no or minimal affinity for other potential interfering components in the sample. Therefore, the assay sensitivity can be significantly improved by utilizing appropriate IC compared to traditional non-IC assay. Additional bioanalytical selectivity can be obtained by choosing a capture reagent that specifically binds to a unique region of the analyte (e.g. complementarity-determining region of the therapeutic antibody) or a specific form of the analyte (e.g. specific posttranslational modification of a biomarker). For example, a specific capture reagent that recognizes the binding site for soluble target of a monoclonal antibody (mAb) can be used to extract the free form of the antibody. This approach is usually considered as a specific capture instead of generic capture. To serve this purpose, careful characterization of the capture reagent is required. Biotinylation of reagent followed by immobilization on streptavidin-coated support is usually used for IC using specific capture reagent. In addition, surface-activated support such as tosylactivated magnetic beads can also be used to couple specific capture reagent by covalent binding of amino or sulfhydryl groups.

High-quality reagents with high-binding affinity and selectivity are paramount to development of a highly sensitive and selective LBA assay. However, the generation and characterization of the specific reagents for LBA usually takes time and resources, which may not be ideal in drug discovery and early development. In contract, the selectivity of capture reagent can be sacrificed in the LC-MS/MS-based assay due to the high selectivity via chromatographic separation and mass spectrometric detection. For example, the protein analyte eluted from IC is usually digested to generate surrogate peptide by endoproteases. Although a small amount of the other proteins with various affinities for the less-specific capture reagent could be co-captured and digested, the unique surrogate peptide of the analyte can be isolated by LC separation and selectively monitored by MS/MS. Highly selective MS systems enable use of generic or less-specific reagents for IC.

Generic capture reagent usually captures a common sequence of amino acids or a common region of different proteins. For example, protein A and protein G, known as bacterial cell wall proteins, can strongly bind to the Fc region of immunoglobulin G (IgG). Binding specificities and affinities differ among source species and antibody subclass (Table 7.2). Protein L binds to all classes of immunoglobulin (e.g. IgG, IgM, IgA, IgE, and IgD) with kappa light chain, thus, offering alternative to protein A and G (Eliasson et al. 1989). A comparison of different immunoglobulin-binding proteins is summarized in Table 7.3. Another approach for IC utilizes an antibody specifically against the Fc region of human IgGs, whereas not against IgGs from animal species. As a result, the anti-human IgG Fc antibody allows for differentiation

Table 7.2 Affinity of protein A and protein G to IgG subclass from different species.

	Immunoglobulin	Protein A	Protein G
Human	IgG1	++	++
	IgG2	++	++
	IgG3	–	++
	IgG4	++	++
Rat	IgG1	–	+
	IgG2a	–	++
	IgG2b	–	+
	IgG2c	+	+
Mouse	IgG1	+	++
	IgG2a	++	++
	IgG2b	++	+
	IgG3	+	+
Rabbit	IgG	++	++
Goat	IgG	+/–	++
Sheep	IgG	+/–	++
Guinea Pig	IgG	++	+
Bovine	IgG	+	++

++, strong affinity; +, moderate/week affinity; –, no affinity; +/–, require evaluation.

between therapeutic human mAbs and endogenous IgGs in animal matrices, which offers a higher degree of selectivity compared to the approach using protein A and protein G. This type of antibodies is especially useful for preclinical studies during drug discovery and early development. However, due to cross-reactivity with endogenous IgGs in human matrix, this approach may not be useful for human samples. For more specific capture of therapeutic humanized mAb in human samples, an anti-idiotype antibody that binds to the unique epitope on the variable region of the target antibody is required.

Protein A, protein G, protein A/G, and protein L are commercially available in immobilized and high-throughput format, while commercially available biotinylated anti-human IgG Fc antibody can be readily coated on streptavidin support. Cartoons of different types of IC reagents are shown in Figure 7.2.

In addition, the generic approach can also be applicable for the extraction of PEGylated peptides and proteins. Polyethylene glycol (PEG) can be covalently conjugated to the target drug to protect the drug from *in vivo* immune system, therefore reducing immunogenicity. PEGylation is also used to improve drug solubility and to reduce renal excretion of the drug. Various anti-PEG antibodies can be obtained commercially for IC of different PEG chains. For example, a rabbit IgG anti-PEG mAb was found to be

Table 7.3 A comparison of immunoglobulin binding proteins.

	Protein A	Protein G	Protein A/G	Protein L
Production source	*E. coli*	*E. coli*	*E. coli*	*E. coli*
Molecular weight (kDa)	44.6	21.6	50.5	35.8
Number of binding sites for Ig	5	2	4+2	4
Optimal binding pH	8	5	5–8	7.5
Ig binding target	Fc	Fc	Fc	VL-kappa

VL-kappa, the kappa light chain variable domain.

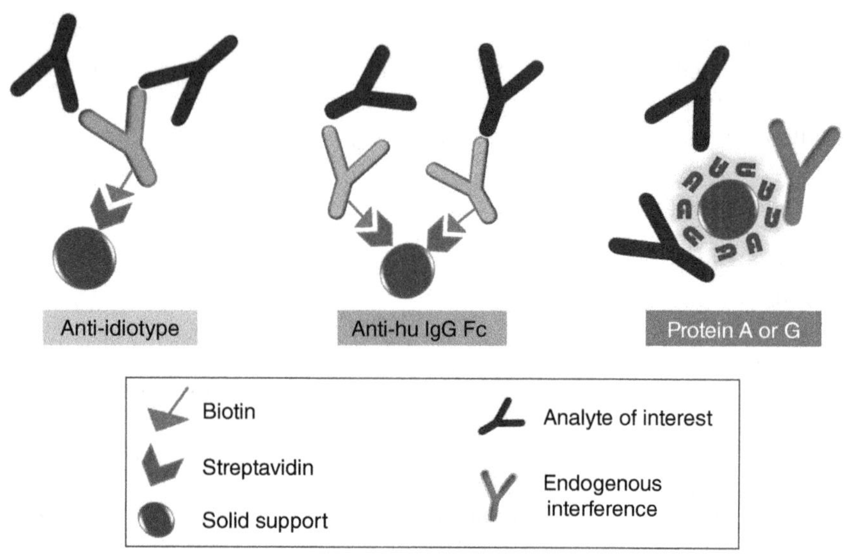

Figure 7.2 Examples of different capture reagents.

strongly bound to a 40 kDa branched PEG conjugated to C-terminus of a therapeutic peptide while the surrogate peptide at the N-terminus was monitored for the intact PEGylated drug after anti-PEG IC (Xu et al. 2010).

IC at the protein level can effectively simplify biological samples and therefore largely improve the efficiency of the subsequent digestion. However, protein capture may not be suitable in some situations. For example, in biomarker analysis, the difference between the recombinant protein standard and the endogenous protein biomarker in posttranslational modifications may lead to discrepancy in the step of protein capture. IC can be employed at the peptide level using anti-peptide antibody (APA), known as the Stable Isotope Standards and Capture by Anti-Peptide Antibodies (SISCAPA) technique (Anderson et al. 2004, 2009). Unlike proteins, peptides are generally too small to be immunogenic, which presents major challenge for the production of APA. To activate an immune response, peptides can be attached to a carrier protein to make an immunogen which can stimulate the host system to produce antibodies. The SISCAPA process typically begins with enzyme digestion of whole plasma or serum sample. A stable isotope-labeled (SIL) peptide is used as an internal standard and is added into the digestion mixture prior to the IC step. The digested sample with spiked SIL peptide is incubated with specific APA bound to a solid support. The captured target peptide and the corresponding SIL peptide can be eluted and quantified using LC-MS/MS. This technique is able to achieve high sensitivity similar to immunoassays (Kuhn et al. 2009; Neubert et al. 2010; Schoenherr et al. 2010). There are some additional advantages, namely the ability to multiplex analytes and the potential of sequential IC. Multiplexed capture and simultaneous quantitation of multiple drugs can be achieved by incubating the sample with multiple APAs targeting different peptides (Kuhn et al. 2009; Kuhn and Carr 2016; Lee et al. 2016; Schoenherr et al. 2012; Whiteaker et al. 2010). Sequential IC at the protein and tryptic peptide levels can highly enrich the analyte, therefore significantly improve the sensitivity (Neubert et al. 2013; Palandra et al. 2013; Schultz et al. 2016). Figure 7.3 illustrates the workflows of different IC formats.

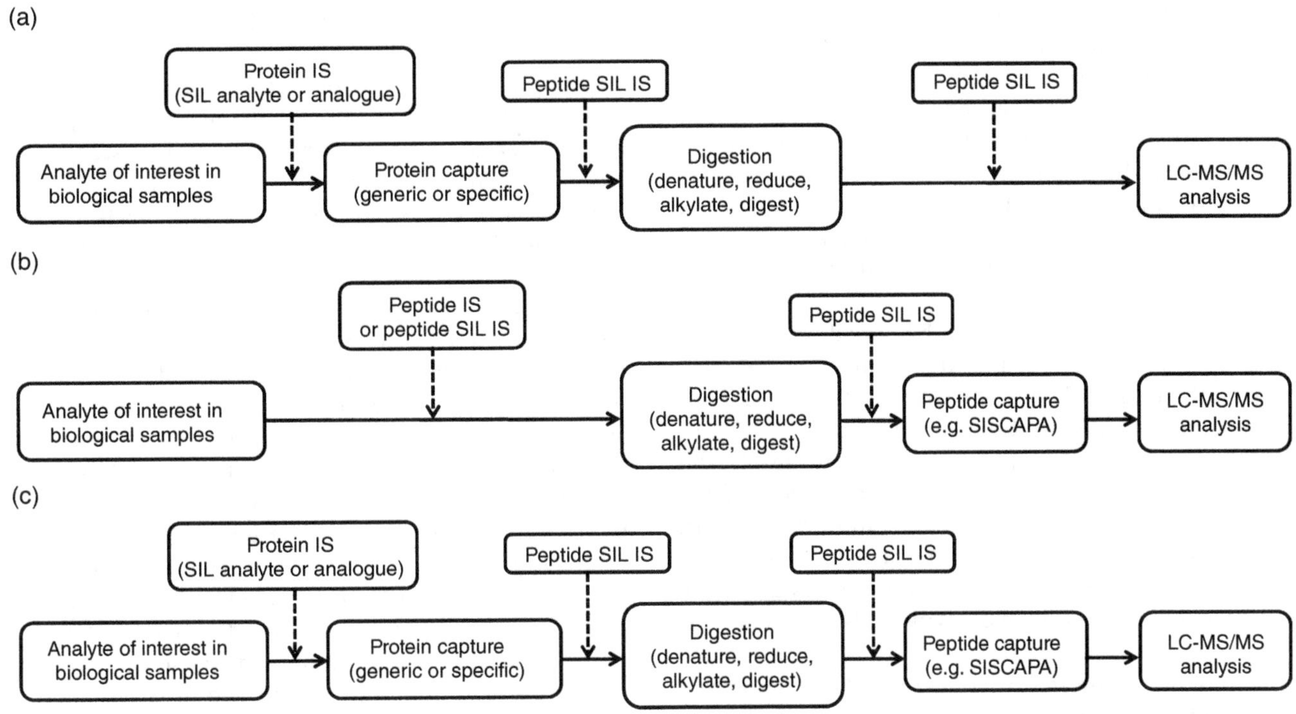

Figure 7.3 Workflows of typical immunocapture formats. Protein (a), peptide (b), and sequential (c) immunocapture.

7.4 Platforms for Immunocapture

Bead-based platform has been widely used as the solid support to immobilize the capture reagents. Magnetic beads are small, usually at the micrometer level in diameter. The small size presents a large surface area per milligram beads and a correspondingly high capacity for the target molecule. The beads are superparamagnetic with high iron content, which ensures easy and fast magnetic separation by magnet stands. Bead-based platform is popular since it is flexible during method development and it has potential of adapting to automated processes for method validation and production work. For example, magnetic beads in 96-well format can be processed automatically on the KingFisher purification system from Thermo Fisher Scientific (Waltham, MA). Alternatively, microplates can be used as the solid support for IC. A recent publication evaluated enzyme-linked immunosorbent assay (ELISA) microplates versus commonly used bead format for IC for protein quantitation by LC-MS/MS (Yang et al. 2015). The microplates from Greiner Bio-One (Monroe, NC) and Thermo Fisher Scientific (Waltham, MA) provided comparable sensitivity and linear range compared to tosylactivated Dynabeads from Life Technologies (Camarillo, CA), which suggested microplates as a viable IC platform. Lately, cartridge and column-based 96-well platforms are gaining popularity due to high capture capacity, hands-free workflow, and high sample throughput. A few recent examples are AssayMap cartridges from Agilent Technologies (Santa Clara, CA), MISA microcolumns from Thermo Fisher Scientific (Waltham, MA), and PhyTip columns from PhyNexus (San Jose, CA). The automation of complex immobilization/capture/washing/elution procedure makes it possible to implement IC technology to support routine and large-scale bioanalysis.

7.5 Internal Standard Selection

In quantitative LC-MS/MS bioanalysis, the selection of internal standard for protein analyte is more complex (Bronsema et al. 2012, 2013) than that for small molecules. For the latter, either a SIL version of analyte or a structural analogue can be readily employed. The extracted proteins usually need to be digested to a mixture of peptides, one or more of which are used for quantitation, and therefore there are more options for internal standard selection. Examples are: (i) the SIL version of signature peptide, (ii) the modified peptide with additional amino acid residues flanking SIL signature peptide, (iii) the intact protein with SIL amino acid residues, and (iv) analogue protein. Peptide internal standard has been widely used due to the relative ease of peptide synthesis. While it might do well in normalizing the sample-to-sample variation in the surrogate peptide of the protein analyte during LC-MS/MS analysis (e.g. matrix effect), it does not compensate for the variations in sample extraction and actual release of the surrogate peptide from the protein analyte during digestion (Figure 7.3). Therefore, if optimal and consistent extraction and digestion are not achieved, the accuracy of absolute quantitation of the protein analyte might be subject to error associated with extraction and digestion that is applied to the protein analyte but not the SIL surrogate peptide.

Protein internal standard can be added into biological samples at the very beginning of sample processing, prior to the IC step. This approach can compensate for the variations throughout the entire procedure, including IC, digestion, LC separation, MS ionization, and therefore enhance the accuracy and precision of the bioanalytical method. The most ideal internal standard for a protein analyte is the same protein that is uniformly incorporated with SIL amino acid residues. SIL protein could be produced by living cells or by cell-free systems in the presence of only heavy isotope-labeled version of one or more essential amino acids. However, the subsequent purification and extensive characterization are required before it can be used as an internal standard. Usually substantial time and resources are needed to obtain a SIL protein of each protein drug candidate, therefore, it may not be readily available at drug discovery and early development stage. Alternatively, a structural analogue of the target protein may serve as an internal standard, but it could only be applicable for generic capture rather than specific capture due to the low specificity of binding of the analogue protein. In addition, a general approach using common intact antibody internal standard was reported to support preclinical studies (Li et al. 2012). A human antidinitrophenol (DNP) IgG_2 antibody is unlikely to exist in preclinical species but possesses common peptide sequences with other human antibodies. A SIL version of this antibody was used as a universal whole protein internal standard and captured by anti-human IgG Fc antibody that recognizes human biotherapeutics but not the endogenous IgG in animal samples.

7.6 Performance Evaluation

Capture efficiency/recovery, matrix effect, selectivity, and tolerance to soluble target are the important parameters to demonstrate the performance of IC step in LC-MS/MS bioanalysis of proteins.

Recovery of the analyte of interest pertains to the efficiency of IC within the limits of variability. Differences in the capture reagent specificity, binding strength, capture platform, and experimental conditions could result in different extents of loss of the analyte in the IC process. The recovery is not necessary to be 100% and is usually low for generic capture, but it should be consistent, precise, and reproducible at different concentrations across the intended dynamic range. The quality control (QC) samples at low, medium, and high concentration levels should be used for the evaluation. The recovery of the IC step can be evaluated using two sets of samples (in triplicates per set) as shown in Figure 7.4. Set 1: analyte spiked in plasma (pre-IC matrix); Set 2: analyte spiked in digestion buffer (post-IC/pre-digestion). Samples in set 1 are processed through immuno-capture and enzyme digestion while samples in set 2 are processed only through the digestion step. The recovery of IC step can be determined by comparing the responses of set 1 versus those of set 2.

Matrix effect typically refers to the matrix-related suppression or enhancement of analyte ionization during LC-MS/MS analysis and can be quantified by measuring the matrix factor (MF). MF is defined as the ratio of the response of analyte in the presence of matrix (spiked post-extraction) to the response in the absence of matrix (spiked in neat solution). Measurement of MF in a protein assay can be a challenge because the analyte is not measured directly. The determination of matrix effect for IC-digestion-LC-MS/MS assay is recommended to be conducted at the peptide level since the surrogate peptide is the final analyte for MS detection. Blank matrix is processed through IC, digestion, and necessary extraction steps. Then the surrogate peptide and the internal standard peptide are spiked in the post-extraction samples and in neat solutions. It is recommended that at least six individual lots of matrix be evaluated at

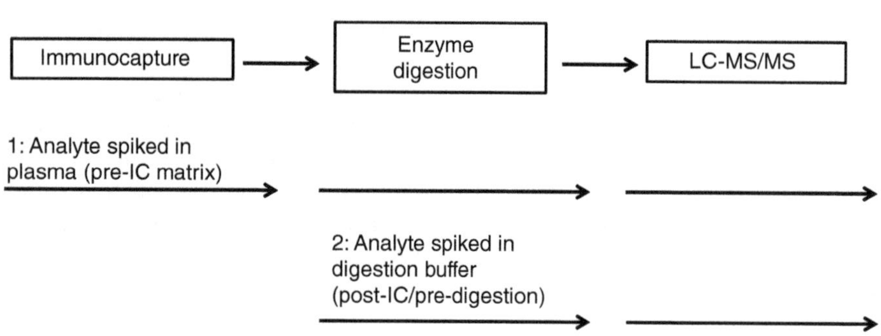

Figure 7.4 Evaluation of immunocapture recovery.

both low and high concentrations. The % coefficient of variation (%CV) of internal standard normalized MF calculated from the six matrix lots is used to determine the acceptance of matrix effect, which should not be >20%.

Selectivity is the ability of a method to measure and differentiate the analyte of interest in the presence of other matrix components, such as catabolites, degradants, concomitant medication and associated metabolite(s) in the study samples. A minimum of six lots of blank matrix should be tested and the selectivity should be ensured at the lower limit of quantification (LLOQ) level. Since IC is based on affinity binding interactions, it is desirable to evaluate the potential interference from analyte-related compounds that may bind to the capture reagent, for example, anti-drug antibodies (ADAs), isoforms, endogenous analogs, and soluble target. Special interfering compounds may not be available during method development and validation. Therefore, the evaluation of selectivity should be revisited when study samples become available.

Assay tolerance to soluble target is usually evaluated when a specific capture reagent is used for IC such as anti-idiotype mAb capture. Soluble target in circulation may block the binding sites on the target protein and thus affect the ability of the capture antibody to capture the intended analyte (bound and unbound). The evaluation may not be needed for animal species due to no binding between humanized antibody drug and animal target or low circulating level of animal target. However, for clinical application, if the levels of soluble target in the clinical population are projected to be high and variable, the intended human assay should be characterized for the possible interference due to soluble target.

7.7 Applications and Representative Protocols

IC technology provides the cleanest sample for LC-MS/MS analysis compared to other available extraction and enrichment methods. A large number of research papers have been published illustrating the various applications of IC in LC-MS/MS bioanalysis of endogenous peptides/proteins such as biomarkers and targets (Lin et al. 2013; Neubert et al. 2016; Torsetnes et al. 2013), protein-based biotherapeutics (Becher et al. 2013; Fernandez Ocana et al. 2012; Liu et al. 2015; Wang and Heilig 2012), and unwanted immunogenicity (Chen et al. 2016; Jiang et al. 2014).

7.7.1 Endogenous Peptides/Proteins

Quantitation of peptides/proteins with key physiological functions in human body is critical to understand the diseases and system pharmacology. However, endogenous peptides/proteins are often present in low abundance in highly complex matrices such as plasma, urine, cerebrospinal fluid (CSF), and tissues. IC LC-MS/MS for quantification of endogenous peptides/proteins has been a subject of great interest in recent years due to its ability to attain good sensitivity and high specificity which leads to a better understanding of what is being measured (Chappell et al. 2016; Jones and Schultz 2016; Lee et al. 2016; Schultz et al. 2016; Torsetnes et al. 2013, 2014).

Glucagon-like peptide-1 (GLP-1) is a 30 amino acid long incretin hormone peptide secreted by intestinal enteroendocrine and certain neurons (Holst 2007). It is a potent regulator of glucose homeostasis by enhancing glucose-dependent insulin secretion. Circulating GLP-1 is present at very low level and the half-life of its active form is very short, which contributes to bioanalytical challenges. An early publication described an IC method using primary antibody coupled with secondary antibody on magnetic bead format to determine the plasma levels of GLP-1 (Wolf et al. 2004). This was a great early application of IC LC-MS; however, the sensitivity (11 pM) was not sufficient to measure endogenous levels. Recently, an ultrasensitive assay with an LLOQ of 0.5 pM was reported, which is ~20-fold more sensitive (Chappell et al. 2014b). One of the key elements that result in better sensitivity and selectivity is the enrichment of the analytes from a complex matrix through the use of mAbs specific to GLP-1. Two mAbs (HYB147-06 and HYB147-13) with high affinity and low dissociation were selected to capture both the active form (7-36 amide) and inactive form (9-36 amide) of GLP-1. This assay was sensitive enough to measure the typical changes in endogenous GLP-1 levels after food intake.

There have been excellent reviews on quantitative LC-MS methods for circulating human proteins in recent years (Chappell et al. 2014a; Ewles and Goodwin 2011; Gillette and Carr 2013). Sample preparation workflows that incorporate IC at the protein or peptide level have been increasingly used to achieve the desired bioanalytical sensitivity that is required to measure low-abundance endogenous proteins. The examples are APA enrichment for the measurements of total salivary pepsin/pepsinogen (Neubert et al. 2010), thyroglobulin (Tg) in circulation (Kushnir et al. 2014) and circulating D-dimer (Wang et al. 2012b), specific protein IC LC-MS/MS analyses of neuron specific enolase (NSE) in serum (Torsetnes et al. 2013), stromal cell-derived factor 1α (SDF-1α) in plasma (Wang et al. 2012a) and Tau in CSF (McAvoy et al. 2014), sequential protein and tryptic peptide IC LC-MS/MS of interleukin 21 (IL-21) (Palandra et al. 2013), and beta-nerve growth factor (β-NGF) (Neubert et al. 2013; Schultz et al. 2016). Those hybrid approaches can robustly quantify endogenous proteins at the sub $ng\,ml^{-1}$ to low $pg\,ml^{-1}$ range. The integrity of capture reagents and the reproducibility of IC steps are critical to a successful assay. The robustness of a recently reported assay using sequential protein and peptide IC has been demonstrated in support of large interventional clinical trials with ~19 000 clinical samples analyzed (Schultz et al. 2016).

Representative protocol (endogenous Peptide-X):

1) Resuspend the Dynabeads M-280 tosylactivated (Life Technologies, 30 mg beads per ml) in the original vial by vortexing for at least 30 seconds.
2) Transfer 1.67 ml of the beads (~50 mg beads) to a tube, and place the tube in a magnetic device for one minute and then remove the supernatant.
3) Wash the beads three times with 2 ml of PBS buffer.
4) Resuspend the washed beads in 1 mg of anti-Peptide-X antibody (ratio of 100 µg antibody to 5 mg beads) and add PBS buffer to a total volume of 0.75 ml. Mix thoroughly by vortexing.
5) Add 0.5 ml of 3 M ammonium sulphate in PBS (pH 7.4) and mix.
6) Incubate overnight at 37 °C with end-over-end rotation.
7) Place the incubation tube in a magnetic device for two minutes and then remove the supernatant.
8) Wash the beads three times with 2 ml of 150 mM NaCl in PBS with 0.05% Tween-20, and then resuspend in 2.4 ml of the same buffer to a final concentration of the Ab-conjugated beads at $20\,mg\,ml^{-1}$. Store at 4 °C until use.
9) Transfer 0.5 ml of human plasma samples to a 2 ml protein LoBind Plate and dilute with 0.5 ml of 1xRIPA buffer.
10) Add 50 µl of IS (SIL version of Peptide-X) solution to each sample.
11) Add 50 µl of Ab-conjugated beads to each sample, and incubate at room temperature for one hour with end-over-end rotation.
12) Wash the beads twice with 2 ml of 1xRIPA buffer and twice with 2 ml of water.
13) Elute the captured Peptide-X from the beads by shaking with 30 µl of elution buffer ($400\,\mu g\,ml^{-1}$ BSA in 20% acetic acid, 10% methanol) for 15 minutes at room temperature.
14) Filtrate the eluted sample using an Ultrafree-MC 0.22 µm PVDF filter (Millipore) and then analyze on LC-MS/MS.

7.7.2 Protein-based Biotherapeutics

LBAs are the most commonly used approaches for PK measurement of protein-based biotherapeutics. The hybrid IC LC-MS/MS approach complements the traditional LBA and provides the specificity necessary for overcoming limitations of LBA (An et al. 2014). For example, a therapeutic protein, recombinant-methionyl human leptin (16 155 Da) shares an identical amino acid sequence with endogenous leptin with the addition of a methionine at the N-terminal. LBA was not able to differentiate the drug from its endogenous counterpart due to cross-reactivity (Wang and Heilig 2012). A monoclonal mouse anti-r-metHu-Leptin antibody was used to capture both recombinant and endogenous leptin from human plasma while the matrix complexity was significantly reduced. After sample cleanup by IC, a top-down assay using Q-Trap mass spectrometry was able to distinguish and specifically quantify both forms due to a 131 Da mass difference.

Therapeutic mAbs represent the majority of biotherapeutics now on the market. Bioanalytical assay capable of determining systemic concentration of the therapeutic antibody is critical to the characterization of its pharmacokinetic property (Ewles et al. 2016). Becher et al. (2013) presented a specific IC method to measure cetuximab, a human:murine chimeric mAb in human serum. Magnetic beads functionalized with soluble epidermal growth factor receptor (sEGFR), the pharmacological target of cetuximab, were used for the extraction of the mAb from human serum. The LLOQ of $20\,ng\,ml^{-1}$ was similar to that of ELISA methods. An advantage of the IC-based method is its capability to quantify specific analytes, such as free or total antibody. Fernandez Ocana et al. (2012) developed a LC-MS/MS-based strategy to measure free and total anti-MadCAM mAb in human serum. The free assay used a biotinylated anti-idiotype antibody immobilized on streptavidin magnetic beads as IC reagent while the total assay used protein G capture.

A recent publication described a hybrid assay for the quantification of the payload molecules of an antibody–drug conjugate (ADC) (Liu et al. 2015). In this approach, the ADCs are immunocaptured followed by enzymatical release of the payload, which is then measured by LC-MS/MS for quantitation of conjugated payload. The capture reagent is a key component for the selectivity of an ADC assay. Different capture reagents may be associated with different bioanalytical outcomes because of the complex nature of the ADC molecule and the potential *in vivo* change. This hybrid assay employed anti-idiotype antibody to capture ADC antibody that was not obscured by soluble target binding. Therefore, the molecules available to bind tumor cell target and thus contributing to bioactivity were measured. Most recently, a comprehensive methodology using hybrid ligand binding (IC) LC-MS/MS approaches was developed in this laboratory to understand and characterize the overall PK properties of the ADC (Wang et al. 2016).

Representative protocol (ADC-X):

1) Reconstitute the lyophilized biotin (EZ-link Sulfo-NHS-LC-Biotin, Thermo Scientific) with deionized water immediately prior to use.
2) Mix the reconstituted biotin with a mouse anti-ADC-X mAb (anti-idiotype) at a molar challenge ratio of 12 : 1, incubate at room temperature for 30 minutes.
3) Purify the biotinylated capture antibody using Zeba Spin Desalting Column, 7 K MWCO (Thermo Scientific).
4) Determine the protein concentration of the purified sample based on its absorbance at 280 nm using NanoDrop 2000, and store at −70 °C until use.
5) Equilibrate AssayMAP streptavidin (SA-W) cartridges using 50 µl of PBS at a flow rate of 25 $\mu l\ min^{-1}$ on an Agilent AssayMAP Bravo system (Agilent Technologies).
6) Immobilize 50 µl of biotinylated capture antibody (at a protein concentration of 600 $\mu g\ ml^{-1}$) on the streptavidin cartridge at a flow rate of 5 $\mu l\ min^{-1}$.
7) Dilute monkey plasma samples with PBS at a ratio of 1 : 1.
8) Load 25 µl of diluted plasma samples on the cartridges with immobilized capture antibody at a slow flow rate of 2 $\mu l\ min^{-1}$.
9) Wash the cup of the cartridge twice with 50 µl of washing solution (500 mM NaCl and 0.1% BSA in PBS).
10) Wash the resin of the cartridge by passing the same washing solution over the cartridges at a flow rate of 25 $\mu l\ min^{-1}$.
11) Wash the syringe three times with 50 µl of elution buffer (12 mM HCl with 0.5% BSA in water).
12) Elute the captured ADC-X from the cartridges using 36 µl of elution buffer at a flow rate of 5 $\mu l\ min^{-1}$.
13) Neutralize the eluted samples with 250 mM ammonium bicarbonate to an optimized pH for the subsequent enzyme digestion.

7.7.3 Immunogenicity

Biotherapeutics can potentially activate immune responses, resulting in the production of ADAs that may affect the efficacy and pharmacokinetics of the drug. Characterization and measurement of ADA are critical to immunogenicity assessment. Immunoassay is the most widely used platform for ADA detection due to its high sensitivity. However, high levels of drug in study samples can interfere with traditional detection of ADA, especially in bridging assay format. A methodology was reported to overcome this drug tolerance issue (Neubert et al. 2008). The available ADA binding sites were completely saturated by adding excessive amount of drug to the samples. Drug–ADA complexes were then isolated using protein G magnetic beads, followed by digestion and LC-MALDI MS analysis (MALDI: matrix-assisted laser desorption ionization) of the protein drug, the concentration of which is used to infer that of ADA. Recently, an IC LC-MS/MS approach was described for simultaneous isotyping and semiquantitation of ADA in human plasma (Chen et al. 2016). Biotinylated drug or ADA was used to capture ADAs or drug–ADA complexes, respectively. The resulting complexes were immobilized on streptavidin magnetic beads and separated from matrix by a magnet. The released ADA was digested and the specific universal peptides of different isotype of immunoglobulins were then analyzed for each ADA isotype using LC-MS/MS.

Bead extraction and acid dissociation (BEAD) is an effective approach to purify neutralizing antibody (NAb) from serum samples. NAb is a type of antidrug antibody that specifically blocks the biological activity of therapeutic proteins, thus resulting in the reduced efficacy. Typically, serum samples are treated with acid to dissociate the binding between the NAb and the protein drug, and then treated with a capture solution which contains biotinylated drug to bind the released NAb. The complexes containing biotinylated drug can be captured by streptavidin-coated magnetic beads and the captured NAb is eluted for quantitation in a cell-based NAb assay. Optimization of the BEAD is critical to the method development of the cell-based NAb assay. Particularly, certain amounts of the drug and endogenous serum components from samples could be carried over to final BEAD eluates due to the formation of protein complexes with ADA or nonspecific binding with the beads, and therefore could cause interference in the subsequent cell-based assay. A recent publication described an LC-MS/MS approach to simultaneously measure the residual drug, endogenous IgG, and NAb-positive control in BEAD eluates (Jiang et al. 2014). The results provided useful information for immunogenicity assay development in regards to the efficiency of removing the drug and endogenous components by the BEAD and the recovery of the NAb from the BEAD.

Representative protocol (ADA measurement using anti-Drug Ab as capture reagent):

1) Label the mouse anti-Drug-X mAb using an EZ-Link Sulfo-NHS-LC biotinlylation kit (Thermo Scientific).
2) Dilute the biotinylated antibody to a concentration of 100 μg ml^{-1} in water and stored at −70 °C until use.
3) Transfer 144 μl of human plasma sample into a 96-well plate, and mix each sample with 6 μl of 5 mg ml^{-1} Drug-X solution.
4) Incubate the samples in the plate at 37 °C with gentle mixing for one hour and then store at −70 °C overnight.
5) Add 100 μl of 100 μg ml^{-1} biotinylated anti-Drug-X mAb to each sample, and incubate at room temperature for two hours to form biotin–mAb–Drug–ADA complex.
6) Add 100 μl of freshly prepared 5 mg ml^{-1} streptavidin magnetic beads and 475 μl of Tris-buffered saline containing 0.1% Tween-20 (TBS-T) to each sample.
7) Mix the samples gently for one hour at room temperature.
8) Separate the beads on a magnetic device and wash the beads three times with 300 μl of TBS-T and once with 300 μl of water.
9) Elute the captured Drug–ADA complex from the beads with 150 μl of 0.1 M glycine, pH 2.5.
10) Neutralize the eluent immediately with 45 μl of 1 M Tris-HCl, pH 8, and add 10 μl of 0.1% BSA.

7.8 Validation Criteria and Regulatory Considerations

Since the IC step is similar to the capture step of LBA, the assay is generally considered as a hybrid approach which involves an initial or middle step of ligand binding and final LC-MS/MS analysis. Currently there is no regulatory guidance that specifically applies to this type of hybrid assay. The requirements for both LBA and chromatographic methods should be considered and combined to develop a fit-for-purpose approach for assay validation (Ackermann et al. 2015; Dufield et al. 2014; Jenkins et al. 2015). There are a small but growing number of the assays that have been subjected to rigorous validation in regulated environment.

For assay acceptance, an industry consensus was to use the applicable LBA validation criteria for the hybrid assay, which is typically wider than the criteria for LC-MS/MS-based assay (Jenkins et al. 2015). The back-calculated standard results and measured QC values should be within ±20% of the nominal concentrations for accuracy with the precision of multiple replicates within 20% CV, while 25% accuracy and precision are applied for the LLOQ level. Dilution linearity should be evaluated when the concentration above ULOQ is expected in study samples. For more complex IC LC-MS/MS assay, it is recommended to serially dilute the Dilution QC to prepare a set of samples with different concentrations within the calibration range, since the capture used at IC step might be saturated at high analyte concentration. The accuracy should be within ±20% of nominal concentration, while the precision of the measured values across the dilution series should not exceed 20%. The calibration range and regression model for LBA and chromatographic methods are different. Linear regression over a wide calibration range is usually applicable for LC-MS/MS assay, while nonlinear with four or five parameter logistic is typically for LBA. For LC-MS/MS-based assay when IC step is employed, linear curve fitting is recommended when possible; however, nonlinear model may be acceptable with demonstrated calibration curve performance.

Critical reagents are defined as the reagents that are critical in LBA, which could be extended to specific capture antibody used in the IC step for hybrid assay. For an IC reagent that is key element of a validated assay, its stability and maintenance procedures should be clearly established and routinely evaluated. Due to the high selectivity of the subsequent LC-MS/MS analysis, the specificity of capture reagent for hybrid assay may not be as critical as that for LBA. In such case, the demonstration of an acceptable accuracy and precision run may be sufficient to indicate reagent suitability.

During the early phase of drug discovery and development (e.g. candidate selection, go vs. no-go decisions), a scientifically sound assay with limited evaluations may be sufficient to provide the meaningful answer. However, at the development phase, most bioanalytical assays are required to be validated for regulatory filing and in support of clinical studies. AlthoughIC-LC-MS/MS approach has been increasingly applied in quantitative bioanalysis of large molecules for years, it is still at a relatively early stage. On the FDA's Guidance to the Industry on Bioanalytical Method Validation (US FDA 2013), it states that FDA encourages the development and use of new bioanalytical technologies, but the validity should be supported by data and scientific evidences. FDA also suggests early dialog with the regulators when new technologies are used.

7.9 Summary

IC in combination with LC-MS/MS has proven to be highly successful for the quantitative bioanalysis of protein or peptide analytes. As the IC technique continues to evolve, a growing number of applications are expected to emerge. The technique can utilize a variety of capture reagents to selectively bind the analyte of interest from biological matrix, enabling wash away of the unbound matrix interference. Typically a clean sample can be achieved by sample preparation with IC, which is well suited for micro-flow or nano-flow LC-MS/MS, and hence, an increase in assay sensitivity can be achieved. IC as a sample preparation approach plays a vital role in the continued evolution of highly specific and highly sensitive LC-MS bioanalysis.

References

Ackermann, B., Neubert, H., Hughes, N. et al. (2015). 2015 White Paper on recent issues in bioanalysis: focus on new technologies and biomarkers (part 2 – hybrid LBA/LCMS and input from regulatory agencies). *Bioanalysis* 7: 3019–3034.

An, B., Zhang, M., and Qu, J. (2014). Toward sensitive and accurate analysis of antibody biotherapeutics by liquid chromatography coupled with mass spectrometry. *Drug Metab. Dispos.* 42: 1858–1866.

Anderson, N.L., Anderson, N.G., Haines, L.R. et al. (2004). Mass spectrometric quantitation of peptides and proteins using stable isotope standards and capture by anti-peptide antibodies (SISCAPA). *J. Proteome Res.* 3: 235–244.

Anderson, N.L., Jackson, A., Smith, D. et al. (2009). SISCAPA peptide enrichment on magnetic beads using an in-line bead trap device. *Mol. Cell. Proteomics* 8: 995–1005.

Becher, F., Dubois, M., Fenaille, F., and Ezan, E. (2013). Mass spectrometry protocol for the absolute quantification of a monoclonal antibody in serum with immunopurification. *Methods Mol. Biol.* 988: 345–352.

Bronsema, K.J., Bischoff, R., and van de Merbel, N.C. (2012). Internal standards in the quantitative determination of protein biopharmaceuticals using liquid chromatography coupled to mass spectrometry. *J. Chromatogr. B Analyt. Technol. Biomed. Life Sci.* 893–894: 1–14.

Bronsema, K.J., Bischoff, R., and van de Merbel, N.C. (2013). High-sensitivity LC-MS/MS quantification of peptides and proteins in complex biological samples: the impact of enzymatic digestion and internal standard selection on method performance. *Anal. Chem.* 85: 9528–9535.

Chappell, D.L., Lassman, M.E., McAvoy, T. et al. (2014a). Quantitation of human peptides and proteins via MS: review of analytically validated assays. *Bioanalysis* 6: 1843–1857.

Chappell, D.L., Lee, A.Y., Bernstein, H.S. et al. (2016). Development and validation of an IA-LC/MS method to quantitate active and total B-type natriuretic peptide in human plasma. *Bioanalysis* 8: 2341–2349.

Chappell, D.L., Lee, A.Y., Castro-Perez, J. et al. (2014b). An ultrasensitive method for the quantitation of active and inactive GLP-1 in human plasma via immunoaffinity LC-MS/MS. *Bioanalysis* 6: 33–42.

Chen, L.Z., Roos, D., and Philip, E. (2016). Development of immunocapture-LC/MS assay for simultaneous ADA isotyping and semiquantitation. *J. Immunol. Res.* 4: 1–14.

Dufield, D., Neubert, H., Garofolo, F. et al. (2014). 2014 White Paper on recent issues in bioanalysis: a full immersion in bioanalysis (part 2 – hybrid LBA/LCMS, ELN & regulatory agencies' input). *Bioanalysis* 6: 3237–3249.

Eliasson, M., Andersson, R., Olsson, A. et al. (1989). Differential IgG binding characteristics of staphylococcal protein A, streptococcal protein G, and a chimeric protein AG. *J. Biol. Chem.* 142: 575–581.

Ewles, M. and Goodwin, L. (2011). Bioanalytical approaches to analyzing peptides and proteins by LC-MS/MS. *Bioanalysis* 3: 1379–1397.

Ewles, M., Mannu, R., Fox, C. et al. (2016). LC-MS/MS strategies for therapeutic antibodies and investigation into the quantitative impact of antidrug-antibodies. *Bioanalysis* 8: 2565–2579.

Fernandez Ocana, M., James, I.T., Kabir, M. et al. (2012). Clinical pharmacokinetic assessment of an anti-MAdCAM monoclonal antibody therapeutic by LC-MS/MS. *Anal. Chem.* 84: 5959–5967.

Gillette, M.A. and Carr, S.A. (2013). Quantitative analysis of peptides and proteins in biomedicine by targeted mass spectrometry. *Nat. Methods* 10: 28–34.

Holst, J.J. (2007). The physiology of glucagon-like peptide 1. *Physiol. Rev.* 87: 1409–1439.

Jenkins, R., Duggan, J.X., Aubry, A.F. et al. (2015). Recommendations for validation of LC-MS/MS bioanalytical methods for protein biotherapeutics. *AAPS J.* 17: 1–16.

Jiang, H., Xu, W., Titsch, C.A. et al. (2014). Innovative use of LC-MS/MS for simultaneous quantitation of neutralizing antibody, residual drug, and human immunoglobulin G in immunogenicity assay development. *Anal. Chem.* 86: 2673–2680.

Jones, B.R. and Schultz, G.A. (2016). Adaptation of hybrid immunoaffinity LC-MS methods for protein bioanalysis in a contract research organization. *Bioanalysis* 8: 1545–1549.

Kim, H. and Hage, D.S. (2006). Immobilization methods for affinity chromatography. In: *Handbook of Affinity Chromatography* (ed. D.S. Hage). New York: Taylor & Francis.

Kuhn, E., Addona, T., Keshishian, H. et al. (2009). Developing multiplexed assays for troponin I and interleukin-33 in plasma by peptide immunoaffinity enrichment and targeted mass spectrometry. *Clin. Chem.* 55: 1108–1117.

Kuhn, E. and Carr, S.A. (2016). Multiplexed immunoaffinity enrichment of peptides with anti-peptide antibodies and quantification by stable isotope dilution multiple reaction monitoring mass spectrometry. *Methods Mol. Biol.* 1410: 135–167.

Kushnir, M.M., Rockwood, A.L., Straseski, J.A., and Meikle, A.W. (2014). Comparison of LC-MS/MS to immunoassay for measurement of thyroglobulin in fine-needle aspiration samples. *Clin. Chem.* 60: 1452–1453.

Lee, A.Y., Chappell, D.L., Bak, M.J. et al. (2016). Multiplexed quantification of proglucagon-derived peptides by immunoaffinity enrichment and tandem mass spectrometry after a meal tolerance test. *Clin. Chem.* 62: 227–235.

Li, H., Ortiz, R., Tran, L. et al. (2012). General LC-MS/MS method approach to quantify therapeutic monoclonal antibodies using a common whole antibody internal standard with application to preclinical studies. *Anal. Chem.* 84: 1267–1273.

Lin, D., Alborn, W.E., Slebos, R.J., and Liebler, D.C. (2013). Comparison of protein immunoprecipitation-multiple reaction monitoring with ELISA for assay of biomarker candidates in plasma. *J. Proteome Res.* 12: 5996–6003.

Liu, A., Kozhich, A., Passmore, D. et al. (2015). Quantitative bioanalysis of antibody-conjugated payload in monkey plasma using a hybrid immuno-capture LC-MS/MS approach: assay development, validation, and a case study. *J. Chromatogr. B Analyt. Technol. Biomed. Life Sci.* 1002: 54–62.

McAvoy, T., Lassman, M.E., Spellman, D.S. et al. (2014). Quantification of tau in cerebrospinal fluid by immunoaffinity enrichment and tandem mass spectrometry. *Clin. Chem.* 60: 683–689.

Moser, A.C. and Hage, D.S. (2010). Immunoaffinity chromatography: an introduction to applications and recent developments. *Bioanalysis* 2: 769–790.

Neubert, H., Fan, Y.Y., and Fernandez Ocana, M. (2016). Quantification of protein biomarkers in tissues: new capabilities with pellet digestion peptide immunoaffinity LC-MS/MS. *Bioanalysis* 8: 1551–1555.

Neubert, H., Gale, J., and Muirhead, D. (2010). Online high-flow peptide immunoaffinity enrichment and nanoflow LC-MS/MS: assay development for total salivary pepsin/pepsinogen. *Clin. Chem.* 56: 1413–1423.

Neubert, H., Grace, C., Rumpel, K., and James, I. (2008). Assessing immunogenicity in the presence of excess protein therapeutic using immunoprecipitation and quantitative mass spectrometry. *Anal. Chem.* 80: 6907–6914.

Neubert, H., Muirhead, D., Kabir, M. et al. (2013). Sequential protein and peptide immunoaffinity capture for mass spectrometry-based quantification of total human beta-nerve growth factor. *Anal. Chem.* 85: 1719–1726.

Palandra, J., Finelli, A., Zhu, M. et al. (2013). Highly specific and sensitive measurements of human and monkey interleukin 21 using sequential protein and tryptic peptide immunoaffinity LC-MS/MS. *Anal. Chem.* 85: 5522–5529.

Schoenherr, R.M., Whiteaker, J.R., Zhao, L. et al. (2012). Multiplexed quantification of estrogen receptor and HER2/Neu in tissue and cell lysates by peptide immunoaffinity enrichment mass spectrometry. *Proteomics* 12: 1253–1260.

Schoenherr, R.M., Zhao, L., Whiteaker, J.R. et al. (2010). Automated screening of monoclonal antibodies for SISCAPA assays using a magnetic bead processor and liquid chromatography-selected reaction monitoring-mass spectrometry. *J. Immunol. Methods* 353: 49–61.

Schultz, G.A., McCardle, K., and Neubert, H. (2016). Large-scale implementation of sequential protein and peptide immunoaffinity enrichment LC/nanoLC-MS/MS for human beta-nerve growth factor. *Bioanalysis* 8: 753–764.

Torsetnes, S.B., Levernaes, M.S., Broughton, M.N. et al. (2014). Multiplexing determination of small cell lung cancer biomarkers and their isovariants in serum by immunocapture LC-MS/MS. *Anal. Chem.* 86: 6983–6992.

Torsetnes, S.B., Lovbak, S.G., Claus, C. et al. (2013). Immunocapture and LC-MS/MS for selective quantification and differentiation of the isozymes of the biomarker neuron-specific enolase in serum. *J. Chromatogr. B Analyt. Technol. Biomed. Life Sci.* 929: 125–132.

US FDA (2013). *Draft Guidance to the Industry on Bioanalytical Method Validation* (May 2013). https://www.fda.gov/downloads/drugs/guidances/ucm070107.Pdf (accessed July 2017).

Wang, W., Choi, B.K., Li, W. et al. (2012a). Quantification of intact and truncated stromal cell-derived factor-1α in circulation by immunoaffinity enrichment and tandem mass spectrometry. *J. Am. Soc. Mass Spectrom.* 25: 614–625.

Wang, J., Gu, H., Liu, A. et al. (2016). Antibody-drug conjugate bioanalysis using LB-LC-MS/MS hybrid assays: strategies, methodology and correlation to ligand-binding assays. *Bioanalysis* 8: 1383–1401.

Wang, Y. and Heilig, J.S. (2012). Differentiation and quantification of endogenous and recombinant-methionyl human leptin in clinical plasma samples by immunocapture/mass spectrometry. *J. Pharm. Biomed. Anal.* 70: 440–446.

Wang, W., Walker, N.D., Zhu, L.J. et al. (2012b). Quantification of circulating D-dimer by peptide immunoaffinity enrichment and tandem mass spectrometry. *Anal. Chem.* 84: 6891–6898.

Whiteaker, J.R., Zhao, L., Anderson, L., and Paulovich, A.G. (2010). An automated and multiplexed method for high throughput peptide immunoaffinity enrichment and multiple reaction monitoring mass spectrometry-based quantification of protein biomarkers. *Mol. Cell. Proteomics* 9: 184–196.

Wolf, R., Hoffmann, T., Rosche, F., and Demuth, H.U. (2004). Simultaneous determination of incretin hormones and their truncated forms from human plasma by immunoprecipitation and liquid chromatography-mass spectrometry. *J. Chromatogr. B Analyt. Technol. Biomed. Life Sci.* 803: 91–99.

Xu, Y., Mehl, J.T., Bakhtiar, R., and Woolf, E.J. (2010). Immunoaffinity purification using anti-PEG antibody followed by two-dimensional liquid chromatography/tandem mass spectrometry for the quantification of a PEGylated therapeutic peptide in human plasma. *Anal. Chem.* 82: 6877–6886.

Yang, W., Kernstock, R., Simmons, N., and Alak, A. (2015). ELISA microplate: a viable immunocapture platform over magnetic beads for immunoaffinity-LC-MS/MS quantitation of protein therapeutics? *Bioanalysis* 7: 307–318.

8

Microextraction Techniques in LC-MS Bioanalysis

Márcio Rodrigues, Ana Fortuna, Amílcar Falcão, and Gilberto Alves

8.1 Introduction

Accurate, precise, and simultaneously fast, economical, and environmental-friendly procedures to extract and quantify drugs and/or their metabolites in biological samples is undeniably important for understanding the therapeutic and toxic effects of drug molecules and development of more selective and effective therapeutics (Kataoka et al. 2016). Knowing the temporal changes in drug and/or its metabolite(s) concentrations in biological samples is required in drug discovery and development programs, as well in routine therapeutic drug monitoring in order to select and optimize drug candidates, investigate patient compliance, assess the influence of co-medication, and so on (Fu et al. 2005; Musteata and Pawliszyn 2007).

Due to the high sensitivity and selectivity, liquid chromatography (LC) coupled with mass spectrometry (LC-MS) or tandem mass spectrometry (LC-MS/MS) are analytical methods of choice often employed to quantify/identify drugs and their metabolites in biological samples. However, despite the advances in the development of analytical instrumentation for the determination of analytes in biological fluids, most of them cannot deal with the sample matrices directly because the various acids, bases, proteins, salts, and other organic or inorganic compounds in the biological samples may cause matrix effect and interfere with the detection of the targeted analytes. Therefore, sample preparation to extract, isolate, and concentrate the analytes of interest becomes a crucial part of the bioanalysis procedure (Fu et al. 2005; Kataoka 2010; Musteata and Pawliszyn 2007). In addition to the complexity of matrices, limited sample volumes and low analyte concentrations also present challenges to bioanalysis (Ocaña-González et al. 2016). An optimized sample preparation procedure is required to effectively remove endogenous components, reduce matrix effects, while providing good recovery with high-throughput.

The most commonly used sample preparation procedures include protein precipitation (PPT), liquid–liquid extraction (LLE), and solid-phase extraction (SPE). With PPT, all small endogenous molecules remain in the processed samples (no selectivity) and those molecules may interfere with the MS ionization process and decrease the performance of the LC column, etc. PPT is often associated with high matrix effect, which, if not addressed properly, may significantly impact the data quality. Compared with PPT, LLE is a more selective sample cleanup technique. However, LLE has its own disadvantages: it is not suitable for hydrophilic compounds, and method development can be challenging when trying to extract multiple analytes with different lipophilicity. In general, LLE requires large amounts of hazardous organic solvents (Ocaña-González et al. 2016; Psillakis and Kalogerakis 2003). Due to the availability of different types of packing material, SPE is suitable for a wide variety of compounds in various sample matrices. When optimized, it can achieve both high recovery and good selectivity of an analyte with very good reproducibility. However, SPE is considered to be very expensive when used to analyze a large set of samples. To overcome or reduce the above drawbacks of LLE and SPE, their corresponding miniaturized extraction procedures, liquid-phase microextraction (LPME) and solid-phase microextraction (SPME) respectively, have been developed as alternative sample preparation techniques for drug analysis (Farhadi et al. 2012). The microextraction techniques are defined as the extraction techniques in which the volume of the extracting phase is very small in relation to the sample volume under defined extraction conditions (Lord and Pawliszyn 2000). To date, a number of different microextraction techniques have been developed and their application to a variety of analytical fields is continuously increasing. In general, the aim of developing and employing these techniques in bioanalysis is to improve compatibility of sample extraction with modern analytical instrumentation, to avoid the use of

Sample Preparation in LC-MS Bioanalysis, First Edition. Edited by Wenkui Li, Wenying Jian, and Yunlin Fu.

hazardous chemicals, and to be capable of dealing with lower sample volume and solvent consumption. The advances achieved in this field are discussed in the present chapter with a particular emphasis on the various SPME and LPME techniques (Figure 8.1) that have been widely applied in LC-MS bioanalysis.

8.2 Solid-Phase Microextraction

SPME was introduced in 1989 by Arthur and Pawliszyn in an attempt to reduce the operation time and reagent consumption associated with the traditional SPE methods (Arthur and Pawliszyn 1990; Płotka-Wasylka et al. 2015; Vas and Vékey 2004). SPME is currently roughly divided into two types, static in-vessel microextraction (sample stir microextraction) and dynamic in-flow microextraction (sample flow microextraction) (Figure 8.1). The mostly commonly used formats of SPME are fibers (fiber SPME) and capillary tubes (in-tube SPME) coated with an appropriate stationary phase (Kataoka 2011).

SPME was first introduced in rod fiber geometry (fiber SPME) and it was invented to extract organic compounds in combination with gas chromatography (GC) analysis (Mirnaghi et al. 2013). Indeed, the most widely used technique of SPME is fiber SPME, in which a fiber-coated polymeric stationary phase is used as an extraction device, and the absorption or adsorption of analytes occurs on the outer surface of the fiber. Fiber SPME has been specially used to extract volatile and semi-volatile organic compounds from samples (Risticevic et al. 2009). It has also been coupled directly with LC to analyze weakly volatile or thermally labile compounds not amenable to GC analysis (Nerín et al. 2009; Risticevic et al. 2009; Vas and Vékey 2004). SPME/LC interface equipped with a special desorption chamber or solvent desorption prior to LC analysis can be used instead of thermal desorption in the injection port of the GC system (Kataoka and Saito 2011; Vas and Vékey 2004).

On the other hand, in-tube SPME was primarily developed to extend SPME to high-throughput applications and automated analytical systems, in which analytes are absorbed or adsorbed onto the inner surface of a capillary tube (Kataoka and Saito 2011). Since 1999, several other developments have been achieved in SPME, such as the emerging new alternative microextraction techniques using stirring bars (stir-bar sorptive extraction, which is addressed in detail in Chapter 13), thin films (thin-film microextraction), microsyringes (in-needle SPME), and pipette tips (in-tip SPME) (Figure 8.1) (Farhadi et al. 2012; Kataoka 2010).

8.2.1 Conventional Fiber SPME

The fiber SPME is a sample preparation technique in which fused-silica fibers coated on the surface of an appropriate stationary phase are employed. More specifically, two types of fiber SPME techniques have been used to extract analytes: headspace (HS) for the analysis of highly volatile compounds and direct immersion (DI) for the analysis of compounds with lower volatility (Figure 8.2). The fibers can be coated with several types of commercially available polymeric coatings, such as polydimethylsiloxane (PDMS), divinylbenzene, carboxen, polyethylene glycol, and carbowax (Carasek and Merib 2015; Kataoka and Saito 2011; Płotka-Wasylka

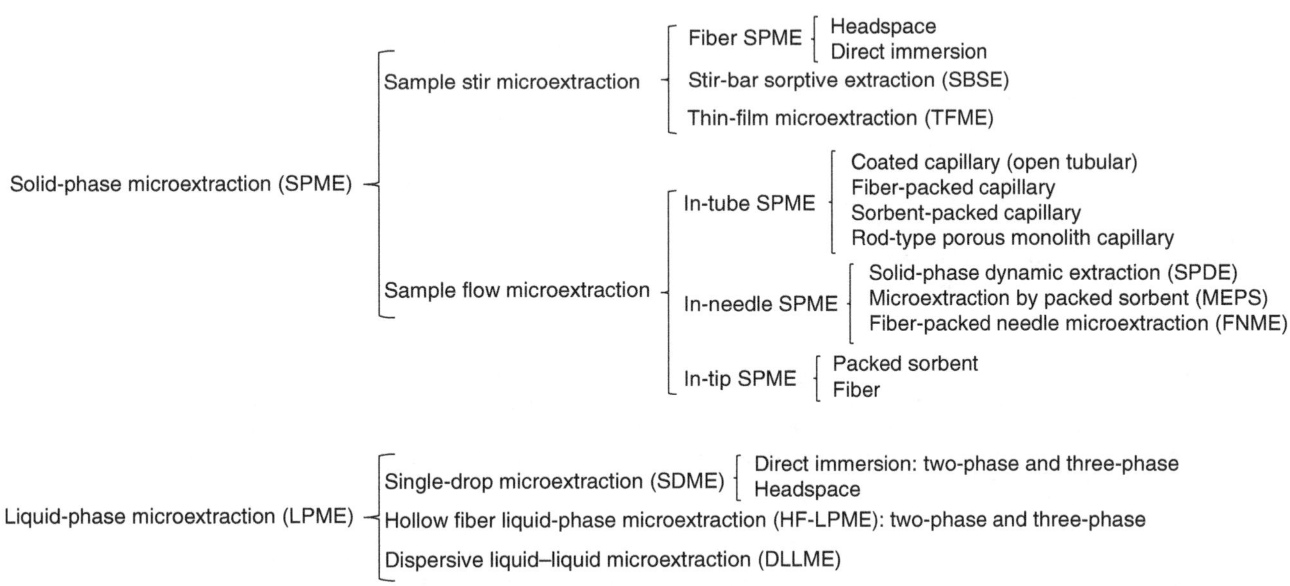

Figure 8.1 Classification of the main microextraction techniques.

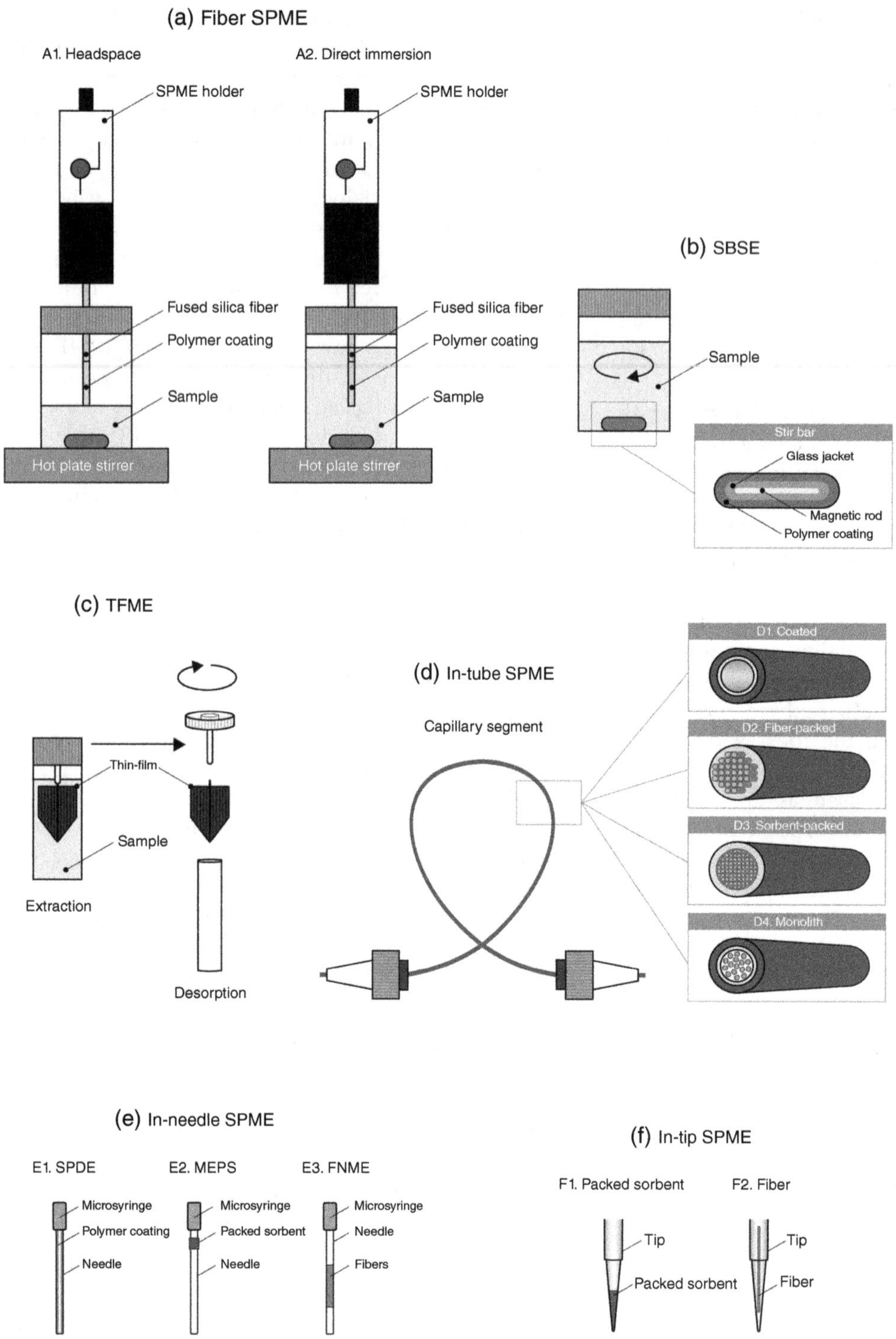

Figure 8.2 Main principles of solid-phase microextraction (SPME) techniques: (a) fiber SPME (A1. headspace and A2. direct immersion), (b) stir-bar sorptive extraction (SBSE), (c) thin-film microextraction (TFME), (d) in-tube SPME (D1. coated capillary, D2. fiber-packed capillary, D3. Sorbent-packed capillary, and D4. rod-type porous monolith capillary), (e) in-needle SPME (E1. solid-phase dynamic extraction [SPDE], E2. microextraction by packed sorbent [MEPS], E3. fiber-packed needle microextraction [FNME]), and (f) in-tip SPME (F1. packed sorbent and F2. fiber).

et al. 2015). Often, in order to extract a mixture of compounds with different polarities, these polymeric coatings are used in combination. Emerging PDMS/divinylbenzene, PDMS/carboxen, and carbowax/divinylbenzene are the most commonly used combinations (Płotka-Wasylka et al. 2015).

The general step-by-step protocol of a fiber SPME procedure is described below. Firstly, the sample is placed into a vial, which is sealed with a septum type cap. Then, the SPME needle pierces the septum and the fiber is extended through the needle into the sample for analyte extraction. Next, the sample is stirred with a small stirring bar to increase the rate of extraction and shorten the equilibration time. After an appropriate extraction time, the fiber is withdrawn into the needle which is removed from the septum. Afterwards, the analyte is desorbed from the fiber coating by heating the fiber in the injection port of a GC system, or by loading solvent into the desorption chamber at the SPME/LC interface. The extracted analyte can be transferred directly to the chromatographic column for analysis. Recently, the online desorption has been employed, with an LC interface composed of a six-port injection valve and a special desorption chamber that enables analyte desorption with a specific solvent or mobile phase prior to LC-MS analysis. During sample extraction, the desorption chamber is placed in the "injection" mode and the mobile phase comes into contact with the fiber, desorbs the analytes, and delivers them directly to the chromatographic column. In contrast, the analyte can be desorbed from the fiber coating by off-line solvent desorption and then this solvent solution with the desorbed analytes is injected into a GC or LC system (Kataoka and Saito 2011; Mirnaghi et al. 2013).

Before analyzing any sample, the fibers should be cleaned in order to remove interfering compounds that may give a high background noise in the analysis. Fibers can be cleaned by inserting them into a heated auxiliary injection port or into a syringe needle immersed in a flow of clean gas. When the SPME fiber is extended into the sample, the partition of the target analytes from the sample matrix into the extractive stationary phase occurs. Although SPME has a maximum sensitivity at the partition equilibrium, there is a proportional relationship between the amount of analyte adsorbed by the SPME fiber and its initial concentration in the sample even before partition equilibrium is reached. Therefore, full equilibration is not required for quantitative analysis by SPME (Kataoka and Saito 2011).

This microextraction technique has been used for many applications in different analytical fields, including environmental, food analysis, and bioanalysis. In the bioanalytical field, for example, a fiber SPME method has been developed to determine tranexamic acid in plasma with a good recovery (Bojko et al. 2011). However, in other methods the recovery of the analytes was shown to be low. The fiber SPME technique is characterized by several advantages: it is a rapid, simple, and solvent-free technique compatible with LC apparatus; it could be used to *in situ* and *in vivo* sampling and for polar and nonpolar analytes in a wide range of matrices; and it is easily automated (Kataoka and Saito 2011; Spietelun et al. 2013). However, the fiber SPME is also characterized by some disadvantages, including the batch-to-batch variation of fiber coatings, the time-consuming preparation of fiber coating, the limited selections of stationary phases, the lack of robustness of fiber coatings, and its low recovery (Płotka-Wasylka et al. 2015; Spietelun et al. 2013; Vas and Vékey 2004). Nevertheless, in recent years, a series of modifications to the original technique have been proposed in order to improve the quality of the results and to adapt fiber SPME procedures to new and more challenging applications, for instance, by using ionic liquids, carbon nanotubes (CNTs), graphene, carbon, and silica materials as fiber coating.

Several experimental conditions can influence the extraction efficiency in fiber SPME, namely agitation, pH and ionic strength of the sample, extraction time, and temperature. Sample agitation accelerates the transfer of analytes from the matrix to the coating, which is important to ensure a fast and efficient extraction of analytes. There are different agitation methods that can be used, such as fast sample flow, rapid fiber movement, stirring, and sonication (Risticevic et al. 2010; Theodoridis et al. 2000).

The sample pH is also an important parameter because it is preferable to have analytes in neutral forms for maximum extraction (Carasek and Merib 2015; Risticevic et al. 2010; Theodoridis et al. 2000). The adjustment of sample pH can be done by adding buffer to the sample, promoting the conversion of the analytes to the neutral forms. To improve the extraction efficiency of basic and acidic compounds, it is necessary to adjust the sample pH to high or low pH values, respectively. On the other hand, the optimum pH must be empirically determined for the extraction of analytes that possesses both acidic and basic functionalities (Risticevic et al. 2010). It should also be emphasized that the determination of the optimum sample pH should be evaluated within the stability ranges of the coatings. For instance, an extreme pH value should only be used in HS-SPME mode due to the potential fiber degradation when DI-SPME is employed (Risticevic et al. 2010; Theodoridis et al. 2000).

The addition of salts to the sample solution leads to an increase in the ionic strength and a decrease in the aqueous solubility of the analytes of interest, a process known as salting out that facilitates the transfer of analytes to the coating of SPME (Carasek and Merib 2015). For a specific target analyte and sample matrix, experiments should be performed to evaluate the effect of salt addition on the assay performance since this strategy can increase the extraction of not only the target analytes but also the interfering compounds (Risticevic et al. 2010). The salts commonly used are sodium chloride, sodium sulphate, sodium carbonate, potassium carbonate, and ammonium sulphate (Risticevic et al. 2010; Theodoridis et al. 2000). Generally, the addition of salts is preferred for HS-SPME because fiber coating of DI-SPME is prone to deteriorate during agitation with salts (Pereira et al. 2014a).

The extraction time is considered the time-limiting step and is one of the most crucial steps during fiber DI-SPME and HS-SPME methods development. For HS-SPME, the time period during which the sample is incubated before coating (incubation time) should also be considered. The incubation time is important as it allows the equilibration of sample components and the transfer of analytes to the HS above the sample, being critical for HS-SPME procedures. The optimum extraction time is always a compromise between the length of time, sensitivity, and reproducibility of the analytical method (Risticevic et al. 2010; Theodoridis et al. 2000).

Extraction temperature is another determinant factor that should be considered during fiber DI-SPME and HS-SPME method development. Increasing the temperature is associated with an increased diffusion coefficient of the analytes and/or increase the HS capacity, which leads to an enhancement in the extraction rate or the mass transfer rate onto the fiber coating. This enables shorter equilibration times and enables faster SPME procedures. However, as the temperature increases, the fiber coating loses its ability to adsorb analytes and, consequently, the analyte recovery at equilibrium conditions decreases. Furthermore, it is worth noting that extremely high sample temperatures may result in decomposition of some compounds of interest and formation of other artifacts (Risticevic et al. 2010; Theodoridis et al. 2000).

8.2.2 Stir-Bar Sorptive Extraction

Stir-bar sorptive extraction (SBSE) was introduced by Baltussen et al. (1999) in order to overcome some of the aforementioned limitations of fiber SPME. In this SPME microextraction technique, the organic compounds are enriched from aqueous samples by sorption onto a thick film of PDMS coated onto glass-enveloped magnetic bar (Figure 8.2). The stirring during a predetermined period of time enables the extraction of the organic compounds. Then, the magnetic bar is removed from the sample, dried, and introduced into a thermal desorption unit. Alternatively, a solvent desorption system can be used. Finally, the recovered compounds could be analyzed by LC systems (Sánchez-Rojas et al. 2009). This topic is extensively discussed in Chapter 13.

8.2.3 Thin-Film Microextraction

Similar to fiber SPME and SBSE, thin-film microextraction (TFME) is also a sample stir SPME technique. TFME is based on the use of a flat film sheet as the extraction phase (Saraji and Keykavooci 2016). The sorbents used in TFME have a larger volume of extraction phase and a higher surface-to-volume ratio without increased thickness and, thus, exhibit improved extraction efficiency and loadability as compared to the traditional SPME fibers (Duan et al. 2011; Mirnaghi et al. 2013). The main advantages of TFME are its high selectivity and high extraction recovery, and its applicability to large sample volumes. The improved extraction selectivity results in a high degree of cleanup of interferences from biological samples, minimizing the matrix effect (Mirnaghi et al. 2013).

The protocols of TFME encompass, in general, three steps. Firstly, the thin film, typically PDMS or its derivatives, is coated with a liquid organic solvent (the acceptor phase). The next step is to submerge the thin film into the sample for the extraction process. Finally, the extracted analytes are desorbed from the acceptor phase for analysis (Figure 8.2). The desorption methods include thermal desorption for GC analysis or solvent desorption through redissolution into an organic solvent for LC analysis (Ocaña-González et al. 2016). The TFME methodology has been applied for the quantification of several endogenous and exogenous compounds by LC-MS/MS. Examples include bile acids from bronchoalveolar lavage (Bessonneau et al. 2014), illicit substances (cannabinoids, steroids, narcotics, β-blockers, stimulants, β_2-agonists, glucocorticoids, hormones, and other anabolic compounds) from human saliva (Bessonneau et al. 2015a), nonconjugated and deconjugated testosterone, and epitestosterone from human urine (Zhan et al. 2011), eicosanoids from rat plasma (Bessonneau et al. 2015b), and diazepam, oxazepam, caffeine, riboflavin, and sucrose from human plasma (Mirnaghi and Pawliszyn 2012).

8.2.4 In-Tube SPME

In-tube SPME is an efficient sample preparation technique using fused-silica capillary column of either an inner surface coating or a sorbent bed. This extraction phase can be coupled online with LC and it is suitable for automation. It overcomes fiber-related drawbacks such as fragility, low sorption capacity, and bleeding from thick-film coatings. In-tube SPME typically uses an inner wall-coated fused-silica open tubular capillary but other types have also been developed, namely fiber-packed, sorbent-packed, and rod-type porous monolith capillaries (Figure 8.2). In in-tube SPME the analytes are directly extracted from the sample into the fused-silica capillary column. Then, the analytes could be desorbed by the flow of mobile phase and could be directly injected into the LC apparatus. It should be highlighted that the extraction efficiency is in general low but reproducible. The main factors that influence the extraction of analytes are the polarity of the capillary coating, the number and volume of draw/eject cycles, and the pH of the sample. Similar to the other microextraction techniques, several commercially available capillary columns are currently available (Kataoka 2002, 2010; Kataoka et al. 2016). Some analytical methods employing in-tube SPME have been developed to determine compounds present in biological samples (Table 8.1).

8.2.5 In-Needle SPME

8.2.5.1 Solid-Phase Dynamic Extraction

Solid-phase dynamic extraction (SPDE) is an in-needle technique for gas and liquid sampling (Kataoka 2010). In SPDE, the analytes are concentrated onto PDMS and active carbon on the inside of a stainless steel needle of a HS syringe (Figure 8.2) (Jochmann et al. 2006; Kataoka 2010). During the extraction, the needle can be directly immersed into the sample or in the HS of sample and then the syringe plunger is moved up and down several times in order to increase the analytes sorption in the coating. Finally, the trapped analytes are recovered by thermal desorption from the coating directly in the GC injector (Jochmann et al. 2006; Kataoka 2010, 2011). The main advantages of SPDE compared with SPME are the larger coating volume employed and the shorter extraction time required. However, the carryover is a drawback that should be evaluated because the analytes tend to remain on the inner wall of the needle after heat desorption in the GC injection port (Kataoka 2010).

8.2.5.2 Microextraction by Packed Sorbent

Microextraction by packed sorbent (MEPS) is a miniaturized version of conventional SPE in a more efficient and sophisticated format. This technique was originally developed by Abdel-Rehim in 2003 (Abdel-Rehim 2004).

The MEPS involves enclosing few milligrams (1–4 mg) of a solid sorbent material into a liquid handling gas-tight syringe (100–250 μl) as a plug or as a cartridge (Alves et al. 2013; Moein et al. 2015). While in the conventional SPE technique the packing material is incorporated into a separate column/cartridge, in the MEPS technique the packing material is coupled to the syringe (Figure 8.2). Most commonly used stationary phases in SPE are also commercially available in MEPS, including silica, ethyl silica (C2), octyl silica (C8), octadecyl silica (C18), mixed-mode, and strong anion exchange (SAX). In both techniques (SPE and MEPS), analyte extraction occurs on the packed solid sorbent and the underlying principles are similar. However, MEPS requires shorter time of extraction and a reduced sample volume than conventional SPE. In addition, as the solid-phase sorbent of MEPS is contained within injection syringes, the technique is amenable to online sample preparation. Nevertheless, the main advantage of MEPS is that the sorbent is frequently reusable and, therefore, the costs of the analysis per sample can be pronouncedly reduced in comparison with conventional SPE (Płotka-Wasylka et al. 2015).

The main steps involved in MEPS are sorbent conditioning, sample loading and retention, washing, elution, and cleaning/reconditioning to minimize the carryover and to reuse the device. The sorbent is conditioned by using an appropriate organic solvent such as methanol or acetonitrile and then the excess of organic phase is removed by passing a more polar solvent through the solid sorbent (e.g. water). After this conditioning step, the sample is loaded through the sorbent but, before that, a sample pretreatment step is usually required to avoid clogging of the sorbent (e.g. PPT, centrifugation, and dilution). When the sample is applied onto the MEPS sorbent (sample loading), it is critical to achieve a good adsorption of the analytes by the solid sorbent in order to attain a good recovery. Hence, the passage of the sample through the sorbent should be slowly performed (~5–20 $\mu l\, s^{-1}$). In addition, the number of MEPS loading cycles is also important for increasing the adsorption of the analytes, as the multiple cycles of pulling/pushing of the sample enable concentration of the analytes on the sorbent. Specifically, the pH and the ionic strength of the samples may also affect the MEPS performance. Following sample loading, the washing step is applied to increase the selectivity by removing endogenous compounds that may interfere with the analysis of the analytes. Pure water, 0.1% formic acid, water/methanol (95 : 5, v/v), or water/methanol (90 : 10, v/v) are frequently used in this step. The composition and the volume of washing solutions need to be adequately optimized to reduce the leakage of the adsorbed analytes. The elution step is performed to release the analytes from the sorbent. The elution solutions could be constituted by an organic solvent such as methanol or acetonitrile. In comparison to conventional SPE, MEPS usually requires much smaller sample volume, which is ideal to be integrated into online sample preparation of various analysis workflows. Lastly, a key factor that makes MEPS more advantageous in comparison to conventional SPE is the possibility of reusing the MEPS

Table 8.1 Bioanalytical methods for analysis of drugs and/or metabolites and other compounds from biological matrices by liquid chromatography–(tandem) mass spectrometry applying microextraction as the sample preparation procedure.

Compound class/ compound	Matrix (sample volume)	Type of extraction	LLOQ	Recovery	Comments	References
Anabolic steroids						
Boldenone	Human urine (100 μl for unconjugated anabolic steroid drugs/300 μl for conjugated anabolic steroid drugs)	In-tube SPME	$0.5\,ng\,ml^{-1}$	93.1–93.6%	Supel-Q PLOT coating (divinylbenzene polymer)	Saito et al. (2010)
Nandrolone				85.7–87.0%		
Testosterone				96.1–99.3%		
Methyltestosterone				103.1–104.7%		
Epiandrosterone				98.9–112.0%		
Androsterone				92.2–106.4%		
Stanozolol				99.7–117.3%		
Antibiotics						
Linezolid	Human plasma (50 μl)	MEPS	$0.3814\,ng\,ml^{-1}$	>70%	Dilution with water before MEPS C18 sorbent MEPS used off-line	Szultka et al. (2013)
Amoxicillin			$0.4249\,ng\,ml^{-1}$			
Anticancer drugs						
Busulphan	Human plasma (50 μl)	MEPS	$5\,ng\,ml^{-1}$	~60%	Polystyrene sorbent MEPS used online ~100–150 samples/MEPS sorbent	Abdel-Rehim et al. (2007)
Antidepressants and antipsycothic						
Amitriptyline	Human plasma and urine (500 μl)	In-tube SPME	$0.28/0.23\,ng\,ml^{-1}$[a]	95.7/89.9%[b]	Hybrid silica monolith coating	Zheng et al. (2010)
Trazodone			$0.19/0.34\,ng\,ml^{-1}$[a]	58.1/108.8%[b]		
Citalopram			$0.49/0.19\,ng\,ml^{-1}$[a]	40.5/112.5%[b]		
Doxepin			$0.53/0.77\,ng\,ml^{-1}$[a]	61.1/109.1%[b]		
Paroxetine			$0.26/5.00\,ng\,ml^{-1}$[a]	67.6/97.0%[b]		
Fluvoxamine			$9.45/2.19\,ng\,ml^{-1}$[a]	111.2/125.3%[b]		
Fluoxetine			$0.33/0.30\,ng\,ml^{-1}$[a]	75.9/80.8%[b]		
Sertraline			$8.79/9.56\,ng\,ml^{-1}$[a]	94.7/71.0%[b]		
Clomipramine			$0.56/0.35\,ng\,ml^{-1}$[a]	97.7/75.8%[b]		
Clozapine			$8.66/9.83\,ng\,ml^{-1}$[a]	94.4/111.5%[b]		
Antidyslipidemics						
Pravastatin Pravastatin lactone	Rat plasma and urine (50 μl)	MEPS	5/5 nM[a]	97–109%/ 92–101%[b]	Plasma: PPT with acetonitrile before MEPS C8 sorbent MEPS used off-line ~30 samples/MEPS sorbent (plasma) ~60 samples/MEPS sorbent (urine)	Vlčková et al. (2012)

Antihypertensives and antithrombotics						
Aliskiren Prasugrel Rivaroxaban	Human urine (200 μl)	MEPS	0.005 ng ml^{-1} 0.0005 ng ml^{-1} 0.005 ng ml^{-1}	95.1–99.8% 98.6–101.8% 99.3–101.9%	pH adjustment to 2.5 with formic acid before MEPS C8 sorbent MEPS used off-line	Magiera (2013)
Acebutolol Metoprolol	Human plasma (500 μl)	MEPS	10 nM	86–108%	Dilution with milli-Q water before MEPS Carbon-XCOS sorbent MEPS used online	Abuzooda et al. (2015)
Aliskiren Enalapril Enalaprilat	Human plasma and urine (50 μl)	MEPS	0.01/0.01 ng ml^{-1}[a]	90.3–93.2/89.5–91.8%[b] 81.2–85.5/84–86.2%[b] 75.4–77.8/75.5–79.5%[b]	Plasma: PPT with acetonitrile before MEPS Urine: only pH adjustment before MEPS C8 sorbent MEPS used off-line	Magiera and Kusa (2015)
(*R*)-Metoprolol (*S*)-Metoprolol	Human plasma and saliva (100 μl)	MEPS	2.5 ng ml^{-1}	96.44–98.61%	Dilution with water before MEPS C18 sorbent MEPS used off-line	Elmongy et al. (2016)
Antimicrobial preservatives						
Methylparaben Ethylparaben Propylparaben Butylparaben	Human serum (1000 μl)	DLLME	0.6 ng ml^{-1} 0.2 ng ml^{-1} 0.7 ng ml^{-1} 0.6 ng ml^{-1}	96–102% 101–102% 98–103% 97–103%	PPT with acetone before DLLME Extraction solvent: chloroform Disperser: acetone	Vela-Soria et al. (2013)
Antivirals						
Atazanavir Amprenavir	Human plasma (100 μl)	In-tip SPME	2.77 ng ml^{-1} 4.50 ng ml^{-1}	~100%	PPT in a propylene 96-well plate + Online extraction with a short monolithic C18	Naxing Xu et al. (2006)
Local anesthetics						
Lidocaine Ropivacaine Bupivacaine	Human blood (25 μl)	MEPS	10 nM	~90%	C18 sorbent MEPS used online ~50 samples/MEPS sorbent	Said et al. (2010)
Lidocaine Ropivacaine Mepivacaine Bupivacaine	Plasma and urine (100 μl)	MEPS	5.0 nM	60–80%	Dilution with formic acid 0.1% before MEPS MIP sorbent MEPS used off-line ~100 samples/MEPS sorbent	Daryanavard et al. (2013)

(*Continued*)

Table 8.1 (Continued)

Compound class/ compound	Matrix (sample volume)	Type of extraction	LLOQ	Recovery	Comments	References
Opioid analgesics						
Naloxone Buprenorphine Norbuprenorphine	Human plasma (5000 µl)	HF-LPME	$0.1\,ng\,ml^{-1}$ $0.1\,ng\,ml^{-1}$ $0.05\,ng\,ml^{-1}$	92.1–106.0%	Polyvinylidene fluoride hollow fiber Organic solvent: 1-octanol/ chloroform/toluene (2/4/4)	Sun et al. (2014)
Several class drugs						
Amphetamine Phenylpropanolamine Cimetidine Morphine Sotalol Atenolol Practolol	Human plasma (500 µl)	HF-LPME	$25\,ng\,ml^{-1}$ (atenolol) $50\,ng\,ml^{-1}$ (cimetidine)	78% 65% 73% 59% 29% 16% 39%	Q3/2 Accurel KM polypropylene hollow fiber Acceptor phase: 0.05 M hydrochloric acid	Ho et al. (2005)

C8, octyl silica; C18, octadecyl silica; DLLME, dispersive liquid–liquid microextraction; HF-LPME, hollow fiber liquid-phase microextraction; In-tip SPME, In-tip solid-phase microextraction; In-tube SPME, In-tube solid-phase microextraction; LLOQ, lower limit of quantification; MEPS, microextraction by packed sorbent; MIP, molecular imprinted polymer; PPT, protein precipitation.

[a] LLOQ values in plasma and urine, respectively.

[b] Recovery values in plasma and urine, respectively.

devices. Due to this aspect, the carryover becomes one of the major concerns in the application of MEPS, particularly when this technique is coupled to more sensitive instrumentation such as MS. Thus, evaluation of carryover is mandatory during the development and optimization of MEPS protocols in order to establish the appropriate conditions that effectively clean the MEPS sorbent between each extraction. This step is important not only for cleaning the sorbent to eliminate memory effects but also acting as the conditioning step for the next extraction cycle (Alves et al. 2013; Pereira et al. 2014a, 2014b).

Many research papers have reported the use of MEPS for the extraction of a wide range of analytes from biological samples in different matrices such as plasma, blood, urine, and saliva. Several examples of LC-MS methods that employed MEPS as the sample extraction procedure are presented in Table 8.1.

8.2.5.3 Fiber-Packed Needle Microextraction

In contrast with MEPS, fiber-packed needle microextraction (FNME) uses fiber instead of particle materials (Figure 8.2). However, the main use of FNME has been with GC applications, and few publications can be found with other analytical techniques (Saito et al. 2009).

8.2.6 In-Tip SPME

In-tip SPME is a recently developed technique for sample preparation. The in-tip SPME constitutes a miniaturized format of SPE, in which a solid packing particle material is inserted into a pipette or alternatively a fiber-packed tip could be used (Figure 8.2) (Kataoka 2010).

In a 96-well in-tip SPME extraction system, the sorbents are situated in a polypropylene pipette tip and the analytes are extracted by repeated cycles of aspiration and desorption of the sample. The sorbent types used in pipette tips and the mechanism of extraction of the analytes are the same as those of the conventional SPE technique. The most common sorbents are those based on silica and monolith particles with large through-pores (Kataoka 2010; Namera et al. 2011).

In a typical protocol of in-tip pipette extraction, firstly the sorbent is preactivated and then the sample is passed through the pipette by repeated cycles of aspiration and desorption. After the extraction, the adsorbed analytes are eluted with a solvent and could be directly injected into a LC apparatus (Namera et al. 2011). These steps could be done in the off-line mode or could be automated using a robot (online) (Kataoka 2010). Table 8.1 summarizes an example of a method employing in-tip SPME as the sample extraction procedure.

8.2.7 New Sorbents for SPME

8.2.7.1 Monolithic Sorbent

Over the last 20 years, the monoliths are used as materials of various chromatographic columns (Namera and Saito 2013). However, it was not until recently that their application as sorbents in sample preparation has been extensively explored (Nema et al. 2011).

In analytical chemistry, monolith refers to a solid piece of stationary phase that is characterized with high permeability and that presents a binary porous structure consisting of uniform distribution of macropores and mesopores (Bunch and Wang 2013; Nema et al. 2014; Xu et al. 2011). The larger macropores enable solvents and proteins to pass through the column at an elevated flow rate, whereas smaller mesopores enable analytes' separation (Bunch and Wang 2013; Nema et al. 2014). These structures result in several advantages of monolith materials when compared to conventional particle materials. These advantages include but are not limited to high permeability for biological samples, no void volume, low backpressure, separation capabilities and chemical modifications similar to those of the particle-based technique, and mechanical robustness. The main disadvantages are the small surface area, narrow column dimensions, and limited commercial suppliers (Namera and Saito 2013).

Monoliths are broadly classified on the basis of the nature of materials used for preparation. Many types of monoliths exist but generally they are categorized into porous polymer and silica. All other types of monoliths are derived from the chemistry of these two types of monoliths, either with certain modifications or by using combination of monomers. To prepare porous polymer monoliths, the typically used monomers are acrylamides, methacrylates, and styrene, and they are polymerized *in situ* in a tube by mixing a suitable composition of monomers. In contrast, tetraethoxysilane and polyethylene glycol are used in the preparation of a silica monolith. The synthesis procedure for both types of monoliths is quite similar, with both the polymerization speed and phase separation playing an important role in the construction of a continuous monolith structure (Namera et al. 2011; Namera and Saito 2013).

Monoliths can be prepared *in situ* within narrow capillaries and applied to microscale separation without channeling in contrast to the most conventional materials used in SPE, in which the intrinsic problem is their inability to fill completely the available space. Consequently, the channeling between particles reduces the extraction efficiency in sorbent materials. In summary, monoliths can be packed or embedded in LC columns, SPE plates, and pipette tips. The mechanism of mass transfer for monolith in SPE is hydrophobic adsorption on the surface of the sorbent and partition into bonded alkyl chains, which is similar to that of LC. The physical properties such as surface area, average pore diameter, and pore volume of the adsorbents are factors that determine the transferring of analytes to the adsorption phase.

Monolithic materials offer a great potential for the extraction of analytes from complex matrices. Some variants of pipette-tip-mounted monoliths have been developed, whereas OMIXs, MonoTips, ZipTips, and NuTips are commercially available. The amount of sorbent that can be packed into OMIXs and MonoTips pipette tips is higher than that for ZipTips or NuTips pipette tips (Namera and Saito 2013). Consequently, a larger sample volume can be used to concentrate the analytes on the former type of tips. The monolith tips could be used for extraction of drugs in biological samples as it can be seen in Table 8.1.

8.2.7.2 Carbon Nanotubes

CNTs consist exclusively of carbon atoms arranged in a series of condensed benzene rings rolled up to form tube-like structures. The structures of CNTs are classified into two types based on the number of layers: single-walled carbon nanotubes and multi-walled carbon nanotubes (He et al. 2013; Jiménez-Soto et al. 2010; Liang et al. 2014). The use of CNTs in SPE and in SPME is related to their adsorption capabilities, such as their ability to establish non-covalent forces, hydrogen bonding, π–π stacking, electrostatic forces, van der Waals forces, and hydrophobic interactions (Ghaemi et al. 2014; Herrera-Herrera et al. 2012; Lasarte-Aragonés et al. 2011). The CNTs are also characterized by a large surface area, especially on the outside, interstitial spaces within nanotubes bundles, and excellent chemical, mechanical, and thermal stability (Herrera-Herrera et al. 2012; Lasarte-Aragonés et al. 2011). The CNTs constitute an attractive material for either nonpolar (in the case of non-functionalized CNTs) or polar compounds (in the case of functionalized CNTs) (Liang et al. 2014). In addition, due to their diversification regarding the retention mechanisms and characteristics, CNTs can be used in the extraction of inorganic and/or organic analytes (Herrera-Herrera et al. 2012; Liang et al. 2014). The CNTs have been immobilized on the SPME fiber by physical deposition technique or dipping technique. In fact, the CNTs are largely used as SPME fiber coatings in different fields of application such as environmental and biological analysis. However, despite the several techniques developed, most of them have been coupled with GC similarly to what happens with fiber SPME (Ghaemi et al. 2014; Jiménez-Soto et al. 2010; Ravelo-Pérez et al. 2010). Overall, the application of SPME to biological samples has been challenging due to the complexity of the matrices and the low amount of extracting phase regardless of the type of SPME device.

8.3 Liquid-Phase Microextraction

LPME is a miniaturized format of LLE, in which few μl of solvents are required to concentrate analytes rather than the large volume (ml) of solvent used in conventional LLE (Lee et al. 2008). The LPME provides advantages over conventional LLE by giving high analyte enrichment, excellent sample cleanup, and low consumption of organic solvent. It also overcomes issues associated with SPME such as lack of commercial supplier and sample carryover (Lee et al. 2008; Pedersen-Bjergaard and Rasmussen 2008). LPME technique is simple to implement and use. It is, generally, fast and characterized by its affordability and compatibility with widely available materials and apparatus such as capillary GC or LC systems (Lee et al. 2008). LPME can be divided into three principal categories (Figure 8.1), which will be extensively addressed in the following sections.

8.3.1 Single-Drop Microextraction

Single-drop microextraction (SDME) is a type of LPME in which the extraction medium is in the form of a drop (Lee et al. 2008). SDME is considered the simplest form of LPME, which only involves the use of 1–3 μl of an organic solvent at the tip of a microsyringe needle (Lee et al. 2008; Pedersen-Bjergaard and Rasmussen 2008). After extraction, the microdrop is retracted back into the syringe and transferred for further analysis.

In practice, two main approaches can be used to perform SDME: the direct immersion (DI)-SDME (two-phase and three-phase) and the headspace (HS)-SDME (Figure 8.3) (Kataoka 2010). In two-phase DI-SPME the analytes are extracted from an aqueous sample to an organic phase. Commonly used water-immiscible solvents for extraction include 1-octanol, toluene, xylene, and chloroform. The DI-SPME can, however, involve a three-phase extraction system, where the analytes are extracted from an aqueous sample to an organic phase, and then back-extracted into a separate aqueous phase enabling the final analysis by a LC system (Choi et al. 2011; Lee et al. 2008). The HS-SDME uses a microdrop of an appropriate solvent in the HS of the sample solution or in a flowing air sample stream that is more suitable to extract volatile or semi-volatile analytes (Kataoka 2010; Sarafraz-Yazdi and Amiri 2010; Xu et al. 2007). In contrast to two-phase DI-SDME, it is not necessary for the extracting solvent to be water-immiscible, since it is not in direct contact with the sample. In HS-SDME,

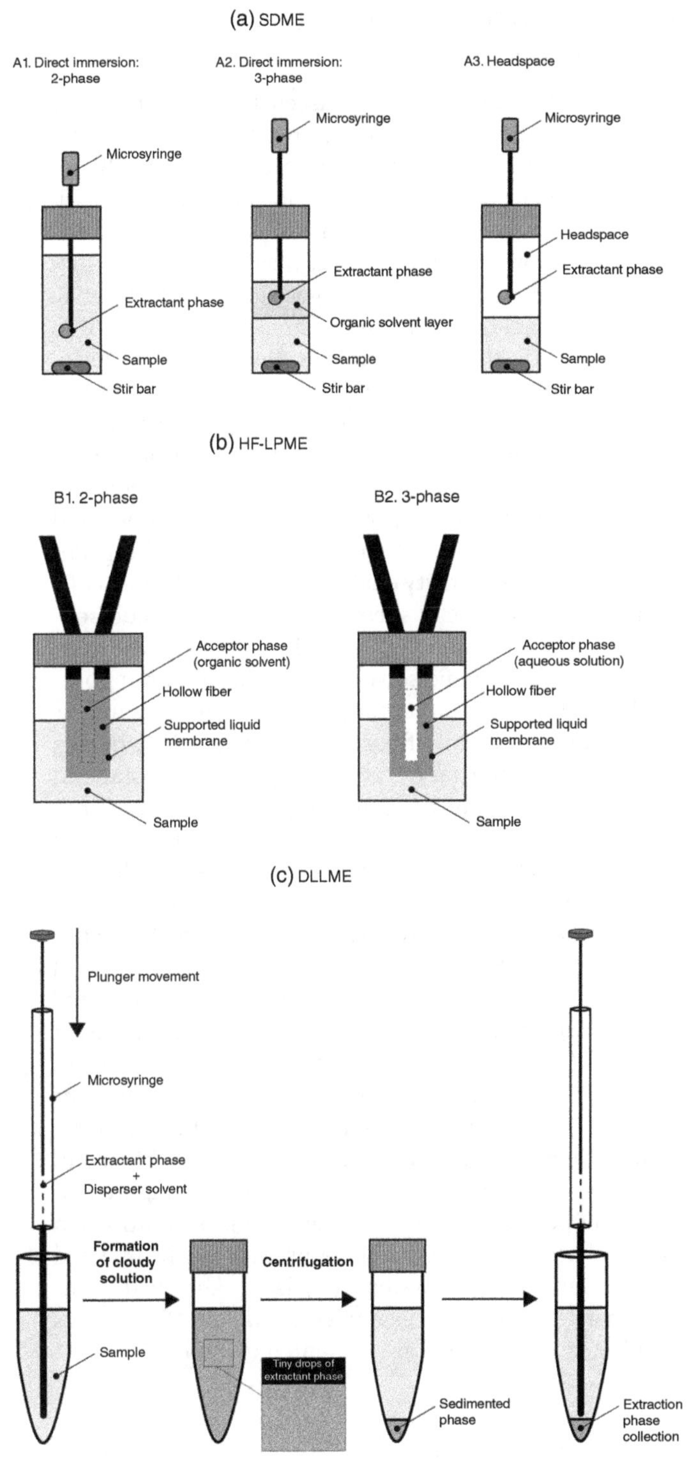

Figure 8.3 Main principles of liquid-phase microextraction (LPME) techniques: (a) single-drop microextraction (SDME) (A1. two-phase direct immersion, A2. three-phase direct immersion, and A3. headspace), (b) hollow fiber liquid-phase microextraction (HF-LPME) (B1. two-phase and B2. three-phase), and (c) dispersive liquid–liquid microextraction (DLLME).

water can also be used as the solvent to extract water-soluble analytes. This significantly enhances the range of extractable analytes as well as the range of analytical methods that can be coupled to SDME. In addition, HS-SDME is found to provide excellent cleanup for samples of complex matrices (Sarafraz-Yazdi and Amiri 2010).

SDME has been used for the quantification of drugs in different matrices and it has been coupled with LC systems. For example, Stege et al. (2011) developed a SDME method for the quantification of ephedrine, cocaine, strychnine, morphine, caffeine, theophylline, and piroxicam in horse urine by LC-MS/MS.

8.3.2 Hollow Fiber Liquid-Phase Microextraction

Pedersen-Bjergaard and Rasmussen introduced hollow fiber liquid-phase microextraction (HF-LPME) in 1999 as a solution to improve the stability and the reliability of LPME (Pedersen-Bjergaard and Rasmussen 1999). In this concept, the extracting phase was placed inside the lumen of a porous hollow fiber typically made of propylene (Gjelstad and Pedersen-Bjergaard 2013; Lucena et al. 2009; Pedersen-Bjergaard and Rasmussen 2008). Prior to the HF-LPME, the hollow fiber is dipped for a few seconds into a water-immiscible organic solvent. For example, 1-octanol, *n*-hexane, *o*-xylene, and toluene can be used as organic solvents in HF-LPME (Han and Row 2012). The organic solvent penetrates into the pores of the hollow fiber by capillary forces, forming a supported liquid membrane (Gjelstad and Pedersen-Bjergaard 2013). The high porosity enables the immobilization of a considerable volume of solvent as a thin film (Pedersen-Bjergaard and Rasmussen 2008). In the two-phase configuration, the organic solvent immobilized in the pores of the hollow fiber (acceptor phase) can be the same as that inside the hollow fiber lumen (Figure 8.3). In this configuration, the target analytes (neutral substances) are extracted from an aqueous sample (donor phase) into an organic solvent based on classical partition. Since the acceptor solution is organic, it can be injected directly into GC or alternatively into LC after the evaporation of solvent and reconstitution in an aqueous medium (Bello-López et al. 2012; Gjelstad and Pedersen-Bjergaard 2013). In the three-phase HF-LPME configuration, the acceptor solution can be an aqueous solution. The target analytes (basic or acidic substances) are extracted from an aqueous phase (donor phase) to another aqueous phase (acceptor phase) present inside the lumen of the hollow fiber through the organic solvent immobilized in the pores of the hollow fiber (organic phase) (Figure 8.3) (Bello-López et al. 2012). Since the acceptor solutions in three-phase HF-LPME are aqueous, they can be injected directly into a LC system (Gjelstad and Pedersen-Bjergaard 2013). Some examples of HF-LPME methods combined with LC-MS are presented in Table 8.1.

8.3.3 Dispersive Liquid–Liquid Microextraction

Rezaee and colleagues developed dispersive liquid–liquid microextraction (DLLME) as a novel LLE technique in which small volumes (μl) of an extraction solvent was used (Rezaee et al. 2006). DLLME is a powerful preconcentration technique that uses a ternary component solvent system. The solvent system consists of an aqueous sample, a polar water-miscible solvent (disperser solvent), and a nonpolar water-immiscible solvent (extracting solvent) (Saraji and Boroujeni 2014). A dispersive solution, containing the disperser solvent (miscible with both aqueous solvent and extraction solvent) and the water-immiscible extraction solvent, is rapidly injected into a sample containing the analytes of interest (Figure 8.3). In general, the disperser solvent can be methanol, acetonitrile, or acetone and the water-immiscible extraction solvent could be chlorobenzene, chloroform, carbon disulfide, undecanol, 1-dodecanol, 2-dodecanol, and *n*-hexadecane (Ojeda and Rojas 2009; Rezaee et al. 2010; Saraji and Boroujeni 2014). After this step, the mixture is agitated to form in the test tube a cloudy solution containing fine droplets of the extraction solvent by dispersion in the sample solution, a process facilitated by the disperser solvent (Ojeda and Rojas 2009; Rezaee et al. 2010). This step allows the analytes to be rapidly extracted into the extraction phase because a very high contact area is generated between the aqueous phase and the extraction solvent (Saraji and Boroujeni 2014). After centrifugation or solidification, the quantification of the enriched analytes in the settled phase (with or without further treatment) can be performed by conventional analytical techniques (Han and Row 2012; Rezaee et al. 2010;Sarafraz-Yazdi and Amiri 2010; Saraji and Boroujeni 2014). The preferred technique for the analysis of extracts is GC since water-immiscible solvents are generally used in the final step of DLLME. However, DLLME also could be used in combination with LC. The halogenated hydrocarbons such as chlorobenzene, carbon tetrachloride, chloroform, and tetrachloroethylene, usually selected as extracting solvents in DLLME, are not compatible with the LC mobile phase because they have high density and are not soluble in the mobile phase and an extra step is needed to evaporate them before final analysis in LC apparatus (Rezaee et al. 2010).

An appropriate extraction solvent that enables high extraction efficiency should have the following characteristics: capability to extract the compounds of interest, low solubility in water, formation of tiny droplets in the presence of the disperser solvent, suitable chromatographic behavior, and a density different from the aqueous phase to enable phase separation (Saraji and Boroujeni 2014). On the other hand, the dispersive solvent plays a key role that helps extraction solvent to form fine droplets in aqueous samples and the selection of the disperser solvent should be based on the miscibility of the solvent in both the aqueous phase (sample solution) and the extraction solvent (Sarafraz-Yazdi and Amiri 2010; Saraji and Boroujeni 2014).

In general, DLLME is a technique of choice for the analysis of samples with a simple matrix. The technique has low selectivity and low sample cleanup efficiency and, therefore, samples in water are the commonly studied. In order to overcome those disadvantages of DLLME, several pretreatment steps can be applied to complex matrices, including dilution, homogenization, sonication, precipitation, centrifugation, and filtration. In more complex matrices such as serum, PPT with a suitable reagent is usually necessary before extraction. These pretreatment steps have expanded the use of DLLME for the determination of the drugs and metabolites and other compounds present in biological samples (Table 8.1).

8.3.4 Influence Factors on LPME Efficiency

There are some factors affecting the method optimization and extraction efficiency (recovery and enrichment) in LPME techniques. These factors are but not limited to: the organic solvent, the volume of donor and acceptor solution, extraction time, pH, agitation of the sample, and salt addition (Han and Row 2012; Pawliszyn and Pedersen-Bjergaard 2006; Psillakis and Kalogerakis 2003).

The choice of organic solvent is a critical factor. The solubility of the organic solvent in water should be low in order to prevent its dissolution into the aqueous phase; its volatility should be low in order to avoid evaporation during extraction; its polarity should match that of the polypropylene fiber so that it can be easily immobilized within the pores of the hollow fiber (Han and Row 2012; Psillakis and Kalogerakis 2003). In general, several water-immiscible solvents with different solubility and polarity may be used as extraction solvents. For DI-SDME and DLLME, the density of organic solvent plays an important role in the extraction process. In DI-SDME, the density of organic solvent should be lower than that of the water. On the other hand, DLLME requires an organic solvent with higher density than the water (Han and Row 2012).

The volume of donor and acceptor solutions directly affects the extraction efficiency. In general, as the ratio of acceptor-to-donor phase increases, the extraction efficiency and enrichment factor also increase. However, enrichment factor decreases when it exceeds a certain limit. Keeping a low extraction volume is also necessary to obtain high selectivity (Han and Row 2012; Psillakis and Kalogerakis 2003).

Extraction in LPME is a time-dependent process, in which the extraction efficiency is attained at the equilibrium condition. In SDME and HF-LPME, the equilibrium time ranges usually between 30 and 60 minutes (Han and Row 2012). Longer extraction times generally result in an increased extraction efficiency but such conditions are usually not used because it could be unfeasible (Psillakis and Kalogerakis 2003). In contrast to SDME and HF-LPME, the extraction time in DLLME is less critical because of the large surface area between extraction solvent and aqueous phase generated after the formation of cloudy solution resulting in a short extraction time (Han and Row 2012; Saraji and Boroujeni 2014).

The pH adjustment can impact the extraction efficiency as it influences the solubility of the target acidic/basic analytes. The pH in the donor and acceptor phases, particularly in three-phase (HF-LPME and also in DI-SDME), is critical, since it influences the distribution ratio, enrichment factor, and recoveries of target analytes. In three-phase HF-LPME, the pH of donor solution should be adjusted so that the analytes of interest are in their unionized form in order to obtain high enrichment factor and high recoveries. In this form, the solubility of analyte in the donor solution decreases and an efficient transfer into the organic phase is obtained. On the other hand, the pH of the acceptor solution should be adjusted to make the analytes of interest in their ionized state, ensuring efficient extraction of analytes into the acceptor solution and preventing them being trapped in the organic phase (Psillakis and Kalogerakis 2003). Thus, for basic substances, the pH of the sample should be adjusted to alkaline to suppress ionization of target analytes, while the aqueous acceptor solution should be acidified. In contrast, for the extraction of acidic substances, the pH-gradient is reversed. It is necessary to acidify the sample, while the acceptor solution should be alkaline (Gjelstad and Pedersen-Bjergaard 2013). The same principles of pH adjustment of the sample are applied in the case of three-phase DI-SDME in which the analytes are extracted from an aqueous sample to an organic phase, and then back-extracted into a separate aqueous phase enabling the final analysis by a LC system.

The agitation of the sample is important to accelerate the extraction kinetics and enhance the extraction efficiency, since stirring allows the continuous exposure of the extraction surface to the aqueous sample and the thermodynamic equilibrium can be achieved in a shorter time by inducing the convection in the membrane phase. Sample agitation can be done in several ways, such as stirring or vibrating the sample. However, in HF-LPME and SDME, the use of magnetic stirrer in high stirring rate promotes formation of bubbles that tends to adhere to the surface of the hollow fiber, leads to solvent evaporation, and instability of microdrops, resulting in decreased extraction efficiency (Han and Row 2012; Psillakis and Kalogerakis 2003).

Salt addition is widely used in microextraction to improve the analyte partitioning into the organic phase by the salting-out effect. However, the presence of higher concentrations of salt could change the physical properties of the extraction film and thus reduce the diffusion rates of the analytes in the organic phase (Han and Row 2012; Psillakis and Kalogerakis 2003). Hence, the effect of salt addition may influence the extraction efficiency of the target analytes depending on their nature (Psillakis and Kalogerakis 2003).

The disadvantages and advantages of the most popular LPME techniques (i.e. SDME, HF-LPME, and DLLME) as well as the main developments related to each of these techniques are compiled in Table 8.2.

Table 8.2 Advantages, disadvantages, and trends of liquid-phase microextraction (LPME) techniques (Sarafraz-Yazdi and Amiri 2010; Spietelun et al. 2014).

Technique	Advantages	Disadvantages
SDME	Inexpensive, rapid, and simple Minimal exposure to organic solvents No special equipment required Combined with many techniques for determination of the analytes Easy to operate Versatile (numerous solvents can be used) Possibility of *in situ* derivatization or complexation Variety of extraction modes High enrichment factor obtained	Instability of the drop Restrictions on the selection of extraction solvent Ease of dislodgement of the microdrop Limited drop volume Limited rate of agitation of the sample solution Air bubble formation Time consuming Average precision Limited drop surface (slow kinetics) Special equipment required (solidified floating organic drop microextraction)
Development of SDME technique	Application of ionic liquids and coacervates as an acceptor phase Modifications of the SDME technique: dynamic-LPME, dynamic liquid-phase nanoextraction, droplet-membrane-droplet-LPME, directly suspended droplet microextraction, solidification of a floating organic drop microextraction, single-drop coacervative microextraction, and solidified floating vesicular coacervative drop microextraction	
HF-LPME	Inexpensive, simple, cleanup Supported of solvent on membrane pores Possibility of automation and miniaturization Combined with many techniques for determination of the analytes High versatility and selectivity Headspace and immersion modes Possibility of *in situ* derivatization	Memory effects when reusing membranes Preconditioning of membranes Average precision performed in manual mode Most of studies carried out in static mode Higher sampling time and temperature compared to SDME (lower evaporation rate)
Development of HF-LPME technique	Modifications of the HF-LPME technique: hollow fiber membrane liquid–liquid–liquid microextraction, hollow fiber solid–liquid phase microextraction, solid-phase membrane tip extraction, solvent bar microextraction, dual solvent-stir-bar microextraction, solvent cooling-assisted dynamic hollow-fiber-supported headspace LPME, dynamic-HF-LPME, electro membrane isolation, electro membrane extraction, and on-chip electro membrane extraction	
DLLME	Inexpensive, simple, and fast (a few seconds) Easy to operate Possibility of automation Enormous contact area between the acceptor phase and sample Combined with many techniques for determination of the analytes Fast extraction kinetics High recovery and high enrichment factor obtained	Three solvents are needed Restrictions on the selection of extraction solvent Centrifugation/freezing/auxiliary solvent/demulsifier must be applied
Development of DLLME technique	Modification of the DLLME technique: ultrasound-assisted emulsification-microextraction, ultrasound dispersive liquid–liquid microextraction, vortex-assisted liquid–liquid microextraction, in-syringe DLLME, surfactant-assisted-DLLME, ion pair-based surfactant-assisted microextraction, ultrasound-assisted surfactant-enhanced emulsification microextraction, vortex-assisted surfactant-enhanced-emulsification liquid–liquid microextraction, solvent terminated-DLLME, solvent demulsification dispersive-DLLME, sequential injection-DLLME, supramolecular-based dispersive liquid–liquid microextraction, air-assisted liquid–liquid microextraction, DLLME variants with solidification of a floating organic, cold-induced aggregation microextraction, *in situ* solvent-formation microextraction, and temperature-controlled ionic liquid dispersive liquid-phase microextraction, totally organic solvent-free emulsification microextraction procedure, ionic liquid-linked dual magnetic microextraction, supramolecular-based dispersive liquid–liquid microextraction, supramolecular solvent-based vortex-mixed microextraction, supramolecular solvent-based microextraction, and microwave-assisted DLLME. Application of ionic liquids, coacervates, supramolecular systems, and solvents with a density lower than water as extraction solvents.	

SDME, Single drop microextraction; HF-LPME, hollow-fiber liquid-phase microextraction; DLLME, dispersive liquid–liquid microextraction.

8.4 Summary

In recent years, the interest in miniaturized new extraction techniques with lower sample volume requirement, simpler equipment and handling, and lower solvent consumption has led to the development of a series of microextraction methods. These techniques have been developed to address the issues associated with low recovery and high matrix effects in an effort of increasing the reliability and robustness of the LC-MS-based bioanalytical methods. The two main types of microextraction techniques are SPME and LPME.

With regard to SPME, the recent development of MEPS and its wide application to several drugs and metabolites and other compounds in biological samples should be emphasized. This technique affords several advantages that enable its wide application, most prominent of which is the reutilization of the MEPS sorbent for a large set of samples. The in-tip SPME is a promising format for extraction of compounds. On the other hand, the HS-SPME, DI-SPME, and TFME are also used but to less extent because such techniques are more suitable to be coupled to GC systems. The new sorbents, such as monolithic sorbents and CNTs, are being used in some SPME techniques, representing important alternatives to the more conventional sorbents. The LPME has been extensively employed, mainly through HF-LPME and DLLME formats. Particularly, DLLME is a LPME technique with special advantageous as it enables the extraction of the analytes using few microliters of extraction and dispersive solvent, while it usually requires a pretreatment step.

In summary, in this chapter, the main advantages and recent developments of well-established microextraction techniques have been discussed with a particular focus on the SPME and LPME techniques for the analysis of drugs from biological samples.

Acknowledgements

This work was supported by FEDER funds through the POCI – COMPETE 2020 – Operational Programme Competitiveness and Internationalization in Axis I – Strengthening research, technological development, and innovation (Project POCI-01-0145-FEDER-00157491) and National Funds by FCT – Foundation for Science and Technology (Project UID/Multi/00709/2013). The authors are also grateful to Dr. Ricardo Relvas (Faculty of Health Sciences, University of Beira Interior, Covilhã, Portugal) for the assistance in preparing the figures presented in this chapter.

References

Abdel-Rehim, M. (2004). New trend in sample preparation: on-line microextraction in packed syringe for liquid and gas chromatography applications: I. Determination of local anaesthetics in human plasma samples using gas chromatography-mass spectrometry. *J. Chromatogr. B* 801 (2): 317–321.

Abdel-Rehim, M., Hassan, Z., Skansem, P., and Hassan, M. (2007). Simultaneous determination of busulphan in plasma samples by liquid chromatography-electrospray ionization mass spectrometry utilizing microextraction in packed syringe (MEPS) as on-line sample preparation method. *J. Liq. Chromatogr. Related Technol.* 30 (20): 3029–3041.

Abuzooda, T., Amini, A., and Abdel-Rehim, M. (2015). Graphite-based microextraction by packed sorbent for online extraction of β-blockers from human plasma samples. *J. Chromatogr. B* 992: 86–90.

Alves, G., Rodrigues, M., Fortuna, A. et al. (2013). A critical review of microextraction by packed sorbent as a sample preparation approach in drug bioanalysis. *Bioanalysis* 5 (11): 1409–1442.

Arthur, C.L. and Pawliszyn, J. (1990). Solid phase microextraction with thermal desorption using fused silica optical fibers. *Anal. Chem.* 62 (19): 2145–2148.

Baltussen, E., Sandra, P., David, F., and Cramers, C. (1999). Stir bar sorptive extraction (SBSE), a novel extraction technique for aqueous samples: theory and principles. *J. Microcolumn Separations.* 11 (10): 737–747.

Bello-López, M.A., Ramos-Payán, M., Ocaña-González, J.A. et al. (2012). Analytical applications of hollow fiber liquid phase microextraction (HF-LPME): a review. *Anal. Lett.* 45 (8): 804–830.

Bessonneau, V., Bojko, B., Azad, A. et al. (2014). Determination of bronchoalveolar lavage bile acids by solid phase microextraction liquid chromatography-tandem mass spectrometry in combination with metabolite profiling: comparison with enzymatic assay. *J. Chromatogr. A* 1367: 33–38.

Bessonneau, V., Boyaci, E., Maciazek-Jurczyk, M., and Pawliszyn, J. (2015a). *In vivo* solid phase microextraction sampling of human saliva for non-invasive and on-site monitoring. *Anal. Chim. Acta* 856: 35–45.

Bessonneau, V., Zhan, Y., De Lannoy, I.A. et al. (2015b). *In vivo* solid-phase microextraction liquid chromatography-tandem mass spectrometry for monitoring blood eicosanoids time profile after lipopolysaccharide-induced inflammation in Sprague-Dawley rats. *J. Chromatogr. A* 1424: 134–138.

Bojko, B., Vuckovic, D., Cudjoe, E. et al. (2011). Determination of tranexamic acid concentration by solid phase microextraction and liquid chromatography-tandem mass spectrometry: first step to *in vivo* analysis. *J. Chromatogr. B* 879 (32): 3781–3787.

Bunch, D.R. and Wang, S. (2013). Applications of monolithic solid-phase extraction in chromatography-based clinical chemistry assays. *Anal. Bioanal. Chem.* 405 (10): 3021–3033.

Carasek, E. and Merib, J. (2015). Membrane-based microextraction techniques in analytical chemistry: a review. *Anal. Chim. Acta* 880: 8–25.

Choi, K., Kim, J., and Chung, D.S. (2011). Single-drop microextraction in bioanalysis. *Bioanalysis* 3 (7): 799–815.

Daryanavard, S.M., Jeppsson-Dadoun, A., Andersson, L.I. et al. (2013). Molecularly imprinted polymer in microextraction by packed sorbent for the simultaneous determination of local anesthetics: lidocaine, ropivacaine, mepivacaine and bupivacaine in plasma and urine samples. *Biomed. Chromatogr.* 27 (11): 1481–1488.

Duan, C., Shen, Z., Wu, D., and Guan, Y. (2011). Recent developments in solid-phase microextraction for on-site sampling and sample preparation. *TRAC Trend Anal. Chem.* 30 (10): 1568–1574.

Elmongy, H., Ahmed, H., Wahbi, A.A. et al. (2016). Determination of metoprolol enantiomers in human plasma and saliva samples utilizing microextraction by packed sorbent and liquid chromatography-tandem mass spectrometry. *Biomed. Chromatogr.* 30 (8): 1309–1317.

Farhadi, K., Hatami, M., and Matin, A.A. (2012). Microextraction techniques in therapeutic drug monitoring. *Biomed. Chromatogr.* 26 (8): 972–989.

Fu, X., Liao, Y., and Liu, H. (2005). Sample preparation for pharmaceutical analysis. *Anal. Bioanal. Chem.* 381 (1): 75–77.

Ghaemi, F., Amiri, A., and Yunus, R. (2014). Methods for coating solid-phase microextraction fibers with carbon nanotubes. *TrAC Trends Anal. Chem.* 59: 133–143.

Gjelstad, A. and Pedersen-Bjergaard, S. (2013). Perspective: hollow fibre liquid-phase microextraction – principles, performance, applicability, and future directions. *Sci. Chromatogr.* 5 (3): 181–189.

Han, D. and Row, K.H. (2012). Trends in liquid-phase microextraction, and its application to environmental and biological samples. *Microchim. Acta* 176 (1): 1–22.

He, H., Pham-Huy, L.A., Dramou, P. et al. (2013). Carbon nanotubes: applications in pharmacy and medicine. *Biomed. Res. Int.* 2013: 1–12.

Herrera-Herrera, A.V., González-Curbelo, M.Á., Hernández-Borges, J., and Rodríguez-Delgado, M.Á. (2012). Carbon nanotubes applications in separation science: a review. *Anal. Chim. Acta* 734: 1–30.

Ho, T.S., Reubsaet, J.L., Anthonsen, H.S. et al. (2005). Liquid-phase microextraction based on carrier mediated transport combined with liquid chromatography-mass spectrometry. New concept for the determination of polar drugs in a single drop of human plasma. *J. Chromatogr. A* 1072 (1): 29–36.

Jiménez-Soto, J.M., Lucena, R., Cárdenas, S., and Valcárcel, M. (2010). Solid phase (micro) extraction tools based on carbon nanotubes and related nanostructures. In: *Handbook of Carbon Nanotubes* (ed. J.M. Marulanda), 409–428. Rijeka: InTech.

Jochmann, M.A., Kmiecik, M.P., and Schmidt, T.C. (2006). Solid-phase dynamic extraction for the enrichment of polar volatile organic compounds from water. *J. Chromatogr. A* 1115 (1–2): 208–216.

Kataoka, H. (2002). Automated sample preparation using in-tube solid-phase microextraction and its application – a review. *Anal. Bioanal. Chem.* 373 (1–2): 31–45.

Kataoka, H. (2010). Recent developments and applications of microextraction techniques in drug analysis. *Anal. Bioanal. Chem.* 396 (1): 339–364.

Kataoka, H. (2011). Current developments and future trends in solid-phase microextraction techniques for pharmaceutical and biomedical analyses. *Anal. Sci.* 27 (9): 893–905.

Kataoka, H., Ishizaki, A., and Saito, K. (2016). Recent progress in solid-phase microextraction and its pharmaceutical and biomedical applications. *Anal. Methods* 8: 5773–5788.

Kataoka, H. and Saito, K. (2011). Recent advances in SPME techniques in biomedical analysis. *J. Pharm. Biomed. Anal.* 54 (5): 926–950.

Lasarte-Aragonés, G., Lucena, R., Cárdenas, S., and Valcárcel, M. (2011). Nanoparticle-based microextraction techniques in bioanalysis. *Bioanalysis* 3 (22): 2533–2548.

Lee, J., Lee, H.K., Rasmussen, K.E., and Pedersen-Bjergaard, S. (2008). Environmental and bioanalytical applications of hollow fiber membrane liquid-phase microextraction: a review. *Anal. Chim. Acta* 624 (2): 253–268.

Liang, X., Liu, S., Wang, S. et al. (2014). Carbon-based sorbents: carbon nanotubes. *J. Chromatogr. A* 1357: 53–67.

Lord, H. and Pawliszyn, J. (2000). Microextraction of drugs. *J Chromatogr A.* 902 (1): 17–63.

Lucena, R., Cruz-Vera, M., Cárdenas, S., and Valcárcel, M. (2009). Liquid-phase microextraction in bioanalytical sample preparation. *Bioanalysis* 1 (1): 135–149.

Magiera, S. (2013). Fast, simultaneous quantification of three novel cardiac drugs in human urine by MEPS-UHPLC-MS/MS for therapeutic drug monitoring. *J. Chromatogr. B* 938: 86–95.

Magiera, S. and Kusa, J. (2015). Evaluation of a rapid method for the therapeutic drug monitoring of aliskiren, enalapril and its active metabolite in plasma and urine by UHPLC-MS/MS. *J. Chromatogr. B* 980: 79–87.

Mirnaghi, F.S., Hein, D., and Pawliszyn, J. (2013). Thin-film microextraction coupled with mass spectrometry and liquid chromatography-mass spectrometry. *Chromatographia* 76 (19): 1215–1223.

Mirnaghi, F.S. and Pawliszyn, J. (2012). Development of coatings for automated 96-blade solid phase microextraction-liquid chromatography–tandem mass spectrometry system, capable of extracting a wide polarity range of analytes from biological fluids. *J. Chromatogr. A* 1261: 91–98.

Moein, M.M., Abdel-Rehim, A., and Abdel-Rehim, M. (2015). Microextraction by packed sorbent (MEPS). *TRAC Trend Anal. Chem.* 67: 34–44.

Musteata, F.M. and Pawliszyn, J. (2007). Bioanalytical applications of solid-phase microextraction. *TRAC Trend Anal. Chem.* 26 (1): 36–45.

Namera, A., Nakamoto, A., Saito, T., and Miyazaki, S. (2011). Monolith as a new sample preparation material: recent devices and applications. *J. Sep. Sci.* 34 (8): 901–924.

Namera, A. and Saito, T. (2013). Advances in monolithic materials for sample preparation in drug and pharmaceutical analysis. *TRAC Trend Anal. Chem.* 45: 182–196.

Naxing Xu, R., Fan, L., Kim, G.E., and El-Shourbagy, T.A. (2006). A monolithic-phase based on-line extraction approach for determination of pharmaceutical components in human plasma by HPLC-MS/MS and a comparison with liquid-liquid extraction. *J. Pharm. Biomed. Anal.* 40 (3): 728–736.

Nema, T., Chan, E.C., and Ho, P.C. (2011). Efficiency of a miniaturized silica monolithic cartridge in reducing matrix ions as demonstrated in the simultaneous extraction of morphine and codeine from urine samples for quantification with liquid chromatography-tandem mass spectrometry (LC-MS/MS). *J. Mass Spectrom.* 46 (9): 891–900.

Nema, T., Chan, E.C., and Ho, P.C. (2014). Applications of monolithic materials for sample preparation. *J. Pharm. Biomed. Anal.* 87: 130–141.

Nerín, C., Salafranca, J., Aznar, M., and Batlle, R. (2009). Critical review on recent developments in solventless techniques for extraction of analytes. *Anal. Bioanal. Chem.* 393 (3): 809–833.

Ocaña-González, J.A., Fernández-Torres, R., Bello-López, M.Á., and Ramos-Payán, M. (2016). New developments in microextraction techniques in bioanalysis. A review. *Anal. Chim. Acta* 905: 8–23.

Ojeda, C.B. and Rojas, F.S. (2009). Separation and preconcentration by dispersive liquid–liquid microextraction procedure: a review. *Chromatographia* 69 (11): 1149–1159.

Pawliszyn, J. and Pedersen-Bjergaard, S. (2006). Analytical microextraction: current status and future trends. *J. Chromatogr. Sci.* 44 (6): 291–307.

Pedersen-Bjergaard, S. and Rasmussen, K.E. (1999). Liquid-liquid-liquid microextraction for sample preparation of biological fluids prior to capillary electrophoresis. *Anal. Chem.* 71 (14): 2650–2656.

Pedersen-Bjergaard, S. and Rasmussen, K.E. (2008). Liquid-phase microextraction with porous hollow fibers, a miniaturized and highly flexible format for liquid-liquid extraction. *J. Chromatogr. A* 1184 (1–2): 132–142.

Pereira, J., Câmara, J.S., Colmsjö, A., and Abdel-Rehim, M. (2014a). Microextraction by packed sorbent: an emerging, selective and high-throughput extraction technique in bioanalysis. *Biomed. Chromatogr.* 28 (6): 839–847.

Pereira, J., Silva, C.L., Perestrelo, R. et al. (2014b). Re-exploring the high-throughput potential of microextraction techniques, SPME and MEPS, as powerful strategies for medical diagnostic purposes. Innovative approaches, recent applications and future trends. *Anal. Bioanal. Chem.* 406 (8): 2101–2122.

Płotka-Wasylka, J., Szczepańska, N., de la Guardia, M., and Namieśnik, J. (2015). Miniaturized solid-phase extraction techniques. *TRAC Trend Anal. Chem.* 73: 19–38.

Psillakis, E. and Kalogerakis, N. (2003). Developments in liquid-phase microextraction. *TRAC Trend Anal. Chem.* 22 (9): 565–574.

Ravelo-Pérez, L.M., Herrera-Herrera, A.V., Hernández-Borges, J., and Rodríguez-Delgado, M.A. (2010). Carbon nanotubes: solid-phase extraction. *J. Chromatogr. A* 1217 (16): 2618–2641.

Rezaee, M., Assadi, Y., Milani Hosseini, M.R. et al. (2006). Determination of organic compounds in water using dispersive liquid-liquid microextraction. *J. Chromatogr. A* 1116 (1–2): 1–9.

Rezaee, M., Yamini, Y., and Faraji, M. (2010). Evolution of dispersive liquid–liquid microextraction method. *J. Chromatogr. A* 1217 (16): 2342–2357.

Risticevic, S., Lord, H., Górecki, T. et al. (2010). Protocol for solid-phase microextraction method development. *Nat. Protoc.* 5: 122–139.

Risticevic, S., Niri, V.H., Vuckovic, D., and Pawliszyn, J. (2009). Recent developments in solid-phase microextraction. *Anal. Bioanal. Chem.* 393 (3): 781–795.

Said, R., Kamel, M., El-Beqqali, A., and Abdel-Rehim, M. (2010). Microextraction by packed sorbent for LC-MS/MS determination of drugs in whole blood samples. *Bioanalysis* 2 (2): 197–205.

Saito, Y., Ueta, I., Ogawa, M. et al. (2009). Fiber-packed needle-type sample preparation device designed for gas chromatographic analysis. *Anal. Bioanal. Chem.* 393 (3): 861–869.

Saito, K., Yagi, K., Ishizaki, A., and Kataoka, H. (2010). Determination of anabolic steroids in human urine by automated in-tube solid-phase microextraction coupled with liquid chromatography-mass spectrometry. *J. Pharm. Biomed. Anal.* 52 (5): 727–733.

Sánchez-Rojas, F., Bosch-Ojeda, C., and Cano-Pavón, J.M. (2009). A review of stir bar sorptive extraction. *Chromatographia* 69: S79–S94.

Sarafraz-Yazdi, A. and Amiri, A. (2010). Liquid-phase microextraction. *TRAC Trend Anal. Chem.* 29 (1): 1–14.

Saraji, M. and Boroujeni, M.K. (2014). Recent developments in dispersive liquid-liquid microextraction. *Anal. Bioanal. Chem.* 406 (8): 2027–2066.

Saraji, M. and Keykavooci, A. (2016). Octadecylsilane/Nylon-6 composite as a thin-film microextraction sorbent for the determination of bisphenol a in water samples. *J. Sep. Sci.* 39 (18): 3616–3623.

Spietelun, A., Marcinkowski, Ł., de la Guardia, M., and Namieśnik, J. (2013). Recent developments and future trends in solid phase microextraction techniques towards green analytical chemistry. *J. Chromatogr. A* 1321: 1–13.

Spietelun, A., Marcinkowski, Ł., de la Guardia, M., and Namieśnik, J. (2014). Green aspects, developments and perspectives of liquid phase microextraction techniques. *Talanta* 119: 34–45.

Stege, P.W., Lapierre, A.V., Martinez, L.D. et al. (2011). A combination of single-drop microextraction and open tubular capillary electrochromatography with carbon nanotubes as stationary phase for the determination of low concentration of illicit drugs in horse urine. *Talanta* 86: 278–283.

Sun, W., Qu, S., and Du, Z. (2014). Hollow fiber liquid-phase microextraction combined with ultra-high performance liquid chromatography-tandem mass spectrometry for the simultaneous determination of naloxone, buprenorphine and norbuprenorphine in human plasma. *J. Chromatogr. B* 951–952: 157–163.

Szultka, M., Krzeminski, R., Szeliga, J. et al. (2013). A new approach for antibiotic drugs determination in human plasma by liquid chromatography-mass spectrometry. *J. Chromatogr. A* 1272: 41–49.

Theodoridis, G., Koster, E.H., and de Jong, G.J. (2000). Solid-phase microextraction for the analysis of biological samples. *J. Chromatogr. B* 745 (1): 49–82.

Vas, G. and Vékey, K. (2004). Solid-phase microextraction: a powerful sample preparation tool prior to mass spectrometric analysis. *J. Mass Spectrom.* 39 (3): 233–254.

Vela-Soria, F., Ballesteros, O., Rodríguez, I. et al. (2013). A new treatment by dispersive liquid-liquid microextraction for the determination of parabens in human serum samples. *Anal. Bioanal. Chem.* 405 (23): 7259–7267.

Vlčková, H., Rabatinová, M., Mikšová, A. et al. (2012). Determination of pravastatin and pravastatin lactone in rat plasma and urine using UHPLC-MS/MS and microextraction by packed sorbent. *Talanta* 90: 22–29.

Xu, L., Basheer, C., and Lee, H.K. (2007). Developments in single-drop microextraction. *J. Chromatogr. A* 1152 (1–2): 184–192.

Xu, L., Shi, Z.G., and Feng, Y.Q. (2011). Porous monoliths: sorbents for miniaturized extraction in biological analysis. *Anal. Bioanal. Chem.* 399 (10): 3345–3357.

Zhan, Y., Musteata, F.M., Basset, F.A., and Pawliszyn, J. (2011). Determination of free and deconjugated testosterone and epitestosterone in urine using SPME and LC-MS/MS. *Bioanalysis* 3 (1): 23–30.

Zheng, M.M., Wang, S.T., Hu, W.K., and Feng, Y.Q. (2010). In-tube solid-phase microextraction based on hybrid silica monolith coupled to liquid chromatography-mass spectrometry for automated analysis of ten antidepressants in human urine and plasma. *J. Chromatogr. A* 1217 (48): 7493–7501.

9

Microsampling Applications with LC-MS Bioanalysis

Chester L. Bowen and Matthew Barfield

9.1 Introduction

Historically, plasma has been the preferred matrix for the bioanalytical laboratory supporting pharmacokinetic (PK) preclinical and clinical *in vivo* studies due to its relative cleanliness, and ease of collection and handling. Advancing the 3R's principles (reduction, refinement, and replacement) within the pharmaceutical industry has been undertaken by many companies and organizations over the past several years in an effort to reduce blood volumes and/or use fewer rodent animals in studies (Jonsson et al. 2012). The greatest impact of the 3R's and microsampling is within rodent species in terms of collection procedures as well as the vast number of rodent studies performed during the drug development process.

Standard parameters are well understood for liquid samples, but not for novel sample formats, such as dried blood. In regard to microsampling, several consortia have convened and are actively meeting to discuss the scientific hurdles involved in developing and validating microsampling (dried and liquid) methods. Within the International Consortium for Innovation and Quality in Pharmaceutical Development (IQ), the Microsampling Working Group focuses on clinical study design, regulatory acceptance, and matrix bridging design to actively advance microsampling. This group is able to harness a wealth of scientific knowledge, share both the financial and scientific burden of the investigations, collaborate to advance dried blood spot (DBS) and microsampling technology, and influence regulators more effectively than individual companies.

While local Institutional Animal Care and Use Committee (IACUC) guidelines vary, a general practice allows for 7.5% of the total blood volume to be collected over 24 hours with a 1-week recovery period (Diehl et al. 2001). Following those guidelines, ~0.800 ml of blood would be available from a female, 6-week old Wistar Han rat, weighing ~136 g, for sampling within a 24-hour period. For transgenic mice the available blood is much less. For a 6-week old female, weighing ~15 g, <100 μl of blood would be available for collection following these guidelines. In order to obtain reproducible and high-quality data, the sampling methodology is as important as the bioanalytical method; accordingly, approaches that interfere minimally with the intended species and produce the least amount of stress and discomfort to the animal are highly desirable (Musteata 2009). The implementation of any of these micro-sampling techniques allows for expanded pharmacokinetic/toxicokinetic (PK/TK) data collection from a minimal number of animals. Due to the lower blood volume required, the researcher has the ability to remove satellite animals, therefore reducing animal numbers and having a direct correlation with PK/TK and safety endpoints.

Various techniques are available for microsampling blood and plasma, and the bioanalytical considerations have been discussed around each of the techniques. Depending on the collection technique, method validation must be designed to investigate and test all possible situations that the biological samples could be subjected to. This includes sample collection, processing, transportation, and finally bioanalysis with a micro-volume of matrix. This validation process ensures that an assay to be used for regulatory submissions is accurate, precise, and robust. Most of these parameters are well understood for liquid samples (independent of volume), but not for novel sample formats, such as dried blood.

As we move to micro-volumes of biological matrix (either wet or dried), there is a need to ensure that there is sufficient volume to complete all needed analysis, including possible repeats and incurred sample reanalysis (ISR). In regard to dried blood (DBS or MITRA), the sampling is achieved through taking multiple (three to four) samples per time point, allowing for repeat or additional analysis if needed. In the case of capillary plasma separation, the assay volume must be adequate to allow for multiple analyses from the expected plasma yield

Sample Preparation in LC-MS Bioanalysis, First Edition. Edited by Wenkui Li, Wenying Jian, and Yunlin Fu.

(~30 μl). When micro-sampling is employed, the potential for additional bioanalytical work (metabolite identification) is extremely limited; therefore, careful consideration in regard to the scope of bioanalytical work needs to be performed prior to study initiation.

Advance in LC-MS/MS has allowed for a decrease in sample volume without a deleterious impact on the lower limit of quantification, thus facilitating implementation of microsampling in bioanalysis. Although detection methods are now amenable to relatively small sample volumes, the discussion regarding method validation and sample handling of these dried or capillary collected, micro-volumes has grown.

Several years ago, the first investigation was published using both microsampling (DBS) coupled with enzymatic digestion LC-MS/MS workflow and immunoassay methodology for the quantification of large (>70 kDa) protein therapeutic (Kehler et al. 2011). In this work, a microsampling method in dried rat blood was developed and qualified for a protein biologic at a range of 25–1000 ng ml^{-1} and 500–50 000 ng ml^{-1} for the immunoassay and LC-MS/MS method, respectively. This work showed that the DBS microsampling technique is not limited to small molecules. For the optimized DBS method followed by immunoassay detection, a punch was taken from a dried 20 μl blood spot on a DMPK-C (GE Healthcare) untreated card. The blood spot was combined with buffer (10 mM tris-buffered saline [TBS, pH7.5], 150 nM sodium chloride, 0.1% bovine serum albumin, and 0.1% Tween 20), followed by mixing for two hours. A polystyrene plate was coated overnight with capture reagent followed by washing with buffer. The plate was then blocked with Superblock in TBS (ThermoScientific) and incubated for two hours, followed by plate washing. Standards, QCs, and samples were then added to the plate, incubated for two hours and washed. A biotinylated detection antibody, which is specific for an epitope of the therapeutic, was added to plate, incubated for one hour and washed. Streptavidin conjugated to horseradish peroxidase was added, incubated for one hour and washed. The signal was developed using SuperSignal ELISA Femto Development Reagent (ThermoScientific) and allowed to shake for one minute followed by chemo luminescent detection (Kehler et al. 2011).

9.2 Plasma Microsampling Considerations

9.2.1 Sample Collection

Following several years of optimization, a novel blood collection and plasma isolation technique has been developed using a specially designed micro-capillary separation tube containing EDTA and a thixotropic gel (Drummond Scientific) (Bowen et al. 2013). To summarize; one end of a micro-capillary tube, mylar wrapped and containing thixotropic gel, was fitted with a self-sealing plug that enable it to be centrifuged following contact with the blood. Figure 9.1 illustrates (a) the capillary tubes when empty, (b) filled with blood, (c) plasma/red blood cells separated after centrifugation, and (d) the labeling container following centrifugation. Inverting the blood-filled capillary tube, allowing for liquid contact (30 seconds) with the porous plug is essential for proper separation. Upon centrifugation, the thixotropic gel migrates within the tube based on density to form a stable barrier between the erythrocytes and plasma. All of the isolated plasma was then dispensed using a Wiretrol® device or Captrol® dispenser (Drummond Scientific), and stored as warranted based on compound stability. Using the technique, isolation of ~30 μl of plasma is possible from 75 μl of blood. Once the plasma is isolated, it is stored in a standard polypropylene container and frozen at −20 °C (or colder is warranted through stability assessment) and shipped to bioanalytical laboratory for analysis.

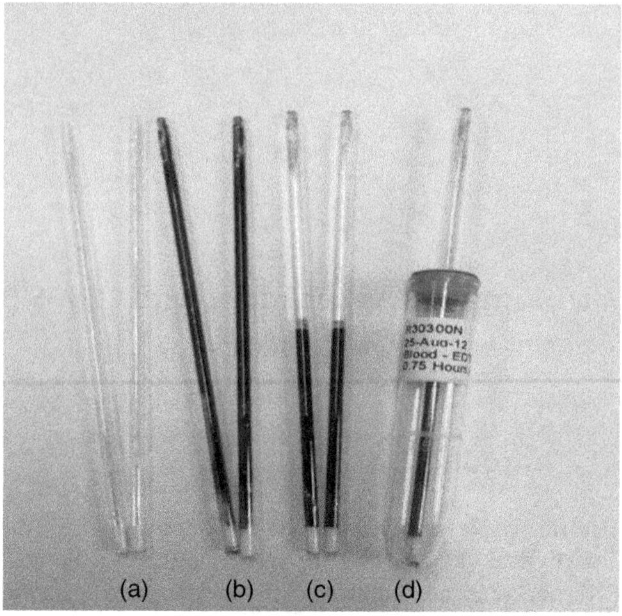

Figure 9.1 Illustration of Drummond Capillary Device for plasma collection and isolation, (a) the capillary tubes when empty, (b) filled with blood, (c) plasma/red blood cells separated after centrifugation, and (d) the labeling container following centrifugation. *Source:* Reprinted with permission from Bowen et al. (2013). Copyright 2013. Future Sciences Ltd.

9.2.2 Interaction with Thixotropic Gel

The use of capillary collection allows for near-optimal yield of isolated plasma due to the incorporation of the thixotropic gel. In theory, besides minimization of the blood volume, this is the only modification when compared to the conventional isolation technique of placing blood into a container, centrifuging, and then manually collecting the plasma. The lack of efficiency with the conventional, tube-based plasma separation technique is caused by the manual removal of the plasma with a pipette and ensuring that no material from the blood–plasma interface is removed. Transferring portions of the interface during plasma isolation could lead to added interferences in the assay. The thixotropic gel has been used in clinical applications since approximately 1975 for serum isolation. The gel is basically an inert material that migrates to the blood–plasma interface upon centrifugation and forms a barrier between the two liquid layers. The density of the gel is between that of the red blood cells and the plasma. When blood in the tube is centrifuged, the gel undergoes a change in viscosity and moves to form a physical barrier between the cells and plasma (Sun et al. 2012).

To compare isolation techniques as part of the validation for a GSK proprietary compound (GSKA), quality control (QC) samples were prepared in blood at two different concentrations (150 and 20 000 ng ml^{-1}). Each QC level was then split into two aliquots. Plasma was isolated from the first aliquot utilizing the capillary tubes (77 μl blood), while plasma from the second aliquot was isolated using the conventional technique (250 μl blood). Table 9.1 illustrates the percent difference (based on peak area ratio of the analyte to internal standard, IS) of GSKA response based on the two different methods of plasma isolation. Response is ~10% lower for the plasma isolated using the capillary technique. Since this was our first full comparison, and results were within the ±15% acceptance window that is used for all other validation assessments, no further investigation into this difference was made. The difference was not felt to be inherent to the technique but rather the preparation. However, this test must be performed with each compound to ensure that the thixotropic gel does not have a deleterious effect on the assay or cause any interference with the mass spectrometer and selected ion transitions.

As further investigations were undertaken and procedures optimized, six different over-the-counter drugs were chosen to compare plasma levels utilizing the standard or capillary plasma isolation techniques. Fresh rat blood (used within 24 hours of draw) was independently spiked with atorvastatin, minoxidil, omeprazole, caffeine, sitamaquine, or ketaconazole at three different concentrations. The concentrations chosen were based on the estimated clinically relevant concentrations at the efficacious dose, 10-fold lower (constituting mid-level QC), and 100-fold lower (representing the low level) (Schulz et al. 2012). As shown in Table 9.2, plasma isolated utilizing the capillary tube consistently has a lower peak response (−3.1 to −21%) for the compounds when compared to the plasma isolated with the conventional technique. The discrepancy in results, both on compound and concentrations, illustrates the need to thoroughly investigate the effect of the thixotropic gel and the general process on intended bioanalytical assay.

Table 9.1 Effect of the different technique used to derive plasma for quantification of GSKA in rat plasma.

	Peak area ratio		Peak area ratio	
	150 ng ml^{-1}		20 000 ng ml^{-1}	
	Conventional plasma isolation	Capillary plasma isolation	Conventional plasma isolation	Capillary plasma isolation
Rep 1	0.102	0.098	14.6	13.1
Rep 2	0.116	0.090	14.9	13.2
Rep 3	0.109	0.097	14.6	13.1
Rep 4	0.103	0.097	14.9	12.8
Rep 5	0.115	0.096	14.6	13.2
Rep 6	0.101	0.098	14.8	13.4
Mean	0.108	0.096	14.733	13.133
SD	0.007	0.003	0.151	0.197
% CV	6.5	3.1	1.0	1.5
% Difference	−11.1		−10.9	

Table 9.2 Plasma concentrations of six over-the-counter drugs measured in the plasma samples generated utilizing conventional and capillary separation techniques.

	Low concentration		Mid concentration		Cmax	
	Conventional	Capillary	Conventional	Capillary	Conventional	Capillary
Atorvastatin	0.138	0.130	1.46	1.38	13.1	12.7
% Difference		−5.8		−5.7		−3.1
Minoxidil	0.016	0.015	0.162	0.155	1.47	1.42
% Difference		−6.8		−4.1		−3.4
Omeprazole	0.222	0.198	2.28	2.058	19.7	18.4
% Difference		−10.8		−9.7		−6.6
Caffeine	0.012	0.011	0.103	0.100	1.08	1.03
% Difference		−6.0		−2.9		−4.6
Sitamaquine	0.011	0.009	0.126	0.106	1.40	1.28
% Difference		−21.1		−15.7		−8.6
Ketaconazole	0.045	0.038	0.399	0.352	1.99	1.90
% Difference		−15.3		−11.9		−4.5

9.2.3 Sample Manipulation

Once these micro-volumes of plasma are isolated, accurate pipetting is essential prior to extraction and analysis. Numerous pipetting tools are available to the bioanalytical scientist. Both air and positive displacement pipettes are most common in a laboratory setting, with performance strictly regulated by the international regulation ISO 8665 (Clark and Shull 2003). For either type of pipette, the use of the correct fitting tips and proper attachment, to ensure a good seal, is essential. Pre-wetting the pipette tip has benefits for both techniques. In air displacement, pre-wetting increases the tip humidity and decreases sample evaporation, while in positive displacement it works to eliminate dead air space.

Positive displacement pipettes offer several advantages over air displacement pipettes. When using positive displacement, the liquid does not come into contact with the pipette. This prevents the creation of aerosols during dispensing, and the risk for cross-contamination is greatly reduced. Also, because there is no air head space, accuracy and reproducibility are not dependent on the density, viscosity, or vapor pressure of the liquid (*Liquid Handling Notebook* n.d.). Air displacement pipettes operate under precise draining of the liquid from the tip due to an increase in air pressure by depressing the pipette tip. Sample delivery is impacted by surface tension, which is affected by temperature and viscosity. Working with all liquids at ambient temperature minimizes the variation.

In a comparison of air and positive displacement pipettes within our lab (Figure 9.2), a marked and statistically relevant difference was noted between them. It is difficult to say which type of pipette is "more accurate" for aliquoting micro-volumes of plasma; all pipettes are calibrated quarterly and both types are considered acceptable for use in our bioanalytical laboratory. The issue arises when one step in the bioanalytical process, such as QC samples are prepared with one type of pipette and stored for future use or shipped to other laboratories for cross-validation. Standard curves could then be prepared at a later time with the other type of pipette, adding variability to the process. This can add additional complexity in cases of cross-validation, where QCs are prepared in one laboratory, and sent to another laboratory for comparison to independently prepared standards. Environmental conditions, such as temperature and relative humidity, could vary greatly affecting the air pressure within the tips and lead to greater variability.

Other assessments that need to be performed with the analyte of interest include nonspecific binding, temperature, and light sensitivity, etc. Variation between pipetting styles of individual analysts and introduction of automation could lead to high variations when dealing with these micro-volumes and should be investigated.

9.2.4 Extraction, LC-MS/MS Analysis, and Detection

To account for small sample volumes (5–10 μl per analytical assay), most sample preparation methods can be

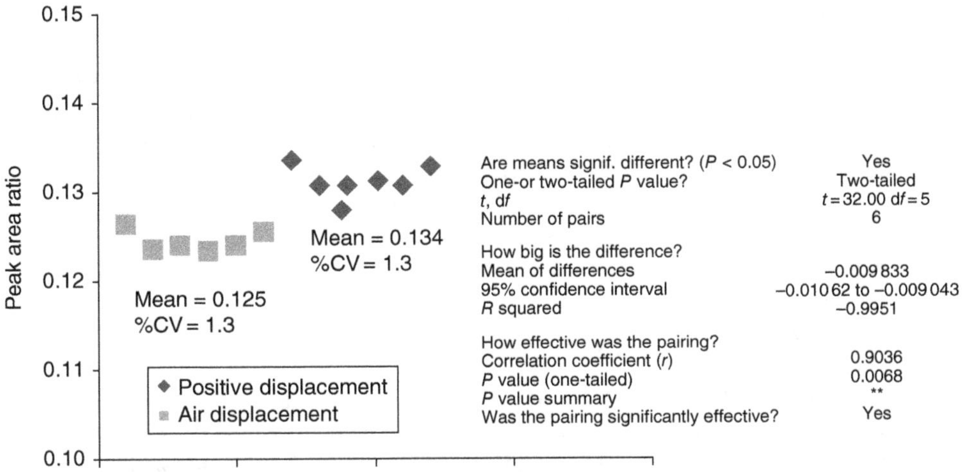

Figure 9.2 Statistical comparisons of air and positive displacement pipettes for 5 µl of plasma.

scaled down in terms of consumable size. Protein precipitation (PPT), solid-phase extraction (SPE), or liquid–liquid extraction (LLE) are all still viable extraction options, often with better recovery and less interferences due to the decrease in matrix volume (Chambers et al. 2007; Van Eeckhaut et al. 2009).

When considering the method development approach, plasma is accurately aliquoted into a 1.2 ml tube; IS working solution is added. This is followed by extraction using PPT, LLE, or SPE. The use of an extraction technique helps remove some of the nonspecific and endogenous interferences from the biological sample and keep the detector clean and capable of analysis of a large number of samples without losing sensitivity or requiring cleaning.

Even with decreased sample volume, similar detection limits can often be reached with proper instrument tuning, utilizing the new advances in mass spectrometer detection (including software advances, high-resolution mass spectrometry, and microfluidics) and the benefit of decrease in matrix effects with smaller volumes (Waters Corporation 2011; Bowen et al. 2016; Stravs et al. 2016).

9.3 Dried Blood (Matrix) Spot (DBS) Considerations

9.3.1 Sample Collection

DBS sampling has been introduced as a method for the collection, storage, transport, and analysis of samples for new chemical entities and biological compounds (Spooner et al. 2009; Kehler et al. 2011). A small volume of blood (typically 15–30 µl) is spotted onto specialty cellulose paper. Multiple spots are placed for a given time point to allow for PK and analytical repeats as well as ISR. Based on previously published data (Spooner et al. 2009), there was no notable difference in compound distribution in blood across spots derived from blood volumes between 10 and 20 µl, when a punch was taken from the center of the spot. This indicates that blood volumes within several microliters (µl) of each other will diffuse across the cellulose media evenly based on the applied volume. In theory, a precise spotting volume is not needed, and normalization of the sampled volume can be achieved using a fixed-sized punch. However, research has also indicated that the varying levels of hematocrit in blood will affect the spot size and therefore analyte distribution.

A common question posed when using DBS is when to add the IS. With a liquid sample the IS is usually added before or during the extraction process; however, with DBS there are four possible options (Abu-Rabie et al. 2011).

1) Addition to the extraction solvent: The gold standard due to its simplicity and has historically shown to work reproducibly. However, in reality, this approach may not be contributing significantly to assay performance since the IS is not incorporated into the sample or medium.
2) Treatment of the paper before use: Possible the best approach scientifically as the IS is incorporated onto the substrate prior to spotting. However, in reality, very hard to achieve practically and can remove the advantages of DBS.
3) Addition of the IS to the sample before spotting: As above in item 2.
4) Application of the IS on the spot after spotting: Fairly simple to perform in the bioanalytical laboratory and the IS is incorporated into the sample and medium. Possibly the best approach for the future as long as the addition of IS does not impact the integrity of blood spot.

One major consideration with the commercially available cellulose-based spotting media is the difference in the blood diffusion across the media for the blood with varying hematocrit levels. The varying hematocrit levels also have an effect on analyte extraction recovery from cellulose media prior to analysis (Denniff and Spooner 2010). The above phenomenon is commonly called "hematocrit effect." To overcome the issue due to the hematocrit effect, spotting accurate blood volume followed by cutting the entire spot for subsequent analysis is an option. In this regard, Aqua-Caps (Drummond Scientific) and positive displacement pipettes are two options for accurately spotting precise volumes of blood onto the DBS media. Accurate spotting not only takes away some of the simplicity of the DBS technique, but also takes significantly more time to complete the entire process from sample collection to analysis. Based on the potential issues regarding DBS and varying levels of hematocrit (sample size and extraction), a careful assessment of the technique in regard to validity and accuracy of the data should be performed prior to utilization of this technique. Recently, the regulatory agencies have been requesting a comparison of PK data obtained using "wet" (plasma, serum, or blood) versus "dry" (DBS) for each study to proceed with the "dry" method in drug development. The quality of the DBS samples is key. However, the dual sampling approach can be very expensive and resource-draining for any program.

Although not investigated by the author, at-home monitoring is also being explored using the DBS technology with spotting directly from a finger onto the collection device for dosed pharmaceutical entities. Collection of DBS through at-home monitoring is already in place for monitoring of Phenylketonuria (PKU) which enables constant monitoring without visits to the hospital. Phenylalanine is a natural part of all protein foods, but it accumulates in children and adults with PKU. The level of phenylalanine can be controlled by a special diet, but blood samples need to be taken to make sure the level of phenylalanine is not too high or too low. The finger or heel is simply pricked with a lancet, blood allowed to pool on area and then spotted, directly from finger, onto filter paper. The spotted paper is then placed into a waxy cover sleeve and mailed to the laboratory for analysis (Alder Hey Children's NHS Foundation Trust 2015).

A recently published report (Jager et al. 2015) investigated the at-home monitoring of tamoxifen during a self-sampling and at-home feasibility study for the individualization of tamoxifen therapy based on (*Z*)-endoxifen concentrations. Each patient received a short training for sample collection and a package of DBS collection cards, lancets, zip-closure bags, written instruction guide, and prepaid envelope. A questionnaire was completed by the patient regarding their experience with the DBS sampling experience. On arrival at the laboratory the spots were inspected for proper spotting. In regard to the patient's experiences, 97% found the training adequate and 92% stated the self-sampling went well. The results of this study showed that the majority of patients included in this study were capable of performing the at-home self-sampling, and preferred this over venous sampling.

9.3.2 Sample Manipulation

Parameters that have impact on the DBS method include utilization of untreated or chemically treated cellulose card, drying temperature and conditions, shipping and storage conditions, and laboratory additives used to enhance analyte or metabolite stability.

The choice of spotting media can affect compound stability, extraction efficiency and recovery, and potential interaction between analyte and media additives. On several occasions, the use of DBS has led to a pronounced increase in analyte stability when compared to a standard wet assay (Bowen et al. 2012). In most cases, this increase in stability can be attributed to removal of esterase activity or minimizing hydrolysis. However, the chemical card treatments are necessarily not "inert," and should be investigated particularly when new chemical treatments are explored. In rare cases, the compound may interact with the additives on the chemically treated card. For instance, use of the DMPK B card with primary amines can lead to on-card derivatization of the amine.

The simplest procedure for DBS card drying is to place the spotted cards on a drying rack for approximately two hours under ambient conditions. This is followed by placement into a zip-closure plastic bag containing desiccant. These spotted cards are then stored and shipped under ambient conditions until time of analysis. Procedures of "speed" drying have been investigated that include blow down with nitrogen and heated drying. All of these procedures are acceptable assuming there is no detrimental effect on the compound stability. However, in multisite studies it is difficult to ensure that all locations have access to the same drying capabilities. Therefore, while it is a possibility, accelerated drying is not recommended, and could have a deleterious effect on unstable metabolites.

Phase II metabolites often require stabilization before sample storage and transport, prior to bioanalysis. In a liquid form (blood or plasma), stabilization agent (acid, base, or other stabilization agent) is often added to prevent conversion of any phase II metabolite back to the parent compound. If not controlled, this conversion can lead to an overestimation of the parent compound.

Pretreatment of DBS media prior to spotting using ascorbic acid has been shown to minimize conversion (Bowen et al. 2012). The ascorbic acid can be added to the card, which was allowed to dry prior to blood spotting. These spotted cards can then be stored under ambient conditions prior to analysis.

9.3.3 Extraction, LC-MS/MS Analysis, and Detection

Blood is inherently more complex than plasma. This complexity could lead to additional interferences and challenges during the development of an LC-MS/MS assay for DBS. In addition to the plasma constituents, blood also contains red blood cells (erythrocytes), white blood cells (leukocytes), and platelets (thromobocytes).

Independent of the extraction method used, the DBS is punched and placed into a 1.2 ml tube. Due to the need to desorb the analyte from the card materials, the IS is often added in an aqueous solution (pH adjusted if needed) and allowed to remain in contact with the punch for up to 30 minutes with general mixing. This will ensure optimal recovery of the analyte into the aqueous phase for further manipulation and cleanup.

The use of an cleanup or extraction technique (PPT, LLE, or SPE) is then evoked to remove the nonspecific and endogenous interferences from the biological sample and helps to keep the detector clean and capable of analysis of a large number of samples without losing sensitivity or requiring cleaning.

Even with decreased sample volume, similar detection limits can often be reached by utilizing the advanced mass spectrometer with proper sample cleanup. With the potential increase in matrix effects with blood samples, a high-resolution mass spectrometer may be an option.

9.4 Volumetric Absorptive Microsampling (VAMS)

Volumetric Absorptive Microsampling (VAMS) device, MITRA, is relatively new for blood or other liquid sampling. The device is designed to deliver the benefits of DBS while tackling the issues of hematocrit. In utilization of the MITRA technique, a liquid sample is absorbed onto a porous, polymeric tip through capillary action. Through the manufacturing process, the size of the polymeric tip is controlled to only collect an accurate volume of blood (~10 µl) by capillary action. The internal volume of the tip is designed and precisely controlled to collect an accurate sample. The collection device has been designed to be complimentary to the common 96-well blocks for ease with automation and to minimize the possibility of the sampler coming into contact with other surfaces during shipment and storage. The current, ergonomic design is shown in Figure 9.3, and is similar to a pipette tip.

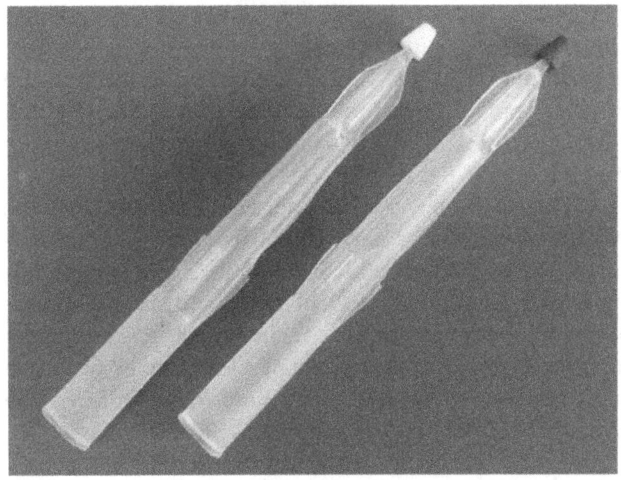

Figure 9.3 Illustration of MITRA Device for storage and collection of dried blood. *Source:* Reprinted with permission from Denniff and Spooner (2014). Copyright 2014. American Chemical Society.

As reported by Denniff and Spooner (2014), the VAMS apparently overcomes the issue due to hematocrit effect in conventional DBS LC-MS/MS bioanalysis, which is based on the size rather than the blood volume of the DBS punch. They reported a minimal difference in blood absorbed due to varying levels of hematocrit. Eight pools of fresh EDTA rat blood were prepared with hematocrit levels ranging from 20 to 70%. The average volume of blood absorbed onto the tips was 10.5 ± 01 µl (CV 3.6%) and the difference in blood volume across the entire hematocrit range is 5% (from 20 to 70%). In a separate investigation, caffeine and paraxanthine recovery in human blood with varying levels of hematocrit were recently compared (De Kesel et al. 2015) in utilizing the MITRA sampling device. Caffeine recovery was decreased from 101 to 92%, while parananthine decreased from 87 to 73%, when hematocrit in blood was increased from 21 to 62%. This data indicated a need to optimize recovery for the analyte of interest when developing a VAMS-related bioanalysis workflow.

9.4.1 Sample Collection

The device is filled by dipping the polymeric tip into a pool of blood and allowing it to fill until the tip is completely red (two to five seconds). Over filling for an additional five seconds was shown to have no impact on the analytical result, indicating that minimal additional blood is absorbed in this situation (Denniff and Spooner 2014).

The porous tip, pipette design allows direct sampling from the animal, bleeding site, or even at-home monitoring with no risk of clotting on the device. The tip can be directly applied to the pooled blood on either the skin surface or tube. With the fixed tip volume, these collectors are easy to use; the key is to ensure the tip is in contact with the blood to ensure adequate wicking onto the porous material. Once collected, the pipette-tip apparatus is stored under ambient conditions in a clamshell-type cartridge, which can be labeled and shipped under ambient conditions to the bioanalytical laboratory for analysis. Using the pipette design, blood can be placed in tubes, and analysis can be automated and performed in a high-throughput environment.

9.4.2 Sample Manipulation

Labeling of the individual MITRA devices are difficult, so care must be taken when preparing the tips for extraction. Minimal manipulation is possible with the MITRA samplers as there is no variation in the choice of the polymeric tip or sampler. The samplers are easily amendable to automation techniques and should be investigated for high-throughput workflows. Since the porous tips do not contain an anticoagulant, one consideration when using these samplers is the potential difference between the *in vivo* blood samples compared to the blood used for preparation of standards and quality controls as the latter often contains anticoagulant. An investigation should be conducted during method development and/or validation to ensure that the presence or lack of anticoagulant will not lead to varied analytical results.

9.4.3 Extraction, LC-MS/MS Analysis, and Detection

Dried blood on the MITRA tips would have similar complexities as mentioned earlier in regard to DBS. It is recommended to allow the blood to dry on the tips for up to two weeks (could be compound dependent) prior to starting method development in regard to sample extraction. The clamshell transport container (Figure 9.4) for the tips is designed to facilitate sampling, and allow drying while in transit through the incorporation of desiccant that can be added to the zip lock-type plastic storage container.

To authors' experience, the characteristics of the dried blood on the tip might change over time (e.g. two week period of time). As a result of this change, the optimized extraction conditions (solvent, pH, and time) that yields a high recovery with the overnight-dried blood on the polymeric tips may not deliver the same recovery for the blood that have been dried for a week. In reality, it has been often the case where the duration of sample collection and analysis varies from sample to samples and some of the difference can be significant. Therefore, during method development one should test the extraction conditions on samples of various ages (overnight to one month). This will ensure that the validated conditions are rugged enough to extract the analyte of interest from the tips independent of sample age. Alternatively, it may be necessary (compound-specific and needs to be investigated during method development) to have the QC samples prepared at various time points to mimic sample collection times.

Recently, we validated a method for quantitative analysis of proprietary compound in human blood using the MITRA collection devices. For sample extraction, the samples were soaked for 1 hour in 1% aqueous formic acid while sonication, which was followed by 30 minutes of vigorous vortexing. After this step, an LLE extraction with ethyl acetate was employed. The method was validated according to the EMA and FDA guidelines (EMA 2011; US FDA 2001). The key to developing this method was the addition of the 1% formic acid and the one-hour sonication. A major difference between the dried blood on paper (DBS) and the blood on polymeric tips is the three-dimensional shape of the tip. Taking this into

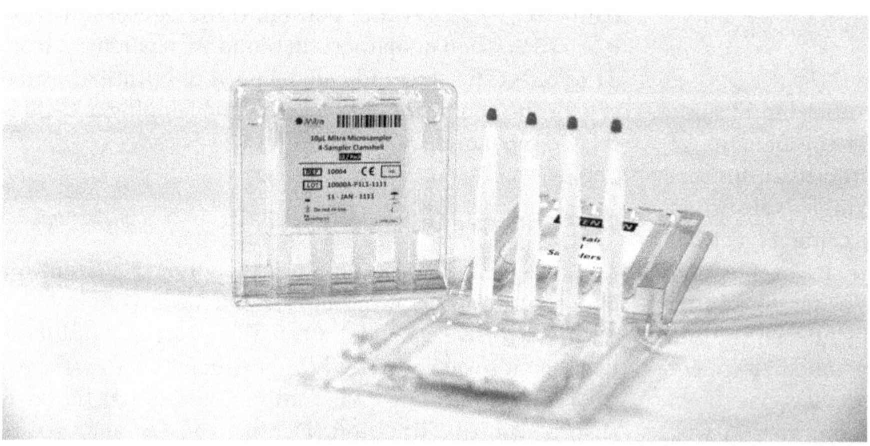

Figure 9.4 Illustration of MITRA Clamshell Device for storage and shipment of dried blood. *Source:* Reproduced with permission from Neoteryx, 2016.

Table 9.3 Validation statistics for a GSK proprietary compound extracted from VAMS samplers.

Items	Criteria
Calibration model	Linear-weighted $1/(x * x)$
Validated range	50–25 000 ng ml^{-1}
Precision (% CV)	≤9.6%
Within-run	
Precision (% CV)	≤10.4%
Between-run	
Accuracy (% bias)	−8.1% ≤ Bias ≤ 13.0%
Stability in 50/50/1% acetonitrile/water/formic acid (1.0 mg ml^{-1})	At least 59 d at −20 °C
Stability in human whole blood (liquid)	At least 4 h at 37 °C
	At least 24 h at ambient temperature
	At least 9 h at 4 °C
	At least 9 h on wet ice
	At least 10 d at −20 °C
	At least 307 d at −80 °C
Effect of temperature extremes on dried blood	At least 72 h at 40 °C
	At least 72 h −20 °C
Long-term storage stability	At least 28 d in dried human whole blood on VAMS tips at ambient temperature
Processed sample reinjectability	At least 48 h at 4 °C
Matrix dilution	Tenfold in 20/80 acetonitrile/water reconstituted extract from dried blank human blood on MITRA microsampling tips
Hematocrit effect	Suitability demonstrated in human blood at HCT levels between 0.30 and 0.60

consideration, additional time and harsher conditions (chemically and mixing) are necessary to ensure that the analyte of interests are extracted from the porous materials with consistently high recovery across the dynamic range (Table 9.3). It should be noted that in liquid blood the proprietary compound is not stable at −20 °C for longer than 10 days, where in the dry form it is stable for at least 28 days. As with DBS, this increase in stability can be attributed to removal of esterase activity or minimizing hydrolysis.

9.5 Emerging Techniques

The popularity of microsampling within bioanalysis has rapidly increased over the last 10 years due to the benefits it affords. These heightened levels of interest make developing new devices cost-effective for manufacturers. From lessons previously learned, the rules for new emerging sampling techniques are simple:

1) The quality of the sample is key – Samples cannot be lost due to the sample collection or storage device.
2) The quality of the sample is often dependent on the ease of sample collection – Ideally, minimal or no training is needed for using the new device as problems often occur in late-stage clinical development with large multisite studies where adequate training with new devices is often difficult to ensure.

One interesting development that is radically different from the existing techniques is solid-phase microextraction (SPME) coupled directly to LC/MS/MS (Ahmad et al. 2015). SPME uses coated fibers that are directly inserted into an animal or human via a cannula, left for less than a minute for the drug to equilibrate with the coated fiber. The fibers are then allowed to dry and can be shipped at ambient conditions for analysis. Conventional analysis can be performed by mixing the fiber with a suitable extraction solvent, to remove the analyte of interest, and then injecting onto a LC/MS/MS for analysis. Also a direct analysis approach can be taken whereby the fiber is inserted into a mass spectrometer source and a high voltage applied. This removes the need to extract from the fiber, and is under further development. The amount of blood used is minimal, i.e. 1 or 2 µl. The technique combines the attractive features of SPME microsampling using minimal sample volumes with the speed, selectivity, and sensitivity capabilities of MS detection. DBS still has a future and perhaps the biggest impact will be with drug monitoring (Wilhelm et al. 2014) especially with the added advantage of home sampling. The ability to use simple finger pricks opens up this avenue of science.

9.6 Summary

The main aim of a quantitative bioanalytical assay is to accurately measure the concentration of the circulating drug at a given time. Sampling and preservation of the sample until analysis is pivotal. Microsampling offers many scientific and logistical advantages and these have been well recognized within the industry. Preclinically the 3R savings are implicit and the ability to sample from pediatrics opens avenue to clinicians that previously could not be imagined.

The largest lessons that have been learned are the ability for the many to be able to use microsampling devices. However, one must effectively review the pros and cons of the technique chosen to ensure that the correct choices are made depending on the stage of compound development. Microsampling may not always be the ideal option. Late-phase clinical studies are inherently complex with many sampling sites and occasions. Specialist training in these situations to use what maybe a simple device to a trained analyst is extremely impractical and this can dramatically impact the quality of the sample, as simplicity is key. Microsampling is an extremely important area and the array of new devices and technologies that are being developed will continue to grow and enhance this area of research.

Acknowledgements

The authors would like to acknowledge Maria Busz for her analysis of the various pipetting options and Kasie Fang for her work on the MITRA method development and validation. We would also like to thank Lee Abberley and Scott Summerfield for their thorough review of this work. All of the above are part of the Biomarker, Immunology, and Bioanalysis (BIB) group at GlaxoSmithKline Pharmaceuticals.

References

Abu-Rabie, P., Denniff, P., Spooner, N. et al. (2011). Method of applying internal standard to dried matrix spot samples for use in quantitative bioanalysis. *Anal. Chem.* 83 (22): 8779–8786.

Ahmad, S., Tucker, M., Spooner, N. et al. (2015). Direct ionization of solid-phase microextraction fibers for quantitative drug bioanalysis: from peripheral circulation to mass spectrometry detection. *Anal. Chem.* 87 (1): 754–759.

Alder Hey Children's NHS Foundation Trust, Department of Clinical Biochemistry (2015). http://www.alderhey.nhs.uk/wp-content/uploads/Collection-of-blood-spots-PKU-PIAG-75.pdf (accessed December 2016).

Bowen, C., Kehler, J., Boram, S. et al. (2016). Modify of the fly: triple quad to high resolution in support of a dermal clinical study requiring an ultra low LLOQ. *Bioanalysis* 8 (3): 205–214.

Bowen, C., Licea-Perez, H., Karlinsey, M. et al. (2013). A novel approach to capillary plasma microsampling for quantitative bioanalysis. *Bioanalysis* 5 (9): 1131–1135.

Bowen, C., Volpatti, J., Cades, J. et al. (2012). Evaluation of glucuornide metabolite stability in dried blood spots. *Bioanalysis* 4 (23): 2823–2832.

Chambers, E., Wagrowki-Diehl, D., Lu, Z., and Mazzeo, J. (2007). Systematic and comprehensive strategy for reducing matrix effect in LC/MS/MS analysis. *J. Chromatogr. B* 852 (1–2): 22–34.

Clark, J. and Shull, A. (2003). Gravimetric and spectrophotometer errors impact on pipette calibration certainty. *Cal. Lab.* (January–March): 31–38.

De Kesel, P., Lambert, W., and Stove, C. (2015). Does volumetric absorptive microsampling eliminate the hematocrit bias for caffeine and paraxanthine in dried blood spots? A Comparative Study. *Anal. Chim. Acta* 881: 65–73.

Denniff, P. and Spooner, N. (2010). The effect of hematocrit on assay bias when using DBS samples for the bioanalysis of drugs. *Bioanalysis* 2 (8): 1385–1395.

Denniff, P. and Spooner, N. (2014). Volumetric absorptive microsampling: a dried sample collections technique for quantitative bioanalysis. *Anal. Chem.* 86 (16): 8489–8495.

Diehl, K., Hull, R., Morton, D. et al. (2001). A good practice guide to the administration of substances and removal of blood, including routes and volumes. *J. Appl. Toxicol.* 21 (1): 15–23.

European Medicines Agency (2011). *Guideline on Bioanalytical Method Validation.* http://www.ema.europa.eu/docs/en_GB/document_library/Scientific_guideline/2011/08/WC500109686.pdf (accessed December 2016).

Jager, N., Rosing, H., Linn, S. et al. (2015). Dried blood spot self-sampling at home for the individualization of tamoxifen treatment: a feasibility study. *Ther. Drug Monitor.* 37 (6): 83–36.

Jonsson, O., Villar, R., Nilsson, L. et al. (2012). Capillary microsampling of 25 μL blood for the determination of toxicokinetic parameters in regulatory studies in animals. *Bioanalysis* 4 (6): 661–674.

Kehler, J., Akella, N., Citerone, D., and Szapacs, M. (2011). Application of DBS for the quantitative assessment of a protein biologic using on-card digestion LC-MS/MS or immunoassay. *Bioanalysis* 3 (20): 2283–2290.

Liquid Handling Notebook (n.d.). https://www.fws.gov/aah/PDF/LiquidHandlingNotebook.pdf (accessed December 2016).

Musteata, F. (2009). Pharmacokinetic applications of microdevices and microsampling techniques. *Bioanalysis* 1 (1): 171–185.

Schulz, M., Iwersen-Bergmann, S., Andresen, H., and Schmoldt, A. (2012). Therapeutic and toxic blood concentrations of nearly 1000 drugs and other xenobiotics. *Crit. Care* 16 (4): R136.

Spooner, N., Lad, R., and Barfield, M. (2009). Dried blood spot as a sample collection technique for the determination of pharmacokinetics clinical studies: considerations for the validation of a quantitative bioanalytical method. *Anal. Chem.* 81 (4): 1557–1563.

Stravs, M., Mechelke, J., Ferguson, P. et al. (2016). Microvolume trace environmental analysis using peak-focusing online solid-phase extraction–nano-liquid chromatography–high-resolution mass spectrometry. *Anal. Bioanal. Chem.* 4–8 (7): 1879–1890.

Sun, K., Oh, H., Emerson, J.F., and Raghavan, S.R. (2012). A new method for centrifugal separation of blood components: creating a rigid barrier between density-stratified layers using a UV-curable thixotropic gel. *J. Mater. Chem.* 22: 2378–2382.

US FDA (2001). *Guidance for Drug Evaluation and Research. Guidance for Industry: Bioanalytical Method Validation* (May 2001). http://www.fda.gov/downloads/Drugs/GuidanceComplianceRegulatoryInformation/Guidances/UCM070107.pdf (accesed December 2016).

Van Eeckhaut, A., Lanckmans, K., Sarre, S. et al. (2009). Validation of bioanalytical LC-MS/MS assays: evaluation of matrix effects. *J. Chromatogr. B* 877 (23): 2198–2207.

Waters Corporation (2011). *Monitoring the Matrix Background Through the Sample Preparation Process in Complex Matrices Using the Xevo TQD with RADAR Functionality.* Water Application Note 72000397EN. http://www.waters.com/webassets/cms/library/docs/720003979en.pdf (accessed December 2016).

Wilhelm, A.J., den Burger, J.C.G., and Swart, E.L. (2014). Therapeutic drug monitoring by dried blood spot: progress to date and future directions. *Clin. Pharmacokinet.* 53 (11): 961–973.

10

Nanomaterials for Sample Preparation in LC-MS Bioanalysis

Rodrigo A. González-Fuenzalida, Neus Jornet-Martinez, Rosa Herráez-Hernández, and Pilar Campíns-Falcó

10.1 Introduction

Sample preparation is a critical step of many qualitative/quantitative analytical methods. The main goals of sample preparation are selective isolation of the target analytes from matrix, and/or pre-concentration of analytes and/or their transformation into chemical forms capable of providing better analytical parameters (e.g. selectivity, sensitivity, and accuracy). In **bioanalysis**, this step is particularly important due to the complexity of samples owing to their biochemical nature, the diversity of biological samples and their variability, as well as the difficulties to preserve their original properties (González-Fuenzalida et al. 2015). Sample preparation is needed even when a powerful technique such as liquid chromatography (LC) coupled to mass spectrometry (MS) is used. The most widely used techniques for sample preparation in bioanalysis are protein precipitation (PPT), liquid–liquid extraction (LLE), and solid-phase extraction (SPE), often combined with a previous treatment such as hydrolysis.

The increasing demand for the reduction in consumption of solvents, reagents, and materials as well as the need for more efficient and selective sample treatments have led to the development of new forms of LLE and SPE, such as liquid-phase microextraction (LPME) or micro-SPE (μ-SPE). New formats of sample extraction have also emerged such as hollow fiber LPME (HF-LPME), solid-phase microextraction (SPME), and magnetic SPE (MSPE), among others. At the same time, new phases with improved extraction capabilities, such as those containing ionic liquids, polymers, and **nanomaterials**, have been introduced.

Nanomaterials have received much attention because of their specific and unique properties. These materials have large surface area, providing excellent adsorption capacities when used as sorbents for sample preparation. Nanomaterials can also improve the selectivity through specific interaction with the target analytes. In addition, they can be synthetized in easy and cost-effective ways, and often show long-term life and reusability. Nanomaterials are increasingly used for sample preparation in environmental analysis, food analysis, and many other fields including bioanalysis. However, the number of articles that report nanomaterial-based sample treatment followed by LC-MS is still limited.

On the other hand, nanomaterials have been used to improve the performance of the main forms of extraction and microextraction, and they have contributed to the development and popularity reached by some of those techniques (e.g. MSPE and magnetic in-tube SPME) (Moliner-Martínez et al. 2012, 2014, 2015).

In this chapter, the main nanomaterials used for sample preparation in bioanalysis will be discussed. Several aspects of these materials, such as their physical and chemical characteristics relevant to bioanalysis, uses sorbents in different extraction techniques, and potential advantages and disadvantages, will also be discussed with some examples. Relevant properties and experimental details of such methods are summarized in Tables 10.1 and 10.2. Finally, future perspectives of these materials in the context of sample preparation for bioanalysis will be discussed.

10.2 Carbon Nanomaterials

Carbon-based nanomaterials, especially carbon nanotubes (CNTs) and graphene, have become popular sorbents for sample preparation because of the easiness in preparation, functionalization, and manipulation of these materials, as well as their favorable surface-to-volume ratios. Graphene is a single layer of carbon atoms forming a honeycomb (hexagonal) lattice while CNTs are hollow cylinders that consist of one (single-wall carbon nanotubes, SWCNTs) or several (multiwall carbon nanotubes, MWCNTs) graphene layers. Both graphene and CNTs have surfaces with hydrophobic properties

Sample Preparation in LC-MS Bioanalysis, First Edition. Edited by Wenkui Li, Wenying Jian, and Yunlin Fu.

Table 10.1 A summary of characteristics of the different sample treatments using nanomaterials in LC-MS.

Nanomaterial	Analyte	Sample	Extraction technique	LC technique	Use (or purpose)	LOD / recovery / Ext. eff.	Ref
MWCNTs	Peptides (BSA and ovalbumin)	Plasma	dSPE	1D- and 2D-nano-LC-MS/MS	Peptidome analysis	— / high recovery / —	Li et al. (2007)
MWCNTs	β_2-Agonists	Swine urine	dSPE	UPLC-ESI-MS/MS	Method validation	0.07–0.17 $\mu g l^{-1}$ (as CCβ) / 86.4–110.3% / high adsorption rates	Du et al. (2012)
Immobilized MWCNTs	Diuretics	Urine	HF-LPME	LC-ESI-MS/MS	Trace analysis	0.09–0.51 $ng\,ml^{-1}$ / 89–96% / —	Ho et al. (2013)
Oxidized MWCNTs with Fe_3O_4 NPs	NAaPs	Cell lysates	MSPE	1D- and 2D-nano-LC-MS/MS	*In vitro* studies	— / — / 95.8 ± 5.0%	Zhang et al. (2012)
Oxidized MWCNTs with Fe_3O_4 NPs	NAaPs	Living cells	MSPE	1D- and 2D-nano-LC-MS/MS	*In vitro*/*in vivo* studies	— / — / 93.5 ± 2.3% (*in vitro*); 95.3 ± 1.0% (*in vivo*)	Zhang et al. (2013)
AuNPs conjugated with antibody	OxPCs	Plasma	dSPE	LC-ESI-MS/MS	Analysis of potential biomarkers	— / — / high extraction yields	Hinterwirth et al. (2013)
AuNPs conjugated with antibody	MDA-LDL	Plasma	dSPE	LC-ESI-MS/MS	Analysis of potential biomarkers	— / 80% / —	Haller et al. (2015)
Ultrasmall AuNPs	N-linked glycopeptides	Rat kidney	SPE	LC-ESI-MS/MS	Analysis of potential biomarkers	— / — / up to 97% (as capturing selectivity)	Tran et al. (2012)
Magnetic supraparticle core with ZrO_2 shell	Phosphorylated peptides (β-casein and BSA)	Saliva	MSPE	Nano-LC-MS/MS	Phosphopeptide analysis	— / — / high effectiveness	Ma et al. (2014)
TiO_2 NPs layer in a capillary column	Phosphorylated peptides (α-casein, β-casein and BSA)	Digest mixtures of proteins	IT-SPME	Nano-LC-ESI-MS/MS	Phosphopeptide analysis	50 fmol / — / —	Lin et al. (2008)
TEPA-MP with Fe_3O_4 NPs	Estrogens	Blood	Magnetic d-μ-SPE	UPLC-MS/MS	Analysis of EDCs	0.022–0.14 $\mu g l^{-1}$ / 85–105% / —	Zhao et al. (2013)
Core-sheath PS/PNIPAAm nanofiber	Drugs	Plasma	μ-SPE	LC-MS/MS	Drug analysis	0.1–1 $ng\,ml^{-1}$ (as lower limit of the linear range) / — / up to 90.7% (as plasma protein binding)	Wu et al. (2013)
Graphene/mesoporous silica composites	Endogenous peptides	Mouse brain	SPE	Nano-LC-ESI-MS/MS	Nanoproteomics	— / — / good results (high number of peptides)	Yin et al. (2012)
CTAB-MCM-41 (mesoporous silica)	PFCAs	Plasma	μ-SPE	LC-ESI-MS/MS	Potential routine analysis	21.23–65.07 $ng\,l^{-1}$ / 87.6–102.5% / —	Lashgari and Lee (2016)
Ordered mesoporous (4.8 nm) carbon	Endogenous peptides	Human serum	SPE	1D- and 2D-nano-LC-MS/MS	Analysis of potential biomarkers	— / — / good extraction results (high number of unique peptides)	Qin et al. (2011)

(*Continued*)

Table 10.1 (Continued)

Nanomaterial	Analyte	Sample	Extraction technique	LC technique	Use (or purpose)	LOD / recovery / Ext. eff.	Ref
Magnetic mesoporous (3.4 nm) carbon	Endogenous peptides	Human serum	MSPE	LC-ESI-MS/MS	Analysis of LMW peptides	— / — / good extraction results (high number of unique peptides)	Wan et al. (2013)
Magnetic mesoporous (3.7 nm) RAM	Diazepam	Rat plasma	MSPE	LC-ESI-MS	Drug analysis	$3\,ng\,ml^{-1}$ / 92.8–95.2% / —	Liu et al. (2013)
Magnetic mesoporous (20 nm) titania	Phosphorylated peptides (β-casein, ovalbumin, and BSA)	Mouse brain (digest)	MSPE	Nano-LC-ESI-MS/MS	Phosphopeptide analysis	0.4 nM (β-casein) / — / good results (high number of peptides)	Lu et al. (2013)
Carboxyl-functionalized magnetic-CNT-MIPs	Catecholamines (DA, E, NE)	Plasma	Magnetic dispersive μ-SPE	UFLC-MS/MS	Potential routine analysis	$10\,ng\,l^{-1}$ (DA); $18\,ng\,l^{-1}$ (E); $76\,ng\,l^{-1}$ (NE) / 87.5–110% / —	Ma et al. (2016)
Magnetic MIPs with Fe_3O_4 NPs	Roxithromycin	Plasma	MSPE	LC-ESI-MS/MS	Trace analysis	$3.8\,ng\,ml^{-1}$ / 86.5–91.5% / —	Ding et al. (2016)

LC, liquid chromatography; LOD, limit of detection; Ext. Eff., extraction efficiency; Ref, reference; MWCNTs, multi-walled carbon nanotubes; BSA, bovine serum albumin; dSPE, dispersive solid-phase extraction; 1D- and 2D-nano-LC, one-dimensional and two-dimensional liquid nanochromatography; MS/MS, tandem mass spectrometry; UPLC, ultra performance liquid chromatography; ESI, electrospray ionization; CCβ, detection capability; HF-LPME, hollow fiber-based liquid-phase microextraction; Fe_3O_4 NPs, iron-oxide nanoparticles; NAaPs, nucleic acid associated proteins; MSPE, magnetic solid-phase extraction; AuNPs, gold nanoparticles; OxPCs, oxidized phosphatidylcholines; MDA-LDL, malondialdehyde-modified low-density lipoprotein; nano-LC, liquid nanochromatography; TiO_2 NPs, titanium dioxide nanoparticles; IT-SPME, in-tube solid-phase microextraction; TEPA-MP, tetraethylenepentamine-functionalized magnetic polymer; d-μ-SPE, dispersive micro-solid-phase extraction; EDCs, endocrine disrupting chemicals; PS/PNIPAAm, polystyrene/poly(*N*-isopropylacrylamide); SPE, solid-phase extraction; μ-SPE, micro-solid-phase extraction; CTAB, cetrimonium bromide; MCM-41, mobil composition of matter number 41; PFCAs, perfluorinated carboxylic acids; LMW, low molecular weight; MS, mass spectrometry; RAM, restricted access material; MIPs, molecularly imprinted polymers; DA, dopamine; E, epinephrine; NE, norepinephrine; UFLC, ultrafast liquid chromatography.

Table 10.2 Number of steps of the different sample treatments using nanomaterials in LC-MS.

Nanomaterial	Sample preparation steps	Details of the procedure (operational time)	Ref
MWCNTs	dSPE[a]	Mixing, **incubation** (15 min), centrifugation (10 min), washes, elution, centrifugation, filtration	Li et al. (2007)
MWCNTs	dSPE	Mixing (1 min), **incubation** (2 h), agitation (5 min), centrifugation (3 min), decantation, elution (1 min), centrifugation (2 min), filtration	Du et al. (2012)
Immobilized MWCNTs	HF-LPME	Sample-device contact, agitation (30 min), desorption (water bath, 5 min)	Ho et al. (2013)
Oxidized MWCNTs with Fe_3O_4 NPs	MSPE protein digestion	Mixing on ice (1 min), centrifugation (5 min), three washes (using a magnet), resuspension, **incubation** (16 h), SPE, lyophilization	Zhang et al. (2012)
Oxidized MWCNTs with Fe_3O_4 NPs	MPSE protein digestion	On ice: Three washes, mixing, **incubation** (9 h), three washes, cell lysis, centrifugation (10 min), washes (using a magnet), resuspension, **sonication (5 s) incubation** (16 h), SPE, lyophilization	Zhang et al. (2013)
AuNPs conjugated with antibody	dSPE	**Incubation** (overnight), four washes (centrifugation, resuspension), US resuspension (5 min), centrifugation	Hinterwirth et al. (2013)
AuNPs conjugated with antibody	dSPE	**Incubation** (overnight), three washes (centrifugation, resuspension), US resuspension, centrifugation (30 min)	Haller et al. (2015)
Ultrasmall AuNPs	Sample digestion SPE solvent change	Cell lysis, FASP, elution, SPE, **Incubation** (1 h), SPE, mixing, sonication (5 min), **incubation** (overnight), three washes, enzymatic cleavage (5 h), cleanup[b] centrifugation, evaporation, resuspension	Tran et al. (2012)
Magnetic supraparticle core with ZrO_2 shell	MSPE[a]	Mixing, **incubation** (30 min), separation (using a magnet), wash, elution	Ma et al. (2014)
TiO_2 NPs layer in a capillary column	Protein digestion IT-SPME	**Incubation** (overnight, a 45 min additional pretreatment for BSA), SPE, centrifugation Sample load, wash (25 min), elution	Lin et al. (2008)
TEPA-MP with Fe_3O_4 NPs	Sample preprocessing magnetic d-μ-SPE solvent change	Three SLE (agitation, 1 min, and centrifugation, 5 min), mixing (1 min), separation (under a magnetic field, 2 min), evaporation, resuspension, filtration	Zhao et al. (2013)
Core-sheath PS/PNIPAAm nanofiber	μ-SPE	Sample-device contact, agitation (2 min), wash, centrifugation (1 min), desorption (under agitation, 10 min)	Wu et al. (2013)
Graphene/mesoporous silica composites	SPE[c]	Sample preprocessing (not detailed), extraction-enrichment, centrifugal separation, elution	Yin et al. (2012)
CTAB-MCM-41 (mesoporous silica)	Sample preprocessing μ-SPE	Anticoagulation, freeze-drying, protein precipitation, centrifugation, device conditioning, sample-device contact, agitation (25 min), desorption (under US, 15–17 min)	Lashgari and Lee (2016)
Ordered mesoporous (4.8 nm) carbon	SPE[a]	Denaturation (5 min), mixing, agitation (0.5 h), decantation, two washes, elution, centrifugation (3 min), lyophilization	Qin et al. (2011)
Magnetic mesoporous (3.4 nm) carbon	MSPE[a]	Denaturation (5 min), mixing, agitation (0.5 h), decantation, two washes, elution (using a magnet), centrifugation (5 min), lyophilization	Wan et al. (2013)
Magnetic mesoporous (3.7 nm) RAM	MSPE	Centrifugation (10 min), mixing, agitation (3 min), separation (using a magnet), two washes, decantation (using a magnet), elution (5 min)	Liu et al. (2013)
Magnetic mesoporous (20 nm) titania	Sample preprocessing protein digestion MSPE	Homogenization on ice, mincing, wash, homogenization (2 min), agitation (30 min), centrifugation (1.5 h), lyophilization, resuspension, mixing, **three incubation** (1.5 h and the latter overnight), lyophilization, resuspension, mixing (30 min), separation (using a magnet), three washes, elution (30 min), lyophilization, resuspension	Lu et al. (2013)

(Continued)

Table 10.2 (Continued)

Nanomaterial	Sample preparation steps	Details of the procedure (operational time)	Ref
Carboxyl-functionalized magnetic-CNT-MIPs	(Magnetic) d-μ-SPE solvent change	Homogenization (1 min), centrifugation (2 min), mixing, agitation (5 min), separation (under magnetic field, 1 min), wash, elution, evaporation, resuspension, filtration	Ma et al. (2016)
Magnetic MIPs with Fe_3O_4 NPs	MSPE solvent change	MIP conditioning, mixing, agitation (5 min), separation (using a magnet), wash, elution (three times under US, 30 s each), evaporation, resuspension, filtration	Ding et al. (2016)

Note:
[a] Protein digestion is performed for the preparation of standards, procedure including an **incubation** step overnight or that lasts more than 12 hours;
[b] using a spin column;
[c] procedure schematically explained.
Ref, reference; MWCNTs, multi-walled carbon nanotubes; dSPE, dispersive solid-phase extraction; HF-LPME, hollow fiber-based liquid-phase microextraction; Fe_3O_4 NPs, iron-oxide nanoparticles; MSPE, magnetic solid-phase extraction; AuNPs, gold nanoparticles; FASP, filter-aided sample preparation; TiO_2 NPs, titanium dioxide nanoparticles; IT-SPME, in-tube solid-phase microextraction; TEPA-MP, tetraethylenepentamine-functionalized magnetic polymer; d-μ-SPE, dispersive micro-solid-phase extraction; SLE, solid–liquid extraction; PS/PNIPAAm, polystyrene/poly(*N*-isopropylacrylamide); μ-SPE, micro-solid-phase extraction; SPE, solid-phase extraction; CTAB, cetrimonium bromide; MCM-41, mobil composition of matter number 41; RAM, restricted access material; CNTs, carbon nanotubes; MIPs, molecularly imprinted polymers.

and thus exhibit high affinities for organic and inorganic molecules. In addition, because of the existence of six-membered rings and delocalized electrons in their structure, they show strong affinity toward aromatic compounds. As a result, they can interact with the analytes by different adsorption mechanisms, including π-π stacking, Van der Waals forces, hydrophobic interactions, and electrostatic forces (Ravelo-Pérez et al. 2010). CNTs are the most commonly used nanomaterials in bioanalysis (see Table 10.1).

CNTs have been extensively applied for the extraction of analytes from biological matrices such as urine, saliva, plasma, blood, and tissues. They are generally used as the sorbent in a dispersive SPE format (D-SPE) (see Figure 10.1). The process involves the dispersion of CNTs into the (liquid) sample, followed by the separation of CNTs with the retained analytes by centrifugation or filtration (many protocols also incorporated one or more washing steps), and finally, the desorption of the trapped analytes with an appropriate solvent which has to be also suitable for LC-MS. Table 10.2 summarizes relevant details of nanomaterial-based sample preparation procedures in bioanalysis. For example, Li et al. (2007) reported the use of MWCNTs for the D-SPE of peptides from digested human plasma as an alternative to SPE cartridges packed with conventional sorbents (silica, C_2, C_8, and C_{18}). The authors concluded that MWCNTs were superior to C18 and C8 sorbents for capturing the smallest peptides. MWCNTs yielded 2521 peptides in the digest sample, while with C18 and C8 sorbents only 2280 and 1724 peptides were found, respectively. According to the authors, the employment of MWCNTs for sample treatment followed by separation and detection by LC/MS provided better sensitivity than previously reported methods, especially for peptides with low molecular masses (600–1200 Da range) which are of particular interest as disease-related biomarkers. MWCNTs have been also used for the determination of 10 β-agonists in urine by D-SPE and UHPLC-ESI–MS/MS (Du et al. 2012).

The utility of MWCNTs as sorbents for HF-LPME followed by LC-MS/MS analysis has been also proved. This technique employs a membrane (HF) between sample solution (donor) and extractant (acceptor) which avoids mixing of the phases (see Figure 10.1). The analytes are extracted from aqueous sample through the organic solvent immobilized in the wall pores of the membrane and into the extractant solution inside the lumen of the hollow fiber. The advantages of this technique are very low solvent consumption and separation of the analytes from high molecular mass compounds which cannot pass through the membrane. Thus, HF-LPME is of special interest for quantitative analysis of analytes in biological samples. Ho et al. (2013) employed the technique for analysis of diuretics in urine. In this study, MWCNTs were dispersed in *n*-octanol and trapped within the pores of the membrane. The nanotubes were weakly bonded to the diuretics in samples by π-π interactions, leading to higher extraction efficiencies than those obtained with the same membranes without MWCNTs.

CNTs can be modified through chemical reactions in order to improve their selectivity and/or efficiency. For example, oxidation of CNTs leads to the formation of carboxyl, hydroxyl, and epoxy groups, which in return improves the dispersity of CNTs in aqueous media, and makes possible new types of adsorption mechanisms

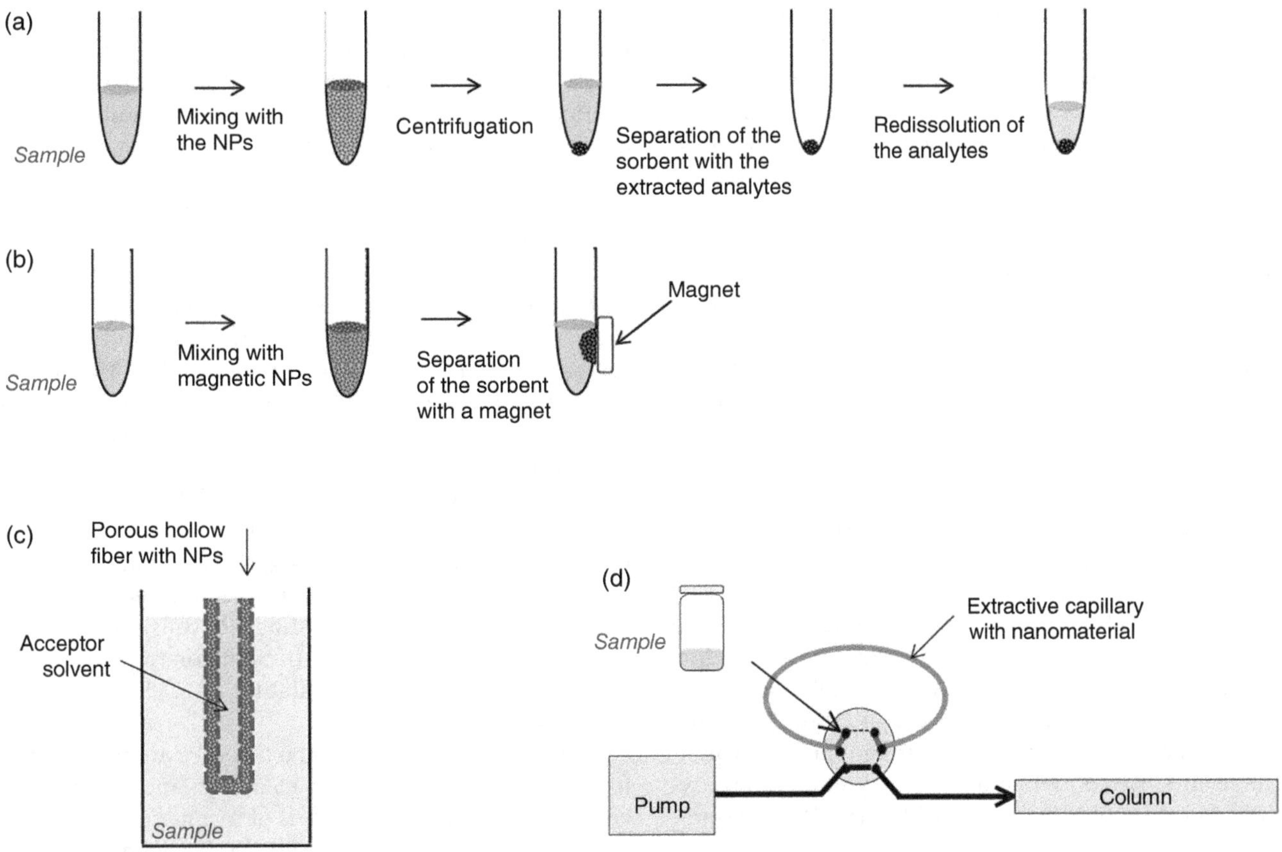

Figure 10.1 Basic principles of the main extraction techniques used with nanomaterials for sample preparation in bioanalysis: (a) dSPE, (b) MSPE, (c) HF-LPME, and (d) IT-SPME. In (b) after the separation of sorbent with the aid of a magnet, the extracted analytes are separated from sample matrix and redissolved like in (a); in (d) samples are passed through the capillary for extraction of the analytes with the valve in load position (continuous line) and then transferred to the chromatographic column by mobile phase when the valve changed to the inject position (dotted line).

(Ravelo-Pérez et al. 2010). On the other hand, CNTs can be immobilized onto magnetic nanoparticles (MNPs) that can be used for MSPE (see Figure 10.1). MNPs are a kind of metal oxide nanoparticles (see Section 10.3.2) that have been increasingly used for D-SPE due to the increased speed of collection and separation of the analytes as a result of the application of an external magnetic field. Zhang et al. (2012) prepared magnetic oxidized CNTs composite oMWCNTs@Fe_3O_4 for the extraction of nucleic acid associated proteins (NAaPs) from cell lysates prior to their analysis by nano-LC-MS/MS. For this purpose oMWCNTs@Fe_3O_4 were mixed with cell lysates and, because of the strong interaction between CNTs and nucleic acids (NAs), NAaPs could be efficiently extracted and separated with a magnet. Once isolated, the oMWCNTs@Fe_3O_4@NAs@NAsPs composite was directly treated for protein digestion and the digested peptides were purified with C18 SPE and chromatographed. The authors reported that, compared with the traditional method (end-circled capillary-based strategy), the use of oMWCNTs@Fe_3O_4 provided higher extraction efficiency and better reproducibility. The same nanoparticles were used in a latter study to extract NAaPs from living cells taking advantage of the fact that oMWCNTs@Fe_3O_4 can be ingested by cells naturally (Zhang et al. 2013). In this study the nanoparticles were incubated with cells, and the cells were collected, washed, and cracked for the isolation of NAaPs. Although the application of MSPE with CNTs offers some advantages, achieving homogeneous dispersions may be sometimes difficult. Unwanted interactions between matrix components and MNPs is another limitation that needs to be considered when implementing the technique.

10.3 Metallic NPs

10.3.1 Metal Nanoparticles

Gold nanoparticles (AuNPs) are biocompatible materials with high surface which can be easily functionalized for specific recognition of target molecules. AuNPs have strong affinity for thiol groups, and they also show some affinity for amine groups (Tables 10.1 and 10.2). They have been applied to surface functionalization via Au—S for selective biomolecule retention in bioseparations. Because of their high surface-to-volume ratio as well as easy and low cost synthesis with narrow size distribution, AuNPs are a popular sorbent for trapping protein/peptide, lipids, and other organic compounds from complex matrices. This was the strategy applied by Hinterwirth et al. (2013) for the selective extraction of oxidized low-density lipoproteins (OxLDL) from plasma, as these compounds are potential biomarkers. For this purpose anti-OxLDL antibodies were immobilized onto AuNPs, and the resulting material was used for the selective trapping of OxLDL following a D-SPE scheme. The extracts with the oxidized lipoproteins were directly processed by LC-MS/MS. Haller et al. (2015) studied the usefulness of modified AuNPs for the selective capture of OxLDL like malondialdehyde-modified low-density lipoprotein (MDA-LDL), which are used as biomarkers of oxidative stress. In this study different approaches for the immobilization of anti-MDA-LDL antibodies were tested. Under optimized conditions, up to three antibodies per nanoparticle could be immobilized, which resulted in suitable extraction recovery and selectivity for the detection of MDA-LDL by LC-MS/MS in plasma samples.

Ultrasmall AuNPs contain a few hundreds core atoms and are very useful for extraction purposes due to their biocompatibility and high surface-to-volume ratio. The presence of thiolate ligands prevents the aggregation of this kind of NPs giving stable and well-defined nanostructures. This was illustrated by Tran et al. (2012), who functionalized glutathione-protected ultrasmall AuNPs with hydrazide groups. The obtained NPs were used for the isolation/enrichment of N-glycosylated peptides prior to their analysis by LC-MS. The authors reported that the synthetized NPs were very stable in biological samples (rat kidney tissue) and had large capacity for peptide capturing. The analysis of the extracted samples showed that about 90% of captured analytes was glycopeptide.

10.3.2 Metal Oxide Nanoparticles

Metal oxide nanoparticles such as TiO_2, ZnO, ZrO_2, SnO_2, and others are relevant materials for the enrichment of compounds of interest in bioanalysis. Several TiO_2-based methods have been developed for the enrichment of traces of phosphopeptides from complex matrices prior to analysis by LC-MS. It is known that molecules with phosphate groups can self-assemble onto the surface of TiO_2 particles, whereas their elution can be achieved at alkaline pHs. An important parameter to consider is the size of particles because the surface area active for interaction increases as the particle size decreases; particles of few nanometer size are preferable as demonstrated for several peptides (Tran et al. 2012).

Metal oxide NPs have also been used for in-tube solid phase microextraction (IT-SPME). In this technique, the extractive phase is packed in a capillary column or deposited in its inner surface, so the analytes are trapped when the sample is passed through the capillary. The main advantage of IT-SPME is that the extractive capillary column can be connected to the LC system, so that sample treatment can be carried out online (see Figure 10.1). Using this technique, Lin et al. (2008) developed a procedure for the analysis of phosphopeptide from proteolytic digestion of biological samples by employing a TiO_2 nanoparticle-coated capillary column which was coupled to an LC-ESI-MS/MS system. According to the authors, the extractive phase was very efficient for the extraction of phosphopeptides and stable after hundreds of extractions.

A remarkable kind of metal oxide NPs are those made from ferromagnetic metal elements including iron, cobalt, and nickel. Applications of magnetic metal oxide can be divided into two types: (i) MNPs act as the sorbent material and (ii) MNPs act as a magnetic support for other nanomaterials responsible for the interaction with the analytes. Most applications in bioanalysis correspond to the latter type. The most utilized MNPs are ferrite (Fe_3O_4) nanoparticles.

Different combinations of TiO_2 with other nanomaterials have been developed in order to improve the selective enrichment of phospholipids from complex biological samples. Lu et al. (2013) synthesized a Fe_3O_4@ mesoporous TiO_2 composite with a core-shell structure (Fe_3O_4 spherical nucleus coated with TiO_2 NPs). This material was successfully applied to the selective enrichment of phosphoproteins under a M-SPE scheme. The measured specific surface area of this composite was higher of that of Fe_3O_4@mesoporousTiO_2 microspheres previously developed for the selective enrichment of phosphopeptides from tryptic digestion of mouse brain for LC–ESI-MS analysis.

Besides TiO_2-based sorbents, other metal oxide-based-nanoparticles, such as ZrO_2, have been utilized for similar purposes. This is the case of the multifunctional nanomaterial with a yolk-shell structure based on magnetic supraparticles (MSP) core and a crystal ZrO_2 shell developed by Ma et al. (2014). The magnetic yolk-shell MSP@ZnO_2 microspheres were applied to the selective enrichment of low-abundant phosphopeptides followed by nanoflow LC-MS/MS analysis of human saliva sample.

10.4 Nanoporous Materials

10.4.1 Polymeric Nanomaterials

Polymeric nanomaterials consist of a polymer or mixture of polymers with different shapes (e.g. platelets, fibers, and spheroids). For the latter, at least one dimension of the different shapes must be in the range of 1–50 nm. The use of biocompatible polymeric nanomaterials has proved to be very useful for bioanalysis (Reyes-Gallardo et al. 2015). In the context of sample preparation, these materials minimize chemical changes in sample during its treatment. They also avoid temporal resolution which is a problem commonly encountered when analyzing biological matrices because nanoscale fiber structures offer high mass transfer rate of analytes between liquid and solid phases. This was demonstrated in a study with core-sheath biocompatible electrospun nanofibers with a polystyrene core and protein-resistant poly(*N*-isopropylacrilamide) polymer as sheath. The described material was tested for different drugs in plasma samples by μ-SPE and LC-MS/MS (Wu et al. 2013).

Ferrite NPs in combination with tetraethylenepentamine (TEPA) magnetic polymer has been synthetized and applied for dispersive μ-SPE of phenolic estrogens in blood by Zhao et al. (2013). According to the authors, the designed adsorbent was inexpensive and showed strong adsorption capabilities, facilitating the removal of phospholipids through magnetic separation with acceptable recoveries (85–105%). In addition, the proposed method eliminated the matrix effects caused by phospholipids.

10.4.2 Mesoporous Materials

Mesoporous materials contain nanopores in 2–50 nm range. Because of their highly specific surface, these sorbents are very well suited for the retention of low-molecular weight analytes. For example, MCM-41 is an aluminosilicate-derived material which possesses a hexagonal array of uniform, regular, one-dimensional pores of diameters in the range of 2–6.5 nm (Stein and Melde 2001). Surfactant templated MCM-41 (CTAB-MCM-41) has been used as sorbent under a μ-SPE format. The sorbent was packed in a sealed porous polypropylene membrane envelope. Compared to SPE with cartridges, the bags did not suffer from high backpressure or blockage. The developed μ-SPE method was used for the determination of trace levels of perfluorinated carboxylic acids in human plasma (Lashgari and Lee 2016) by LC-MS/MS.

Mesoporous materials are often combined with NPs in order to improve the extraction performance. For example, mesoporous silica-based material ($mSiO_2$) has been used to prepare a C_8-modified graphene@$mSiO_2$ composite (Yin et al. 2012). The material was successfully employed for the highly specific enrichment of the endogenous peptides from a mouse brain tissue prior to LC-MS analysis.

Recently, different studies have shown that ordered mesoporous carbon materials can provide better extraction efficiency for some compounds than the silica mesoporous materials because the former are more hydrophobic (Qin et al. 2011). This has been illustrated for the extraction and analysis of low-molecular weight peptides as potential biomarkers demonstrated that with an ordered mesoporous carbon-based sorbent the efficiency for the extraction of peptides from a small amount of serum was advantageous over silica mesoporous materials, ferrite NPs, and C_{18}-based SPE procedures, as well as ultrafiltration. Carbonaceous materials can be also combined with magnetic NPs (Wan et al. 2013). The material showed better performance in the enrichment of endogenous peptides than magnetic mesoporous silica due to the mentioned hydrophobicity.

A particular interest in the analysis of biofluids is restricted access material (RAM). This kind of materials possess an inner surface favorable to the retention of low-molecular weight compounds by hydrophilic, hydrophobic, ionic or affinity interactions, while the outer surface prevents the retention of high-molecular weight compounds by a size exclusion mechanism. In such a way, RAM enables the isolation and enrichment of analytes at trace levels in the presence of macromolecules. This is a clear advantage over SPE, in which the macromolecules such as proteins compete with small-molecule analytes for the binding sites, resulting in lower recoveries and increased interferences. The inner and outer surfaces of RAM can be chemically modified with groups of different hydrophobicity in order to enhance the selectivity of the extraction. RAM-based sorbents have been used for some decades, but additional benefits have been achieved through combination with other nanomaterials. As an example, the successful extraction of diazepam from rat plasma has been reported using a C_8-Fe_3O_4@$mSiO_2$ composite RAM sorbent (Liu et al. 2013). In the sample preparation process, these materials captured the small analyte molecules by hydrophobic interaction with the C_8 groups located at the inner walls of the mesoporous channels, whereas macromolecules were excluded. Moreover, the magnetic properties of the sorbent facilitated the separation of the analytes from the sample matrix.

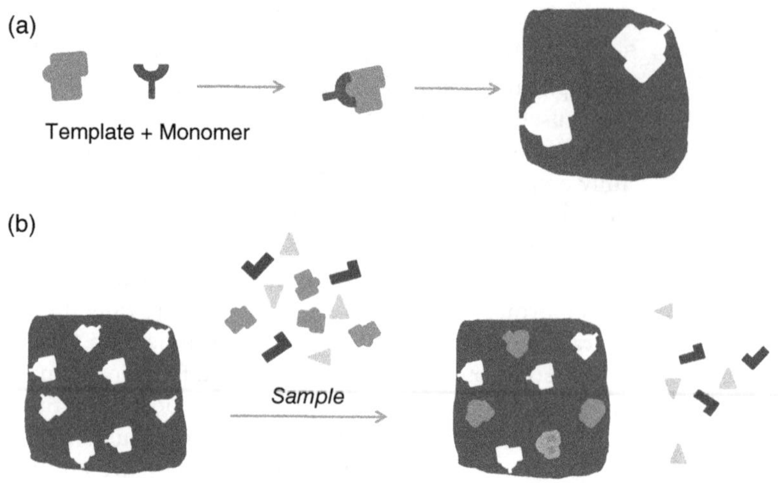

Figure 10.2 Extraction with MIPs: (a) preparation and (b) principle of MIPs for sample preparation.

10.4.3 Molecular Imprinted Polymers (MIPs)

MIPs are polymers synthetized using the target analyte as a template and an appropriate functional monomer (for example, methacrylic acid or 4-vinylpyridine) in the presence of a porogenic solvent and a cross-linker agent. Polymerization takes place around the analyte, and once it is completed, the analyte is removed, leaving a rigid 3D cavity that is complementary to the analyte (see Figure 10.2). The synthesis of these sorbent materials is often inexpensive, fast, and reproducible. For these reasons, MIPs have been extensively used for the specific extraction of the target compound from complex matrices, including biological samples.

Moreover, MIPs can be used for the preparation of different types of nanocomposites with magnetic NPs, CNTs, and other materials by incorporating the NPs to the polymerization mixture. For example, polyvinylpyrrolidines and acrylic acid-based MIP polymers have been synthesized in the presence of ferrite magnetic NPs. The sorbent was used for M-SPE of roxithromycin from human plasma (Ding et al. 2016). MIPs have been also combined with magnetic-functionalized carboxyl-CNTs (Ma et al. 2016) for dispersive μ-SPE combined with ultrafast liquid chromatography–tandem quadrupole mass spectrometry for the analysis of catecholamines in human plasma.

10.5 Future Perspectives

This chapter reports the variety and the high potential of nanomaterials for sample preparation in LC-MS/MS analysis. Table 10.3 summarizes advantages or disadvantages of the several kinds of nanomaterials in reference to sample treatment. Future perspectives of this field lie in miniaturization (SPME, LPME) and automation of sample treatment. While currently most widely used procedures based on NPs are off-line approaches, the development of online procedures is also being explored. Online microextraction is a powerful tool to analyze samples with speed and sensitivity with integrated sample extraction, pre-concentration, and subsequent LC-MS/MS. There is no doubt that the use of nanomaterials online can contribute to improve the analyte recoveries with increased sensitivity and selectivity (González-Fuenzalida et al. 2015, 2016; Moliner-Martínez et al. 2011, 2012, 2014, 2015).

Some nanomaterials extensively used in other areas remain relatively unexplored for LC-MS bioanalysis. This is particularly the case for silica NPs and metal organic frameworks (MOFs). For example, silica NPs bind to proteins, which have been exploited in some studies although not specifically for the preparation of biological samples (Sarparanta et al. 2012). These NPs show good biocompatibility, and they can be easily synthesized and functionalized at very low cost. Thus, future applications in the area of bioanalysis can be expected.

Acknowledgements

The authors are grateful to the Spanish Ministerio de Economía y Competitividad (MINECO)/FEDER (project CTQ2014-53916-P) and to Generalitat Valenciana (PROMETEO 2016/109). NJ and RAG express their gratitude to MINECO/FEDER for their postdoctoral grant.

Table 10.3 General advantages and disadvantages of the different types of nanomaterials in sample preparation.

Type of nanomaterials	Advantages	Disadvantages	Sample preparation technique
Carbon nanomaterials: CNTs	• High adsorption capacity • Easy modification/ functionalization	• High hydrophobicity: incompatible in aqueous media • Spontaneous aggregation • High pressure in packed columns and cartridges • Nonbiodegradable	• SPE • dSPE/d-μ-SPE • HF-LPME • MSPE
Metal nanoparticles: AuNPs	• Excellent biocompatibility using appropriate ligands • Highly stability • Easy surface modification • Good biomolecule recognition via Au—S bonds	• Nonbiodegradable • High cost for large-scale production • Few application as sorbent material	• SPE • d-SPE
Metal oxide nanoparticles: TiO_2, ZnO_2	• High stability • Low cost • Nontoxic • Selective adsorption of organic phosphate groups	• Nonbiodegradable • Application limited: enrichment of phospholipids/phosphopeptides	• IT-SPME
Magnetic nanoparticles: Fe_3O_4	• Quick and easy isolation from the samples • Controlled clustering	• Limited biocompatibility • Poor colloidal stability (formation of aggregates) • Nonspecific adsorption • Nonbiodegradability	• MSPE
Polymeric materials	• Good biocompatibility • Good biodegradation • Easy functionalization • Tunable composition	• Instability: colloidal fouling and biofouling • High-cost production	• SPE/μ-SPE • MSPE
Mesoposorus materials	• Controllable porosity • Easy functionalization • Biodegradability	• Surface modification produces a decrease in pore size • Large particles size limits the accessibility and reduces the desorption and elution efficiency	• SPE/μ-SPE • MSPE

References

Ding, J., Zhang, F., Zhang, X. et al. (2016). Determination of roxithromycin from human plasma samples based on magnetic surface molecularly imprinted polymers followed by liquid chromatography-tandem mass spectromer. *J. Chromatogr. B* 1021: 221–228.

Du, X.D., Wu, Y.L., Yang, H.J., and Yang, T. (2012). Simultaneous determination of 10 β_2-agonists in swine urine using liquid chromatography-tandem mass spectrometry and multi-walled carbon nanotubes as a reversed dispersive solid phase extraction sorbent. *J. Chromatogr. A* 1260: 25–32.

González-Fuenzalida, R.A., López-García, E., Moliner-Martínez, Y., and Campíns-Falcó, P. (2016). Adsorbent phases with nanomaterials for in-tube solid-phase microextraction coupled on-line to liquid nanochromatography. *J. Chromatogr. A* 1432: 17–25.

González-Fuenzalida, R.A., Moliner-Martínez, Y., Verdú-Andrés, J. et al. (2015). Microextraction with phases containing nanoparticles. *Bioanalysis* 7: 2163–2170.

Haller, E., Lindner, W., and Lämmerhofer, M. (2015). Gold nanoparticle-antibody conjugates for specific extraction and subsequent analysis by liquid chromatography-tandem mass spectrometry of malondialdehyde-modified low density lipoprotein as biomarker for cardiovascular risk. *Anal. Chim. Acta* 857: 53–63.

Hinterwirth, H., Stuebiger, G., Lindner, W., and Lämmerhofer, M. (2013). Gold nanoparticle anti-oxidized low-density-conjugated lipoprotein antibodies for targeted lipidomics of oxidative stress biomarkers. *Anal. Chem.* 85 (17): 8376–8384.

Ho, T.T., Li, Z.G., Lin, H.Y., and Lee, M.R. (2013). Determination of diuretics in urine using immobilized multi-walled carbon nanotubes in hollow fiber liquid-phase microextraction combined with liquid chromatography-tandem mass spectrometry. *J. Chin. Chem. Soc.* 60 (8): 1033–1042.

Lashgari, M. and Lee, H.K. (2016). Micro-solid phase extraction of perfluorinated carboxylic acids from human plasma. *J. Chromatogr. A* 1432: 7–16.

Li, X., Xu, S., Pan, C. et al. (2007). Enrichment of peptides from plasma for peptidome analysis using multiwalled carbon nanotubes. *J. Sep. Sci.* 30 (6): 930–943.

Lin, B., Li, T., Zhao, Y. et al. (2008). Preparation of a TiO_2. NPs-deposited capillary column by liquid phase deposition and its application in phosphopeptide analysis. *J. Chromatogr. A* 1192 (1): 95–102.

Liu, X., Yu, Y., Li, Y. et al. (2013). Restricted access magnetic core-mesoporous shell microspheres with C8-modified interior pore-walls for the determination of diazepam in rat plasma by LC-MS. *Talanta* 106: 321–327.

Lu, J., Wang, M., Deng, C., and Zhang, X. (2013). Facile synthesis of Fe_3O_4@mesoporous TiO_2 microspheres for selective enrichment of phosphopeptides for phosphoproteomics analysis. *Talanta* 105: 20–27.

Ma, J.B., Qiu, H.W., Rui, Q.H. et al. (2016). Fast determination of catecholamines in human plasma using carboxyl-functionalized magnetic-carbon nanotube molecularly imprinted polymer followed by liquid chromatography-tandem quadrupole mass spectrometry. *J. Chromatogr. A* 1429: 86–96.

Ma, W.F., Zhang, C., Zhang, Y.T. et al. (2014). Magnetic MSP@ZrO_2 microspheres with yolk-shell structure: designed synthesis and application in highly selective enrichment of phosphopeptides. *Langmuir* 30 (22): 6602–6611.

Moliner-Martínez, Y., Herráez-Hernández, R., Verdu-Andres, J. et al. (2015). Recent advances of in-tube solid phase microextraction. *TrAC* 71: 205–213.

Moliner-Martínez, Y., Prima-Garcia, H., Ribera, A. et al. (2012). Magnetic in-tube solid phase microextraction. *Anal. Chem.* 84: 7233–7240.

Moliner-Martínez, Y., Ribera, A., Coronado, E., and Campíns-Falcó, P. (2011). Preconcentration of emerging contaminants in environmental water samples by using silica supported Fe_3O_4 magnetic nanoparticles for improving mass detection in capillary liquid chromatography. *J. Chromatogr. A* 1218: 2276–2283.

Moliner-Martínez, Y., Vitta, Y., Prima-Garcia, H. et al. (2014). Silica supported Fe_3O_4 magnetic nanoparticles for magnetic solid-phase extraction and magnetic in-tube solid-phase microextraction: application to organophosphorous compounds. *Anal. Bioanal. Chem.* 406: 2211–2215.

Qin, H., Gao, P., Wang, F. et al. (2011). Highly efficient extraction of serum peptides by ordered mesoporous carbon. *Angew. Chem. Int. Ed.* 50 (51): 12218–12221.

Ravelo-Pérez, L.M., Herrera-Herrera, V., Hernández-Borges, J., and Rodríguez-Delgado, M.A. (2010). Carbon nanotubes: solid-phase extraction. *J. Chromatogr. A* 1217: 2618–2641.

Reyes-Gallardo, E.M., Lucena, R., Cárdenas, S., and Valcárcel, M. (2015). Polymer–nanoparticles composites in bioanalytical sample preparation. *Bioanalysis* 7 (14): 1723–1730.

Sarparanta, M., Bimbo, L.M., Rytkonen, J. et al. (2012). Intravenous delivery of hydrophobin-functionalized porous silicon nanoparticles: stability, plasma protein adsorption and biodistribution. *Mol. Pharm.* 9 (3): 654–663.

Stein, A. and Melde, B. (2001). The role of surfactants and amphiphiles in the synthesis of porous inorganic solids. In: *Reactions and Synthesis in Surfactant Systems* (ed. J. Texter), 819–851. New York: Marcel Dekker.

Tran, T.H., Park, S.Y., Lee, H. et al. (2012). Ultrasmall gold nanoparticles for highly specific isolation/enrichment of *N*-linked glycosylated peptides. *Analyst* 137 (4): 991–998.

Wan, H., Qin, H., Xiong, Z. et al. (2013). Facile synthesis of yolk-shell magnetic mesoporous carbon microspheres for efficient enrichment of low abundance peptides. *Nanoscale* 5 (22): 10936–10944.

Wu, Q., Wu, D., and Guan, Y. (2013). Fast equilibrium micro-extraction from biological fluids with biocompatible core-sheath electrospun nanofibers. *Anal. Chem.* 85 (12): 5924–5932.

Yin, P., Wang, Y., Li, Y. et al. (2012). Preparation of sandwich-structured graphene/mesoporous silica composites with C8-modified pore wall for highly efficient selective enrichment of endogenous peptides for mass spectrometry analysis. *Proteomics* 12 (18): 2784–2791.

Zhang, Y., Hu, Z., Qin, H. et al. (2013). Cell nucleus targeting for living cell extraction of nucleic acid associated proteins with intracellular nanoprobes of magnetic carbon nanotubes. *Anal. Chem.* 85 (15): 7038–7043.

Zhang, Y., Hu, Z., Qin, H. et al. (2012). Highly efficient extraction of cellular nucleic acid associated proteins in vitro with magnetic oxidized carbon nanotubes. *Anal. Chem.* 84 (23): 10454–10462.

Zhao, Y.G., Chen, X.H., Pan, S.D. et al. (2013). Simultaneous analysis of eight phenolic environmental estrogens in blood using dispersive micro-solid-phase extraction combined with ultra fast liquid chromatography-tandem mass spectrometry. *Talanta* 115: 787–797.

11

Sample Preparation via Molecularly Imprinted Polymers (MIPs) in LC-MS Bioanalysis

Myriam Díaz-Álvarez and Antonio Martín-Esteban

11.1 Introduction

The main objectives of sample preparation are the removal of unwanted matrix components which might eventually affect extraction, pre-concentration, and detection of target analytes or conversion of (if needed) the analytes to a more suitable form for detection. More recently, new objectives such as improvement of selectivity in extraction, miniaturization, and reduction in use of organic solvents have been set (Smith 2003). In order to fulfill such requirements, sample treatment techniques such as solid-phase extraction (SPE) and solid-phase microextraction (SPME), among others, have been incorporated to bioanalytical laboratories. However, those techniques still suffer from lack of selectivity even with extensive optimization. As an alternative, antibodies immobilized onto an adequate support, called immunosorbents, were proposed some years ago in order to overcome the issue of assay selectivity. The inherent selectivity of antibodies provides an excellent cleanup of the analytes of interest from the sample matrix. Several analytical methods based on immobilized antibody have been developed for the determination of a variety of analytes in food, environmental, and biological samples (Hennion and Pichon 2003). However, generation of suitable antibodies is difficult, time-consuming, and expensive. Furthermore, it is often difficult to guarantee its success. Besides, it is important to point out that once the antibodies have been obtained, they have to be immobilized onto an adequate support, which may result in poor antibody orientation or even complete denaturation. Because of these limitations, an alternative approach to synthesize host molecules, which can recognize targeted guest species, has been developed. This approach is called "molecular imprinting."

Molecular imprinting, shown schematically in Figure 11.1, is based on the preparation of a highly cross-linked polymer around a template (the analyte) in the presence of a suitable monomer. The template and monomer(s) are first mixed in order to form a stable pre-polymerization complex in a selected solvent. Subsequently, the polymerization is initiated in the presence of a suitable cross-linker. After polymerization, the polymer is ground and sieved to an appropriate particle size, and the template is removed, leaving cavities complementary in shape, size, and functionality to the template. These cavities are able to selectively rebind, in given conditions, the analyte (the template) from a complex mixture. The preparation of molecularly imprinted polymers (MIPs) is easy and inexpensive and is a powerful tool for providing selectivity to the analytical process (Tamayo et al. 2007; Bui and Haupt 2010; Martín-Esteban and Sellergren 2010; Hu et al. 2013; Martin-Esteban 2013).

11.2 Preparation of MIPs

As mentioned above, the first step in the preparation of MIPs involves the prearrangement of the template and monomer(s). The template molecule associates with the functional monomer(s) to form a covalent or a non-covalent bonded complex. The covalent approach is attributed to Wulff and Sarhan (1972), who described polymer preparation with chiral cavities for the separation of racemic mixtures. This first example of imprinted polymer was based on the reversible formation of ester linkages between a sugar and phenylboronic acid, which was derivatized with a vinyl group. Imprinting with covalent interactions allows the cavity structures to be probed in detail but this system is not very flexible when choosing the functional monomer(s) and the template species, restricting its application to only a few molecules.

A more flexible approach, known as non-covalent imprinting, was introduced by Arshady and Mosbach (1981). In this approach, the template–monomer interactions are based on hydrogen bonding, electrostatic interactions, and/or metal ion coordination. Therefore,

Sample Preparation in LC-MS Bioanalysis, First Edition. Edited by Wenkui Li, Wenying Jian, and Yunlin Fu.

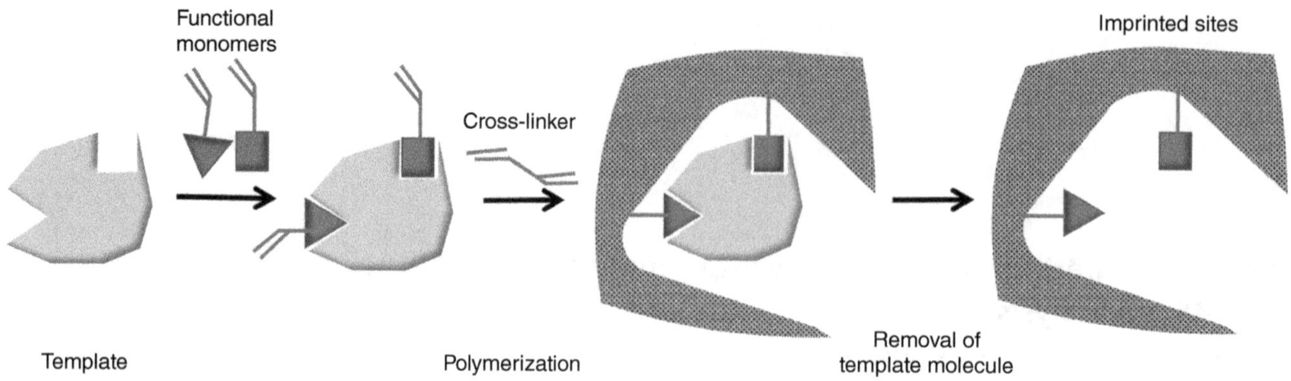

Figure 11.1 Principle of molecular imprinting.

it covers a wide range of monomers and templates, thus increasing its range of application in chemical analysis. This flexibility has led to a spectacular increase in the number of applications published during past years. This chapter will, therefore, focus on the variables that affect the performance of non-covalent imprinted polymers and their applications.

The template, monomer(s), cross-linker, and solvent (porogen) used are obviously the key parameters for the obtainment of a successful selective MIP. All the parameters mentioned have a strong influence on the overall performance of MIPs in terms of affinity, selectivity, loading capacity, etc. The proper selection of these parameters will ensure the appropriate properties of the MIPs for the intended application.

11.2.1 Template and Monomer(s)

The first step in the preparation of MIPs is the prearrangement of the template and the monomer(s) in a solvent. As the selection of the monomer depends upon the characteristics of the template, the template should contain in its structure functional chemical groups capable of interacting with the monomer(s) with sufficient strength to form a stable complex (Sellergren 1999). Thus, for templates containing acidic groups, Brönsted basic functional monomers (e.g. 2- or 4-vinylpyridine [VPy]) are preferably choice, whereas for basic templates, acidic functional monomers (e.g. methacrylic acid [MAA]) are used. In the case of weakly polar to apolar templates, with few polar interaction sites, it may be beneficial to use amphiphilic monomers, stabilizing the monomer-template assemblies by hydrophobic and van der Waals forces. Although such rule of thumb is valid, for some systems monomer selection is more complex than expected. The selection is best carried out by using high-throughput polymer synthesis techniques, proper experimental design, and combinatorial techniques. Typically, a combinatorial library of MIPs in small-scale (mini-MIPs) is prepared using different monomers, cross-linkers, and porogens. The ability of obtained mini-MIPs to recognize target analyte is evaluated by equilibrium rebinding-elution experiments. Subsequently, a scaled-up version of the optimum polymer is prepared (Lanza and Sellergren 2004; Villoslada and Takeuchi 2005; Martin-Esteban and Tadeo 2006). The selection of suitable monomers may also be carried out by computational methods. This approach relies on molecular modeling software to design and screen a virtual library of monomers against the desired template. Through this approach it is possible to calculate binding energies and to predict template–monomer interaction positions, making it easier to select the best functional monomer (Piletsky et al. 2001; Chianella et al. 2006).

Besides, it has to be pointed out that since the template–monomer interactions are governed by an equilibrium process, a large amount of monomer should be used in order to drive the equilibrium toward the formation of the template–monomer complex. In general, a template–monomer molar ratio of 1 : 4 should provide the needed stability for the formed complex. However, the excess of free monomers leads to the formation of nonspecific binding sites, therefore, when using such MIP materials to conduct SPE, the loading, washing, and elution conditions must be correctly selected, as described later.

The solvent used in the pre-polymerization step is also of prime importance since it has a direct influence on the strength of the template–monomer interaction. In general, solvents with a low dielectric constant, such as chloroform and toluene, offer an adequate medium to stabilize hydrogen bonding and/or electrostatic interactions between monomer(s) and templates. Solvents with higher dielectric constants (i.e. acetonitrile) have also been used but the polymers obtained usually show a lower affinity to rebind the template. Protic solvents, such as water and methanol, are not recommended since they disrupt the template–monomer hydrogen-bonding interactions. Polymerizable host monomers have been designed to form a strong complex with a given template in solution and, in some cases, this has resulted in a quantitative yield of high-energy binding sites (stoichiometric imprinting), which can recognize the template also under aqueous conditions (Wulff and Knorr 2001; Hall et al. 2005).

Finally, the size and shape of the template has a remarkable influence on the selectivity of the obtained polymers. In general, slight difference in the structure of the template near the functional group responsible for the interaction with the monomer may have an impact on the selectivity of the imprinted polymers. In some cases, this difference prevented the polymers from binding to structurally related compounds (Sellergren et al. 1988; Ramstrom et al. 1996). However, in other cases, the impact seemed negligible (Tamayo et al. 2003).

11.2.2 Cross-linker

In order to preserve the stability of the template–monomer complex during the polymerization process and to increase polymer porosity, a high degree of cross-linking is necessary. The most commonly used cross-linkers are ethylene glycol dimethacrylate (EGDMA), trimethylolpropane trimethacrylate (TRIM), and divinylbencene (DVB). It has been reported that at least 50% of the total monomer in a MAA-system has to be EGDMA, otherwise no recognition can take place (Sellergren 1989). The presence of a cross-linker not only preserves the binding sites but also has a direct influence on the physical and chemical properties of the polymeric matrix. EGDMA is the cross-linker often used in methacrylate-based systems, because it provides mechanical and thermal stability, good wettability, and rapid mass transfer.

11.2.3 Porogen

Apart from its influence on the strength of template–monomer interaction during the prearrangement step, the solvent (porogen) plays an important role in the morphology of the formed polymer in terms of specific surface area and pore diameter. In general, low surface area and low macroporosity are associated with low template recognition in the subsequent rebinding experiments owing to the slow analyte diffusion to the binding sites located in micropores.

As stated earlier, the polymer obtained by bulk polymerization has to be ground and sieved to the desired particle size (25–50 μm). This process is tedious and time-consuming, and the particles obtained are often irregular in size and shape. In general, only 50% or less of the total amount of polymer is useful for analytical purposes and some binding sites of the polymers are partially destroyed during grinding, which results in a considerable loss of loading capacity of the imprinted polymer versus the theoretical loading capacity considering the amount of template used in preparation.

To overcome the challenges mentioned above, different polymerization strategies have been proposed by several authors for the direct preparation of imprinted polymers, which enables spherical particles to be obtained in the desired particle size. Among those strategies, the most commonly used ones are: (i) imprinting in the pores of preformed beaded silica (Yilmaz et al. 2002; Abou-Hany et al. 2015), (ii) a two-step swelling technique using water as a suspension medium (Haginaka and Sakai 2000; Sambe et al. 2007), and (iii) precipitation polymerization (Wang et al. 2003; Turiel et al. 2005). More recently, the synthesis of core-shell particles has received special attention. In the core-shell particles, since the MIP is located in the outer surface of the particle ("shell"), the mass-transfer of target analytes to the binding sites is improved (Sellergren et al. 2002; Tamayo et al. 2005; Barahona et al. 2010).

Once the polymer has been prepared, it is necessary to remove the template from the polymer in order to free the binding sites. This step is usually carried out by washing the polymer repeatedly with a solvent capable of disrupting the template–monomer interactions. However, it is known that, even after exhaustive washing by different methodologies (Ellwanger et al. 2001), trace templates remain in the polymer network. On the other hand, it is also true that in the vast majority of publications, no template bleeding has been reported at a concentration level that might interfere with the intended applications.

In any case, the best approach to address the bleeding issue is the use of an analogue of the target analyte as a template. By this approach, the bleeding of the template, if any, does not interfere with the quantification of the target analyte. This approach was originally proposed by Andersson et al. (1997) for the selective extraction of sameridine on a MIP synthesized using a template compound closely related to the target analyte. However, the use of template-analogue imprinted polymers often leads to an inferior molecular recognition of the MIPs as compared to those prepared using the analyte itself as the template. A very interesting alternative, proposed by Kawaguchi et al. (2005) is the use of a stable isotope-labeled compound as the template molecule. Isotope imprinting has been further exploited for the determination of bisphenol A (BPA) in water samples. In this case, BPA-d_{16} was used as a template and BPA-$^{13}C_{12}$ was added to the sample as surrogate standard before SPE. Subsequently, the extracted sample was analyzed by LC-MS. Figure 11.2 shows the LC-MS single ion monitoring chromatograms of BPA, BPA-$^{13}C_{12}$, and BPA-d_{16} in river water sample using SPE as the sample preparation method. As can be seen, leaking of template BPA-d_{16} occurs but it does not affect the quantification of analyte BPA.

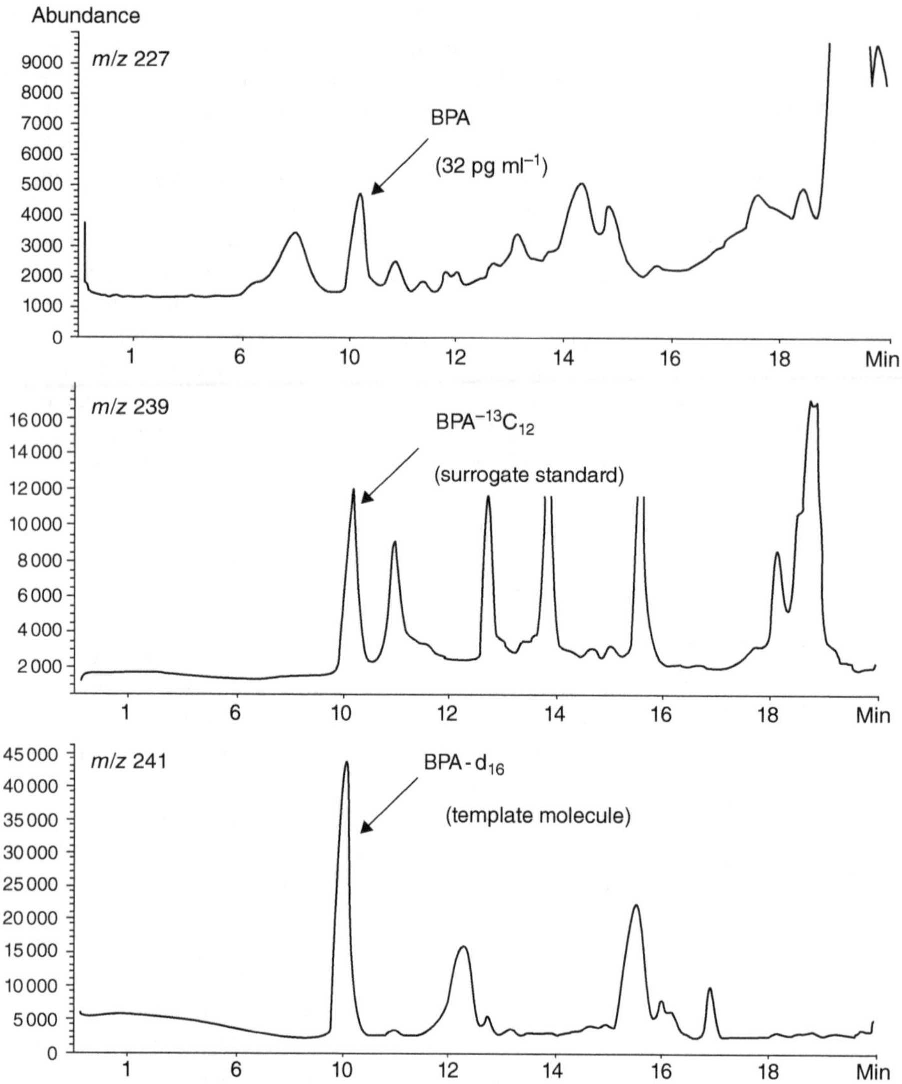

Figure 11.2 Single ion monitoring chromatograms of BPA, BPA-$^{13}C_{12}$, and BPA-d_{16} in river water samples subjected to MISPE by LC-MS. *Source:* Reprinted from Kawaguchi et al. (2005) with permission from Elsevier.

11.3 MIPs for Sample Preparation in Bioanalysis

11.3.1 Molecularly Imprinted Solid-phase Extraction (MISPE)

The use of MIPs as selective sorbent materials allows for a customized sample preparation. Clean extracts using MIP in SPE, known as molecularly imprinted SPE (MISPE), can be obtained with minimal sample pretreatment and is by far the most advanced MIP technique (Bui and Haupt 2010; Hu et al. 2013; Martin-Esteban 2013) with many successful applications as summarized in Table 11.1.

There are several configurations of MISPE, including (i) batch SPE, where the MIP is incubated with the sample, (ii) conventional off-line SPE, where the MIP is packed into cartridges, and (iii) in-line or online SPE.

11.3.1.1 Off-line Protocols

Typically, a small amount (15–500 mg) of imprinted polymer is packed into polyethylene cartridges. Then, as it is shown in Figure 11.3, the sample is commonly loaded onto the MIP cartridge in a low-polarity solvent, in which the specific interactions are maximized. After a washing step to remove unwanted components that are nonspecifically bound to the polymeric matrix, a solvent is added to disrupt the non-covalent interactions between the analyte and the imprinted polymer to elute the analytes, ideally free of interfering matrix components, and the final extract is further analyzed by chromatographic techniques. Samples such urine, serum, or plasma are usually

Table 11.1 Examples of MISPE in pharmaceutical analysis and bioanalysis.

Analytes	Sample and matrix	Mode	Reference
Pentamidine	Urine	In-line	Sellergren (1994)
Sameridine	Plasma	Batch	Andersson et al. (1997)
Theophylline	Serum	In-line, PE	Mullett and Lai (1998)
Theophylline	Plasma	In-line, DPE	Mullett and Lai (1999)
Theophylline	Serum	Off-line	Li et al. (2009)
β-Agonists	Urine	Off-line	Van Hoof et al. (2005)
Clenbuterol	Urine	Off-line	van Hout et al. (2003)
Cephalexin	Serum	In-line, DPE	Lai and Wu (2003)
Metformin	Plasma, urine	In-line, DPE	Lai and Feng (2003)
Verapamil and its metabolites	Urine, plasma, and cell cultures	Online, RAM	Mullett et al. (2004)
Diphenylphosphate	Flame retardants in urine	Off-line	Moller et al. (2004)
Ropivacaine	Plasma	Off-line	Cobb and Andersson (2005)
Ropivacaine	Plasma	Online	Abdel-Rehim et al. (2006)
Ropivacaine and bupivacaine	Plasma	Online	Cobb et al. (2007)
Trimethoprim	Urine	Off-line	Hu et al. (2005)
MNAL (tobacco-specific nitrosamine)	Urine	Off-line	Xia et al. (2005)
Cocaine and metabolites	Urine	Off-line, Mag	Sánchez-González et al. (2016)
Mycophenolic acid	Plasma of transplant patients	Off-line	Yin et al. (2006)
Alfuzosin	Plasma and soil	Off-line	Chapuis et al. (2006)
Diazepam and metabolites	Hair	Off-line	Ariffin et al. (2007)
Carbamazepine	Urine and waste water	Off-line	Beltran et al. (2007)
Benzodiazepines	Hair	Off-line	Anderson et al. (2008)
Zidovudine and stavudine	Serum	Off-line	Duy et al. (2009)
Methamphetamine, amphetamine, and methylenedioxyphenylalkylamine designer drugs	Blood	Off-line	Kumazawa et al. (2012)
Diethyl thiophosphate and diethyl dithiophosphate	Urine	Off-line	Santos et al. (2012)
Cannabinoids	Urine, oral fluid	Off-line	Cela-Pérez et al. (2016)
Urinary benzene metabolites	Urine	Off-line	Chauhan et al. (2015)
Biochanin A, daidzein, and genistein	Urine	Off-line	Chrzanowska et al. (2015)
Morphine	Urine	Off-line, MWCNTs	Kolaei et al. (2016)
Venlafaxine and metabolites	Plasma	Off-line	Miranda et al. (2016)

PE, pulsed elusion; DPE, differential pulsed elution; RAM, restricted access material; Mag, magnetic; MWCNTs, multiwalled carbon nanotubes.

pre-diluted with a low-polarity solvent (i.e. acetonitrile) to maximize the non-covalent interactions between the analyte and the binding sites of MIPs during the loading step. However, aqueous samples can also be directly loaded onto MIP cartridges. In this case, MIPs behave like a reverse-phase sorbent and thus both target analytes and matrix components are retained trough nonspecific interactions. Then, the washing solvent has to be suitable for removing the matrix components and, more important, of redistributing nonspecifically bound analytes to the selective imprints. Unfortunately, the success of such procedure is not always achieved. Lately, efforts have been directed toward the synthesis of water-compatible MIPs by incorporating hydrophilic surface properties to the polymer in order to reduce nonspecific hydrophobic interactions (Wan et al. 2015).

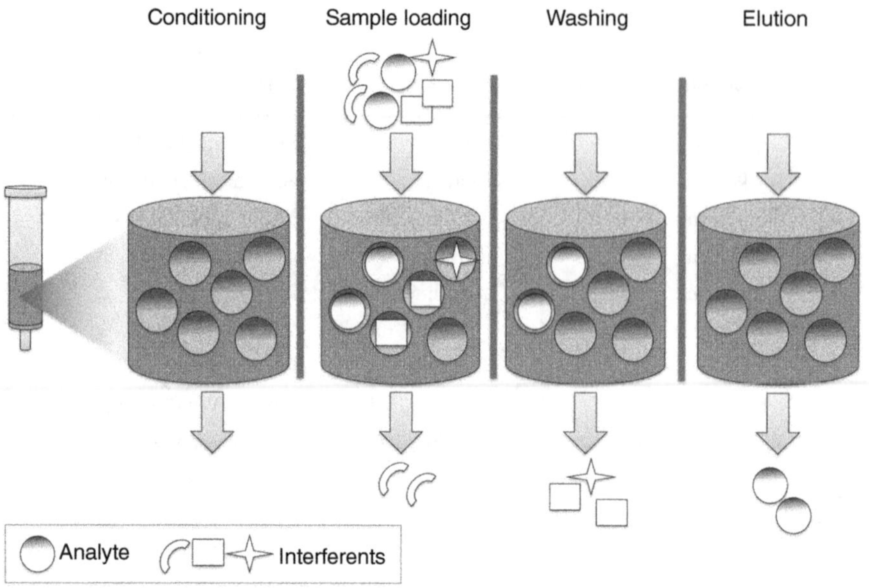

Figure 11.3 Molecularly imprinted solid-phase extraction (MISPE) procedure.

In order to improve mass transfer of target analytes, some in situ polymerization strategies have also been proposed. In this regard, Du et al. (2012) described the synthesis of kinetin molecularly imprinted monolith in a syringe by an in-situ polymerization technique. The synthesis procedure is relatively simple and can be easily performed in any laboratory. Similarly, in-situ monolithic molecularly imprinted polymer sol–gel packed tips (MMSTs) have been prepared and evaluated for the extraction of lung cancer biomarker l-tyrosine (Tyr) from human plasma and urine samples by LC-MS/MS (Moein et al. 2014).

The in-situ synthesis of MIPs on the surface of microfiltration glass-fiber membranes in multi-well filter plates has been also proposed (Ceolin et al. 2009; Renkecz et al. 2011, 2012). Briefly, the synthesis procedure is as follows: firstly, membranes of a 24-well filter plate are washed with methanol and dried before use. Then, 30–50 μl aliquots of polymerization mixture are transferred onto the filter plate membranes under oxygen-free argon atmosphere. The plate is covered with UV transparent cling film and placed under a UV lamp for plate irradiation for three hours. After polymerization, the template is removed by successive washing and ready for SPE experiments. The observed MIP on the surface of the modified membranes forms a veil-like web between adjacent fibers, whereas inside the membrane the polymer is partly deposited in clusters on the glass fibers. This approach is a cost-effective, one-step procedure for preparing MIP-composite membranes and offers a viable alternative to existing MISPE cartridges.

11.3.1.2 Online Protocols

In this format, a small precolumn packed with the imprinted polymer (typically about 50 mg) is placed in the loop of a six-port injection valve. After sample loading and washing out interfering components, the analytes are eluted by the mobile phase and then separated on the analytical column. However, the lack of compatibility of the elution solvent necessary to desorb analytes from the MIP precolumn and the mobile phase for chromatography on the analytical column make it difficult to develop such online MISPE methods. This problem can be circumvented by online directing the eluent from the MIP precolumn to the injection loop and subsequently injected onto the chromatographic system (Theodoridis et al. 2004), or by mixing the organic elution solvent with an aqueous-rich solvent before reaching the analytical column (Ou et al. 2006). However, such approaches increase the complexity of the online system and, therefore, require the incorporation of extra-instrumentation (i.e. pumps).

On the other hand, in order to prevent a large amount of matrix components reaching the MIP precolumn, Mullett et al. (2004) reported a two-step online SPE-LC-MS procedure for the determination of verapamil and its metabolites in urine, plasma, and cell cultures. The procedure involved a restricted access material (RAM) cartridge and a verapamil-imprinted polymer (Figure 11.4). In brief, the raw sample was loaded onto the RAM cartridge where the analytes were retained by the alkyl hydrophobic-bonded phase, while hydrophilic electroneutral diol groups bound to the external surface of the spherical particles protected the sorbent from binding to proteins in the sample. After this first cleanup, the analytes were desorbed from the RAM precolumn using acetonitrile, and transferred to the MIP precolumn. Finally, the analytes retained by the MIP were desorbed by the mobile phase and transferred to the analytical column for chromatographic separation. Using this approach, verapamil and its metabolites were easily quantified by LC-MS in <35 minutes, including sample preparation, selective fractionation, chromatographic separation, and detection.

Combining the RAM and the MIP properties into one single material has been proposed to eliminate the fractionation step using the RAM precolumn. Haginaka et al. (1999) reported that hydrophilized MIPs, prepared by postgrafting a hydrophilic polymer, providing RAM properties, on the surface of a preformed MIP, allowed for the repeated direct injection of plasma samples for the determination of (*S*)-naproxen with minor fouling by sticky proteins. Alternatively, the direct inclusion of a hydrophilic comonomer in the imprinting polymerization was proposed (Dirion et al. 2003). It was anticipated that this could contribute to defining the imprinted site in addition to creating a more hydrophilic polymer matrix. Such approach was successfully applied to the determination of ropivacaine and bupivacaine in plasma samples (Cobb et al. 2007).

11.3.1.3 In-line Protocols

Due to the high selectivity of MIPs, the direct coupling of a MIP column in-line with the detection system might allow the extraction, enrichment, separation, and determination of target analytes in single step. MISPE with direct in-line UV-detection was first described by Sellergren (1994) for the determination of pentamidine in urine. In that work, after loading 100 ml of diluted urine sample, the column was washed with 100 ml of a buffer solution (pH 9) and pentamidine was eluted with ~1.5 ml of buffer at pH 3. The selective elution allowed for the successful detection of pentamidine free of co-extractives by UV at 270 nm. However, since a large volume of sample was loaded, a high amount of matrix components were retained on MIP columns by hydrophobic interactions, which made it necessary to use a large volume of washing solutions.

MISPE with pulsed elution (MISPE-PE) was proposed as an alternative, which was exemplified in the determination of theophylline in chloroform-diluted serum samples (Mullett and Lai 1998). After sample loading and washing off the interfering components, a 20 µl of methanol was injected to elute theophylline allowing its detection by UV at 270 nm (Mullett and Lai 1998). MISPE-PE has been subsequently improved by applying successive pulses of 20 µl of different solvents, known as MISPE

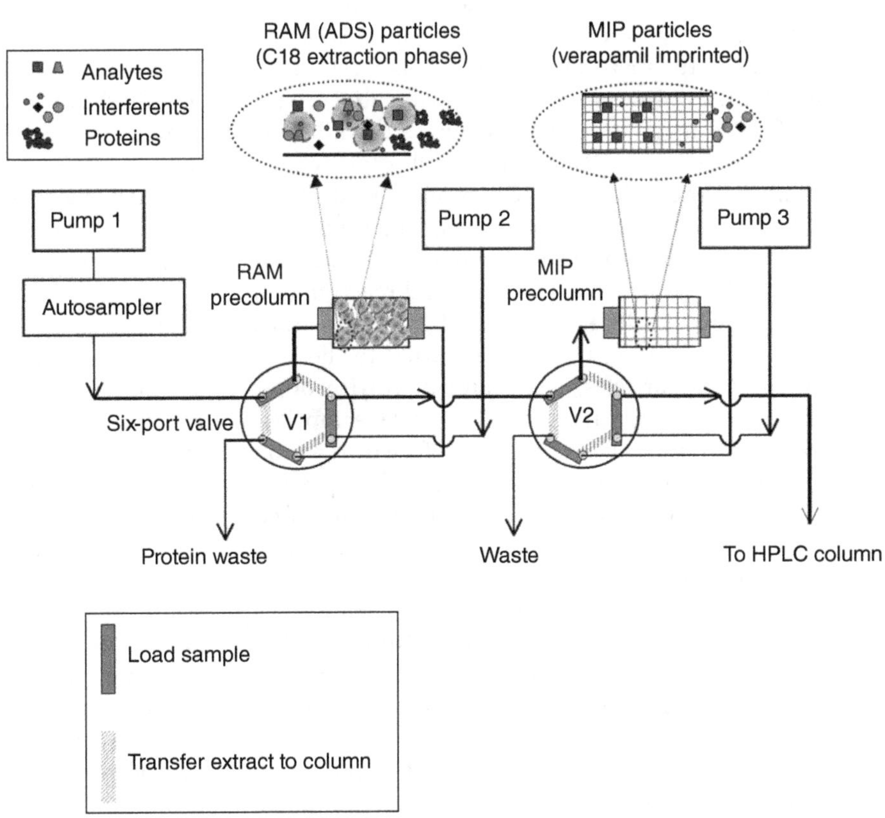

Figure 11.4 Experimental configuration of the online two-step SPE-LC procedure for the determination of verapamil and its metabolites using a restricted access material (RAM) cartridge and a verapamil-imprinted polymer. *Source:* Reprinted from Mullett et al. (2004) with permission from Elsevier.

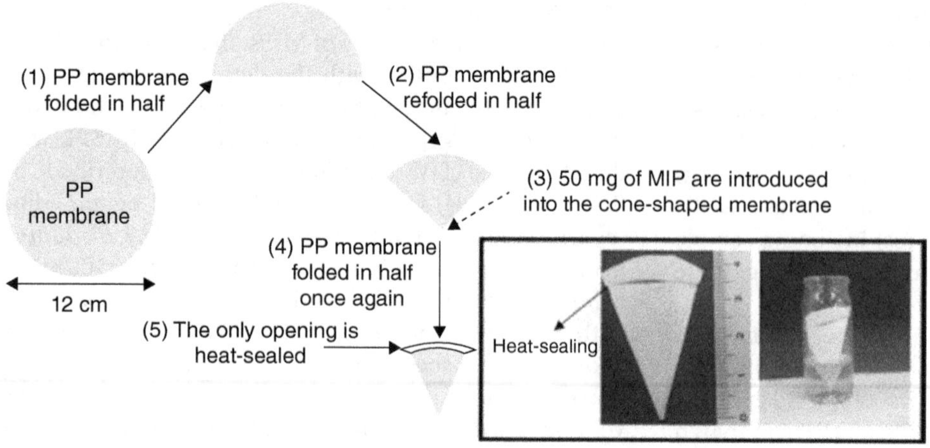

Figure 11.5 Schematic of preparation of molecularly imprinted microSPE device. *Source:* Reprinted from Sánchez-González et al. (2015) with permission from Elsevier.

with differential pulsed elution (MISPE-DPE), in order to wash out remaining interfering components and the nonspecifically retained analyte fraction (Mullett and Lai 1999). Both approaches allow for analyte enrichment through injection of large volumes of sample, leading to a high analytical sensitivity.

11.3.1.4 Other Protocols

The use of magnetic nanoparticles as support for different stationary phases in batch SPE is one of the most active research area in sample preparation. Upon sample extraction, such modified magnetic particles can be separated from the matrix by a simple magnet, thereby avoiding tedious filtration and/or centrifugation steps. The surface of magnetic particles can be modified by the incorporation of MIPs, which provides the needed selectivity for the extraction (Chen and Li 2012; Hu et al. 2013). One of the simplest ways to synthesize magnetic MIPs is the encapsulation of magnetite nanoparticles in silica through a typical sol–gel reaction (Hu et al. 2013). Silica shells are easily modifiable and, thus, desirable functional groups can be grafted to the surface of the nanoparticles. For example, vinyl groups, which are necessary for the subsequent surface imprinting, can be introduced onto particle surface by reaction with 3-methacryloyloxypropyltrimethoxysilane (MPS). Finally, the MIPs are formed on the surface of modified magnetic particles by the copolymerization of vinyl groups with typical imprinting monomers and cross-linkers in the presence of the template molecule. This procedure has been successfully used for the extraction of sterols in serum (Zhang et al. 2011) and cocaine and its metabolites in urine (Sánchez-González et al. 2016), among others.

A rather simple approach, so-called molecularly imprinted micro-solid-phase extraction (MI-MSPE) has been proposed. The MI-MSPE (Figure 11.5) device comprises MIP particles enclosed within a porous polypropylene (PP) flat-sheet membrane envelope. Once the MI-MSPE device is prepared, it is conditioned and placed in the sample. The device tumbles freely within the sample solution during extraction for a certain period of time under stirring. After extraction, the device is manually removed with the help of tweezers and placed in a vial for elution of analytes. This approach has been successfully applied to the selective extraction of cocaine and its metabolites from human urine (Sánchez-González et al. 2015). Alternatively, MIP pills have been used for the extraction of cannabinoids from urine and oral fluids. In this approach, the MIP pill is added to the sample and sonicated for a certain period of time. After that, the MIP pill is transferred to a clean tube for elution of target analytes (Lendoiro et al. 2014).

11.3.2 MIPs in Other Sample Preparation Techniques

Other sample preparation techniques, such as SPME or stir-bar sorptive extraction (SBSE), have been incorporated to analytical laboratories as an alternative to SPE. However, the variety of commercially available fibers (or stir bars) is limited and extraction is based on nonselective absorption (i.e. polar analytes are attracted to polar phases). In this regard, the incorporation of MIPs to these techniques would ideally provide a powerful analytical tool with the characteristics of both technologies (simplicity, flexibility, and selectivity).

The preparation of the first MIP-coated silica fiber was proposed by Koster et al. (2001) for the SPME of brombuterol from human urine. Silica fibers were activated by silylation and subsequently immersed in a polymerization solution composed of clenbuterol, MAA, and

EGDMA in acetonitrile to carry out polymerization during 12 hours at 4°C under irradiation at 350 nm. According to authors, fibers with a polymeric film thickness of ~75 μm were obtained in a reproducible manner. Specific extraction of brombuterol from urine has been obtained on the MIP-coated silica fiber. Recently, key factors influencing both the morphology and the thickness of the obtained MIP coating has been systematically studied in a series of papers dealing with the preparation of different MIP-coated silica fibers. The results obtained are rather promising and further development is expected in the near future (Hu et al. 2008, 2009).

A completely different approach for the direct preparation of MIP fibers (monoliths), using fused silica capillaries as molds, has been proposed (Djozan and Baheri 2007; Turiel et al. 2007). Following a simple experimental procedure, MIP monoliths of 1 cm length are obtained with thickness dependent on the inner diameter of the silica capillary used. Moreover, the obtained fibers were flexible and can be bended to a certain extent.

The use of MIP-coated stir bars prepared by chemical bonding of the MIP to silylated stir bars has been proposed. This procedure is similar to that described above for the preparation of MIP-coated fibers for SPME and has been applied for the determination of β_2-agonists (Xu et al. 2010) and sulfa drugs (Xu et al. 2011) in different samples.

Finally, the use of water-compatible MIPs in matrix solid-phase dispersion (MSPD) has been demonstrated for the selective extraction of fluoroquinolones from eggs and tissue (Yan et al. 2007), with average recoveries 72.2–114.1%, depending upon the analyte and the sample, and relative standard deviations of <7.0%. The superior performance of MIPs compared to conventional sorbents (C18 and Florisil, recoveries were 24.9–84.6%) was clearly demonstrated. More recently, similar methodologies have been applied to the MSPD of steroids in goat milk (Ganan et al. 2012, 2014).

11.4 Fragment Imprinting

Traditional imprinting is well-suited for small molecules but it fails when the targets are complex, poorly soluble in organic solvents, or simply unavailable in sufficient quantities for imprinting. In this regard, the imprinting of most biomolecules (i.e. proteins, nucleic acids, and polysaccharides) is a challenging task. In these cases, polymers may need to be synthesized in aqueous media, allowing for solubilization of biomolecules without affecting their native conformation. However, the interactions between the target and the functional monomers commonly used in molecular imprinting in aqueous solution are not optimal.

In order to overcome this problem, fragments of the target protein may be used as templates (Quaglia et al. 2001; Rachkov and Minoura 2001). This approach uses available templates that are compatible with the imprinting, resulting in binding sites, complementary to the protein substructure but able to retain the larger biomolecule. Following this so-called epitope imprinting approach, short peptides (epitope) have been used as templates to generate MIP material for binding larger target containing the epitope sequence. Besides, the combination of epitope imprinting with surface imprinting techniques has led to the obtainment of MIPs that are able to recognize peptides (Titirici and Sellergren 2004) or even proteins (Nishino et al. 2005, 2006) based on just a short terminal peptide sequences.

11.5 Summary

MIPs are well-suited to be incorporated to sample preparation techniques to obtain cleaner sample extracts as compared to some conventional methods. Such cleaner extracts are often accompanied with reduced ion suppression in MS detection. As shown by the numerous recent examples, the quantitative analysis of pharmaceutical drugs, drugs of abuse, or endogenous biomarkers in various biomatrices (urine, serum, plasma, hair, and tissue samples) has been improved by the application of MIPs.

During the last few years, refinement of the use of MIPs in sample preparation has been observed. Further development in the areas of molecular recognition in aqueous media and protein imprinting is expected. Commercialization of certain MIPs will more or less represent a breakthrough of sample preparation in the bioanalysis field.

References

Abdel-Rehim, M., Andersson, L.I., Altun, Z., and Blomberg, L.G. (2006). Microextraction in packed syringe online with liquid chromatography-tandem mass spectrometry: molecularly imprinted polymer as packing material for MEPS in selective extraction of ropivacaine from plasma. *J. Liq. Chromatogr. Relat. Technol.* 29: 1725–1736.

Abou-Hany, R., Urraca, J., Descalzo, A. et al. (2015). Tailoring molecularly imprinted polymer beads for alternariol recognition and analysis by a screening with mycotoxin surrogates. *J. Chromatogr. A* 1425: 231–239.

Anderson, R.A., Ariffin, M.M., Cormack, P.A.G., and Miller, E.I. (2008). Comparison of molecularly imprinted solid-phase extraction (MISPE) with classical solid-phase extraction (SPE) for the detection of benzodiazepines in post-mortem hair samples. *Forensic Sci. Int.* 174: 40–46.

Andersson, L., Paprica, A., and Arvidsson, T. (1997). A highly selective solid phase extraction sorbent for pre-concentration of sameridine made by molecular imprinting. *Chromatographia* 46: 57–62.

Ariffin, M.M., Miller, E.I., Cormack, P.A.G., and Anderson, R.A. (2007). Molecularly imprinted solid-phase extraction of diazepam and its metabolites from hair samples. *Anal. Chem.* 79: 256–262.

Arshady, R. and Mosbach, K. (1981). Synthesis of substrate-selective polymers by host-guest polymerization. *Makromol. Chem.* 182: 687.

Barahona, F., Turiel, E., Cormack, P.A.G., and Martin-Esteban, A. (2010). Chromatographic performance of molecularly imprinted polymers: core-shell microspheres by precipitation polymerization and grafted MIP films via iniferter-modified silica beads. *J. Polym. Sci. A Polym. Chem.* 48: 1058–1066.

Beltran, A., Caro, E., Marce, R.M. et al. (2007). Synthesis and application of a carbamazepine-imprinted polymer for solid-phase extraction from urine and wastewater. *Anal. Chim. Acta* 597: 6–11.

Bui, B. and Haupt, K. (2010). Molecularly imprinted polymers: synthetic receptors in bioanalysis. *Anal. Bioanal. Chem.* 398: 2481–2492.

Cela-Pérez, M.C., Bates, F., Jiménez-Morigosa, C. et al. (2016). Water-compatible imprinted pills for sensitive determination of cannabinoids in urine and oral fluid. *J. Chromatogr. A* 1429: 53–64.

Ceolin, G., Navarro-Villoslada, F., Moreno-Bondi, M. et al. (2009). Accelerated development procedure for molecularly imprinted polymers using membrane filterplates. *J. Comb. Chem.* 11: 645–652.

Chapuis, F., Mullot, J.-U., Pichon, V. et al. (2006). Molecularly imprinted polymers for the clean-up of a basic drug from environmental and biological samples. *J. Chromatogr. A* 1135: 127–134.

Chauhan, A., Bhatia, T., Gupta, M.K. et al. (2015). Imprinted nanospheres based on precipitation polymerization for the simultaneous extraction of six urinary benzene metabolites from urine followed by injector port silylation and gas chromatography-tandem mass spectrometric analysis. *J. Chromatogr. B Analyt. Technol. Biomed. Life Sci.* 1001: 66–74.

Chen, L. and Li, B. (2012). Application of magnetic molecularly imprinted polymers in analytical chemistry. *Anal. Methods* 4: 2613–2621.

Chianella, I., Karim, K., Piletska, E. et al. (2006). Computational design and synthesis of molecularly imprinted polymers with high binding capacity for pharmaceutical applications-model case: adsorbent for abacavir. *Anal. Chim. Acta* 559: 73–78.

Chrzanowska, A.M., Poliwoda, A., and Wieczorek, P.P. (2015). Surface molecularly imprinted silica for selective solid-phase extraction of biochanin A, daidzein and genistein from urine samples. *J. Chromatogr. A* 1392: 1–9.

Cobb, Z. and Andersson, L.I. (2005). Determination of ropivacaine in human plasma using highly selective molecular imprint-based solid phase extraction and fast LC-MS analysis. *Anal. Bioanal. Chem.* 383: 645–650.

Cobb, Z., Sellergren, B., and Andersson, L. (2007). Water-compatible molecularly imprinted polymers for efficient direct injection on-line solid-phase extraction of ropivacaine and bupivacaine from human plasma. *Analyst* 132: 1262–1271.

Dirion, B., Cobb, Z., Schillinger, E. et al. (2003). Water-compatible molecularly imprinted polymers obtained via high-throughput synthesis and experimental design. *J. Am. Chem. Soc.* 125: 15101–15109.

Djozan, D. and Baheri, T. (2007). Preparation and evaluation of solid-phase microextraction fibers based on monolithic molecularly imprinted polymers for selective extraction of diacetylmorphine and analogous compounds. *J. Chromatogr. A* 1166: 16–23.

Du, F., Ruan, G., Liang, S. et al. (2012). Monolithic molecularly imprinted solid-phase extraction for the selective determination of trace cytokinins in plant samples with liquid chromatography-electrospray tandem mass spectrometry. *Anal. Bioanal. Chem.* 404: 489–501.

Duy, S.V., Lefebvre-Tournier, I., Pichon, V. et al. (2009). Molecularly imprinted polymer for analysis of zidovudine and stavudine in human serum by liquid chromatography-mass spectrometry. *J. Chromatogr. B Analyt. Technol. Biomed. Life Sci.* 877: 1101–1108.

Ellwanger, A., Berggren, C., Bayoudh, S. et al. (2001). Evaluation of methods aimed at complete removal of template from molecularly imprinted polymers. *Analyst* 126: 784–792.

Ganan, J., Gallego-Pico, A., Garcinuno, R. et al. (2012). Development of a molecularly imprinted polymer-matrix solid-phase dispersion method for selective determination of beta-estradiol as anabolic growth promoter in goat milk. *Anal. Bioanal. Chem.* 403: 3025–3029.

Ganan, J., Morante-Zarcero, S., Gallego-Pico, A. et al. (2014). Evaluation of a molecularly imprinted polymer for determination of steroids in goat milk by matrix solid phase dispersion. *Talanta* 126: 157–162.

Haginaka, J. and Sakai, Y. (2000). Uniform-sized molecularly imprinted polymer material for (S)-propranolol. *J. Pharm. Biomed. Anal.* 22: 899–907.

Haginaka, J., Takekira, H., Hosoya, K., and Tanaka, N. (1999). Uniform-sized molecularly imprinted polymer for (S)-naproxen selectively modified with hydrophilic external layer. *J. Chromatogr. A* 849: 331–339.

Hall, A., Manesiotis, P., Emgenbroich, M. et al. (2005). Urea host monomers for stoichiometric molecular imprinting of oxyanions. *J. Org. Chem.* 70: 1732–1736.

Hennion, M. and Pichon, V. (2003). Immuno-based sample preparation for trace analysis. *J. Chromatogr. A* 1000: 29–52.

van Hout, M.W.J., Niederlander, H.A.G., de Zeeuw, R.A., and de Jong, G.J. (2003). Ion suppression in the determination of clenbuterol in urine by solid-phase extraction atmospheric pressure chemical ionisation ion-trap mass spectrometry. *Rapid Commun. Mass Spectrom.* 17: 245–250.

Hu, S.G., Li, L., and He, X.W. (2005). Comparison of trimethoprim molecularly imprinted polymers in bulk and in sphere as the sorbent for solid-phase extraction and extraction of trimethoprim from human urine and pharmaceutical tablet and their determination by high-performance liquid chromatography. *Anal. Chim. Acta* 537: 215–222.

Hu, X., Pan, J., Hu, Y. et al. (2008). Preparation and evaluation of solid-phase microextraction fiber based on molecularly imprinted polymers for trace analysis of tetracyclines in complicated samples. *J. Chromatogr. A* 1188: 97–107.

Hu, X., Pan, J., Hu, Y., and Li, G. (2009). Preparation and evaluation of propranolol molecularly imprinted solid-phase microextraction fiber for trace analysis of beta-blockers in urine and plasma samples. *J. Chromatogr. A* 1216: 190–197.

Hu, Y., Pan, J., Zhang, K. et al. (2013). Novel applications of molecularly-imprinted polymers in sample preparation. *Trends Analyt. Chem.* 43: 37–52.

Kawaguchi, M., Hayatsu, Y., Nakata, H. et al. (2005). Molecularly imprinted solid phase extraction using stable isotope labeled compounds as template and liquid chromatography-mass spectrometry for trace analysis of bisphenol A in water sample. *Anal. Chim. Acta* 539: 83–89.

Kolaei, M., Dashtian, K., Rafiee, Z., and Ghaedi, M. (2016). Ultrasonic-assisted magnetic solid phase extraction of morphine in urine samples by new imprinted polymer-supported on MWCNT-Fe3O4-NPs: central composite design optimization. *Ultrason Sonochem.* 33: 240–248.

Koster, E., Crescenzi, C., den Hoedt, W. et al. (2001). Fibers coated with molecularly imprinted polymers for solid-phase microextraction. *Anal. Chem.* 73: 3140–3145.

Kumazawa, T., Hasegawa, C., Hara, K. et al. (2012). Molecularly imprinted solid-phase extraction for the selective determination of methamphetamine, amphetamine, and methylenedioxyphenylalkylamine designer drugs in human whole blood by gas chromatography-mass spectrometry. *J. Sep. Sci.* 35: 726–733.

Lai, E.P.C. and Feng, S.Y. (2003). Molecularly imprinted solid phase extraction for rapid screening of metformin. *Microchem. J.* 75: 159–168.

Lai, E.P.C. and Wu, S.G. (2003). Molecularly imprinted solid phase extraction for rapid screening of cephalexin in human plasma and serum. *Anal. Chim. Acta* 481: 165–174.

Lanza, F. and Sellergren, B. (2004). Molecularly imprinted polymers via high-throughput and combinatorial techniques. *Macromol. Rapid Commun.* 25: 59–68.

Lendoiro, E., de Castro, A., Fernandez-Vega, H. et al. (2014). Molecularly imprinted polymer for selective determination of delta(9)-tetrahydrocannabinol and 11-nor-delta(9)-tetrahydrocannabinol carboxylic acid using LC-MS/MS in urine and oral fluid. *Anal. Bioanal. Chem.* 406: 3589–3597.

Li, Y., Zhou, W.H., Yang, H.H., and Wang, X.R. (2009). Grafting of molecularly imprinted polymers from the surface of silica gel particles via reversible addition-fragmentation chain transfer polymerization: a selective sorbent for theophylline. *Talanta* 79 (2): 141–145.

Martin-Esteban, A. (2013). Molecularly-imprinted polymers as a versatile, highly selective tool in sample preparation. *Trends Analyt. Chem.* 45: 169–181.

Martín-Esteban, A. and Sellergren, B. (2010). The use of molecularly imprinted polymers for sampling and sample preparation. In: *Handbook of Sample Preparation* (ed. J. Pawliszyn and H.L. Lord). Hodoken, NJ: Wiley.

Martin-Esteban, A. and Tadeo, J.L. (2006). Selective molecularly imprinted polymer obtained from a combinatorial library for the extraction of bisphenol A. *Comb. Chem. High Throughput Screen.* 9: 747–751.

Miranda, L.F.C., Domingues, D.S., and Queiroz, M.E.C. (2016). Selective solid-phase extraction using molecularly imprinted polymers for analysis of venlafaxine, O-desmethylvenlafaxine, and N-desmethylvenlafaxine in plasma samples by LC-MS/MS. *J. Chromatogr. A* 1458: 46–53.

Moein, M., El-Beqqali, A., Abdel-Rehim, A. et al. (2014). Preparation of monolithic molecularly imprinted polymer sol-gel packed tips for high-throughput bioanalysis: extraction and quantification of L-tyrosine in human plasma and urine samples utilizing liquid chromatography and tandem mass spectrometry. *J. Chromatogr. B Analyt. Technol. Biomed. Life Sci.* 967: 168–173.

Moller, K., Crescenzi, C., and Nilsson, U. (2004). Determination of a flame retardant hydrolysis product in human urine by SPE and LC-MS. Comparison of molecularly imprinted solid-phase extraction with a mixed-mode anion exchanger. *Anal. Bioanal. Chem.* 378: 197–204.

Mullett, W.M. and Lai, E.P.C. (1998). Determination of theophylline in serum by molecularly imprinted solid-phase extraction with pulsed elution. *Anal. Chem.* 70: 3636–3641.

Mullett, W.M. and Lai, E.P.C. (1999). Molecularly imprinted solid phase extraction micro-column with differential pulsed elution for theophylline determination. *Microchem. J.* 61: 143–155.

Mullett, W., Walles, M., Levsen, K. et al. (2004). Multidimensional on-line sample preparation of verapamil and its metabolites by a molecularly imprinted polymer coupled to liquid chromatography-mass spectrometry. *J. Chromatogr. B Analyt. Technol. Biomed. Life Sci.* 801: 297–306.

Nishino, H., Gruber, M., Ellson, R., and Shea, K. (2005). Fabrication of novel molecularly imprinted polymer film for specific protein recognition. *Abstr. Pap. Am. Chem. Soc.* 229: U1139–U1140.

Nishino, H., Huang, C., and Shea, K. (2006). Selective protein capture by epitope imprinting. *Angew. Chem. Int. Ed.* 45: 2392–2396.

Ou, J., Hu, L., Hu, L. et al. (2006). Determination of phenolic compounds in river water with on-line coupling bisphenol A imprinted monolithic precolumn with high performance liquid chromatography. *Talanta* 69: 1001–1006.

Piletsky, S., Karim, K., Piletska, E. et al. (2001). Recognition of ephedrine enantiomers by molecularly imprinted polymers designed using a computational approach. *Analyst* 126: 1826–1830.

Quaglia, M., Chenon, K., Hall, A. et al. (2001). Target analogue imprinted polymers with affinity for folic acid and related compounds. *J. Am. Chem. Soc.* 123: 2146–2154.

Rachkov, A. and Minoura, N. (2001). Towards molecularly imprinted polymers selective to peptides and proteins. The epitope approach. *Biochim. Biophys. Acta* 1544: 255–266.

Ramstrom, O., Yu, C., and Mosbach, K. (1996). Chiral recognition in adrenergic receptor binding mimics prepared by molecular imprinting. *J. Mol. Recognit.* 9: 691–696.

Renkecz, T., Ceolin, G., and Horvath, V. (2011). Selective solid phase extraction of propranolol on multiwell membrane filter plates modified with molecularly imprinted polymer. *Analyst* 136: 2175–2182.

Renkecz, T., Laszlo, K., and Horvath, V. (2012). In situ synthesis of molecularly imprinted nanoparticles in porous support membranes using high-viscosity polymerization solvents. *J. Mol. Recognit.* 25: 320–329.

Sambe, H., Hoshina, K., and Haginaka, J. (2007). Molecularly imprinted polymers for triazine herbicides prepared by multi-step swelling and polymerization method – their application to the determination of methylthiotriazine herbicides in river water. *J. Chromatogr. A* 1152: 130–137.

Sánchez-González, J., Tabernero, M., Bermejo, A. et al. (2015). Porous membrane-protected molecularly imprinted polymer micro-solid-phase extraction for analysis of urinary cocaine and its metabolites using liquid chromatography – tandem mass spectrometry. *Anal. Chim. Acta* 898: 50–59.

Sánchez-González, J., Tabernero, M., Bermejo, A. et al. (2016). Development of magnetic molecularly imprinted polymers for solid phase extraction of cocaine and metabolites in urine before high performance liquid chromatography – tandem mass spectrometry. *Talanta* 147: 641–649.

Santos, M.G., Vitor, R.V., Andrade, F.L. et al. (2012). Molecularly imprinted solid phase extraction of urinary diethyl thiophosphate and diethyl dithiophosphate and their analysis by gas chromatography-mass spectrometry. *J. Chromatogr. B Analyt. Technol. Biomed. Life Sci.* 909: 70–76.

Sellergren, B. (1989). Molecular imprinting by noncovalent interactions – enantioselectivity and binding-capacity of polymers prepared under conditions favoring the formation of template complexes. *Macromol. Chem. Phys.* 190: 2703–2711.

Sellergren, B. (1994). Direct drug determination by selective sample enrichment on an imprinted polymer. *Anal. Chem.* 66: 1578–1582.

Sellergren, B. (1999). Polymer- and template-related factors influencing the efficiency in molecularly imprinted solid-phase extractions. *Trends Analyt. Chem.* 18: 164–174.

Sellergren, B., Lepisto, M., and Mosbach, K. (1988). Highly enantioselective and substrate-selective polymers obtained by molecular imprinting utilizing noncovalent interactions – NMR and chromatographic studies on the nature of recognition. *J. Am. Chem. Soc.* 110: 5853–5860.

Sellergren, B., Ruckert, B., and Hall, A. (2002). Layer-by-layer grafting of molecularly imprinted polymers via iniferter modified supports. *Adv. Mater.* 14: 1204–1208.

Smith, R.M. (2003). Before the injection – modern methods of sample preparation for separation techniques. *J. Chromatogr. A.* 1000 (1–2): 3–27.

Tamayo, F.G., Casillas, J.L., and Martin-Esteban, A. (2003). Highly selective fenuron-imprinted polymer with a homogeneous binding site distribution prepared by precipitation polymerisation and its application to the clean-up of fenuron in plant samples. *Anal. Chim. Acta* 482: 165–173.

Tamayo, F.G., Titirici, M.M., Martin-Esteban, A., and Sellergren, B. (2005). Synthesis and evaluation of new propazine-imprinted polymer formats for use as stationary phases in liquid chromatography. *Anal. Chim. Acta* 542: 38–46.

Tamayo, F.G., Turiel, E., and Martin-Esteban, A. (2007). Molecularly imprinted polymers for solid-phase extraction and solid-phase microextraction: recent developments and future trends. *J. Chromatogr. A* 1152: 32–40.

Theodoridis, G., Zacharis, C., Tzanavaras, P. et al. (2004). Automated sample preparation based on the sequential injection principle – solid-phase extraction on a molecularly imprinted polymer coupled on-line to high-performance liquid chromatography. *J. Chromatogr. A* 1030: 69–76.

Titirici, M. and Sellergren, B. (2004). Peptide recognition via hierarchical imprinting. *Anal. Bioanal. Chem.* 378: 1913–1921.

Turiel, E., Tadeo, J.L., Cormack, P.A.G., and Martin-Esteban, A. (2005). HPLC imprinted-stationary phase prepared by precipitation polymerisation for the determination of thiabendazole in fruit. *Analyst* 130: 1601–1607.

Turiel, E., Tadeo, J.L., and Martin-Esteban, A. (2007). Molecularly imprinted polymeric fibers for solid-phase microextraction. *Anal. Chem.* 79: 3099–3104.

Van Hoof, N., Courtheyn, D., Antignac, J.P. et al. (2005). Multi-residue liquid chromatography/tandem mass spectrometric analysis of beta-agonists in urine using molecular imprinted polymers. *Rapid Commun. Mass Spectrom.* 19: 2801–2808.

Villoslada, F. and Takeuchi, T. (2005). Multivariate analysis and experimental design in the screening of combinatorial libraries of molecular imprinted polymers. *Bull. Chem. Soc. Jpn.* 78: 1354–1361.

Wan, Y., Ma, H., Lu, B. et al. (2015). MIPs in aqueous environments. *Mol. Imprinted Polym. Biotechnol.* 150: 131–166.

Wang, J., Cormack, P., Sherrington, D., and Khoshdel, E. (2003). Monodisperse, molecularly imprinted polymer microspheres prepared by precipitation polymerization for affinity separation applications. *Angew. Chem. Int. Ed.* 42: 5336–5338.

Wulff, G. and Knorr, K. (2001). Stoichiometric noncovalent interaction in molecular imprinting. *Bioseparation* 10: 257–276.

Wulff, G. and Sarhan, A. (1972). Über die Anwendung von enzymanalog gebauten Polymeren zur Racemattrennung. *Angew. Chem.* 84: 364.

Xia, Y., McGuffey, J.E., Bhattacharyya, S. et al. (2005). Analysis of the tobacco-specific nitrosamine 4-(methylnitrosamino)-1-(3-pyridyl)-1-butanol in urine by extraction on a molecularly imprinted polymer column and liquid chromatography/atmospheric pressure ionization tandem mass spectrometry. *Anal. Chem.* 77: 7639–7645.

Xu, Z., Hu, Y., Hu, Y., and Li, G. (2010). Investigation of ractopamine molecularly imprinted stir bar sorptive extraction and its application for trace analysis of beta(2)-agonists in complex samples. *J. Chromatogr. A* 1217: 3612–3618.

Xu, Z., Song, C., Hu, Y., and Li, G. (2011). Molecularly imprinted stir bar sorptive extraction coupled with high performance liquid chromatography for trace analysis of sulfa drugs in complex samples. *Talanta* 85: 97–103.

Yan, H., Qiao, F., and Row, K. (2007). Molecularly imprinted-matrix solid-phase dispersion for selective extraction of five fluoroquinolones in eggs and tissue. *Anal. Chem.* 79: 8242–8248.

Yilmaz, E., Ramstrom, O., Moller, P. et al. (2002). A facile method for preparing molecularly imprinted polymer spheres using spherical silica templates. *J. Mater. Chem.* 12: 1577–1581.

Yin, J., Wang, S., Yang, G. et al. (2006). Molecularly imprinted solid-phase extraction for rapid screening of mycophenolic acid in human plasma. *J. Chromatogr. B Analyt. Technol. Biomed. Life Sci.* 844 (1): 142–147.

Zhang, Z., Tan, W., Hu, Y., and Li, G. (2011). Simultaneous determination of trace sterols in complicated biological samples by gas chromatography-mass spectrometry coupled with extraction using beta-sitosterol magnetic molecularly imprinted polymer beads. *J. Chromatogr. A* 1218: 4275–4283.

12

Stir-bar Sorptive Extraction for Sample Preparation in LC-MS Bioanalysis

Maria Kechagia, Maria Kissoudi, and Victoria F. Samanidou

12.1 Introduction

Chemical analysis for the determination (quantification and identification) of compounds of interest is achieved in three basic steps: sampling, sample preparation, and measurement (Almeida and Nogueira 2006). Among these steps, sample preparation is the most time-consuming one, as it usually requires more than two-thirds of total analysis time. Moreover, since it is often error prone, it is considered as the most determining step for the accuracy and precision of final results.

Sample preparation aims to transform the sample to a suitable form for analysis, without significant loss of the analytes. As it is well known, there are many sample preparation techniques, which are used for different types of matrices and analytes. In recent years the trend of making analytical chemistry more environmental-friendly has been established and the principles of Green chemistry have been already accepted by many analytical laboratories. By taking into account that large quantities of organic solvents are usually used in sample preparation, the contribution of this step in pollution is of utmost significance (Tobiszewski et al. 2009; Nogueira 2015).

In light of green analytical chemistry (GAC), trends in sample preparation include miniaturization, automation, and simplification. Also, the elimination or reduction of the amount of organic solvents involved in the procedure is required. The techniques that promise all the aforesaid include solid-phase microextraction (SPME), micro-liquid–liquid extraction (MLLE), single-drop microextraction (SDME), dispersive liquid–liquid extraction (DLLE), stir-bar sorptive extraction (SBSE), gum phase extraction (GPE), open tubular trapping (OTT), and more recently, microextraction by packed sorbent (MEPS). These techniques have replaced traditionally used ones, such as Liquid–Liquid Extraction (LLE) and Solid-Phase Extraction (SPE), which are more solvent consuming. These environmentally friendly approaches are well-established techniques, which use active, exhaustive, or dynamic isolating modes. In the meantime, the passive, non-exhaustive, or static isolating modes have appeared to be more acceptable in all scientific areas due to the cost-effectiveness, and much easier manipulation, features that are especially found in SPME, in-tube SPME, and SBSE. All these sample preparation techniques are considered as solvent-free for simultaneous extraction and preconcentration of the analytes of interest. Moreover, this opens the possibility of using immersion or headspace (HS) sampling modes in a single step. A summary of the main techniques used widely for isolation volatiles and semi-volatile compounds from liquid, solid, or gas matrices is shown in the diagram illustrated in Figure 12.1 (Baltussen et al. 2002; Nogueira 2012; Camino-Sánchez et al. 2014).

SBSE was initially introduced in 1999 by Baltussen and co-workers (Baltussen et al. 1999). This technique is a modification of SPME using a stir bar consisting of a magnet covered with glass which in turn is coated by a layer (typically 0.5–1 mm) of sorptive material (usually polydimethylsiloxane – PDMS) for the extraction. The bar is subsequently inserted into a vial, which contains the sample. The bar is stirred until equilibrium of analyte concentration between sorbent and sample matrix is reached. After the extraction, the bar is removed and transferred to a clean vial, where the target compounds are desorbed either by thermal or liquid mode. Finally, the target compounds are analyzed by liquid or gas chromatography interfaced with different modes of detection such as ultraviolet, mass spectrometry, or fluorescence, and so on (Prieto et al. 2010; Camino-Sánchez et al. 2014; Nazyropoulou and Samanidou 2015; Nogueira 2015).

A variety of coating materials exist. Some are commercially available, while others have already been or can be developed in-house. However, the most commonly used one is the commercial PDMS as it overcomes the main drawback of SBSE, i.e. low compatibility with more polar

Sample Preparation in LC-MS Bioanalysis, First Edition. Edited by Wenkui Li, Wenying Jian, and Yunlin Fu.

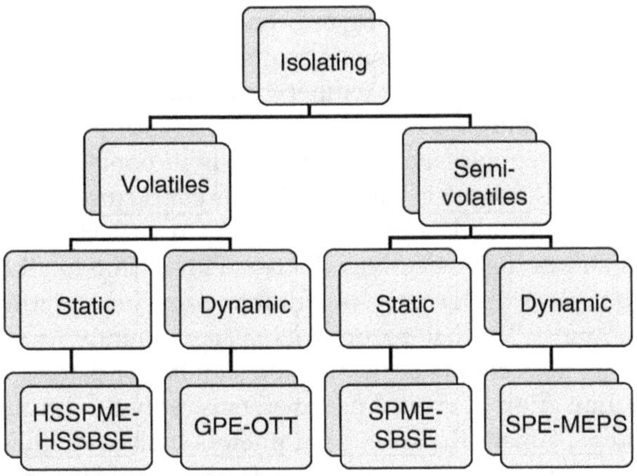

Figure 12.1 Diagram summarizing the most used methods for trace analysis of compounds (volatile and semi-volatile) based on sorption.

compounds. The number of publications using SBSE has been increased in the last decade, due to the fact that SBSE can be applied for environmental, food, pharmaceutical, and clinical samples (Tobiszewski et al. 2009; Prieto et al. 2010; Nogueira 2015).

The principle of SBSE is briefly discussed in this chapter along with its application to LC-MS bioanalysis.

12.2 SBSE Principle

Sorptive extraction is, by nature, an equilibrium technique. For aqueous samples, the extraction of solute from the aqueous phase into the extraction phase is controlled by the partitioning coefficient of the solute between the extraction (silicone) phase and the aqueous phase. Several studies have correlated this partitioning coefficient with the octanol–water distribution coefficient ($K_{o/w}$). Apparently, $K_{o/w}$ gives a good indication of whether and how well a given solute can be extracted with SPME or SBSE.

Notwithstanding, it is very important to know that sorptive equilibrium is also dependent upon the phase ratio (β), and thus the amount of PDMS applied. This relationship is shown in Eq. (12.1) (Kawaguchi et al. 2006a; David and Sandra 2007; Nogueira 2012; Camino-Sánchez et al. 2014; Nazyropoulou and Samanidou 2015).

$$K_{o/w} \approx K_{PDMS/w} = \frac{C_{PDMS}}{C_w} = \frac{m_{PDMS}}{m_w} = \frac{V_w}{V_{PDMS}} = \frac{m_{PDMS}}{m_w}\beta \tag{12.1}$$

It is noted that C_{PDMS} is the concentration of the analyte in the PDMS phase, C_w is the concentration of the analyte in water, m_{PDMS} is the mass of the analyte in the PDMS phase, m_w is the mass of the analyte in the aqueous phase, and β is the phase ratio (V_w/V_{PDMS}).

In Eq. (12.2), the initial mass of the target analyte in the aqueous phase ($m_{w,0}$) is given by the mass-balance:

$$m_{w,0} = m_{PDMS} + m_w \tag{12.2}$$

where m_w is the mass of the analyte in water (Kawaguchi et al. 2006a; Nazyropoulou and Samanidou 2015).

The theoretical recovery R (%), which is expressed as the ratio of the extracted amount of solute in the PDMS phase (m_{PDMS}) to the initial amount of solute in water ($m_{w,0}$) is thus dependent upon $K_{PDMS/w}$ and β, as described in Eq. (12.3).

$$R = \frac{m_{PDMS}}{m_{w,0}} \times 100 = \frac{K_{PDMS/w}}{K_{PDMS} + \beta} \times 100 \tag{12.3}$$

Using Eq. (12.3), the theoretical recovery for a solute with a known partition coefficient and a given β can be calculated. Taking Eq. (12.3) into consideration, it is obvious that the extraction efficiency is increased with increasing $K_{PDMS/w}$. Since $K_{PDMS/w}$ is approximately similar to $K_{o/w}$, the extraction efficiency can be predicted.

Apart from $K_{PDMS/w}$, $\beta(V_w/V_{PDMS})$ is also an essential quantity. The larger the amount of PDMS is, the larger β is and the higher extraction efficiency is achieved. In addition, some researchers have reported that β is a significant parameter for the recovery and the equilibrium time. When β is large, the results are the decrease of recovery and the extension of the equilibrium time. On the contrary, when β is small, the recovery is higher and the equilibrium time is shorter. Accordingly, in SBSE, it is crucial to increase the amount of PDMS as much as possible. Furthermore, from this expression, one can realize that the higher the hydrophobicity of the target analytes or the PDMS volume (lower β), the higher is the extraction efficiency obtained by SBSE (David and Sandra 2007; Kawaguchi et al. 2013; Camino-Sánchez et al. 2014).

A software program called KowWIN (Syracuse Research Corp., Syracuse, NY, USA) is available for the calculation of theoretical recoveries for a specific sample volume and dimensions of a given stir bar. The operation of that program is based on a log $K_{o/w}$ calculator (Kawaguchi et al. 2006a; Nazyropoulou and Samanidou 2015; Nogueira 2015).

Figure 12.2 shows the influence of the theoretical efficiency of the sorptive technique, SBSE, against $\log K_{o/w}$. It is obvious that when nonpolar target compounds are extracted, the efficiency is increased. The figure depicts the way that the constant β influences the theoretical recovery by SBSE. When the sample volume is low, the plots move toward the left side and the extraction capacity is higher (Nazyropoulou and Samanidou 2015; Nogueira 2015).

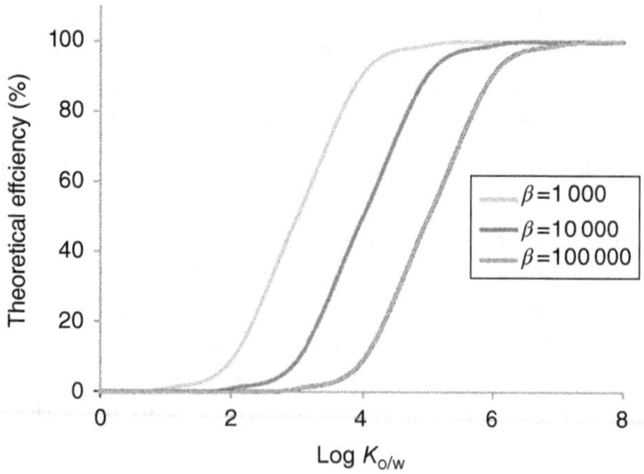

Figure 12.2 The increase of β influences the theoretical efficiency of SBSE (PDMS) toward log $K_{o/w}$.

The required volume of the PDMS coating material ranges between 24 and 100 μl. It is noted that the equilibrium state on the PDMS phase is obtained while the sorption is taking place, so that the concentration of the target compound is kept stable and reproducible, making the analysis accurate and precise (Kawaguchi et al. 2006b; Nazyropoulou and Samanidou 2015).

12.3 SBSE Steps

In SBSE, two distinguished procedures are taking place: sorption/extraction and desorption. The principles of these steps are described in the following sections.

12.3.1 Extraction Procedure

During the extraction, the polymer-coated stir bar comes in contact with analytes, in two modes, either by immersion or by HS sampling. The latter is applied for volatile compounds.

In the immersion mode, the polymer-coated stir bar is deposited on a HS vial containing the liquid sample. The bar is immersed in the sample and it is stirred under controlled conditions. After extraction, the stir bar is removed and, if necessary, is rinsed by distilled water to remove any potential interference derived from the sample matrix, such as salts, sugars, and proteins. The rinsing step is very important because this can avoid the formation of nonvolatile material that can clog the desorption unit, during thermal desorption of the analytes (Prieto et al. 2010; Nogueira 2015). Then the stir bar is transferred carefully to a clean paper and is placed to another clean vial for desorption.

The HS extraction is used for sampling in the vapor phase. In headspace SBSE (HSSBSE), a vial is used, which contains a magnetic stirring rod. The sample is placed in the same vial and the polymer-coated stir bar is placed in the HS. By increasing the temperature and stirring of the sample, the solutes are transferred in the vapor phase. In this way the polymer-coated stir bar is in contact with the vapors. After HS extraction, the stir bar is rinsed with distilled water and wiped with a clean paper. HSSBSE mode shows high selectivity, increased life-time for the stir bar, and apparently is suitable only for volatile compounds. To analyze nonvolatile compounds using HSSBSE, a preceding derivatization step is mandatory.

Optimization of critical parameters is required during this step. Namely, the kinetic parameters such as equilibrium time and agitation speed, and the thermodynamic parameters such as pH adjustment, polarity, and ionic strength of the matrix are the most important. Additionally, sample and PDMS volumes as well as dilution factor are crucial to achieve higher sensitivity (Prieto et al. 2010; Nogueira 2012, 2015).

12.3.2 Desorption

As mentioned above, after the extraction step (immersion or HS mode), the stir bar is removed, rinsed with distilled water, wiped with a paper tissue, and submitted to desorption procedure. This can be performed either thermally (called thermal desorption – TD) or by means of a suitable solvent system (called liquid desorption – LD). TD is followed by Gas Chromatographic analysis while LD is followed by High-Performance Liquid Chromatography (HPLC), Capillary Electrophoresis (CE), or Gas Chromatography (GC). TD is usually used for thermally stable volatile or semi-volatile compounds, whereas LD is preferred for semi-volatile, nonvolatile, and thermo-labile compounds (Figure 12.3) (Tobiszewski et al. 2009; Prieto et al. 2010).

12.3.2.1 Thermal Desorption

TD is performed without organic solvents and allows for direct introduction of the extracted analytes in the chromatographic system. The TD operates in temperatures between 150 and 300 °C and the duration of the desorption is estimated up to 15 minutes. To achieve the high sensitivity, the use of an expensive unit on the GC setup is required, which is the thermal desorption unit (TDU). The TDU consists of two programmable temperature vaporizes (PTVs). The first PTV is headed to desorb the analytes from the polymeric-coated stir bar, while the second one is kept cool, at temperature between −150 and 40 °C in order to cryofocus the desorbed analytes before their entrance into the GC. The advantages of the TD mode include the fact that it is solvent-free, provides high sensitivity, and allows the complete introduction of the sample in the chromatographic system.

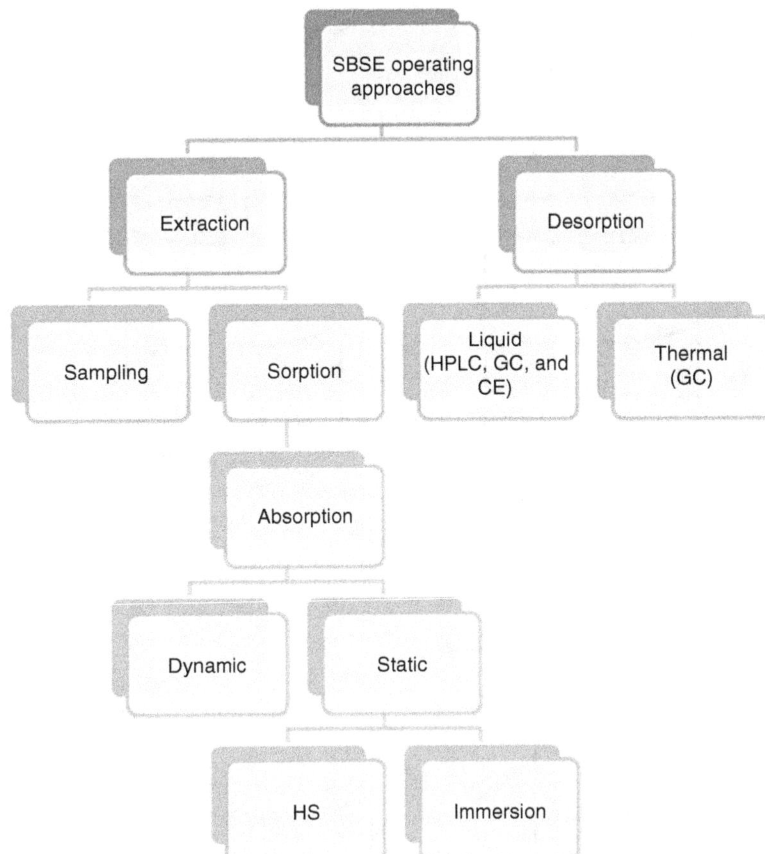

Figure 12.3 Diagram summarizing the SBSE's operating modes.

12.3.2.2 Liquid Desorption

On the other hand, in the LD mode the stir bar is transferred into a glass vial which contains the suitable organic solvent, usually methanol, acetonitrile, water, aqueous buffers, or a mixture of those. The amount of solvents to be used is <1.5 ml. Actually, the ideal amount ranges from 100 to 200 μl. The smaller the amount of solvent the more the process complies with the principles of GAC. Five to 10 minutes are required for complete desorption. Mechanical or sonication treatment can improve the desorption process. In comparison with TD, the LD mode does not need any especially expensive unit, offers the opportunity of sample reanalysis, and allows the coupling with GC, HPLC, and CE. Furthermore, LD is an off-line approach, while TD operates online (Prieto et al. 2010; Spietelun et al. 2013; Nogueira 2015).

12.4 Derivatization

Derivatization of polar and thermally labile compounds is one of the most used alternatives of the SBSE implementation. Different derivatization approaches can be applied: in situ, on-stir-bar, or post-extraction. These modes are described in Figure 12.4. Several specific derivatization reactions can take place such as alkylation, acetylation, acylation, esterification, and silylation (Prieto et al. 2010; Nogueira 2012).

12.4.1 In Situ Derivatization

In situ derivatization is the simplest and the most widely applied approach. It takes place in the aqueous sample before, or simultaneously, with the extraction step. So, the desired derivatives are firstly formed and then the extraction into the polymer-coating material phase, which is followed by immersion or HS sampling. This approach boosts both the affinity of the analyte for the polymer-coating material phase and the subsequent GC separation.

12.4.2 On-Stir-Bar Derivatization

The **on-stir-bar derivatization** mode can be performed by preloading the polymeric-coated material phase with the derivatization agent (e.g. acetic anhydride). In this way, the reaction occurs as soon as the analytes are incorporated in the PDMS phase (simultaneous extraction and derivatization). Another option is by prior concentration of the analytes in the PDMS phase and

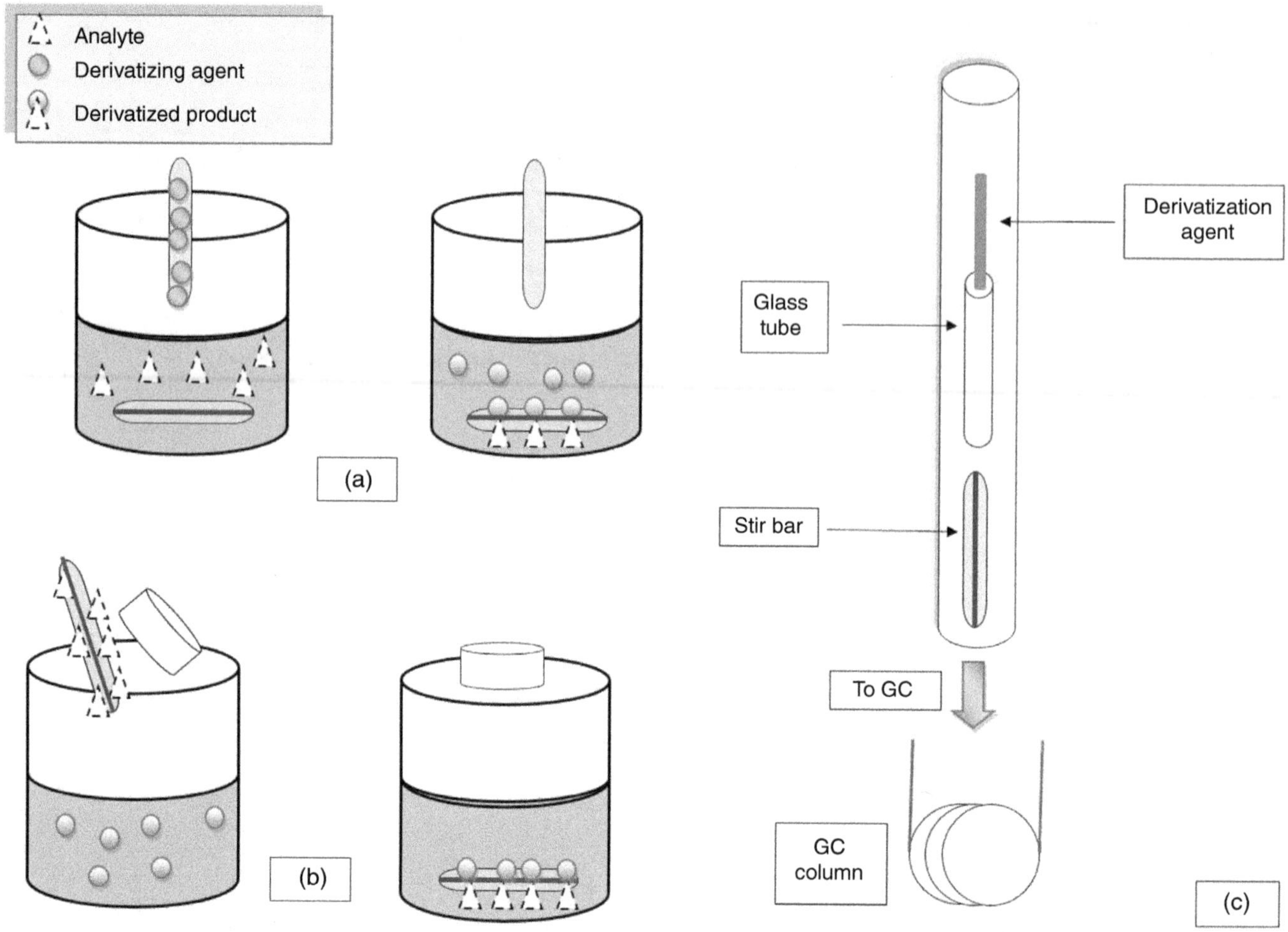

Figure 12.4 Illustration of derivatization modes. (a) In situ derivatization, (b) on-stir-bar derivatization, and (c) post-extraction mode.

then exposing the stir bar to the vapor of the derivatization agent (extraction followed by derivatization).

12.4.3 Post-Extraction Mode

The **post-extraction mode** is mainly used for silylating agents, since they are very susceptible to traces of water and other protic solvents, and it can be achieved with both back-extraction modes such as TD (in-tube) and LD (in-extract). In the first case, the derivatization agent, *N,O*-bis(trimethylsilyl)trifluoroacetamide (BSTFA), is added with the stir bar into the glass tube inside the TD chamber after extraction. Whereas in the second case the derivatization agent, *N*-(tert-butyldimethylsilyl)-*N*-methyl trifluoroacetamide (MTBSTFA), is added to the resulting organic extract after back-extraction (Prieto et al. 2010; Nogueira 2012, 2015). Mainly, it has been applied for in-extract (i.e. LD) derivatization of phenolic compounds and acidic pharmaceuticals and herbicides.

Both derivatization approaches, acetylation and silylation, have improved the sensitivity as compared to approaches without derivatization. Between the two derivatization approaches, in situ acetylation provides better sensitivity for the compounds with hydrophilic properties due to the increased affinity of the compounds for the PDMS phase, whereas in-tube silylation provides better results for the analytes with hydrophobic properties.

Although the derivatization procedures are used to overcome the lack of efficiency of SBSE (PDMS) for the most polar analytes, it is limited to some special derivatization agents, depending on the chemical properties of the analytes and on the type of reaction to be carried out. From the analytical point of view, derivatization procedures are always the last choice, due to the significant complexity usually involved (Kassem 2011; Nogueira 2012).

12.5 Coating Materials

In SBSE, the analytes are sorbed in homogenous and nonporous sorbents. Specifically, the sorption is a concomitant phenomenon of both absorption and adsorption. On one hand, absorption is the process leading to extraction where analytes are partitioned from their matrix into a liquid phase. On the other hand, adsorption is the process where there is an increase in the concentration of a dissolved substance at the interface of a condensed phase due to the operation of surface forces. Adsorption can also occur at the interface between a condensed phase and a gaseous phase. Therefore, in SBSE, the target compounds are retained by the sorptive materials without forming temporary bonding with them.

Sorptive material is a group of polymeric sorbents with a glass transition temperature (T_g). The sorptive material has to be kept below T_g during the sampling–storage–desorption process. At temperatures above the T_g, the polymeric sorbents have a gum-like, or even liquid-like behavior, which is not characteristic of solid materials. On the other hand, at cool conditions, sorbents lose their sorptive properties with a low specific surface area (Baltussen et al. 2002).

The most widely used sorbent in sorptive sample preparation is the nonpolar PDMS, which is a 100% methyl-substituted siloxane polymer. PDMS is used as a coating material to the 1–4 cm-long magnetic stir bars, which are commercially available under the trademark Twister™ (Gerstel GmbH, Mullheiman der Ruhr, Germany). PDMS is widely used because it is an inert material, exhibiting less loss of unstable and/or polar target compounds by irreversible adsorption or by catalytic (surface) reaction. Furthermore, the synthesis of PDMS is not a demanding procedure. Figure 12.5 depicts the structure of PDMS which has a glass transition temperature of −125 °C (Baltussen et al. 2002; Kawaguchi et al. 2013). The usual amount of PDMS used as a coating material is about 24–126 μl. The role of PDMS is to have hydrophobic interactions with the analyte molecules, where the van der Waals forces exist as the retention mechanism. Also, the oxygen atoms of PDMS can form hydrogen bonds with analyte molecules depending on their structures. The SBSE using PDMS as the coating material is a very promising technique with a great capacity and quantitative recoveries, mainly for apolar compounds, because PDMS requires a lower phase ratio between the sorbent phase and the sample media (David and Sandra 2007; Nogueira 2012; Płotka-Wasylka et al. 2015).

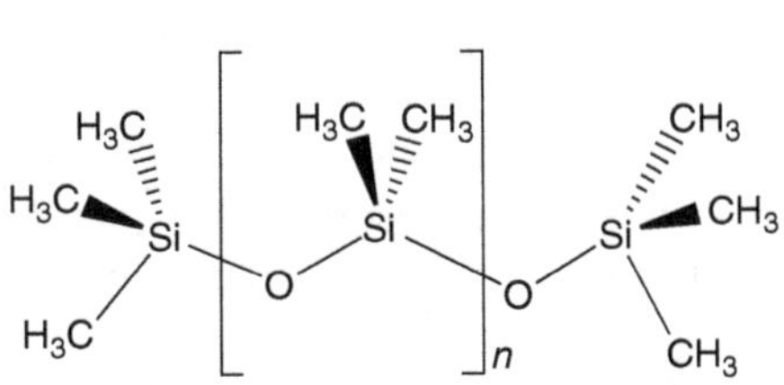

Figure 12.5 The structure of PDMS polymeric phase which has a glass transition temperature of −125 °C.

Stir bars that have PDMS as a coating material consist of three main parts. The first part, also the inner part, is a magnetic stirring rod that is used for the transformation of the rotating movement of a stirring plate to a liquid sample. The rod is covered by a thin glass jacket (the second part), and the third and also the outer part is a layer of PDMS sorbent, which is used for the extraction of target molecules. The glass part plays a role in lowering the risk of decomposition of the PDMS layer, which is catalyzed by the metals in the magnetic bar (Kawaguchi et al. 2013; Płotka-Wasylka et al. 2015).

In addition, SBSE can be used in an automated mode through the GERSTEL® Multi-Purpose Sampler. The SBSE extraction procedure can be applied in many analytical fields. Recent developments in sample preparation include devices for the desorption of target molecules in multi-shot mode (David and Sandra 2007; Spietelun et al. 2013).

It needs to be pointed out that PDMS is a nonpolar material and, therefore, its application for the extraction of polar compounds is limited due to the low recovery. Recently, new coating materials have been developed to improve the recovery, flexibility, and selectivity of the SBSE. Improvement in coating materials expands the applicability of the SBSE from nonpolar analytes to semi-polar and polar analytes. New coating materials for magnetic stir bars include:

- polyurethane foams (PU)
- silicone materials
- poly(ethylene glycol)-modified silicone (EG Silicone Twister)
- poly(dimethylsiloxane)/polypyrrole
- poly(phthalazine ether sulfone ketone) (PPESK)
- polyvinyl alcohol
- polyacrylate (Acrylate Twister)
- carbon nanotube–poly(dimethylsiloxane) (CNT-PDMS)
- alkyl-diol-silica (ADS) restricted access materials
- Molecularly Imprinted Polymers (MIPs)
- sorbents obtained with sol–gel techniques
- monolithic materials such as poly(methacrylic acid-3-sulfopropyl ester potassium salt-co-divinylbenzene) (MADB), poly(vinylimidazole–divinylbenzene) (VIDB), poly(acrylic acid–ethylene dimethacrylate)
- cyclodextrinalkyl-diol-silica (ADS) restricted access materials (RAM)
- low cost and mechanically durable silicon materials
- divinylbenzene (DVB)
- sol–gel amino modified multi-walled carbon nanotubes–PDMS

The above-mentioned materials are applied for extraction in various analytical fields and some of them are very specific.

Another important improvement is the two-phase sorptive elements, such as PDMS/polypyrrole (PPY), which are a combination of two materials (or more in some cases) with different concentration capabilities. These stir bars have a modified structure compared to the traditional SBSE device. They consist of a PDMS rod with two magnetic stoppers at both ends, which is packed with various types of adsorbents. The most commonly used dual-phase stir bar is the one with activated carbon as the adsorbent material with an increased efficiency in extracting semipolar and polar analytes (Nogueira 2012; Kawaguchi et al. 2013; Camino-Sánchez et al. 2014; Nogueira 2015; Płotka-Wasylka et al. 2015).

It is worth mentioning that the only commercially available coating materials are PDMS, polyacrylate (PA), and ethylene glycol/silicone (EG/silicone). The research that concerns the development of new coatings focuses on two technologies: Molecular Imprinting Technology (MIT) and sol–gel procedures. MIT aims to the design of artificial receptors or ligands with a predefined selectivity and specificity for a specific target compound. MIPs are usually antibodies or biological receptors useful for the extraction and the analysis of complex samples such as biological fluids. The extraction of the target compounds displays great selectivity, because it relies on recognition of the three-dimensional structure or functional groups. In contrast, the sol–gel procedure aims to the production of improved solid materials from small molecules. Generally, the sol–gel technique concerns the transition of a solution system from a liquid "sol" (mostly colloidal) into a solid "gel" phase. This novel process allows for the design of improved materials with different forms: ultrafine or spherical-shaped powders, thin film coatings, fibers, and porous or dense materials. These materials can be placed on magnetic stir bars to give them different physical and chemical properties (David and Sandra 2007; Nogueira 2012; Camino-Sánchez et al. 2014; Płotka-Wasylka et al. 2015).

12.6 Applications

The applications of SBSE in environmental and biomedical analysis have recently been reviewed by Kawaguchi et al. (2006b) and by David and Sandra (2007).

For bioanalytical purposes, SBSE is a suitable technique among other well-established techniques and has found many applications (Novakova and Vlčkova 2009). In the field of bioanalysis, different matrices including urine, plasma, and saliva can be analyzed. Particular attention should be given to the possibilities of in situ derivatization, combined with SBSE, since target compounds are frequently quite polar, e.g. metabolites.

Table 12.1 showed the applications of SBSE in bioanalysis.

In a recent application, barbiturates and benzodiazepines as well as their metabolites in urine and blood have been determined after extraction by SBSE. Furthermore, RAM sorbent was used for SBSE of caffeine and its metabolites from rat plasma and SBSE with derivatization was employed for the determination of oxidative stress marker 4-hydroxynoneal (Novakova and Vlčkova 2009).

Melo et al. developed a novel two-phase polymeric coating consisting of PDMS and PPY for the SBSE of antidepressants (mirtazapine, citalopram, paroxetine, duloxetine, fluoxetine, and sertraline) from plasma samples, followed by liquid chromatographic analysis (SBSE/LC-UV). The mechanism of the extraction was adsorption by PPY and sorption by PDMS. SBSE parameters, such as extraction time, temperature, pH of the matrix, and desorption time, were optimized in order to achieve suitable analytical sensitivity in a short time period. The analysis showed that the extraction using the PDMS/PPY-coated stir bar was very efficient with high sensitivity and selectivity toward the target compounds (Melo et al. 2009).

Another biological application of SBSE was proposed by Unceta et al. The aim of the research was the development of SBSE method coupled with HPLC-fluorescence detection (FLD) for the quantification of serotonin reuptake inhibitors in plasma, urine, and brain tissue samples. Also, optimization of important variables of SBSE technique was applied for crucial parameters, such as pH, temperature, volume, extraction time, organic modifier addition, desorption solvent composition, and desorption mode. This method showed many advantages, such as simplicity of the extraction procedure, high sensitivity, accuracy, and high selectivity because of the coupling of SBSE with the fluorescence detector (Unceta et al. 2010).

Talebpour et al. described an SBSE-HPLC-UV method for the analysis of carvedilol in human serum. The innovation of the method was the coating material of the stir bar, which was a film of poly(methyl methacrylate/ethyleneglycoldimethacrylate) (PA-EG) and PDMS, coupled with LD. Furthermore, this novel method used imidazolium-based ionic liquid as a modifier in desorption solvent. The optimization of the crucial parameters was the same as the other applications described above. The result was high sensitivity and short sample preparation time. The developed method was simple, sensitive, precise, and accurate and it can be applied for the determination of carvedilol in the human serum. Consequently, the proposed method can be used for therapeutic drug monitoring (Talebpour et al. 2012).

Chaves et al. reported an SBSE/LC-UV method for the quantification of antidepressants, such as sertraline, mirtazapine, fluoxetine, citalopram, paroxetine, imipramine,

Table 12.1 Examples of SBSE's applications in bioanalysis.

Analyte	Matrix	Extraction phase	Method	Desorption	Analysis	Reference
Antidepressants	Plasma	PDMS/PPY	SBSE	LD	LC-UV	Melo et al. (2009)
Serotonin reuptake inhibitors	Plasma, urine and brain tissue	PDMS	SBSE	LD	HPLC-FLD	Unceta et al. (2010)
Carvedilol	Human serum	PA-EG/PDMS	SBSE	LD with ionic liquid	HPLC-UV	Talebpour et al. (2012)
Antidepressants	Plasma	PDMS	SBSE	LD	LC-UV	Chaves et al. (2007)
Thyroxine	Urine	PEG/OH-PDMS/γ-MPTS	SBSE	LD	HPLC-UV/ICP-MS	Fan et al. (2013)
Steroid sex hormones	Urine	Poly (methacrylic acid stearyl ester–ethylene dimethacrylate)	SBSE	LD	HPLC-DAD	Huang et al. (2008)
Dopamine	Urine	MIP	MIP-SB	LD	HPLC-FL	Lei et al. (2014)
Diclofenac	Pediatric urine	PDMS	SBSE	LD	HPLC-UV	Kole et al. (2011)
Polycyclic aromatic hydrocarbons	Urine	PDMS	SBSE	LD	LC-MS/MS	Zhao et al. (2013)
SCIT, SDCIT, and SDDCIT	Urine	CIP	IP-SBSE	LD	LC-ITMS	Unceta et al. (2013)
Fluoxetine	Plasma	PDMS	SBSE	LD	LC-MS	Fernandes et al. (2006)
THCA	Urine	PDMS	SBSE	LD	LC-MS/MS	Goto et al. (2010)

nortriptyline, amitriptyne, and desipramine, in plasma samples. Important parameters of the SBSE technique, such as extraction time, pH, ionic strength, influence of plasma proteins, and desorption conditions: solvents, modes (magnetic stir, ultrasonic), time, and number of desorption steps, were optimized. The SBSE media was a 10 mm length glass-encapsulated magnetic stir bar, externally coated with 22 μg of PDMS. The developed method was applied for the determination of antidepressants in plasma samples from patients receiving therapeutic dosages and the evaluation of plasma levels in urgent toxicological analyses after the accidental or suicidal intake of higher doses (Chaves et al. 2007).

Fan et al. described a SBSE combined with HPLC-UV/inductively coupled plasma mass spectrometry (ICP-MS) for analysis of thyroxine in urine samples. In this study, a polyethyleneglycol (PEG)/hydroxyl polydimethylsiloxane (OH-PDMS)/γ-mercaptopropyltrimethoxysilane (γ-MPTS) coated stir bar was prepared the by sol–gel process and its extraction performance for the extraction of amphoteric thyroxines (3,3′,5,5′-tetraiodothyronin, T4; 3,3′,5-triiodothyronine, T3; reversed-3,3′,5-triiodothyronine, rT3) and their metabolite (3,5-diiodothyronine, T2) was studied. The prepared PEG/OH-PDMS/γ-MPTS coated stir bar, by the sol–gel process, could be reused for more than 20 times. In comparison with well-known pretreatment techniques, such as SPE and LLE, the PEG/OH-PDMS/γ-MPTS coating SBSE provided higher enrichment factor and good sensitivity for the analysis of thyroxines. The developed method of SBSE-HPLC-UV/ICP-MS showed a great application potential in the determination of amphiprotic thyroxine and its metabolites in urine samples (Fan et al. 2013).

Huang et al. developed a plain, very quick, and sensitive method for simultaneous determination of six steroid sex hormones (SSHs) in urine matrix by the combination of SBSE with HPLC and diode array detection (DAD). A poly(methacrylic acid stearyl ester–ethylene dimethacrylate) was prepared and used as the SBSE extraction phase. To optimize the extraction mode, several parameters, including agitation speed, desorption solvent, extraction and desorption time, pH value, inorganic salt, and organic solvent content of the sample matrix were investigated. Compared to the other extraction methods for SSH determination, the suggested method was simple, rapid, and precise. Consequently, it can be a useful tool for the determination of SSHs in biological and environmental samples (Huang et al. 2008).

Lei and co-workers prepared a stir bar coated with molecularly imprinted polymer (MIP-SB) and it was applied in analysis of dopamine in urine. The adsorptive capacity of the MIP-SB was almost four times over that of non-imprinted stir bar. The MIP-SB showed good extracting selectivity for dopamine when it was used to adsorb dopamine and its analogues. The final method combining MIP-SB sorptive extraction with FLD was characterized as simple and suitable for the determination of dopamine in urine samples (Lei et al. 2014).

Kole et al. developed and validated a novel SBSE-HPLC-UV method for the extraction of diclofenac (DIC) from pediatric urine. An optimized SBSE protocol was developed that selectively extracted DIC among various concurrently administered drugs, with selectivity and sensitivity being the prime objectives of the bioanalytical method for clinical samples. The results showed that, for the present study, stir bars which were used 150 times were still functional based on in-house acceptance criteria and extraction efficiency (Kole et al. 2011).

A simple and sensitive method for the simultaneous determination of 11 monohydroxylated polycyclic aromatic hydrocarbons (OH-PAHs) in human urine samples (smokers and nonsmokers) based on SBSE-HPLC-tandem mass spectrometry (MS/MS) was introduced by Zhao and co-workers. Factors, such as pH, value of matrix, extraction time, desorption solvents, and desorption time were optimized to improve the extraction efficiency. The proposed method has many advantages, such as easy operation, decreased interference, and less organic solvent consumption (Zhao et al. 2013).

Unceta et al. developed a chiral imprinted polymer (CIP)-coated stir bar for the selective extraction of (+)-(*S*)-citalopram (SCIT) and its main metabolites, (+)-(*S*)-desmethylcitalopram (SDCIT) and (+)-(*S*)-didesmethylcitalopram (SDDCIT), from urine samples. The aim of the device was to be capable of selectively extracting the three target analytes from urine samples without saturating the imprinted sites. Thus, a chiral imprinted material was used for the SBSE, followed by analysis using LC-ion trap mass spectrometry (ITMS) (Unceta et al. 2013).

Fernandes and co-workers reported an SBSE-LC-MS method for the analysis of fluoxetine in plasma samples. The first step was the protein precipitation, followed by SBSE and LC-MS analysis. Methanol was used as desorption solvent. The analysis showed that the method was specific, accurate, sensitive, and selective and it presented linearity over the concentration range of 10–500 ng ml^{-1} (Fernandes et al. 2006).

Goto et al. described an SBSE-LC-MS/MS method for the determination of 11-nor-9-carboxy-Δ^9-tetrahydrocannabinol (THCA), the main metabolite of Δ^9-tetrahydrocannabinol, in human urine. The coating material was PDMS and LD was used. The method presented linearity in the range of 10–200 ng ml^{-1} and it had very adequate lower limit of detection (1.3 ng ml^{-1}). Also, the method was accurate, precise, and sensitive (Goto et al. 2010).

Magi et al. proposed a simple, sensitive, and selective method for the determination of the most commonly used UV filter molecules in water that have endocrine-disrupting potential, namely benzophenone-3 (BP-3), octocrylene (OC), ethylhexyl dimethyl p-aminobenzoate (OD-PABA), ethylhexylmethoxycinnamate, ethylhexyl salicylate (EHS), and homosalate (HMS). The UV filter molecules in water samples were extracted by SBSE, followed by LC-MS/MS analysis with good sensitivity and selectivity. The method showed satisfactory linearity and recovery of the analytes (Magi et al. 2012).

Margoum et al. used an SBSE method followed by LD in combined with HPLC-MS/MS for the determination of 15 pesticides or selected metabolites from different families (herbicides, insecticides, and fungicides) in surface water samples. The stir bars were coated with a 1.0 mm film of PDMS. Furthermore, the sample preparation procedure was optimized for all the crucial parameters. The results showed satisfying recoveries (93–101%) with the evaluated measurement uncertainty ranging between 13 and 51% (Margoum et al. 2013).

Nguyen et al. developed an SBSE-LC-MS/MS method for the determination of six UV filter molecules in seawater. The sample preparation included SBSE extraction procedure and LD. The coating material was a 0.5 mm film of PDMS. The combination of SBSE and LC-MS/MS proved to be a simple, accurate, and sensitive tool for the analysis of the examined UV filter molecules. Also, the method was very rapid, ~five minutes, economical, and eco-friendly due to the small amount of solvents that were used (Nguyen et al. 2011).

Blasco and co-workers presented two sorptive extraction-based approaches, SPME and SBSE, in combination with LC-MS for analyzing pesticides, such as chlorpyriphos methyl, diazinon, fonofos, phenthoate, phosalone, and pirimiphos ethyl in honey. Focusing on the SBSE method, the coating material was a 1 mm PDMS layer. The significant parameters were optimized in order to gain high recoveries and linearity. The two procedures were simple, rapid, without extra preliminary steps of sample preparation, and economical with reduced volumes of toxic solvents. Also, both of them had great linearity and similar precision. They concluded that SBSE had higher accuracy that SPME (Blasco et al. 2004).

12.7 Summary

SBSE is one of the most used sorptive methods for sample preparation. Despite the fact that it was first introduced to the field of environmental analysis, with the development of new coating materials for stir bars, its applicability has been expanded to food and biological analysis. However, there are not many sorbents commercially available for the analysis of nonpolar and semipolar compounds. This is the major limitation of SBSE technique. For this reason, researchers work on the improvement of coating materials and analytical approaches, such as in situ derivatization, on-stir-bar derivatization, and post-extraction derivatization, in order to make SBSE a significant micro sample preparation technique. Applications have demonstrated that SBSE not only has great sensitivity and selectivity for trace analysis of complex matrices but also fulfills the requirements of GAC with automation capability. Furthermore, SBSE is simple in operation and requires relatively low cost in instrumentation. Another advantage of this technique is its versatility for extraction of different compounds. The future expectation of the technique is the expansion of its applicability in all analytical fields via overcoming the limitations and challenges that are associated with the technique (Kawaguchi et al. 2006a; David and Sandra 2007; Tobiszewski et al. 2009; Camino-Sánchez et al. 2014; Nogueira 2015; Płotka-Wasylka et al. 2015).

References

Almeida, C. and Nogueira, J.M.F. (2006). Determination of steroid sex hormones in water and urine matrices by stir bar sorptive extraction and liquid chromatography with diode array detection. *J. Pharm. Biomed. Anal.* 41 (8): 1303–1311.

Baltussen, E., Cramers, C.A., and Sandra, P.J.F. (2002). Sorptive sample preparation – a review. *Anal. Bioanal. Chem.* 373 (19): 3–22.

Baltussen, E., Sandra, P., David, F., and Cramers, C. (1999). Stir bar sorptive extraction (SBSE), a novel extraction technique for aqueous samples: theory and principles. *J. Microcolumn Sep.* 11 (11): 737–747.

Blasco, C., Fernández, M., Picó, Y., and Font, G. (2004). Comparison of solid-phase microextraction and stir bar sorptive extraction for determining six organophosphorus insecticides in honey by liquid chromatography–mass spectrometry. *J. Chromatogr. A* 1030 (8): 77–85.

Camino-Sánchez, F.J., Rodríguez-Gómez, R., Zafra-Gómez, A. et al. (2014). Stir bar sorptive extraction: recent applications, limitations and future trends. *Talanta* 130 (11): 388–399.

Chaves, A.R., Silva, S.M., Queiroz, R.H.C. et al. (2007). Stir bar sorptive extraction and liquid chromatography with UV detection for determination of antidepressants in plasma samples. *J. Chromatogr. B* 850 (7): 295–302.

David, F. and Sandra, P. (2007). Stir bar sorptive extraction for trace analysis. *J. Chromatogr. A* 1152 (16): 54–69.

Fan, W., Mao, X., He, M. et al. (2013). Stir bar sorptive extraction combined with high performance liquid chromatography-ultraviolet/inductively coupled plasma mass spectrometry for analysis of thyroxine in urine samples. *J. Chromatogr. A* 1318 (8): 49–57.

Fernandes, C., Jiayu, P., Sandra, P., and Lanças, F.M. (2006). Stir bar sorptive extraction-LC-MS for the analysis of fluoxetine in plasma. *Chromatographia* 64 (4): 517–521.

Goto, Y., Araki, T., Fuchigami, T., and Arizono, K. (2010). Analysis of 11-nor-9-carboxy-Δ^9-tetrahydrocannabinol in urine by LC-MS-MS after stir-bar sorptive extraction and liquid desorption. *Forensic Toxicol.* 28 (1): 38–42.

Huang, X., Yuan, D., and Huang, B. (2008). Determination of steroid sex hormones in urine matrix by stir bar sorptive extraction based on monolithic material and liquid chromatography with diode array detection. *Talanta* 75 (5): 172–177.

Kassem, G.M. (2011). Stir bar sorptive extraction for central nervous system drugs from biological fluids. *Arabian J. Chem.* 4 (10): 25–35.

Kawaguchi, M., Ito, R., Hayatsu, Y. et al. (2006a). Stir bar sorptive extraction with in situ de-conjugation and thermal desorption gas chromatography-mass spectrometry for measurement of 4-nonylphenol glucuronide in human urine sample. *J. Pharm. Biomed. Anal.* 40 (5): 82–87.

Kawaguchi, M., Ito, R., Nakazawa, H., and Takatsu, A. (2013). Applications of stir-bar sorptive extraction to food analysis. *Trends Anal. Chem.* 45 (14): 280–293.

Kawaguchi, M., Ito, R., Saito, K., and Nakazawa, H. (2006b). Novel stir bar sorptive extraction methods for environmental and biomedical analysis. *J. Pharm. Biomed. Anal.* 40 (8): 500–508.

Kole, P.L., Millership, J., and McElnay, C.J. (2011). Determination of diclofenac from paediatric urine samples by stir bar sorptive extraction (SBSE)–HPLC–UV technique. *Talanta* 85 (10): 1948–1958.

Lei, Y., Xu, G., Wei, F. et al. (2014). Preparation of a stir bar coated with molecularly imprinted polymer and its application in analysis of dopamine in urine. *J. Pharm. Biomed. Anal.* 94 (6): 118–124.

Magi, E., Di Carro, M., Scapolla, C., and Nguyen, K.T.N. (2012). Stir bar sorptive extraction and LC–MS/MS for trace analysis of UV filters in different water matrices. *Chromatographia* 75 (9): 973–982.

Margoum, C., Guillemain, C., Yang, X., and Coquery, M. (2013). Stir bar sorptive extraction coupled to liquid chromatography-tandem mass spectrometry for the determination of pesticides in water samples: method validation and measurement uncertainty. *Talanta* 116 (7): 1–7.

Melo, L.P., Nogueira, A.M., Lancas, F.M., and Queiroz, M.E.C. (2009). Polydimethylsiloxane/poly-pyrrole stir bar sorptive extraction and liquid chromatography (SBSE/LC-UV) analysis of antidepressants in plasma samples. *Anal. Chim. Acta* 633 (7): 57–64.

Nazyropoulou, C. and Samanidou, V. (2015). Stir bar sorptive extraction applied to the analysis of biological fluids. *Bioanalysis* 7 (17): 2241–2250.

Nguyen, K.T.N., Di Carro, M., Scapolla, C., and Magi, E. (2011). Rapid and selective determination of UV filters in seawater by liquid chromatography–tandem mass spectrometry combined with stir bar sorptive extraction. *Talanta* 85 (9): 2375–2384.

Nogueira, J.M.F. (2012). Novel sorption-based methodologies for static microextraction analysis: a review on SBSE and related techniques. *Anal. Chim. Acta* 757 (9): 1–10.

Nogueira, J.M.F. (2015). Stir-bar sorptive extraction: 15 years making sample preparation more environment-friendly. *Trends Anal. Chem.* 71 (9): 214–223.

Novakova, L. and Vlčkova, H. (2009). A review of current trends and advances in modern bio-analytical methods: chromatography and sample preparation. *Anal. Chim. Acta* 656 (27): 8–35.

Płotka-Wasylka, J., Szczepańska, N., de la Guardia, M., and Namieśnik, J. (2015). Miniaturized solid-phase extraction techniques. *Trends Anal. Chem.* 73 (19): 19–38.

Prieto, A., Basauri, O., Rodil, R. et al. (2010). Stir-bar sorptive extraction: a view on method optimisation, novel applications, limitations and potential solutions. *J. Chromatogr. A* 1217 (24): 2642–2666.

Spietelun, A., Marcinkowski, L., de la Guardia, M., and Namieśnik, J. (2013). Recent developments and future trends in solid phase microextraction techniques towards green analytical chemistry. *J. Chromatogr. A* 1321 (13): 1–13.

Talebpour, Z., Taraji, M., and Adib, N. (2012). Stir bar sorptive extraction and high performance liquid chromatographic determination of carvedilol in human serum using two different polymeric phases and an ionic liquid as desorption solvent. *J. Chromatogr. A* 1236 (5): 1–6.

Tobiszewski, M., Mechlińska, A., Zygmunt, B., and Namieśnik, J. (2009). Green analytical chemistry in sample preparation for determination of trace organic pollutants. *TrAC – Trends Anal. Chem.* 28 (8): 943–951.

Unceta, N., Gómez-Caballero, A., García, D. et al. (2013). Enantioselective extraction of (+)-(*S*)-citalopram and its main metabolites using a tailor-made stir bar chiral imprinted polymer for their LC-ESI-MS/MS quantitation in urine samples. *Talanta* 116 (5): 448–453.

Unceta, N., Ugarte, A., Sαnchez, A. et al. (2010). Development of a stir bar sorptive extraction based HPLC-FLD method for the quantification of serotonin reuptake inhibitors in plasma, urine and brain tissue samples. *J. Pharm. Biomed. Anal.* 51 (7): 178–185.

Zhao, G., Chen, Y., Wang, S. et al. (2013). Simultaneous determination of 11 monohydroxylated PAHs in human urine by stir bar sorptive extraction and liquid chromatography/tandem mass spectrometry. *Talanta* 116 (4): 822–826.

13

Monolithic Spin Column Extraction in LC-MS Bioanalysis

Akira Namera and Takeshi Saito

13.1 Introduction

Extraction and enrichment of target analytes of interest in complex sample matrices constitute one of the most important and challenging tasks in analytical chemistry. Liquid–liquid extraction (LLE) or solid-phase extraction (SPE) has been widely used to achieve the extraction, purification, and enrichment of target analytes from complex matrices, such as biological, environmental, and food samples. However, these methods involve laborious, intensive, repetitive, and tedious procedures. In addition, the organic solvents used are generally toxic to both humans and the environment. Therefore, recent investigations in analytical sciences have focused on developing new techniques to reduce sample volume, sample preparation time, and the associated costs and to minimize/eliminate the use of toxic organic solvents. Microextractions, such as solid-phase microextraction (SPME) and liquid-phase microextraction (LPME), have been reported in the last two decades as novel extraction techniques that offer many advantages, i.e. reduced sample size, decreased sample volume, easy operation, rapid detection, minimization of the volume of solvent, and miniaturization of the extraction device. These techniques have been used to extract analytes from biological samples. A major disadvantage of these techniques is the relatively long extraction time (30 minutes or more) of the analytes due to the partition equilibrium between the sample matrix and the extraction phase. A new technique is needed to overcome this problem. Recently, microextraction by packed sorbent (MEPS) or sorbent packed tip has been developed and used for extraction of analytes from biological samples. Although this microextraction technique is effective to extract analytes from small sample volumes, the amount of the adsorbed analytes are dependent on the amount of sorbent on the devices, therefore, limiting the use of the devices. A risk of contamination also exists due to the repeated use of the packed sorbent in sample extraction. Therefore, a simple technique that is suitable for the extraction of analyte(s) of interest with a wide concentration range in complex matrices is desirable. In this regard, monoliths has been one of the most interesting materials for biological sample preparation because the analytes can be enriched with a small amount of the sorbent materials and eluted with a small volume of a solvent.

13.2 History of Monoliths

Liquid chromatography (LC) typically has lower resolution in comparison to gas chromatography (GC) and capillary electrophoresis (CE). To achieve a higher level of resolution by LC, a high flow rate of the mobile phase run on a stationary phase with reduced particle size is required. As a result, the pressure of the LC system could become very high, which often leads to malfunction and other issues of the system. Many studies have been carried out to produce new sorbent materials that can help achieve high-resolution in chromatography separation at fast flow rate without high pressure. In the early 1990s, some researchers focused on a highly porous polymer materials (Hjertén et al. 1989; Minakuchi et al. 1996; Tennikova et al. 1990). The material is made of continuous porous polymer and the overall structure is a single piece of three-dimensional network. Therefore, it is named "monolith," meaning a single huge rock or stone (ancient Greek words, "monos" [one or single] and "lithos" [stone]" [Gvishi 2015; Tomašić and Jović 2006]). The polymeric material can be used to separate analytes using a high number of theoretical plates but with a low backpressure. Monoliths are mainly classified into two categories: polymer- and silica-based monoliths.

Sample Preparation in LC-MS Bioanalysis, First Edition. Edited by Wenkui Li, Wenying Jian, and Yunlin Fu.

Polymer-based monoliths are suitable for the separation of large molecules such as proteins, while silica-based monoliths are best suited for small molecules. Thus, these two monoliths are complementary. Recently, an organo-silica hybrid-based third type of monolith has been developed (Acquah et al. 2016).

The polymer-based monoliths were initiatively studied by Svec et al. (2003). In the preparation (Svec et al. 2003; Wang et al. 1993), a mixture of monomers, cross-linkers, and porogens placed in a glass or stainless steel tube is subjected to heat or UV irradiation for the formation of the polymeric skeleton. The pore size is controlled by the polymerization temperature, composition of the porogen, and percentage of the cross-linking monomer. The surface area of the polymer-based monoliths was low in the early stage of development. An approach was later introduced to fabricate monoliths with larger surface areas. This involves extending a new cross-linking monomer with previously formed polystyrene copolymers by a hyper-cross-linking reaction (Urban et al. 2010).

The silica-based monoliths were initiatively studied by Tanaka, Nakanishi, and other researchers (Unger et al. 2011). Alkoxysilane and polyethylene oxide (PEO) are used in the preparation of silica monoliths using sol–gel technology (Nakanishi 1997; Tanaka et al. 2001). In silica monoliths, through-pores and mesopores are controlled independently by the composition of the monomeric mixtures. Moreover, the chemically active surface of the silica allows for the facile insertion of many functional groups, such as octadecyl (ODS) and cation-exchange resin directly onto the monolith surface.

13.3 The Use of Monolith as Sorbent in Solid-Phase Extraction

In SPE, both the area and the chemical nature of the sorbent's surface are the parameters that determine the nature of the interaction between the target analyte and the sorbent. Practically, the surface area affects the analyte recovery and the reproducibility of the extraction, while the nature of the surface affects the selectivity of the analytes adsorbed onto the sorbent. The surface area of commercial SPE cartridges is over $300\,m^2\,g^{-1}$; for example, SepPak C_{18} and OASIS HLB have surface areas of 325 and $830\,m^2\,g^{-1}$, respectively, as shown in Table 13.1 (Namera and Saito 2013). Monolithic silica with a surface area over $300\,m^2\,g^{-1}$ is suitable for the extraction of analytes. However, early monolithic polymers had surface areas of a few tens $m^2\,g^{-1}$. Efforts have been made to increase the surface area of monolithic polymers. The achieved areas were over $600\,m^2\,g^{-1}$, which is similar to those of commercial SPE (Poole 2003). However, monolithic polymers are difficult to be embedded into a syringe barrel because they either swell or shrink when in contact with solvent, creating a gap between the wall of the barrel and the polymer. As a result, the analyte and the solvent flow without retention onto the sorbent. This is a major drawback and also a main reason why there have been hardly any commercially available polymeric-based monolithic products. Many research groups have used in-house monolithic polymers, however, the only commercially available polymer-based monolithic column is CIM monolith, which has been used for the puri-

Table 13.1 Physicochemical properties of SPE columns.

	SepPak® C18	Oasis® HLB	Bond Elut® C18	Bond Elut Certify	InertSep® C18	InertSep® Pharm	MonoSpin C18
Supplier	Waters	Waters	Agilent	Agilent	GL Sciences	GL Sciences	GL Sciences
Sorbent bed material	Silica	Polymer	Silica	Silica	Silica	Polymer	Silica
Sorbent bed mass (mg)	100	30	50	50	100	30	8.5
Cartridge volume (ml)	1	1	1	1	1	1	1
Particle size (μm)	55–105	30	40, 120	40, 120	60	30	—
Particle shape	Irregular	Spherical	Irregular	Irregular	Irregular	Spherical	Monolithic
Through-pore size (μm)	—	—	—	—	—	—	5
Mesopore size (A)	125	80	60	60	60	70	100
Surface area ($m^2\,g^{-1}$)	325	830	500	500	450	600	350
Total carbon (%)	12	—	17.4	9	19	—	23
Total porosity	—	—	—	—	—	—	>0.8
Minimum elution volume (μl)	500	400	300	600	250	500	100

fication of large molecules such as proteins and DNA (Prasanna and Vijayalakshmi 2010).

In contrast, silica-based monolithic materials shrink less when compared to the polymer-based ones. Therefore, they can be more easily embedded in the syringe barrel, which makes them more suitable for the extraction of analytes. Finally, although both silica- and polymer-based monolithic materials are useful to extract analytes of interest in biological samples, intellectual property seems to be one of the biggest barriers that prevent more monolithic-based products from getting into market swiftly.

13.4 Monolithic Spin Column for Sample Preparation

Monolithic silica has been developed as a new type of stationary phase for HPLC as described above. Initially, monolithic silica was formed directly inside the capillary tube, which was used for the extraction of phenols in environmental samples (Shintani et al. 2003). Amphetamines in urine could also be extracted using the monolithic silica packed into a capillary glass tube (0.2-mm internal diameter [ID]) (Nakamoto et al. 2006). However, sample filtration was required to avoid blockage within the tube. In order to overcome this problem, a monolithic silica disk (4.3-mm ID, 1.5-mm thickness) was packed into a spin column (Figure 13.1). This column is commercially available from GL Sciences (Tokyo, Japan) under the trade name MonoFas® (Ota et al. 2006). The surface of MonoFas mainly contains free silanol groups. Various functional groups, such as octadecyl, amino, and trimethylaminopropyl, can be easily bound to the silanol on the surface, which leads to the production of a wide range of columns commercially available under the trade name MonoSpin (Table 13.2).

13.4.1 Extraction Procedure

The representative sample handling procedure is shown in Figure 13.2. The procedure involves sample loading, washing, and elution of the target analytes by centrifugation of the column. The flow rate can be easily controlled by adjusting the rotation speed and time. In addition, multiple samples can be processed simultaneously.

13.4.2 Extraction Mechanism

The mass transfer mechanism in monoliths depends on the hydrophobic adsorption to the surface of the sorbent and the partition coefficient of the analyte to the functional surface moiety, such as bonded alkyl chains or ion exchange resins. This mechanism is similar to that in the conventional SPE. Physical properties, such as surface area, average pore diameter, and pore volume of the sorbents, are also important in the adsorption and desorption process of the analytes. In general, the adsorption capacity of the sorbent is proportional to its surface area, while the pore size affects the molecules adsorbed in the diffusion process. Therefore, the sorbent's surface area is one of the most important factors that affect the adsorption capacity (extraction efficiency). The surface area of polymeric monoliths developed in early days was in a range of only 5–20 $m^2\ g^{-1}$, which was too small to be useful in SPE (Svec 2006). Polymeric monoliths, with surface areas up to 1000 $m^2\ g^{-1}$, have been developed by some research groups by increasing the degree of cross-linking in the structures. The capacity of the surface area is sufficient to routinely extract the analytes of interest. However, polymeric monoliths have some critical defects, i.e. their weakness to the mechanical force applied by the high flow rate of solvent, and their swelling and shrink-

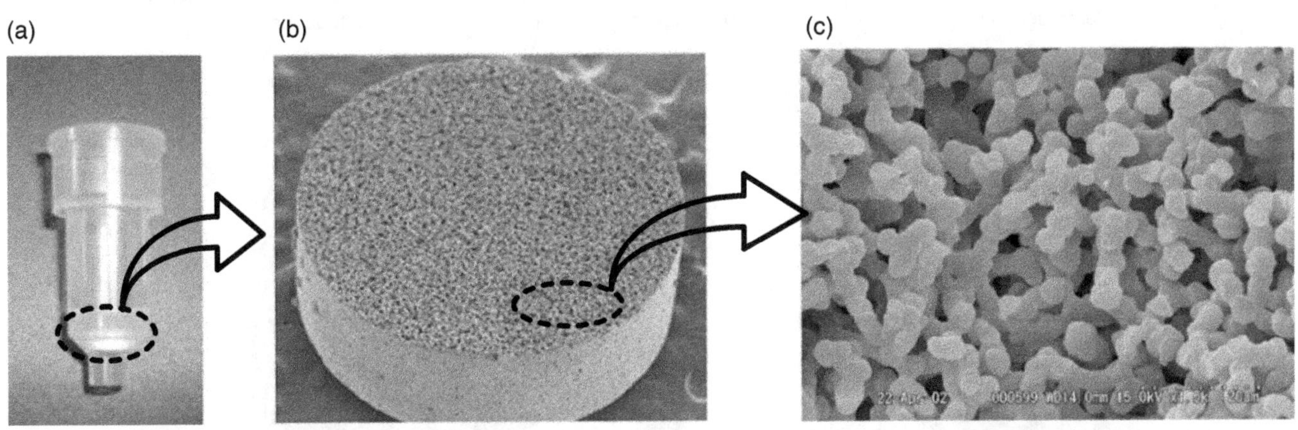

Figure 13.1 Photograph of MonoSpin column (a), and electromicrograph of the disk (b) and monolithic silica (c), prepared with mixtures of TEOS and PEO in MonoSpin (mesopore size: 10 nm, through-pore size: 5 lm, surface area: 350 $m^2\ g^{-1}$). *Source:* Courtesy of Shigenori Ota of GL Sciences, Japan.

Table 13.2 Physicochemical properties of commercial MonoSpin columns.

MonoSpin brands	Functional structures	Extraction modes	Targets	Purpose	Through-pore (μm)	Meso pore (nm)
C18	$-(CH_2)_{17}-CH_3$	Reversed phase	Hydrophobic (nonpolar)		5	10
			Peptide	desalting		
C18 FF	$-(CH_2)_{17}-CH_3$	Reversed phase			20	15
Ph	$-C_6H_5$	Reversed phase	Moderately nonpolar		5	10
SAX	$-(CH_2)_3-N^+-(CH_3)_3$	Ion exchange	Strong anion (hydrophobic)		5	10
SCX	$-(CH_2)_3-C_6H_4-SO_3^-$	Ion exchange	Strong cation (hydrophobic)		5	10
C18-AX	$-(CH_2)_{17}-CH_3$	Reversed phase	Strong anion hydrophobic		5	10
	$-(CH_2)_3-N^+-(CH_3)_3$	Ion exchange				
C18-CX	$-(CH_2)_{17}-CH_3$	Reversed phase	Strong cation, hydrophobic		5	10
	$-(CH_2)_3-C_6H_4-SO_3^-$	Ion exchange				
NH2	$-(CH_2)_3-NH_2$	HILIC	Hydrophilic (polar)		5	10
CBA	$-CH_2-COOH$	Ion exchange	Strong basic		5	10
Amide	$-CH_2-CONH_2$	hydrophilics	Hydrophilic (polar)		5	10
PBA	$-NH-C_6H_4-B(OH)_2$		*cis*-Diol		5	10
TiO	TiO_2		Phosphate		20	15
ME	$-CH_2-N(COOH)_2$	Chelating	Metal		5	10
Phospholipid	TiO_2, ZrO_2			de-phospholipids	5	10
Trypsin	Trypsin	Enzyme hydrolysis			5	10
ProA	Protein A	Affinity			2	60
ProG	Protein G	Affinity			2	60

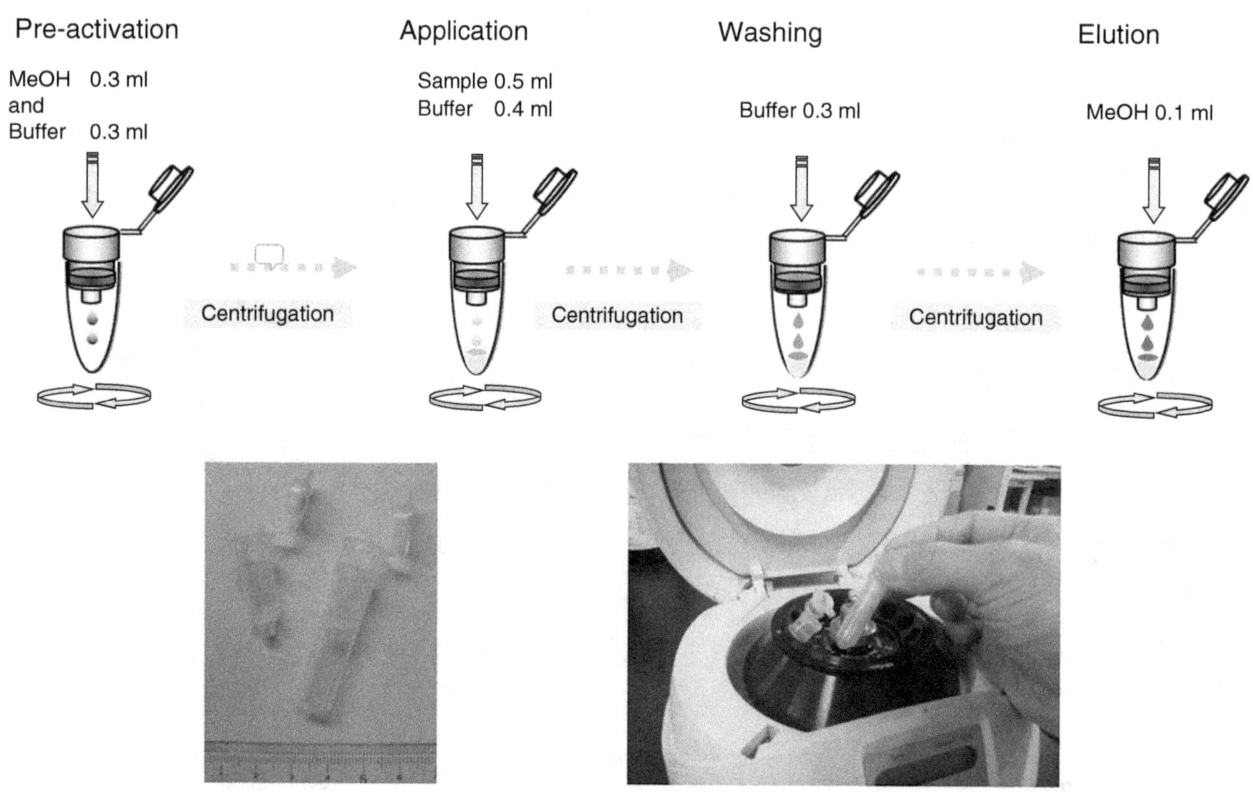

Figure 13.2 General extraction procedure using MonoSpin columns.

ing influenced by the ratio of the organic solvent and water. Therefore, it is difficult to pack the polymeric monolith into the cartridge or syringe barrel.

The surface area of silica monoliths is generally over $300\,m^2\,g^{-1}$. Although this number is smaller than that of Oasis HLB® (Namera and Saito 2013), the effective or accessible surface area of silica monoliths is larger than that of the packed column. In other word, larger amounts of analyte of interest can be adsorbed with a smaller amount of sorbent and, therefore, the amount of sorbent used can be reduced.

The height equivalent to a theoretical plate (HETP) is an index used to measure the performance of an HPLC column. HETP is dependent on the particle diameter. The shorter each theoretical plate, the more plates are "contained" in given length of column/cartridge. Particles with a small diameter are associated with an increased HETP numbers and capacity; however, small particles are also associated with an increased backpressure. Although monolithic silica has similar characteristics to smaller particle sorbent, i.e. particle diameters of $<2\,\mu m$, the backpressure is dramatically lower than that of the particle sorbent. Therefore, the monolithic silica adsorbs the analyte with a narrow band on the column. As a result, the analyte can be easily eluted from the column with a small volume of solvent. The elution volume usually required to provide good extraction efficiencies in conventional SPE column/cartridge is two or three times the volume of the sorbent bed. However, with monolithic silica, the analyte adsorbed can be eluted with as little as $100\,\mu l$ of an elution solvent.

13.4.3 Advantages and Limitations

Compared to conventional particle sorbents, monolith sorbents have some advantages. These advantages include but are not limited to: zero dead volume, no channeling, ease of operation, and ease of shape modification.

13.4.3.1 Zero Dead Volume

A zero dead volume reduces the amount of solvent required for the elution of the analyte. In conventional SPE, the stationary phase is packed into the space between the two frits to avoid spilling of the sorbent from the syringe barrel or cartridge. The volume of solvent that fills up the frits is called "dead volume." Monolithic silica does not need to use these frits because the one-piece structure of monolithic silica supports its body. Therefore, zero dead volume is possible, and the amount of eluting solvent can be reduced and the possible interaction of analyte with frit is avoided.

13.4.3.2 No Channeling

Most sorbents in conventional SPE are spherical or irregular particles in shape and they are tightly packed into a cartridge or column. Extra instruments such as mechanical pumps or syringes are inevitably needed to aid the flow of the sample solution through the sorbent under high or reduced pressure. However, an intrinsic problem with all particles is their inability to fill completely all the available space. Consequently, a crack (channel) forms between the particles and the wall of the cartridge. This leads to a reduced interaction of the analytes of interest with the extraction sorbents and, consequently, reduced extraction efficiency. In contrast, monoliths can be prepared in situ within narrow capillaries and applied to microscale separations without channeling.

13.4.3.3 Ease of Operation

When doing sample preparation using the monolithic silica spin column, all of the extraction steps such as sample loading, washing, and elution are achieved by centrifugation. A low-volume elution is possible to desorb the analyte from the column and no solvent evaporation is needed. Moreover, many samples can be processed simultaneously without the use of a manifold. Therefore, the sample preparation time is reduced and the productivity is increased, although the first-time users need to pay attention to flow rate, especially when using a manifold, to avoid drying the SPE column completely which may impact the recovery of the analyte from the column.

13.4.3.4 Ease of Shape Modification

In most applications, the surface of the monolith is bonded to C_{18} groups. Multifunctional moieties can be attached to the silica surface to meet the study needs. Various types of spin columns have been developed and commercialized as shown in Table 13.2. Simultaneous extractions of acidic and basic drugs can be performed with high recovery using monolithic spin column without interference from endogenous substances in the sample matrix. In order to meet more challenging objectives, it is possible to extract both hydrophobic and hydrophilic compounds in a single extraction by adding activated carbon to the sorbent (Saito et al. 2013a). Moreover, highly selective single extraction of both hydrophobic and hydrophilic compounds can be achieved by adding titania (titanium oxide) to the sorbent (Saito et al. 2011).

Of course, there are some obvious disadvantages associated with monolith sorbents as compared to conventional SPE sorbent materials:

13.4.3.5 Loading Sample Volume

The acceptable loading volume for MonoSpin® columns is about 0.8 ml, which limits the ability to enrich analytes from samples with larger volumes. Particularly, larger loading volumes are required for certain applications in environmental analysis and bioanalysis. For those applications, a larger capacity of MonoSpin column packed with a 9-mm ID disk is available.

13.4.3.6 High Viscosity Sample

Forensic samples, such as high viscosity spatters and blood clots, need to be diluted with water or buffer, and submitted to either centrifugation or filtration to remove the solids. After centrifugation, the resulting samples are still difficult to pass through the conventional MonoSpin columns with a pore size of 5 μm. Fast-flow MonoSpin columns, packed with a larger through-pore size (20 μm) monolithic silica, have been developed for extraction of such high viscosity samples.

13.4.3.7 Commercial Suppliers

Nowadays, only a few suppliers provide reliable monolithic silica columns. The MonoSpin series are available from GL Sciences in Japan. SOLA, available from Thermo Scientific, contains a similar packing materials as MonoSpin columns. Smaller extraction devices exist. For example, MonoTip, NuTip, and ZipTip are commercially available from GL Science, Glygen, and Millipore, respectively. These are suited for low sample volume applications due to the small amount of sorbent packed in the tips.

13.4.4 Applications

Spin-column techniques have been applied for the extraction of various compounds, including amphetamines (Namera et al. 2008), a-PVP (Namera et al. 2013), MAM-2201 (Saito et al. 2013b), benzodiazepines and its metabolites (Namera et al. 2011a), dibucaine (Saito et al. 2008a), eperisone (Miura et al. 2014), naphazoline (Saito et al. 2008a), and opiates (Namera et al. 2011a). Common applications are listed in Table 13.3. Among the many applications, use of monolith silica with C_{18} moieties bonded to its surface is the most popular one in the same way as conventional SPE. The extraction procedure is the same as a conventional SPE as

Table 13.3 Representative applications of MonoSpin® extraction in LC-MS bioanalysis.

Material	Target	Sample	LOD (ng ml^{-1})	Recovery (%)	Detection	Reference
C18	Amitraz, metabolites	Serum	5	95.5, 92.2	LC-MS	Saito et al. (2008b)
	Dibudcaine, naphazoline	Serum	5–10	70.2–78.6	LC-MS	Saito et al. (2008a)
	Amphetamines	Urine	100	96–111	LC-UV	Namera et al. (2008)
	Dopamine, epinephrine, norepinephrine	Urine	0.2	95–101	LC-ECD	Tsunoda et al. (2011)
	α-PVP	Hair	0.2 ng mg^{-1}	73.1–101.5	LC-MS	Namera et al. (2013)
	MAM-2201	Blood	1		LC-MS/MS	Saito et al. (2013b)
	Eperison, tolperison, tizanidine	Serum	0.25	94.1–99.8	LC-MS/MS	Miura et al. (2014)
	Aconitine, colchicine	Urine, serum	0.05–0.1	96.6–101.9	LC-MS/MS	Saito et al. (2015)
	Amino acids	Plasma		19.0–87.5	MS/MS	Tamashima et al. (2015)
	11 medicinal toxicants	Serum		11.0–130.8	LC-MS/MS	Usui et al. (2017)
	Cortisol, cortisone	Urine	2.5 (LLOQ)	46.5, 50.8	LC-MS/MS	Matsumoto et al. (2017)
C18 FF	10 medicines	blood	0.5–2.0	76–108	LC-MS	Namera et al. (2017)
SCX	Opiates Benzodiazepines, metabolites	Urine Serum	10 1–10	69.2–98.9 83.3–112.3	LC-MS	Namera et al. (2011a)
	Arginine derivatives	Plasma	3.75–9.0 (fmol)	89–104	LC-FL	Nonaka et al. (2014)
SAX	Alendronate	Plasma	2.0	41.1–51.2	LC-MS/MS	Yamada et al. (2018)
CBA	Tetrodotoxin	Serum	1	86.5–90.4	LC-MS/MS	Saito et al. (2014)
Amide	Tetrodotoxin	Urine	0.5	90.7–92.7	LC-MS/MS	Saito et al. (2014)
C18-C	Organophosphates, carbamates	Serum	0.1–50	11.9–99.2	LC-MS/MS	Saito et al. (2013a)
C18-CX	β-blockers, calcium blockers	Serum			LC-MS	Namera, personal data
Trypsin	Cetuximab	Serum	0.1		LC-MS/MS	Shibata et al. (2017)

shown in the following protocols. In brief, 0.2–1.0 ml volume of the biological sample or the diluted sample is applied to an activated spin column. After washing the column with distilled water or suitable buffer, the analyte is then eluted with solvent such as methanol or acetonitrile. The eluate can be directly injected to LC without evaporation.

Recently, new spin column in which multifunctional moieties attached to the silica surface has been developed to simultaneously extract analytes that have different chemical properties. For example, the simultaneous extraction of acidic (e.g. phenobarbital and secobarbital) and basic drugs (e.g. amitriptyline, chlorpromazine, clotiazepam, diazepam, imipramine, lidocaine, and phenacetin) in urine can be achieved with no interference due to endogenous substances using MonoSpin C18-CX (Namera et al. 2011b). In brief, the urine (0.2 ml) was acidified with citrate buffer (0.1 M, pH 3) and then loaded onto the column. After washing, the analytes were eluted with 0.1 ml of methanol containing 2% ammonium hydroxide. By this procedure, all the tested drugs were simultaneously extracted from urine with recovery between 65 and 123%. Concentrations of up to 2500 ng ml^{-1} of the targeted drugs in urine were analyzed with linear curves ($r^2 > 0.996$) and limits of detection at 1–25 ng ml^{-1}. The intra- and inter-day precision at three different concentrations ranged from 2.1 to 14.7%. The proposed method was successfully applied for clinical and forensic samples. The same procedure was applied for the extraction of β-blockers (bisoprolol, carazolol, carvedilol, pindolol, and propranolol) and calcium blockers (amlodipine, felodipine, nicardipine, nifedipine, and nitrendipine) in human serum and the typical chromatogram using LC–MS is shown in Figure 13.3. The results obtained were in good agreement with those obtained by conventional methods and confirmed that this sorbent is suitable for LC-MS bioanalysis.

Typical protocol 1 (reversed-phase extraction mode)

1) Borate buffer (0.4 ml, 0.1 M, pH 10)[a] and the internal standard are added to the urine sample (0.5 ml)[b], and the mixture is then vortex-mixed for several seconds.
2) The MonoSpin C18[c] column is connected to a waste tube and pre-activated by washing with 0.3 ml of methanol and water (or buffer) with the aid of centrifugal force[d] (centrifugation).
3) The solution used for the pre-activation in the waste tube is discarded.
4) The pre-activated column is attached to the empty waste tube again. The vortex-mixed (pH-adjusted) sample mixture is poured onto the column.
5) The analyte is adsorbed on the column and the column is centrifuged[e] at 3000 rpm for two minutes.
6) The interferences in the column are eliminated by washing the column with 0.3 ml of buffer or methanolic solution[f] and centrifuged at 3000 rpm for one minute.
7) The column is then attached to a new tube, the analyte is eluted with 0.1 ml of methanol, and the column is then centrifuged at 3000 rpm for one minute.
8) The eluate[g] is directly injected into the LC without evaporation of the solvent.

Notes

[a]The buffer is added to the sample to adjust the pH of the sample. The pH of the buffer is usually $pK_a \pm 2$ of each analyte in reversed-phase extraction mode. If an acidic analyte is to be extracted by MonoSpin C18, then the employment of an acidic buffer, such as a citrate buffer (pH 3), is recommended.

[b]Low viscosity samples, such as urine and its solutions, can be used directly on the MonoSpin columns. For high viscosity samples, such as whole blood, FF-type MonoSpin columns should be used (Namera et al. 2017). When insoluble particles are present, the sample should be first filtered or centrifuged.

[c]The recommended maximum column capacity is 0.8 ml. For sample volumes of 1.0 ml, the pinhole at the head cap of the column should be adjusted.

[d]The column is centrifuged at 3000 rpm for 30 seconds.

[e]The rotation speed is between 2500 and 3000 rpm, and the recovery of the analyte is dependent on rotation speed (flow rate). When FF-type MonoSpin column is used, the suitable rotation speed is around 1000–1500 rpm.

[f]The adsorption profile should be monitored for each analyte to confirm the retention of the target analyte to the sorbent.

[g]No peak or broad peak is associated with highly water-soluble analyte(s) when methanol content is high in the sample eluate. In such case, the eluate is twice or three times diluted with the mobile phase.

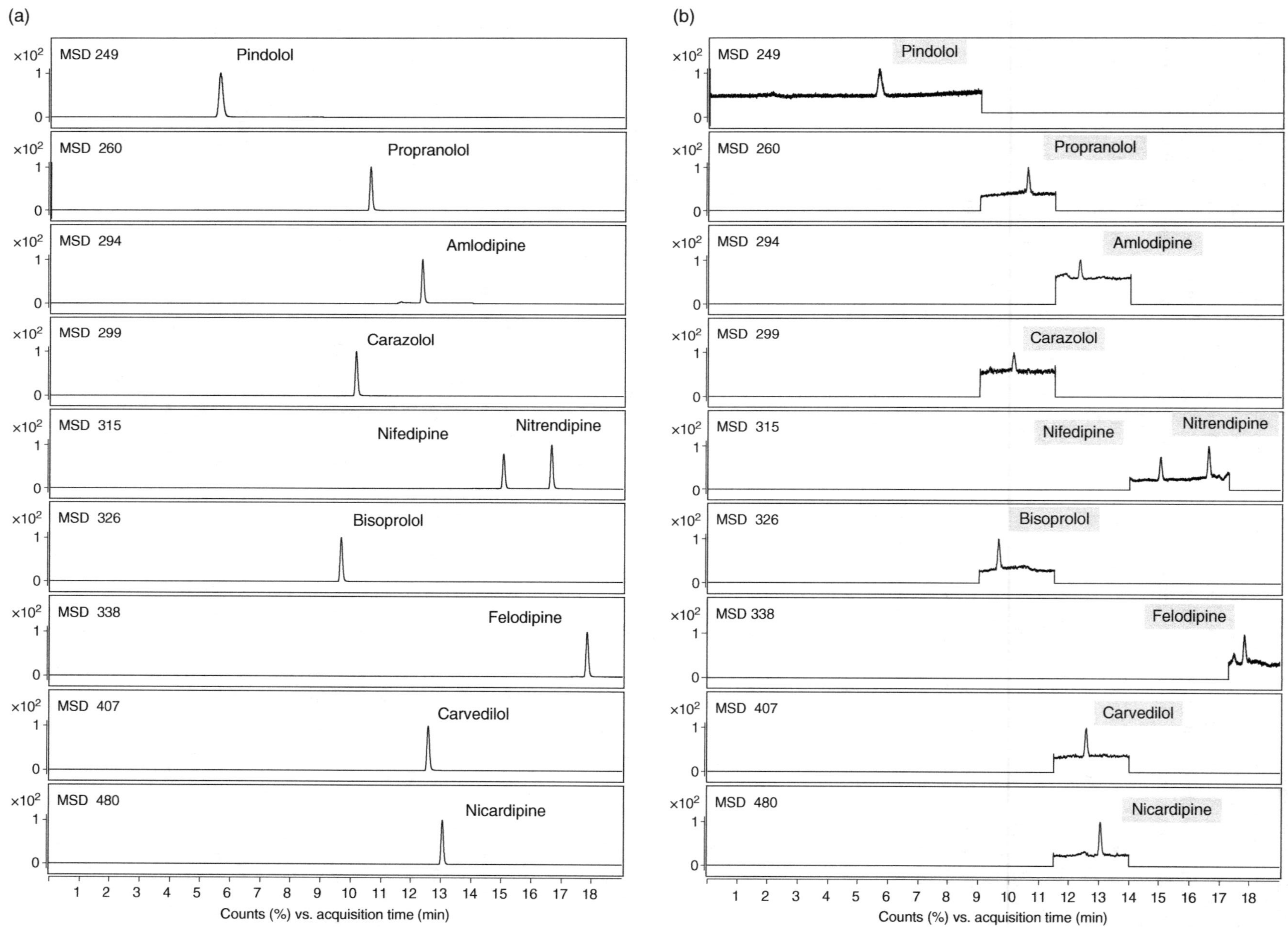

Figure 13.3 Representative chromatograms of β-blockers and calcium blockers obtained using LC-MS analysis of human serum. (a) Serum spiked at 250 ng ml^{-1} and (b) serum spiked at the concentration of LLOQ (5 ng ml^{-1}).

Typical protocol 2 (mixed-mode extraction (reversed phase and cationic exchange)

1) Citrate buffer (0.4 ml, 0.1 M, pH 3)[a] and the internal standard are added to the urine sample (0.2 ml), and the mixture is vortex-mixed for several seconds.
2) The MonoSpin C18-CX[b] column is connected to a waste tube and is preactivated by washing with 0.5 ml of methanol and citrate buffer with the aid of centrifugal force[c] (centrifugation).
3) The solution used for preactivation in the waste tube is discarded.
4) The preactivated column is attached to the empty waste tube again. The vortex-mixed (pH-adjusted) sample mixture is poured onto the column.
5) The analyte is adsorbed on the column and the column is centrifuged[d] at a suitable rotation for two minutes.
6) The passed solution in the waste tube is discarded.
7) The interferences in the column are eliminated by washing the column with 0.3 ml of buffer.[e]
8) The column is then attached to a new tube, the analyte is eluted with 0.1 ml of methanol containing a 2% aqueous ammonium solution.[f]
9) The eluate[g] is directly injected into the LC without evaporation of the solvent.

Notes

[a]The concentration and pH of the citrate buffer used are 10 mM and 3, respectively.
[b]The recommended maximum column capacity is 0.8 ml. For sample volumes of 1.0 ml the pinhole at the head of the column should be adjusted.
[c]The column is centrifuged at 3000 rpm for 30 seconds.
[d]The recovery of the analyte is dependent on the rotation speed (flow rate). The rotation speed is usually 5000 rpm.
[e]The adsorption profile should be monitored for each analyte. The water-soluble analyte is eluted with high content of methanol in the mobile phase.
[f]The basic and acidic analytes are simultaneously eluted with the basic solution.
[g]Highly water-soluble analytes may show broad or no peaks when methanol content is high in the sample eluate. In this case, the eluate is twice or three times diluted with the mobile phase.

References

Acquah, C., Moy, C.K., Danquah, M.K., and Ongkudon, C.M. (2016). Development and characteristics of polymer monoliths for advanced LC bioscreening applications: a review. *J. Chromatogr. B Analyt. Technol. Biomed. Life Sci.* 1015–1016: 121–134.

Gvishi, R. (2015). Monolithic sol–gel materials. In: *The Sol–Gel Handbook – Synthesis, Characterization, and Applications: Synthesis, Characterization and Applications* (ed. D. Levy and M. Zayat). Weinheim, Germany: Wiley-VCH.

Hjertén, S., Liao, J.L., and Zhang, R. (1989). High-performance liquid chromatography on continuous polymer beds. *J. Chromatogr. A* 473: 273–275.

Matsumoto, A., Shimanoe, C., Tanaka, K. et al. (2017). Development of suitable method for large-scale urinary glucocorticoid analysis by liquid chromatography-mass spectrometry. *J. Chromatogr. B* 1057: 62–69.

Minakuchi, H., Nakanishi, K., Soga, N. et al. (1996). Octadecylsilylated porous silica rods as separation media for reversed-phase liquid chromatography. *Anal. Chem.* 68 (19): 3498–3501.

Miura, N., Saito, T., Taira, T. et al. (2014). Rapid simultaneous determination of eperisone, tolperisone, and tizanidine in human serum by using a MonoSpin® C18 extraction column and liquid chromatography/tandem mass spectrometry. *J. AOAC Int.* 97 (6): 1546–1551.

Nakamoto, A., Namera, A., Nishida, M. et al. (2006). Monolithic silica capillary column extraction of methamphetamine and amphetamine in urine coupled with thin-layer chromatographic detection. *Forensic Toxicol.* 24 (2): 75–79.

Nakanishi, K. (1997). Pore structure control of silica gels based on phase separation. *J. Porous Mater.* 4: 67–112.

Namera, A., Konuma, K., Saito, T. et al. (2013). Simple segmental hair analysis for α-pyrrolidinophenone-type designer drugs by MonoSpin extraction for evaluation of abuse history. *J. Chromatogr. B Analyt. Technol. Biomed. Life Sci.* 942–943: 15–20.

Namera, A., Nagao, M., Nakamoto, A. et al. (2011a). Monolithic spin column: a new extraction device for analysis of drugs in urine and serum by GC/MS and HPLC/MS. *J. AOAC Int.* 94 (3): 765–774.

Namera, A., Nakamoto, A., Nishida, M. et al. (2008). Extraction of amphetamines and methylenedioxyamphetamines from urine using a monolithic silica disk-packed spin column and high-performance liquid chromatography-diode array detection. *J. Chromatogr. A* 1208 (1–2): 71–75.

Namera, A. and Saito, T. (2013). Advances in monolithic materials for sample preparation in drug and pharmaceutical analysis. *Trends. Analyt. Chem.* 45: 182–196.

Namera, A., Saito, T., Ota, S. et al. (2017). Optimization and application of octadecyl-modified monolithic silica for solid-phase extraction of drugs in whole blood samples. *J. Chromatogr. A* 1517: 9–17.

Namera, A., Yamamoto, S., Saito, T. et al. (2011b). Simultaneous extraction of acidic and basic drugs from urine using mixed-mode monolithic silica spin column bonded with octadecyl and cation-exchange group. *J. Sep. Sci.* 34 (16–17): 2232–2239.

Nonaka, S., Sekine, M., Tsunoda, M. et al. (2014). Simultaneous determination of N(G)-monomethyl-L-arginine, N(G),N(G)-dimethyl-L-arginine, N(G),N(G')-dimethyl-L-arginine, and L-arginine using monolithic silica disk-packed spin columns and a monolithic silica column. *J. Sep. Sci.* 37 (16): 2087–2094.

Ota, Y., Tokuda, S., Buchanan, P.K., and Hattori, T. (2006). Phylogenetic relationships of Japanese species of *Heterobasidion – H. annosum* sensu lato and an undetermined *Heterobasidion* sp. *Mycologia* 98 (5): 717–725.

Poole, C.F. (2003). New trends in solid-phase extraction. *TrAC Trend. Anal. Chem.* 22 (6): 362–373.

Prasanna, R.R. and Vijayalakshmi, M.A. (2010). Characterization of metal chelate methacrylate monolithic disk for purification of polyclonal and monoclonal immunoglobulin G. *J. Chromatogr. A* 1217 (23): 3660–3667.

Saito, T., Aoki, H., Namera, A. et al. (2011). Mix-mode TiO-C18 monolith spin column extraction and GC-MS for the simultaneous assay of organophosphorus compounds and glufosinate, and glyphosate in human serum and urine. *Anal. Sci.* 27 (10): 999–1005.

Saito, T., Miura, N., Namera, A. et al. (2013a). Rapid determination of polar and non-polar pesticides in human serum, using mixed-mode C-C18 monolithic spin column extraction and LC–MS/MS. *Chromatographia* 76 (13): 781–789.

Saito, T., Miura, N., Namera, A. et al. (2014). A rapid sample preparation procedure using MonoSpin CBA and amide columns for tetrodotoxin detection in serum and urine using LC–MS/MS analysis. *Chromatographia* 77 (9): 687–693.

Saito, T., Morita, S., Kishiyama, I. et al. (2008a). Simultaneous determination of dibucaine and naphazoline in human serum by monolithic silica spin column extraction and liquid chromatography-mass spectrometry. *J. Chromatogr. B Analyt. Technol. Biomed. Life Sci.* 872 (1–2): 186–190.

Saito, T., Namera, A., Miura, N. et al. (2013b). A fatal case of MAM-2201 poisoning. *Forensic Toxicol.* 31 (2): 333–337.

Saito, T., Umebachi, R., Namera, A. et al. (2015). Identification and quantification of aconitines and colchicine in serum, urine, and plants using MonoSpin C18 and LC-MSMS. *Chromatographia* 78 (15): 1041–1048.

Saito, T., Yamamoto, R., Inoue, S. et al. (2008b). Simultaneous determination of amitraz and its metabolite in human serum by monolithic silica spin column extraction and liquid chromatography-mass spectrometry. *J. Chromatogr. B Analyt. Technol. Biomed. Life Sci.* 867 (1): 99–104.

Shibata, K., Naito, T., Okamura, J. et al. (2017). Simple and rapid LC-MS/MS method for the absolute determination of cetuximab in human serum using an immobilized trypsin. *J. Pharm. Biomed. Anal.* 146: 266–272.

Shintani, Y., Zhou, X., Furuno, M. et al. (2003). Monolithic silica column for in-tube solid-phase microextraction coupled to high-performance liquid chromatography. *J. Chromatogr. A* 985 (1–2): 351–357.

Svec, F. (2006). Less common applications of monoliths: preconcentration and solid-phase extraction. *J. Chromatogr. B Analyt. Technol. Biomed. Life Sci.* 841 (1–2): 52–64.

Svec, F., Tennikova, T.B., and Deyl, Z. (2003). *Monolithic Materials: Preparation, Properties, and Applications.* Amsterdam, The Netherlands: Elsevier.

Tamashima, E., Hayama, T., Yoshida, H. et al. (2015). Direct tandem mass spectrometric analysis of amino acids in plasma using fluorous derivatization and monolithic solid-phase purification. *J. Pharm. Biomed. Anal.* 115: 201–207.

Tanaka, N., Kobayashi, H., Nakanishi, K. et al. (2001). Monolithic LC columns. *Anal. Chem.* 73 (15): 420A–429A.

Tennikova, T.B., Belenkii, B.G., and Svec, F. (1990). High-performance membrane chromatography. A novel method of protein separation. *J. Liq. Chromatogr.* 13 (1): 63–70.

Tomašić, V. and Jović, F. (2006). State-of-the-art in the monolithic catalysts/reactors. *Appl. Catal. Gen.* 311: 112–121.

Tsunoda, M., Aoyama, C., Ota, S. et al. (2011). Extraction of catecholamines from urine using a monolithic silica disk-packed spin column and high-performance liquid chromatography-electrochemical detection. *Anal. Methods* 3: 582–585.

Unger, K.K., Tanaka, N., and Machtejevas, E. (2011). *Monolithic Silicas in Separation Science.* Weinheim, Germany: Wiley-VCH.

Urban, J., Svec, F., and Fréchet, J.M. (2010). Hypercrosslinking: new approach to porous polymer monolithic capillary columns with large surface area for the highly efficient separation of small molecules. *J. Chromatogr. A* 1217 (52): 8212–8221.

Usui, K., Mori, H., Tachi, T. et al. (2017). A rapid method to screen poisoning causative agents in an acute care hospital in Japan. *J. Clin. Pharm. Ther.* 42: 454–460.

Wang, Q.C., Svec, F., and Fréchet, J.M. (1993). Macroporous polymeric stationary-phase rod as continuous separation medium for reversed-phase chromatography. *Anal. Chem.* 65 (17): 2243–2248.

Yamada, M., Lee, X.P., Fujishiro, M. et al. (2018). Highly sensitive determination of alendronate in human plasma and dialysate using metal-free HPLC-MS/MS. *Leg. Med.* 30: 14–20.

14

Aptamer-based Sample Preparation in LC-MS Bioanalysis

Fuyou Du, Zongyu Wei, Qiulian Zeng, and Guihua Ruan

14.1 Introduction

Due to its excellent sensitivity, specificity, and capability for continuous high-throughput measurement, LC-MS has been considered as an important breakthrough in the field of bioanalysis. Since the 1990s, LC-MS has been generally accepted as the method-of-choice for the identification and quantification of various molecules, from small to large ones, in biological matrices, including cells, tissues, blood, plasma, serum, and so on (Li et al. 2013a). However, challenges remain in the quantification of many analytes, especially those at trace and ultra-trace levels, due to the limit of instrumental detection and interferences of abundant endogenous and exogenous components from complex matrix. In this regard, proper sample preparation has to be introduced to isolate and enrich the analytes of interest while removing unwanted and interfering components from the complex biological samples prior to LC-MS analysis.

Sample preparation is considered pivotal for the entire process of LC-MS bioanalysis of trace molecules. Many sample preparation techniques, including protein precipitation (PPT), liquid–liquid extraction (LLE), solid-phase extraction (SPE), solid-phase microextraction (SPME), micro SPE (MSPE), stir-bar sorptive extraction (SBSE), microextraction by packed sorbent (MEPS), single-drop microextraction (SDME), hollow-fiber liquid-phase microextraction (HF-LPME), dispersive liquid–liquid microextraction (DLLME), etc. have been introduced (Moein et al. 2017). However, selectivity remains a challenge for those techniques. Recently, some more selective sample preparation techniques have been developed and introduced to the LC-MS bioanalysis filed. These techniques are based on highly selective sorbents, including restricted access materials (RAMs), molecularly imprinted polymers (MIPs), antibodies, and aptamers (Du et al. 2010, 2015; Iliuk et al. 2011; Martín-Esteban 2013; Wang et al. 2012). Compared with RAMs, MIPs, and antibodies, aptamers possess some unique advantages, such as high specificity and binding affinity, good stability, inexpensive, nontoxic, easy to synthesis, and controllable modification. These features make aptamers very promising sorbents in sample preparation in the field of environmental and biomedical research and monitoring (Meng et al. 2016; Tan et al. 2013).

As single-stranded DNA or RNA molecules, aptamers bind to their target molecules with high affinity and specificity by folding into distinct secondary and tertiary structures. Aptamers can be artificially generated via a procedure known as Systematic Evolution of Ligands by Exponential Enrichment (SELEX), including *in vitro* SELEX, cell-SELEX, non-SELEX, automated-SELEX, microfluidic SELEX, and so on (Yüce et al. 2015). The modification makes aptamers more suitable for recognition of the intended targets such as metal ions (Du et al. 2013; Qu et al. 2016), small molecules (Iliuk et al. 2011), proteins (Deng et al. 2014; Du et al. 2014; Lin et al. 2016a), viruses, bacteria, whole cells (Li et al. 2014; Meng et al. 2016; Tan et al. 2013), and even tissues (Li et al. 2014; Liu et al. 2012). The published research has shown that the aptamers generated by SELEX techniques can also be chemically modified, which allows researchers to improve the binding properties of the aptamers, e.g. capturing a much broader range of target analytes (Du et al. 2015; Meng et al. 2016; Tan et al. 2013).

Aptamer-functionalized materials (AFMs) are prepared by immobilizing aptamers onto a solid support through a linking reagent (linker). Besides the inherent high selectivity and strong affinity, AFMs possesses better chemical, biochemical, and mechanical stability as compared to the free aptamers. This makes AFMs more attractive in a variety of applications. Aptamer-based sample preparation, therefore, has been developed by using AFMs as sorbents to extract, isolate, and purify analytes from complex biological samples for LC-MS bioanalysis. In this chapter, we focused on the developments and

Sample Preparation in LC-MS Bioanalysis, First Edition. Edited by Wenkui Li, Wenying Jian, and Yunlin Fu.

applications of aptamer-based sample preparation techniques, including aptamer-based solid-phase extraction (Apt-SPE), aptamer-based solid-phase microextraction (Apt-SPME), aptamer-based microfluidic sample preparation, and other techniques. Each of the technique was accompanied by typical protocols for LC-MS bioanalysis.

14.2 Aptamer-based Sample Preparation

14.2.1 Aptamer-based Solid-phase Extraction

To improve the selectivity and affinity of SPE, Apt-SPE was developed by integrating SPE technique with AFMs. Apt-SPE possesses the advantages of SPE, including simplicity, versatility, low consumption of reagents, and convenience in combination with different workflow (in online or off-line).

14.2.1.1 Aptamer-based Affinity Columns for SPE

Aptamer-based affinity columns (Apt-ACs), including aptamer-functionalized material-packed column (AFMPC), aptamer-functionalized open tubular capillary column (AFOTCC), aptamer-functionalized monolithic column (AFMC), and aptamer-functionalized spin column (AFSC), and aptamer-functionalized stir-bar sorptive extraction (Apt-SBSE), have been developed for the selective SPE of target analytes from complex samples.

AFMPC is prepared by packing AFMs into a commercial cartridge, a stainless steel chromatographic column, a small plastic column, or a capillary column. The preparation of AFMs is considered as a critical step in the development of an AFMPC for SPE. Generally, biotinylated aptamers were immobilized onto a streptavidin-sepharose or streptavidin-coated magnetic porous glass beads, while 5′-amino-modified aptamers were immobilized onto a thiol-activated sepharose, CNBr-activated sepharose, NHS-activated sepharose, polystyrene porous particles, or carboxyLink coupling gel (immobilized diaminodipropylamine, DADPA) (Deng et al. 2001; Madru et al. 2009). Then, the prepared AFMs were packed into empty SPE cartridges.

AFOTCC is prepared by covalently immobilizing aptamers onto the inner surface of a fused-silica capillary column. However, the coverage of aptamers on the AFOTCC surface is generally low, leading to a need of improving affinity capture efficiency and reducing background interference due to nonspecific adsorption to the silica surfaces that are not covered by aptamers (Connor and McGown 2006). The limited surface area of the inner wall and the small loading capacity restricts the applications of AFOTC column in sample preparation.

AFMC is prepared by immobilizing a biotinylated DNA aptamer onto streptavidin-modified polymer monolithic capillary column (Zhao et al. 2008a, 2008b). When using this non-covalent approach, a risk of leakage exists, especially when using organic modifiers that may affect the biotin–streptavidin affinity during the SPE procedure. Furthermore, this non-covalently immobilized support could be fragile and the biotinylated DNA aptamer on monolith could be deteriorated when applying elution solutions with high organic content (Madru et al. 2011). To overcome these issues, more robust covalent immobilization methods have been used to prepare aptamer-functionalized poly(glycidyl methacrylate-coethylene dimethacrylate) (poly(GMA-co-EDMA)) monolith (Han et al. 2012) and aptamer-functionalized organic–inorganic hybrid monolith (Deng et al. 2012; Jiang et al. 2014; Zhao et al. 2016). These AFMCs exhibited excellent selectivity and enrichment capability for capturing and pre-concentrating thrombin from human serum samples for analysis with high recovery (Deng et al. 2012).

AFSC is prepared by immobilizing biotinylated aptamer onto streptavidin agarose resin in Handee™ spin column, to which small polyethylene frits are pre-inserted. AFSC was reported with good selectivity and highly specific enrichment capability for extraction of histone proteins directly from complex whole-cell lysates (Shao et al. 2012).

For Apt-SBSE, aptamer was first immobilized on the surface of metal–organic frameworks ($Zn_4O(BDC)_3$, MOF-5), and then the prepared aptamer-functionalized MOF-5 was applied to selectively enrich the analytes of interest, such as polychlorinated biphenyls (PCBs), at low levels (Lin et al. 2016b).

Several typical preparation schemes for Apt-ACs are shown in Figure 14.1.

14.2.1.2 Aptamer-based Magnetic SPE

Magnetic materials, including magnetic nanoparticles, magnetic microspheres, and magnetic beads, have received considerable attention, owing to their good biocompatibility, paramagnetic property, easy surface modification, easy operation, and high binding capacity (Li et al. 2013b). Due to the high selectivity of aptamers and the advantages of magnetic materials mentioned above, aptamer-functionalized magnetic materials (Apt-MMs) are believed to be better candidates than others in developing highly selective magnetic SPE methods for extracting and enriching various targets from complex samples.

The preparation of an Apt-MM usually involves three steps, including (i) synthesis of the magnetic particle (Fe_3O_4 or γ-Fe_2O_3), (ii) coating of the

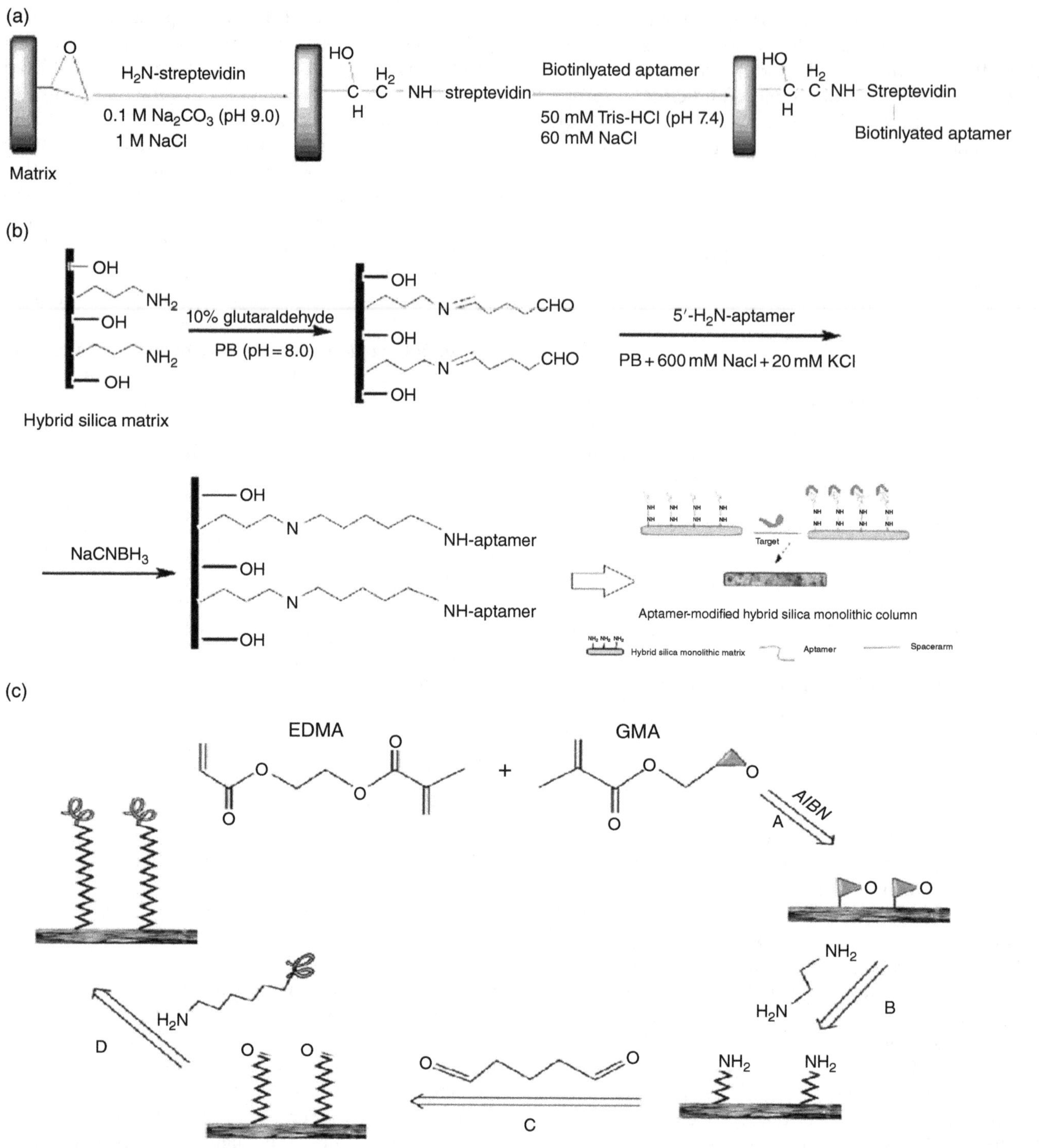

Figure 14.1 Typical schemes of the preparation of aptamer-based affinity columns, including: (a) Immobilization of biotinylated aptamers onto a streptavidin-sepharose for polymer monolithic (Zhao et al. 2008a). *Source:* Reproduced with permission of American Chemical Society. (b) Aptamer-functionalized poly(GMA-co-EDMA) affinity monolithic column (Han et al. 2012). *Source:* Reproduced with permission of Elsevier. (c) Aptamer-functionalized organic–inorganic hybrid monolith (Deng et al. 2012). *Source:* Reproduced with permission of American Chemical Society. (d) and (e) Aptamer-functionalized silica hybrid monolithic capillary column (Jiang et al. 2014; Zhao et al. 2016). *Source:* Reproduced with permission of Royal Society of Chemistry.

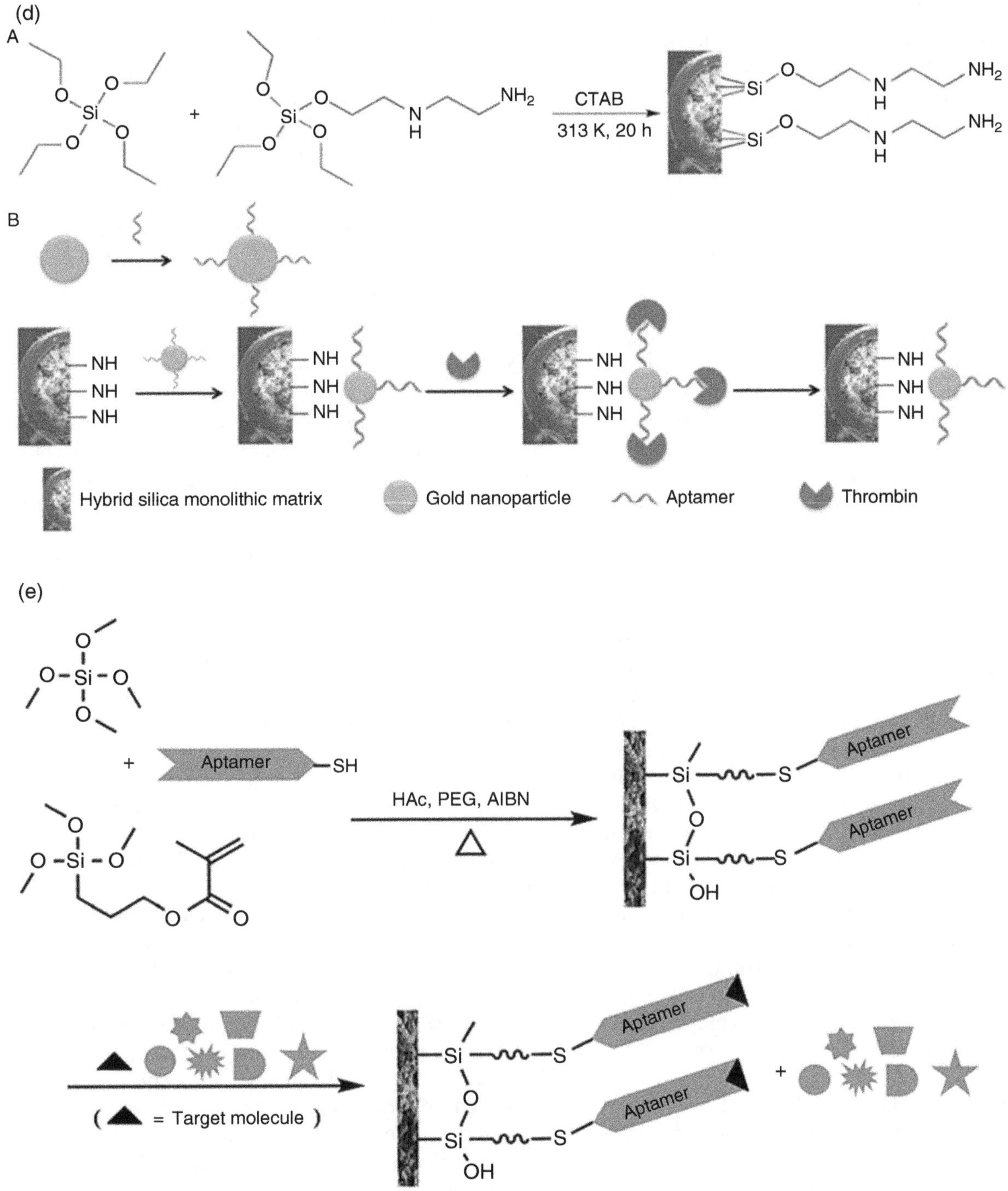

Figure 14.1 (Continued)

magnetic core with silica, styrene/acrylamide copolymer, or gold particles, and (iii) immobilization of aptamer onto the resultant magnetic particles. The two immobilization methods, i.e. streptavidin–biotin sandwich immobilization and *N*-(3-dimethylamnopropyl)-*N*-ethylcarbodiimide hydrochloride/*N*-hydroxysuccinimide (EDC/NHS) coupling immobilization, are the most common coupling techniques in the preparation of Apt-MMs (Lin et al. 2016c; Richards et al. 2016; Zhang et al. 2012; Zhao et al. 2011). In order to increase the efficiency of specific enrichment by the aptamer-functionalized magnetic beads, the magnetic beads are first prepared via suspension polymerization and then grafted with glycidylmethacrylate (GMA) brushes, which results in the high quantities of active groups on bead surface (Ozalp et al. 2014). The resulting magnetic beads could be immobilized with different aptamers by multiple-amino groups. Several typical methods for preparation of aptamer-immobilized magnetic materials are shown in Figure 14.2. For some commercial magnetic materials, the detailed immobilization methods can be found in the respective manufacturer's instruction.

Many applications of AFMMs have been reported for magnetic SPE of small molecules, proteins, pathogen, and cancer cells from complex biological samples (Li et al. 2013b; Lin et al. 2016c; Richards et al. 2016; Smith et al. 2007; Zhang et al. 2012). For example, aptamer-modified magnetic beads were used to selectively extract proteins in wide concentration ranges (even at the femtomolar level) with a high recovery (about 90% for human-α thrombin and about 110% for human neutrophil elastase (HNE)) from diluted serum samples without additional sample preparation. This resulted in the highly sensitive detection of the proteins (LOD of 2 fM for thrombin and 100 fM for HNE, respectively) in human serum without much matrix interference (Zhao et al. 2011). The aptamer-functionalized magnetic nanoparticles (Apt-MNPs) were also used to extract and enrich cancer cells, including CCRF-CEM acute leukemia cells, Ramos cells, and Toledo cells, from complex biological samples. This makes it simple and easy to accurately identify various target cancer cells present in biological samples without additional sample preparation (Smith et al. 2007). The aptamer-based magnetic SPE is an attractive alternative to antibody-based magnetic SPE in the enrichment of target analytes, the removal of interfering compounds from the sample matrix, and the improvement of the subsequent detection. A comparable limit of detection (1 pg ml^{-1}) and quantification (2.5 pg ml^{-1}) were achieved for determination of gonadotropin-releasing hormone (GnRH) in equine urine (Richards et al. 2016).

14.2.1.3 Aptamer-based Surface Affinity SPE

The common substrates for immobilization of aptamer onto aptamer-based surface affinity solid-phase extraction (Apt-SA-SPE) are carbon nanotubes, graphene oxide, gold, silica, etc. The aptamer-functionalized affinity materials can (i) specifically capture target analytes, (ii) remove interfering components, and (iii) act as an ionization substrate for mass spectrometric detection (Ocsoy et al. 2013; Zhang et al. 2013). Aptamers are covalently attached to the surface of a gold layer that serves as the MS ionization substrate. The resultant aptamer-functionalized surface substrate was demonstrated to have high affinity and high specificity in capturing thrombin from human plasma, resulting in high efficient removal of unwanted interfering proteins and significant improvement for the subsequent TOF-MS analysis (Zhang et al. 2013). The aptamer-functionalized Au@MnO nanoflower surfaces also can act as an ionization substrate and a targeting and sensing platform for selective extraction adenosine triphosphate (ATP) from cancer cells followed by mass spectral analysis, and exhibited high specific capture capability for the targets. The proposed Apt-SA-SPE could be reused repeatedly for capture of the targets after regeneration without loss of specificity and extraction performance (Ocsoy et al. 2013).

14.2.1.4 Aptamer-functionalized Materials for Other SPE Format

In addition to the above mentioned, other SPE methods have been successfully developed and applied to selectively extract and enrich trace target analytes from complex samples by incorporating precipitation or/and centrifugation steps (Rouah-Martin et al. 2014; Song et al. 2014; Yasun et al. 2012). In one example, aptamer-functionalized gold nanorods (Apt-AuNRs) were first dropped into samples for extraction of low-abundance proteins, and then centrifuged and washed for the subsequent detection (Yasun et al. 2012). The results showed that thrombin at 1 ppm level in buffer could be enriched with ~90% capturing efficiency using a low concentration (0.19 nM) of aptamer-conjugated gold nanorods. A similar capturing efficiency was observed at 1 ppb level in plasma. This demonstrated that the combination of Apt-AuNRs with centrifugation was efficient to extract and enrich low-abundance proteins from biological fluids. Aptamer-based precipitation assay (or simply aptamoprecipitation) has been reported for the purification of His-tagged proteins from complex *E. coli* cell lysates and extraction of thrombin from a mixture of thrombin and bovine serum albumin (BSA) by using RNA aptamer-functionalized polystyrene resins as SPE materials (Song et al. 2014). Although the protein recoveries achieved by the proposed aptamoprecipitation method were not high (about 50%), the purity of the purified His-tagged protein and thrombin was more than 95%, which was superior or at least comparable to the conventional immobilized metal affinity precipitation or immunoprecipitation (Song et al. 2014).

In addition, Lee et al. (2014) developed a novel aptamer-based SPE method by combining the selectivity and specificity of ssDNA aptamer-functionalized microbeads with the advantages of the integrated selective enrichment target (ISET) platform. The method exhibited specific extraction and high enrichment efficiency as demonstrated by the reached detection limit of 1 fmol for thrombin, which is considered a

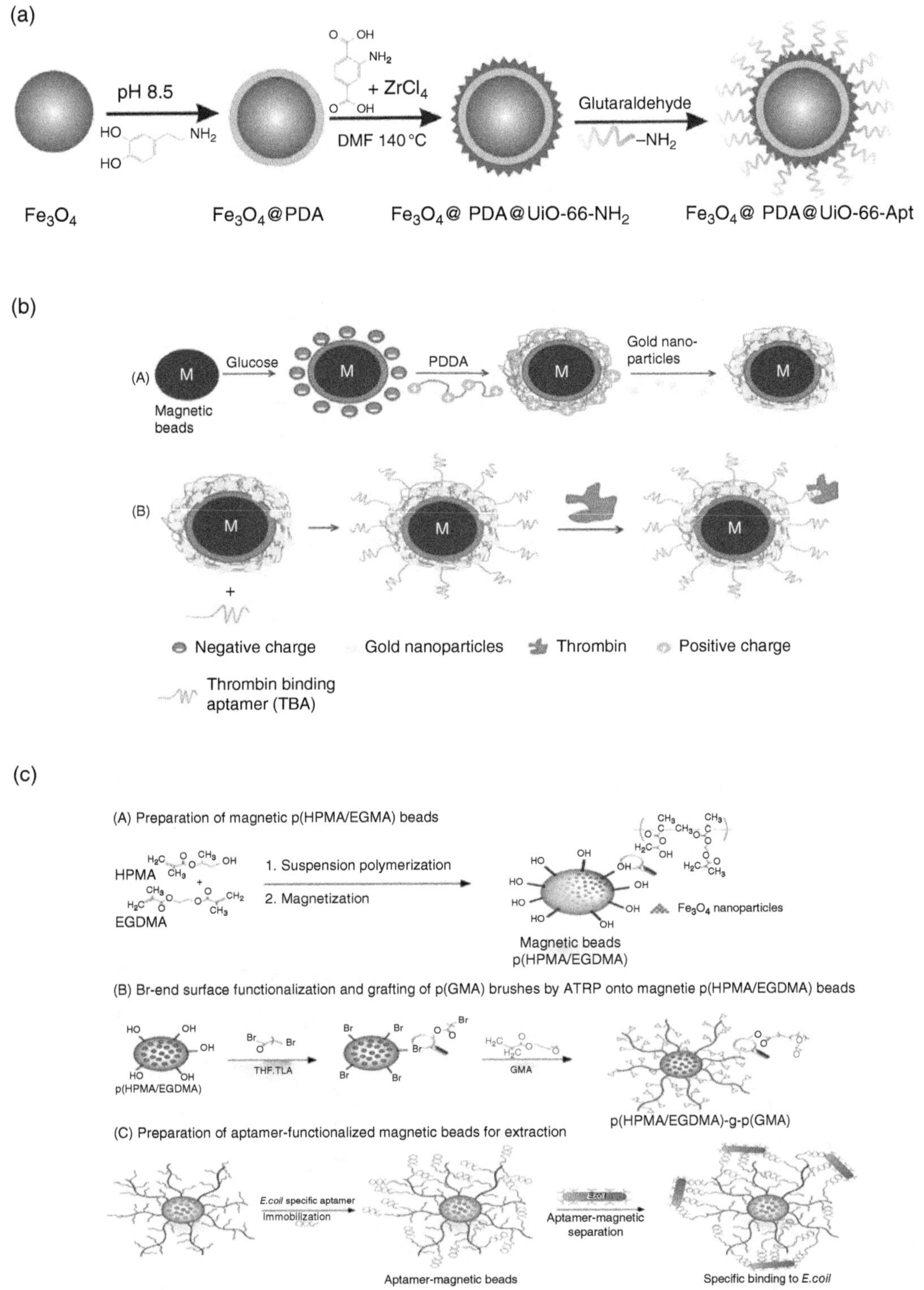

Figure 14.2 Typical schemes of the preparation of aptamer-immobilized magnetic materials, including: (a) Aptamer-functionalized Fe_3O_4@ polydopamine@UiO-66-Apt (Lin et al. 2016a), (b) aptamer-functionalized gold-coated magnetic microspheres (Zhang et al. 2012), and (c) aptamer-functionalized p(HPMA/EGDMA)-g-p(GMA) magnetic beads (Ozalp et al. 2014). *Source:* Reproduced with permission of Elsevier.

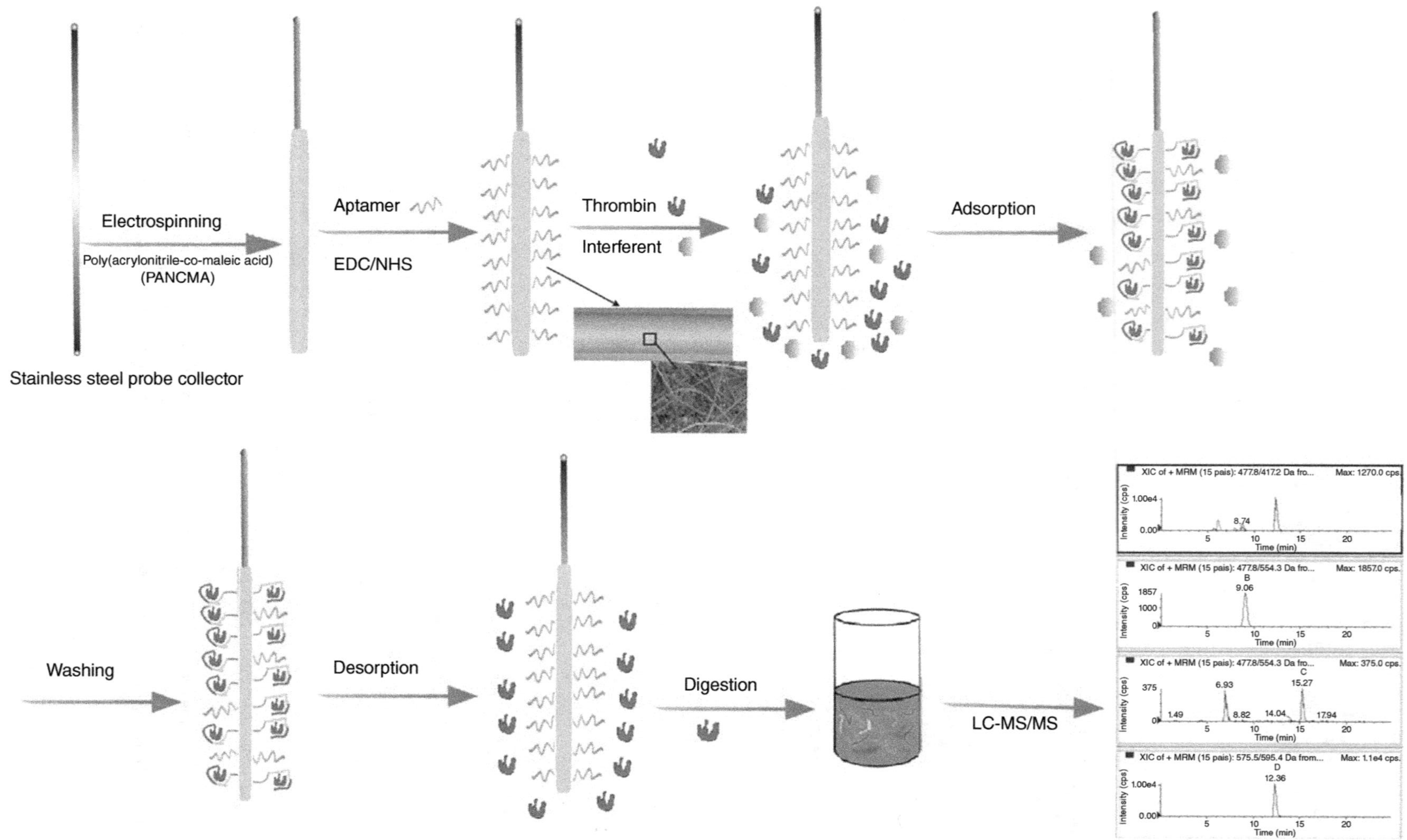

Figure 14.3 Scheme of the fabrication process and application of electrospun enhanced aptamer-functionalized SPME.

significant increase as compared to the corresponding results obtained using antibody-based ISET.

14.2.2 Aptamer-based Solid-phase Microextraction

SPME is a simple, fast, and environmental-friendly sample preparation technique that integrates sampling, extraction, concentration, and elution to analysis into a single step. This technique was initially developed by Pawliszyn and co-workers in the early 1990s (Arthur and Pawliszyn 1990), and it has attracted much attention in the field of environment, food, drug research, and development due to its potential in various applications.

Incorporating aptamer into SPME has some well-demonstrated advantages as compared to conventional SPME. For example, in order to improve the selectivity and sensitivity of SPME for the extraction of very polar analytes such as adenosine, aptamer-functionalized solid-phase microextraction (Apt-SPME) was developed and applied to human plasma samples (Guo et al. 2016; Mu et al. 2013). Compared with commercial SPME fibers, the adenosine aptamer-functionalized sol–gel SPME fiber exhibited specific affinity and excellent selectivity for adenosine with extraction efficiency more than 20-fold. In a separate study, Du et al. (2014) covalently immobilized thrombin aptamer ligand onto electrospun microfibers made of hydrophilic polymer poly(acrylonitrile-co-maleic acid) (PANCMA) on stainless steel rods. The resultant SPME fibers exhibited high selectivity with good binding capacity for the extraction of thrombin from 20-fold diluted human plasma samples without a need of any other purification. By combining with LC-MS/MS, the developed aptamer-functionalized SPME was successfully applied for the determination of thrombin in diluted and undiluted human plasma samples (Figure 14.3). The limit of detection for thrombin was 0.3 nM in 20-fold diluted human plasma samples. Although the extraction efficiency was somewhat reduced in undiluted human plasma as compared to the standard solution, this new method opened up the possibility of selective enrichment of a given targeted protein using aptamer affinity-based SPME from complex sample either *in vivo* or *ex vivo*.

14.2.3 Aptamer-based Microfluidic Sample Preparation

AFMs have been applied to improve the selectivity and specificity of the microfluidic sample preparation workflows for the enrichment of various low-abundance analytes from complex samples. In this process, aptamers were first modified with biotin and then the biotinylated aptamers were immobilized onto streptavidin-coated poly(dimethylsiloxane) (PDMS) on the microfluidic channel surface (Xu et al. 2009), or onto the channel surface of the other modified materials, including glass (Sheng et al. 2012), silicon nanowire arrays (SiNWAs) (Shen et al. 2013; Srinivas et al. 2011), microgel particles (Srinivas et al. 2011), polymerized microtubules (Kim and Kim 2014), bis-acrylamide/azlactone microbeads (Hilton et al. 2011; Nguyen et al. 2011), polyethylene glycol-silane-polystyrene microbeads (Koh et al. 2012), or gold film (Wang et al. 2013). The resultant aptamer-functionalized microfluidic materials possess high binding capacity and high specific recognition toward target analytes, including proteins and cells from biological samples (Kim and Kim 2014; Koh et al. 2012; Nguyen et al. 2011; Shen et al. 2013; Srinivas et al. 2011).

Thermally controlled aptamer-functionalized microfluidic (TCAFMF) devices were proposed for online sample preparation (Hilton et al. 2011; Nguyen et al. 2011). The device consisted of a microchamber packed with aptamer-functionalized microbeads, a microheater, and temperature sensor. For the designed TCAFMF devices, specific analyte capture was achieved using surface-immobilized aptamers in the microchamber, and an enrichment factor of 566- to 686-fold for the extraction of adenosine monophosphate (AMP) was achieved (Nguyen et al. 2011). The release of the captured analytes was realized by a moderate temperature change to reversibly disrupt the analyte–aptamer interaction, which was controlled via an integrated heater with a temperature sensor. Importantly, the aptamer-based sensor response was consistent after repeated regeneration when the temperature returned to the optimal condition, which indicated that the TCAFMF devices can be reused many times without losing specific affinity and enrichment ability (Hilton et al. 2011).

14.3 Representative Protocols

Protocol 1 describes an aptamer-based magnetic SPE method for quantitative analysis of GnRH in equine urine by LC-MS/MS (Richards et al. 2016).

In this work, an aptamer-based magnetic SPE method was developed and applied to enrich a model compound GnRH from equine urine in order to assess the viability of an aptamer-based enrichment technique as an alternative to antibody-based sample preparation. In comparison to the antibody enrichment protocol, the aptamer-based enrichment method exhibited greater specificity, affinity, and stability at ambient temperature.

Equipment and Reagents

- Oasiss HLB cartridges (Holliston, MA, USA)
- Dionexs Ultimate 3000 RSLC nanoUHPLC (LC Packings, Amsterdam, The Netherlands)
- TSQ Vantage triple quadrupole mass spectrometer (ThermoScientific, Australia)
- Biotinylated GnRH aptamer (NOX-1255) (Noxxon Pharmaceuticals, Germany)
- Biotinylated stereoisomer of the GnRH aptamer, using D-DNA (Geneworks, Australia)
- Streptavidin magnetic beads (M-280, 10 mg ml^{-1}) (Dynabeads®, Life Technologies, Australia)
- DNA selection buffer (20 mM Tris.HCl, pH 7.4/137 mM NaCl/5 mM KCl/2 mM $CaCl_2$/1 mM $MgCl_2$/0.005% TritonX-100)
- DNA wash buffer (20 mM Tris.HCl, pH 7.4/137 mM NaCl/5 mM KCl/2 mM $CaCl_2$/1 mM $MgCl_2$)
- Citric acid solution (60 ml, 100 mM, pH 2 containing 15% acetonitrile, v/v)

Method

1) Preparation of aptamer-based magnetic beads by immobilizing the biotinylated GnRH aptamer onto the magnetic beads. Briefly, magnetic beads (300 ml) were twice washed with DNA selection buffer (1 ml) and then resuspended in DNA selection buffer (600 ml). The biotinylated GnRH aptamer (1 pmol ml^{-1}) was immobilized on the magnetic beads (1/1, v/v), by incubation at ambient temperature for 15 minutes with gentle rotation. The aptamer-coated beads were washed twice with DNA selection buffer (1 ml), resuspended in DNA selection buffer (1 ml), incubated at 90 °C for 2 minutes, and then allowed to refold through cooling to ambient temperature (15 minutes).
2) Treatment using Oasiss HLB cartridges. In order to reduce the nonspecific binding and to improve the specific enrichment of the target analytes in equine urine samples, Oasiss HLB cartridges were used to pre-treat the urine samples prior to using the aptamer-based enrichment process. Firstly, Oasiss HLB cartridges were preconditioned with methanol (2 ml) and water (2 ml). Following loading of the urine samples (3 ml), the cartridges were washed with water (2 ml) and air dried for one minute. Samples were eluted with 1 ml methanol/water (80:20, v:v). The obtained eluent was collected and evaporated to dryness at 40 °C using a vacuum concentrator.
3) Extraction using aptamer-based magnetic beads. Briefly, the sample residuals were redissolved in 400 ml of DNA selection buffer, and then 100 ml of aptamer-coated magnetic beads were added, which was followed by incubation at 4 °C for 12–16 hours with gentle mixing. After incubation, the mixture was centrifuged (1000 × g, 30 seconds) prior to separation on the magnetic separator. The supernatant was discarded. The collected aptamer-based magnetic beads were washed twice with DNA wash buffer (500 ml). The GnRH–aptamer complex was dissociated by addition of acetic acid (250 ml, 2% v/v), heated (75 °C, 5 minutes), and mixed (ambient temperature, 30 minutes). The supernatant was collected, and the aptamer-based magnetic beads were regenerated by successive washes with DNA selection buffer. The supernatant was transferred into clean polypropylene microtubes, evaporated to dryness under vacuum (40 °C), reconstituted in citric acid solution, centrifuged (19 000 g, 20 minutes), and transferred into LC vials for analysis by LC-MS/MS.

Protocol 2 describes an aptamer-based SPME method for LC-MS bioanalysis of thrombin from biological samples (Du et al. 2014).

A highly selective and biocompatible aptamer-based SPME method is expected to extend the application of the SPME in biological samples. In this work, aptamer-based SPME for the extraction of human thrombin in plasma samples for LC-MS/MS analysis was developed.

Equipment and Reagents

- A high-voltage power supply (UW-SYS E2047, University of Waterloo, Canada)
- A stainless steel rod-collector (55 mm × 1.5 mm, i.d.)
- A Kd Scientific syringe pump (Holliston, MA, USA)
- SK-300 SHAKER (JEIO TECH, Ontario, Canada)
- Shimadzu LC-10AD HPLC system
- API 4000 mass spectrometer (AB Sciex, Concord, Ontario, Canada)
- Analytical HPLC column (100 mm × 1.0 mm i.d., particle size 5 μm, BioBasic-C8 column, Thermo Scientific, USA)
- Chemical-grade maleic anhydride (≥99.0%; Sigma-Aldrich, St. Louis, MO, USA)
- Chemical-grade acrylonitrile (≥99.0%; Sigma-Aldrich, St. Louis, MO, USA)

- Chemical-grade 1-ethyl-3-(3-dimethylaminopropyl) carbodiimide hydrochloride (EDC·HCl, ≥99.0% Sigma-Aldrich, St. Louis, MO, USA)
- Anti-thrombin DNA aptamer (Apt) (5′/5AmMC6/-AGT CCG TGG TAG GGC AGG TTG GGG TGA CT-3′, Integrated DNA Technologies Inc., Canada)

Method

1) Preparation of electrospun poly(acrylonitrile-co-maleic acid) (PANCMA) probe. Briefly, 4.0 g PANCMA was dissolved in 50 ml of *N,N*-dimethylformamide (DMF) at room temperature with gentle stirring for 12 hours. The obtained homogeneous solution was placed in a syringe bearing a 1.0 mm inner diameter metal needle connected to a high-voltage power supply. A grounded counter electrode was connected to the stainless steel rod-collector (55 mm × 1.5 mm, i.d.). In this work, the optimal electrospinning voltage was 10.0 kV, the distance between the needle tip and the collector was 130 mm, the flow rate of the solution was 0.30 ml h^{-1}, the coating length of probe-collector was 30 mm, and the collection time was 50 minutes. The obtained electrospun membrane on the probe was dried for at least three hours at 60 °C in a vacuum oven before it was used.
2) Immobilization of thrombin aptamer on the electrospun PANCMA Probe.[a] Briefly, four PANCMA probes were thoroughly washed with water and rinsed with a 50 mM PBS solution containing 20 mM KCl and 600 mM NaCl (pH 8.0). After this, the pretreated probes were submerged into 2.0 ml of EDC/NHS solution (100 mM EDC and 100 mM NHS in 50 mM PBS buffer, pH 8.0) and shaken gently at room temperature to activate the —COOH group of PANCMA. After incubation for two hours, 240 nmol aptamer was added and incubated overnight. The prepared aptamer-functionalized poly (acrylonitrile-co-maleic acid) (Apt-PANCMA) probes were taken out and washed several times with a 10 mM PBS buffer (pH 7.4), then dipped into the same PBS solution for future use.
3) The aptamer-based electrospun PANCMA probe was used for the extraction of thrombin from human plasma.[b] The Apt-PANCMA probes were first washed with 10 ml of PBS (10 mM, pH 7.4), then dipped into 2.0 ml of sample. The incubation was held for 1.0 hour at room temperature under shaking (130 rpm, SK-300 SHAKER). The Apt-PANCMA probes were taken out from the sample solution, and first washed three times with 2 ml of PBS, then consecutively washed three times with 10 ml of pure water to remove any nonspecifically bound or weakly bound species. The captured thrombin was eluted from the Apt-PANCMA probes by using 2 ml of eluting solution (acetonitrile/water, 80/20, v/v), which was followed by drying at ambient temperature with N_2 gas. The resulting residuals were dissolved in 146 μl of NH_4HCO_3 buffer (25 mM). After addition of 30 μl of trypsin solution (1.0 mg ml^{-1} in NH_4HCO_3 buffer) was added, the sample solutions were incubated overnight at 37 °C. To the digestion mixture, 4 μl of the internal peptide SSIIHIER solution (500 nM in formic acid/acetonitrile/water [0.1/10/90, v/v/v] solution) and 20 μl acetonitrile/formic acid (80/20, v/v) was added. The mixture was then filtered through a 0.22 μm Supor® membrane (Pall Corporation, USA) prior to LC-MS/MS analysis.[c]
4) After extraction, the aptamer-based electrospun PANCMA probe should be regenerated by successive washes with 10 ml of PBS.

Notes

[a]The EDC/NHS protocol was preferred for immobilization of thrombin aptamer on the electrospun PANCMA probe. The immobilization conditions should be optimized by investigating the aptamer coupling efficiency and the specificity of the probe for thrombin.

[b]In the SPME process, appropriate dilution of the plasma sample was necessary in order to improve the selectivity and sensitivity of the method. The unspecific adsorption of other proteins on the probe could happen, which results in low recovery of the target protein. In human plasma samples, the unspecific adsorption of prothrombin and the specific adsorption of thrombin are competitive. Although the recovery of prothrombin recovery was very low, its unspecific adsorption was obvious without dilution of the plasma sample, owing to its high concentration. Therefore, an appropriate dilution was necessary for selective binding thrombin and at the same time decreasing other proteins.

[c]Internal peptide SSIIHIER was necessary for the accurate quantitation of thrombin by LC-MS/MS analysis.

Protocol 3 describes an aptamer-based SPME method for LC-MS bioanalysis of adenosine phosphates from biological samples (Guo et al. 2016).

Adenosine phosphates are important biomolecules that represent energy currency of living cells. Therefore, there is always a demand for sensitive and reliable analysis of adenosine phosphates in bio-samples. In this work, the aptamer-based SPME was developed for monitoring the trace levels of ATP, adenosine diphosphate (ADP), and AMP in human serum by LC-MS/MS.

Equipment and Reagents

- Silica fibers (4.0 cm × 140 μm, i.d.)
- Agilent 1200 series HPLC system (Agilent, San Francisco, CA, USA)
- Applied Biosystems API 5000 (MDS Sciex, Toronto, Canada)
- Chemical-grade maleic anhydride (≥99.0%; Sigma-Aldrich, St. Louis, MO, USA)
- Chemical-grade acrylonitrile (≥99.0%; Sigma-Aldrich, St. Louis, MO, USA)
- Chemical-grade 1-ethyl-3-(3-dimethylaminopropyl) carbodiimide hydrochloride (EDC·HCl, ≥99.0% Sigma-Aldrich, St. Louis, MO, USA)
- Chemical-grade acrylamide (AA) (Sigma-Aldrich, St. Louis, MO, USA)
- Chemical-grade trimethylolpropane trimethacrylate (TRIM) (Sigma-Aldrich, St. Louis, MO, USA)
- Chemical-grade carbonyl diimidazole (CDI) (Sigma-Aldrich, St. Louis, MO, USA)
- Chemical-grade azo(bis) isobutyronitrile (AIBN) (Damao Reagent Plant, Tianjin, China)
- 5′-NH_2-$(CH_2)_6$-modified DNA oligonucleotides (ATP aptamer: 5′-TTT TTT TTT TTT ACC TGG GGG AGT ATT GCG GAG GAA GGT-3)
- Tris-selective buffer (20 mmol l^{-1} Tris, 140 mmol l^{-1} NaCl, 5 mmol l^{-1} KCl, 10 mmol l^{-1} $MgCl_2$, pH 7.4)
- TE buffer (10 mmol l^{-1} Tris hydrochloride, 1 mmol l^{-1} ethylenediaminetetraacetic acid [EDTA], pH 7.4)

Method

1) Preparation of porous polymer-coated fiber (PP-fiber).[a] Briefly, 172 mg of AA, 3 ml of TRIM, and 45 mg of AIBN were dissolved adequately in 27 ml of acetone, and then 1.5 ml of the mixture was transferred into a small glass tube, degassed in a sonicating bath for five minutes, and then deoxygenized with a stream of nitrogen for five minutes. Subsequently, the silylated fibers were inserted into the mixture in the tubes, which were sealed immediately by a rubber stopper and inserted into an HGC-12 nitrogen evaporator for polymerization at 60 °C for 1.5 hours. The polymerization process was repeated five times. The obtained porous polymer coating was scraped to ensure a uniform length of 10 mm on the fiber top.
2) Immobilization of aptamer on the PP-fiber. The PP-fiber was first immersed into carbonyl diimidazole solution (1.5 ml, 0.7 mol l^{-1}, in dehydrated acetonitrile) for reaction for 1.5 hours, and then rinsed with dehydrated acetonitrile three times and dried with nitrogen gas. After 30 minutes conditioning in TE buffer, the fiber was immersed in aptamer solution (1.0 ml) for 24 hours at room temperature to obtain the aptamer-based PP fiber (Apt-PP-fiber).
3) Apply Apt-PP-fiber for the extraction of adenosine phosphates from human serum samples.[b]

   Before extraction, an Apt-PP-fiber was rinsed with 500 μl of TE buffer three times, and then the Apt-PP-fiber was immersed into 5 ml of sample solution for 60 minutes at the stirring rate of 300 rpm. Subsequently, the Apt-PP-fiber was pulled out and rinsed with 5 ml of Tris-selective buffer for one minute in order to remove any nonspecifically bound or weakly bound species, and then 100 μl of eluting solution (acetonitrile/water, 40/60, v/v) to elute the bound adenosine. The collected elution solution was subjected to LC-MS/MS analysis.
4) After desorption, the Apt-PP-fiber was conditioned in Tris-selective buffer for 10 minutes, and then stocked in a sealed glass tube in a refrigerator at 4 °C.

Notes

[a]The pretreatment of silicon fiber, including solvent cleaning, alkaline-washing, acid-washing, silylation, and coating of porous polymer onto the silicon fiber surface was according to the reported method.

[b]For real samples, 1 ml of human serum was mixed with 3% v/v perchloric acid solution (1 ml) and 100 mg l^{-1} ATP, ADP, and AMP mixed standard solution (10 or 1 μl) in a vortex mixer for 1 minute, and then centrifuged at 10 000 rpm for 30 minutes. Subsequently, the supernatant was filtered through a 0.45 μm millipore filter, and 1 ml of filtrate was transferred into a glass vial and diluted with Tris-selective buffer to a volume of 5.00 ml.

14.4 Summary

Sample preparation continues to be an important part of LC-MS bioanalysis. Many advances have been achieved in the last few decades for the improvement of the reliability and robustness of the bioanalytical method. Aptamer-based sample preparation has attracted a considerable interest with a variety of applications in bioanalysis due to the high affinity and selectivity associated with aptamer. In this chapter, we introduced several aptamer-based sample preparation techniques, including Apt-SPE, Apt-SPME, and aptamer-based microfluidic sample preparation. We also introduced three typical protocols about the applications of aptamer-based sample preparation in the LC-MS bioanalysis of model compounds. All these techniques are very promising for LC-MS bioanalysis but their developments and applications are still limited at present, resulting from a limited number of aptamers available, the biostability and non-specific adsorption of aptamer-based materials, the high cost of the large-scale aptamer production, and the LC-MS instrument. With the development of the more efficient and low-cost aptamer selection methods, the more multifunctional support materials for aptamer, and the more requirements of highly selective sample preparation techniques, aptamer-based sample preparation is attracting more and more attention in the field of LC-MS bioanalysis.

Acknowledgements

Financial support is gratefully acknowledged from the National Natural Science Foundation of China (Grant No. 21265004 and 21465008), the Natural Science Foundation from Guangxi Zhuang Autonomous Region (No.2014GXNSFAA118063,and2015GXNSFAA139024), the Guangxi Colleges and Universities Key Laboratory of Food Safety and Detection (Guijiaoren [2014] 49), and the project of high-level innovation team and outstanding scholar in Guangxi colleges and universities, respectively.

References

Arthur, C.L. and Pawliszyn, J. (1990). Solid phase microextraction with thermal desorption using fused silica optical fibers. *Anal. Chem.* 62 (19): 2145–2148.

Connor, A.C. and McGown, L.B. (2006). Aptamer stationary phase for protein capture in affinity capillary chromatography. *J. Chromatogr. A* 1111 (2): 115–119.

Deng, Q., German, I., Buchanan, D., and Kennedy, R.T. (2001). Retention and separation of adenosine and analogues by affinity chromatography with an aptamer stationary phase. *Anal. Chem.* 73 (22): 5415–5421.

Deng, N., Liang, Z., Liang, Y. et al. (2012). Aptamer modified organic-inorganic hybrid silica monolithic capillary columns for highly selective recognition of thrombin. *Anal. Chem.* 84 (23): 10186–10190.

Deng, B., Lin, Y., Wang, C. et al. (2014). Aptamer binding assays for proteins: the thrombin example – a review. *Anal. Chim. Acta* 837: 1–15.

Du, F., Alam, M.N., and Pawliszyn, J. (2014). Aptamer-functionalized solid phase microextraction–liquid chromatography/tandem mass spectrometry for selective enrichment and determination of thrombin. *Anal. Chim. Acta* 845: 45–52.

Du, F., Bai, Y., Bai, Y., and Liu, H. (2010). Quantitative detection of trace systemins in solanaceous plants by immunoaffinity purification combined with liquid chromatography/electrospray quadrupole time-of-flight mass spectrometry. *Anal. Chem.* 82 (22): 9374–9383.

Du, F., Guo, L., Qin, Q. et al. (2015). Recent advances in aptamer-functionalized materials in sample preparation. *Trends Analyt. Chem.* 67: 134–146.

Du, Y., Li, B., and Wang, E. (2013). "Fitting" makes "sensing" simple: label-free detection strategies based on nucleic acid aptamers. *Acc. Chem. Res.* 46 (2): 203–213.

Guo, X., Ye, T.T., Liu, L.Y., and Hu, X.G. (2016). Preparation and characterization of an aptamer-functionalized solid-phase microextraction fiber and its application in the selective monitoring of adenosine phosphates with liquid chromatography and tandem mass spectrometry. *J. Sep. Sci.* 39 (8): 1533–1541.

Han, B., Zhao, C., Yin, J., and Wang, H. (2012). High performance aptamer affinity chromatography for single-step selective extraction and screening of basic protein lysozyme. *J. Chromatogr. B Analyt. Technol. Biomed. Life Sci.* 903: 112–117.

Hilton, J.P., Nguyen, T.H., Pei, R. et al. (2011). A microfluidic affinity sensor for the detection of cocaine. *Sens. Actuators A* 166 (2): 241–246.

Iliuk, A.B., Hu, L., and Tao, W.A. (2011). Aptamer in bioanalytical applications. *Anal. Chem.* 83 (12): 4440–4452.

Jiang, H.P., Zhu, J.X., Peng, C.Y. et al. (2014). Facile one-pot synthesis of a aptamer-based organic–silica hybrid monolithic capillary column by "thiol–ene" click chemistry for detection of enantiomers of chemotherapeutic anthracyclines. *Analyst* 139: 4940–4946.

Kim, M. and Kim, T. (2014). Aptamer-functionalized microtubules for continuous and selective concentration of target analytes. *Sens. Actuators B* 202: 1229–1236.

Koh, Y., Lee, B.R., Yoon, H.J. et al. (2012). Bead affinity chromatography in a temperature-controllable microsystem for biomarker detection. *Anal. Bioanal. Chem.* 404 (8): 2267–2275.

Lee, S.J., Adler, B., Ekström, S. et al. (2014). Aptamer/ISET-MS: a new affinity-based MALDI technique for improved detection of biomarkers. *Anal. Chem.* 86 (15): 7627–7634.

Li, X., Zhang, W., Liu, L. et al. (2014). In vitro selection of DNA aptamers for metastatic breast cancer cell recognition and tissue imaging. *Anal. Chem.* 86 (13): 6596–6603.

Li, W., Zhang, J., and Tse, F. (eds.) (2013a). *Handbook of LC-MS Bioanlysis, Best Practices, Experimental Protocols, and Regulations*. Hoboken, NJ: Wiley.

Li, X.S., Zhu, G.T., Luo, Y.B. et al. (2013b). Synthesis and applications of functionalized magnetic materials in sample preparation. *Trends Analyt. Chem.* 45: 233–247.

Lin, S., Gan, N., Cao, Y. et al. (2016a). Selective dispersive solid phase extraction-chromatography tandem mass spectrometry based on aptamer-functionalized UiO-66-NH_2 for determination of polychlorinated biphenyls. *J. Chromatogr. A* 1446: 34–40.

Lin, S., Gan, N., Zhang, J. et al. (2016b). Aptamer-functionalized stir bar sorptive extraction coupled with gas chromatography–mass spectrometry for selective enrichment and determination of polychlorinated biphenyls in fish samples. *Talanta* 149: 266–274.

Lin, X., Sun, X., Luo, S. et al. (2016c). Development of DNA-based signal amplification and microfluidic technology for protein assay: a review. *Trends Analyt. Chem.* 80: 132–148.

Liu, J., Liu, H., Sefah, K. et al. (2012). Selection of aptamers specific for adipose tissue. *PLoS One* 7 (5): e37789.

Madru, B., Chapuis-Hugon, F., Peyrin, E., and Pichon, V. (2009). Determination of cocaine in human plasma by selective solid-phase extraction using an aptamer-based sorbent. *Anal. Chem.* 81 (16): 7081–7086.

Madru, B., Chapuis-Hugon, F., and Pichon, V. (2011). Novel extraction supports based on immobilised aptamers: evaluation for the selective extraction of cocaine. *Talanta* 85 (1): 616–624.

Martín-Esteban, A. (2013). Molecularly-imprinted polymers as a versatile, highly selective tool in sample preparation. *Trends Analyt. Chem.* 45: 169–181.

Meng, H.M., Liu, H., Kuai, H. et al. (2016). Aptamer-integrated DNA nanostructures for biosensing, bioimaging and cancer therapy. *Chem. Soc. Rev.* 45 (9): 2583–2602.

Moein MM, El Beqqali A, Abdel-Rehim M. Bioanalytical method development and validation: critical concepts and strategies. *J. Chromatogr. B Analyt. Technol. Biomed. Life Sci.* 2017; 1043:3–11.

Mu, L., Hu, X., Wen, J., and Zhou, Q. (2013). Robust aptamer sol-gel solid phase microextraction of very polar adenosine from human plasma. *J. Chromatogr. A* 1279: 7–12.

Nguyen, T.H., Pei, R., Stojanovic, M., and Lin, Q. (2011). Demonstration and characterization of biomolecular enrichment on microfluidic aptamer-functionalized surfaces. *Sens. Actuators B* 155: 58–66.

Ocsoy, I., Gulbakan, B., Shukoor, M.I. et al. (2013). Aptamer-conjugated multifunctional nanoflowers as a platform for targeting, capture, and detection in laser desorption ionization mass spectrometry. *ACS Nano* 7 (1): 417–427.

Ozalp, V.C., Bayramoglu, G., Kavruk, M. et al. (2014). Pathogen detection by core–shell type aptamer-magnetic preconcentration coupled to real-time PCR. *Anal. Biochem.* 447: 119–125.

Qu, H., Csordas, A.T., Wang, J. et al. (2016). Rapid and label-free strategy to isolate aptamers for metal ions. *ACS Nano* 10 (8): 7558–7565.

Richards, S.L., Cawley, A.T., Cavicchioli, R. et al. (2016). Aptamer based peptide enrichment for quantitative analysis of gonadotropin-releasing hormone by LC-MS/MS. *Talanta* 150: 671–680.

Rouah-Martin, E., Maho, W., Mehta, J. et al. (2014). Aptamer-based extraction of ergot alkaloids from ergot contaminated rye feed. *Adv. Biosci. Biotechnol.* 5 (8): 692–698.

Shao, N., Zhang, K., Chen, Y. et al. (2012). Preparation and characterization of DNA aptamer based spin column for enrichment and separation of histones. *Chem. Commun.* 48 (53): 6684–6686.

Shen, Q., Xu, L., Zhao, L. et al. (2013). Specific capture and release of circulating tumor cells using aptamer-modified nanosubstrates. *Adv. Mater.* 25 (16): 2368–2373.

Sheng, W., Chen, T., Kamath, R. et al. (2012). Aptamer-enabled efficient isolation of cancer cells from whole blood using a microfluidic device. *Anal. Chem.* 84 (9): 4199–4206.

Smith, J.E., Medley, C.D., Tang, Z. et al. (2007). Aptamer-conjugated nanoparticles for the collection and detection of multiple cancer cells. *Anal. Chem.* 79 (8): 3075–3082.

Song, S., Cho, Y.S., Lee, S.J., and Hah, S.S. (2014). Aptamer-based precipitation as an alternative to the conventional immunoprecipitation for purification of target proteins. *Bull. Korean Chem. Soc.* 35 (9): 2665–2668.

Srinivas, R.L., Chapin, S.C., and Doyle, P.S. (2011). Aptamer-functionalized microgel particles for protein detection. *Anal. Chem.* 83 (23): 9138–9145.

Tan, W., Donovan, M.J., and Jiang, J. (2013). Aptamers from cell-based selection for bioanalytical applications. *Chem. Rev.* 113 (4): 2842–2862.

Wang, F., Guan, Y., Zhang, S., and Xia, Y. (2012). Hydrophilic modification of silica-titania mesoporous materials as restricted-access matrix adsorbents for enrichment of phosphopeptides. *J. Chromatogr. A* 1246: 76–83.

Wang, B., Huang, F., Nguyen, T.H. et al. (2013). Microcantilever-based label-free characterization of temperature-dependent biomolecular affinity binding. *Sens. Actuators B* 176: 653–659.

Xu, Y., Phillips, J.A., Yan, J. et al. (2009). Aptamer-based microfluidic device for enrichment, sorting, and detection of multiple cancer cells. *Anal. Chem.* 81 (17): 7436–7442.

Yasun, E., Gulbakan, B., Ocsoy, I. et al. (2012). Enrichment and detection of rare proteins with aptamer-conjugated gold nanorods. *Anal. Chem.* 84 (14): 6008–6015.

Yüce, M., Ullah, N., and Budak, H. (2015). Trends in aptamer selection methods and applications. *Analyst* 140: 5379–5399.

Zhang, X., Zhu, S., Deng, C., and Zhang, X. (2012). Highly sensitive thrombin detection by matrix assisted laser desorption ionization-time of flight mass spectrometry with aptamer functionalized core-shell Fe_3O_4@C@Au magnetic microspheres. *Talanta* 88: 295–302.

Zhang, X., Zhu, S., Xiong, Y. et al. (2013). Development of a MALDI-TOF MS strategy for the high-throughput analysis of biomarkers: on-target aptamer immobilization and laser-accelerated proteolysis. *Angew. Chem. Int. Ed.* 52 (23): 6055–6058.

Zhao, Q., Li, X.F., and Le, X.C. (2008a). Aptamer-modified monolithic capillary chromatography for protein separation and detection. *Anal. Chem.* 80 (10): 3915–3920.

Zhao, Q., Li, X.F., and Le, X.C. (2011). Aptamer capturing of enzymes on magnetic beads to enhance assay specificity and sensitivity. *Anal. Chem.* 83 (24): 9234–9236.

Zhao, Q., Li, X.F., Shao, Y., and Le, X.C. (2008b). Aptamer-based affinity chromatographic assays for thrombin. *Anal. Chem.* 80 (19): 7586–7593.

Zhao, J., Zhu, Q., Zhao, L. et al. (2016). Preparation of an aptamer based organic-inorganic hybrid monolithic column with gold nanoparticles as an intermediary for the enrichment of proteins. *Analyst* 141: 4961–4967.

15

Sample Extraction via Electromembrane in LC-MS Bioanalysis

María Ramos-Payán, Rut Fernández-Torres, and Miguel Ángel Bello-López

15.1 Introduction

Electromembrane extraction (EME) is a sample preparation technique based on the miniaturization of the traditional liquid–liquid extraction for aqueous samples prior to instrumental analysis such as chromatography, electrophoresis, or mass spectrometry, among others. It was introduced as a rapid alternative to liquid-phase microextraction (LPME) for the first time in 2006 and presented a significant advance in the analysis of chemical and biochemical substances from biological material. The application of an electrical potential across the supported liquid membrane (SLM) was discovered to increase the mass transfer significantly and consequently the extraction times were highly decreased compared to LPME. EME also provides high pre-concentration and efficient sample cleanup. In recent years, EME has been one of the sample pretreatment techniques widely used due to its various advantages. Nowadays, research and developments of EME have been focused on automation, miniaturization, and new materials as well.

Although EME is closely related to LPME, the principle of extraction is completely different. EME is based on electrokinetic migration of analyte of interest into an electrical field sustained across the SLM, while the traditional LPME is carried out by passive diffusion due to a pH gradient used as a driving force. The original setup is very simple and consists of two aqueous solutions separated by a membrane which immobilizes an organic solvent. In addition, two electrodes are located at both sides of the membrane (one in the sample or donor phase and the other in the acceptor solution). Figure 15.1 shows the typical EME setups for the extraction of acidic compounds. EME has been demonstrated either for basic, acid compounds, and zwitterionic drugs. For extraction of acid analytes, the cathode (negative electrode) is located in the sample (donor phase) and the anode is located in the acceptor phase and vice versa for the extraction of basic analytes. The pH is adjusted in both phases to ensure that the analytes are negatively charged or positively charged for the extraction of acid drugs or basic drugs, respectively. This is mandatory in order for the analytes to migrate in the electrical field. When the electrical field is turned on by an external power supply, the extraction process is facilitated by an electric field across the SLM, and subsequently the analytes are selectively extracted from the aqueous sample, through a microliter-volume of organic solvent sustained as a thin SLM into the acceptor solution. The analytes are effectively trapped in the acceptor solution, and are prevented from reentering the SLM due to their charge and due to the direction of the electrical field. Once the extraction is completed, the electrical field is turned off, and the acceptor solution is collected and analyzed by chromatography, MS, or electrophoresis.

As seen in Figure 15.1, there are mainly two different supports (membranes) for EME: hollow fibers and flat membranes. Both have been successfully applied for biological sample preparation. Regardless of the type of membrane employed, the EME can operate in two different configurations: two phases and three-phase configuration. The first EME was configured in a three-phase system using hollow fibers. In a three-phase configuration, the SLM (organic solvent) is sandwiched between two aqueous solutions (the sample and the acceptor phase, respectively), providing an aqueous extract directly compatible with high-performance liquid chromatography (HPLC), liquid chromatography–mass spectrometry (LC-MS), and capillary electrophoresis (CE). However, in a two-phase configuration, the acceptor phase is an organic solvent, providing an organic extract directly compatible with gas chromatography.

Three phases can be also configured as carrier-assisted EME. That variant consists in adding some carriers (such

Sample Preparation in LC-MS Bioanalysis, First Edition. Edited by Wenkui Li, Wenying Jian, and Yunlin Fu.

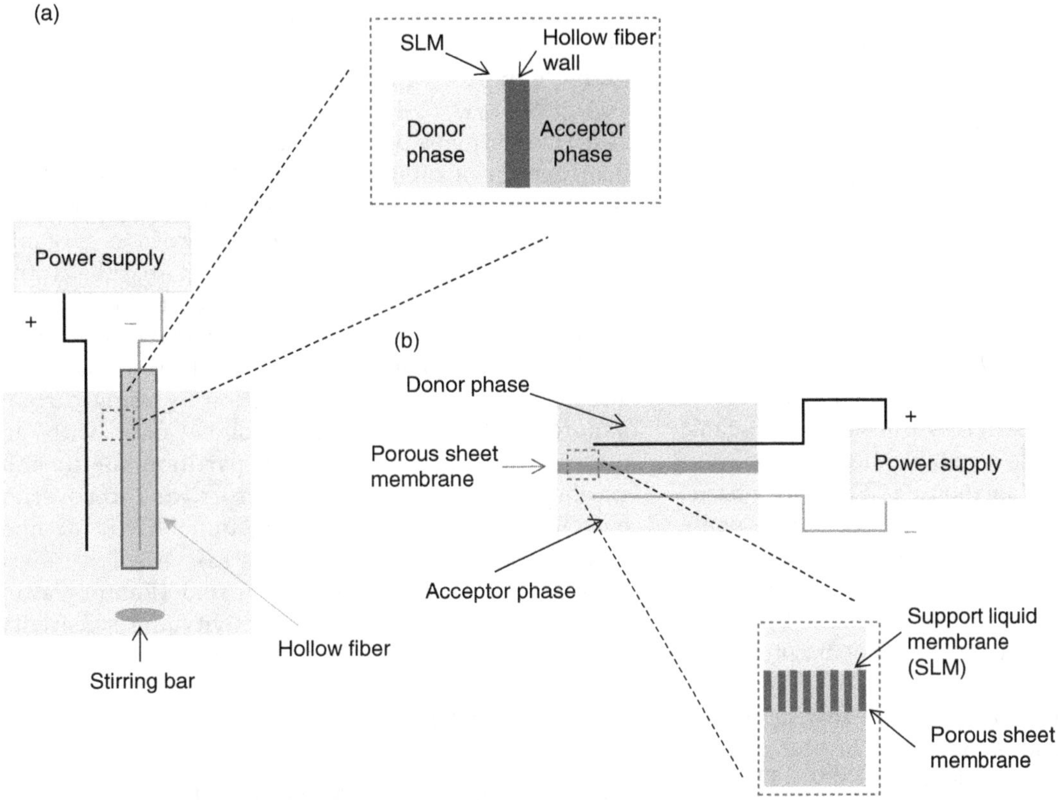

Figure 15.1 Typical electromembrane extraction setups for the extraction of acidic compounds. (a) Hollow fiber configuration and (b) Porous sheet membrane.

as di-(2-ethylhexyl) phosphate [DEHP] and tris-(2-ethylhexyl) phosphate [TEHP]) in the composition of the SLM in order to improve the electrokinetic migration of the analytes through the SLM into the acceptor solution. In the contact region of the liquid membrane and the acceptor solution, the analytes are released from the ion-pair complex into the acceptor solution, whereas counterions (protons) present in a very high concentration in the acceptor solution ion-pair with the carrier in the contact area. Carrier concentration must be optimized for EME since high concentrations of carrier decrease SLM electrical resistance and consequently increase the current level, which create bubbles around the membrane resulting in a system unstable with changes in the pH in both phases. As a result, the achieved extraction efficiency is lower compared with the pure organic solvent. It has been reported that several carriers can be used in organic solvents for better extraction of heavy metal cations, which improve the general transport of the analytes through an SLM into an acceptor solution (Hosseiny Davarani et al. 2015). The transport mechanism for heavy metal cations is proton driven, which is based on proton/analyte ion exchange between the carrier and the donor solution, and subsequently analyte/proton ion exchange between the carrier and the acceptor solution. A sufficiently large excess of protons in the acceptor solution is necessary both to provide extraction and to prevent back extraction of analytes as ion-pairs with the carrier.

In the selection of a sample preparation technique, not only the effectiveness needs to be considered but also the cost of the equipment, cost of operation, the complexity of method development, the amount of organic solvent required, and the level of automation. In the last few years, modern requirements in the field of bioanalysis often involve miniaturized and high-throughput sample preparation techniques that consume low amounts of both sample and potentially hazardous organic solvents. For this reason the reduction of hazardous organic solvent (owing to environmental or costs perspectives) as well as the small amounts of available samples or the reduction in time of processing analysis have been the major incentives for scientists leading to miniaturization of the existing sample preparation methods. EME meets several of these requirements, so it has risen as a clear alternative to liquid–liquid extraction. Additionally, it is also directly compatible with modern and future miniaturized analytical instruments, as well as easily automated.

15.2 Factors Affecting the Extraction Efficiency of EME

EME offers unique possibilities to adjust the extraction selectivity. This depends on the acceptor and sample pH, stirring speed, organic solvent for SLM, temperature, and voltage applied.

15.2.1 Composition of Supported Liquid Membrane and Nature of the Supports

A liquid membrane is formed by a thin layer of organic phase between two phases. This thin layer of organic phase is immobilized onto a suitable inert microporous support (polypropylene membrane). The membrane provides the structural support for the membrane phase (organic extracting solvent) which is the active component in the separation. The extracting solvent used for SLM is basically an organic solvent which has been chosen based on selectivity of the components present in both phases.

The chemical nature of the SLM is highly significant to success in EME, and should thus be the primary parameter to be optimized (Seidi et al. 2011a). The analyte flow through the SLM is influenced by different aspects. The flow is partially determined by the distribution ratio of the analyte of interest to SLM, which depends on the type of solvent used. Since the type of organic solvent also influences the diffusion coefficient of the analyte, a good selection of the organic solvent is critical in order to increase selectivity and to obtain good cleanup during extraction. Unlike LPME, the organic solvent used for EME must meet certain requirements, i.e. organic phase must have a certain dipole moment or electrical conductivity to withstand a flow of relatively low current in the system, and it also needs certain chemical properties to allow the phase transfer electrokinetic migration of the analytes. The mechanical robustness is a major challenge in EME processes, as part of the SLM leaks into the sample due to agitation and application of the electric field when the water-solubility of the organic solvent is extremely low (Pabby and Sastre 2013). As a consequence, the extraction efficiency of the compounds decreases. For this reason, the membrane is replaced after extraction for good reproducibility and a minimum carry-over. Furthermore, a low viscosity is preferred since a thicker membrane results in an increased diffusion path and theoretically reduces the extraction recovery. Most of the reported applications use 1-octanol as SLM for the extraction of acidic analytes and 2-nitrophenyl octyl ether (NPOE) for basic ones. A detailed explanation of organic solvents used for SLM can be found in Section 15.4.

To date, different types of porous materials have been evaluated as support for liquid membranes. For example, tailored membranes as a mixture of NPOE and 25% (w/w) DEHP have been shown to be very selective for hydrophobic and more hydrophilic basic drug extractions (Gjelstad et al. 2006). However, there is still a lack of theoretical knowledge about the role of the support in the extraction process. Organic solvents with additives as mono- or di-substituted phosphate groups were demonstrated to be effective carrier molecules in combination with both primary alcohols and ketones (Seip et al. 2011). EME has been also tested for acid drug extraction by using carbon nanotube (CNT)-reinforced hollow fiber with an increase of the overall analytes' partition coefficient, excellent pre-concentration factors, and recoveries over 90% compared to a conventional EME method (Hasheminasab and Fakhari 2013). Also, a novel approach of EME has been reported (Ramos-Payán et al. 2014) with increased selectivity and sensitivity thanks to the use of silver nanometallic-decorated hollow fibers or agar films containing silver nanoparticles (Román-Hidalgo et al. 2015).

15.2.2 Composition of Acceptor and Donor (Sample) Phase

15.2.2.1 For Extraction of Basic Drugs

For the extraction of basic drugs, the analytes are positively charged in both phases. The pH of the sample is lower than the pK_a value of the analytes of interest. The analytes are extracted across the SLM with a positive electrode placed in the sample and a negative electrode placed in the acceptor solution. In order to determinate the most suitable organic solvent to be immobilized in the pores of the membrane, the polarity of the basic drugs has to be taken into account. For extraction of nonpolar basic drugs, nitro-aromatic solvents are efficient for the electrokinetic migration of the molecules through the SLM. In some cases, the addition of hydrophobic alkylated phosphate reagents to SLM can improve the phase transfer and electrokinetic migration of basic analytes. On the other hand, since the high polarity of basic drugs counteract the influence of the electrical field, addition of an ion-pair reagent to the organic solvent, such as DEHP or TEHP is necessary in order to form ion pairs with the analytes and to facilitate the transport through SLM.

15.2.2.2 For Extraction of Acidic Drugs

For acid drug extraction, the analytes are negatively charged in both phases so that the pH of the sample and acceptor solution should be higher than the pK_a values of the analytes. In this case, the acid drugs are extracted from the donor to the acceptor solution due to an electrical field promoted by the application of a potential difference between the cathode and the anode, located in the sample and acceptor solution, respectively. Aliphatic alcohols are necessary for the high efficiency extraction of acid drugs. 1-Octanol and 1-heptanol have been reported as good organic solvents for SLM for acidic analytes (Gjelstad 2010). They are insoluble in water, easily impregnate the membrane, and show an appropriate electrical resistance to the applied voltage.

In both cases (acidic and basic drugs), it is critical to maintain the ion balance. The ion balance in the system can be determined by the pH values of donor and acceptor phases. The total ionic concentration on the donor phase in relation to the acceptor phase has also an effect on the flux over the membrane. An increase in this ratio would result in a decrease in the flux.

15.2.3 Electrical Field Application: Voltage and Time

The selectivity of the EME depends significantly on the direction and magnitude of the electric field applied. The direction controls if cations or anions are to be extracted, whereas the magnitude of the electrical field controls the type of compounds to be extracted. The applied voltage must be optimized. A very high current is often associated with a decrease in the extraction efficiency due to bubble formation at the electrodes and undesired pH changes due to electrolysis.

The extraction time is a parameter directly related to the voltage applied, as both affect the flux of analytes through the SLM. It is conceivable that an increase in both extraction time and voltage could improve the extraction efficiency; however, there is an antagonist effect. For very long extraction times, the balance is reversed due to the instability of the electrical power system, electrolysis that results in changes in the phase pH, or reversibility of distribution equilibria.

15.2.4 Agitation/Stirring Speed

The stirring speed is one of the essential parameters for sample extraction and must be optimized. The extraction device is agitated during extraction to ensure efficient convection in the sample and to maintain high extraction efficiency. An increase in the stirring speed promotes the kinetics and improves the extraction efficiency. This occurs by increasing the mass transfer and reducing the thickness of the double layer around the SLM. However, an excessive increase in the stirring speed produces bubbles in both phases and may produce leakage of the organic solvent. However, stirring speed is not necessary when EME is downscaled to a microfluidic device since the channels are very small and the low sample and acceptor volumes minimize the need of stirring to ensure the adequate transport of the analytes to the SLM.

15.2.5 Presence of Salt/Salt Effect

The effect produced by adding ionic substances has not been well studied. On one hand, the presence of ionic substances causes an increase in the value of the ion balance in the system and therefore decreases the flux of the analytes through the SLM. On the other hand, some studies have shown that the presence of ionic substances could benefit extraction efficiency due to an increase in viscosity and a change in the conductivity of the donor solution (Basheer et al. 2010).

15.2.6 Temperature

The temperature of the extraction must be controlled since it can affect the stability of the analytes and the SLM when temperature is over 40 °C. On the other hand, an increase in the temperature can speedup the extraction process.

15.3 Recent Developments in EME

The first EME approach was developed using a hollow fiber as a membrane. By this approach, an electrode was placed in the sample solution and the other electrode inside the lumen of the hollow fiber (acceptor solution). Lately, flat membranes have also been employed with good EME extractions by applying a potential difference at both sides of the membrane (sample and acceptor phase, respectively). In the past few years, other innovations have been made for EME.

15.3.1 On-chip EME

An interesting development is the downscaling of EME into a microfluidic device (Petersen et al. 2010), which allows for analyses with very small sample volumes, low consumption of chemicals and reagents, high extraction efficiency, and rapid extraction due to a very short diffusion path. Transport phenomena are faster in microsystems than in systems with regular size and, therefore, extractions can be accomplished rapidly. The device is composed of planar porous polypropylene membranes located between two poly(methyl methacrylate) (PMMA) plates, each with channel structures facing the membrane. Each channel has one electrode. A differential potential is applied between both electrodes, generating an electrical field. The sample and the acceptor solutions

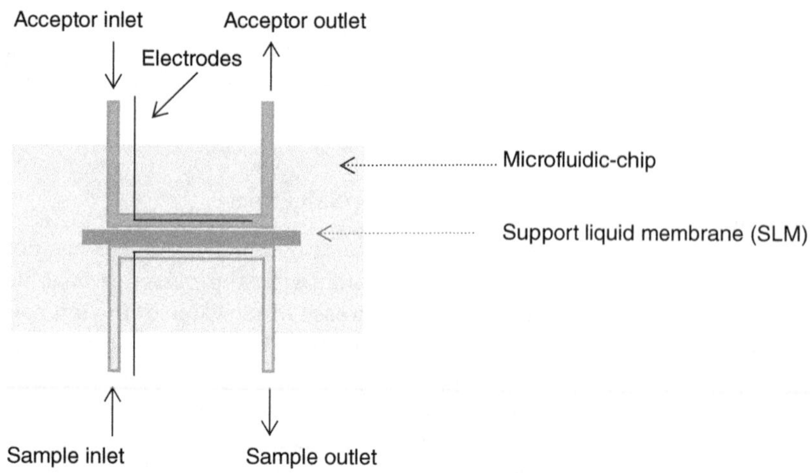

Figure 15.2 Schematic illustration of on-chip EME.

are pumped to the chip and the analytes are extracted from the sample solution (channel 1) through the SLM and into an acceptor phase (channel 2) located on the other side of the membrane. A schematic illustration of these microdevices can be seen in Figure 15.2.

The principle of EME may be envisaged as a combination of electrophoresis and three-phase liquid extraction. Compared with existing technologies for sample preparation, on-chip EME has features of: (i) rapid isolation (in one to three minutes), (ii) high selectivity and sample cleanup, (iii) limited consumption of organic solvent, (iv) compatible with modern and future miniaturized analytical instruments, and (v) easily automated. The technique has been successfully employed for the EME of different basic and acidic drugs from human blood plasma, whole blood, and urine for analysis. In the original design, the acceptor solution is collected after extraction and analyzed off-line, but the device can be also configured for online analysis (Petersen et al. 2012). The microfluidic device allows working in two different modes: double-flow and stopped flow in the acceptor solution. For online analysis, the acceptor solution is continuously pumped into an ultraviolet (UV) detector or mass spectrometer (MS). Its application for online monitoring of metabolic reactions showed a new attractive approach as the sample preparation procedure. One of the main advantages of this miniaturized EME is that it is rapid, simple-handling, requires very small sample volume, and is possible to make consecutive extractions without the necessity to change the membrane.

15.3.2 Low-voltage EME

The aim of using lower voltages in EME is to minimize or eliminate sample/analyte degradation in the extraction process. As an example, this type of EME has been used for the extraction of acid and basic drugs (Seidi et al. 2012), providing high recoveries (over 90%) by applying 40 V for only 15 minutes. However, it has also been demonstrated that this low-voltage EME does not work for certain drug molecules (Domínguez et al. 2012).

15.3.3 Drop-to-drop EME

Drop-to-drop EME is a miniaturized EME developed by Petersen et al. This modality uses flat membranes and works under stagnant conditions (Petersen et al. 2009). It uses aluminum foil as a sample reservoir which is connected to the anode. The membrane is placed over the small aluminum well (keeping contact with the sample) and subsequently the organic solvent is delivered into a membrane for rapid immobilization. Finally, a drop of acceptor solution is applied on the top of the membrane and a negative electrode is inserted into the top of the acceptor droplet. A scheme for this miniaturized EME device is shown in Figure 15.3. The extraction is completed in a short period of time with an extraction efficiency of over 33%. Drop-to-drop EME is a selective extraction from small sample volumes with no pre-concentration. This system is inexpensive and does not require agitation. However, it can be used only for a single extraction due to carryover observed in biological samples.

15.3.4 Pulsed EME

Another variant of EME combines the use of pulsed voltage and common DC constant power, which is named pulsed EME (PEME) (Rezazadeh et al. 2012). This design eliminates the mass transfer barrier with an improved extractability as result of a decrease in the thickness of the double layer at the interfaces. It works by applying a potential difference for a short period of time during extraction, but the sample solution is still being stirred despite the voltage is turned off. This variant demonstrated higher pre-concentration factors

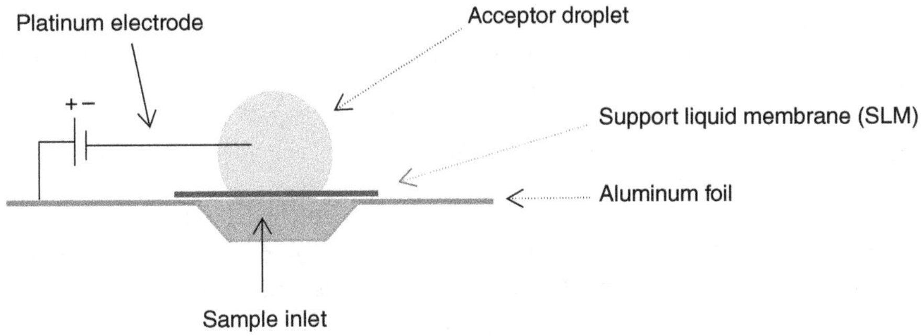

Figure 15.3 Schematic illustration for drop-to-drop EME.

than conventional EME and it has been successfully applied for the analysis of biological fluids.

15.3.5 EME Combined with Other Preparation Techniques

A combination of EME with other techniques (CE) is also possible for the pretreatment of the samples (Ramos-Payán et al. 2013).

A two-step method combining EME with a low-density solvent based ultrasound-assisted emulsification electromembrane microextraction (EME-LDS-USAEME) has been reported with enrichment factors of up to 2198, low limits of detection of 0.005 $mg\,l^{-1}$, and good linearity (Guo and Lee 2012).

Another very sensitive method is based on a combination of EME with dispersive liquid–liquid microextraction (DLLME) for biological samples (Seidi et al. 2013).

It should be also mentioned a variant named electromembrane-surrounded solid-phase microextraction (EME-SPME) was developed. In EME-SPME, a carbonaceous pencil lead is used as the cathode, which simultaneously acted as a SPME sorbent (Rezazadeh et al. 2013a).

15.4 Bioanalytical Applications

EME has been used for drug analysis and also for the analysis of more complex molecules such as amino acids and peptides. Drugs have been determinated most frequently in urine samples, while the extraction of more complex biomolecules have been determinated mostly from human serum, plasma, and whole blood samples.

15.4.1 Drug Analysis

As mentioned earlier, EME is a powerful pre-concentration tool and cleanup procedure that allows for direct analysis using diverse techniques, including LC-MS. Although many of applications have involved analytical techniques other than LC-MS, in general, it can be assumed that most of the obtained extracts can be directly analyzed by LC-MS. So this section includes most of the bioanalytical applications of EME, regardless of the analytical method used for final analyses of the interested compounds.

Particular attention should be paid to the latter applications, including some conceptual and configuration changes that improves the applicability and efficiency of EME.

Urine applications are the most frequent ones and the literature covers a wide range of compounds, including basic drugs such as narcotics (Seidi et al. 2011a), mebendazole, tricyclic antidepressants, amphetamine-type stimulants, phenazopyridine (Fotouhi et al. 2015b), or acidic drugs like diclofenac (Seidi et al. 2012) among others. Determinations in plasma and blood are also numerous by EME, as, for example, the extraction of six drugs of abuse in five minutes resulting in recoveries within the range of 10–30%. In general, recoveries by EME in such samples are low, between 10 and 60% (Jamt et al. 2012). Saliva applications are less common and an example is the determination of propanolol and betaxolol by EME.

Typical EME configurations were also used for the determination of enantiomers (trimipramine, amlodipine, and tolterodine) from plasma and urine samples (Nojavan and Fakhari 2010; Fakhari et al. 2013) where NPOE was the used SLM and detection was by means of maltodextrin-modified CE.

Low-voltage EME was first reported by Kjelsen et al. (2008) for the analysis of five basic drugs in human plasma, urine, and breast milk. A 10 V voltage was applied to a 1-isopropyl-4nitrobenzene SLM, using as donor/acceptor phase an aqueous solution of 10 nM HCl with recoveries between 50 and 93%. The authors additionally used a common 9 V battery as power supply obtaining similar enrichments. Other application of low-voltage EME was described for the determination of diclofenac (acid drug) and nalmefene (basic drug) at 40 V obtaining recoveries higher than 90% from urine samples. In both cases the time of extraction was <15 minutes. Additionally, drop-by-drop application was reported by Petersen et al. (2009) for basic drugs in human urine and plasma using NPOE as SLM. A two-phase low-voltage

EME was developed for the determination of tricyclic antidepressants (imipramine, amitriptyline, and chlorpromazine) in human urine samples (Sanagi et al. 2014). Extraction was carried out on porous polypropylene membrane filters impregnated with NPOE.

With respect to the SLM used, most basic dugs were extracted by means of NPOE (Basheer et al. 2010; Kubáň and Boček 2012); in some cases recoveries could be enhanced by modifying NPOE with additives like TEHP (Seidi et al. 2011b). On the contrary, acidic drugs are usually extracted with 1-octanol. Other SLM less frequently used were ethyl nitrobenzene (ENB) for the analysis, among others, of pethidine, nortriptyline, tramadol, methadone, haloperidol, and loperamide (Gjelstad et al. 2009).

As previously mentioned, in the last few years, several modifications of the traditional EME configuration have been reported and applied to the analysis of diverse drugs in biological matrices. The coming part will briefly describe these modified configurations.

An application based on PEME for the extraction of naltrexone and nalmefene from biological fluids was presented for the first time by Rezazadeh et al. (2012). Likewise, the performance of a PEME coupled with HPLC with UV detection was compared to a previously described HF-LPME procedure for the extraction of olanzapine and fluoxetine from human plasma and urine samples. SLM used in this PEME procedure was 2-nitrophenyl pentyl ether (NPPE) and 1-octanol for the HF-LPME. PEME provides higher extraction efficiencies in a shorter time. Another PEME coupled with HPLC was also described for the extraction and determination of diclofenac and mefenamic acid from urine and plasma samples (Fotouhi et al. 2015a).

A device consisting of a dual hollow fiber EME (Tabani et al. 2013) allows the simultaneous extraction and preconcentration of acidic (ibuprofen) and basic (thebaine) drugs. The procedure was applied to plasma and urine samples and the extracts were analyzed by CE. Exhaustive and stable EME of acidic drugs (flurbiprofen, gemfibrozil, ibuprofen, ketorolac, ketoprofen, probenecid, and quinaldic acid) from human plasma was proposed (Huang et al. 2015b). 1-Heptanol was mixed with NPOE in order to improve the SLM stability and recoveries higher than 91% were obtained after 10 minutes. A configuration that includes two separate cells has been also used for the simultaneous extraction of both acidic and basic analytes (Nojavan and Asadi 2016); using this device, naproxen, ibuprofen, and ketamine were extracted from urine samples using 2-ethylhexanol and 1-octanol as SLMs.

The use of carbon nanotube-reinforced hollow fiber in EME devices (CNTs/EME) has been an interesting topic. Thus, they have been employed for the extraction of a basic drug (buprenorphine) from urine samples (Hasheminasab and Fakhari 2013) using NPOE as the SLM. Additionally, acidic drugs (ibuprofen and naproxen) were also extracted from plasma, urine, and breast milk samples using 1-octanol as the SLM. In a two-phase configuration, this approach was also used for the extraction of tramadol and methadone from plasma and urine samples (Hasheminasab et al. 2014) being analyzed by GC. Graphene oxide-reinforced two-phase EME coupled with GC was applied for the determination of methamphetamine as a model analyte in biological samples (urine and hair) with high recoveries (95–98.5%).

In some cases two or more successive extraction procedures were applied in order to enhance the extraction performance or selectivity. Thus, an EME and DLLME procedure followed by gas chromatography–flame ionization detection (GC-FID) (Seidi et al. 2013) was developed for the determination of tricyclic antidepressants (amitriptyline, trimipramine, and doxepin) in untreated human plasma and urine samples, but the results obtained were not interesting in terms of selectivity.

Pedersen-Bjergaard et al. described in 2014 the concept of parallel EME (Pa-EME) using flat membranes in a multiwall format (Eibak et al. 2014). This approach was applied to the extraction of four antidepressant drugs (quetiapine, sertraline, amitriptyline, and fluoxetine) from undiluted human plasma samples. The selected SLM was NPOE and they were isolated into formic acid. The extracts were analyzed by LC-MS/MS showing to be free from interfering matrix components. Using this concept, amitriptyline, fluoxetine, and haloperidol were isolated from samples of undiluted human plasma and human urine.

Kubáň and Boček (2014) described a novel approach to EME that uses free liquid membranes (FLMs) in a micro-electromembrane extraction (μ-EME) configuration. An FLM was formed as a plug of a selected organic solvent, 1-ethyl-2-nitrobenezene (ENB) or 2-nitrophenyloctyl ether (NPOE), in a narrow-bore polymeric tubing and was sandwiched between a plug of aqueous donor and aqueous acceptor solutions. The FLM acted as a phase interface that enabled selective transfer of analytes from the donor into the acceptor solution. Then, the μ-EME acceptor solution was analyzed by CE. The procedure was applied to the determination of nortriptyline, haloperidol, and loperamide in urine and blood serum samples. μ-EME was also tested for the simultaneous extraction of inorganic ions (K^+ and ClO_4^-) and acidic (ibuprofen) and basic (procaine) drugs. Good extraction recoveries for inorganic ions were obtained (91–94%) and satisfactory recovery for the tested drugs (35–63%). The procedure was applied to the extraction and later CE drug analysis from undiluted human urine samples.

In 2014, a flat porous membrane of polypropylene supporting a SLM of NPOE was used for the development of an EME procedure (Huang et al. 2014) for the exhaustive extraction of basic drugs (quetiapine, citalopram, amitriptyline, methadone, and sertraline) from human plasma with recoveries in the the range of 83–105% .

A carrier-assisted EME was proposed for the extraction of morphine at urine concentration levels encountered in toxicology and doping (Ahmar et al. 2014). The EME device consisted of a PP hollow fiber supporting a SLM of NPOE containing 10% TEHP and 10% DEHP. Determination was carried out by differential pulse voltammetry using screen-printed carbon electrode. High nitrogen-doped graphene (HND-G) was proposed as a new cationic carrier for the enhancement of EME performance. Using HND-G into 1-octanol as the SLM, the EME procedure was tested using naproxen and sodium diclofenac as model acidic drugs. The applicability of this method was evaluated by the extraction of real urine samples.

Some procedures called surfactant-assisted electromembrane extraction (SA-EME) have been developed. Using conventional EME configurations, Triton X-100 has been added to the donor phase to extract naproxen and diclofenac from plasma and urine samples (Fakhari et al. 2015a). Using 1-octanol as the SLM, recoveries in the 88–92% range were obtained using CE with UV detector as the analytical instrument. Basic drugs (methamphetamine, ephedrine, and methadone) were also extracted using this configuration but with NPOE as the SLM from hair and urine samples. Using pulsed electric DC as the driving force and combined to GC analysis, a procedure for the extraction and analysis of alfentanil, sufentanil, and methadone was proposed (Zahedi et al. 2016) with 0.02% (w/v) sodium dodecyl sulphate added to the samples. Recoveries in the 70.5–95.2% range were obtained for the extraction from plasma, breast milk, and urine samples.

A homemade graphene/polyaniline composite as coating of the electrode has been used in an EME configuration called electromembrane-surrounded solid-phase microextraction (EM-SPME (Abedi and Ebrahimzadeh 2015). Mefenamic acid and ibuprofen were used as model analytes and extracted human urine and plasma samples were analyzed by ion mobility spectrometry with satisfactory results. Amphetamine and methamphetamine were also extracted from human urine and whole blood samples. Tramadol was extracted from urine samples (Fakhari et al. 2015b), but in this case, reduced graphene oxide spin-coated pencil lead was used not only as a solid sorbent for SPME and reinforced cathode of EME, but also as the working electrode in the later linear sweep voltammetry analysis. A self-assembled monolayer of benzyl mercaptan on a copper wire was proposed for EM-SPME and applied for the extraction of some antihistamines (naphazoline and antazoline) from human urine and plasma samples with acceptable results (Shamsayei et al. 2016). Electrochemically synthesized nanostructured polypyrrole coating was utilized as both cathode and SPME sorbent and the procedure was used for the analysis of amitriptyline and doxepin as model analytes from human whole blood and urine samples (Mohammadkhani et al. 2016).

Prof. Yaminis' group proposed the use of external cylindrical electrodes (surrounded by the hollow fiber) in order to improve the extraction EME efficiency (Asl et al. 2015a). Diclofenac and mefenamic acid as model analytes were extracted from urine and plasma samples but the obtained results were similar to typical EME configurations. This group also proposed dynamic electromembrane extraction (DEME) (Asl et al. 2015b) in that acceptor solution is repeatedly pumped into the lumen of hollow fiber by a syringe pump whereas a peristaltic pump is used to move sample solution around the mounted hollow fiber into the flow cell. Amitriptyline and nortriptyline were extracted from biological samples using this configuration.

Ionic liquids have been used as the SLM in some EME procedures. Thus, some of them ([C_4MIm][PF6], [C_6MIm][PF6] and [C_8MIm][PF6]) were tested for the EME extraction from pig kidney of four chlorophenoxy acid herbicides, namely, 2,4-dichlorophenoxy acetic acid, 2-(2,4-dichlorophenox)propionic acid, 4-chloromethylphenoxyacetic acid, and 2,4,5-trichlorophenoxyacetic acid with a significant increase in the extraction efficiency. The same ionic liquids were also used for the extraction of strychnine and brucine in human urine samples and the results compared with those obtained using 2-ethylnitrobenzene (ENB) as the organic SLM showed improvement.

A homemade polymer inclusion membrane (HPIM) was used to develop an EME extraction (Mamat and See 2015). HPIM was prepared by casting a solution of the desired proportions of cellulose acetate, tris(2-ethylhexyl)phosphate (TEHP), and di-(2-ethylhexyl)phosphoric acid (D2EHPA) in dichloromethane on glass capillary tubing. Amphetamine, methamphetamine, and 3,4-methylenedioxy-*N*-methylamphetamine (MDMA) were selected as model analytes for their extraction from human plasma samples. Excellent recoveries near to 100% were reported.

An on-chip EME coupled with HPLC was developed for the efficient extraction of nortriptyline and amitriptyline as basic model analytes from urine samples (Abdossalami Asl et al. 2015). This device provides effective and reproducible extractions using a low volume of sample. Other on-chip EME device was described for the simultaneous extraction of mefenamic acid and diclofenac as acidic model analytes, and betaxolol, as a basic model analyte, followed by HPLC-UV determination. The microfluidic procedure was applied for the analysis of urine and plasma samples. On the other hand, an on-chip EME coupled with HPLC was proposed for the consecutive extraction of drugs with different chemical properties (Asl et al. 2016). Betaxolol, naltrexone, and nalmefene were extracted from urine and plasma samples.

A two-step voltage dual electromembrane extraction (TSV-DEME) using 1-octanol and 2-ethylhexanol as SLMs for the extraction of acidic and basic analytes, respectively, was described (Asadi and Nojavan 2016) and its efficiency compared to simple DEME, showed higher recoveries (38–68%). Codeine, naproxen, and ketamine were extracted from breast milk and plasma.

A virtually rotating SLM EME device was proposed (Hosseiny Davarani et al. 2016). The virtual (non-mechanical) rotating of the SLM was achieved using a novel electrode assembly including a central electrode immersed inside the lumen of the SLM and five counter electrodes surrounding the SLM. A particular electronic circuit was designed to distribute the potential among five counter electrodes in a rotating pattern. Extraction of verapamil, trimipramine, and clomipramine as model analytes and 2-ethylhexanol as the SLM was tested on human urine samples.

Polar ($\log P$: −0.40 to 1.32) basic drugs (metaraminol, benzamidine, sotalol, phenylpropanolamine, ephedrine, and trimethoprim) were extracted from plasma using pure bis(2-ethylhexyl) phosphate as the SLM (Huang et al. 2016) and electrolysis in the sample and acceptor solutions was kept at an acceptable level. Recoveries in the range 25–91% were obtained.

Clozapine was extracted from human plasma samples by combination of EME with differential pulse voltammetry. NPOE was used as the SLM into the pores of the hollow fiber and three microelectrodes, an Ag/AgCl, a platinum wire, and a graphite pencil lead as the reference, counter, and working electrodes, respectively.

15.4.2 Biomolecule Analysis

EME extraction was also used for the analysis of more complex molecules such as amino acids and peptides. In this case, the extraction of compounds from human serum, plasma, and whole blood samples is performed well with ENB, or 1-octanol modified with DEHP, or diisobutyl ketone using voltages not exceeding 50 V as the driving force by means of a traditional electromembrane configuration. The extraction times are relatively short, around the order of 10 minutes. In this case, low recoveries are usually obtained for the studied analytes, the best ones being nearly 40%. Four polyamines mainly present in saliva samples (putrescine, cadaverine, spermidine, and spermine) were directly determined by CZE with capacitively coupled contactless conductivity detection (CZE-C^4D) after EME extraction with recoveries in the 78–97% range (Liu et al. 2014) using a 1 : 30 (v/v) mixture of ENB/DEHP.

EME coupled with LC-MS/MS was developed for the determination of 10 volatile organic compounds metabolites: (*N*-acetyl-*S*-(phenyl)-l-cysteine, phenylglyoxylic acid, mandelic acid, *N*-acetyl-*S*-(benzyl)-l-cysteine, hippuric acid, 2-methylhippuric acid, 3-methylhippuric acid, 4-methylhippuric acid, *N*-acetyl-*S*-(1,2-dichlorovinyl)-l-cysteine, and *N*-acetyl-*S*-(2,2-dichlorovinyl)-l-cysteine), in dried urine spot samples (Suh et al. 2015a). Aliquots of urine were spotted onto Bond Elut DMS cards and dried at room temperature. After drying, the punched-out dried urine spot was eluted with water. Compounds were extracted through a typical HF-EME configuration using 1-octanol as the SLM.

A typical hollow fiber EME configuration was applied, using 95% of 1-octanol and 5% of di-(2-ethylhexyl)-phosphate as the SLM, for the extraction of four gonadotropin-releasing hormone agonist anticancer peptides (alarelin, leuprolide, buserelin, and triptorelin) in human plasma samples (Nojavan et al. 2016).

More recently, some EME modifications have been applied to the analysis of biomolecules. Thus, has been proposed the use of pulsed electrical DC current in EME devices (PEME) (Rezazadeh et al. 2013b) in order to enhance their performance. A mixture of NPOE, DEHP, and TEHP was immobilized in the pores of hollow fiber as the SLM and the procedure was applied to the extraction and determination of three aminoacids (histidine, phenylalanine, and tryptophan) in plasma, saliva, milk, and whey samples with recoveries in the 7.1–21.6% range.

EME was coupled to dispersive liquid–liquid microextraction (EME-DLLME) for the extraction of some pyridine derivatives (3-methylpyridine, 2,4-lutidine, quinolone, and 4-dimethylaminopyridine) (Arjomandi-Behzad et al. 2014). A mixture of 90% NPOE and 10% DEHP was immobilized in the pores of the hollow fiber as the SLM and the extraction procedure was applied to the analysis of smokers' urines.

Urinary benzene, toluene, ethylbenzene, and xylene metabolites (*S*)-phenylmercapturic acid, hippuric acid, phenylglyoxylic acid, and methylhippuric acid isomers were extracted from human urine samples (Suh et al. 2015b) using a typical hollow fiber EME configuration with 1-octanol as the SLM. Analytes were measured by HPLC-MS/MS with recoveries between 52.8 and 79%.

Pedersen-Bjergaard et al. illustrated for the first time the possibility of exhaustive extraction of peptides using EME under low system-current conditions (<50 μA) (Huang et al. 2015a). Selected target analytes were bradykinin acetate, angiotensin II antipeptide, angiotensin II acetate, neurotensin, angiotensin I trifluoroacetate, and leu-enkephalin. A flat membrane supporting a 1 : 1 (v/v) mixture of 1-nonanol diluted with 2-decanone containing 15% (v/v) DEHP as the suitable SLM allowed recoveries for the model peptides in the range of 77–94%. A two-step approach of this configuration permitted the selective isolation of some of the above-mentioned peptides based on their isoelectric points. The observations indicated that the two-step EME may have a future potential for the fractionation of peptides and other ampholytic compounds.

A procedure called dynamic EME which consists in the continual renewal of the acceptor compartment has been developed for the extraction of lipophilic neuropeptides (Drouin et al. 2016). This new setup minimized the effect of electrolysis on extraction performance while enabling high extraction yield (up to 72%).

15.5 Summary

EME has been demonstrated to be a powerful technique which provides high pre-concentration and efficient sample cleanup. In addition, since the extraction is performed under the influence of an electrical field, the extraction selectivity can be controlled by the direction and the magnitude of the electrical field and several other experimental parameters. EME has been successfully applied to environmental and biological samples. Due to its exceptional selectivity and cleanup offered, it allows analysis of biological samples with the minimum sample treatment. The latest development of this technique toward miniaturization systems, new supports, and its compatibility online to instrumental techniques makes EME a very attractive technique in the future. Also, there are already new theories that help to thoroughly understand the different mechanisms occurring in EME in order to improve the extraction efficiencies.

EME is undoubtedly a very competitive technique that offers a lot of potential compared to other modern extraction techniques as well as the more traditional ones used in the treatment of samples. It is expected that EME will offer great advances in the future and can be placed among the top techniques used for sample treatment.

References

Abdossalami Asl, Y., Yamini, Y., Seidi, S., and Ebrahimpour, B. (2015). A new effective on chip electromembrane extraction coupled with high performance liquid chromatography for enhancement of extraction efficiency. *Anal. Chim. Acta* 898: 42–49.

Abedi, H. and Ebrahimzadeh, H. (2015). Electromembrane-surrounded solid-phase microextraction coupled to ion mobility spectrometry for the determination of nonsteroidal anti-inflammatory drugs: a rapid screening method in complicated matrices. *J. Sep. Sci.* 38 (8): 1358–1364.

Ahmar, H., Tabani, H., Hossein Koruni, M. et al. (2014). A new platform for sensing urinary morphine based on carrier assisted electromembrane extraction followed by adsorptive stripping voltammetric detection on screen-printed electrode. *Biosens. Bioelectron.* 54: 189–194.

Arjomandi-Behzad, L., Yamini, Y., and Rezazadeh, M. (2014). Extraction of pyridine derivatives from human urine using electromembrane extraction coupled to dispersive liquid-liquid microextraction followed by gas chromatography determination. *Talanta* 126: 73–81.

Asadi, S. and Nojavan, S. (2016). Two-step voltage dual electromembrane extraction: a new approach to simultaneous extraction of acidic and basic drugs. *Anal. Chim. Acta* 923: 24–32.

Asl, Y.A., Yamini, Y., Rezazadeh, M., and Seidi, S. (2015a). Electromembrane extraction using a cylindrical electrode: a new view for the augmentation of extraction efficiency. *Anal. Methods* 7 (1): 197–204.

Asl, Y.A., Yamini, Y., and Seidi, S. (2016). A novel approach to the consecutive extraction of drugs with different properties via on chip electromembrane extraction. *Analyst* 141 (1): 311–318.

Asl, Y.A., Yamini, Y., Seidi, S., and Amanzadeh, H. (2015b). Dynamic electromembrane extraction: automated movement of donor and acceptor phases to improve extraction efficiency. *J. Chromatogr. A* 1419: 10–18.

Basheer, C., Lee, J., Pedersen-Bjergaard, S. et al. (2010). Simultaneous extraction of acidic and basic drugs at neutral sample pH: a novel electro-mediated microextraction approach. *J. Chromatogr. A* 1217 (43): 6661–6667.

Domínguez, N.C., Gjelstad, A., Nadal, A.M. et al. (2012). Selective electromembrane extraction at low voltages based on analyte polarity and charge. *J. Chromatogr. A* 1248: 48–54.

Drouin, N., Rudaz, S., and Schappler, J. (2016). Dynamic-electromembrane extraction: a technical development for the extraction of neuropeptides. *Anal. Chem.* 88 (10): 5308–5315.

Eibak, L.E.E., Lars, E.E., Parmer, M.P. et al. (2014). Parallel electromembrane extraction a multiwall plate. *Anal. Bioanal. Chem.* 406 (2): 431–440.

Fakhari, A.R., Hasheminasab, K.S., Aladaghlo, Z., and Koruni, M.H. (2015a). Surfactant-assisted electromembrane extraction combined with capillary electrophoresis as a novel technique for the determination of acidic drugs in biological fluids. *Electrophoresis* 36 (24): 3034–3041.

Fakhari, A.R., Sahragard, A., Ahmar, H., and Tabani, H. (2015b). A novel platform sensing based on combination of electromembrane-assisted solid phase microextraction with linear sweep voltammetry for the determination of tramadol. *J. Electroanal. Chem.* 747: 12–19.

Fakhari, A.R., Tabani, H., Behdad, H. et al. (2013). Electrically-enhanced microextraction combined with maltodextrin-modified capillary electrophoresis for quantification of tolterodine enantiomers in biological samples. *Microchem. J.* 106: 186–193.

Fotouhi, L., Seidi, S., Yamini, Y., and Hosseini, E. (2015a). Evaluation of pulsed electromembrane extraction for the analysis of diclofenac and mefenamic acid in biological fluids. *Anal. Methods* 7 (6): 2848–2854.

Fotouhi, L., Yamini, Y., Hosseini, R., and Rezazadeh, M. (2015b). Determination of phenazopyridine in biological fluids using electromembrane extraction followed by high-performance liquid chromatography. *Can. J. Chem.* 93 (7): 702–707.

Gjelstad, A. (2010). Electromembrane extraction: the use of electrical potential for isolation of charged substances from biological matrices. *LC-GC North America* 28: 92–112.

Gjelstad, A., Rasmussen, K.E., and Pedersen-Bjergaard, S. (2006). Electrokinetic migration across artificial liquid membranes. Tuning the membrane chemistry to different types of drug substances. *J. Chromatogr. A* 1124 (1–2): 29–34.

Gjelstad, A., Rasmussen, K.E., and Pedersen-Bjergaard, S. (2009). Electromembrame extraction of basic drugs from untreated human plasma and whole blood under physiological pH conditions. *Anal. Bioanal. Chem.* 393 (3): 921–928.

Guo, L. and Lee, H.K. (2012). Electro membrane extraction followed by low-density solvent based ultrasound-assisted emulsification microextraction combined with derivatization for determining chlorophenols and analysis by gas chromatography-mass spectrometry. *J. Chromatogr. A* 1243: 14–22.

Hasheminasab, K.S. and Fakhari, A.R. (2013). Development and application of carbon nanotubes assisted electromembrane extraction (CNTs/EME) for the determination of buprenorphine as a model of basic drugs from urine samples. *Anal. Chim. Acta* 767 (1): 75–80.

Hasheminasab, K.S., Fakhari, A.R., and Koruni, M.H. (2014). Development of carbon-nanotube-assisted electromembrane extraction in the two-phase mode combined with GC for the determination of basic drugs. *J. Sep. Sci.* 37 (1–2): 85–91.

Hosseiny Davarani, S.S., Moazami, H.R., Memarian, E., and Nojavan, S. (2016). Electromembrane extraction through a virtually rotating supported liquid membrane. *Electrophoresis* 37 (2): 339–346.

Hosseiny Davarani, S.S., Sheikhi, N., Nojavan, S. et al. (2015). Electromembrane extraction of heavy metal cations from aqueous media based on flat membrane: method transfer from hollow fiber to flat membrane. *Anal. Methods* 7: 2680–2687.

Huang, C., Eibak, L.E.E., Gjelstad, A. et al. (2014). Development of a flat membrane based device for electromembrane extraction: a new approach for exhaustive extraction of basic drugs from human plasma. *J. Chromatogr. A* 1326: 7–12.

Huang, C., Gjelstad, A., and Pedersen-Bjergaard, S. (2015a). Exhaustive extraction of peptides by electromembrane extraction. *Anal. Chim. Acta* 853 (1): 328–334.

Huang, C., Gjelstad, A., Seip, K.F. et al. (2015b). Exhaustive and stable electromembrane extraction of acidic drugs from human plasma. *J. Chromatogr. A* 1425: 81–87.

Huang, C., Seip, K.F., Gjelstad, A., and Pedersen-Bjergaard, S. (2016). Electromembrane extraction of polar basic drugs from plasma with pure bis(2-ethylhexyl) phosphite as supported liquid membrane. *Anal. Chim. Acta* 934: 80–87.

Jamt, R.E.G., Gjelstad, A., Eibak, L.E.E. et al. (2012). Electromembrane extraction of stimulating drugs from undiluted whole blood. *J. Chromatogr. A* 1232: 27–36.

Kjelsen, I.J.Ø., Gjelstad, A., Rasmussen, K.E., and Pedersen-Bjergaard, S. (2008). Low-voltage electromembrane extraction of basic drugs from biological samples. *J. Chromatogr. A* 1180 (1–2): 1–9.

Kubáň, P. and Boček, P. (2012). Capillary electrophoresis with capacitively coupled contactless conductivity detection: a universal tool for the determination of supported liquid membrane selectivity in electromembrane extraction of complex samples. *J. Chromatogr. A* 1267: 96–101.

Kubáň, P. and Boček, P. (2014). Micro-electromembrane extraction across free liquid membranes. Extractions of basic drugs from undiluted biological samples. *J. Chromatogr. A* 1337: 32–39.

Liu, Y., Zhang, X., Guo, L. et al. (2014). Electromembrane extraction of salivary polyamines followed by capillary zone electrophoresis with capacitively coupled contactless conductivity detection. *Talanta* 128: 386–392.

Mamat, N.A. and See, H.H. (2015). Development and evaluation of electromembrane extraction across a hollow polymer inclusion membrane. *J. Chromatogr. A* 1406: 34–39.

Mohammadkhani, E., Yamini, Y., Rezazadeh, M., and Seidi, S. (2016). Electromembrane surrounded solid phase microextraction using electrochemically synthesized nanostructured polypyrrole fiber. *J. Chromatogr. A* 1443: 75–82.

Nojavan, S. and Asadi, S. (2016). Electromembrane extraction using two separate cells: a new design for simultaneous extraction of acidic and basic compounds. *Electrophoresis* 37 (4): 587–594.

Nojavan, S., Bidarmanesh, T., Mohammadi, A., and Yaripour, S. (2016). Electromembrane extraction of gonadotropin-releasing hormone agonists from plasma and wastewater samples. *Electrophoresis* 37 (5–6): 826–833.

Nojavan, S. and Fakhari, A.R. (2010). Electro membrane extraction combined with capillary electrophoresis for the determination of amlodipine enantiomers in biological samples. *J. Sep. Sci.* 33 (20): 3231–3238.

Pabby, A.K. and Sastre, A.M. (2013). State-of-the-art review on hollow fibre contactor technology and membrane-based extraction processes. *J. Membr. Sci.* 430: 263–303.

Petersen, N.J., Jensen, H., Hansen, S.H. et al. (2010). On-chip electro membrane extraction. *Microfluid. Nanofluidics* 9 (4–5): 881–888.

Petersen, N.J., Jensen, H., Hansen, S.H. et al. (2009). Drop-to-drop microextraction across a supported liquid membrane by an electrical field under stagnant conditions. *J. Chromatogr. A* 1216 (9): 1496–1502.

Petersen, N.J., Pedersen, J.S., Poulsen, N.N. et al. (2012). On-chip electromembrane extraction for monitoring drug metabolism in real time by electrospray ionization mass spectrometry. *Analyst* 137 (14): 3321–3327.

Ramos-Payán, M., Fernández-Torres, R., Pérez Bernal, J.L. et al. (2014). A novel approach for electromembrane extraction based on the use of silver nanometallic-decorated hollow fibers. *Anal. Chim. Acta* 849: 7–11.

Ramos-Payán, M., Li, B., Jensen, H. et al. (2013). Nano-electromembrane extraction. *Anal. Chim. Acta* 785: 60–66.

Rezazadeh, M., Yamini, Y., Seidi, S., and Ebrahimpour, B. (2013a). Electromembrane surrounded solid phase microextraction: a novel approach for efficient extraction from complicated matrices. *J. Chromatogr. A* 1280: 16–22.

Rezazadeh, M., Yamini, Y., Seidi, S., and Esrafili, A. (2012). Pulsed electromembrane extraction: a new concept of electrically enhanced extraction. *J. Chromatogr. A* 1262: 214–218.

Rezazadeh, M., Yamini, Y., Seidi, S., and Esrafili, A. (2013b). One-way and two-way pulsed electromembrane extraction for trace analysis of amino acids in foods and biological samples. *Anal. Chim. Acta* 773: 52–59.

Román-Hidalgo, C., Ramos-Payán, M., Ocaña-González, J.A. et al. (2015). Agar films containing silver nanoparticles as new supports for electromembrane extraction. *Anal. Bioanal. Chem.* 407: 1519–1525.

Sanagi, M.M., Suhaila, N., Hanapi, M. et al. (2014). Two-phase electrodriven membrane extraction combined with liquid chromatography for the determination of tricyclic antidepressants in aqueous matrices. *Anal. Methods* 6: 8802–8809.

Seidi, S., Yamini, Y., Heydari, A. et al. (2011a). Determination of thebaine in water samples, biological fluids, poppy capsule, and narcotic drugs, using electromembrane extraction followed by high-performance liquid chromatography analysis. *Anal. Chim. Acta* 701 (2): 181–188.

Seidi, S., Yamini, Y., and Rezazadeh, M. (2013). Combination of electromembrane extraction with dispersive liquid-liquid microextraction followed by gas chromatographic analysis as a fast and sensitive technique for determination of tricyclic antidepressants. *J. Chromatogr. B Analyt. Technol. Biomed. Life Sci.* 913-914: 138–146.

Seidi, S., Yamini, Y., Rezazadeh, M., and Esrafili, A. (2012). Low-voltage electrically-enhanced microextraction as a novel technique for simultaneous extraction of acidic and basic drugs from biological fluids. *J. Chromatogr. A* 1243: 6–13.

Seidi, S., Yamini, Y., Saleh, A., and Moradi, M. (2011b). Electromembrane extraction of levamisole from human biological fluids. *J. Sep. Sci.* 34 (5): 585–593.

Seip, K.F., Stigsson, J., Gjelstad, A. et al. (2011). Electromembrane extraction of peptides – fundamental studies on the supported liquid membrane. *J. Sep. Sci.* 34 (23): 3410–3417.

Shamsayei, M., Yamini, Y., Rezazadeh, M. et al. (2016). Self-assembled benzyl mercaptan monolayer as a coating in electromembrane surrounded solid-phase microextraction of antihistamines in urine and plasma samples. *New J. Chem.* 40 (6): 5268–5276.

Suh, J.H., Eom, H.Y., Kim, U. et al. (2015a). Highly sensitive electromembrane extraction for the determination of volatile organic compound metabolites in dried urine spot. *J. Chromatogr. A* 1416: 1–9.

Suh, J.H., Lee, H.Y., Kim, U. et al. (2015b). Simultaneous determination of benzene, toluene, ethylbenzene, and xylene metabolites in human urine using electromembrane extraction combined with liquid chromatography and tandem mass spectrometry. *J. Sep. Sci.* 38 (24): 4276–4285.

Tabani, H., Fakhari, A.R., and Shahsavani, A. (2013). Simultaneous determination of acidic and basic drugs using dual hollow fibre electromembrane extraction combined with CE. *Electrophoresis* 34 (2): 269–276.

Zahedi, P., Davarani, S.S.H., Moazami, H.R., and Nojavan, S. (2016). Surfactant assisted pulsed two-phase electromembrane extraction followed by GC analysis for quantification of basic drugs in biological samples. *J. Pharm. Biomed. Anal.* 117: 485–491.

Part II

Matrix-specific Sample Preparation Techniques in LC-MS Bioanalysis

16

Tissue Sample Preparation in LC-MS Bioanalysis

Hong Gao and John Williams

16.1 Introduction

Tissue sample bioanalysis is required at all stages of drug discovery and development for pharmacokinetic/pharmacodynamic (PK/PD) and toxicokinetic/toxicodynamic (TK/TD) assessment in addition to diagnostic screening. The goal of quantitative tissue bioanalysis using LC/MS is to determine the concentrations of parent drug and/or its metabolites in the tissue samples of interest. The most critical step in developing a robust LC/MS method for tissue sample analysis is the preparation of the solid tissue into a form from which the analyte(s) of interest can be readily and reproducibly extracted for analysis. The process of liquefying a solid tissue is called homogenization. This chapter describes the different homogenization techniques and their application for quantitative analysis of different types of tissues. Most importantly, this chapter provides detailed homogenization protocols for the tissue types that are common to the bioanalysis community as well as those less common, such as skin, hair, nail, and bone.

The physical and histological properties of the solid tissue samples determine the appropriate homogenization technique for bioanalysis (Gao et al. 2013a). The four most commonly used techniques for tissue homogenization are (i) mechanical shearing, (ii) ultrasonic disruption, (iii) chemical digestion, and (iv) enzymatic digestion. Depending on the type of tissue, analyte stability, assay throughput, the level of accuracy and precision requirement, the tissue sample preparation strategy varies. Therefore, an assay protocol should be defined prior to the intended work (Xue et al. 2012).

16.2 Selection of Homogenization Method

The method selected for tissue homogenization will directly impact the analyte recovery (extraction efficiency), analyte stability under homogenization conditions, and assay quality. Understanding the properties of both the tissue and the analyte is critical for selecting the appropriate homogenization technique that will deliver reliable results.

16.2.1 Tissue Properties

The physical and histological properties of the tissue are the key factors for consideration of LC/MS bioanalysis of tissue samples. These properties determine the parameters to be employed in the homogenization process of the tissues. Histologically, tissues are grouped as epithelial, connective, muscular, and nervous tissues (William 1972). Based upon the physical properties, tissues can be classified as soft, fibrous (tough), and hard as discussed previously in the literature (Gao et al. 2013a). Organs categorized as soft have low levels of connective tissue and can be easily homogenized by many different techniques. These soft tissues include adipose, brain, liver, and spleen. Fibrous tissues contain greater levels of fibrous connective tissue and are elastic and strong. Examples are arteries, heart, intestine, tumor, cartilage, and uterus. Bones, hair, skin, and nails are classified as hard tissues and are difficult to homogenize using mechanical force alone. Alternative techniques such as chemical or enzymatic digestion are often required as part of the homogenization process. The size of the container used for homogenization must

Sample Preparation in LC-MS Bioanalysis, First Edition. Edited by Wenkui Li, Wenying Jian, and Yunlin Fu.

Table 16.1 Animal organ weight.

Species	Mouse	Hamster	Rat	Rabbit	Monkey	Dog
Body weight (kg, male)	0.02–0.06	0.11–0.14	0.25–0.4	2.5–6	5	10
Organ weight (g)						
Adrenals	0.004	0.027	0.05	0.5	1.2	1
Brain	0.36	—	1.8	14	90	80
Heart	0.08	0.46	1	5	18.5	80
Kidneys	0.32	0.96	2	13	25	50
Liver	1.75	6.05	10	77	150	320
Lung	0.12	0.74	1.5	18	33	100
Spleen	0.1	0.54	0.75	1	8	25

Source: From Kwon (2002).

be suited to the size of the organ to be homogenized (Kemper et al. 2012). Table 16.1 summarizes the typical organ weights of different animals often analyzed in bioanalytical laboratories which can guide the selection of sample container and apparatus to be used for tissue homogenization.

16.2.2 Homogenization Techniques

Tissues can be homogenized by different techniques. Mechanical homogenization can be carried out by rotor-stator shearing, bead beating, and ultrasonication to physically disrupt connective tissue and cellular structure to release the analyte from the cells (Table 16.2). Chemical and enzymatic digestion of tissue employs strong acids, bases, or specific proteinases to break down tissue. Chemical and enzymatic digestion techniques avoid the build-up of heat that is common with mechanical disruption but often are at the cost of homogenization time (Table 16.3). The selection of homogenization technique primarily depends on the size, type, and physical properties of the tissue to be processed, and the chemical properties of the analyte (thermal liability, chemical stability, small molecule vs. biological entity) to be analyzed.

16.2.2.1 Rotor-Stator Homogenization

16.2.2.1.1 Suitable Tissues

Rotor-stator homogenizer is an ideal tool for the homogenization of soft tissues and some tough tissues. In general, hard tissues such as bones, hair, skin, and nails are not directly processed using a rotor-stator homogenizer alone. However, the technique can be used as a secondary homogenization following enzymatic digestion to produce a homogenous mixture. Tissues with size <20 mg (e.g. rat lymph node) can be difficult to process using a rotor-stator homogenizer unless adequate homogenization buffer is used. Dilution due to the addition of the buffer can affect the overall assay sensitivity. On the other hand, large organs, such as dog liver, are also challenging due to the size constraints on containers and apparatus. Homogenization of the entire organ followed by subsampling the resulting homogenate for analysis can improve the reproducibility of the measured analyte concentrations by eliminating potential spatial concentration differences within an organ. The strength and elasticity of the connective tissue found in fibrous tissues such as artery, muscle, and uterus make homogenization with a rotor-stator homogenizer difficult as well. Fibrous tissues often require a longer processing time and a higher shearing speed as compared to soft tissues, which can lead to excessive generation of heat. However, homogenization in an ice bath can help minimize sample heating, which is especially important for thermally labile analytes.

16.2.2.1.2 Selection of Apparatus and Probe

Handheld single probe device, such as the Omni Tissue Master-125, is a perfect choice for small numbers of samples. With Omni multi-sample homogenizer, six samples can be processed at once. This offers an improved sample throughput as compared to a single probe device. The Omni Prep 96 Homogenizer offers the highest throughput and is capable of processing 96 samples at once. Rotor-stator probes are made of metal or plastic. Depending on the size of tissue samples, there are a variety of metal probes to choose from. Disposable plastic probes for both multi-sample and 96-tube format is designed for single use to avoid cross-contamination. With fixed size, the disposable plastic probes are ideal for universal application. Furthermore, there is no need for cleaning between samples.

Table 16.2 Commonly used apparatus for tissue homogenization.

Apparatus	Example devices	Mechanism and features
Rotor-stator homogenizer	 Omni Tissue Master-125  Omni Prep Multi-Sample Homogenizer  Omni Prep 96 Homogenizer	• Rotating blades shearing and dispersing tissues • Different size of probe available to accommodate specific needs • Single probe for single sample process. Fixed probe requires thorough cleanup between samples • Six-channel and 96-channel available for multiple sample processing in one batch • Disposable probe is available for single use for single, multi-sample, and 96-sample format. No cross-contamination • Manufactured by Omni, Fisher, and other companies Web site: Omni homogenizers: https://www.omni-inc.com (OMNI International, Products) Fisher Scientific: https://www.fishersci.com (Lab Equipment and Supplies)
Beads beater	 Precellys 24  Omni Bead Ruptor 24 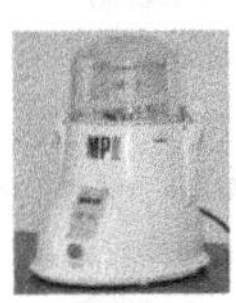 Fast Prep-24	• High-speed agitation of tissue against beads • Beads composed of different materials such as metal, glass, and ceramic • Beads available in different sizes • Sample vials available in different sizes • Programmable for beating power and duration • Can be used for cryogenic sample processing • Cooling system available for some systems Web site: Precellys 24: www.precellys.com (Homogenizers.net – Precellys Homogenizer) Omni Bead Ruptor 24: https://www.omni-inc.com (OMNI International, Products) Fast Prep-24: www.mpbio.com (Sample Preparation Equipment)
Ultrasonicator	 Covaris E220 Focused-ultrasonicator	• High frequency and energy sound waves shear tissue • Adaptive focused acoustics energy • Utilizes a dish-shaped acoustic transducer and computer-controlled electronics • Acoustic energy can be tuned to be delivered precisely and accurately to samples of varying volumes • 96-well format is available for small samples using micro glass vial Web site: Covaris E220: https://covarisinc.com (Covaris Advanced, Pre-Analytical Systems for Life Sciences)

Note: Diluents are typically added to sample prior to performing the homogenization process using the methods described in this table. Pure water, saline, special reagent, buffer, organic solvent, or mixtures of those all can serve as the homogenization buffer.

Table 16.3 Commonly used agents/enzymes for chemical or enzyme digestion of tissues.

Digestion	Agent examples	Mechanism and features
Acid and alkaline	HCl HNO_3 HF H_2SO_4 Hydrogen peroxide KOH NaOH	• Both strong acids and strong bases can hydrolyze peptide bonds to break down tissue structure • Digestion vessels are available to accommodate normal or pressurized digestion • Analyte stability under acidic or basic conditions should guide choice of digestion route
Enzyme	Collagenase Proteinase K DNase Elastase Hyaluronidase Papain Lipase Trypsin Other enzymes	• Collagenases break the peptide bonds in collagen and destroy cellular structures • Proteinase break down proteins into peptides and amino acids • DNase catalyze hydrolytic cleavage of phosphodiester linkages to degrade DNA • Elastase hydrolyze amides and esters in proteins to destroy connective tissues • Hyaluronidase is a polysaccharidase breaking down the bond between 2-acetoamido-2-deoxy-β-D-glucose and D-glucuronate • Papain is a cysteine protease breaking down the peptide bonds • Lipase catalyzes the hydrolysis of fat tissues at the position of glycerol backbone • Trypsin cleaves peptide chains mainly at the carboxyl side of the amino acids lysine or arginine, except when either is followed by proline • Various enzymes available for different types of tissue digestion • The digestion incubation condition can be optimized such as pressure, temperature, and duration

16.2.2.1.3 *Potential Issues*

The rotor-stator probe can clog when processing fibrous tissues. In addition, long sample processing time and heat produced by rotor-stator homogenization of fibrous tissues can be detrimental to thermally labile analytes. Cross-contamination of metal probes should be eliminated by thorough cleaning after each use.

16.2.2.2 Bead Beating

16.2.2.2.1 *Tissue Samples*

By definition, tissue homogenization via bead beating is achieved through high-speed shaking/agitating of a tissue sample against small hard beads inside a sealed sample vial. All types of tissues can be homogenized by bead beaters when the appropriate processing parameters including vial, bead size, and bead materials are considered. Bead beating is very simple, effective, and fast for homogenizing soft tissues. In practice, the sample is homogenized in a homogenization buffer. The presence of the buffer assists in producing a homogeneous suspension of tissue particles of uniform size that can be transferred for extraction prior to analysis.

16.2.2.2.2 *Bead Selection*

Several brands of bead beaters are commercially available (Table 16.3). The hard beads are made of different materials such as metal, glass, and ceramic. The selection of material and size of beads is based on the hardness and the histological properties of the tissues to be pulverized (Table 16.4).

16.2.2.2.3 *Vial Selection*

Vials of different sizes are manufactured to accommodate different sample sizes. Sample vials are generally available with 2-, 7-, 15-, and 30-ml sizes. Several companies (Precellys, Omni, and FastPrep) offer their own line of bead-beating homogenizer products for different applications. From a practical point of view, tissue samples should be collected and weighed directly into the preselected, pre-bead filled, and tared vials at the time of sample collection (Table 16.5).

16.2.2.2.4 *Parameter Selection and Programing*

The homogenization process can be programed by selecting the appropriate beating power (vibrating speed set as rpm), number of cycles, and duration as depicted in Table 16.6. Bead-beating products with cooling systems are, in general, available for homogenization temperature control.

16.2.2.2.5 *Dry Grinding Using Bead Beaters*

Tissues can be ground without addition of any buffer or diluent. The process is called dry grinding. This method can improve the results of wet grinding when applied to certain tough tissues, such as tail, cornea, artery, intestine, tumor, and adipose tissue. For better results, freezing samples prior to dry grinding is recommended. Frozen tissues can be directly ground without addition of buffer. In this case, reinforced sample tubes (refer to Table 16.5) are required due to the extra force required for the homogenization. Usually 2.8 mm ceramic beads

Table 16.4 Bead size and material selection for tissue interruption.

Beads	Size (diameter in mm)	Application
Ceramic	1.4 (Precellys)	For soft tissues like brain, liver, kidneys, mammalian cells, etc.
Ceramic	2.8 (Precellys)	For tough (hard) tissues like ear, lung, heart, tail, artery, etc.
Stainless steel	2.8 (Precellys)	For hard tissue like muscle, skin, bone, teeth, hair, etc.
Mix of glass and ceramic		For animal feces. Sample size from 20 to 200 mg in 2-ml vial
Mix of ceramic in different size	2.8 and 5.0 mm	For wet or dry grinding of tissues like hair, lung, heart, bones, etc. Sample size range: from 20 to 200 mg tissue
Garnet flakes mix with ceramic	0.7 and 2.8 mm	For elastic tissues like skin, cartilage, and retina. Sample size range: from 20 to 200 mg tissue

Note: 1. Information source from Precellys online website: www.precellys.com.
2. Custom design reinforced tubes available for dry grinding.

Table 16.5 Vial selection for beads beater tissue homogenization.

Vial type	Apparatus	Application
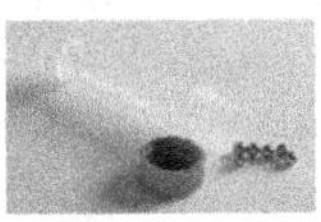 2-ml	Precellys 24 Fast-Prep 24 Omni Bead Ruptor	Suitable for weight from 20 to 200 mg of any types of tissues. Ideal for soft tissues
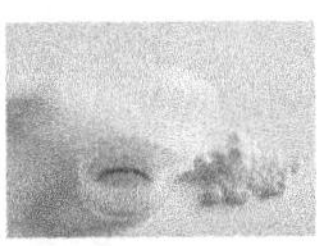 7-ml	Precellys 24-Dual Fast-Prep Omni Bead Ruptor	Suitable for any soft and tough tissues in a weight range of 200 mg to 2 g. Ideal for mouse and rat organs like whole brain, liver, lung, heart, kidneys, and spleen in tissue distribution study
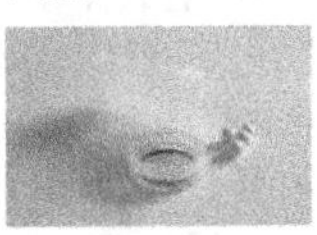 15-ml	Precellys	Suitable for any soft and tough tissues with a weight range of 400 mg to 4 g
 30-ml, 50-ml	Omni Bead Ruptor (30-ml) Fast-Prep (50-ml)	Suitable for soft tissues <15 g in weight
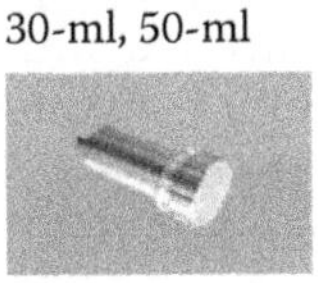 Metal tubes	Precellys Evolution	Suitable for hard tissues like bones, tooth, hair, and others

Note: Information source:
Precellys: www.precellys.com.
Omni: https://www.omni-inc.com.
FastPrep: www.mpbio.com.

Table 16.6 Common parameters for beads beater homogenization (examples).

Tissue	Mouse brain	Rat kidney	Rat heart	Rat lung	Muscle	Skin	Cartilage tumor
Weight	360 mg	1.5 g	1 g	1.4 g	200 mg	200 mg	50 mg
Buffer (μl)	720	3000	2000	2800	400	1000	1000
Vial Beads	2-ml Ceramic	7-ml Ceramic	7-ml Ceramic	7-ml Ceramic	2-ml Ceramic	2-ml Ceramic	7-ml Ceramic
Parameters							
Cycle	1	1	2	1	2	2	3
Speed (rpm)	5000	5500	6000	5000	6500	6800	6500
Duration (s)	25	30	23	25	30	30	15
Pause (s)	0	0	10	0	10	10	10
Dilution (*X*)	3	3	3	3	3	5	20

Note: Buffer is predetermined.
Source: Precellys: www.precellys.com.

in 2-ml reinforced tube is an ideal option. Other tissues like bones, hair, cartilage, and skin are very challenging for homogenization with use of a bead beater alone. Therefore, a combination of multiple techniques is necessary. One example is chemical/enzymatic digestion followed by bead-beating homogenization for difficult tissues. (Please refer to common protocols in Section 16.3.)

16.2.2.3 Ultrasonic Tissue Disruption

Tissue can be disrupted thoroughly when strong acoustic energy is directly focused at the tissue sample. The applied acoustic energy causes pressure disturbances and agitates tissue structure. The focused-ultrasonicators such as Covaris E-Series are versatile instruments engineered for tissue sample homogenization by Adaptive Focused Acoustics (AFA) that is delivered precisely and accurately to tissue samples at different volumes.

16.2.2.3.1 Tissue Samples

Any soft and tough tissues are suitable for acoustic disruption using AFA technology homogenization. However, for better results tissue samples need to be dry pulverized to powder form prior to the sonication. The dry pulverization increases the tissue surface area and breaks up the extracellular matrix to improve the extraction efficiency of the analyte.

16.2.2.3.2 Parameter Selection (Programing)

Covaris E220 series is capable of processing multi-samples with a wide range of sample types and volume. The E220 can be programed to process 1–96 samples in a single batch. With the available racks, the E220 instrument can be readily configured to use all AFA Tubes (volumes from 50 μl to 10 ml), including single AFA microTUBE™ and AFA 96 microTUBE Plate, for processing a single sample or a sample batch. The E220 can be operated as a stand-alone unit, or integrated as part of an automated laboratory system. Parameters setting up for each module may slightly be different and the guidance should be followed.

16.2.2.4 Chemical Digestion

Chemical digestion is used to break down fibrous tissues or hard tissue like hair, bones, and nails for which the outcome via mechanical homogenization is suboptimal. Common reagents include strong acids (HCl, HF, HNO_3, and H_2SO_4), bases (KOH and NaOH), and hydrogen peroxide. The selection of reagents for chemical digestion depends on the tissues type and the analyte stability under the digestion conditions (Henderson et al. 1995). In some cases, temperature and pressurization may be applied for the tissues that are difficult to digest. Bones and hair are the two types of tissues for which chemical digestion is employed for sample preparation. Bones can be digested in concentrated nitric acid and nitric–perchloric acid (AOAC 1990) at elevated temperature. Hair can be digested either by NaOH (Marchei et al. 2005; Favretto et al. 2006) or HCl (Girod and Staub 2000; Heimbuck and Bower 2002) under defined conditions.

16.2.2.5 Enzymatic Digestion

Enzymatic digestion has recently become common for the analysis of drug in tissues like hard skin, hair, and cartilage. A range of enzymes are available with different digestive properties and efficiencies that fit different types of tissue to be digested. Among the many enzymes, collagenase, trypsin, proteinase K, DNase, elastase, glucuronidase, aryl-sulfatase, hyaluronidase, papain, and lipase are the most common. The selection of enzyme depends on tissue type and histological properties (Yu et al. 2004).

16.3 Common Protocols

As discussed above, many techniques can be employed to homogenize a particular type of tissue (e.g. soft vs. fibrous vs. hard tissue). However, the selection of the sample preparation technique depends on what technique is available and, importantly, the stability of the analyte of interest under the intended condition. A survey of recent literature revealed that tissue homogenization protocols can be classified based on the type of tissues, i.e. soft tissues, fibrous tissues, and hard tissues. The following protocols are the most commonly used for the homogenization of soft, fibrous, and hard tissue types in a medium-to-high throughput setting.

16.3.1 For Soft Tissue

Soft tissue is the tissue type that can be easily homogenized because of the absence of dense irregular connective tissue fibers. Complete homogenization can be quickly achieved using mechanical force, most commonly by bead beating.

1) Follow vendor recommendations as captured in Tables 16.4 and 16.5 to select appropriate bead-beating tube volume and bead type.
2) Add pre-weighed tissue sample and homogenization buffer (typically three- to fivefold dilution) to the bead-containing tube.
3) Homogenize the samples using settings as recommended. Typically, one to three 30-second cycles are adequate for complete homogenization.
4) Allow sample foam to disperse, if necessary, prior to aliquoting.
5) Aliquot sample homogenate (e.g. 100 mg) to new vial using positive displacement pipette for subsequent extraction and analysis.

16.3.2 For Fibrous (Tough) Tissue

Fibrous tissue requires additional treatment in addition to mechanical homogenization due to the presence of dense irregular/regular connective tissue fibers that are resistant to mechanical shearing. The most common connective fibers in the extracellular matrix are consisted of the protein collagen. Collagenase, an enzyme responsible for collagen proteolysis, has been used most often to digest fibrous samples prior to mechanical homogenization via bead beating.

1) Following manufacturer recommendations as summarized in Tables 16.4 and 16.5 to select appropriate bead-beating tube volume and bead type.
2) Add pre-weighed tissue sample and digestion buffer (typically three- to fivefold dilution) to bead-containing tube. The digestion buffer contains 5–10 mg ml^{-1} collagenase in the appropriate collagenase buffer specified by vendor.
3) Incubate the tissue samples at 37 °C to allow for digestion of the extracellular matrix. Allow for an approximately two hours incubation period for heart, lung, intestine, and tumor. Skin tissue is especially fibrous and requires an overnight (>12 hours) incubation.
4) Upon completion of incubation, homogenize the samples by bead beating using the recommended settings. Typically, one to three 30-second cycles are adequate for complete homogenization.
5) Allow sample foam to disperse, if necessary, prior to aliquoting.
6) Aliquot sample homogenate to new vial using positive displacement pipette for subsequent extraction and analysis.

16.3.3 For Hard Tissue

Hard tissue like bone is physically hard in texture and difficult to release the analyte for extraction and LC-MS analysis. As described below, hard tissue needs to be broken down into powder or small particles, followed by acid digestion and solvent extraction.

1) Cryogenic Grinding: Place bone sample into liquid nitrogen until it is completely frozen. Use a hammer or mortar and pestle to pulverize the sample into small pieces in a selected container or vial. For consistent results, the commercially available Covaris cryoPREP™ system is a good choice for cryogenic grinding. The system can deliver a calibrated and repeatable mechanical force to cryofracture flash-frozen samples.
2) Direct Grinding: Using bead beating, hard tissue can be directly ground without freezing sample in liquid nitrogen. This saves time and cost. For example, Precellys-24 7-ml metal tube can be used for grinding up to six mammalian bones or teeth samples without any pretreatment. The procedure is simple and straightforward: (i) Place bone samples (about 200 mg per piece) into the 7-ml metal tube prefilled with 3 × 6.8 mm ceramic beads and (ii) setup the program at 7500 rpm with a parameter of 10 × 20 seconds (pause × duration) and agitate for 1 minute.
3) Digestion: Weigh out the pulverized sample into selected vial and add digestion reagent (strong acid or acid/solvent). Stabilizing agent may be added to the

vial prior to addition of digestion reagent if the analyte is liable. Neutralize the incubation buffer prior to aliquoting for analyte extraction.

16.4 Protocols for Special Tissue Sample Preparation

16.4.1 Hair

Bioanalysis of hair is often used for diagnostic purposes to identify the possible usage of illicit drugs (Gottardo et al. 2007), forensically toxicological drugs, and abused drugs (Gaulier et al. 2008). Drugs often measured in hair samples include benzodiazepines (Kim et al. 2011; Morini et al. 2012; Rust et al. 2012), antipsychotics (Patteet et al. 2015), psychoactive agents (Vincenti et al. 2013), steroid hormones (Gao et al. 2016), and alcohol (Crunelle et al. 2014). Many standardized analytical methods are available for the preparation of hair samples. Depending on the analyte to be analyzed, hair sample preparation normally involves the following steps: segmenting (cutting), washing, digestion (incubation) or homogenization, and extraction.

1) Segmenting and Washing: Hair is cut into 3–6 cm in length. To eliminate contamination, the cut hair samples are washed with aqueous (water, 0.3% Tween-20, buffer) and solvent (acetone, methanol, dichloromethane, ethyl acetate, isopropanol, etc.) once or multiple times (Nielsen et al. 2010; Nielsen and Johansen 2012). Hair samples are then dried under nitrogen flow or at room temperature. The selected washing solvent should effectively remove external contamination but without actively extracting analyte from hair.
2) Digestion: After cutting and washing, the hair samples are subjected to digestion in either strong alkaline or acid prior to extraction. Depending on the stability and physicochemical properties of the analyte, digestion solution can be selected from some very common digestion solutions, including NaOH (0.01–0.05 N) at 56 °C (Kintz 2007) and HCl (0.1–0.5 M) (Gottardo et al. 2007). The digestion duration and incubation temperature should be adjusted based on the property of the analyte. Higher temperatures will shorten the incubation duration. For example, digestion with NaOH for 40 minutes at 75 °C was employed for analysis of gamma-hydroxybutyrate (Bertol et al. 2012) while an overnight incubation with 0.01 N NaOH at 56 °C was conducted for the analysis of the same drug (Kintz 2007). Chemical digestion solution is neutralized with either base (if digested with acid) or acid (if digested with base) prior to extraction of the analyte. In addition to alkaline and acid digestion, phosphate buffer (pH 7.6 or 8.4) is also an option that can be used for the overnight digestion of hair samples for the analysis of benzodiazepines (Kintz 2007).
3) Homogenization: Other than chemical digestion, homogenization is an alternate approach to release the analyte from hair. Hair homogenization is mostly performed by ultrasonication (Müller et al. 2000; Thieme and Sachs 2007), and pulverization (Favretto et al. 2011, 2013; Miyaguchi and Inoue 2011; Chen et al. 2012; Shen et al. 2014) to yield high recovery and accurate results prior to analyte extraction when digestion is not performed.
4) Extraction: After digestion or homogenization, the analyte of interest must be extracted into liquid environment for LC/MS analysis. Liquid–liquid extraction (LLE) using organic solvent such as ethyl acetate, dichloromethane or dichloromethane/diethyl ether, dichloromethane/2-propanol/*n*-heptane, Toxi-Tubes A (Agilent Technology), or isopropanol is a popular choice. Methanol is often used as the extraction solvent for fair sample preparation due to its capability of dissolving various compounds. Online (Gao et al. 2013b; Quinete et al. 2015) and off-line (Müller et al. 2000) solid-phase extraction (SPE) has also been used for hair sample preparation.

16.4.2 Nail

The interest in quantitative analysis of drug in human nails has increased recently because the exposure assessment of drug in nails is believed to be more valuable than other conventional matrices in understanding certain forensic toxicity, drug abuse, or diseases (Shen et al. 2014; Patteet et al. 2015). For example, analysis of triclosan and triclocarban, the two antimicrobial agents commonly present in cosmetic and personal care products, in nails helped understand the relationship between exposure and certain diseases (Shi et al. 2013). Nail is a hard tissue which requires special treatment to release the analyte for analysis. Preparation of nail samples follows similar steps as discussed above for other hard tissues:

1) Wash/Decontamination: Nail samples must be decontaminated to remove loose dirt and target chemicals adhering to the nail surface prior to sample preparation. Water and acetone (Kim et al. 2010), detergents (Shi et al. 2013), or organic solvent like dichloromethane (Stepanov et al. 2006), ethyl acetate (Chen et al. 2014), ethanol (Irving and Dickson 2007), acetone (Pufal 2004), and methanol (Kim et al. 2010) were used for the decontamination. Acid solutions such as 0.1–2 mol l^{-1} of hydrochloric acid, nitric acid, and trifluoroacetic acid are also commonly used for nail decontamination.

2) Digestion: Nail digestion with either alkaline or acid is usually performed to ensure drugs or target chemicals are released from nail (Sukumar 2006; Madej 2010). Sodium hydroxide ($0.5–3.0\,mol\,l^{-1}$) is the most used alkaline and $0.1–2\,mol\,l^{-1}$ of hydrochloric acid, nitric acid, and trifluoroacetic acid are commonly used for nail digestion. Temperature selection varies from cabinet to 80 °C depending on the incubation duration and the property of analyte for analysis.
3) Homogenization: Upon decontamination, nail samples can be milled into powder by bead beating or pulverization if digestion is not a desired option.
4) Extraction: Analytes from neutralized nail digestion mixture or pulverized powder can be extracted by LLE or SPE. The method discussed for hair extraction can also be applied for nail samples in most cases. Other than LLE or SPE extraction, ultrasonication with methanol or other solvent has also been used as an efficient extraction method for certain drugs (Fisichella et al. 2014).

16.4.3 Skin

Skin samples are often collected and analyzed for dermatological drug development and therapeutic monitoring in the clinical trials (Sobue et al. 2004; Szekely-Klepser et al. 2005). Skin analysis is also used as a checkpoint in evaluating drug delivery for treating diabetes through skin administration (Getie and Neubert 2004). In some cases animal skin samples are collected and analyzed in addition to other major tissues and organs for investigating drug penetration and distribution in skin (Ebling et al. 1989; Cherlet et al. 2003) or establishment of PK/PD correlation (Neubert et al. 2012). Skin consists of keratinocytes (>90%) and melanocytes, which are fibrous in texture and resistant to common homogenization. Therefore, following mechanical homogenization, additional chemical/enzymatic digestion steps are necessary for optimal results.

1) Pretreatment: Fat tissue under skin is removed and surficial hair is sheared before sample preparation.
2) Multistep Processing: Due to the tough and fibrous nature of skin, multiple steps of processing, such as combination of mechanical homogenization and chemical/enzymatic digestion, are often employed to ensure the fibrous structure is completely disrupted and the analyte is released from the tissue. In one instance, skin samples were cut into small pieces and homogenized in Sorensen's phosphate buffer by rotor-stator homogenizer prior to enzymatic digestion (Ebling et al. 1989). Alternatively, skin can be first digested with acid, followed by mechanical homogenization (Cherlet et al. 2003). Solutions of 1 M HCl and 1–10 M NaOH are typically used for digestion. Chemical methods of digestions using 30% hydrogen peroxide mixed with 2 M ammonium hydroxide (Szekely-Klepser et al. 2005) or 20% TCA (Cherlet et al. 2003) have been reported.
3) Enzyme Digestion: Collagenase and other proteinase enzymes are normally used to digest skin samples (Ebling et al. 1989) when the analyte is not suitable to chemical digestion. For instance, 200 mg of skin piece is incubated with 2 mg of collagenase in 1 ml of Tris buffer at 37 °C for 90 minutes. Additional cycles of incubation can be performed with additional enzyme when needed.
4) Ultrasonication: Some analytes, especially peptide drugs, are sensitive to strong alkaline/acid as well as enzymatic digestion. Therefore, a practical alternative is to incubate skin samples with buffer or solvents (such as methanol) in an ultrasonication device (Covaris E-Seris 200) at a temperature suitable for the analyte of interest (Kassama et al. 2010). For better results it is suggested to pulverize the skin samples by cryofracture prior to homogenization.
5) Extraction: Once analyte is released into the semiliquid form of skin homogenate after digestion or ultrasonication, LLE with organic solvents such as pentane (Ebling et al. 1989), ethyl acetate (Sobue et al. 2004), or hexane (Croubels et al. 2005), SPE (Cherlet et al. 2003), or precipitation with methanol, acetonitrile, or ethanol (Getie and Neubert 2004) can be performed prior to LC/MS analysis.

16.4.4 Bone

Drug penetration into bone is often investigated for its therapeutic effect and toxicity. Analysis of drug exposure in bone requires extensive sample preparation due to the hardness and high density of bone. Typically, bone samples are cut into smaller pieces for further grinding which is followed by digestion in alkaline or acid solution prior to extraction for LC/MS analysis as described below.

1) Grinding: Break bones into small pieces using a hammer or mortar and pestle. Grinding results will be significantly improved if the small bone pieces are initially frozen in liquid nitrogen. For higher throughput, some devices are commercially available, such as Precellys-24 which is equipped with 7-ml metal tube bead beating or the SPEX Freezer/Mill, which is capable of grinding up to four samples at 50 g in weight without need of any pretreatment.
2) Digestion and Extraction: Bone samples are digested either with alkaline or acid solution after grinding. This is followed by LLE using organic solvent(s) or SPE if desired. Direct extraction with methanol without chemical digestion has also been reported when rotor-stator homogenization was used to make grinded bone piece to suspension (Ji et al. 2007, 2008).

16.5 Challenges Associated with Tissue Homogenization

16.5.1 Foaming

Foam is often generated during mechanical homogenization (rotor-stator and bead beating). It may affect the accuracy of transferring the homogenate prior to extraction. Tissue like lung has air in the organ and readily generates foam during homogenization. Therefore, defoaming is highly recommended prior to pipetting tissue homogenate for extraction. There are several ways to get rid of foam. Placing the homogenate on ice-bath until the foam has broken is a simple option but this adds additional time to the assay. The addition of antifoam agents (such as dimethyl polysiloxanes, alcohols, stearates, and glycols) into the sample vial prior to homogenization is another option. However, it is not recommended due to the potential matrix effects caused by the antifoam agents in LC/MS. A quick (<30 seconds) centrifugation of the homogenate will break down the foam but it is not recommended neither as centrifugation may cause heterogeneity of the tissue homogenate that leads to inaccurate and irreproducible sampling (Gao 2012).

16.5.2 Homogenization of Tissue for the Analysis of Labile Analyte

To ensure the stability of labile analyte in tissue, the collected sample should be snap-frozen by placing on dry ice or in liquid nitrogen upon collection. This pretreatment ensures the needed specimen integrity and reduces the risk of contamination. The labile analyte has to be stabilized during the homogenization process. Table 16.7 provides recommendations for stabilization of labile analyte during tissue processing.

Table 16.7 Homogenization of tissue for the analysis of labile analyte.

Liability	Recommendation
Thermal	Applying cooling during homogenization Cryogenic pulverizing in liquid nitrogen (−196 °C) Ultrasonication Chemical or enzymatic digestion
Enzymatic	Add agent to inhibit enzyme activity prior to homogenization
Chemical	Add buffer to maintain the appropriate pH of the sample matrix Acidify or basifying the tissue sample Add stabilizer to homogenization buffer (e.g. NaF/citric acid)

16.5.3 Analyte Recovery

The recovery of analyte of interest from tissue samples can vary from one technique to the other or from one type of tissue to the other. For example, the recovery of compound A is comparable in brain and liver but lower in heart and muscle when comparing polytron homogenization with FAF ultrasonication as shown in Figure 16.1 (Takach et al. 2004). Analyte recovery from enzyme digestion also showed higher results than that from polytron homogenization (Figure 16.2, Yu et al. 2004). The higher analyte recovery yield by means of ultrasonication and enzyme digestion result from a greater level of tissue and cell disruption compared to the mechanical process. To improve analyte recovery, assay accuracy, and data reliability, employment of multiple technologies is usually an effective application. Figure 16.3 shows the significant reduction of tissue particle size when enzymatic digestion followed by bead beating was compared to enzyme digestion alone (Liang et al. 2011). The tissue particle size was more homogenous when a longer enzymatic digestion was conducted prior to bead beating due to the more effective break down of the collagen in the tissue.

16.5.4 Protein Binding

Some compounds exhibit extremely high protein binding in the tissue samples which prevents the compound to be released from tissue during homogenization. For non-covalently bound small molecules, extraction of completely homogenized tissue aliquot in organic solvent will be sufficient to disrupt the protein binding and release the analyte. A more difficult situation can occur when a reactive analyte binds covalently to a protein. For example, ^{14}C-labeled taxane analogue milataxel and its metabolite binds to the xenograft mouse tumor and brain tissues after administration (Safarpour et al. 2009). This covalent binding between the administrated drug and target tissues make the isolation of analyte from tissue impossible by mechanical homogenization only. If the covalent transformation is known to occur, the homogenate can be treated with specific enzymes such as proteinase to release the analyte. Since chemical and enzyme digestion usually can break the connection between compound and tissue proteins, usually no further treatment is required.

16.5.5 Analyte Extraction

Analyte extraction from tissue homogenate faces additional challenges compared to the extraction from a liquid sample such as plasma and urine. Tissue homogenate contains large quantities of cellular membrane particles that may affect the extraction process and efficiency. The extraction techniques such as protein precipitation

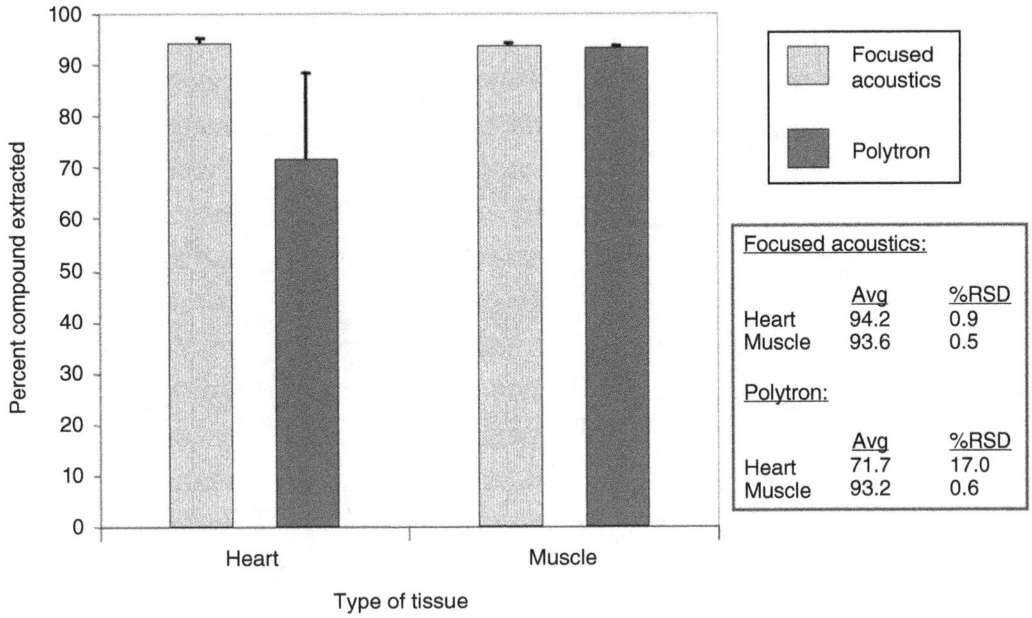

Figure 16.1 Response of test compound A in different tissues after FAF ultrasonication and polytron mechanical homogenization. *Source:* Reproduced with permission from Takach et al. (2004).

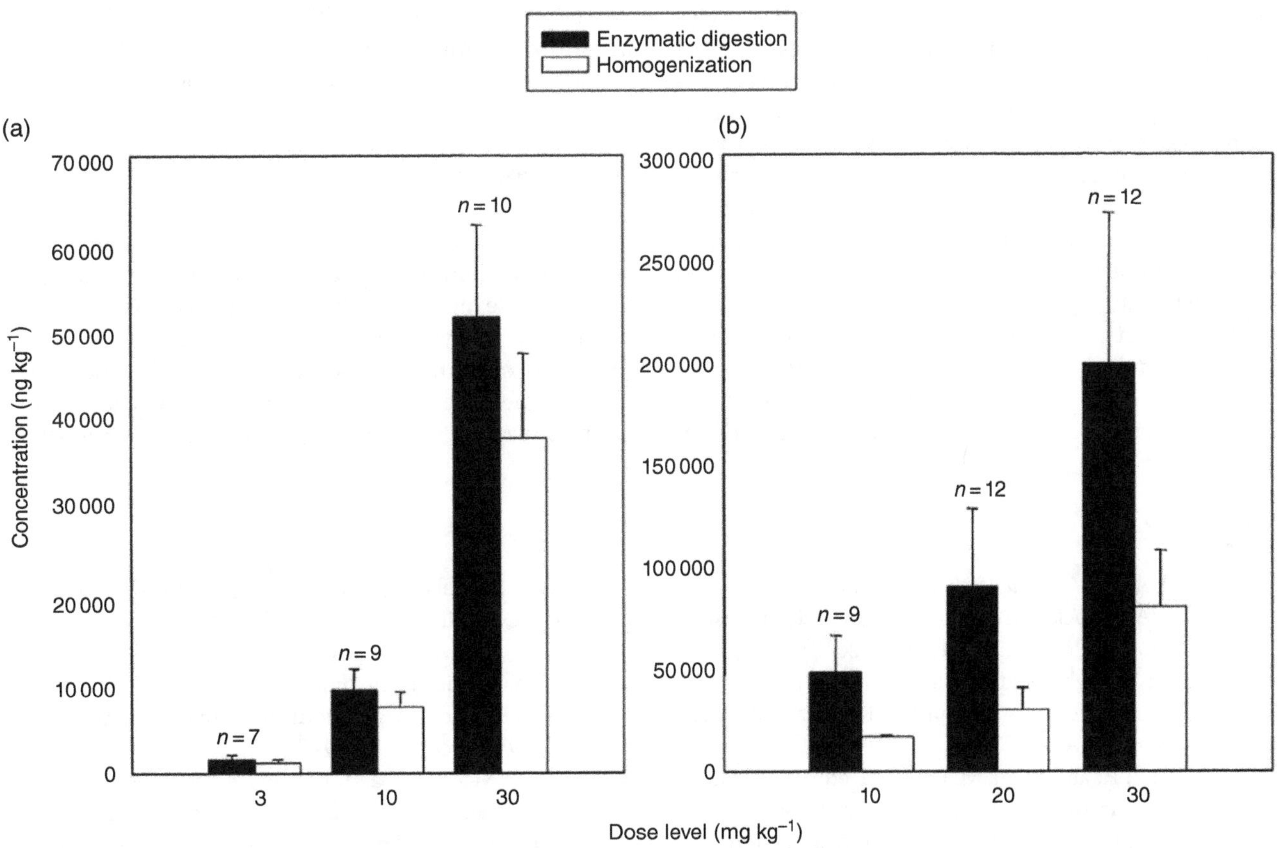

Figure 16.2 Average concentration of desipramine (a) and fluoxetine (b) in rat brain extracted by enzymatic digestion or polytron homogenization. *Source:* Yu et al. (2004). Reproduced with permission of American Chemical Society.

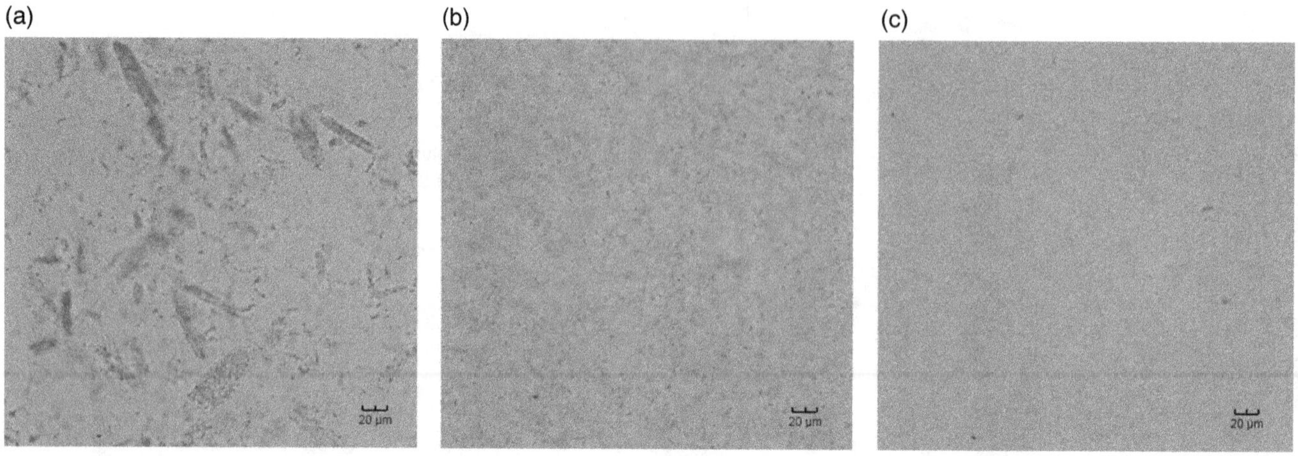

Figure 16.3 Microscopic pictures of homogenate solution of mouse heart tissue. Homogenization after (a) 24 hours incubation in collagenase without bead beater, (b) after 1 hour, and (c) 24 hours incubation in collagenase follow by 120 seconds bead-beater treatment. *Source:* Liang et al. (2011). Reproduced with permission of Future Sciences Ltd.

(PPT), SPE, and LLE used for liquid biological matrix also apply for tissue homogenate samples. However, caution may be needed during the process of each step. A heterogeneous mixture of tissue particles following a poor homogenization can trap analyte in large tissue particles if organic precipitation of the protein is used as an initial extraction step. If LLE is used, tight control over the homogenate pH is required to ensure the analyte remains in a neutral state of ionization for optimal recovery. When SPE is applied for extraction of tissue homogenate, large particles may block the cartridge bed and cause inconsistent extraction efficiency and recovery. If that is the case, multiple homogenization steps may be required to remove the large particles before applying SPE extraction (Korecka et al. 2010). The neutralization of strong acid/base chemical digestion homogenates to stop the reaction can result in salt formation that can suppress ionization during LC/MS analysis. Therefore, additional extraction steps may be required to remove the salts produced during the neutralization step. Some compounds may extensively bind to proteins in tissue and cause low recovery if non-denatured sample extract is analyzed directly (Spoto et al. 2006; Wang et al. 2008). The bound fraction of analyte should be released by protein denaturation prior to LC/MS bioanalysis.

16.6 Summary

Many methodologies are available for tissue sample homogenization prior to extraction for LC-MS analysis. Each of these methodologies has advantages over the other. In practice, the type of tissue, the required throughput, and the physicochemical properties of the analyte of interest have to be taken into consideration when selecting a homogenization method for quantitative LC-MS analysis.

References

AOAC (1990). *AOAC Official Methods of Analysis*, 15e, 84–85. Arlington, VA: Association of Official Analytical Chemists.

Bertol, E., Argo, A., Procaccianti, P. et al. (2012). Detection of gamma-hydroxybutyrate in hair: validation of GC-MS and LC-MS/MS methods and application to a real case. *J. Pharm. Biomed. Anal.* 70: 518–522.

Chen, H., Xiang, P., and Shen, M. (2014). Determination of clozapine in hair and nail: the role of keratinous biological materials in the identification of a bloated cadaver case. *J. Forensic Leg. Med.* 22: 62–67.

Chen, H., Xiang, P., Sun, Q.R., and Shen, M. (2012). Comparison of clozapine in nail and hair of psychiatric patients determined with LC-MS/MS. *Yao Xue Xue Bao* 47: 1193–1199.

Cherlet, M., Schelkens, M., Croubels, S., and De Backer, P. (2003). Quantitative multi-residue analysis of tetracyclines and their 4-epimers in pig tissues by high-performance liquid chromatography combined with positive-ion electrospray ionization mass spectrometry. *Anal. Chim. Acta* 492: 199–213.

Croubels, S., Maes, A., Baert, K., and De Backer, P. (2005). Quantitative determination of salicylic acid and metabolites in animal tissues by liquid chromatography–tandem mass spectrometry. *Anal. Chim. Acta* 529: 179–187.

Crunelle, C., Yegles, M., van Nuijs, A.L.N. et al. (2014). Hair ethyl glucuronide levels as a marker for alcohol use and abuse: a review of the current state of the art. *Drug Alcohol Depend.* 134: 1–11.

Ebling, W., Milis-Williams, L., Harapat, S.R., and Stanski, D. (1989). High-performance liquid chromatographic method for determining thiopental concentrations in twelve rat tissues: application of physiologic modeling of distribution of barbiturate. *J. Chromatogr. Biomed. Appl.* 490: 339–353.

Favretto, D., Frison, G., Vogliardi, S., and Ferrara, S.D. (2006). Potentials of ion trap collisional spectrometry for liquid chromatography/electrospray ionization tandem mass spectrometry determination of buprenorphine and nor-buprenorphine in urine, blood and hair samples. *Rapid Commun. Mass Spectrom.* 20: 1257–1265.

Favretto, D., Stocchero, G., Nalesso, A. et al. (2013). Monitoring haloperidol exposure in body fluids and hair of children by liquid chromatography–high-resolution mass spectrometry. *Ther. Drug Monit.* 35: 493–501.

Favretto, D., Vogliardi, S., Stocchero, G. et al. (2011). High performance liquid chromatography–high resolution mass spectrometry and micropulverized extraction for the quantification of amphetamines, cocaine, opioids, benzodiazepines, antidepressants and hallucinogens in 2.5 mg hair samples. *J. Chromatogr. A* 1218: 6583–6595.

Fisichella, M., Morini, L., Sempio, C., and Groppi, A. (2014). Validation of a multi-analyte LC-MS/MS method for screening and quantification of 87 psychoactive drugs and their metabolites in hair. *Anal. Bioanal. Chem.* 406: 3497–3506.

Gao, H. (2012). Tissue sample LC-MS/MS bioanalysis: method development and considerations. Presentation at 13th annual Land O'Lakes bioanalytical conference in Madison, WI (16–20 July 2012).

Gao, H., Ho, S., and Williams, J. (2013a). LC-MS bioanalysis of drugs in tissue samples. In: *Handbook of LC-MS Bioanalysis, Best Practice, Experimental Protocols, and Regulations* (ed. W. Li, J. Zhang and F. Tse), 297–306. Hoboken, NJ: Wiley.

Gao, W., Kirschbaum, C., Grass, J., and Stalder, T. (2016). LC-MS based analysis of endogenous steroid hormones in human hair. *J. Steroid Biochem. Mol. Biol.* 162: 92–99.

Gao, W., Stalder, T., Foley, P. et al. (2013b). Quantitative analysis of steroid hormones in human hair using a column-switching LCAPCI-MS/MS assay. *J. Chromatogr. B Analyt. Technol. Biomed. Life Sci.* 928: 1–8.

Gaulier, J.M., Sauvage, F.L., Pauthier, H. et al. (2008). Identification of acepromazine in hair: an illustration of the difficulties encountered in investigating drug-facilitated crimes. *J. Forensic Sci.* 53: 755–759.

Getie, M. and Neubert, R. (2004). LC-MS determination of desmopressin acetate in human skin samples. *J. Pharm. Biomed. Anal.* 35: 921–927.

Girod, C. and Staub, C. (2000). Analysis of drugs of abuse in hair by automated solid-phase extraction, GC/EI/MS and GC ion trap/CI/MS. *Forensic Sci. Int.* 107 (1–3): 261–271.

Gottardo, R., Bortolotti, F., De Paoli, G. et al. (2007). Hair analysis for illicit drugs by using capillary zone electrophoresis-electrospray ionization-ion trap mass spectrometry. *J. Chromatogr. A* 1159: 185–189.

Heimbuck, C.A. and Bower, N.W. (2002). Teaching experimental design using a GC-MS analysis of cocaine on money: a cross-disciplinary laboratory. *J. Chem. Educ.* 79 (10): 1254.

Henderson, G.L., Harkey, M.R., and Jones, R.T. (1995). Analysis of hair for cocaine. In: *Hair Testing for Drugs of Abuse: International Research on Standards and Technology*, 91–120, NIH Publication No. 95-3727. Rockville, MD: U.S. Department of Health and Human Services, Public Health Service, National Institutes of Health, National Institute on Drug Abuse.

Irving, R.C. and Dickson, S.J. (2007). The detection of sedatives in hair and nail samples using tandem LC-MS-MS. *Forensic Sci. Int.* 166: 58–67.

Ji, A.J., Saunders, J.P., Amorusi, P. et al. (2008). A sensitive human bone assay for quantitation of tigecycline using LC/MS/MS. *J. Pharm. Biomed. Anal.* 48: 866–875.

Ji, A.J., Saunders, J.P., Wadgaonkar, N.D. et al. (2007). A novel antibiotic bone assay by liquid chromatography/tandem mass spectrometry for quantitation of tigecycline in rat bone. *J. Pharm. Biomed. Anal.* 44: 970–979.

Kassama, Y., Xu, Y., Dunn, W.B. et al. (2010). Assessment of adaptive focused acoustics versus manual vortex/freeze–thaw for intracellular metabolite extraction from *Streptomyces lividans* producing recombinant proteins using GC-MS and multi-block principal component analysis. *Analyst* 135 (5): 934–942.

Kemper, C., Koetzner, L., and Koleto, M. (2012). A primer for best practices in tissue preparation for bioanalysis. *Bioanalysis* 4 (21): 2621–2636.

Kim, J., Lee, S., In, S. et al. (2011). Validation of a simultaneous analytical method for the detection of 27 benzodiazepines and metabolites and zolpidem in hair using LC-MS/MS and its application to human and rat hair. *J. Chromatogr. B* 879: 878–886.

Kim, J.Y., Shin, S.H., and In, M.K. (2010). Determination of amphetamine-type stimulants, ketamine and metabolites in fingernails by gas chromatography-mass spectrometry. *Forensic Sci. Int.* 194: 108–114.

Kintz, P. (2007). Bioanalytical procedures for detection of chemical agents in hair in the case of drug-facilitated crimes. *Anal. Bioanal. Chem.* 388: 1467–1474.

Korecka, M., Clark, C.M., Lee, V.M.Y. et al. (2010). Simultaneous HPLCMS-MS quantification of 8-iso-PGF (2 alpha) and 8,12-iso-iPF(2 alpha) in CSF and brain tissue samples with on-line cleanup. *J. Chromatogr. B Analyt. Technol. Biomed. Life Sci.* 878 (24): 2209–2216.

Kwon, Y. (2002). *Handbook of Essential Pharmacokinetics, Pharmacodynamics, and Drug Metabolism for Industrial Scientists*. New York: Springer.

Liang, X., Ubhayakar, S., Liederer, B.M. et al. (2011). Evaluation of homogenization techniques for the preparation of mouse tissue samples to support drug discovery. *Bioanalysis* 3 (17): 1923–1933.

Madej, K.A. (2010). Analysis of meconium, nails and tears for determination of medicines and drugs of abuse. *Trends Anal. Chem.* 29: 246–259.

Marchei, E., Durgbanshi, A., Rossi, S. et al. (2005). Determination of arecoline (areca nut alkaloid) and nicotine in hair by high-performance liquid chromatography quadrupole mass spectrometry. *Rapid Commun. Mass Spectrom.* 19: 3416–3418.

Miyaguchi, H. and Inoue, H. (2011). Determination of amphetamine-type stimulants, cocaine and ketamine in human hair by liquid chromatography/linear ion-trap-Orbitrap hybrid mass spectrometry. *Analyst* 136: 3503–3511.

Morini, L., Vignali, C., Polla, M. et al. (2012). Comparison of extraction procedures for benzodiazepines determination in hair by LC-MS/MS. *Forensic Sci. Int.* 218: 53–56.

Müller, C., Vogt, S., Goerke, R. et al. (2000). Identification of selected psychopharmaceuticals and their metabolites in hair by LC/ESI-CID and LC/MS/MS. *Forensic Sci. Int.* 113: 415–421.

Neubert, H., Fountain, S., King, L. et al. (2012). Tissue bioanalysis of biotherapeutics and drug targets to support PK/PD. *Bioanalysis* 4 (21): 2589–2604.

Nielsen, M.K. and Johansen, S.S. (2012). Simultaneous determination of 25 common pharmaceuticals in whole blood using ultra-performance liquid chromatography–tandem mass spectrometry. *J. Anal. Toxicol.* 36: 497–506.

Nielsen, M.K., Johansen, S.S., Dalsgaard, P.W., and Linnet, K. (2010). Simultaneous screening and quantification of 52 common pharmaceuticals and drugs of abuse in hair using UPLC–TOF-MS. *Forensic Sci. Int.* 196: 85–92.

Patteet, L., Cappelle, D., Maudens, K.E. et al. (2015). Advances in detection of antipsychotics in biological matrices. *Clin. Chim. Acta* 441: 11–22.

Pufal, E. (2004). Research on the determination of xenobiotics in epidermal products and their usefulness in forensic toxicology. *Probl. Forensic Sci.* 59: 38–49.

Quinete, N., Bertram, J., Reska, M. et al. (2015). Highly selective and automated online SPE LC-MS3 method for determination of cortisol and cortisone in human hair as biomarker for stress related diseases. *Talanta* 134: 310–316.

Rust, K.Y., Baumgartner, M.R., Meggiolaro, N., and Kraemer, T. (2012). Detection and validated quantification of 21 benzodiazepines and 3 "z-drugs" in human hair by LC-MS/MS. *Forensic Sci. Int.* 215: 64–72.

Safarpour, H., Connolly, P., Tong, X. et al. (2009). Overcoming extractability hurdles of a 14C labeled taxane analogue milataxel and its metabolite from xenograft mouse tumor and brain tissues. *J. Pharm. Biomed. Anal.* 49: 774–779.

Shen, M., Chen, H., and Xiang, P. (2014). Determination of opiates in human fingernail-comparison to hair. *J. Chromatogr. B* 967: 84–89.

Shi, Y., Liu, X., Zhang, J., and Shao, B. (2013). Analysis of triclosan and triclocarban in human nails using isotopic dilution liquid chromatography–tandem mass spectrometry. *J. Chromatogr. B* 934: 97–101.

Sobue, S., Sekiguchi, K., and Nabeshima, T. (2004). Intracutaneous distributions of fluconazole, itraconazole, and griseofulvin in guinea pigs and binding to human stratum corneum. *Antimicrob. Agents Chemother.* 48: 216–223.

Spoto, B., Fezza, F., Parlongo, G. et al. (2006). Human adipose tissue binds and metabolizes the endocannabinoids anandamide and 2-arachidonoylglycerol. *Biochimie* 88 (12): 1889–1897.

Stepanov, I., Feuer, R., Jensen, J. et al. (2006). Mass spectrometric quantitation of nicotine, cotinine, and 4-(methylnitrosamino)-1-(3-pyridyl)-1-butanol in human toenails. *Cancer Epidemiol. Biomarkers Prev.* 15: 2378.

Sukumar, A. (2006). Human nails as a biomarker of element exposure. *Rev. Environ. Contam. Toxicol.* 185: 141–177.

Szekely-Klepser, G., Wade, K., Woolson, D. et al. (2005). A validated LC/MS/MS method for the quantification of pyrrole-2,3,5-tricarboxylic acid (PTCA), a eumelanin specific biomarker, in human skin punch biopsies. *J. Chromatogr. B* 826 (1–2): 31–40.

Takach, E.J., Zhu, Q., Yu, S. et al. (2004). New technology in tissue homogenization: using focused acoustic energy to improve extraction efficiency of drug compounds prior to LC/MS/MS analysis. 52nd ASMS conference on mass spectrometry, Nashville, TN (26 May 2004). Poster: WPJ 143.

Thieme, D. and Sachs, H. (2007). Examination of a long-term clozapine administration by high resolution segmental hair analysis. *Forensic Sci. Int.* 166: 110–114.

Vincenti, M., Salomone, A., Gerace, E., and Pirro, V. (2013). Role of LC-MS/MS in hair testing for the determination of common drugs of abuse and other psychoactive drugs. *Bioanalysis* 5 (15): 1919–1938.

Wang, C., Wang, S., Chen, Q., and He, L. (2008). A capillary gas chromatography-selected ion monitoring mass spectrometry method for the analysis of atractylenolide I in rat plasma and tissues, and application in a pharmacokinetic study. *J. Chromatogr. B Analyt. Technol. Biomed. Life Sci.* 863 (2): 215–222.

William, T.K. (1972). *Biological Science*, 2e, 85. New York: W.W. Norton & Company Inc.

Xue, Y.J., Gao, H., Ji, Q.C. et al. (2012). Bioanalysis of drug in tissue: current status and challenges. *Bioanalysis* 4 (21): 2637–2653.

Yu, C., Penn, L.D., Hollembaek, J. et al. (2004). Enzymatic tissue digestion as an alternative sample preparation approach for quantitative analysis using liquid chromatography-tandem mass spectrometry. *Anal. Chem.* 76 (6): 1761–1767.

17

Sample Preparation for LC-MS Bioanalysis of Peripheral Blood Mononuclear Cells

Yunlin Fu and Wenkui Li

17.1 Introduction

Generally, extracellular (plasma and serum) concentrations of a drug and/or its metabolite(s) link well with its pharmacodynamics (PD) or toxicodynamics (TD). However, in some cases, considerable disconnection exists between extracellular concentrations of a drug and its therapeutic efficacy. In such a case, a drug response might be better linked to intracellular, particularly peripheral blood mononuclear cells (PBMCs), rather than extracellular concentrations. Evidence has shown that accumulation of a drug and/or its metabolite(s) intracellularly might contribute to the persistence of its efficacy, despite the drug and/or its metabolite might have been removed from the extracellular medium. For many therapies, failure to maintain adequate intracellular drug concentrations may result in the development of resistance, leading to a decreased probability of response to the current and subsequent therapies (Bazzoli et al. 2010; D'Avolio et al. 2014).

In addition, each individual subject may respond differently to the same drug at the same dose levels. Often, there is a need to individualize therapeutic dosages for a maximized efficacy but minimized toxicity of the treatment. Therefore, in drug development and post-approval therapeutic drug monitoring, in addition to measuring extracellular exposure, obtaining information on intracellular concentrations of the drugs and/or their metabolites can be very important to understand intracellular exposure associated therapeutic failure or toxicity and/or to improve therapeutic efficacy with a reduced toxicity (Bazzoli et al. 2010; D'Avolio et al. 2014).

LC-MS/MS has been widely used for the bioanalysis of intracellular drugs and/or metabolites in support of TK/TD and PK/PD studies in drug development and therapeutic drug monitoring (Cohen et al. 2013; Zhang and Bartels 2013). In the current chapter, a focused effort is on sample preparation workflow for quantitative LC-MS/MS analysis of drugs and/or their metabolite(s) in PBMCs.

17.2 Peripheral Blood Mononuclear Cells (PBMCs)

PBMCs are blood cells with a round-shaped nucleus, including monocytes and lymphocytes, with the lymphocyte population comprised of T cells, B cells, and NK (natural killer) cells. PBMCs make up about one-third of the white blood cells (WBCs), also known as leukocytes, which are essential components in our immune system, and play an integral role in the body's defense such as anti-infection and anti-cancer activities.

PBMCs along with other WBCs are a major source of interest among researchers seeking better therapies for autoimmune disorders and cancer for drug discovery and development and post-approval therapeutic drug monitoring (Zhang and Bartels 2013).

17.3 Sample Preparation Workflow for LC-MS Bioanalysis of PBMC Samples

17.3.1 Cell Separation

Separation of PBMCs from whole blood is considered an important step for the subsequent LC-MS/MS bioanalysis. An ineffective isolation of the cells can significantly affect the bioanalytical results, leading to unreliable data interpretation and subsequent wrong adjustment of dose regimens for the intended therapy. PBMCs can be isolated from whole blood using density gradient centrifugation technique with Ficoll®-Pague®

Sample Preparation in LC-MS Bioanalysis, First Edition. Edited by Wenkui Li, Wenying Jian, and Yunlin Fu.

at a density of 1.0778 g ml^{-1} or equivalent. The whole blood can be treated with heparin, ETDA, or other anticoagulants.

17.3.1.1 Ficoll®/Histopaque® Method

This method was first introduced in 1968. The two most commonly used density gradient media are Ficoll®-Paque® PLUS (FICOLL™ 400 and sodium metrizoate or sodium diatrizoate solution) from GE Healthcare Life Sciences and Histopaque®-1077 from Sigma Aldrich (Farowski et al. 2013). The principle of the method is based on the differential migration of blood cells through the polymer media during the centrifugation process, resulting in separation of different cell types into different layers.

Briefly, either anticoagulant-treated or defibrinated blood samples are diluted with/without a buffered solution (e.g. 1 : 1 dilution) and then carefully placed on top of the Ficoll®/Histopaque® medium in the centrifugation tubes. This is followed by centrifugation, during which blood samples pass through the Ficoll®/Histopaque® medium causing its fractioning into separate layers. The bottom layer contains particles with density higher than the medium. This layer is mostly made up of Ficoll®/Histopaque®-aggregated red blood cells, which sink completely through to the bottom. Immediately above this layer is the diffuse layer that contains mostly granulocytes and unbound Ficoll®/Histopaque® medium. Due to the density of PBMCs being slightly lower than that of the Ficoll®/Histopaque® medium, they sediment at the interface between the Ficoll®/Histopaque® medium and uppermost plasma/platelet layer (Figure 17.1). The cells are then harvested by carefully pipetting them from the liquid interface and washed with a buffer to remove platelets, Ficoll®/Histopaque® medium residue, and plasma, then centrifuged again. Cell isolates are then ready for the other steps prior to LC-MS analysis.

Isolation of PBMCs using Ficoll®/Histopaque® medium is attractive for cost perspectives. However, in practice, the technique is relatively laborious and requires meticulous manual handling to avoid contamination. Furthermore, the overall cell recovery is low (Grievink et al. 2016).

Over the years, alternative gradient isolation methods have been developed aiming to simplify the procedure in terms of time and sample handling to avoid contamination with other cell types.

17.3.1.2 Cell Preparation Tube (CPT) Method

The CPT™ tubes are commercially available. The tubes contain a sodium heparin (or sodium citrate) anticoagulant with a FICOLL®/Hypaque® density medium and a thixotropic polyester gel barrier which separates the layers. This single-tube system for the collection of whole blood and separation of PBMCs reduces the risk of contamination and eliminates the need for additional tubes, pipettes, and reagents. The tube's internal vacuum allows for blood to be easily drawn during venipuncture without dilution. The blood separation media takes advantage of the lower density of PBMCs to isolate them from whole blood. The isolation occurs during centrifugation when the gel portion of the media moves to form a stable barrier under the mononuclear cells and platelets; this separates them from the denser blood components, which sediment in the bottom of the tube (Figure 17.2). The latter washing and centrifugation steps further reduce the quantity of platelets present, resulting in a suspension of concentrated mononuclear cells.

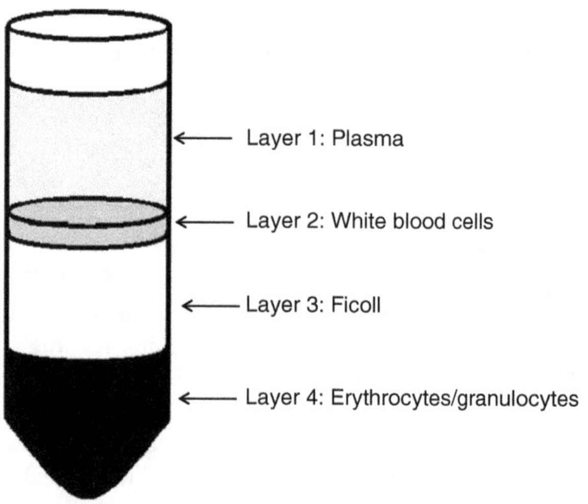

Figure 17.1 Illustration of blood separation for peripheral blood mononuclear cells (PBMCs) using the Ficoll®/Histopaque® method.

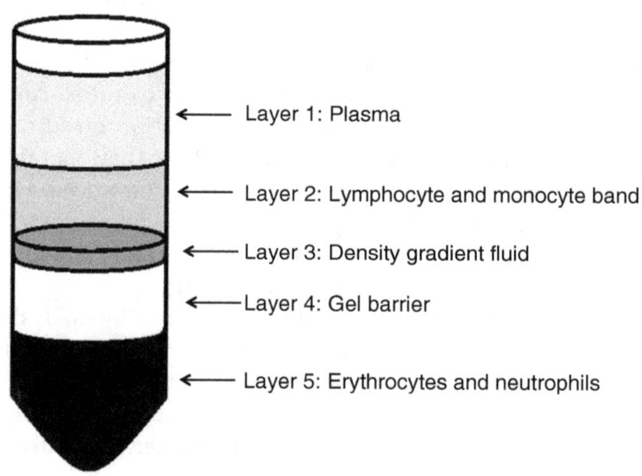

Figure 17.2 Illustration of blood separation for peripheral blood mononuclear cells (PBMCs) using the CPT method.

For a better separation of the cells using the CPT method, centrifugation needs to take place between 20 and 25 °C (Pèlerin et al. 2005) with a short centrifugation time (e.g. 1800 × *g*, 30 minutes with brake). In contrast, with the Ficoll gradient method, separation of cells can occur at 4 °C, but with a slightly longer centrifugation time (e.g. 400 × *g*, 40 minutes with no brake).

Becher et al. (2002) compared the Ficoll and CPT methods by measuring phosphorylated anabolites (stavudine triphosphate and lamivudine triphosphate) in PBMCs, and identical results were obtained. However, compared to the Ficoll method, more erythrocyte contamination was found for the CPT method; although the overall cell recovery via the CPT method is ~70% higher (Grievink et al. 2016).

17.3.1.3 Factors That Need to Be Considered in Cell Separation

In general, the choice of the method for PBMC isolation and the number of washes must be performed on the basis of the analyte of interest as each molecule possesses different physicochemical properties from others. A PBMC separation method for an intended drug molecule must be thoroughly evaluated and/or validated (Cohen et al. 2013; Ter Heine et al. 2010; Zhang and Bartels 2013). In addition to the general precautions mentioned above, some specific precautions should be taken into consideration:

- **Stability**: Enzyme activity should be considered while selecting the separation method. For example, blood sample for the quantification of zidovudine were collected into EDTA and diluted three times before Ficoll®-Hypaque® gradient centrifugation by Durand-Gasselin et al. (2008). The authors indicated that samples should be immediately taken to the laboratory (within five minutes) and all procedures should be performed at 4 °C to minimize enzymatic activity. In this regard, use of CPT is easier, less time consuming, and therefore quicker, which is important for the measurement of the nucleoside's intracellular metabolite(s), the mono-, di-, and triphosphate(s). Those metabolites are generally unstable extracellularly and their stability must be assessed in blood to decide for how long the collected blood sample can be stored in the cell separation tubes before isolation of PBMCs (Pruvost et al. 2001).
- **Contamination**: The pink or red color of the harvested PBMCs is an indication of erythrocyte contamination, which often leads to larger variation in matrix effect during LC-MS/MS. It was reported that the matrix factor can shift from around 0.8 for the PBMCs with regular color to around 0.34 for the PBMCs with reddish color (Shi et al. 2002). To prevent contamination, prior to washing steps, an extra erythrocyte lysis procedure should be applied (Zhang and Bartels 2013). For example, to ensure that PBMC fraction was free of red blood cells, Farowski et al. (2010a, 2010b) employed a hypotonic lysis with phosphate buffer (pH 7.4) just before the PBS washing steps.
- **Passive diffusion vs. efflux of analyte(s) of interest from cells.** While passive diffusion and active transport contribute to xenobiotics' passing from the systemic circulation to the extravascular compartment, membrane transporters (efflux and influx) have an important role in mediating drug absorption and disposition. Taking the above into consideration, the dilution step in the Ficoll®-Hypaque® density gradient method should be theoretically avoided as it may cause disequilibrium of analyte concentration, leading to passive diffusion (Cascorbi 2006; Hartkoorn et al. 2007; Ho and Kim 2005). In this regard, the CPT method appears to be a better fit with less passive diffusion. In addition, the time between blood sample collection and the cell isolation and extraction procedure should be as short as possible to minimize the possible impact due to active drug efflux/influx (Almond et al. 2005; Colombo et al. 2005).

17.3.2 Cell Counting and Recovery

With a given concentration of the analyte of interest inside a given cell, the number of the cells employed in the LC-MS bioanalysis dictates the total intracellular concentrations measured for a given PBMC sample, which is generally expressed as the amount of analyte per 10^6 cells. Therefore, a critical step in the LC-MS bioanalysis of PBMC sample is the determination of the number of cells from which the analyte is to be quantified.

In most cases, cell numbers are determined on a small aliquot using a Coulter counter instrument, a Malassez cell counter, or other devices (Derissen et al. 2016). In many cases, cell viability needs to be examined using the trypan blue method. In such a test, a vital stain (a diazo dye) is used to selectively color dead cells blue while the live cells with intact cell membranes are not colored (Baran et al. 2011). However, this procedure may not be sufficiently accurate or precise, particularly when multiple sites are involved in a single study. In this regard, a biochemical test has been developed for cell counting. The method is based on the linear relationship between the PBMC count and a fluorescence signal generated during detection of DNA in the cell pellet (Benech et al. 2004; Durand-Gasselin et al. 2008). This test could be performed in analytical laboratories where there is no Coulter counter available.

17.3.3 Cellular Volume for Concentration Conversion

To evaluate the intracellular concentration/plasma concentration ratio, it is necessary to convert the amount of drug per 10^6 cells to the amount of drug per volume unit of cells (Derissen et al. 2016). In the case of red blood cells, the measurement of mean cell volume (MCV) is carried out by multiplying a volume of blood by the hematocrit of the blood, and then dividing that product by the number of red blood cells in that volume. Normally, MCV is expressed in picoliters (pl or 10^{-12} l) or femtoliters (fl or 10^{-15} l).

For PBMCs, an arbitrary MCV of 400 fl (or 0.400 pl) or 283 fl (0.283 pl) has been frequently used to convert the measured results (Almond et al. 2005; Colombo et al. 2005; Khoo et al. 2002; Ter Heine et al. 2010). Because of the arbitrary nature of the PBMC volume used for the conversion, the reported intracellular drug concentrations could be potentially biased due to the inter-subject variability in the MCV of the PBMCs, which vary according to the state of the cells (quiescent or stimulated).

17.3.4 Cell Lysis and Extraction

After cells are isolated, a critical step is cell lysis to recover/extract the drug and/or drug metabolites from the cells. Methods for cell lysis and extraction include use of organic solvent or organic acid with/without lysis instrumentation (such as sonicator, dismember device, etc.) (Colombo et al. 2005; Derissen et al. 2016; Ehrhardt et al. 2007; Farowski et al. 2010a; Huang et al. 2004; Jemal et al. 2003). For example, to quantify the cyclosporine level in PBMCs, the cells were lysed with organic solvent (MeOH) containing internal standard (Ansermot et al. 2007); tris-HCl/MeOH lysis was employed to extract delavirdine from PBMCs (Pèlerin et al. 2005); sonication was employed in extraction of atazanavir, an human immunodeficiency virus (HIV) protease inhibitor, in PBMCs (Jemal et al. 2003).

Stability of the analyte of interest during lysis and extraction must be carefully thought through. This is particularly important when analyzing (mono-, di-, and tri-) phosphate metabolites (Cohen et al. 2013). Generally, the extraction of these molecules is carried out at low temperature with a minimized duration and, after extraction, the sample mixture is immediately stored at low temperature prior to LC-MS analysis. For example, for the measurement of ribavirin and ribavirin-mono, di-, and tri-phosphates in PBMCs, the cells were lysed with 500 µl of cold 70 : 30 MeOH:water and stored at −80 °C prior to LC-MS analysis (Jimmerson et al. 2015); For the measurement of fludarabine triphosphate in PBMCs, the cells were lysed with 1 ml of solution containing 70% MeOH and 1.2 mM Tris-Buffer (pH 7.2) followed by vortex-mixing and immediately stored at −20 °C prior to LC-MS analysis (Huang et al. 2013). When using strong acids, such as perchloric acid, for sample treatment, attention must be given to the possible impact of the acid on the stability of the compounds of interest in the sample mixture. Because of this concern, sample treatment using a mixture of organic solvent (proportion from 60 to 80% of ACN or MeOH) and water is favorable.

In method development, the analyte of interest can be spiked at concentrations equivalent to the QC low, mid, and high levels using blank PBMCs (4×10^6 cells). Lysis solvent such as pure MeOH (100%), 75% aqueous MeOH (i.e. MeOH/H_2O 75/25), 50% aqueous MeOH or pure MeOH containing diethylamine (0.1%, 0.25%), or HCl 0.01 N can be evaluated for the extraction recovery of the analyte of interest by comparing LC-MS peak area values or peak response ratio (against internal standard). The solvent that is associated with the best recovery overall across all three concentration levels should be chosen for method validation and subsequent study sample analysis (Colombo et al. 2005).

Depending on the method of lysis used and the nature of the analyte of interest, the lysed sample extract might be directly subjected to LC-MS analysis without any additional treatment (Huang et al. 2013). However, most of the sample extracts usually cannot be assayed directly without a proper pretreatment, such as evaporation of the extraction solvent followed by reconstitution (Pèlerin et al. 2005), neutralization of the acidic sample extract using a basic solution (Cohen et al. 2013), or further processing to remove endogenous compounds prior to LC-MS analysis. For the latter, solid-phase extraction (Huang et al. 2004; Jemal et al. 2003) and liquid–liquid extraction (Ehrhardt et al. 2007) are the common tools.

17.4 Representative Protocols

17.4.1 Ficoll®-Paque® Plus Density Gradient Separation Method

General procedures for PBMC isolation are available from manufacturers for using Ficoll®-Paque® (https://www.sigmaaldrich.com/content/dam/sigma-aldrich/docs/Sigma-Aldrich/General_Information/1/ge-isolation-of-mononuclear-cells.pdf) or Histopaque®-1077 (https://www.sigmaaldrich.com/content/dam/sigma-aldrich/docs/Sigma/Product_Information_Sheet/1/10771pis.pdf). Modification to the baseline procedure can be made to fit the purpose of the intended study. A representative protocol was taken from Derissen et al. (2015) for the quantitative determination of the intracellular 5-fluorouracil nucleotides responsible for the anticancer effect of 5-fluorouracil. Clinical samples were obtained by drawing 16 ml whole blood in lithium heparin tubes. Samples were processed immediately after blood collection, and processing was executed on ice.

- After centrifugation (10 minutes at 1500 × *g* at 4 °C) of the blood, the buffy coat (the fraction of an anticoagulated blood sample that contains most of the WBCs and platelets) was collected and resuspended in cold phosphate-buffered saline pH 7.4 (PBS) to a total volume of 6 ml. This diluted buffy coat was then carefully layered over 4 ml cold Ficoll®-Paque® plus density gradient, avoiding mixing of the layers.
- After centrifugation (20 minutes at 550 × *g* at 4 °C without brake), a two-layer system was formed.
- The PBMCs, forming a white interface between the upper and lower layers, were collected and washed with 35 ml of cold PBS.
- After centrifugation (5 minutes at 1500 × *g* at 4 °C), the supernatant was thoroughly removed. The cell pellet was resuspended in 70 μl PBS, resulting in a homogeneous cell suspension with a total volume of ∼100 μl.
- A 30 μl aliquot of this cell suspension was 10 times diluted with PBS and counted for cell numbers using a hematology analyzer (Cell-Dyn Sapphire; Abbott Diagnostics, Abbott Park, IL, USA).
- A 60 μl aliquot of the cell suspension was used for determination of the 5-FU nucleotide concentrations. To this end, cells were lysed by the addition of 100 μl methanol, followed by extensive vortex-mixing for one minute. After centrifugation (5 minutes at 3000 × *g* at 4 °C), the supernatant (PBMC lysate) was collected and stored at −70 °C until analysis.
- Directly prior to the LC-MS/MS analysis, after vortex-mixing, a 100 μl sample was processed by the addition of 5 μl of internal standard working solution. Then, the samples were again vortex-mixed, centrifuged (5 minutes at 3000 × *g* at 4 °C), and transferred to autosampler vials.

17.4.2 CPT Method

In general, recommended procedures for PBMC isolation are available from the manufacturer for using the Vacutainer™ CPT tubes (www.bdj.co.jp/pas/products/mekkin/1f3pro00000r5drz-att/bd-cpt-manual-362760-362761.pdf; https://www.bdbiosciences.com/ds/ab/others/PI_CPT_heparin_March_2016_VDP40105-07_web_500010323.pdf). Modification to the baseline procedure can be made to fit the purpose of the intended study (Colombo et al. 2005).

- One tube per patient (∼8 ml of blood) was collected, gently inverted 8–10 times for careful homogenization with the anticoagulant, and centrifuged within 5 minutes at 1650 × *g* for 20 minutes at RT, without break.
- The thin mononuclear cell layer was collected at the interface with a pipette and transferred into a polypropylene tube (15 ml) pre-chilled at 4 °C on an ice bath.
- The entire cell washing procedure was performed using pre-chilled PBS solutions kept at 4 °C and the centrifuge was maintained at 4 °C to inhibit enzymatic activity and to prevent active drug efflux out of cells.
- The collected cells were added with 10 ml of cold PBS solution containing 2% fetal bovine serum (FBS), and after careful mixing, the tube was centrifuged for 10 minutes at 650 × *g* (4 °C) and the supernatant was discarded.
- The above washing procedure was repeated twice.
- The third (final) supernatant wash solution was completely aspirated off and an exact volume of 1.0 ml of PBS was added to the cell pellet.
- Cells were carefully put in suspension, from which an aliquot (20 μl) was taken out and diluted 1 : 10 with 180 μl of PBS in a 0.5-ml Eppendorf tube. This aliquot was used for cell counting within 10 minutes, enabling the total number of PBMCs contained in each patient's pellet sample to be calculated, and used for normalizing the intracellular drug concentration determined by LC-MS/MS.
- The remaining 980 μl cell suspension was centrifuged at 650 × *g* for 10 minutes at 4 °C. The supernatant was aspirated off, leaving the PBMC pellet at the bottom of the vial which was immediately frozen and stored at −20 °C until the day of the analysis.
- At the initial stage of the method validation, each PBS solution wash was retained and subsequently checked by LC-MS/MS for the presence of the analyte of interest. These analyses confirmed that no drug was detectable in the third wash, indicating not only that the washing procedure was efficient in removing any extracellular medium possibly associated with the cells but also that the washing procedure at 4 °C prevented drug efflux from the cell.

17.4.3 Additional Consideration in LC-MS Bioanalysis for PBMC Samples

17.4.3.1 Matrix Blank

In general, PBMCs prepared from healthy donors or control group animals should be used as control blanks for the preparation of calibration standards and QC samples. The process of generating control PBMCs should be the same as for study samples. However, relatively large quantities are needed to support study sample analysis and, particularly, long-term stability assessment for the analyte in QC samples to cover the duration between study sample collection and analysis. Often, filters containing leucocytes collected from blood donation can be used for the preparation of human PBMC blanks (Colombo et al. 2005; Derissen et al. 2015). In addition, the blanks PBMCs are commonly used for assessing interference and matrix effect.

17.4.3.2 STDs and QCs

Depending on the nature of the analytes of interest, different approaches can be employed for the preparation of the calibration standards and/or QCs for the analysis of PBMCs. For xenobiotics, the blank PBMCs can be directly spiked with the compounds of interest for the preparation of the calibration standards and QCs (Becher et al. 2002; Ehrhardt et al. 2007). This approach has been employed in many studies, including the one reported by Pèlerin et al. (2005) who spiked into blank PBMCs (10×10^6 cells) with 500 µl of the standard solutions at concentrations of 1, 2, 10, 50, 100, 300, and 500 ng ml^{-1} of each antiretroviral drug (in methanol) for calibration standards, or 5, 75, and 250 ng ml^{-1} of each drug (in methanol) for QCs. The final amounts of each drug were 0.5, 1, 5, 25, 50, 150, and 250 ng per sample containing around 10×10^6 cells for calibration standards and 2.5, 37.5, and 125 ng per sample containing around 10×10^6 cells for QCs.

However, for endogenous compounds, the pure blank matrix does not exist, which renders accurate quantification of the analytes of interest difficult. An approach to overcome this challenge is to prepare the calibration standards and/or QCs in neat solutions (Laverdière et al. 2012). However, the difference in matrix effect between the neat calibration standards and/or QC solutions and study samples can be an issue. In this regard, use of stable isotope-labeled internal standard is expected to compensate for such a difference (Cohen et al. 2013; Zhang and Bartels 2013). The other approach is to build the standard curve with the stable isotope-labeled analogs using PMBCs. For example, stable isotope-labeled dATP was used as surrogate for the quantification of dATP. It was considered that the LC-MS/MS response factor was identical for both labeled and non-labeled molecules. Of course, attention should be given concerning the purity of the isotope labeling. With this approach unlabeled internal standard such as bromo- or chloro-nucleotide can be used (Cohen et al. 2009).

17.5 Summary

For many drugs, there is a correlation between the measured PBMC concentrations and efficacy and/or toxicity of the drugs. Therefore, quantitative analysis of the relevant study of PBMC samples is necessary to facilitate the assessment of the correlations between PBMC concentrations and study endpoints. As a highly selective and sensitive method, LC-MS/MS has been widely used for quantitative analysis of PBMC concentrations of drugs and/or drug metabolites. However, in order to carry out a successful LC-MS/MS analysis of analyte(s) of interest in PBMCs, attention should be given to some important details in the PBMC sample preparation workflow, including cell separation, cell counting, cell lysing, and sample extraction.

References

Almond, L.M., Edirisinghe, D., Dalton, M. et al. (2005). Intracellular and plasma pharmacokinetics of nevirapine in human immunodeficiency virus-infected individuals. *Clin. Pharmacol. Ther.* 78 (2): 132–142.

Ansermot, N., Fathi, M., Veuthey, J.L. et al. (2007). Quantification of cyclosporine A in peripheral blood mononuclear cells by liquid chromatography-electrospray mass spectrometry using a column-switching approach. *J. Chromatogr. B Analyt. Technol. Biomed. Life Sci.* 857: 92–99.

Baran, Y., Bielawski, J., Gunduz, U., and Ogretmen, B. (2011). Targeting glucosylceramide synthase sensitizes imatinib-resistant chronic myeloid leukemia cells via endogenous ceramide accumulation. *J. Cancer Res. Clin. Oncol.* 137: 1535–1544.

Bazzoli, C., Jullien, V., Le Tiec, C. et al. (2010). Intracellular pharmacokinetics of antiretroviral drugs in HIV-infected patients, and their correlation with drug action. *Clin. Pharmacokinet.* 49 (1): 17–45.

Becher, F., Pruvost, A., Goujard, C. et al. (2002). Improved method for the simultaneous determination of d4T, 3TC and ddl intracellular phosphorylated anabolites in human peripheral-blood mononuclear cells using high-performance liquid chromatography/tandem mass spectrometry. *Rapid Commun. Mass Spectrom.* 16 (6): 555–565. Erratum in: *Rapid Commun. Mass Spectrom.* 2007;21(13):2168.

Benech, H., Théodoro, F., Herbet, A. et al. (2004). Peripheral blood mononuclear cell counting using a DNA-detection-based method. *Anal. Biochem.* 330 (1): 172–174.

Cascorbi, I. (2006). Role of pharmacogenetics of ATP-binding cassette transporters in the pharmacokinetics of drugs. *Pharmacol. Ther.* 112 (2): 457–473.

Cohen, S., Gagnieu, M.G., Lefebvre, I., and Guitton, J. (2013). LC-MS bioanalysis of nucleotides. In: *Handbook of LC-MS Bioanalysis: Best Practices, Experimental Protocols and Regulations* (ed. W. Li, J. Zhang and F.L.S. Tse), 559–572. Hoboken, NJ: Wiley.

Cohen, S., Megherbi, M., Jordheim, L.P. et al. (2009). Simultaneous analysis of eight nucleoside triphosphates in cell lines by liquid chromatography coupled with tandem mass spectrometry. *J. Chromatogr. B Analyt. Technol. Biomed. Life Sci.* 877 (30): 3831–3840.

Colombo, S., Beguin, A., Telenti, A. et al. (2005). Intracellular measurements of anti-HIV drugs indinavir, amprenavir, saquinavir, ritonavir, nelfinavir, lopinavir, atazanavir, efavirenz and nevirapine in peripheral blood mononuclear cells by liquid chromatography coupled to tandem mass spectrometry. *J. Chromatogr. B Analyt. Technol. Biomed. Life Sci.* 819 (2): 259–276.

D'Avolio, A., Pensi, D., Baietto, L., and Di Perri, G. (2014). Therapeutic drug monitoring of intracellular anti-infective agents. *J. Pharm. Biomed. Anal.* 101: 183–193.

Derissen, E.J.B., Hillebrand, M.J.X., Rosing, H. et al. (2015). Development of an LC-MS/MS assay for the quantitative determination of the intracellular 5-fluorouracil nucleotides responsible for the anticancer effect of 5-fluorouracil. *J. Pharm. Biomed. Anal.* 110: 58–66.

Derissen, E.J., Jacobs, B.A., Huitema, A.D. et al. (2016). Exploring the intracellular pharmacokinetics of the 5-fluorouracil nucleotides during capecitabine treatment. *Br. J. Clin. Pharmacol.* 81 (5): 949–957.

Durand-Gasselin, L., Pruvost, A., Dehée, A. et al. (2008). High levels of zidovudine (AZT) and its intracellular phosphate metabolites in AZT- and AZT-lamivudine-treated newborns of human immunodeficiency virus-infected mothers. *Antimicrob. Agents Chemother.* 52 (7): 2555–2563.

Ehrhardt, M., Möck, M., Haefeli, W.E. et al. (2007). Monitoring of lopinavir and ritonavir in peripheral blood mononuclear cells, plasma, and ultrafiltrate using a selective and highly sensitive LC/MS/MS assay. *J. Chromatogr. B Analyt. Technol. Biomed. Life Sci.* 850 (1–2): 249–258.

Farowski, F., Cornely, O.A., Vehreschild, J.J. et al. (2010a). Intracellular concentrations of posaconazole in different compartments of peripheral blood. *Antimicrob. Agents Chemother.* 54 (7): 2928–2931.

Farowski, F., Cornely, O.A., Vehreschild, J.J. et al. (2010b). Quantitation of azoles and echinocandins in compartments of peripheral blood by liquid chromatography-tandem mass spectrometry. *Antimicrob. Agents Chemother.* 54 (5): 1815–1819.

Farowski, F., Cornely, O.A., Vehreschild, J.J. et al. (2013). Intracellular concentrations of anidulafungin in different compartments of the peripheral blood. *Int. J. Antimicrob. Agents* 41 (4): 379–382.

Grievink, H.W., Luisman, T., Kluft, C. et al. (2016). Comparison of three isolation techniques for human peripheral blood mononuclear cells: cell recovery and viability, population composition, and cell functionality. *Biopreserv. Biobank.* 14 (5): 410–415.

Hartkoorn, R.C., Chandler, B., Owen, A. et al. (2007). Differential drug susceptibility of intracellular and extracellular tuberculosis, and the impact of P-glycoprotein. *Tuberculosis (Edinb.)* 87 (3): 248–255.

Ho, R.H. and Kim, R.B. (2005). Transporters and drug therapy: implications for drug disposition and disease. *Clin. Pharmacol. Ther.* 78 (3): 260–277.

Huang, L., Lizak, P., Aweeka, F., and Long-Boyle, J. (2013). Determination of intracellular fludarabine triphosphate in human peripheral blood mononuclear cells by LC-MS/MS. *J. Pharm. Biomed. Anal.* 86: 198–203.

Huang, Y., Zurlinden, E., Lin, E. et al. (2004). Liquid chromatographic-tandem mass spectrometric assay for the simultaneous determination of didanosine and stavudine in human plasma, bronchoalveolar lavage fluid, alveolar cells, peripheral blood mononuclear cells, seminal plasma, cerebrospinal fluid and tonsil tissue. *J. Chromatogr. B Analyt. Technol. Biomed. Life Sci.* 799 (1): 51–61.

Jemal, M., Rao, S., Gatz, M., and Whigan, D. (2003). Liquid chromatography-tandem mass spectrometric quantitative determination of the HIV protease inhibitor atazanavir (BMS-232632) in human peripheral blood mononuclear cells (PBMC): practical approaches to PBMC preparation and PBMC assay design for high-throughput analysis. *J. Chromatogr. B Analyt. Technol. Biomed. Life Sci.* 795: 273–289.

Jimmerson, L.C., Ray, M.L., Bushman, L.R. et al. (2015). Measurement of intracellular ribavirin mono-, di- and triphosphate using solid phase extraction and LC-MS/MS quantification. *J. Chromatogr. B Analyt. Technol. Biomed. Life Sci.* 978–979: 163–172.

Khoo, S.H., Hoggard, P.G., Williams, I. et al. (2002). Intracellular accumulation of human immunodeficiency virus protease inhibitors. *Antimicrob. Agents Chemother.* 46 (10): 3228–3235.

Laverdière, I., Caron, P., Couture, F. et al. (2012). Liquid chromatography-coupled tandem mass spectrometry based assay to evaluate inosine-5′-monophosphate dehydrogenase activity in peripheral blood mononuclear cells from stem cell transplant recipients. *Anal. Chem.* 84 (1): 216–223.

Pèlerin, H., Compain, S., Duval, X. et al. (2005). Development of an assay method for the detection and quantification of protease and non-nucleoside reverse transcriptase inhibitors in plasma and in peripherical blood mononuclear cells by liquid chromatography coupled with ultraviolet or tandem mass spectrometry detection. *J. Chromatogr. B* 819: 47–57.

Pruvost, A., Becher, F., Bardouille, P. et al. (2001). Direct determination of phosphorylated intracellular anabolites of stavudine (d4T) by liquid chromatography/tandem mass spectrometry. *Rapid Commun. Mass Spectrom.* 15 (16): 1401–1408. Erratum in: *Rapid Commun. Mass Spectrom.* 2007;21(13):2167.

Shi, G., Wu, J.T., Li, Y. et al. (2002). Novel direct detection method for quantitative determination of intracellular nucleoside triphosphates using weak anion exchange liquid chromatography/tandem mass spectrometry. *Rapid Commun. Mass Spectrom.* 16 (11): 1092–1099.

Ter Heine, R., Mulder, J.W., van Gorp, E.C. et al. (2010). Intracellular and plasma steady-state pharmacokinetics of raltegravir, darunavir, etravirine and ritonavir in heavily pre-treated HIV-infected patients. *Br. J. Clin. Pharmacol.* 69 (5): 475–483.

Zhang, F. and Bartels, M.J. (2013). LC-MS bioanalysis of intracellular drugs. In: *Handbook of LC-MS Bioanalysis: Best Practices, Experimental Protocols and Regulations* (ed. W. Li, J. Zhang and F.L.S. Tse), 337–352. Hoboken, NJ: Wiley.

18

Sample Preparation for LC-MS Bioanalysis of Urine, Cerebrospinal Fluid, Synovial Fluid, Sweat, Tears, and Aqueous Humor Samples

Allena J. Ji

18.1 Introduction

Quantitative LC/MS bioanalysis has been applied in pharmacokinetics, biomarkers, clinical diagnostics, and cosmetics industry for decades. With new technologies rapidly emerging in bioanalytical fields, the scientists are required to possess the skills of liquid chromatography and mass spectrometry, as well as the skills of various sample preparation prior to LC-MS/MS analysis. Most common sample preparation techniques used in various matrices include protein precipitation (PPT), liquid–liquid extraction (LLE), supported liquid extraction (SLE), solid-phase extraction (SPE), ultrafiltration, and direct dilution. This chapter summarizes commonly used sample preparation procedures for urine, cerebrospinal fluid (CSF), synovial fluid, sweat, tears, and aqueous humor to suit for LC-MS/MS analysis, and also includes a brief introduction of sample collection methods for several uncommon matrices.

18.2 Sample Preparation Methods for Urine

Urine is a yellow solution that contains >95% water with various by-products of metabolic constituents like urea, chloride, sodium, potassium, and creatinine as well as other organic and inorganic compounds. The pH of urine can vary in a normal range of 5.7–7.0 (Rose et al. 2015). A normal urine sample does not have a red or reddish-brown color due to hemoglobin from red blood cells, myoglobin from muscle cells, certain medications, or food dyes. If a urine sample is cloudy or foamy, this indicates higher levels of protein. One most important factor is nonspecific binding during quantitative analysis. Urine is lack of protein and lipids which can be associated with nonspecific binding to its collection container, called adsorption. Adsorption can lead to a significant loss of the analyte. The methodologies to identify analyte loss due to nonspecific binding and its resolution have been described in a previous publication (Ji et al. 2010). Scientists should test nonspecific binding of the analyte before any urine extraction method development. Another important factor is sample mixing. Urine sample tubes are often in much bigger volume than plasma sample, which can prevent thorough mixing when the containers are full. Therefore, to obtain reproducible urine results, mixing the sample is a critical step. The commonly used urine sample extraction methods are discussed.

18.2.1 Direct Dilution for Urine Samples

The simplest urine preparation method is direct dilution. This preparation method is suitable for small polar analyte(s) in urine, which are often present in high concentrations, insensitive to pH change, and are not suitable for LLE and SPE methods. Because there is a large variation of salt concentration and pH in urine, the urine samples need to be treated with a strong buffer to adjust the pH, then add an internal standard working solution. The percent of organic solvent in the diluted urine sample needs to be similar to the initial mobile phase of the LC-MS/MS analysis. The major advantages of the dilution method are (i) fast and simple sample preparation and (ii) high extraction recovery, since nothing is lost during sample dilution. The major disadvantages of the dilution method are (i) the sensitivity of the assay is limited due to the dilution; (ii) it requires a longer chromatographic run time in order to separate the analyte from interfering components; (iii) it may have a strong matrix effect which affects assay reproducibility; and (iv) it may lead to a shorter HPLC column life time since the diluted

Sample Preparation in LC-MS Bioanalysis, First Edition. Edited by Wenkui Li, Wenying Jian, and Yunlin Fu.

Table 18.1 An example of urine sample preparation procedure using direct dilution.

1) Thaw-frozen urine samples (~5 ml conical tubes or 20–50 ml bottles), and STDs and QC samples in a 37 ± 5 °C water bath or at room temperature for 1–1.5 h
2) Mix tubes by vortexing for 2 min, and invert tubes four to five times if urine tubes are almost full
3) Centrifuge all tubes at ~3000 rpm for 10 min at room temperature to separate precipitate if any
4) Pipette 100 μl of urine standards, QCs, and unknown samples into a 96-well plate per plate map
5) Add 50 μl of working internal standard solution to all wells. For urine blank without IS, add 50 μl IS solvent
6) Add 500 μl of acetonitrile to each well using a pipette repeater
7) Cap the plate and gently mix at a setting 3 (low speed) in a multi-vortex for 5 min
8) Submit the plate to autosampler of the LC-MS/MS system
9) Inject 5 μl of each well

Note: (i) The analyte in this protocol can be a water-soluble compound at $\mu g\,ml^{-1}$ level; (ii) the initial HPLC gradient program can be 20% water/80% acetonitrile with 0.1% formic acid.

sample is nonspecific removal of unwanted interfering compounds. Table 18.1 is an example of the dilution preparation protocol.

18.2.2 Liquid–Liquid Extraction for Urine Samples

LLE is most commonly used for urine extraction (Ji 2013). The principle of LLE is to extract analyte(s) based on their relative solubility in two different immiscible liquids, usually an aqueous buffer and an organic solvent. In LLE, the first step is to adjust the pH of the urine sample to ensure the analyte and internal standard becomes uncharged for a better extraction recovery. In general, the pH of the urine samples should be adjusted to 2 units below its pK_a when analyzing acidic drugs/biomarkers or 2 units above the pK_a when analyzing the basic drugs/biomarkers. The commonly used organic LLE solvents are methyl *tert*-butyl ether (MTBE), ethyl acetate, methylene chloride, hexane, and chloroform. Among these solvents, MTBE is the most commonly used organic solvent. The advantages of MTBE are: (i) it is suitable to extract many acidic and basic analytes containing hydrophobic groups; (ii) the extraction can remove the majority of the phospholipids that causes matrix effect in LC/MS/MS analysis; (iii) its drying time is short (only 10–15 minutes); and (iv) it is easy to separate the MTBE organic level from the aqueous phase. MTBE is lighter than water, so the bottom aqueous phase can be frozen using a dry ice bath and can allow the top organic unfrozen layer to be poured into another clean tube. A typical LLE procedure using tube-type or 96-well plate format (Ji 2013) is summarized in Table 18.2. It has to be pointed out that the 96-well format procedure has 10–15% lower extraction efficiency compared to the tube format when using a similar procedure due to the efficiency of mixing in the two-phase partition.

18.2.3 Solid-Phase Extraction for Urine Samples

SPE is the commonly used sample cleanup procedure for urine drugs/biomarker analysis prior to LC-MS/MS analysis for several decades. There are three major types of SPE columns based on their packing materials. They are reverse-phase-SPE (RP-SPE), normal-phase-SPE (NP-SPE), and ion exchange-SPE (IX-SPE). The description of each type of SPE has been summarized in the LC-MS/MS bioanalysis handbook (Liu and Aubry 2013). Majority of SPE methods involve steps of activation of SPE sorbent, loading sample, washing off unbound components, eluting analyte from SPE sorbent, concentrating sample extracts using evaporation, and reconstitution in less volume. In general, a pH or a buffer treatment to the urine sample is necessary before loading treated urine samples on SPE cartridges/sorbent plate in order to increase suitable extraction recovery. An SPE product usually has a standard extraction procedure from the vendor. These SPE procedures should be a good starting point for sample preparation. An example of Waters Oasis SPE generic sample preparation protocols (Danaceau 2013) is displayed in Figure 18.1. The generic SPE procedures are suitable to many acidic, neutral, and basic compounds for their initial SPE steps. The advantages of SPE are (i) it is relatively highly selective to a variety of compounds in various matrices; (ii) it is easy to be automated for high throughput; (iii) it has a wide variety of stationary-phase types including reverse phase, normal phase, ion exchange, and mixed modes, allowing the analyst to fine-tune the selectivity of their extraction methods for a wide range of analytes; and (iv) it is the best technique for minimizing matrix interferences including proteins, phospholipids, salts, and other endogenous compounds. Disadvantages of SPE methods are: (i) high cost of SPE cartridge or plates; (ii) it has many extraction steps so that it takes longer than LLE or

Table 18.2 An example of urine sample preparation procedure using LLE.

A: Manual procedure format using conical tubes
1) Thaw-frozen urine samples (~5 ml tubes), and STDs and QC samples in a 37 ± 5 °C water bath or room temperature for 1–1.5 h
2) Invert tubes upside down for four to five times if urine tubes are almost full, then vortex tubes for 2 min
3) Centrifuge all tubes at ~3000 rpm for 10 min at room temperature to separate precipitate if any
4) Pipette 100 μl of urine standards, QCs, and unknown samples into 10-ml glass conical tubes with screw cap
5) Add 50 μl of working internal standard solution to each tube. For urine blank without IS, add 50 μl IS solvent
6) Add 50 μl 100 mM sodium acetate buffer (pH 5.0) into each tube
7) Add 5.0 ml of methyl *tert*-butyl ether (MTBE) to all tubes
8) Cap all tubes and vortex tubes using a multi-tube vortexer for 10 min
9) Centrifuge the tubes at ~3000 rpm for 5 min at room temperature
10) Place conical tubes into a dry ice bath for 1 min, which should freeze the aqueous layer at the bottom
11) Uncap the tubes and pour the top MTBE layer into another set of clean tubes (10 × 13 mm glass tubes)
12) Evaporate the MTBE extracts at 37 °C under nitrogen flow in a TurboVap for ~10 min
13) Reconstitute dried sample residues with 200 μl of reconstitution solution
14) Vortex the tubes for 3 min
15) Transfer sample extract into 2-ml HPLC vials and cap them tightly
16) Submit the sample vials for LC-MS/MS analysis
B: 96-well plate format with a liquid handler
1) Thaw-frozen urine samples (~5 ml tubes), and STDs and QC samples in a 37 ± 5 °C water bath or room temperature for 1–1.5 h
2) If urine tubes are almost full, invert tubes four to five times, then mix tubes by vortexing for 2 min
3) Centrifuge all tubes at ~3000 rpm for 10 min at room temperature
4) Pipette 100 μl of urine standards, QCs, and unknown samples into a 96-well plate per plate map
5) Add 50 μl of working internal standard solution to all wells. For urine blank without IS, add 50 μl IS solvent
6) Add 50 μl of 100 mM sodium acetate (pH 5.0) into each well of the urine sample using a liquid handler
7) Add 800 μl (400 μl × 2 times) of MTBE to each well using a liquid handler
8) Cap the plate with a plate map tightly (or use caps to seal 96-well format microtubes)
9) Vortex the plate at low speed for 10 min (or using tips to aspirate and dispense 600 μl for four to five times)
10) Centrifuge the plate at ~300 rpm for 10 min at room temperature
11) Open the plate mat (or uncap the microtubes in a 96-well plate)
12) Transfer 600 μl (300 μl × 2 times) of MTBE layer to another clean plate
13) Evaporate the sample extracts in the plate at 37 ± 5 °C under nitrogen flow using Turbo Vap96 or equivalent device for ~5–10 min to dryness
14) Add 200 μl of reconstitution solution to each well using a liquid handler
15) Cap the plate with a plate map and vortex the plate at a low speed in a multi-tube vortexer for ~3 min
16) Submit the sample plate for LC-MS/MS analysis

Note: The analyte can be pH insensitive at pK_a around 7 with small portion of polar and multiple nonpolar groups.
Source: Ji (2013). Reproduced with permission of John Wiley & Sons.

direct dilution; (iii) it takes much longer time for method optimization due to many steps of extraction; (iv) it requires skillful and experienced scientist(s) for the method development. Table 18.3 lists an example of basic drug extraction using the mix-mode strong cation extraction method (Danaceau 2013).

It must be noted that the tables are for urine sample preparation; the scientists need to make their own adjustments according to the characteristics of analyte and SPE type used. The goal of SPE is to make the analyte of interest bind on SPE sorbent, and wash off interfering components, and then elute the analyte for the final collection. The final volume of the reconstitution solution and component need to be designed to suit the assay quantitation limits and initial chromatographic conditions.

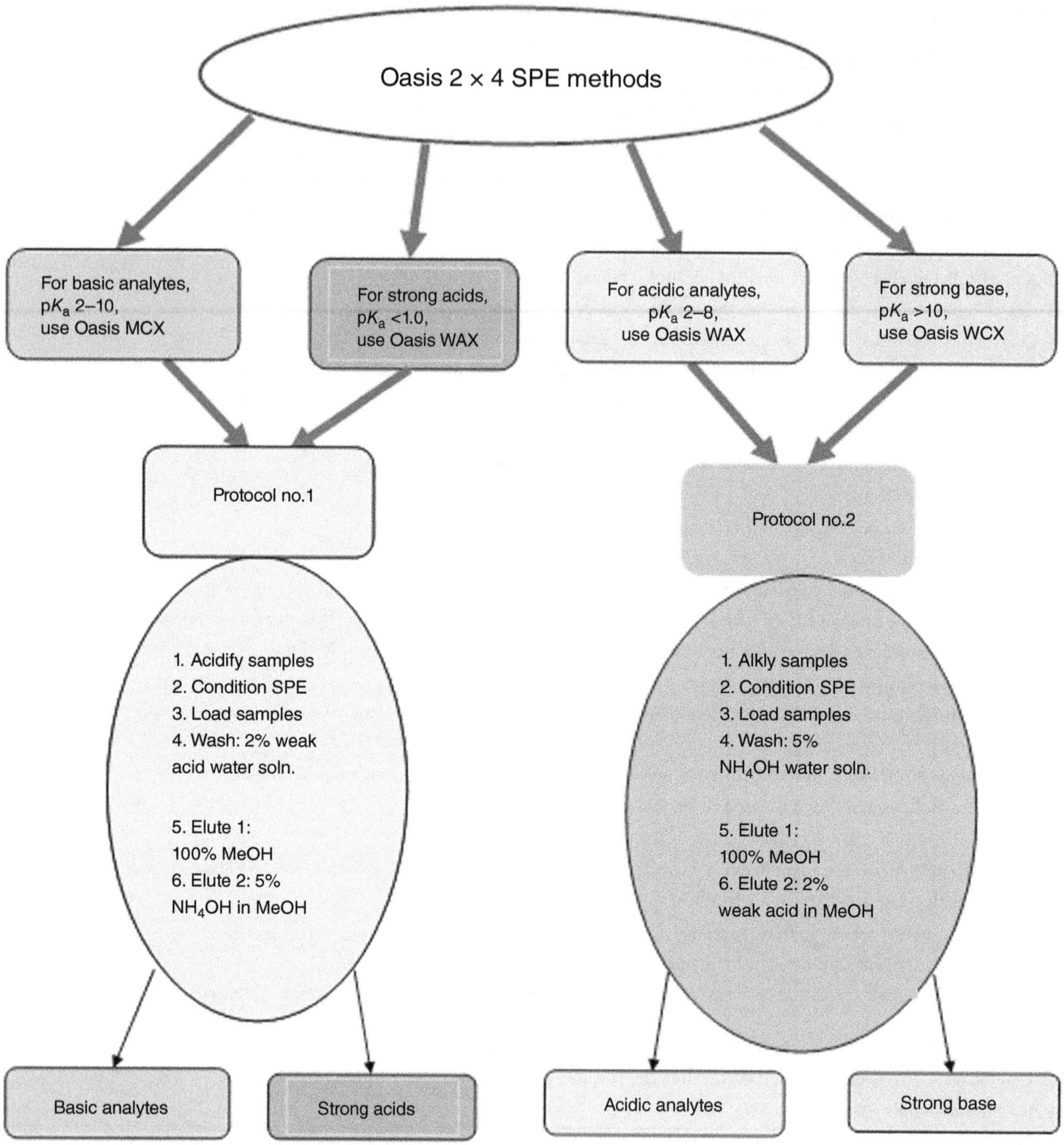

Figure 18.1 Generic SPE procedures for biological sample preparation. MCX, mixed-mode cation exchange; MAX, mixed-mode anion exchange; WAX, weak anion exchange; WCX, weak cation exchange. *Source:* From Danaceau (2013).

Table 18.3 An example of urine sample preparation procedure using oasis MCX SPE.

- *Pretreatment of urine samples*

1) Thaw-frozen urine samples (~5 ml conical tubes), and STDs and QC samples in a 37 ± 5 °C water bath or at room temperature for 1–1.5 h
2) Invert tubes for four to five times if urine tubes are almost full, then mix tubes by vortexing for 2 min
3) Centrifuge all tubes at ~3000 rpm for 10 min at room temperature
4) To a 96-well plate (Plate No. 1), pipette 100 µl of urine standards, QCs, and unknown samples into each well of a 2-ml 96-well plate
5) Add 100 µl of 4% phosphoric acid water solution to each well
6) Add 100 µl of working internal standard solution to all wells. For urine blank without IS, add 100 µl IS solvent (water)
7) Cap the plate with a plate mat
8) Vortex the plate at a low speed for 5 min using a multi-tube vortexer

- *Condition of SPE plate*

9) To a 96-well Oasis MCX SPE plate (Plate No. 2, a waste collection plate underneath of the SPE plate), add 200 µl of methanol to each well using a pipette repeater or eight-channel pipette or a liquid handler. Let the methanol drip by gravity for 5 min. Then apply vacuum to make solvent slightly dry
10) Add 200 µl of water to each well, apply vacuum to dryness for ~10 s (Note: Do not leave this plate dry for longer than 5 min)

- *Load pretreated urine samples on SPE plate*

11) Uncap Plate No. 1
12) Using a multichannel pipette or a liquid handler, aspirate 250 µl of pretreated urine sample from Plate No. 1, and dispense the volume to preconditioned 96-well SPE plate (Plate No. 2)
13) Let the urine sample drip by gravity for 2–3 min. Then gently apply vacuum to pull the sample through the SPE sorbent dropwise, and vacuum to dryness at the end for 10 s

- *Wash off interferences from the samples on the SPE plate*

14) Add 200 µl of water to each well, gently apply vacuum to pull water through the sorbent dropwise
15) Add 200 µl of methanol to each well, gently apply vacuum to pull the solvent through the sorbent dropwise, then to dryness at the end
16) Discard waste solution in the waste collection plate underneath the SPE plate

- *Elute analyte from the SPE plate*

17) Put a clean 2-ml 96-well plate (Plate No. 3) underneath the SPE plate (Plate No. 2)
18) Add the first 50 µl of elution solvent (95% [ACN : MeOH = 6 : 4, v:v]/5%NH_4OH, v/v) to each well of the SPE plate, gently apply vacuum to make the solution elute dropwise, then allow a brief dry. Collect the eluent in Plate No. 3
19) Add the second 50 µl of elution solvent (95% [ACN : MeOH = 6 : 4, v:v]/5%NH_4OH, v/v) to each well of the SPE plate, gently apply vacuum to elute the solution dropwise, then increase vacuum intensity for 10–15 s to induce dryness. Accumulate the eluent in Plate No. 3

- *Concentrate extracted analyte(s)*

20) Evaporate the sample extracts in Plate No. 3 at 37 ± 5 °C under nitrogen flow using Turbo Vap 96 or equivalent device to dryness for 10–20 min
21) Add 100 µl of reconstitution solution, 2% ACN/98% of 0.1% formic acid water (the mobile-phase component at the initial chromatographic condition or lower organic % for reverse phase) to each well using a pipette repeater or liquid handler
22) Cap the plate with a plate mat and vortex the plate at a low speed in a multi-tube vortexer for 3–5 min
23) Submit the sample Plate No. 3 to the HPLC autosampler
24) Inject 5 µl to the LC-MS/MS system

Note: The above protocol is suitable to most basic compounds in toxicology lab such as morphine, codeine, oxycodone, fentanyl, methadone, bupremephine, and their related metabolites. The liquid chromatographic condition is reverse-phase chromatography. ACN, acetonitrile; MeOH, methanol; NH_4OH, ammonium hydroxide.
Source: From Danaceau (2013).

18.3 Sample Preparation Methods for Cerebrospinal Fluid

CSF is a clear, colorless body fluid produced by the chord plexus of the ventricles of the brain. It acts as a cushion or buffer for the brain's cortex, providing basic mechanical and immunological protection to the brain inside the skull. The CSF also serves a vital function in cerebral autoregulation of cerebral blood flow. In normal adults, the total CSF volume is 90–150 ml (Williamson and Snyder 2011). Eighty percent of the CSF is contained in the arachnoid space in the cranium and spinal cord, where a small amount can be drawn for CSF tests through a lumbar puncture.

CSF is a biological sample type used to assess biomarkers related to neurological diseases and therapeutic/investigational drug exposure in the central nervous system. CSF's normal protein concentration is 15–45 mg dl^{-1} (Williamson and Snyder 2011), which is about 400–500 times less than plasma total protein. Therefore, adsorption test for an analyte adhered to the container wall which results in the loss of analyte must be tested in the first step of method development. The detailed sequential tube transfer test for diagnosis of adsorption is described in the previous publication on urine bioanalysis (Ji 2013; Ji et al. 2010). Briefly, prepare a QC low-level solution of CSF, transfer the low QC solution from one tube to another tube for five consecutive tubes (Tube 1 through Tube 5), for each transfer, the tube needs to be incubated for 10–30 minutes at room temperature to allow adsorption occurring and leave a minimum volume of the solution in each tube after each transfer. At the end, analyze each tube. If the concentration decreased with an increase of tube number, then the analyte in CSF has adsorbed to the container wall. An appropriate anti-adsorptive agent at its optimal concentration should be added to CSF samples at the sample collection stage and when CSF QCs are prepared. The removal of interference from added anti-adsorptive agent should be considered in CSF sample extraction or LC-MS/MS analysis.

CSF is relatively clean compared to plasma and urine samples. Therefore, the most common extraction methods for small-molecule analyte(s) in CSF are the PPT method and dilution method. Extraction methods of LLE and SPE are also commonly used based on the analyte's characteristics. The extraction procedures of LLE or SPE are similar to those used for plasma and urine matrices. For protein and peptide analysis, the CSF sample preparation method is similar to plasma protein analysis. Table 18.4 lists an example of CSF sample preparation using the PPT method for small-molecule compounds (Cox et al. 2015) and Table 18.5 lists an example of CSF sample preparation using the direct dilution method (Arning and Bottiglieri 2014).

Table 18.4 An example of cerebrospinal fluid sample preparation procedure using protein precipitation.

1) Thaw-frozen CSF samples (in ~2 ml cryotubes), and STDs and QC samples at room temperature for 1 h
2) Mix each tube by vortexing at a medium speed for ~10 s
3) Pipette 100 μl of CSF standards, QCs, and unknown samples into a 96-well plate per plate map
4) Add 50 μl of working internal standard solution to all wells. For blank without IS, add 50 μl IS solvent (50% ACN/50% water)
5) Add 400 μl of ACN to each well using a pipette repeater
6) Cap the plate and gently mix at a medium speed in a multi-tube vortexer for 5 min
7) Centrifuge the plate at 3000 rpm for 5 min
8) Transfer 400 μl of supernatant to another clean plate
9) Evaporate the sample extracts at 37 ± 5 °C under nitrogen flow using Turbo Vap 96 or equivalent device to dryness for ~30 min
10) Add 100 μl of reconstitution solution (e.g. 20% ACN/80% water/0.1% formic acid, v/v/v) to each well
11) Cap the plate with a plate mat and vortex the plate at a low speed in a multi-tube vortexer for ~3–5 min
12) Submit the plate to autosampler of the LC-MS/MS system
13) Inject 5 μl of each well

Note: The analyte must be soluble in the reconstitution solution which is similar to mobile phase initial condition in its gradient program. ACN, acetonitrile; MeOH, methanol; NH_4OH, ammonium hydroxide.
Source: Cox et al. (2015). Reproduced with permission of Future Sciences Ltd.

Table 18.5 An example of CSF sample preparation procedure using direct dilution.

1) Thaw-frozen CSF samples (in ~2 ml cryotubes), and STDs and QC samples at room temperature for 1 h.
2) Mix each tube by vortexing at a medium speed for ~10 s
3) Pipette 50 μl of CSF standards, QCs, and unknown samples into a 96-well plate per plate map
4) Add 250 μl of working internal standard solution containing 0.2%DTT water solution. For blank without IS, add 250 μl 0.2%DTT water only
5) Cap the plate and gently mix the plate for 10 min
6) Submit the plate to autosampler of the LC-MS/MS system
7) Inject 5 μl of each well

Note: Analyte can be pterin metabolite-type of compounds. The LC condition is a reverse-phase chromatography with a gradient program which uses mobile phase A (water containing 0.1% formic acid and 0.1% heptafluorobutyric acid) and mobile phase B (methanol containing 0.1% formic acid). DTT, 1,4-dithiothreitol.
Source: Arning and Bottiglieri (2014). Reproduced with permission of Springer.

18.4 Sample Preparation Methods for Synovial Fluid

Synovial fluid, often referred to as "joint fluid," is a transparent viscid lubricating fluid secreted by a membrane of an articulation, bursa, or tendon sheath. Normal synovial fluid is straw-colored, clear, and slightly sticky or stringy. Common clinical tests for synovial fluid are glucose, proteins, uric acid, lactic dehydrogenase, and cell culture for bacterial identification (El-Gabalawy 2013). In recent years, it has been identified 113 proteins in healthy synovial fluid that appear to be primarily derived from plasma transudates, 37 proteins primarily derived from synovium, and 11 proteins primarily derived from cartilage (Bennike et al. 2014). The collection of synovial fluid is drawn by inserting a sterile needle through the skin and into the joint space and then drawing the fluid through the needle into a sterile syringe.

The sample preparation of synovial fluid is usually simple. PPT plus ultrafiltration or PPT only is used for sample preparation. Table 18.6 shows an example of PPT and ultrafiltration for the sample extraction (Montesano et al. 2016). Synovial fluid is collected by aseptic arthrocentesis from knee joint into heparinized tubes. Samples were then centrifuged to remove any dispersed solid matter, and frozen at −20 °C in polypropylene tubes.

Table 18.6 An example of synovial fluid sample preparation procedure using protein precipitation and ultrafiltration.

1) Thaw-frozen synovial fluid (SF) samples (in 2-ml cryotubes), and STDs and QC samples at room temperature for 30 min
2) Mix each tube by vortexing at a medium speed for ~10 s two to three times
3) Pipette 175 μl of SF standards, QCs, and unknown samples into 1.5-ml microcentrifuge tubes
4) Add 85 μl of working internal standard methanol solution to each tube. For blank without IS, add 85 μl methanol only
5) Transfer ~250 μl of above solution into an ultrafiltration tube (top part has a filter, bottom part is a small conical tube for collection)
6) Centrifuge the tubes at 13 000 rpm for 15 min
7) Remove top filter tube portion and discard them
8) Transfer 100 μl of the bottom filtrated solution to 200-μl glass insert in 2-ml HPLC vial
9) Submit the samples tray to the LC-MS/MS system
10) Inject 5 μl

Note: The analyte can be an antibiotic agent. Mobile phase A is 5 mM formic acid in methanol and mobile phase B is and 5 mM formic acid in water.
Source: Montesano et al. (2016). Reproduced with permission of Elsevier.

Table 18.7 An example of synovial fluid sample preparation procedure using TCA protein precipitation.

1) Thaw-frozen synovial fluid samples (in 2-ml cryotubes), and STDs and QC samples at room temperature for 30 min
2) Mix each tube by vortexing at a medium speed for ~10 s twice
3) Pipette 500 μl of synovial fluid standards, QCs, and unknown samples into 1.5-ml centrifuge tubes
4) Add 20 μl of working internal standard solution to each tube. For blank without IS, add 20 μl IS solvent, 10 mM ammonium acetate buffer
5) Add 250 μl of 1.2 M TCA water solution
6) Cap and vortex the above tubes at high speed
7) Centrifuge the tubes at 13 000 rpm for 10 min
8) Transfer 600 μl of the supernatant into 2-ml HPLC vials
9) Add 30 μl of 30% ammonia to neutralize the strong acid
10) Submit the samples to the LC-MS/MS system
11) Inject 5 μl

Note: The analyte can be glucosamine. Mobile phases were 10 mM acetic acid in water adjusted to pH 7.5 with ammonia (solvent A) and acetonitrile (solvent B). The analytical column was a Shodex® Asahipak NH2P-50 2D column (5 m, 150 mm × 2.0 mm i.d.). TCA, trichloroacetic acid.
Source: Pastorini et al. (2009). Reproduced with permission of Elsevier.

Another PPT method was reported to use trichloroacetic acid (TCA). It was confirmed to be the best choice according to the composition and protein content of synovial fluid. In addition, compared to other PPT procedures (e.g. addition of acetonitrile or methanol) it has a relatively low sample dilution. Table 18.7 shows the TCA PPT method (Pastorini et al. 2009).

18.5 Sample Preparation Methods for Sweat

Sweat has recently gained popularity as a noninvasive biofluid that can be used in diagnostic and biomarker investigations. The analyte composition of sweat can help indicate underlying pathologies, as is the case with cystic fibrosis, which is characterized in part by increased chloride levels in sweat. Sweat is a clear, hypotonic biofluid which is slightly acidic at a pH range of 4.0–6.8 and consists of water (99%), electrolytes (e.g. sodium, chloride, and potassium), urea, pyruvate, and lactate. Proteins, peptides, amines, amino acids, metal ions, a variety of drugs, cosmetics, and ethanol in very low concentrations are also found in this biofluid (Sato et al. 1989). Except for some high-molecular weight proteins, most sweat components are small molecules (<1000 Da) resulting from metabolic pathways (Mena-Bravo and Luque de Castro 2014).

Sweat sampling is usually stimulated by heat or chemicals, such as pilocarpine, to obtain enough volume of this biofluid for the analytical procedures. The advantage of sweat is noninvasive, less complex than urine, relatively free from interferences, and sample preparation is much simpler than other matrices. The main limitations of sweat as a clinical sample are the difficulty to produce enough sweat for analysis, lack of appropriate sampling devices, and normalization of the sampled volume. However, the advances both in sweat collection devices and highly sensitive analytical technologies increased the interest on sweat testing of drugs over the past few years (De Giovanni and Fucci 2013; Kintz et al. 1996; Mena-Bravo and Luque de Castro 2014). Sweat sampling can be performed with the aid of sweat wipes as liquid perspiration or over time using sweat patches (Jadoon et al. 2015). The experience gained in sweat collection procedures for the analysis of xenobiotics could be used to implement this biofluid in clinical diagnostics (Papanas et al. 2013). A sweat collection device, called Macroduct™ Sweat Stimulation and Sweat Collection System (Figure 18.2), is clinically approved for the collection of eccrine sweat for the diagnosis of cystic fibrosis through measurement of osmolarity and sodium concentrations (Carter et al. 1984). The Macroduct system (ELITechGroup Solutions 2016) consists of the Webster Sweat Inducer, the Macroduct Sweat Collector, and the Sweat-Chek Analyzer. Sweat is stimulated using pilocarpine. On average, 50–60 μl of sweat were collected from each subject during the 30-minute collection interval. After a sufficient volume of sweat had accumulated, a sweat dispenser, or blunt needle on a tuberculin syringe, was connected to the open end of the tubing. The tubing was uncoiled from the Macroduct® Sweat Collector and severed at the point of attachment. The sweat was then transferred from the tubing to microcentrifuge tubes, which were placed immediately on dry ice.

Table 18.8 Sweat sample preparation procedure using direct dilution and filtration.

- *Direct dilution*
  1) Thaw-frozen sweat samples, and STDs and QC samples at room temperature for 10 min. Vortex each tube at a high speed for ~10 s twice
  2) Pipette 10 μl sweat sample to 200-μl HPLC glass insert in 2-ml vial
  3) Add 20 μl of 0.1% (v/v) formic acid in water containing internal standard. (Note: 1 : 1 to 1 : 50 [v:v, sweat : diluent] can be used based on the concentration of the analyte tested)
  4) Mix well at low speed
  5) Submit the sample to LC-MS/MS for analysis
- *Sweat patch extraction and filtration*
  1) Add 1.5 ml of 20 $mmol l^{-1}$ ammonium formate, pH 7, and methanol (50:50, v:v) to a 5-ml conical tube with screw cap
  2) Add sweat patch into the above tubes
  3) Cap the tubes and shake the tubes in a shaker at 300 rpm for 10 min
  4) Centrifuge at 3000 rpm for 5 min
  5) Filter supernatant through Whatman mini-Uniprep syringeless filter vials
  6) Transfer filtered solution to a HPLC vial
  7) Submit sample vials for LC-MS/MS analysis

Note: Analytes in the above protocol are small-molecule drugs, amino acids, and/or metabolites.
Source: Calderón-Santiago et al. (2014). Reproduced with permission of Elsevier.

The sweat sampling and its LC-MS/MS analysis have been used for human drug analysis (Koster et al. 2014)

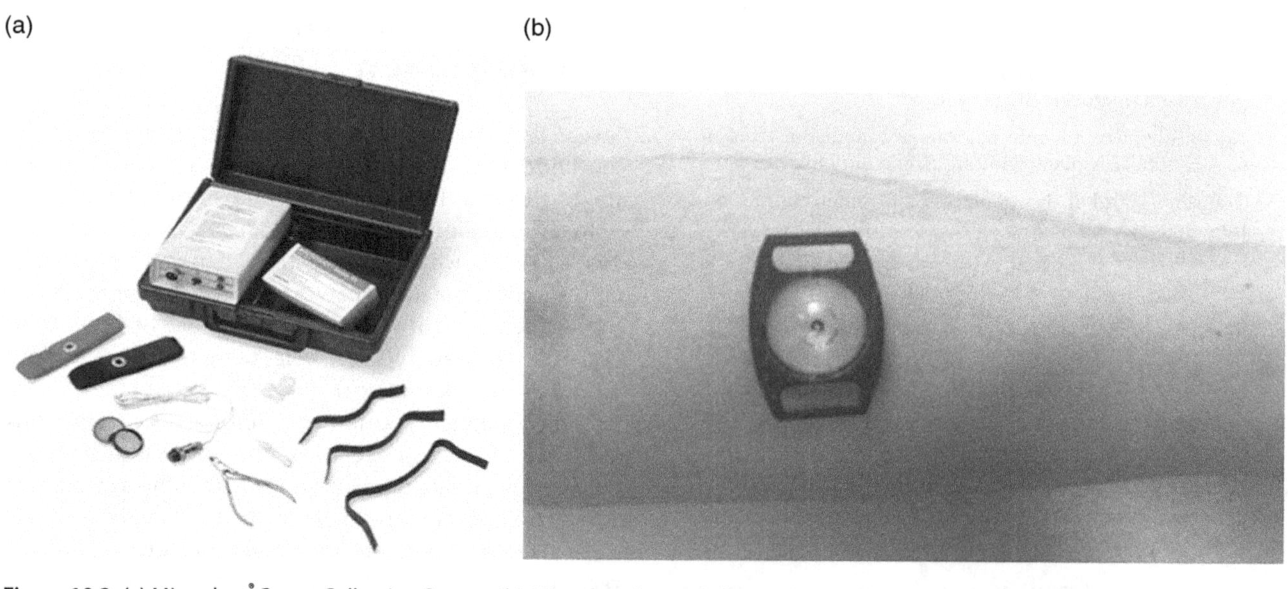

Figure 18.2 (a) Microduct® Sweat Collection System; (b) Microduct Sweat Collector. *Source:* From ELITechGroup Solutions (2016) and Delgado-Povedano et al. (2016).

and metabolomics analysis (Calderón-Santiago et al. 2014; Hooton et al. 2016) in recent years. The common sweat sample preparation methods for small-molecule extraction are direct dilution, patch soaking followed by filtration, and SPE coupled with the Micro SpinColumn™ system (protocol is provided by the vendor with particular sorbent). Table 18.8 lists an example protocol for direct dilution and filtration. Table 18.9 summarizes the procedures for SPE with Micro SpinColumn system procedures for small-molecule sweat analysis (Calderón-Santiago et al. 2014; Delgado-Povedano et al. 2016; Koster et al. 2014). For proteomics analysis (Raiszadeh et al. 2012), sweat sample preparation is similar to most other matrices which involves (i) reduction with 10 mM dithiothreitol (DTT) in 8 M urea at 50 °C for 20 minutes, (ii) alkylation with 50 mM iodoacetamide at room tem-

Table 18.9 Sweat sample preparation procedure using SPE with Micro SpinColumn system.

- *Use C18 centrifugal Micro SpinColumn systems (Harvard Apparatus, MA, USA)*
  1) Add 150 μl of water for solvation of sorbent
  2) Centrifuge at 1000 × *g* for 25 s
  3) Add 150 μl of 50% ACN water solution (v/v) for sorbent conditioning
  4) Centrifuge at 1000 × *g* for 25 s
  5) Add 150 μl of 5% ACN water solution (v/v) for sorbent equilibration
  6) Centrifuge at 1000 × *g* for 25 s
  7) Mix 25 μl of sweat sample with 5 μl of working internal standard
  8) Load 25 μl of treated sweat sample to the SPE column
  9) Centrifuge at 1000 × *g* for 25 s
  10) Wash 150 μl of 5% ACN water solution (v/v) to obtain the sample free from the undesirable retained compounds
  11) Centrifuge at 1000 × *g* for 25 s
  12) Elute with 150 μl of 50% ACN water solution (v/v)
  13) Centrifuge at 1000 × *g* for 25 s
  14) Evaporate all extract to dryness
  15) Reconstitute the dry sample with 20 μl of 0.1% (v/v) formic acid in water
  16) Vortex tubes and transfer the final solution to HPLC vials or a 96-well plate
  17) Submit the samples to the LC-MS/MS system
- *Use hydrophilic centrifugal Micro SpinColumn™ systems (Harvard Apparatus, MA, USA)*
  1) Add 150 μl of water for solvation of sorbent
  2) Centrifuge at 1000 × *g* for 25 s
  3) Add 150 μl of 50% (v/v) ACN water solution (v/v) for sorbent conditioning
  4) Centrifuge at 1000 × *g* for 25 s
  5) Add 150 μl of 5% ACN water solution (v/v) for sorbent equilibration
  6) Centrifuge at 1000 × *g* for 25 s
  7) Mix 25 μl of sweat sample with 5 μl of working internal standard
  8) Load 25 μl of treated sweat sample to the SPE column
  9) Centrifuge at 1000 × *g* for 25 s
  10) Wash 150 μl of 100% ACN to remove undesirable compounds
  11) Centrifuge at 1000 × *g* for 25 s
  12) Elute with 150 μl of 50% ACN water solution (v/v)
  13) Centrifuge at 1000 × *g* for 25 s
  14) Evaporate all extract to dryness
  15) Reconstitute the dry sample with 20 μl of 0.1% (v/v) formic acid in water
  16) Vortex tubes and transfer the final solution to HPLC vials or a 96-well plate
  17) Submit the samples to the LC-MS/MS system

Note: The analytes in the above protocol are small-molecule drugs, amino acids, and/or metabolites. ACN, acetonitrile.
Source: Calderón-Santiago et al. (2014). Reproduced with permission of Elsevier.

perature in the dark for 20 minutes, and (iii) overnight enzymatic digestion with 0.5 µg sequencing grade in 50 mM ammonium bicarbonate (NH_4HCO_3), pH 8 at 37 °C. Digestions were quenched with the addition of 2 µl glacial acetic acid, followed by the addition of 500 fmol of angiotensin II. Digest samples were then desalted with C-18 Zip Tips (SPE extraction) and dried in a vacuum concentrator.

In summary, sweat sample preparation includes direct dilution and SPE with Micro SpinColumn system due to its small volume. Once it is known how the sweat sample is collected and how much the available volume is, and then the sample preparation method can be designed.

18.6 Sample Preparation Methods for Tears

Tear is a clear fluid secreted from lacrimal glands and often serves to clean and lubricate the eyes in response to an irritation of the eyes. Human tears, composed of electrolytes, water, proteins (e.g. antibodies, lysozymes), and lipids, function to moisturize the ocular surface and minimize damage to the corneal epithelium (Moshirfar et al. 2014). Some of the substances in lacrimal fluid (such as lysozyme) fight against bacterial infection as a part of the immune system.

The most common application in tears analysis is ocular antibiotics pharmacokinetic studies such as fluoroquinolones (Arnold et al. 2008), a class of synthetic antibacterial agents which are known for their broad spectrum of antibacterial activity and/or drugs for treatment of bacterial conjunctivitis (Gu et al. 2016). The most commonly used tears collection method is to use Schirmer tear test strips displayed in Figure 18.3 and/or direct collection (Bhatta et al. 2011; Tian et al. 2008). For small-molecule drug analysis, the sample preparation methods can be directly diluted or by using SPE. Due to scarcity of tear sample volume, the LLE method is difficult to use. Tables 18.10 and 18.11 show an example of tear sample preparation procedure using direct dilution (Byrro et al. 2012) and SPE (Bhatta et al. 2011), respectively. It is worth mentioning that all example sample preparation needs adjustments to suit analyte's property and chromatographic conditions.

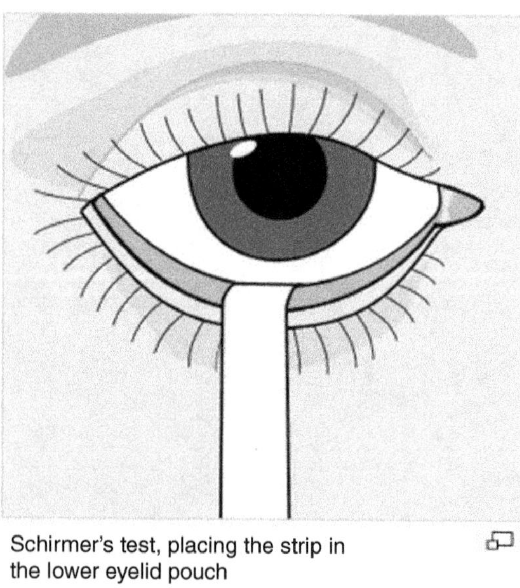

Figure 18.3 Schirmer's tear test strips. *Source:* https://en.wikipedia.org/wiki/Schirmer's_test. Licensed under CC BY-SA 3.0.

Table 18.10 Examples of tears sample preparation procedure using direct dilution.

- *Tears from direct collection in tubes*
  1) Thaw tears samples at room temperature for 30 min and spike STDs and QC samples (prepared in artificial tears)
  2) Pipette 5 µl tears samples, STDs, and QCs to 1-ml polypropylene microcentrifuge tubes
  3) Add 10 µl of working internal standard in methanol to each tube
  4) Add 100 µl of acetonitrile to the same tube
  5) Vortex at low speed (Note: do not mix at high speed in such a low volume)
  6) Transfer sample solution to 200-µl glass insert in HPLC vials
  7) Submit sample vials for LC-MS/MS analysis
- *Tears collected in Schirmer tear test strips*
  1) Preparation of STDs and QCs in tear strips
  2) Spike appropriate amounts of the working solution of analyte in blank tear strips to form eight standards and three levels of QCs
  3) Transfer each tears strip sample, STDs, QCs to 2-ml polypropylene tubes and store the tubes in a −70 °C freezer until bioanalysis
  4) On the day of sample analysis, thaw strip tubes at ambient temperature for 30 min
  5) Open the tube and dry the strips with an air stream
  6) Add 200 µl aliquot of 1 : 1 (v:v) solution of 0.15% aqueous formic acid : methanol to each tube
  7) Vortex at medium speed for 1 min
  8) Place the tubes at upright position on the bench at room temperature for 30 min
  9) Transfer 100 µl of sample solutions to 200-µl glass insert in 2-ml HPLC vials
  10) Submit the vials for LC-MS/MS analysis
  11) Inject 50 µl

Note: Analytes can be small-molecule antibiotics.
Source: Byrro et al. (2012). Reproduced with permission of Elsevier.

Table 18.11 An example of tears sample preparation procedure using SPE.

- *Pretreatment of tear samples*
  1) Thaw tears samples and spiked STDs and QC samples (prepared in artificial tears) at room temperature for 30 min
  2) Pipette 10 μl tears samples to 1-ml polypropylene centrifuge tube
  3) Add 10 μl of working internal standard in methanol
  4) Add 80 μl of MeOH to the same tube
  5) Vortex at low speed (Note: do not mix at high speed in such a low volume)
- *Conditioning the SPE column*
  6) Collect waste by placing a tube underneath of each SPE cartridge
  7) To Oasis HBL (3 cc, 60 mg) cartridge, add 2.0 ml of MeOH. Add 2.0 ml of 0.1% aqueous ammonium hydroxide solution, apply vacuum
- *Load the sample on the SPE column*
  8) Add pretreated sample (~100 μl) and 250 μl of 0.1% aqueous ammonium hydroxide solution on to SPE almost at one step, fast follow-up step
  9) Let solution drip by gravity and then apply vacuum gently to induce a dropwise flow rate
- *Washing off interfering components*
  10) Add 1.5 ml of 4% MeOH water solution to wash the SPE column, apply vacuum
  11) Add second 1.5 ml of 4% MeOH water solution to wash the SPE column, apply vacuum to dryness
  12) Remove and discard waste tubes underneath the SPE column
- *Eluting analyte*
  13) Place a set of clean collection tubes (5 ml 75 × 100 mm glass tubes) underneath of each SPE column
  14) Add 2.0 ml of methanol to each SPE cartridge
  15) Apply low vacuum first, then apply stronger vacuum to induce dryness at the end
- *Concentrating the analyte in the final solution*
  16) Evaporate methanol extracts to dryness under nitrogen flow at 37 °C for 15–20 min
  17) Reconstitute dried sample with 100 μl of 50% MeOH water solution (v/v, or mobile-phase component at initial chromatographic condition)
  18) Vortex each tube at high speed
- *Preparation for LC-MS/MS injection*
  19) Transfer sample solution to HPLC vials with a 200-μl glass insert
  20) Submit the vials for LC-MS/MS analysis
  21) Inject 25 μl into the LC-MS/MS system

Note: Analytes in the above protocol can be small-molecule antibiotics such as ofloxacin and natamycin. MeOH, methanol.
Source: Bhatta et al. (2011). Reproduced with permission of Elsevier.

Table 18.12 A typical aqueous humor sample preparation procedure using protein precipitation.

1) Thaw frozen aqueous humor samples (in 1-ml cryotubes), and STDs and QC samples (spiked analyte in blank porcine aqueous humor) at room temperature for 30 min
2) Mix each tube by vortexing at a medium speed for ~10 s twice
3) Pipette 50 μl of aqueous humor standards, QCs, and unknown samples into 1.5-ml microcentrifuge tubes
4) Add 150 μl of working internal standard solution containing ACN (or MeOH) to each tube. For blank without IS, add 150 μl ACN (or MeOH)
5) Centrifuge the tubes at 13000 rpm for 5 min
6) Transfer 150 μl of supernatant to V-shaped bottom HPLC glass vials
7) Dry the supernatant fully in the HPLC vials
8) Reconstitute the samples with 100 μl of mobile phase A (or reconstitution solution)
9) Submit the samples tray to the LC-MS/MS system
10) Inject 10 μl

Note: The analyte can be a fluoroquinolone type of antibacterial agents. Mobile phase A is 20% water/80% acetonitrile/0.2% formic acid (v/v/v). ACN, acetonitrile; MeOH, methanol.
Source: Domingos et al. (2017). Reproduced with permission of Elsevier.

Table 18.13 An example of aqueous humor sample preparation procedure for protein analysis and peptide identification.

1) Aliquot protein standard (6 –10 μg) and aqueous humor sample to microfuge tubes (~0.5-ml tube)
2) Dry the above tube in a Christ RVC 2–25 vacuum concentrator (Christ)
3) Add 16 μl of 6 M urea to reconstitute the protein
4) Add 20 μl of dithiothreitol (DTT, 200 mM in 50 mM ammonium bicarbonate)
5) Incubate the tube at room temperature for 45 min
6) Add 25 μl of 1 M iodoacetamide (1 M IA in 50 mM ammonium bicarbonate)
7) Incubate the tube at room temperature for another 45 min
8) Add 100 μl with 50 mM ammonium bicarbonate, and trypsin (Trypsin Gold, Promega Corporation, Madison, WI, USA) at a final trypsin/protein ratio of 1:20
9) Vortex the tube and incubate the tubes overnight at 37 °C
10) Dry the tube (peptides) in an RVC2 25 Speedvac concentrator (Christ)
11) Add 10 μl of 0.1% formic acid (FA) for resuspension
12) Sonicate the tubes for 5 min
13) Aliquots (2 μl) of each sample using C18 ZipTip pipette tips (Millipore) following the manufacturer's protocol (desalting)
14) Dry each aliquot in an RVC2 25 Speedvac concentrator (Christ)
15) Add 3 μl of 0.1% FA and sonicate the sample for 5 min
16) Dilute the sample with 2 μl of 0.1% FA to a final volume of 5 μl
17) Submit the sample for LC-MS/MS analysis

Note: Peptide mixtures can be separated (in triplicate) by online nanoLC and analyzed using electrospray tandem mass spectrometry.
Source: From Soria et al. (2015).

18.7 Sample Preparation Methods for Aqueous Humor

Aqueous humor is a transparent, watery fluid that is secreted from the ciliary epithelium, a structure supporting the lens. The fluid contains 98% water, amino acid, electrolytes, ascorbic acid, glutathione, immunoglobulins, organic solute, growth factors, cytokines, and other proteins involved in the metabolism of the avascular tissues of the anterior eye segment (To et al. 2002). Protein profiling has demonstrated the importance of human aqueous humor in the regulation of many functions of the eye. Changes in human aqueous humor proteomes correlate with various stages of disease progression. Using LC-MS/MS methods to quantify protein concentrations in its profile (Soria et al. 2015) and fluoroquinolones as ophthalmic antimicrobials for eye surgery and treatment of bacteria conjunctivitis have been reported (Donnenfeld et al. 2011). Aqueous humor samples are usually obtained immediately before surgery (Donnenfeld et al. 2011). Approximately 150 μl of aqueous humor was drawn from the anterior chamber of the study eye using a 30-gauge needle attached to a 1-ml syringe. Samples are transferred to pre-labeled cryogenic tubes, frozen immediately at −20 °C or −70 °C, then shipped while frozen to bioanalytical site for bioanalysis. Sample preparation methods for aqueous humor include PPT for small-molecule analytes (Bucci et al. 2016; Domingos et al. 2017; Donnenfeld et al. 2011) and protein digestion (Soria et al. 2015). Tables 18.12 and 18.13 summarize typical sample preparation procedures for aqueous humor.

18.8 Summary

Sample preparation or sample cleanup plays a critical role in successful LC-MS quantification of analyte(s) of interest in urine, CSF, synovial fluid, sweat, tears, and aqueous humor samples. Considering the presence of various forms of interfering substances such as salts, acids, bases, proteins, cells, lipids, and enzymes in these matrices, proper sample preparation procedures should be developed and implemented to minimize or eliminate the interference, if any, in the LC-MS quantification of the analyte(s) with suitable assay's sensitivity.

References

Arning, E. and Bottiglieri, T. (2014). LC-MS/MS analysis of cerebrospinal fluid metabolites in the Pterin biosynthetic pathway. *JIMD Rep.* 29: 1–9.

Arnold, D.R., Granvil, C.P., Ward, K.W., and Proksch, K.W. (2008). Quantitative determination of besifloxacin, a novel fluoroquinolone antimicrobial agent in human tears by liquid chromatography–tandem mass spectrometry. *J. Chromatogr. B* 867: 105–110.

Bennike, T., Ayturk, U., Haslauer, C.M. et al. (2014). A Normative Study of the synovial fluid proteome from healthy porcine knee joints. *J. Proteome Res.* 13 (10): 4377–4387.

Bhatta, R.S., Chandasana, H., Rathi, C. et al. (2011). Bioanalytical method development and validation of natamycin in rabbit tears and its application to ocular pharmacokinetic studies. *J. Pharm. Biomed. Anal.* 54: 1096–1100.

Bucci, F.A. Jr., Nguimfack, I.T., and Fluet, A.T. (2016). Pharmacokinetics and aqueous humor penetration of levofloxacin 1.5% and moxifloxacin 0.5% in patients undergoing cataract surgery. *Clin. Ophthalmol.* 10: 783–789.

Byrro, R.M.D., Fulgêncio, G.D.O., Cunha, A.D.S. Jr. et al. (2012). Determination of ofloxacin in tear by HPLC–ESI-MS/MS method: comparison of ophthalmic drug release between a new mucoadhesive chitosan films and a conventional eye drop formulation in rabbit model. *J. Pharm. Biomed. Anal.* 70: 544–548.

Calderón-Santiago, M., Priego-Capote, F., Jurado-Gámez, B., and Luque de Castro, M.D. (2014). Optimization study for metabolomics analysis of human sweat by liquid chromatography–tandem mass spectrometry in high resolution mode. *J. Chromatogr. A* 1333: 70–78.

Carter, E.P., Barrett, A.D., Heeley, A.F., and Kuzemko, J.A. (1984). Improved sweat test method for the diagnosis of cystic fibrosis. *Arch. Dis. Child.* 59 (10): 919–922.

Cox, J.M., Butler, J.P., Lutzke, B.S. et al. (2015). A validated LC-MS/MS method for neurotransmitter metabolite analysis in human cerebrospinal fluid using benzoyl chloride derivatization. *Bioanalysis* 7 (19): 2461–2475.

Danaceau, J. (2013). Successful sample preparation strategies for LC/MS/MS analysis of drugs in complex biological matrices for forensic toxicology applications. Waters Powerpoint presentation. http://www.waters.com/webassets/cms/events/docs/FINALSamplePrepStrategiesClinical_Forensic11_5_13.pdf (accessed 9 October 2018).

De Giovanni, N. and Fucci, N. (2013). The current status of sweat testing for drugs of abuse: a review. *Curr. Med. Chem.* 20 (4): 545–561.

Delgado-Povedano, M.M., Calderón-Santiago, M., Priego-Capote, F., and Luquede Castro, M.D. (2016). Study of sample preparation for quantitative analysis of amino acids in human sweat by liquid chromatography–tandem mass spectrometry. *Talanta* 146: 310–317.

Domingos, L.C., Moreir, M.V., Keller, K.M. et al. (2017). Simultaneous quantification of gatifloxacin, moxifloxacin, and besifloxacin concentrations in cornea and aqueous humor by LC-QTOF/MS after topical ocular dosing. *J. Pharmacol. Toxicol. Methods* 83: 87–93.

Donnenfeld, E.D., Comstock, T.L., and Proksch, J.W. (2011). Human aqueous humor concentrations of besifloxacin, moxifloxacin, and gatifloxacin after topical ocular application. *J. Cataract. Refract. Surg.* 37: 1082–1089.

El-Gabalawy, H.S. (2013). Synovial fluid analysis, synovial biopsy, and synovial pathology. In: *Kelly's Textbook of Rheumatology*, 9e (ed. G.S. Firestein, R.C. Budd, S.E. Gabriel, et al.), 753–769. Philadelphia, PA: Elsevier Saunders.

ELITechGroup Solutions (2016). *Macroduct® Sweat Collection System*. http://www.elitechgroup.com/product/macroduct-sweat-collection-system (accessed 9 October 2018).

Gu, X.F., Mao, B.Y., Xia, M. et al. (2016). Rapid, sensitive and selective HPLC–MS/MS method for the quantification of topically applied besifloxacin in rabbit plasma and ocular tissues: application to a pharmacokinetic study. *J. Pharm. Biomed. Anal.* 117: 37–46.

Hooton, K., Han, W., and Li, L. (2016). Comprehensive and quantitative profiling of the human sweat submetabolome using high-performance chemical isotope labeling LC-MS. *Anal. Chem.* 88: 7378–7386.

Jadoon, S., Karim, S., Akram, M.R. et al. (2015). Recent developments in sweat analysis and its applications. *Int. J. Anal. Chem.* 2015: 164974. 7 pages.

Ji, A. (2013). LC-MS bioanalysis of drugs in urine. In: *Handbook of LC-MS Bioanalysis, Best Practices, Experimental Protocols, and Regulations* (ed. W. Li, J. Zhang and F. Tse), 307–316. Hoboken, NJ: Wiley.

Ji, A.J., Jiang, Z., Livson, Y. et al. (2010). Challenges in urine bioanalytical assays: overcoming nonspecific binding. *Bioanalysis* 2 (9): 1573–1586.

Kintz, P., Tracqui, A., Mangin, P., and Edel, Y. (1996). Sweat testing in opioid users with a sweat patch. *J. Anal. Toxicol.* 20: 393–397.

Koster, R.A., Alffenaar, J.W.C., Greijdanus, B. et al. (2014). Application of sweat patch screening for 16 drugs and metabolites using a fast and highly selective LC-MS/MS method. *Ther. Drug Monit.* 36: 35–45.

Liu, G. and Aubry, A. (2013). Best practices in biological sample preparation for LC-MS bioanalysis. In: *Handbook of LC-MS Bioanalysis, Best Practices, Experimental Protocols, and Regulations* (ed. W. Li, J. Zhang and F. Tse), 165–184. Hoboken, NJ: Wiley.

Mena-Bravo, A. and Luque de Castro, M.D. (2014). Sweat: a sample with limited present applications and promising future in metabolomics. *J. Pharm. Biomed. Anal.* 90: 139–147.

Montesano, C., Curini, R., Sergi, M. et al. (2016). Determination of marbofloxacin in plasma and synovial fluid by ultrafiltration followed by HPLC-MS/MS. *J. Pharm. Biomed. Anal.* 123: 31–36.

Moshirfar, M., Pierson, K., Hanamaikai, K. et al. (2014). Artificial tears potpourri: a literature review. *Clin. Ophthalmol.* 31 (8): 1419–1433.

Papanas, N., Boulton, A.J.M., Malik, R.A. et al. (2013). A simple new non-invasive sweat indicator test for the diagnosis of diabetic neuropathy. *Diabet. Med.* 30 (5): 525–534.

Pastorini, E., Rotini, R., Guardigli, M. et al. (2009). Development and validation of a HPLC-ES-MS/MS method for the determination of glucosamine in human synovial fluid. *J. Pharm. Biomed. Anal.* 50: 1009–1014.

Raiszadeh, M.M., Ross, M.M., Russo, P.S. et al. (2012). A proteomic analysis of eccrine sweat: implications for the discovery of schizophrenia biomarker proteins. *J. Proteome Res.* 11 (4): 2127–2139.

Rose, C., Parker, A., Jefferson, B., and Cartmell, E. (2015). The characterization of feces and urine: a review of the literature to inform advanced treatment technology. *Crit. Rev. Environ. Sci. Technol.* 45 (17): 1827–1879.

Sato, K., Kang, W.H., Saga, K., and Sato, K.T. (1989). Biology of sweat glands and their disorders. I. Normal sweat gland function. *J. Am. Acad. Dermatol.* 20: 537–563.

Soria, J., Villarrubia, A., Merayo-Lloves, J. et al. (2015). Label-free LC-MS/MS quantitative analysis of aqueous humor from keratoconic and normal eyes. *Mol. Vis.* 21: 451–460.

Tian, D., Feng, F., Li, Y. et al. (2008). Liquid chromatography–mass spectrometry method for the determination of thiamphenicol in rabbit tears. *J. Pharm. Biomed. Anal.* 48: 1015–1019.

To, C.H., Kong, C., Chan, C. et al. (2002). The mechanism of aqueous humor formation. *Clin. Exp. Optom.* 85: 335–349.

Williamson, M.A. and Snyder, L.M. (2011). *Wallach's Interpretation of Diagnostic Tests*, 9e. Philadelphia, PA: Lippincott Williams & Wilkins, a Wolters Kluwer pp. 100, 304.

19

Sample Preparation for LC-MS Bioanalysis of Liposomal Samples

Wenying Jian, Rao N.V.S. Mamidi, and Naidong Weng

19.1 Introduction

Liposomes, according to FDA 2015 draft guidance on liposome drug products, are "microvesicles composed of a bilayer and/or a concentric series of multiple bilayers separated by aqueous compartments formed by amphipathic molecules such as phospholipids that enclose a central aqueous compartment" (FDA 2015). Liposomes have been employed as drug delivery carriers for poorly soluble drugs and/or to provide targeting for cancer treatment (Allen and Cullis 2013; Bobo et al. 2016). There are several advantages of liposome formulation: (i) Controlled and sustained release of the drugs; (ii) Reduced systemic exposure of free drug and therefore reduced toxicity; and (iii) Targeted delivery via the enhanced permeation and retention (EPR) effect due to the leaky vasculature and impaired lymphatic draining of tumor tissue in comparison to normal tissue (Matsumura and Maeda 1986). The classic liposomes typically exhibit short circulating half-life due to recognition by the immune system and clearance by macrophages (Hsu and Juliano 1982; Allen and Hansen 1991). Coating the liposome surface with polyethylene glycol (PEG) results in prolonged *in vivo* half-life due to "steric stabilization" effects, which reduce nonspecific protein–protein interaction and protein–cell interaction (Klibanov et al. 1990; Papahadjopoulos et al. 1991; Senior et al. 1991). Since the first liposome drug product, a PEGylated liposome formulation of doxorubicin was approved by US FDA in 1995 (Northfelt et al. 1996; Barenholz 2012), there have been a growing number of trials and approvals for liposome delivery (Barenholz 2012). Currently, there are approximately a dozen of liposome-based drugs, including both classic and PEGylated liposomes, approved for clinical use (Allen and Cullis 2013; Zylberberg and Matosevic 2016). They include liposomal formulations of doxorubicin, daunorubicin, cytarabine, Amphotericin B, vincristine, and so on, for indications such as cancer and infectious disease.

Most of the approved liposomal drug formulations are for intravenous administration. After being injected to blood stream, liposomal drugs exist in following forms: liposome-encapsulated drugs and nonencapsulated drugs (including protein-bound drug and completely free drug), the sum of which are regarded as total drugs (Figure 19.1). In order for effective assessment and prediction of the efficacy and toxicity of liposomal drugs *in vivo*, it is important to understand the pharmacokinetics (PK) of each of the forms (Jiang et al. 2011). This is particularly important for establishing the equivalency for generic products of liposomal drug products. It has been shown in preclinical studies that bioequivalence of traditional PK (measuring total drug) alone may not be sufficient to provide assurance on biological equivalence (toxicity and efficacy) of liposomal products that vary with minor differences in compositions (Mamidi et al. 2010). According to the 2015 FDA draft guidance, for approval of new liposomal formulations with sustained release, three criteria must be taken into consideration: rate of clearance of liposomal carrier itself, the rate of release of entrapped drugs from the liposomal carrier, and the rate of clearance and metabolism of free drug upon its release (FDA 2015; Zylberberg and Matosevic 2016). This requires validated bioanalytical methods that reliably quantify encapsulated and nonencapsulated drug substance. It should be noted that from regulatory perspective, there is no requirement to separately measure protein-bound and completely free drug. In literature and in FDA guidance, "free drug" is often used to refer to nonencapsulated fractions including both protein-bound and protein-free drugs. For clarification, in this chapter, "nonencapsulated" will be used exclusively to refer to drug outside liposome, while "free" will be specifically used to refer to the fraction free from protein binding.

Sample Preparation in LC-MS Bioanalysis, First Edition. Edited by Wenkui Li, Wenying Jian, and Yunlin Fu.

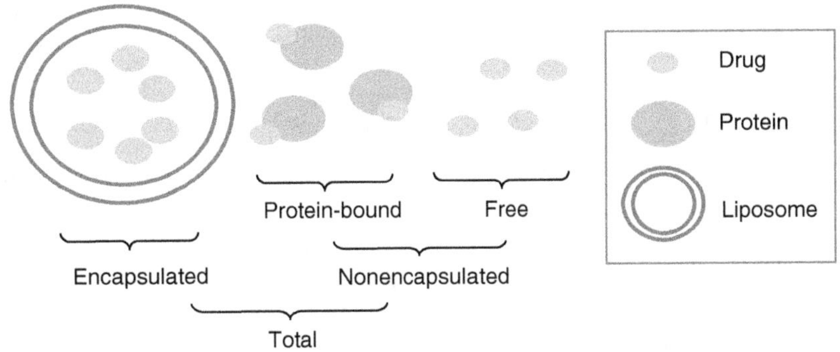

Figure 19.1 Different fractions of the drug in biological samples after administration of liposomal formulation.

It should also be noted that in the FDA guidance, there is differentiation between the terms "encapsulated" and "intercalated," the former of which means drugs being entrapped in the aqueous compartment of liposome as in the cases of hydrophilic drugs, while the latter refers to drugs being contained in the lipid layer(s) as for hydrophobic drugs. In this chapter, both cases are referred to as "encapsulated" since there is no difference between them in terms of bioanalytical quantitation.

Since most of the therapeutic agents used in liposome formulation are small-molecule drugs, the default bioanalytical method is liquid chromatography-mass spectrometry (LC-MS), which provides superior selectivity and sensitivity for analyzing small molecules in biological samples. For a LC-MS-based bioanalysis method, sample preparation to eliminate interference from biological matrix is usually the key step. For liposomal samples, there is additional challenge to separately measure encapsulated and nonencapsulated drugs. Sample preparation to separate the different forms while maintaining integrity of the liposome and avoiding contamination of nonencapsulated drug by liposome is the most critical component of a successful bioanalytical method for liposomal drug. Previous literature studies have shown that majority of the drug in the plasma after liposome formulation administration was liposome-encapsulated (Gabizon et al. 1994; Bekersky et al. 2002; Yu et al. 2013). For example, after administration of liposomal doxorubicin and liposomal vincristine to mice, the nonencapsulated drug levels were less than 0.03 and 0.3% of total drug in the plasma for doxorubicin and vincristine, respectively (Mayer and St-Onge 1995). As shown in Figure 19.2, after a single intravenous administration of liposomal doxorubicin in mouse, the released doxorubicin barely reached 1% of the liposomal drug (Yamamoto et al. 2011). Leakage of drug from liposome or contamination of nonencapsulated drug by liposome during sample preparation, even to a slight extent, could cause overestimation and variability in the measurement of nonencapsulated drug. In the following sections, major types of sample extractions for LC-MS bioanalysis of liposomal drugs and the key consideration to overcome the challenges associated with liposome formulation will be discussed.

19.2 Major Types of Sample Extraction Techniques for Liposomal Samples

To characterize the PK profile of the difference species of a liposomal formulation after being administrated *in vivo*, a proper and reliable sample preparation method to separate nonencapsulated drug from liposome-encapsulated drug is crucial. Published methodologies to separate these two forms include gel filtration (Druckmann et al. 1989), solid-phase extraction (SPE) (Thies et al. 1990; Gabizon et al. 1994; Bellott et al. 2001; Griese et al. 2002; Mross et al. 2004; Deshpande et al. 2010; Wei et al. 2010), ultrafiltration (Mayer and St-Onge 1995; Bekersky et al. 2002; Bedikian et al. 2006), ultracentrifugation (Yamauchi et al. 2008; Yu et al. 2013), and other innovative techniques such as online extraction (Yamamoto et al. 2011) and derivatization (Smits et al. 2013). Among them, SPE is the most commonly used method for *in vivo* sample analysis and it can measure nonencapsulated and encapsulated fractions simultaneously, while ultracentrifugation and ultrafiltration have been shown to measure nonencapsulated concentration. In the cases when the total amount of analyte need to be measured, simple procedure such as protein precipitation (PPT) can be employed to disrupt the liposome and to remove protein (Liu et al. 2008; Wei et al. 2008).

19.2.1 Solid-Phase Extraction

In general, SPE methods utilize electrostatic or hydrophobic interaction between resin and compound to retain nonencapsulated drug while the liposome flows through. Two types of stationary phases are commonly used, reversed-phase C18 and cation exchange, depending on the physiochemical property of the drug. In most of the

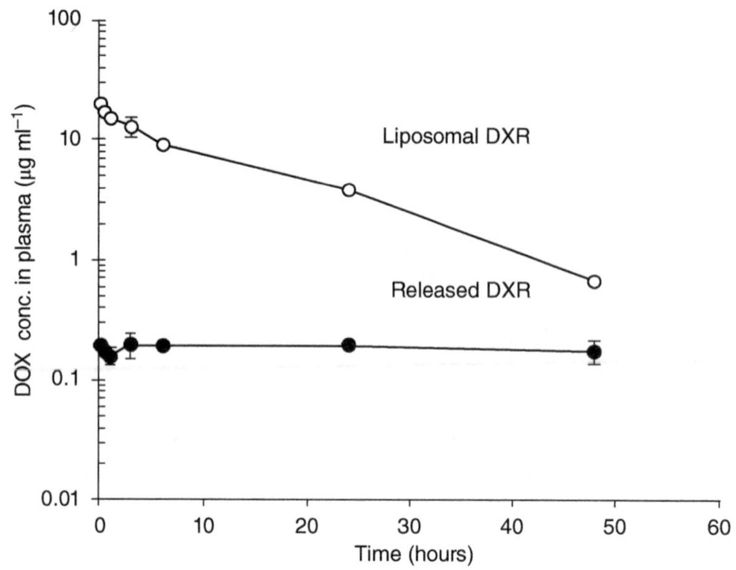

Figure 19.2 Time course of released (nonencapsulated) doxorubicin (DXR) and liposomal encapsulated DXR in mouse plasma after intravenous administration of DXR-containing liposomes (1 mg kg^{-1}, $n = 3$). (●) released (nonencapsulated) DXR and (○) liposomal-encapsulated DXR. *Source*: Reproduced with permission from Yamamoto et al. (2011). Reproduced with permission of Elsevier.

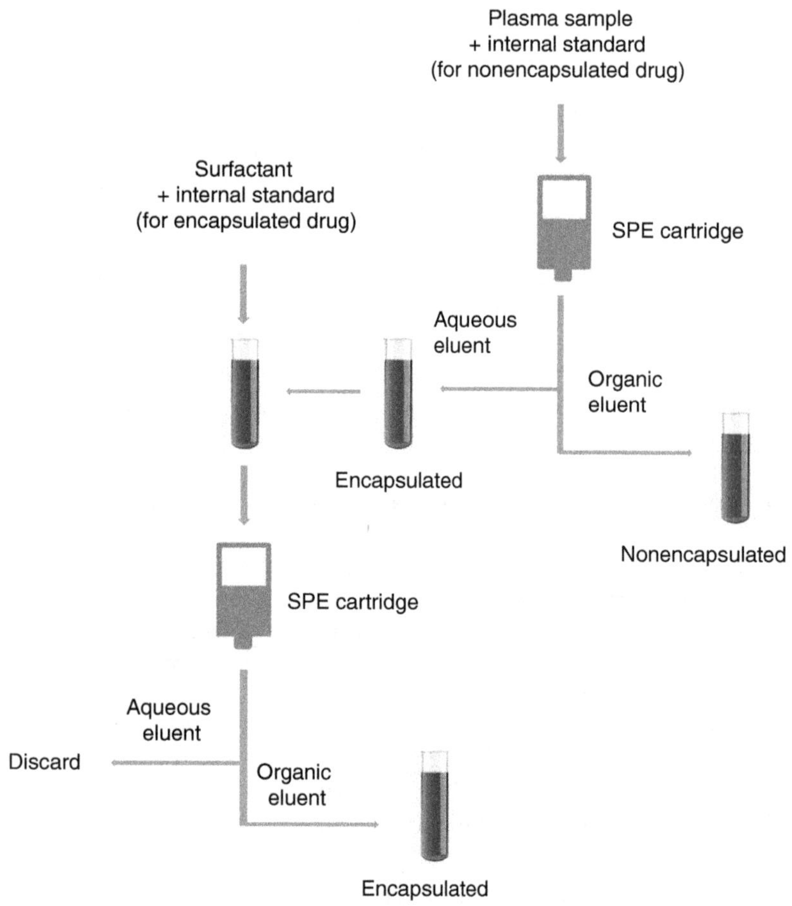

Figure 19.3 Solid-phase extraction procedure for measurement of encapsulated and nonencapsulated liposomal drugs in the plasma sample.

literature studies, C18 SPE column has been employed for various compounds including doxorubicin (Mross et al. 2004; Wei et al. 2010), daunorubicin (Bellott et al. 2001; Griese et al. 2002), and amphotericin B (Deshpande et al. 2010), while there were also reports on using cation exchange resin for analysis of doxorubicin (Gabizon et al. 1994). Figure 19.3 depicts a general workflow of SPE procedure for the separation of nonencapsulated and encapsulated liposomal drugs in plasma samples. When the plasma samples are loaded to the SPE cartridge, nonencapsulated drug is retained on the cartridge and the liposomal components pass through and are collected in the aqueous eluent. The cartridge is then eluted with organic solvent such as methanol to collect the nonencapsulated drug for LC-MS analysis. The liposome in the aqueous collection needs to be further processed to disrupt the liposome and release the encapsulated drug for LC-MS analysis. This is often achieved by a simple PPT procedure using high volume of organic solvent. Alternatively, the aqueous collection can be treated with surfactant to disrupt the liposome membrane followed by a secondary reversed-phase SPE to remove the lipid components and the surfactant before LC-MS analysis.

The SPE method can be conducted in most of the modern biological laboratories without requirement of additional equipment and devices. However, in practice, SPE has its limitations, the major one being drug release from liposomes during sample preparation, leading to artificially elevated concentration of nonencapsulated drug (Mayer and St-Onge 1995). The SPE loading and washing conditions should be carefully optimized to avoid rupturing the liposome. In general, extreme dilution, extreme pH, organic solvent, and vacuum should be avoided during the sample loading step and the aqueous washing step for collection of liposome components. Usually, there is a balance between providing an environment for nonencapsulated analyte to retain on the stationary phase while maintaining the integration of the liposome. For example, for retention of Amphotericin B on a SPE cartridge, 250 μl of plasma sample was buffered with 500 μl of 0.1% aqueous ammonium solution, the concentration of which was limited to 0.1% in such a way that it did not break liposomes in plasma while holding the nonencapsulated drug on the cartridge (Deshpande et al. 2010). In addition, the washing should be thorough to avoid contamination of nonencapsulated drug by residual liposome that might have retained on the stationary phase and co-elute with the nonencapsulated drug. This usually requires washing with copious volume of aqueous buffer.

19.2.2 Ultrafiltration

Ultrafiltration utilizes the molecular weight cutoff (MWCO) mechanism of a filtration device to separate free drug from liposome-encapsulated drug. In an ultrafiltration procedure, plasma samples are loaded to an ultrafiltration filter device and centrifuged (e.g. 3000 g for one hour). The free drug passes through the filter and is collected in the reservoir below, while the liposomal and protein-bound nonencapsulated drug is retained above the filter. The filtrate can be directly subject to LC-MS analysis. There are only a few literature studies showing quantitation of protein-bound free drug concentration in PK studies of liposomal drugs (Mayer and St-Onge 1995; Bekersky et al. 2002; Bedikian et al. 2006). All of them uses filter with MWCO of 30 kDa (pore size <5 nm) which can effectively prevent pass-through of liposome particles (usually ~100 nm) (Mayer et al. 1994). It has been shown that the ultrafiltration method provides lower process-induced leakage of liposome than SPE because it is an equilibrium method with milder conditions. In one study comparing the two techniques, the percentage of liposome contamination into the free fraction has been shown to be 1.29–8.38% for SPE while negligible (<0.2%) lipid could be detected using ultrafiltration devices (Mayer and St-Onge 1995). The procedure of ultrafiltration is simple and straightforward, and does not require additional instrument rather than a benchtop centrifuge machine. Therefore, it can be conducted directly on-site immediately after the plasma is collected, which eliminates complication caused by potential stability issue associated long-term storage and freeze–thaw of the plasma samples, as will be discussed in more detail in Section 19.3.1.

A potential issue associated with ultrafiltration is adsorption of free drug to the filtration device, mainly the filter membrane. It has been shown that adsorption of doxorubicin and vincristine dissolved in saline to the filtration device resulted in recovery ranging from 41 to 96%, for concentrations ranging from 0.04 to 4.0 $\mu g\,ml^{-1}$ (Mayer and St-Onge 1995). The issue can be overcome by treating the device with passivation solution such as 5% PEG-8000. In plasma samples, the adsorption issue is ameliorated and the results obtained for protein binding were similar regardless whether the device was pretreated (Mayer and St-Onge 1995). Nevertheless, adsorption of free drug to the filtration device or the collection container should always be evaluated for the ultrafiltration method because drug at low concentrations in a protein-free environment is highly prone to such issue. Other potential problem of the ultrafiltration method is defected filter that could result in errors. In a PK study of liposomal vincristine in cancer patients, free vincristine concentrations in patients after intravenous infusion were not consistent between two cycles of treatments, which was attributed to potential assay errors due to mechanical failure of the ultrafiltration device (Bedikian et al. 2006). Very importantly, the ultrafiltration method is not useful for recovery of encapsulated drugs and therefore a separate method, usually SPE, has to be established if the encapsulated fraction to be determined.

19.2.3 Ultracentrifugation

Ultracentrifugation method has been used traditionally to determine plasma protein binding. It uses gravitational forces to separate components of different sizes in the samples. There have been only a couple of published examples of using this technique to quantify free drug concentrations for liposomal samples (Yamauchi et al. 2008; Yu et al. 2013). One such study is determination of free eribulin mesylate (E7389) in mice after liposomal dose (Yu et al. 2013). In that study, 1 ml of plasma samples was centrifuged at 10 °C for four hours at approximately 100 000 rpm (355 040 × *g*). Aliquot of 150 µl was removed from the top layer using a syringe carefully to avoid contamination from the bottom layer which contains the liposomal drug and the protein-bound drug. Protein concentration of the ultracentrifuged top layer was determined to evaluate the effectiveness of the separation and was shown to be only 1%, indicating a good separation of free drug from the rest of the components. The advantage of ultracentrifugation is that it is simple and free from adsorption issue associated with the ultrafiltration method. However, it is time-consuming and requires expensive equipment and special operational skills. It may also be subject to errors due to diffusion of the molecules in the centrifuged samples. Overall, it is not considered a method that can be conducted in high-throughput manner for routine sample analysis.

19.2.4 Other Methods

There are other techniques available for separation of nonencapsulated/free drug from encapsulated drugs, such as gel filtration, Ficoll density gradient, dialysis, protamine aggregation, and so on (Dipali et al. 1996). However, they were only used for analysis of liposomal formulation and not applicable for analysis of relatively low concentration of drugs in biological samples. Recently, a new method based on a capillary electrophoresis (CE) separation of doxorubicin and encapsulated doxorubicin followed by laser-induced fluorescence detection was developed for analysis of plasma samples (Kim and Wainer 2010). The CE conditions such as column dimension, run time, and running buffer were optimized to achieve complete separation of free and encapsulated doxorubicin without significant leaching of free drug from liposome. The only pretreatment of the plasma sample before injection to the CE is dilution fourfold by running buffer. Another online SPE-SPE-HPLC-fluorescence method based on restricted access material (RAM) was developed for analysis of plasma samples (Yamamoto et al. 2011). During the online sample preparation process, nonencapsulated doxorubicin was extracted on the first RAM SPE column, whereas the liposomes were eluted and collapsed online, and the released doxorubicin was delivered to the second RAM SPE column for extraction. RAM is an adsorption medium that allows the retention of small molecules in the stationary phase, while larger size molecules are prevented from accessing the inner layer by size exclusion and hydrophilic interaction. It was proposed that by using RAM, leaking of doxorubicin from liposome on conventional reversed-phase C18 SPE can be minimized because the release of drug was attributed to the hydrophobic interaction of the liposome with the C18 stationary phase. Another novel method to distinguish nonencapsulated from encapsulated drug involves transforming the former into a different structure. In this unique example for analyzing liposomal formulation of prodrug prednisolone phosphate, the free prodrug was converted into prednisolone by adding alkaline phosphatase to the sample, while the prodrug encapsulated in the liposome remains unchanged (Smits et al. 2013). The samples were then precipitated using organic solvent and both the prednisolone and prednisolone phosphate were analyzed simultaneously on reversed-phase HPLC-UV for determination of nonencapsulated and encapsulated fractions, respectively. All the above-discussed novel sample preparation methods can potentially be applied for MS-based bioanalysis of plasma samples.

19.3 Key Considerations in Sample Preparation for Liposomal Samples

19.3.1 Stability

The biggest challenge of liposomal sample bioanalysis resides in the nature of the formulation for which majority of the drugs are encapsulated in the liposome while only a small percentage is present outside. For samples containing low level of nonencapsulated drug (e.g. <1%), leakage or contamination of the liposome to the nonencapsulated fraction during sample storage or extraction process would cause significant elevation of its concentration, therefore, giving misleading data. This phenomenon has been well documented and discussed in past literature studies (Mayer and St-Onge 1995; Griese et al. 2002; Zamboni et al. 2007; Deshpande et al. 2010; Kim and Wainer 2010; Yamamoto et al. 2011). Kim and Wainer used the novel CE fluorescence method to quantify free and encapsulated doxorubicin in plasma as shown in the previous section. Measurement of free fraction of liposomal doxorubicin in buffer was accurate while the assay failed to meet the FDA guideline when the liposomal doxorubicin was spiked in plasma (Kim and Wainer 2010). Further investigation showed that the percentage of free doxorubicin derived from a set concentration of liposomal doxorubicin varied with different spiking concentrations as well as different batches of human plasma. It was proposed that leakage from liposomal formulation was due to an active process most likely by hydrolytic enzyme such as lipase which could be impacted by diverse factors such as age, gender, disease status, etc. Kaminskas et al. (2012) argue that there is continual reestablishment of the equilibrium between encapsulated and nonencapsulated fractions during the separation process which dictated that levels of nonencapsulated drug lower than 1% of total plasma drug could not be assayed with acceptable accuracy.

In addition to precaution during sample extraction as discussed in the previous section, storage stability is the key for reproducible measurement of nonencapsulated drug concentration. Particularly, freeze–thaw should be avoided as much as possible. The best approach is to freshly separate free drug from liposomal drug immediately after plasma sample collection, a practice adopted in some previously reported studies (Bedikian et al. 2006; Zamboni et al. 2007; Peyrl et al. 2009). This requires suitable equipment and experienced personnel at the same site of clinical study. From this perspective, ultrafiltration is relatively easier to implement than SPE as it requires less specialized devices and skills to operate, unless there is dedicated bioanalytical facility at the clinical site. If freeze–thaw is unavoidable, cryoprotectant reagent should be added to the sample to protect liposome from rupturing during the freeze–thaw steps (Van Bommel and Crommelin 1984). Wei et al. (2010) have shown that mixing glycerol with plasma (1 : 5, v : v) can effectively reduce the release of doxorubicin from liposome in plasma samples. Mross et al. (2004) used 55% glucose solution mixed with plasma at a ratio of 1 : 10 as a cryoprotectant reagent. It has been also recommended that the freeze–thaw cycle should be minimized to only one but for the analysis (Deshpande et al. 2010). In addition to freeze–thaw, prolonged storage of plasma sample in thawed condition should also be avoided. Yamamoto et al. (2011) showed that plasma liposomal samples contained 4.8% of nonencapsulated doxorubicin stored at 4 °C continued to release doxorubicin to 5.9 and 6.3% over 24 and 48 hours of storage, respectively.

Measurement of metabolite of free drug has been proposed as one approach to circumvent the issue of liposomal rupture or contamination because it can be readily distinguished from the parent drug using MS-based methods. For example, doxorubicinol, a metabolite of doxorubicin, instead of the free parent drug itself, has been quantified in a PK study of liposomal doxorubicin (Mross et al. 2004). However, the level of doxorubicin in patients after liposomal doxorubicin infusion is only 1–20 $ng\,ml^{-1}$ while that of doxorubicinol is even 2- to 100-fold lower, presenting serious challenge of accurate measurement. It has also been argued that measurement of doxorubicinol can provide some supportive PK information but not necessarily direct correlation with the free doxorubicin level because tissue update and processing of liposomal doxorubicin can also release doxorubicinol (Jiang et al. 2011).

Finally, the impact of liposomal rupture and contamination on the measurement of nonencapsulated drug concentration should be considered in the context of overall relative levels of different components in the sample. For example, in the cases where nonencapsulated drug is expected to be >10% of the total concentration, the impact from coexistence of liposome could be negligible, whereas for instances of <1%, accurate measurement of nonencapsulated concentration could be in fact impossible, even with all the precautions in place.

19.3.2 Protein Binding

The nonencapsulated fraction of drug in a liposomal sample is composed of protein-bound portion and truly free fraction. Different sample extraction methods and quantitation strategy dictates which portion is measured. SPE-based methods do not separate protein-bound free drug from actually free drug, because both forms can be retained on the SPE cartridge (Bellott et al. 2001). Free concentration determined by the SPE method therefore is the entire fraction of nonencapsulated drugs, including both protein-bound free drug and completely free drug. Unlike ultrafiltration and ultracentrifugation methods, which have been used to determine plasma protein binding of drugs dosed with liposomal formulation, SPE is not suitable for such purpose.

Ultrafiltration is one of the standard procedures for determination of protein binding. For liposomal samples, during the ultrafiltration process, the protein-bound portion of the nonencapsulated drug is retained above the filter and only the protein-free drugs go through the filter and are collected for LC-MS analysis. If the calibration samples are prepared by directly spiking reference standard in filtrate, the measurement is absolute concentration of free drug in the filtrate and thus concentration of protein-bound free portion of nonencapsulated drugs in the samples. Alternatively, if the calibration standard samples are prepared in plasma and taken through the sample ultrafiltration process before LC-MS analysis, the measurement should be considered as combined nonencapsulated drug concentrations regardless of protein binding. This is because the protein-bound portion of free drug in the calibration samples is also removed during ultrafiltration and therefore the loss in the samples are corrected. An assumption here is that the protein-binding are not significantly different between incurred and spiked plasma samples, which may not always be true.

19.3.3 Quality Control Samples

For a bioanalytical method to quantify accurately and precisely, the calibration standard samples should be prepared in appropriate matrix and quality control

(QC) samples should mimic the composition of the incurred samples. This is particularly important for liposomal samples for which the different components may have impact on each other. Calibration curve for encapsulated and nonencapsulated drug can be prepared by spiking the liposomal and free drug to plasma, respectively. QC sample should contain both encapsulated and nonencapsulated fraction in order to fully evaluate the impact of these two components on each other. This is especially important for the assay of nonencapsulated drug due to the potential rupture or contamination from the liposomal portion. On the other hand, it is less likely that the nonencapsulated drug can significantly affect the measurement of encapsulated fraction, given the low percentage of the former. Nevertheless, QC samples containing both components should be included (Deshpande et al. 2010). For full understanding and evaluation of the assay performance for each individual component and their impact on each other, three sets of QCs are recommended, including QC prepared by spiking liposomal drug only (liposomal QC, to be used in encapsulated assay), free drug only (nonencapsulated QC, to be used in nonencapsulated assay), and both liposomal and free drug (mixed QC, to be used for both assays). The ratio of the nonencapsulated/encapsulated drugs in the mixed QC samples should mimic the expected ratio of the two fractions in the incurred samples in order to fully evaluate the potential impact from liposomal drugs during storage and sample preparation.

19.3.4 Matrix Effect

For measurement of encapsulated drug in a liposomal sample using LC-MS methods, the liposomes need to be ruptured to release drug, after being separated from nonencapsulated fraction and before injection to the instrument. Same is true for determination of the total drug level in the sample for which the liposomes also need to be fully burst. The rupture can be induced by adding organic solvent, detergent, pH modifier, or increasing the temperature. The resulting samples contain the components from liposome such as phospholipids, cholesterol, and PEG, all of which are known to cause matrix effects in ion source of MS and potential analytical variability (Tong et al. 2002; Shou and Naidong 2003). The sample should be further purified by procedures such as SPE to remove the matrix effect-causing components. If no purification is conducted, potential matrix effects caused by the lipid components should be thoroughly investigated during method development by monitoring the elution of phospholipids and/or by post-column infusion (King et al. 2000; Little et al. 2006). If there are significant matrix effects, mitigations such as stable isotope-labeled internal standard, atmospheric pressure chemical ionization (APCI) rather than electrospray (ESI) mode of ionization, or extensive chromatographic gradient to separate the analyte from lipids should be incorporated into the method.

19.3.5 Addition of Internal Standard

Use of internal standard for LC-MS-based bioanalysis can effectively compensate for assay viabilities caused by sample extraction, injection, chromatography, instrument fluctuation, and matrix effects. An ideal internal standard is a stable-isotope labeled analog of the analyte that exhibits exactly identical physicochemical properties as the analyte but only difference in molecular weight. In the cases when no stable isotope-labeled compound is available, structure analog can also serve the purpose with less effectiveness. Internal standard should be added in equal amount to all the samples as early as possible during sample extraction so that the entire bioanalytical process is tracked by the internal standard. However, for bioanalysis of liposomal samples, addition of internal standard depends on the assay procedure and targeted components. In the SPE procedure for measurement of nonencapsulated analyte, internal standard should be mixed with the sample before the sample extraction (Figure 19.3). The internal standard in its free form will bind to the stationary phase as same as the analyte and therefore provide effective tracking of the nonencapsulated fraction of the analyte for the entire bioanalytical process. On the other hand, the internal standard cannot track with the encapsulated analyte, which are not retained on the stationary phase and are collected during the loading and initial aqueous washing steps. Internal standard for the encapsulated drug should be added to the collected samples before further procedures including rupturing of the liposome, purification (if any), and LC-MS analysis so that all the subsequent processes are covered (Figure 19.3). This is particularly important for controlling of matrix effects potentially caused by the lipid components of the encapsulated fraction. For analysis of nonencapsulated analyte using ultrafiltration, internal standard should be mixed with the samples before loading to the device so that potential adsorption loss to the filter membrane and variable collection volumes can be effectively controlled. For measurement of nonencapsulated analyte by using either SPE or ultrafiltration, the internal standard should be prepared in aqueous solution and mixed gently with the samples by low-speed vortexing or inverting capped tubes. For encapsulated analyte, such precautions are not necessary since the liposome will nevertheless be burst and therefore the internal standard can be prepared in the organic solvent being used for rupturing the samples and mixed vigorously. Same is true for measurement of total analyte for which internal standard can be included in the PPT solvent which also serves the purpose of rupturing the liposome.

19.4 Typical Protocols

19.4.1 SPE Procedure for Measurement of Nonencapsulated and Encapsulated Analyte in Liposomal Samples (Modified Based on Bellott et al. (2001), Wei et al. (2010), and Mross et al. (2004))

1) Condition reversed-phase C18 SPE-PAK column (30 mg, 1 ml) using 1 ml of methanol and 1 ml of water.
2) Mix 0.2 ml of plasma sample with 0.05 ml of internal standard working solution in water by gentle vortexing.
3) Place new collection tube under the SPE column.
4) Load the sample to the SPE column. Let the liquid pass through the column under atmospheric pressure.
5) Add 1 ml of water to the column. Let the liquid pass through the column under atmospheric pressure. Collect the follow-through in the same collection tube (Collection 1).
6) Remove the collection tube for Collection 1.
7) Wash the column with 1 ml of water two times.
8) Place new collection tube under the SPE column.
9) Add 1 ml of methanol to elute nonencapsulated analyte and internal standard. Collect the flow-through in the collection tube (Collection 2).
10) Collection 2 is evaporated under nitrogen and reconstituted into appropriate solvent for injection into the LC-MS system for analysis of **nonencapsulated analyte**.
11) To Collection 1, add 0.1 ml internal standard solution and 0.5 ml of Triton X-100 detergent to disrupt the liposome. Vortex to mix thoroughly.
12) Condition a new set of reversed-phase C18 SPE-PAK columns (30 mg, 1 ml) using 1 ml of methanol and 1 ml of water.
13) Load sample from step 11 to the SPE column.
14) Wash the column with 1 ml of water three times.
15) Place new collection tube under the SPE column.
16) Add 1 ml of methanol to elute the analyte and internal standard. Collect the follow-through in the collection tube (Collection 3).
17) Collection 3 is evaporated under nitrogen and reconstituted into appropriate solvent for injection into the LC-MS system for analysis of **encapsulated analyte**.

Notes

1) The internal standard solution used in step 2 should be in aqueous solution to avoid disrupting the liposomes in the samples.
2) At step 2, plasma may need to be mixed with buffer to adjust pH in order to retain nonencapsulated analyte on the SPE. The buffer should be aqueous and the final pH should be optimized to avoid rupture of liposome.
3) Mixing of the sample in step 2 should be done by gently vortexing at low setting.
4) Washing can be done using appropriate aqueous buffer instead of pure water.
5) The SPE should be operated under atmospheric pressure without applied vacuum before step 8 to avoid rupturing the liposome. After that, disruption of liposome is not a concern and vacuum can be applied.

19.4.2 Ultrafiltration Procedure for Measurement of Nonencapsulated Analyte in Liposomal Samples (Modified Based on Bedikian et al. (2006) and Mayer and St-Onge (1995))

1) Mix 0.2 ml of plasma sample with 0.05 ml of internal standard working solution in water by gentle vortexing.
2) Add the sample to a Centrifree (1.0 ml) device with 30 kDa MWCO.
3) Centrifuge at 4 °C at 2000 g for 30 minutes in a centrifuge with fixed-angle rotor.
4) Transfer the ultrafiltrate in the collection reservoir for injection to the LC-MS system for analysis of **nonencapsulated analyte**.

Notes

1) The internal standard solution should be in aqueous solution to avoid disrupting the liposome in the samples.
2) Mixing of the sample in step 1 should be done by gently vortexing at low setting.
3) The volume of the filtrate may vary and prolonged centrifugation may be needed in order to increase the volume of the filtrate.
4) The plasma sample may be too thick to pass through the filter. In this case, the procedure can include a dilution step. The plasma sample can be mixed with aqueous solution before being loaded to the filter. Appropriate buffer and volume should be tested to avoid disrupting the liposome.
5) In order to improve recovery of analyte that is prone to adsorption loss, the filter can be pretreated by passing solutions such as 5% PEG-8000 in water, 1% bovine serum albumin (BSA), or 1% milk powder in water.

19.4.3 Protein Precipitation Procedure for Measurement of Total Analyte in Liposomal Samples (Modified from Wei et al. (2008))

1) Mix 0.2 ml of plasma sample with 0.05 ml of internal standard working solution.
2) Add 0.5 ml of acetonitrile and mixed thoroughly.
3) Centrifuge the sample at 15 000*g* for 10 minutes at room temperature.
4) Transfer 0.6 ml of supernatant to a new tube and evaporate under nitrogen.
5) Reconstituted into appropriate solvent for injection into the LC-MS system for analysis of **total analyte**.

Notes

1) Suitable internal standard, preferably stable isotope-labeled ones, is important for controlling matrix effects caused by the lipids derived from liposome.
2) If the matrix effect is severe, an additional sample purification procedure such as SPE or liquid–liquid extraction can be applied to the protein-precipitated samples for further purification.

19.5 Summary

Liposomal formulation has been clinically applied for targeted delivery of cancer drugs and other poorly soluble drugs. After a liposome drug is administered to patients, the total amount of drug in plasma may exist in different forms. Majority of the drug are being trapped in the liposome, known as the encapsulated fraction, while a small amount of drugs are present outside of liposome as nonencapsulated fraction, which consisted of protein-bound form and completely free form. It is important to determine the concentration of each subpopulation in order to gain a full understanding of the pharmacokinetic behavior of liposomal drug, and its efficacy and toxicity properties and this is required by health authorities for review of liposomal drug product submissions for approval.

In order to reliably quantify the concentrations of encapsulated and nonencapsulated drugs, a sample extraction method to effectively separate these two forms is the key. SPE can be applied for analysis of both fractions, while ultrafiltration can be used for analysis of nonencapsulated fraction. During SPE, liposomes flow through under aqueous condition while nonencapsulated drug are retained, and the two fractions can be collected separately for analysis. SPE can be conducted in most modern bioanalytical laboratories without requirement of additional equipment and devices and has the advantage of simultaneous measurement of both encapsulated and nonencapsulated fractions in one assay. However, it has the limitation of being prone to drug release from liposome during extraction, leading to artificially elevated concentration of nonencapsulated drugs. The SPE conditions should be carefully optimized to avoid such issue. In the ultrafiltration process, a MWCO membrane retains the liposomal drug above the filter while the free drugs pass through and are collected for LC-MS analysis. The ultrafiltration method has the advantage of being simple and straightforward. It can be conducted at the clinical sites immediately after plasma sample collection, therefore minimizing potential complication associated with frozen storage and freeze–thaw of the samples. However, it can only measure nonencapsulated fraction and an assay for quantitation of encapsulated drugs needs to be separately established. Nonspecific adsorption loss of free drug to the filtration device should also be evaluated and managed.

For *in vivo* samples, encapsulated drugs exist as the major form and the nonencapsulated drug is often only a small fraction of total drug. Slight rupture of liposome or contamination from liposome may potentially cause significant overestimation of the concentration of nonencapsulated drugs. Care should be given during the entire process of sample collection, storage, and sample extraction to avoid disruption of the liposomal structure. In addition, appropriate bioanalytical strategies for stabilization of the samples, preparation of QCs, control of matrix effect, addition of internal standard, and so on, should be adopted for establishment of reliable assays for analysis of incurred samples.

References

Allen, T.M. and Cullis, P.R. (2013). Liposomal drug delivery systems: from concept to clinical applications. *Adv. Drug Deliv. Rev.* 65 (1): 36–48.

Allen, T.M. and Hansen, C. (1991). Pharmacokinetics of stealth versus conventional liposomes: effect of dose. *Biochim. Biophys. Acta* 1068 (2): 133–141.

Barenholz, Y. (2012). Doxil-the first FDA-approved nano-drug: lessons learned. *J. Control. Release* 160 (2): 117–134.

Bedikian, A.Y., Vardeleon, A., Smith, T. et al. (2006). Pharmacokinetics and urinary excretion of vincristine sulfate liposomes injection in metastatic melanoma patients. *J. Clin. Pharmacol.* 46 (7): 727–737.

Bekersky, I., Fielding, R.M., Dressler, D.E. et al. (2002). Plasma protein binding of amphotericin B and pharmacokinetics of bound versus unbound amphotericin B after administration of intravenous liposomal amphotericin B (AmBisome) and amphotericin B deoxycholate. *Antimicrob. Agents Chemother.* 46 (3): 834–840.

Bellott, R., Pouna, P., and Robert, J. (2001). Separation and determination of liposomal and non-liposomal daunorubicin from the plasma of patients treated with Daunoxome. *J. Chromatogr. B Biomed. Sci. Appl.* 757 (2): 257–267.

Bobo, D., Robinson, K.J., Islam, J. et al. (2016). Nanoparticle-based medicines: a review of FDA-approved materials and clinical trials to date. *Pharm. Res.* 33 (10): 2373–2387.

Deshpande, N.M., Gangrade, M.G., Kekare, M.B., and Vaidya, V.V. (2010). Determination of free and liposomal amphotericin B in human plasma by liquid chromatography-mass spectroscopy with solid phase extraction and protein precipitation techniques. *J. Chromatogr. B Analyt. Technol. Biomed. Life Sci.* 878 (3–4): 315–326.

Dipali, S.R., Kulkarni, S.B., and Betageri, G.V. (1996). Comparative study of separation of non-encapsulated drug from unilamellar liposomes by various methods. *J. Pharm. Pharmacol.* 48 (11): 1112–1115.

Druckmann, S., Gabizon, A., and Barenholz, Y. (1989). Separation of liposome-associated doxorubicin from non-liposome-associated doxorubicin in human plasma: implications for pharmacokinetic studies. *Biochim. Biophys. Acta* 980 (3): 381–384.

FDA (2015). *Liposome Drug Products: Chemistry, Manufacturing, and Controls; Human Pharmacokinetics and Bioavailability; and Labeling Documentation. Guidance for Industry*. Silver Spring, MD: U.S. Food and Drug Administration.

Gabizon, A., Catane, R., Uziely, B. et al. (1994). Prolonged circulation time and enhanced accumulation in malignant exudates of doxorubicin encapsulated in polyethylene-glycol coated liposomes. *Cancer Res.* 54 (4): 987–992.

Griese, N., Blaschke, G., Boos, J., and Hempel, G. (2002). Determination of free and liposome-associated daunorubicin and daunorubicinol in plasma by capillary electrophoresis. *J. Chromatogr. A* 979 (1–2): 379–388.

Hsu, M.J. and Juliano, R.L. (1982). Interactions of liposomes with the reticuloendothelial system. II: nonspecific and receptor-mediated uptake of liposomes by mouse peritoneal macrophages. *Biochim. Biophys. Acta* 720 (4): 411–419.

Jiang, W., Lionberger, R., and Yu, L.X. (2011). In vitro and in vivo characterizations of PEGylated liposomal doxorubicin. *Bioanalysis* 3 (3): 333–344.

Kaminskas, L.M., McLeod, V.M., Kelly, B.D. et al. (2012). A comparison of changes to doxorubicin pharmacokinetics, antitumor activity, and toxicity mediated by PEGylated dendrimer and PEGylated liposome drug delivery systems. *Nanomedicine* 8 (1): 103–111.

Kim, H.S. and Wainer, I.W. (2010). Simultaneous analysis of liposomal doxorubicin and doxorubicin using capillary electrophoresis and laser induced fluorescence. *J. Pharm. Biomed. Anal.* 52 (3): 372–376.

King, R., Bonfiglio, R., Fernandez-Metzler, C. et al. (2000). Mechanistic investigation of ionization suppression in electrospray ionization. *J. Am. Soc. Mass Spectrom.* 11 (11): 942–950.

Klibanov, A.L., Maruyama, K., Torchilin, V.P., and Huang, L. (1990). Amphipathic polyethyleneglycols effectively prolong the circulation time of liposomes. *FEBS Lett.* 268 (1): 235–237.

Little, J.L., Wempe, M.F., and Buchanan, C.M. (2006). Liquid chromatography-mass spectrometry/mass spectrometry method development for drug metabolism studies: examining lipid matrix ionization effects in plasma. *J. Chromatogr. B Analyt. Technol. Biomed. Life Sci.* 833 (2): 219–230.

Liu, Y., Yang, Y., Liu, X., and Jiang, T. (2008). Quantification of pegylated liposomal doxorubicin and doxorubicinol in rat plasma by liquid chromatography/electrospray tandem mass spectroscopy: application to preclinical pharmacokinetic studies. *Talanta* 74 (4): 887–895.

Mamidi, R.N., Weng, S., Stellar, S. et al. (2010). Pharmacokinetics, efficacy and toxicity of different pegylated liposomal doxorubicin formulations in preclinical models: is a conventional bioequivalence approach sufficient to ensure therapeutic equivalence of pegylated liposomal doxorubicin products? *Cancer Chemother. Pharmacol.* 66 (6): 1173–1184.

Matsumura, Y. and Maeda, H. (1986). A new concept for macromolecular therapeutics in cancer chemotherapy: mechanism of tumoritropic accumulation of proteins and the antitumor agent smancs. *Cancer Res.* 46 (12 Pt 1): 6387–6392.

Mayer, L.D., Cullis, P.R., and Bally, M.B. (1994). The use of transmembrane pH gradient-driven drug encapsulation in the pharmacodynamic evaluation of liposomal doxorubicin. *J. Lipid Res.* 4 (1): 529–553.

Mayer, L.D. and St-Onge, G. (1995). Determination of free and liposome-associated doxorubicin and vincristine levels in plasma under equilibrium conditions employing ultrafiltration techniques. *Anal. Biochem.* 232 (2): 149–157.

Mross, K., Niemann, B., Massing, U. et al. (2004). Pharmacokinetics of liposomal doxorubicin (TLC-D99; Myocet) in patients with solid tumors: an open-label, single-dose study. *Cancer Chemother. Pharmacol.* 54 (6): 514–524.

Northfelt, D.W., Martin, F.J., Working, P. et al. (1996). Doxorubicin encapsulated in liposomes containing surface-bound polyethylene glycol: pharmacokinetics, tumor localization, and safety in patients with AIDS-related Kaposi's sarcoma. *J. Clin. Pharmacol.* 36 (1): 55–63.

Papahadjopoulos, D., Allen, T.M., Gabizon, A. et al. (1991). Sterically stabilized liposomes: improvements in pharmacokinetics and antitumor therapeutic efficacy. *Proc. Natl. Acad. Sci. U. S. A.* 88 (24): 11460–11464.

Peyrl, A., Sauermann, R., Traunmueller, F. et al. (2009). Pharmacokinetics and safety of intrathecal liposomal cytarabine in children aged <3 years. *Clin. Pharmacokinet.* 48 (4): 265–271.

Senior, J., Delgado, C., Fisher, D. et al. (1991). Influence of surface hydrophilicity of liposomes on their interaction with plasma protein and clearance from the circulation: studies with poly(ethylene glycol)-coated vesicles. *Biochim. Biophys. Acta* 1062 (1): 77–82.

Shou, W.Z. and Naidong, W. (2003). Post-column infusion study of the "dosing vehicle effect" in the liquid chromatography/tandem mass spectrometric analysis of discovery pharmacokinetic samples. *Rapid Commun. Mass Spectrom.* 17 (6): 589–597.

Smits, E.A., Smits, C.J., and Vromans, H. (2013). The development of a method to quantify encapsulated and free prednisolone phosphate in liposomal formulations. *J. Pharm. Biomed. Anal.* 75: 47–54.

Thies, R.L., Cowens, D.W., Cullis, P.R. et al. (1990). Method for rapid separation of liposome-associated doxorubicin from free doxorubicin in plasma. *Anal. Biochem.* 188 (1): 65–71.

Tong, X.S., Wang, J., Zheng, S. et al. (2002). Effect of signal interference from dosing excipients on pharmacokinetic screening of drug candidates by liquid chromatography/mass spectrometry. *Anal. Chem.* 74 (24): 6305–6313.

Van Bommel, E.M.G. and Crommelin, D.J.A. (1984). Stability of doxorubicin-liposomes on storage: as an aqueous dispersion, frozen or freeze-dried. *Int. J. Pharm.* 22: 299–310.

Wei, G., Xiao, S., Gu, Y., and Si, D. (2010). Validated HPLC assay of liposome-encapsulated and non-liposomal doxorubicin in plasma and its application to pharmacokinetic study. *Asian J. Pharmacodyn. Pharmacokinet.* 10 (3): 229–236.

Wei, G., Xiao, S., Si, D., and Liu, C. (2008). Improved HPLC method for doxorubicin quantification in rat plasma to study the pharmacokinetics of micelle-encapsulated and liposome-encapsulated doxorubicin formulations. *Biomed. Chromatogr.* 22 (11): 1252–1258.

Yamamoto, E., Hyodo, K., Ohnishi, N. et al. (2011). Direct, simultaneous measurement of liposome-encapsulated and released drugs in plasma by on-line SPE-SPE-HPLC. *J. Chromatogr. B Analyt. Technol. Biomed. Life Sci.* 879 (30): 3620–3625.

Yamauchi, M., Kusano, H., Saito, E. et al. (2008). Controlled release of a protein kinase inhibitor (UCN-01) from liposomes influenced by the particle size. *Int. J. Pharm.* 351: 250–258.

Yu, Y., Desjardins, C., Saxton, P. et al. (2013). Characterization of the pharmacokinetics of a liposomal formulation of eribulin mesylate (E7389) in mice. *Int. J. Pharm.* 443 (1–2): 9–16.

Zamboni, W.C., Strychor, S., Joseph, E. et al. (2007). Plasma, tumor, and tissue disposition of STEALTH liposomal CKD-602 (S-CKD602) and nonliposomal CKD-602 in mice bearing A375 human melanoma xenografts. *Clin. Cancer Res.* 13 (23): 7217–7223.

Zylberberg, C. and Matosevic, S. (2016). Pharmaceutical liposomal drug delivery: a review of new delivery systems and a look at the regulatory landscape. *Drug Deliv.* 23 (9): 3319–3329.

Part III

Sample Preparation Techniques for LC-MS Bioanalysis of Challenging Molecules

20

Key Pre-analytical Considerations in LC-MS Bioanalysis

Fumin Li, Bruce J. Hidy, and Rand G. Jenkins

20.1 Introduction

Experienced bioanalysts recognize that providing effective bioanalytical support to a nonclinical or clinical study requires thorough technical evaluations and logistical considerations, advanced planning, and good communication with the study team or sponsors. In many cases, particularly when new chemical entities (NCEs) (e.g. drugs and their metabolites) need to be measured in biological samples, the above activities should take place very early, even before the study protocol has been finalized. Therefore, it is useful to define the pre-analytical phase of a study, of which the above activities are thoroughly considered. Fundamentally, the pre-analytical phase begins with pre-study planning to establish the study goals and an appropriate design, followed by development, review, and approval of the study protocol. When subject samples are to be obtained and analyzed for a therapeutic and/or its metabolite(s), or for a biomarker molecule, instructions must be provided, often in the form of a "lab manual." The lab manual defines appropriate sample collection, processing, storage and shipping procedure, and/or conditions. When the study is underway, the compliance of study subjects and staff with study protocol, sample collection, handling, and storage are essential aspects of the pre-analytical phase. Figure 20.1 summarizes various intricate processes during pre-analytical phases or planning. Deviation from the above can lead to unexpected study results. Studies have shown that most of the variations and errors encountered in the clinical chemistry (*in vitro* diagnostics testing) would occur in the pre-analytical phase before laboratory analysis (Guder et al. 2014; Schleicher 2006). This chapter is intended to: (i) inform or remind the readers in the LC-MS bioanalysis field of the importance of the pre-analytical phase, (ii) provide a framework for conducting the pre-analytical evaluation/planning of the bioanalysis process, and (iii) discuss some key issues/factors that, if not anticipated and proactively addressed, can significantly impact the integrity of the biological sample and analyte(s) of interest and compromise the validity of the analytical data.

For LC-MS-based bioanalytical assays, achieving the required sensitivity, acceptable selectivity/specificity, or desired assay accuracy and precision can be challenging for some analytes. Molecular properties of the analyte, including size, polarity, stability, ionizability, collision-induced dissociation (CID) fragmentation, as well as potential endogenous interferences, can have an impact on assay development. It is important to recognize that some of the same factors that cause challenges during the laboratory phase of bioanalysis may have already impacted the integrity of an analyte in study samples during the pre-analytical phase if not addressed, sometimes to such an extent that analyzing the sample may be useless for the intended purpose.

20.2 The Pre-analytical Phase

As noted above, the pre-analytical phase of a study occurs in an interconnected sequence of stages or processes ahead of the analytical phase. For the laboratory, the pre-analytical phase often begins with a simple inquiry from the project team or sponsor:

Can you measure analyte X in matrix Y?

In order for this question to be answered (beyond "it depends"), sufficient information must be available to properly evaluate the request and establish an appropriate bioanalysis project scope to support the intended study.

Important information for bioanalysis generally falls into two categories: the Proposed Study Design and the Assay Technical Requirements, as outlined in Tables 20.1 and 20.2, respectively.

Sample Preparation in LC-MS Bioanalysis, First Edition. Edited by Wenkui Li, Wenying Jian, and Yunlin Fu.

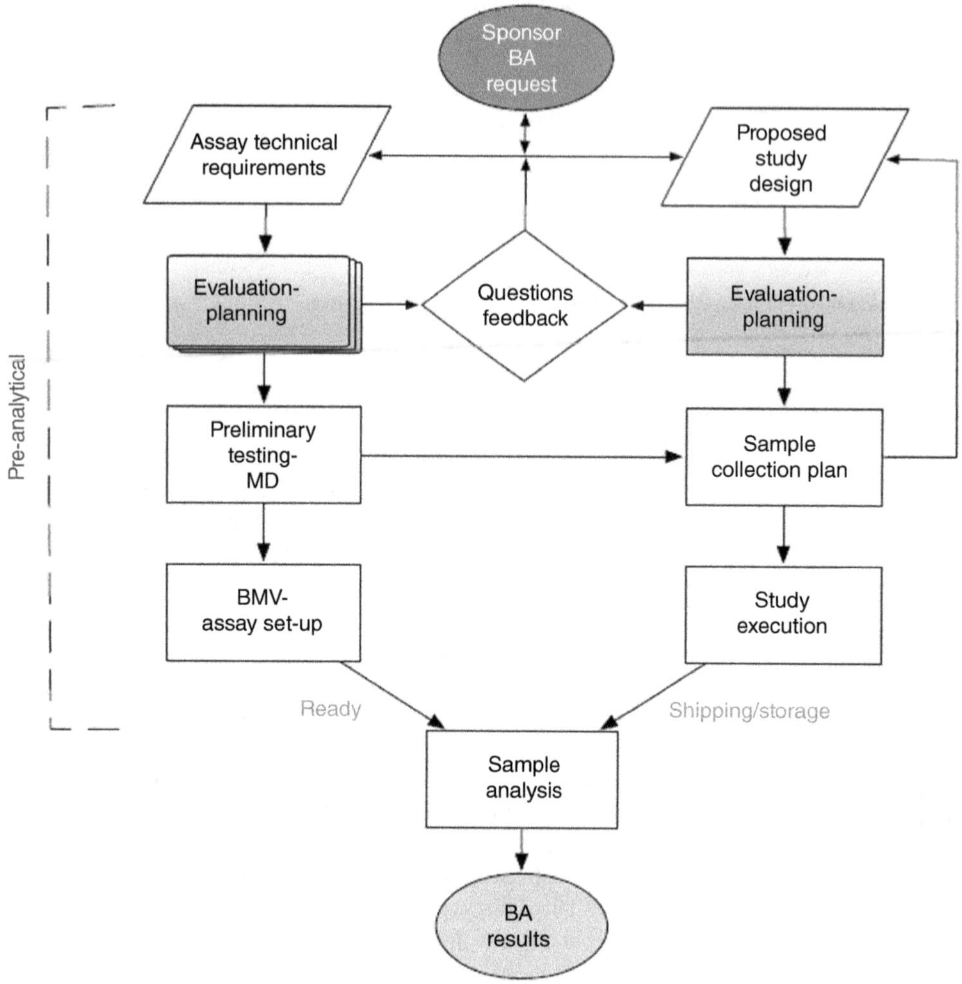

Figure 20.1 The work flow of pertinent planning in the pre-analytical phase before laboratory analysis.

Table 20.1 Proposed study design.

Type	Exploratory, non-regulated or regulated; nonclinical or clinical
Development phase	Preclinical, clinical (phase I–IV)
Therapeutic area	Disease and drug target
Study subject population	Species, age range, sex, normal healthy vs. diseased
Drug under investigation	Small or large molecules
Drug administration	Route, dosage, and schedule, expected concomitant meds
Samples to be collected and analyzed	Matrix, anticoagulant, schedule, procedures
Analyte(s) to be measured	Drug, metabolites, biomarker(s)
Other aspects	Single site, multicenter

Table 20.2 Assay technical requirements.

Analyte(s) and properties	Parent drug, metabolite(s), or biomarker, isomer, isoform, free vs. total, etc. endogenous or an analogue to an endogenous counterpart
Matrix (fluid or solid)	Blood (whole, serum, plasma) Urine or stool Limited volume biofluid (ocular, CSF) Tissue (biopsy sample)
Species (nonclinical or clinical)	Rodent (rat, mouse, special strain) Monkey, minipig, dog, cat, other Human
Sensitivity	LLOQ (must vs. nice to have)
Range	2–3 logs
Technology requested/preferred	LC-MS, LBA, or hybrid

While mostly a paper exercise, a bioanalyst should carefully consider all pertinent information to define the project scope, anticipating – as much as foreseeable – any pre-analytical issues that might impact the conduct or bioanalysis outcome of the study. The relevant study-related details and proposed assay technical requirements must be considered in parallel. In some situations, it may be advisable to conduct preliminary laboratory testing to assess a critical analyte property or to confirm the acceptability of a specific handling procedure. Further communication with the study team or sponsor is needed, either during or following pre-analytical evaluation, to clarify details and provide feedback on assay feasibility and any potential issues from a bioanalysis perspective. Ultimately, the goal of a thorough pre-analytical evaluation is to have a well-thought out (and mutually agreed upon) project scope that clearly defines the bioanalytical requirements/specifications, taking into account anticipated study-related factors and sample collection/handling precautions (Dakappagari et al. 2017; Rai and Vitzthum 2006).

20.3 Bioanalytical Evaluation-planning

20.3.1 Study-related Considerations

There are several key study-related questions, as listed in the following, to be considered during the pre-analytical planning.

- What kind of study is planned?
- What drug is being investigated and by what dosing scheme?
- What matrices will be collected and analyzed for which analytes?
- What decisions will be made based on bioanalytical results?
- Are there bioanalytical data from prior studies to support the establishment of assay range?
- Where and in what kinds of facilities will the study take place?
- What are the regulatory expectations?

20.3.1.1 Study Type and Development Phase

The possible types of studies that require bioanalytical support range from discovery, early exploratory nonclinical, to GLP tox, to late-stage pivotal Phase III clinical trials and beyond. The required assay and study rigor, along with regulatory expectations, can vary considerably across these studies. Regardless, being able to make scientifically valid decisions based on bioanalytical data critically depends on the reproducibility and reliability of the bioanalytical measurements. This means that the analytes being measured are what we think they are and that their integrity in the study sample is well maintained, such that their measured levels reflect the concentrations existing *in vivo* at the time of sample collection. Information related to therapeutic areas, diseases, and drug targets can be helpful to provide context and better understanding of the study needs and how they will be addressed in the scope of the bioanalysis. For example, oncology studies have unique challenges including slow patient enrollment, small study sample counts, and the rapid turn-around need of bioanalytical results once samples are available for analysis. The size and design of a study can determine how samples are managed, either via a central lab or shipped directly from doctors' clinics. There can be lots of direct but small shipments from participating doctors' clinics to a bioanalytical laboratory, creating challenges in maintaining a good chain of custody to ensure sample integrity and managing daunting number of shipments such as little as one sample at a time. It is critical to establish seamless collaboration between sponsors, clinics, central labs, and bioanalytical laboratories on sample collection, storage, shipment, and analysis. One of the biggest logistic challenges is the discrepancy in sample information between clinics and a bioanalytical laboratory. Database reconciliation can be a painful and time-consuming process. Having good communication and information sharing among different organizations/groups are highly encouraged.

20.3.1.2 Study Subject Population

In nonclinical studies, the animal species can create some limitations which influence bioanalytical decisions. With rodent studies, sample volumes will be severely limited (e.g. typically <30 µl in plasma), particularly for serial bleeding as opposed to sparse sampling per time point. Rats are known to have inherently high activity for some enzymes, such as esterase, which can impact stability of some analytes. Blank matrices from some rare species or special strains, such as gene knockout mice, can be difficult to obtain. With primate studies, more stringent safety precautions are expected (e.g. Biosafety Level-2 [BSL-2]) due to simian viruses that may be potentially pathogenic to humans. Clinical study populations may involve adult or pediatric and normal healthy or diseased groups. Sample volumes are typically restricted for pediatric and sometimes also for diseased individuals (e.g. <50 µl). Greater or more variable selectivity/specificity issues may be observed in diseased populations, due to endogenous composition differences and potential concomitant medications. Evaluation with diseased state matrices is a recommended practice for bioanalysis of biologics, and of small molecules to a lesser extent. There is generally a long lead time to procure rare matrices and it is critical to take this timing into consideration for overall project planning.

20.3.1.3 Drug and Administration

Studies requiring bioanalytical support typically involve small- or large-molecule drug candidates in a nonclinical or clinical setting. These compounds are dosed in certain quantities by various routes according to a planned schedule. It is important for bioanalysts to have a basic understanding of the ADME properties expected for a drug, with respect to how it will be administered, where in the body it will be distributed, and how it will be biotransformed (i.e. metabolized or catabolized) and ultimately eliminated. These factors directly impact the TK or PK of the candidates, including AUC_{0-t}, C_{max}, volume of distribution, clearance and half-life, which are determined based on the bioanalysis data. Many of the bioanalytical method requirements are closely tied to these factors, including the appropriateness of the matrix and the assay sensitivity and range.

The route of administration can have a significant effect on the assay requirements. Oral dosing is typical for many small-molecule drugs, while intravenous (IV) dosing is common for large molecules. From an assay perspective, IV bolus dosing will produce a huge concentration spike shortly after administration, which may drop fairly quickly during the distribution phase, while IV infusion will produce a sustained relatively high level, which persists for days or weeks after dosing with many antibody-based therapeutics. A high upper limit of quantitation (ULOQ) or dilutions are often needed to measure peak drug concentrations. Other dosing routes, such as subcutaneous, transdermal, inhalation, or ocular may be applied in various situations. These often result in the need for sensitive assays with much lower limit of quantification (LLOQ) due to slow absorption or low dosing of potent compounds. Drugs with short half-life but long tapering profiles also require sensitive assays to accurately determine the area under the curve (AUC). The planned drug dosage and schedule (e.g. single, multiple, rising, or steady-state) will also have an impact on the required assay sensitivity and range.

20.3.1.4 Samples to Be Collected for Analysis

The biomatrix to be analyzed is chosen based on the purpose of the study and the behavior of the drug, including its absorption, distribution, metabolism, and excretion (ADME) properties. Different tissues may be chosen to evaluate distribution, or to confirm the presence at the intended site of action. Fluid matrices may be chosen for various reasons, with blood being the most common one. An aspect of blood, which is sometimes underappreciated in selecting the actual fraction to analyze (i.e. whole blood, serum, or plasma), is the potential that an analyte may partition into or bind to the surface of red blood cells (RBCs). Choosing the wrong blood-derived fraction can result in an underestimate of the total analyte concentration and exposure. Urine may be evaluated to assess elimination, while other special situations may require cerebrospinal fluid (CSF), saliva, or other fluids like bronchoalveolar lavage (BAL). These materials are generally low in protein contents, which can result in analyte-losses due to nonspecific binding (NSB) (Fourier et al. 2015; Ji et al. 2010) if not addressed up-front, often during collection.

20.3.1.5 Analyte(s) to Be Measured

The drug under investigation is measured in most studies to assess TK/PK. Metabolites may also be of interest to quantify, particularly if they are active and present in significant abundance relative to the parent drug (US FDA 2016). Biomarkers are increasingly of interest in drug development, to the extent that some companies require them for every drug development program.

20.3.1.6 Other Aspects

The configuration and logistics of a study can have significant impacts on bioanalysis considerations (Dakappagari et al. 2017). Single-site studies, common in nonclinical or Phase I clinical, are in general better controlled as compared to others, rendering it easier to manage any special collection or handling requirements. Multi-site studies, common in later-phase studies, are often spread out globally, and it can be difficult to ensure that pre-analytical factors are sufficiently controlled. In some countries or remote areas, lack of basic infrastructure, e.g. centrifugation, refrigeration, or freezing equipment, for conventional PK sample collection and logistics is not uncommon. In such cases, the use of dried matrices, such as dried blood spot (DBS) sampling, can fill the void.

20.3.2 Assay-related Considerations

There are some critical assay-related factors to be considered once study-related factors as highlighted above have been well considered. These key assay-related factors include, but are not limited to, the following:

- Identity and form of the analyte(s) to be measured?
- Any prior knowledge for the bioanalysis of the analyte(s) and/or its analogs?

20.3.2.1 Analyte(s) and Properties – Small vs. Large Molecule

Small-molecule drugs (xenobiotics) and their metabolites, as well as endogenous biomarkers, vary significantly in their chemical and physical properties. For NCEs, such information is especially limited or not readily available. While there may be published or preexisting bioanalytical methods available, it is still a good practice to consider an analyte's basic chemical properties: molecular weight, structure, and functional groups that affect volatility, hydrophobicity, polarity, and solubility. Certain types of functional groups or structural features are known to impact instability under suboptimal conditions. Very small molecules may be difficult for detection with needed selectivity due to the high background noises that are typically observed in the low *m/z* range of most mass spectrometers. Derivatization of a suitable functional group can be applied to increase the molecular weight, improve mass spectrometric response, and/or improve chromatographic behaviors of the analyte(s) as well. On the other hand, large molecules, including peptides, polypeptide proteins, and polynucleotide-based molecules typically generate complex mixtures of multiply charged precursor ions under electrospray ionization (ESI) conditions. This leads to dilution of the overall signal and increased challenges of specific detection of analyte of interest.

20.3.2.2 Matrix and Species

The choice of biological matrix along with the species to be sampled from presents some potential issues related to sample handling. Matrices can be fluids or solid tissues. While plasma, typically containing an anticoagulant (e.g. heparin, EDTA, and citrate), is the most common matrix derived from blood, in some cases, plasma may not be an appropriate matrix. For example, molecules that tend to significantly partition into RBCs (erythrocytes) are generally best analyzed from whole blood to determine total drug exposure. Serum is often the preferred matrix for protein analysis, as it tends to be "cleaner" (particularly for ligand-binding assays, LBAs). Most proteins tend to be reasonably stable during serum processing. However, due to the lengthy coagulation process (preferably at ambient temperature), serum may be a poor choice for quantitative analysis of therapeutic or biomarker peptides or proteins, particularly those susceptible to cleavage by endogenous proteases during serum processing, even when protease inhibitors are present (Percy et al. 2013; van den Broek et al. 2008). Use of EDTA anticoagulant (also inhibiting some metalloproteases) and processing the plasma on ice may be an effective alternative (Evans et al. 2001).

Tissue samples can be as big as the entire organs obtained from study animals or as little as small mg-sized biopsy pieces collected in clinical studies. Sometimes snap freezing is the best practice to ensure stability until analysis. If sophisticated handling is necessary, such as dissection of specific components, mincing, or some form of homogenization, maintaining analyte integrity can be problematic. Keeping tissues on ice is typical, and cryogenic pulverizing can be a useful approach, that can mechanically facilitate subsampling while maintaining stability. Tissue distribution or analyte homogeneity within the same organ can be variable, affecting sample-to-sample reproducibility. A homogenization procedure is often necessary to ensure homogeneity.

20.4 Common Pre-analytical Issues in LC-MS Bioanalysis

20.4.1 Stability in Matrix

The potential analyte losses due to stability-related issues during study sample collection, processing, interim storage and shipping, as well as downstream handling and analysis at the laboratory, must be carefully considered during pre-analytical planning (Li et al. 2011). Appropriate mitigation steps should be identified and incorporated into protocols as needed to ensure analyte stability in the study samples.

Analyte loss due to instability can be generally caused by: (i) chemical or biochemical degradation, such as hydrolysis, oxidation, isomerization, deamidation, or other reactivity; and (ii) physical instability, such as aggregation of protein analytes.

The following examples illustrate a variety of situations where analytes required stabilization in plasma matrix.

Case 1 – Degradation due to light exposure. Rilpivirine (RPV) is a non-nucleoside reverse transcriptase inhibitor that is highly potent against both wild-type and drug-resistant HIV-1 strains. RPV is a *trans*-isomer (E form), which is known to convert to its *cis*-isomer (Z form) under light. RPV in human plasma, following separation from a blood sample, can undergo this photochemical isomerization under light (Else et al. 2014). In a RPV method reported by Kim et al. (2017), the chromatographic separations were optimized to ensure baseline separation between E- and Z-isomer. Yellow light was employed throughout the entire bioanalysis process from sample collection, preparation, to LC-MS analysis. The stability of the RPV in human plasma was successfully established. In addition, the response of the Z-isomer (derived from RPV) was monitored for all known samples, i.e. calibrators and quality control samples (QCs), to evaluate the degree of the photo-conversion during the process of bioanalysis. The peak area ratio (PAR) of Z- to E-isomer was reported as a conversion rate for each sample. The average conversion rate in the control samples was <0.4%, demonstrating that the conversion from E- to Z-isomer during sample handing and analysis, if any, was minimal when samples were processed under yellow light.

Case 2 – Degradation due to chemical reaction. Analytes containing thiols, as a category, are known to react rapidly in untreated blood and plasma to form disulfide conjugates with endogenous thiol-containing molecules, including glutathione and/or other free cysteine-containing species (Giustarini et al. 2016b). These reactions may have already occurred *in vivo*, and it is possible to design a "total" assay format by treating the biomatrix sample with a reducing agent, such as dithiothreotol (DTT) or tris(2-carboxyethyl)-phosphine (TCEP), to cleave all the conjugates to release the free thiol analyte. Whether this reduction step is initially done or not, measuring the thiol analyte requires stabilization, usually by derivatization (Giustarini et al. 2016a). A particularly challenging example for bioanalysis was Omapatrilat, a thiol-containing ACE inhibitor candidate, which also produced two thiol metabolites. Jemal et al. (2001) applied immediate derivatization with methyacrylate (MA) at the time of blood collection to stabilize the three analytes. In another example, Furlong et al. (2013) utilized 2-bromo-3′-methoxyacetophenone (MPB) to stabilize the active thiol-containing metabolites of clopidogrel.

Case 3 – Instability due to oxidation. Levodopa and carbidopa are frequently used in combination to treat patients with Parkinson's disease. Levodopa, as a prodrug, is converted to dopamine by the enzyme dihydroxyphenylalanine (DOPA) decarboxylase in the peripheral circulation and in the central nervous system after penetrating the brain–blood barrier. Chi et al. (2017) described a method where levodopa's ex vivo conversion to dopamine due to oxidation during sample collection and processing was prevented by adding one part of sodium metabisulfite as an antioxidant to every 10 parts of rat plasma samples. By using this approach, stability of both levodopa and carbidopa in rat plasma was established under the following conditions: for 25.7 hours when stored at 2–8 °C, 5 freeze (−80 ± 10 °C)/thaw (2–8 °C) cycles and at least 322 days when stored at −80 ± 10 °C. These stability data clearly demonstrated that sodium metabisulfite is a highly effective additive to stop the conversion of levodopa to dopamine in rat plasma.

Case 4 –Enzymatic hydrolysis. Pitavastatin lactone undergoes interconversion to pitavastatin in biological matrices such as plasma. In a method published by Qi et al. (2013), both rat plasma and dog plasma samples were treated with a buffer upon sample collection so that the final pH of the samples was maintained at around pH 4.2. It was demonstrated that sample stability could be further enhanced when 5 mM of phenylmethylsulfonyl fluoride (PMSF), an enzyme inhibitor, was added. A combined approach of controlling pH to 4.2 and addition of PMSF greatly reduce the conversion of pitavastatin lactone to pitavastatin. Handling samples on wet ice was also necessary. Under these specific sample conditions, pitavastatin lactone was demonstrated to be stable for at least 69 days in dog plasma and at least 2 months in rat plasma when stored at −80 ± 10 °C.

Case 5 – Instability due to deamidation. Decitabine, a deoxycytosine analog, which acts as a nucleic acid synthesis inhibitor, is used for the treatment of myelodysplastic syndromes (MDS) and acute myeloid leukemia (AML). The compound is unstable in plasma due to the action of the enzyme cytidine deaminase (CDA). Hua et al. (2014) described a decitabine assay where human plasma samples were stabilized by the addition of tetrahydrouridine (THU), a competitive inhibitor of CDA. Decitabine was stable in THU-treated human plasma (~25 mcg ml^{-1}) for at least 68 days and after 5 freeze–thaw cycles when stored at −80 ± 10 °C.

As discussed for thiol-containing and protease-sensitive compounds, some analytes can be extremely unstable, with a very short half-life on the order of minutes. In such cases, treatment of serum or plasma samples following blood processing and centrifugation is not considered a viable option. Treatment of the whole blood sample at the time of collection is required to ensure stabilization of the analyte. Additives can be introduced directly into the blood-drawing tube (e.g. BD Vacutainers®) either prior to (Jemal et al. 2001) or immediately after drawing (Furlong et al. 2013). Alternatively, for enzymatically labile small-molecule drugs and/or metabolites of interest, rapid denaturing of proteins in the matrix should be considered. This can be accomplished by processing whole blood samples with an organic solvent, i.e. protein precipitation.

A representative procedure is summarized in Table 20.3 for assessing the stability of analyte(s) at approximately the low and high QC concentrations in whole blood under the following conditions: (i) store the blood samples at room temperature versus on wet ice or equivalent; (ii) separate plasma at room temperature versus at refrigerated condition via centrifugation; and (iii) time interval that mimic sample collection, processing, and storage. In this evaluation, specific instructions need to be considered for the following: type of blood sample collection tube (glass versus plastic); tube size (volume); anticoagulant; mixing of blood and anticoagulant; time and temperature a sample may be stored prior to centrifugation; temperature, speed (rpm or g), and time for centrifugation; time and temperature a sample may be stored prior to separating the plasma layer; type of plasma sample "transfer" tube (polypropylene); tube size (volume); and storage temperature (−20 °C vs. −70 °C). In addition, considerations should be in place for compounds include: reactive/unstable; photosensitive and "sticky" (subject to NSB).

Table 20.3 Whole blood sample collection/handling evaluation procedures.

Fresh drawn into a Vacutainer[a] or use commercially available whole blood tube containing the appropriate anticoagulant[b]					
Add analyte in an appropriate solution and volume (≤5% of the total volume) to whole blood previously incubated to 37 °C. Gently mix the blood to ensure homogeneity, which is followed by incubation at 37 °C for 5–15 min					
Add to a 4- to 10-ml[a] glass or plastic Vacutainer tube (containing no anticoagulant). Process sample immediately		Add to a 4- to 10-ml[a] glass or plastic Vacutainer tube (containing no anticoagulant). Incubate for 30–120 min at room temperature		Add to a 4- to 10-ml[a] glass or plastic Vacutainer tube (containing no anticoagulant). Incubate for 30–120 min on wet ice or Cryoblock	
Control	Test-A	Test-B	Test-C	Test-D	Test-E
Centrifuge blood for 10–15 min at 1500–1700 *g* at 2–8 °C	Centrifuge blood for 10–15 min at 1500–1700 *g* at room temperature	Centrifuge blood for 10–15 min at 1500–1700 *g* at 2–8 °C	Centrifuge blood for 10–15 min at 1500–1700 *g* at room temperature	Centrifuge blood for 10–15 min at 1500–1700 *g* at 2–8 °C	Centrifuge blood for 10–15 min at 1500–1700 *g* at room temperature
Separate plasma, analyze immediately or store at −20 or −70 °C		Separate plasma, analyze immediately or store at −20 or −70 °C		Separate plasma, analyze immediately or store at −20 or −70 °C[c]	
Analyze plasma samples in six replicates		Analyze plasma samples in six replicates		Analyze plasma samples in six replicates	

[a] Use Vacutainer® type/size specified in sample processing instructions.
[b] This procedure to evaluate the stability of the test substances during sample processing is intended only for plasma assays and not serum-based methods.
[c] Plasma stored in a 7- or 10-ml polypropylene screw-cap flat based tube.

In general, the mean peak area/height ratio for the Test Groups-A, B, C, D, and E must be within ±15.0% of the mean peak area/height ratios from the Control Group at each QC concentration level. Precision for each control and test group must be ≤15.0% CV. If results from a certain condition do not meet the acceptance, sample handling at that particular condition is not recommended. Applying 20% criteria is typically applied for large-molecule assays.

20.4.2 Nonspecific Binding

The potential for analyte losses due to nonspecific adsorption to collection devices or storage containers can be a significant issue for some analytes and matrix types. This problem primarily affects hydrophobic molecules, which are prone to sticking to hydrophobic surfaces, such as plastic containers. Polar molecules can also have issues with glass surfaces due to hydrogen bonding or ionic interactions with surface silanols. NSB losses can also occur via strong ionic interactions between charged analytes (typically bases) and glass container surfaces. This phenomenon is rarely noticeable with blood, serum, or plasma matrices. The high protein and lipid content of these matrices prevents the sticky analyte molecules from adhering to the tube surface by direct shielding or by providing something else for a molecule to stick to. On the other hand, low protein content biomatrices, such as urine and CSF, as well as neat solutions, can be very problematic. To prevent/eliminate compound loss due to NSB, different container types (tubes and pipette tips) must be carefully evaluated. One approach is to introduce additives, ranging from organic solvents like methanol or isopropanol, to nonionic surfactants or zwitterionic detergents, such as Tween 20 or CHAPS (3-([3-cholamidopropyl]dimethylammonio)-1-propanesulfonate). The use of special container types, such as Eppendorf LowBind plastic ware, can be of great benefit especially for biomolecules. If glassware is to be used, it can be treated by siliconizing or using silylating reagents to neutralize ionic binding interactions. While collection of blood rarely presents NSB issues, collection of urine into plastic cups or via a catheter into a bag, followed by transfers to larger containers, is often problematic (Ji et al. 2010). Therefore, it is desirable to test all of the devices to be involved in sample collection (i.e. closely simulate the anticipated exposure conditions) for assessing the magnitude of potential NSB losses and associated mitigation.

A simple protocol (Ji et al. 2010) can be used to assess the NSB of analytes due to adsorption to glass and/or plastic (polypropylene) containers.

- Prepare a stock solution at 100 μg ml^{-1}.
- Label a set of glass and/or polypropylene tubes.
- Transfer 10.0 μl of analyte stock solution into one polypropylene or glass tube labeled for "Tube 1." Dilute with 990 μl of water (or other intended generally high aqueous solvent mixture). Vortex to mix.
- Transfer 10.0 μl of the stock solution into one polypropylene or glass tubes labeled for "Control." Dilute with 990 μl of methanol. Vortex briefly. (Negative control)
- Transfer a portion of Tube 1 and the negative control sample to a 96-well polypropylene plate or a 96-well plate with glass inserts. Dilute with an appropriate diluent (including an appropriate internal standard) as needed for the expected instrument conditions.
- Vortex the remaining samples for 20 seconds. Wait for five minutes. Vortex for 20 seconds.
- Pour the contents of Tube 1 into a second tube of the same material. (Tube 2)
- Vortex the transferred samples for 20 seconds. Wait for five minutes. Vortex for 20 seconds.
- From Tube 2, transfer a portion to a 96-well polypropylene plate or a 96-well plate with glass inserts. Dilute with an appropriate diluent (including an appropriate internal standard) for the expected instrument conditions.
- Repeat step 7 to step 9 three times (ending in "Tube 5").
- From Tube 5, transfer a portion to a 96-well polypropylene plate or a 96-well plate with glass inserts. Dilute with an appropriate diluent for the expected instrument conditions.
- Vortex plates briefly.
- Inject all samples onto an LC-MS/MS instrument platform.
- Peak area ratios (PARs) between an analyte and its internal standard are calculated. PARs from transferred samples are compared to the PAR of the Control sample. A negative bias between Tube 1 and the negative control, as well as a trend from Tubes 1 to 5 indicates absorption to the container surfaces has occurred.
- Appropriate additives can be added and the process repeated to assess the effectiveness of an additive in preventing NSB.

CSF collection can present problems associated with sampling needles and catheters (Bjerke et al. 2010). Therefore, it may be important to obtain the same devices to be used in the clinic for evaluating potential NSB from CSF. In this evaluation, using an attached syringe, a portion of the CSF QCs is drawn into and passed through the sampling device, and the resulting transferred CSF samples are then measured against the same QCs that were not exposed to the devices. Analyte response ratios are calculated to estimate the possible loss due to NSB, if any. It should be noted that nonspecific adsorption to these devices usually cannot be prevented by pretreating them with an additive prior to use, as they are typically components of a sterilized lumbar puncture kit for use by the physician. In practice, once the devices have been sufficiently "flushed" by an initial portion of CSF sample (which should generally be discarded), the surface binding sites may become saturated with the analyte or other strongly adsorbed endogenous molecules, minimizing any further loss during continued CSF collection. It is difficult to predict whether this will be an issue unless adequate testing is performed.

Although issues due to NSB are commonly encountered with many kinds of analytes in LC-MS bioanalysis, it is important to recognize that observed losses can sometimes be due to analyte-specific binding issues. For example, the presence of certain endogenous molecules with high binding affinity to the analyte of interest can cause recovery interference in LC-MS-based assays if not mitigated in the method (Cox et al. 2014). As in LBAs, anti-drug antibodies (ADAs) or endogenous carrier proteins can interfere the enrichment/extraction process causing low analyte recovery and negative bias unless steps are taken to break the binding interaction. The possibility of encountering ADAs upon repeat dosing is significant for therapeutic peptides and proteins, while carrier proteins are common for many hormones, such as sex hormone-binding globulin (SHBG) for testosterone (Wang et al. 2014).

20.5 Summary

This chapter summarizes various pre-analytical considerations in LC-MS bioanalysis. Among the many technical, operational and, logistical issues or factors, assessment of stability and NSB of the analyte(s) of interest in the early stage of method development can be a key. These two issues, if not properly assessed and proactively addressed, can significantly impact the integrity of the biological samples and associated bioanalytical results for the analyte(s) of interest.

References

Bjerke, M., Portelius, E., Minthon, L. et al. (2010). Confounding factors influencing amyloid beta concentration in cerebrospinal fluid. *Int. J. Alzheimers Dis.* 2010: https://doi.org/10.4061/2010/986310.

van den Broek, I., Sparidans, R.W., Schellens, J.H., and Beijnen, J.H. (2008). Quantitative bioanalysis of peptides by liquid chromatography coupled to (tandem) mass spectrometry. *J. Chromatogr. B Analyt. Technol. Biomed. Life Sci.* 872 (1–2): 1–22.

Chi, J., Ling, Y., Jenkins, R., and Li, F. (2017). Quantitation of levodopa and carbidopa in rat plasma by LC-MS/MS: the key role of ion-pairing reversed-phase chromatography. *J. Chromatogr. B Analyt. Technol. Biomed. Life Sci.* 1054: 1–9.

Cox, H.D., Lopes, F., Woldemariam, G.A. et al. (2014). Interlaboratory agreement of insulin-like growth factor 1 concentrations measured by mass spectrometry. *Clin. Chem.* 60 (3): 541–548.

Dakappagari, N., Zhang, H., Stephen, L. et al. (2017). Recommendations for clinical biomarker specimen preservation and stability assessments. *Bioanalysis* 9 (8): 643–653.

Else, L.J., Tjia, J., Jackson, A. et al. (2014). Quantification of rilpivirine in human plasma, cervicovaginal fluid, rectal fluid and genital/rectal mucosal tissues using liquid chromatography-tandem mass spectrometry. *Bioanalysis* 6 (14): 1907–1921.

Evans, M.J., Livesey, J.H., Ellis, M.J., and Yandle, T.G. (2001). Effect of anticoagulants and storage temperatures on stability of plasma and serum hormones. *Clin. Biochem.* 34 (2): 107–112.

Fourier, A., Portelius, E., Zetterberg, H. et al. (2015). Pre-analytical and analytical factors influencing Alzheimer's disease cerebrospinal fluid biomarker variability. *Clin. Chim. Acta* 449: 9–15.

Furlong, M.T., Savant, I., Yuan, M. et al. (2013). A validated HPLC-MS/MS assay for quantifying unstable pharmacologically active metabolites of clopidogrel in human plasma: application to a clinical pharmacokinetic study. *J. Chromatogr. B Analyt. Technol. Biomed. Life Sci.* 926: 36–41.

Giustarini, D., Galvagni, F., Orlandini, M. et al. (2016a). Immediate stabilization of human blood for delayed quantification of endogenous thiols and disulfides. *J. Chromatogr. B Analyt. Technol. Biomed. Life Sci.* 1019: 51–58.

Giustarini, D., Tsikas, D., Colombo, G. et al. (2016b). Pitfalls in the analysis of the physiological antioxidant glutathione (GSH) and its disulfide (GSSG) in biological samples: an elephant in the room. *J. Chromatogr. B Analyt. Technol. Biomed. Life Sci.* 1019: 21–28.

Guder, W.G., Narayanan, S., Wisser, H., and Zawta, B. (2014). *Diagnostic Samples: From the Patient to the Laboratory: The Impact of Preanalytical Variables on the Quality of Laboratory Results*, 4th, Updated Edition, 124. Weinheim: Wiley-Blackwell.

Hua, W., Ierardi, T., Lesslie, M. et al. (2014). Development and validation of a HILIC-MS/MS method for quantification of decitabine in human plasma by using lithium adduct detection. *J. Chromatogr. B Analyt. Technol. Biomed. Life Sci.* 969: 117–122.

Jemal, M., Khan, S., Teitz, D.S. et al. (2001). LC/MS/MS determination of omapatrilat, a sulfhydryl-containing vasopeptidase inhibitor, and its sulfhydryl- and thioether-containing metabolites in human plasma. *Anal. Chem.* 73 (22): 5450–5456.

Ji, A.J., Jiang, Z., Livson, Y. et al. (2010). Challenges in urine bioanalytical assays: overcoming nonspecific binding. *Bioanalysis* 2 (9): 1573–1586.

Kim, M.-Y., Cousineau, C., Li, F. (2017). Quantitation of rilpivirine in the presence of its Z-isomer using LC-MS/MS. 65th ASMS Conference on Mass Spectrometry and Allied Topics, Indianapolis, IN, USA (4–8 June 2017).

Li, W., Zhang, J., and Tse, F.L. (2011). Strategies in quantitative LC-MS/MS analysis of unstable small molecules in biological matrices. *Biomed. Chromatogr.* 25 (1–2): 258–277.

Percy, A.J., Parker, C.E., and Borchers, C.H. (2013). Pre-analytical and analytical variability in absolute quantitative MRM-based plasma proteomic studies. *Bioanalysis* 5 (22): 2837–2856.

Qi, X., Ding, L., Wen, A. et al. (2013). Simple LC-MS/MS methods for simultaneous determination of pitavastatin and its lactone metabolite in human plasma and urine involving a procedure for inhibiting the conversion of pitavastatin lactone to pitavastatin in plasma and its application to a pharmacokinetic study. *J. Pharm. Biomed. Anal.* 72: 8–15.

Rai, A.J. and Vitzthum, F. (2006). Effects of preanalytical variables on peptide and protein measurements in human serum and plasma: implications for clinical proteomics. *Expert Rev. Proteomics* 3 (4): 409–426.

Schleicher, E. (2006). The clinical chemistry laboratory: current status, problems and diagnostic prospects. *Anal. Bioanal. Chem.* 384 (1): 124–131.

US FDA (2016). *Safety Testing of Drug Metabolites: Guidance for Industry*, Revision 1. Silver Spring, MD: U.S. Food and Drug Administration.

Wang, Y., Gay, G.D., Botelho, J.C. et al. (2014). Total testosterone quantitative measurement in serum by LC-MS/MS. *Clin. Chim. Acta* 436: 263–267.

21

Derivatization in Sample Preparation for LC-MS Bioanalysis

Dafang Zhong and Yunting Zhu

21.1 Introduction

LC-MS plays an important role in quantitative bioanalysis owing to its inherent selectivity and sensitivity. However, challenges still exist for the analytes with the following properties: (i) poor ionization, (ii) lack of adequate chromatographic retention, and (iii) chemical instability. The complexity of biological matrix further complicates these challenges. To overcome these challenges, chemical derivatization could be employed as an effective method to improve the stability, chromatographic behavior, and MS detection of the analytes of interest (Deng et al. 2012; Niwa 2012). In general, the reactions of chemical derivatization involve the modification of the original chemical structures and introduction of the new tags to the analytes. After derivatization, physicochemical properties of the analytes change according to the structural modifications. Newly introduced tags will serve as functional groups to improve LC-MS performance (Xu et al. 2011) (Figure 21.1). For example, hydrophobic groups are introduced to improve the chromatographic retention; ionizable or pre-charged moieties are integrated to enhance the ionization efficiency (Qi et al. 2014). Over the past few decades, numerous derivatization methods have been reported to facilitate the quantitative bioanalysis of compounds in various matrices (Santa 2013). Although the application of chemical derivatization in bioanalysis may be time-consuming, it has been acknowledged as an indispensable technique to extend the application of LC-MS analysis to a wide range of analytes. Nowadays, an increasing emphasis has been placed on the development of new derivatization methods for LC-MS bioanalysis with high throughput while achieving high sensitivity and specificity.

21.2 Derivatization Strategies

Major derivatization strategies employed for LC-MS bioanalysis include detection-oriented, separation-oriented, and stability-oriented derivatization. Each strategy involves diverse key considerations, derivatization reagents, and various applications (Table 21.1). The most frequently used incorporating tags are MS response-enhancing moieties, which are aimed at improving detection sensitivity (Iwasaki et al. 2011). In this respect, sufficient ionization could be realized by incorporating permanently charged moieties or efficiently ionizable tags (Gao et al. 2005). Typical derivatization reagents containing permanently charged moieties include *p-N,N,N*-trimethylammonioanilyl *N′*-hydroxysuccinimidyl carbamate iodide (TAHS) (Rebane et al. 2012), 2,5-dioxopyrrolidin-1-yl *N*-tri(pyrrolidino)phosphoranylideneamino carbamate (FOSF) (Rebane et al. 2012), and Girard reagents (Juang et al. 2013; Lavrynenko et al. 2013). Girard reagents react with carbonyl groups and introduce quaternary amines to boost MS responses. Frequently adopted derivatization reagents with easily ionizable moieties include dansyl chloride (DNS-Cl) (Rebane et al. 2014), 4-substituted 1,2,4-triazoline-3,5-dione (PTAD) (Ding et al. 2010; Higashi et al. 2011; Lipkie et al. 2013), and 3-nitrophenylhydrazine (3NPH) (Han et al. 2015) (Figure 21.2). DNS-Cl is widely adopted as a derivatization reagent for phenolic compounds, such as amino acids and estrogens; 3NPH targets carboxyl groups and is a useful reagent for hydroxy carbolic acids to improve the detection limits. In general, detection-oriented derivatization affords extra ionization efficiency to permit the desirable LC-MS bioanalysis of the molecules that are not readily ionized via normal sample preparation, and this is highly desirable for bioanalysis in pharmaceutical applications where analytes

Sample Preparation in LC-MS Bioanalysis, First Edition. Edited by Wenkui Li, Wenying Jian, and Yunlin Fu.

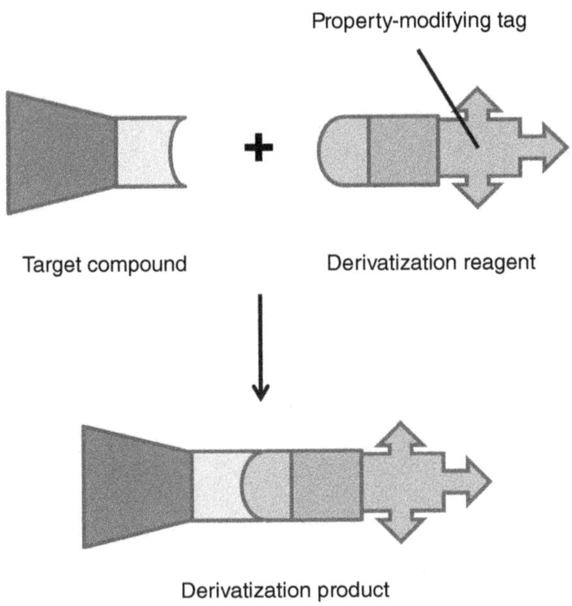

Figure 21.1 General mechanism of derivatization reactions.

of interest are present at low concentrations (Higashi et al. 2005). In addition, sensitive or specific product ions derived from a derivatized compound can also be useful when applying multiple reaction monitoring (MRM) in LC-MS, especially when fragment ions are specific to derivatized compound rather than those merely from derivatizing agent or analyte moiety. The specificity of the method could be significantly magnified due to the production of specific fragments in MS/MS.

Another common derivatization strategy adopted for LC-MS bioanalysis is to integrate hydrophilic or hydrophobic tags to the analytes of interest to improve their chromatographic behavior. The strategy is especially advantageous when developing a LC-MS-based method for simultaneous determination of multiple components with diverse polarities (Lu et al. 2013). A typical class of molecules with undesirable chromatographic behavior is amino acids, as most of those molecules are highly polar with poor retention on reversed-phase columns. Without derivatization, analysis of amino acids by means of classical reversed-phase chromatography in combination with MS is quite challenging (Becker et al. 2012). Recently reported derivatization reagents for amino acids include DNS-Cl, 9-florenylmethoxycarbonyl chloride (Fmoc-Cl) (Ziegler and Abel 2014), FOSF, 1-bromobutane (Sakaguchi et al. 2015), and isobutyl chloroformate (Li et al. 2013) (Figure 21.2). All of the above derivatization reagents contain hydrophobic units such as alkyl chains, ester, or phenyl groups. The integration of these groups renders amino acids extra hydrophobicity for retention on reversed-phase chromatography. Moreover, for the molecules severely affected by the matrix effect, chemical derivatization by incorporating hydrophobic units can serve as an applicable mean to increase the molecular weight and enhance the LC retention to avoid the matrix effect aroused by the low *m/z* co-elution interferences (Santa 2011; Wang et al. 2011).

For chemically fragile molecules, degradation could be easily induced by time-consuming sample preparation, inappropriate pH, or high temperature in the source, all of which will be disastrous to the accuracy of the results. Blocking the labile moieties by derivatization with alternative moieties could be effective to avoid chemical degradation, and also preferable to optimize the chromatographic and mass spectral behavior of the analytes of interest (Vichi et al. 2013). For instance, endogenous thiol compounds including glutathione, cysteine, homocysteine, *N*-acetylcysteine, and cysteinylglycine, fall into the category of bioactive substances with sulphydryl groups, which is acknowledged as unstable compounds with high reactivity (Kusmierek et al. 2011). Derivatization has been applied to prevent the conversion of thiols to oxidized thiols or disulfides via autoxidation, using 4-(hydroxymercuri)benzoate (PHMB) (Liem-Nguyen et al. 2015), 4-chloro-3,5-dinitrobenzotrifloride (CNBF) (Zhang et al. 2014), and methyl or ethyl propiolate (Karakosta et al. 2013) (Figure 21.2). Other examples of unstable functional group include isocyanate, aminophenols, aldehydes, and/or funan ring. To stabilize these compounds, various derivatization strategies have been developed. Special attention should be paid to the timing to perform derivatization. For labile molecules, immediate derivatization after sampling is necessary. For many other molecules, stability-oriented derivatization was performed during sample preparation before submission to LC-MS.

Other than above-mentioned strategies, other useful applications of derivatization can be found in the literature. Cai et al. (2010) presented an LC-MS method for analyzing DNSated neurotransmitters and their metabolites. The most important attribute of this method was the incorporation of electrophilic groups with basic nitrogen atoms, which allows for the simultaneous determination of both monoamines and their acidic metabolites under acidic mobile phase in positive ion mode. The established protocol was fully implemented for the analysis of 10 neurotransmitters and their metabolites in the plasma samples collected from both schizophrenic patients and healthy subjects. The principle of charge-reversal can be realized via derivatization with positive- or negative-charged reagents, so as to enable the simultaneous determination of multiple components (Bollinger et al. 2010). For fatty acids, ionization in negative ion mode is often inefficient in comparison with that in positive ion mode; therefore, multiple charge-reversal chemical derivatization procedures to improve the ionization efficiency have

Table 21.1 Applications of various derivatization strategies.

Strategy	Analyte	Derivatization reagent	Matrix	Sensitivity	Reference
Detection-oriented	Amino acids	TAHS	Solution	22–117 fmol on column	Rebane et al. (2012)
	Amino acids	FOSF	Solution	7–168 fmol on column	Rebane et al. (2012)
	Amino acids	DNS-Cl	Solution	NA	Rebane et al. (2014)
	Alkylphenols	DNS-Cl	Water samples	0.08–0.83 pg per injection	Pernica et al. (2015)
	1,25-Dihydroxyvitamin D_2 and 1,25-dihydroxyvitamin D_3	PTAD	Human plasma	$10\,pg\,ml^{-1}$	Ding et al. (2010)
	Ecdysteroids	Girard reagent P, T, and C	Drosophila pupae	10 pg	Lavrynenko et al. (2013)
	Bisphenol A and estrogens	DNS-Cl	Human plasma	$4.0–43.5\,pg\,ml^{-1}$	Vitku et al. (2015)
	Estrogens and estrogen metabolites	DNS-Cl	Breast cancer cells	$0.05–80\,pg\,ml^{-1}$	Huang et al. (2011a)
	Estradiol	1,2-Dimethylimidazole-5-sulfonyl chloride	Human serum	$0.5\,pg\,ml^{-1}$	Keski-Rahkonen et al. (2015)
	Estrogens	2-Fluoro-1-methylpyridinium-*p*-toluenesulfonate	Human plasma and serum	0.2 pg on column	Faqehi et al. (2016)
	Testosterone	*O*-(3-trimethylammoniumpropyl) hydroxylamine	Human serum	$1\,pg\,ml^{-1}$	Star-Weinstock et al. (2012)
	Allopregnanolone, pregnenolone, and androsterone	1-Amino-4-methylpiperazine	Human serum	$5\,pg\,ml^{-1}$	Ke et al. (2016)
	Menadione (vitamin K_3)	3-Mercaptopropionic acid	Human plasma	$0.03\,ng\,ml^{-1}$	Liu et al. (2014)
	Fatty acids	3NPH	Human feces	0.3–15 fmol	Han et al. (2015)
	Lipid compounds with hydroxy groups	DNS-Cl	Tissue extracts	2.9–109 pg	Tang and Guengerich (2010)
	Phenolic compounds	DNS-Cl	Blood plasma	$0.03–0.5\,ng\,ml^{-1}$	Chang et al. (2010)
	Morphine	DNS-Cl	Human urine	$25\,pg\,ml^{-1}$	Lamshoft et al. (2011)
	Propofol	1,2-Dimethylimidazole-4-sulfonyl chloride	Human serum	$5\,ng\,ml^{-1}$	Maas et al. (2017)

Separation-oriented	Amino acids	Fmoc-Cl	Plant extract	1–15 nM	Ziegler and Abel (2014)
	Amino acids	1-Bromobutane	Hydrolyzed BSA	5.4–91 fmol	Sakaguchi et al. (2015)
	Amino acids	Isobutyl chloroformate	Tobacco sample	0.12–2.82 μg ml^{-1}	Li et al. (2013)
	Ribonucleosides	Acetone	Human urine	0.089–0.55 ng ml^{-1}	Li et al. (2015)
	Biogenic amines	2*H*,2*H*,3*H*,3*H*-perfluoroundecan-1-al	Human plasma	0.01 nM	Hayama et al. (2012)
	Dimethylarginine	Naphthalene-2,3-dicarboxaldehyde	Human plasma	25 nM	Hui et al. (2012)
	Carboxylic acids	Prolylamidepyridines	Human saliva	32–520 amol	Kuwabara et al. (2014)
	3,4-Methylenedioxymethamphetamine	*N*(α)-(5-fluoro-2,4-dinitrophenyl)-D-leucinamide	Human urine	0.1 μg ml^{-1}	Nakanishi et al. (2012)
	Sarcosine	*N*,*N*′-dicyclohexylcarbodiimide	Human urine	1 ng ml^{-1}	Chen et al. (2014)
Stability-oriented	Glutathione, cysteine, homocysteine, and cysteinylglycine	CNBF	Human plasma, urine, and saliva	0.15 μM	Zhang et al. (2014)
	Thiols	PHMB	Yeast	Subpicomole range	Rao et al. (2010)
	Thiols	PHMB	Natural water	0.06–0.5 nM	Liem-Nguyen et al. (2015)
	Cysteine and glutathione	Methyl or ethyl propiolate	Yeast samples	0.6 μmol l^{-1}	Karakosta et al. (2013)
	Thiols	Monobromobimane	Mouse plasma	0.31–4.98 fmol	Sun et al. (2016)
	Volatile thiols	Ebselen	Lipid matrices	0.05–0.5 ng kg^{-1}	Vichi et al. (2013)
	Cysteine and thiol peptides	Ebselen	Methanol/water	NA	Xu et al. (2010)
	Glutathione	Iodoacetic acid	Rat and human hepatocytes	0.05 μM	Sentellas et al. (2014)
	Thiols and oxidized thiols	ω-Bromoacetonylquinolinium bromide (BQB) and d^7-BQB	Human urine	0.833–8.02 nmol l^{-1}	Huang et al. (2011b)

NA, not assayed.

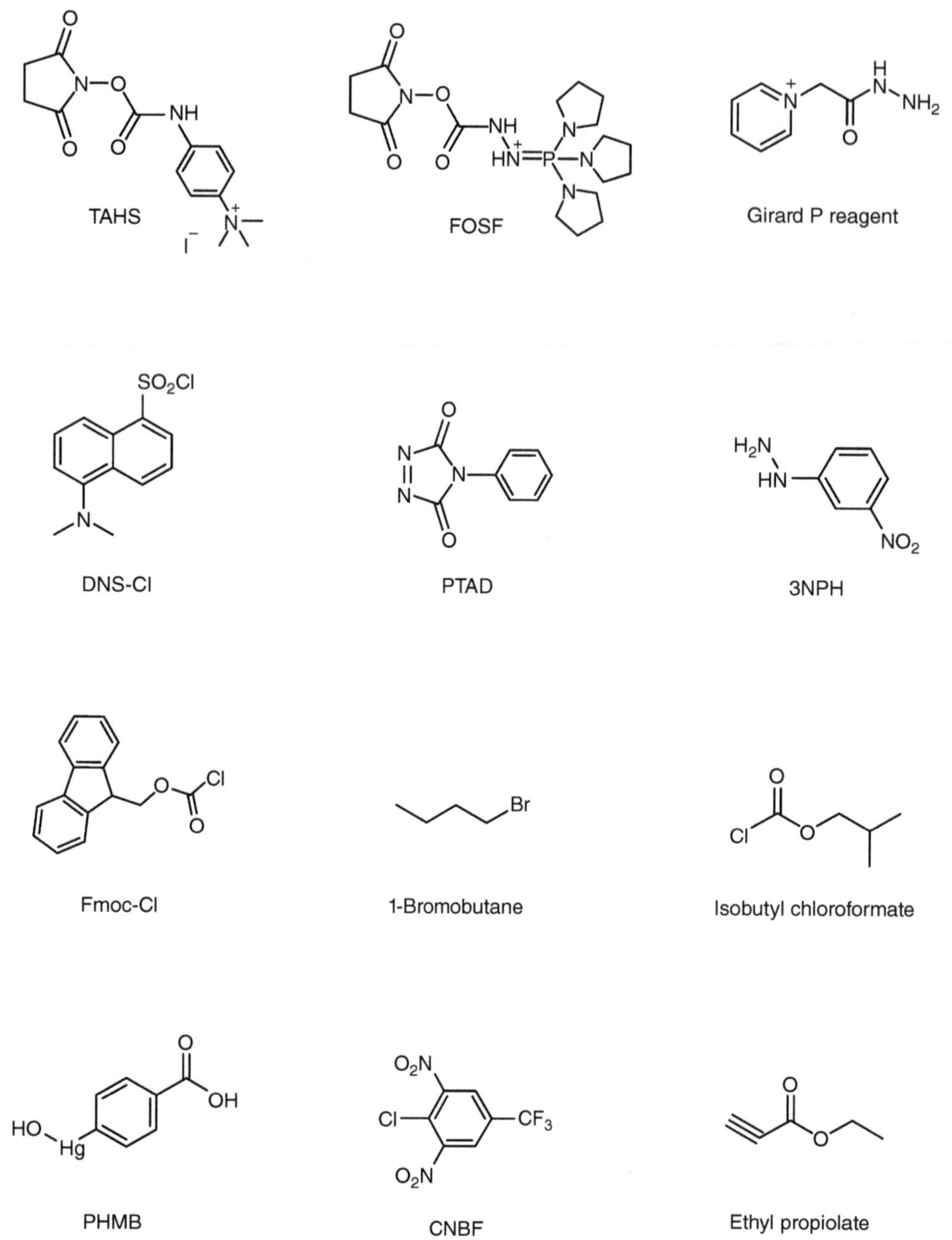

Figure 21.2 Chemical structures of typical derivatization reagents for detection-oriented, separation-oriented, and stability-oriented derivatization.

been established for detection of fatty acids in positive ion mode (Bollinger et al. 2013). Chiral derivatization for the separation of enantiomers is another valuable strategy for LC-MS. Chiral derivatization agents include (*S*)-*N*-(4-nitrophenoxycarbonyl)-L-phenylalanine 2-methoxyethyl ester (Visser et al. 2011), *R*(−)-4-(3-isothiocyanatopyrrolidin-1-yl)-7-(*N*,*N*-dimethylaminosulfonyl)-2,1,3-benzoxadiazole(*R*(−)-DBD-PyNCS) (Min et al. 2011), and (*S*)-1-(4-dimethylaminophenylcarbonyl)-3-aminopyrrolidine (Ogawa et al. 2013).

21.3 Key Considerations for Derivatization

Some key factors should be considered for derivatization of analyte(s) of interest in LC-MS bioanalysis. These factors include, but are not limited to, (i) the nature of the target analyte, (ii) objective of the assay, and (iii) selection of the derivatization reagents. The chemical structures of the analytes exhibit functional groups, which may suggest the reactive function group and potential reaction or reagent suitable for derivatization. For example, chemical derivatization using hydrazine reagents has been proven to be capable of improving the detectability of poorly ionic carbonyl compounds; DNS-Cl contains a sulfonyl group that can readily react with phenols; hydroxylamine can react with ketones to form oximes (Borrey et al. 2007; Kushnir et al. 2006). The aim of the study appears to be another key to the success of derivatization. Derivatization reagents should be carefully evaluated to meet the requirements of the study, such as sensitivity, separation, or stability. In addition, type and availability of the reagent, chromatographic behavior, and fragmentation patterns of the derivatized molecule need to be considered for the method of derivatization. Reagents with potent toxicity and requiring extraordinarily laborious handling are surely not preferred. Side-products in the reaction should be minimal.

When it comes to derivatization procedures, the reaction conditions should be optimized. Multiple factors, such as reaction temperature, substrate concentration, or reaction time, will most likely to affect the final result. The principle of reaction optimization is to reach acceptable productivity with favorable repeatability. For quantitative LC-MS bioanalysis, reaction should proceed in a quantitative manner, which may require strict procedures. Although the optimization of chemical derivatization is time-consuming and labor-intensive, it is generally believed that appropriate derivatization turns out to be the effective and complementary technique to extend the application of LC-MS analysis to a wide range of compounds.

For MS/MS fragmentation, in general, derivatized compounds will generate three types of product ions. First, fragmentation occurs between the analyte moiety and derivatization tag of the derivatized analyte, which produces almost the same fragment ions as the charged intact analyte. The product ions from intact analyte moiety are useful for compound specific detection. The problem is that the production of such ions is analyte-dependent and is not controlled by intention and would not change based on derivatization. Practically, few compounds produce such compound specific ions. The same process also produces the product ions derived from the derivatizing agent moiety. This is very useful for detection of a similar compound class and usually they provide good sensitivity. Neutral loss from the derivatized tags is common for some derivatized compounds, such as dimethylaminoethanol and trimethylaminoethyl ester derivatives. The drawback of the neutral loss approach is that many peaks may appear in the chromatogram, as many compounds with similar functional groups are derivatized and detected in the same manner, and this will lead to the loss of assay specificity. Therefore, efficient separation of the many molecules using ultra performance LC may be required (Inoda et al. 2011). The last type of fragmentation is the one specific to the derivatized compounds rather than that merely from derivatizing agent or analyte moiety. This type of fragmentation is beneficial to the specificity of the method and is most common for derivatized compounds. The requirement for the analyte chemical structure is not as strict as stated above types and many examples of this kind of fragmentation are reported (Kalhorn et al. 2007).

In spite of the valuable analytical benefits of chemical derivatization, it must be recognized that the approach has its limitations, such as by-product formation, interference caused by excess reagents, or analyte loss. Therefore, the benefits, as well as the deficiencies of the certain derivatization reaction should be carefully evaluated prior to implementation. Unfavorable derivatization with severe interference on the target analyte should be avoided.

21.4 Application of Derivatization for Quantitative LC-MS Bioanalysis

Nowadays, quantitative LC-MS bioanalysis is widely employed in various fields for the quantification of diverse analytes. Despite the advancement of instruments and techniques, establishment of a sensitive and robust LC-MS method for quantitative analysis of some molecules is still challenging. Derivatization comes up as alternative means to support routine LC-MS bioanalysis of these molecules.

In the pharmaceutical industry, quantitative LC-MS bioanalysis of metabolites is of great interest and importance to the new drug development, especially when metabolites are pharmacologically active or toxic. However, compared to the parent drugs, most of the metabolites are polar molecules with weak chromatographic retention under conventional reversed-phase chromatographic conditions. Furthermore, some metabolites often present at very low concentrations in the study samples. For the past many years, various derivatization methods have been developed to assist the LC-MS bioanalysis of metabolites. For example, *N*-methyl-nicotinic acid, a pre-ionized derivatization reagent, was used for the sensitive quantification of estrogenic tamoxifen metabolites (Johanning et al. 2015).

For peptide and protein analysis, the most commonly used methods for comparing and quantifying protein rely on the use of differential isotopic labeling (Bantscheff et al. 2012). The frequently used chemical-derivatization includes (i) the isotope-coded affinity tag (iCAT) method, which derivatizes cysteine residues by iCAT reagent

(Gygi et al. 1999); (ii) the isobaric tags for relative and absolute quantification (iTRAQ) method, which derivatizes the *N*-terminus of peptides by iTRAQ reagent (Ross et al. 2004); (iii) the stable-isotope dimethyl labeling method, which derivatizes the *N*-terminal and ε-amino group of lysine through reductive amination by formaldehyde (Hsu et al. 2003).

For metabolomic and biomarker analysis, endogenous compounds, such as amino acids, thiols, and fatty acids, are often the targeted analytes. Most of the endogenous compounds are extremely hydrophilic or present at very low concentrations in biofluids, which adds to the difficulty of LC-MS method development. Derivatization targeting aldehydes, amines, thiols, and carboxylic acids has been frequently employed (Zhu et al. 2015).

For food-safety evaluation, multiple drug residues and toxins are the target analytes for LC-MS bioanalysis. Derivatization-based LC-MS has many advantages to offer better qualitative and quantitative results while increasing sensitivity and specificity. For β-lactam antibiotic residues, aldehyde derivatization was the most frequently adopted strategies; for neutral steroids, DNS-Cl was the common reagent for derivatization.

For many compounds with unfavorable properties for LC-MS bioanalysis, chemical derivatization could be a valuable mean to improve chromatography and/or MS detection of the molecule of interest. In the following examples, representative applications of chemical derivatization for LC-MS bioanalysis of drug substances are to be discussed.

21.4.1 Derivatization for Quantitative LC-MS Bioanalysis of Prasugrel

Prasugrel is a potent antiplatelet aggregation and activation agent for the prevention of atherothrombotic events in patients with acute coronary syndrome. Prasugrel is metabolized *in vivo* to several inactive metabolites and a thiol-containing active metabolite, R-138727 (Ding et al. 2003; Farid et al. 2007b; Rehmel et al. 2006) (Figure 21.3). Since R-138727 is closely related to the antiplatelet activity of prasugrel, it is of great importance to perform quantitative bioanalysis of R-138727 in systemic circulation for pharmacological evaluation. However, R-138727 is a thiol-containing metabolite, and the high reactivity of sulphydryl groups leads to its oxidation in both aqueous solution and plasma. Hence, it is important to stabilize the sulphydryl group in biological samples during sample preparation. The approach of derivatization was chosen to capture and stabilize the reactive thiol group of R-138727 to permit its accurate quantitation (Farid et al. 2007a). In addition, R-138727 possesses two chiral centers (Figure 21.3), and it has been reported that the four stereoisomers possess different antiplatelet activity. Quantitative bioanalysis of the four stereoisomers is of great benefit to assess contribution of each of them to the total R-138727 activity present in humans after administration of prasugrel (Wickremsinhe et al. 2007).

21.4.1.1 Stability of Prasugrel Derivative

Four reagents, methyl acrylate, methacrylamide (MAA), *N*-ethylmaleimide (NEM), and 2-bromo-3′-methoxyacetophenone (MPB), were tested for their ability to rapidly and effectively derivatize R-138727. The highest yield with the least variability was obtained when R-138727 was reacted with MPB (Figure 21.4). It was observed that more than 20% of R-138727 degraded within 15 minutes in plasma kept on ice, which indicated that the analyte must be stabilized by derivatization before separation of plasma from blood. Moreover, more than 30% of R-138727 degraded within 15 minutes in whole blood kept on ice. Therefore, the derivatization step must be carried out immediately after blood is collected. After MPB derivatization, the derivative was proved to be stable for at least 45 minutes on ice or at ambient temperature prior to centrifugation to separate plasma from blood. In human plasma, the derivative was stable up to 24 hours at room temperature, at −20 and −70 °C for up to four freeze/thaw cycles, and for 245 days when stored at −80 °C (Farid et al. 2007a). The ratio of the four stereoisomer derivatives of R-138727 remained constant in human plasma when they were exposed to room temperature for 24 hours, subjected to four freeze/thaw cycles, or held during long-term frozen storage for 31 days at −20 °C (Wickremsinhe et al. 2007).

Lukram et al. reported a method to stabilize R-138727 using NEM as the derivatizing agent with a short sample preparation and analysis cycle time (Figure 21.4). It was observed that the recovery and reproducibility of the NEM derivatization were better than the MPB approach. The derivatized R-138727 was found to be stable in whole blood for one hour on ice and for six hours in plasma after separation from the blood cells (Lukram et al. 2012).

21.4.1.2 Step-by-step Procedure of Derivatization

For standards' preparation, the stock solution of R-138727 was prepared by dissolving the analyte in 0.5 ml acetonitrile. The stock solution was mixed with 4.0 ml of 1 : 4 0.1 M ammonium hydroxide/0.1 M ammonium chloride. The derivatization was carried out by mixing the above solution with 0.5 ml of 0.1 M MPB or NEM in acetonitrile for at least 10 minutes at room temperature. The mixtures were then diluted with acetonitrile to yield stock solutions of ~50 mg ml^{-1}. The working solutions of derivatized R-138727 were prepared by diluting the derivatized stock solution with acetonitrile. All solutions were stored at −20 °C in polypropylene vials.

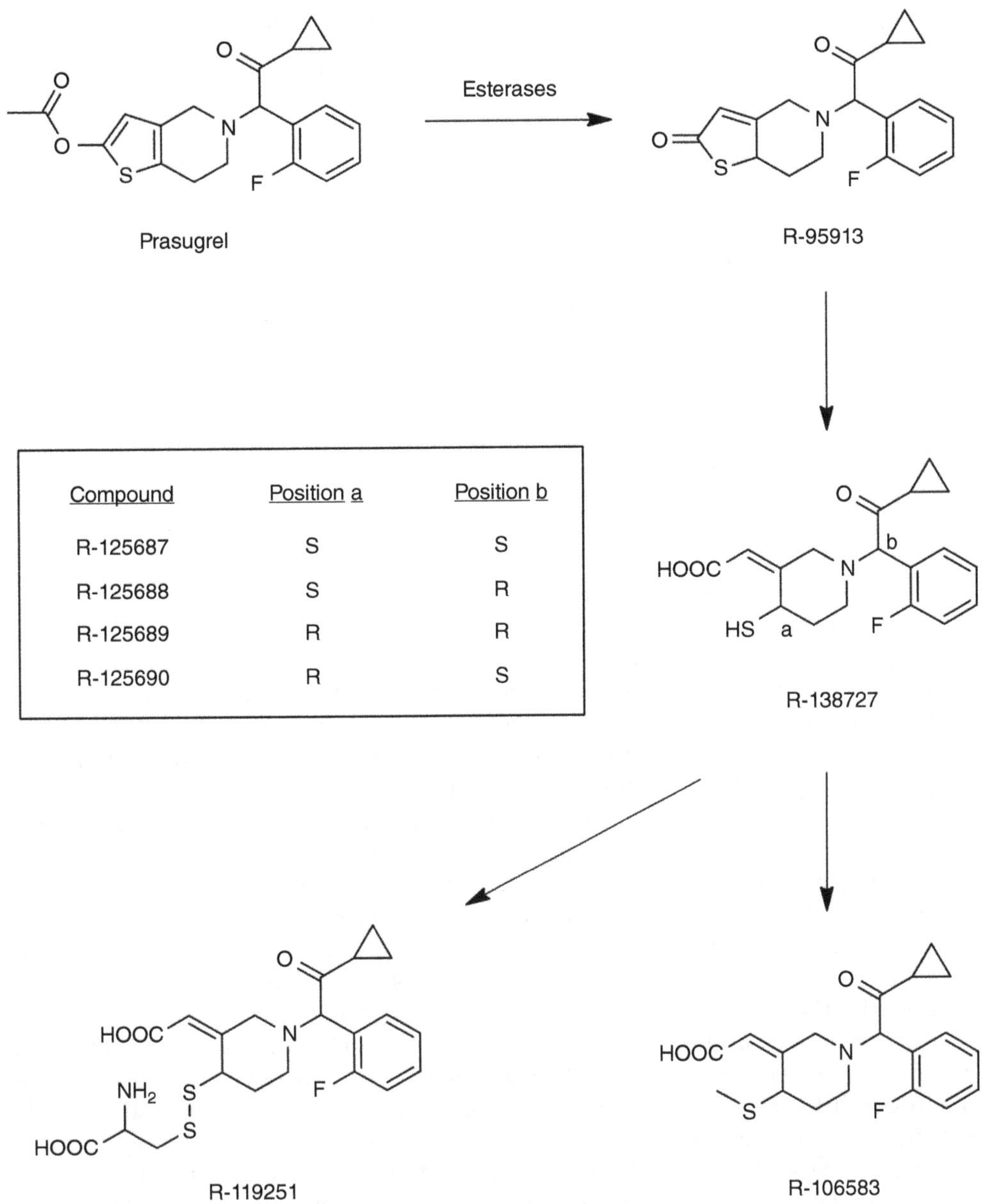

Figure 21.3 Metabolic pathway of prasugrel (Rehmel et al. 2006).

For plasma samples, immediately after blood was drawn, a volume of 25 µl of MPB or NEM (500 mM in acetonitrile) was added to the collected blood, followed by proper mixing. After centrifugation, the harvested plasma was stored at −80 or −20 °C before sample analysis.

21.4.1.3 LC-MS Behavior of Prasugrel Derivative

The protonated ion of MPB-derivatized R-138727 generated multiple fragment ions. The most abundant one was at *m/z* 348, deriving from R-138727 derivative (*m/z* 498). MRM transition was selected for the mass spectrometric specificity to distinguish the analyte from another isobaric metabolite. The validated lower limit of quantification (LLOQ) was 0.5 ng ml^{-1} for R-138727 via MPB derivatization (Farid et al. 2007a).

The protonated ion of NEM-derivatized R-138727 produced the major fragment ion at *m/z* 206, originating from the R-138727. The validated LLOQ was 1 ng ml^{-1} for R-138727 via NEM derivatization (Lukram et al. 2012).

To achieve the chromatographic separation of the stereoisomers, sample extracts (10 µl) were injected onto a 10 × 2 mm BetaBasic C4 Javelin Guard (Thermo Electron Corporation, Waltham, MA) connected to two (150 × 2 mm and 250 × 2 mm) ProntoSIL Chiral AXQN

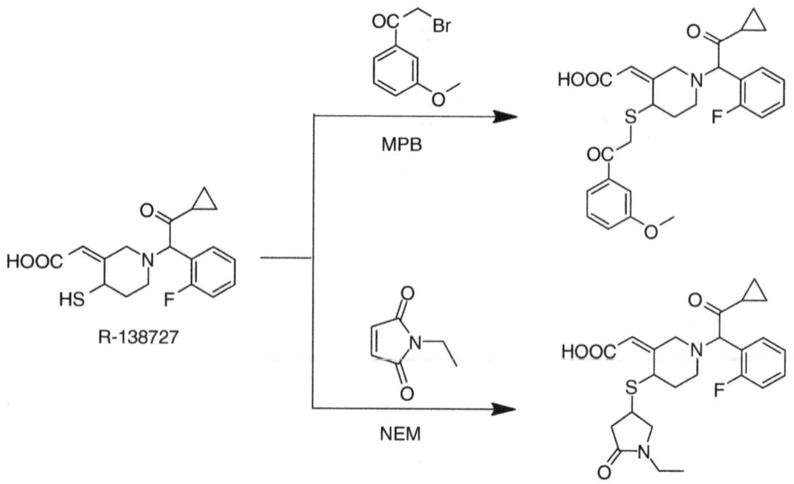

Figure 21.4 Derivatization reaction of the active metabolite of prasugrel (R-138727) (Farid et al. 2007a; Lukram et al. 2012).

columns in series (MAC-MOD Analytical, Chadds Ford, PA). An isocratic mobile phase consisting of 200:1.2:0.6:0.0025 methanol/acetic acid/water/0.1 M ammonium acetate was used, and the total cycle time was 20 minutes per injection (Wickremsinhe et al. 2007).

21.4.2 Derivatization for Quantitative LC-MS Bioanalysis of Minodronic Acid

Minodronic acid (MA) is a potent nitrogen containing bisphosphonate (BP), which has been clinically used for the treatment of osteoporosis in Japan since 2009 (Tanishima and Morio 2013). As MA is more potent than the first- and second-generation BPs, the oral dose of MA is much lower (1 mg once daily) than most of the other BPs, with a C_{max} value in humans of ~200 pg ml^{-1} (Zhou et al. 2015). Therefore, an ultrasensitive analytical method with a LLOQ of 10.0–20.0 pg ml^{-1} is required. However, it was challenging in developing a LC-MS assay with LLOQ at low pictogram per milliliter levels for MA because of the polar nature of the molecule bearing phosphates. In addition, the low molecular weight of MA directly affects the selectivity.

Off-line derivatization in combination with LC-MS detection has proved previously feasible to achieve the needed sensitivity for bioanalysis of BPs. Zhu et al. proposed a sensitive method for the measurement of risedronic acid (RA) and alendronic acid (AA) in human serum and urine using diazomethane as the derivatization reagent with achieved LLOQs of 200 and 50.0 pg, respectively (Zhu et al. 2006). Taking into consideration of the toxicity and the explosiveness of diazomethane, a much safer reagent, trimethylsilyldiazomethane (TMS-DAM), was used for the bioanalysis of ibandronic acid, zoledronic acid, RA, and AA in biological matrix with the achieved LLOQs ranging from 200 pg ml^{-1} to 100 ng ml^{-1} (Chen et al. 2012; Tarcomnicu et al. 2009; Veldboer et al. 2011).

21.4.2.1 Step-by-step Procedure of Derivatization

Internal standard working solutions (100 μl, d4-MA in methanol–water, 50:50, v/v) and water (1 ml) were added to a 1.0 ml aliquot of plasma. The mixture was vortexed and centrifuged for 1 minute at 3500 × *g*. The supernatant was loaded onto an Oasis weak anion exchange (WAX) 6 cc/150 mg cartridge (Waters, Milford, MA, USA) preconditioned with methanol (2.0 ml) and water (2.0 ml). The cartridge was washed with water (2 ml) and methanol (4 ml), and the derivatization reagent (2 ml, TMS-DAM/methanol/water, 25/75/6.25, v/v/v) was added, followed by elution with methanol (1 ml). The eluents from the last two steps were collected and mixed with water (200 μl, used as a catalyst), and the resulting mixture was reloaded onto the same cartridge for a second cycle of derivatization. The eluent from the second cycle of derivatization was collected and incubated in an ultrasonic water bath for two hours at 60 °C. It was then evaporated under a stream of nitrogen at 40 °C to dryness. The residue was reconstituted in water (300 μl), and an aliquot (20 μl) of the resulting solution was injected onto the LC-MS/MS system (Yang et al. 2015) (Figure 21.5).

21.4.2.2 LC-MS Behavior of MA Derivative

The derivatization products of BPs with diazomethane or TMS-DAM vary from case to case. For typical BPs, the products are tetramethyl derivatives. For amino-containing bisphosphonates, such as alendronate, hepta-, and hexa-methyl derivatives were also found (Ranz et al. 2008). For MA derivative, in addition to the reported tetramethyl derivative (TMMA, *m/z* 379), a new penta-methylated derivative (PMMA, *m/z* 393) was detected.

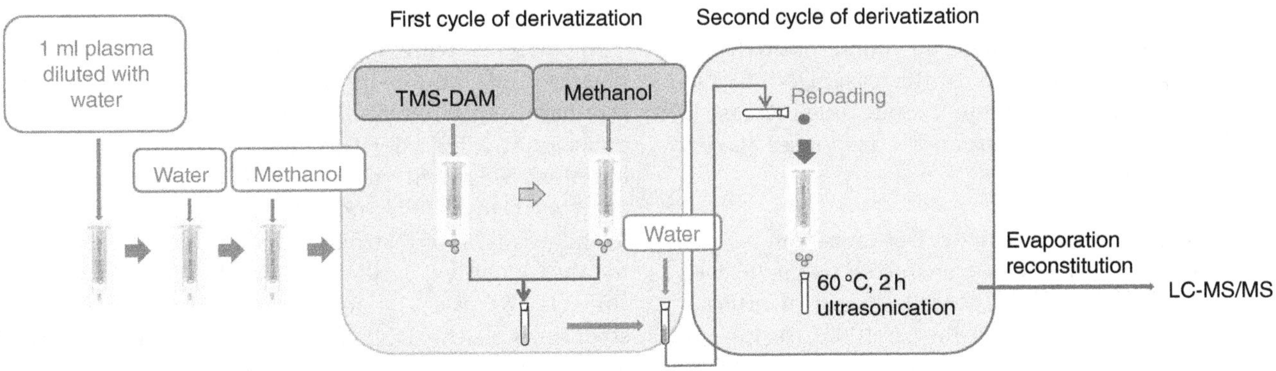

Figure 21.5 Step-by-step procedure of derivatization strategy for minodronic acid (Yang et al. 2015).

PMMA was monitored for quantification because it exhibited a better MS signal-to-noise ratio and fewer interfering peaks than TMMA. The product ion spectra show that similar product ions as PMMA were formed by the loss of the doubly methylated phosphoric acid moiety and water. This fragmentation pattern indicated that the fifth methylation occurred in the nitrogen heterocyclic ring, and it was further presumed to be position 1 or 4 nitrogen, and more likely on position 1 in consideration of the steric hindrance (Chen et al. 2012; Zhu et al. 2006).

The methylation of phosphonic acid enhanced the chromatographic retention of MA on the reverse-phase chromatography system. The optimization of the chromatographic method was focused on separating PMMA from co-eluting interfering peaks. Phenyl and PFP columns showed much better retention of PMMA than C18 columns. Basic mobile phase could offer stronger retention of PMMA than acidic mobile phase, but showed poor separation of PMMA from interfering peaks. Finally, an Eclipse XDB phenyl column combined with an acidic mobile phase successfully separated PMMA from the interfering peaks (Yang et al. 2015).

21.4.2.3 Key Consideration for MA Derivatization

Various factors affect the derivatization yield. Conditioning and washing were performed with HCl solutions of different pH values to evaluate whether the acidity affected the recovery and derivatization yield. The results indicated that the yield of PMMA decreased as the acidity increased, and pure water was optimal. EDTA was used as an anticoagulant instead of heparin because EDTA plasma is more suitable for the SPE procedure than the heparin-treated one. During the on-cartridge derivatization, dripping under gravity achieved a higher yield than keeping the reagent on the cartridge by capping the cartridge. The derivatization yield was further improved by incubating the eluate in a water bath at 40–60 °C or under ultrasonication.

It was also noticed that the yield of d4-PMMA was much more variable and lower than that of PMMA. As a result, the tracking function of d4-MA as an internal standard was less effective, leading to poor linearity and reproducibility. Insufficient on-cartridge exposure time was the reason of insufficient derivatization of internal standard. Therefore, strategies to prolong the exposure time were adopted, including a second cycle of derivatization by reloading the eluent on the cartridge. As a result, the yield of d4-PMMA significantly improved, and the variation in the yield of d4-PMMA decreased, resulting in acceptable linearity.

21.4.3 Derivatization for Quantitative LC-MS Bioanalysis of α-Fluoro-β-alanine, 5-Fluorouracil, and Capecitabine

Capecitabine (Cape), a prodrug of 5-fluorouracil (5-FU), is a novel oral fluoropyrimidine carbamate that is preferentially metabolized to the cytotoxic moiety 5-FU in target tumor tissues. However, 5-FU can be further catabolized to α-fluoro-β-alanine (FBAL), which has been proven to induce cardiotoxicity and neurotoxicity in some patients (Muneoka et al. 2005). Due to these issues, a sensitive and accurate bioanalytical method for the analysis of FBAL, 5-FU, and Cape is required for monitoring chemotherapy treatment in cancer patients. However, both 5-FU and FBAL are hydrophilic compounds with $\text{clog}P < -0.5$, which results in less sufficient retention on reversed-phase column, especially for the metabolite FBAL (Niwa 2012). The lower LLOQ requirements between 1 and 10 ng ml^{-1} is also challenging. Raising $\text{clog}P$ and at the same time gaining ionization efficiency by chemical derivatization is an effective choice for this issue. Due to the remarkable differences in polarity among the three analytes and to difficulties finding an analytical column to retain and separate them, a novel derivatization procedure with DNS-Cl has been established (Licea-Perez et al. 2009).

DNS-Cl is a commonly used derivatization reagent for phenolic compounds to improve detection sensitivity for bioanalysis based on LC-MS. In this case, DNS-Cl targeted amino groups to result in considerable increase in both *c*log*P* and mass spectrometric responses (Licea-Perez et al. 2009).

21.4.3.1 Step-by-step Procedure of Derivatization

Plasma samples (50 μl) were transferred to the 96-well plate, followed by adding a 300 μl aliquot of internal standard solution in acetonitrile (25/100/50 ng ml^{-1} of [$^{13}C_3$]-FBAL/[$^{13}C_1$,$^{15}N_2$]-FU/[$^2H_{11}$]-Cape). Then 0.8 ml of ethyl acetate was added, and the wells were capped with the ArctiSeal mat, vortex-mixed for three minutes, and centrifuged at 3220 × *g* for five minutes. The organic layer was used for analysis of 5-FU and Cape while the remaining aqueous phase was used for analysis of FBAL.

For the organic layer, the ethyl acetate/acetonitrile layer containing 5-FU/Cape was transferred to another ArcticWhite 96-well, 2 ml polypropylene 96-well plate and the solvent evaporated under a stream of nitrogen at 45 °C. After evaporation, 100 μl of 10 mg ml^{-1} DNS-Cl in acetonitrile and 100 μl sodium bicarbonate (100 mM, pH 11) were added to all wells. The plate was sealed and vortex-mixed for 3 minutes, followed by incubation at 60 °C for 30 minutes. After incubation, 1 ml of methyl *tert*-butyl ether (MTBE) was added to all wells to extract the derivatives of 5-FU and Cape. After vortex-mixing and centrifugation, the MTBE layer was transferred to a polypropylene 96-well plate containing 1 ml deactivated (silanized) glass-inserts and evaporated to dryness under a stream of nitrogen at 45 °C. The extract then was reconstituted in 100 μl of 30/70 (v/v) acetonitrile:water.

For the aqueous layer containing FBAL, 100 μl of DNS-Cl (10 mg ml^{-1}) in acetone and 100 μl sodium bicarbonate (100 mM, pH 11) were added and incubated at 60 °C for five minutes. Following addition of 0.5 ml of water, the wells were capped with the ArctiSeal mat, vortex-mixed for three minutes, and centrifuged at 3220 × *g* for five minutes. 300 ml of the supernatant then were loaded onto a 10 mg Oasis MAX, 30 μm, SPE 96-well plate which was previously conditioned with 1 ml methanol and 1 ml water. The loaded SPE plate was then washed with 1 ml of water and 1 ml of methanol. Dansyl-FBAL then was eluted twice with 250 μl volumes of 2% formic acid in methanol (freshly prepared) into a 2 ml 96-well plate containing 1 ml deactivated (silanized) glass-inserts. These sample extracts were evaporated to dryness at 45 °C under a stream of nitrogen, reconstituted with 500 μl of 30/70 (v/v) acetonitrile/water, vortex-mixed and centrifuged before analysis (Licea-Perez et al. 2009).

21.4.3.2 LC-MS Behavior of DNS-Cl Derivatives

The analytes were extracted from 50 μl of human plasma by liquid–liquid extraction (LLE) and incubated with DNS-Cl in the presence of sodium bicarbonate. This derivatization procedure provided excellent MS sensitivity allowing detection of all three analytes. Dansyl-5FU and dansyl-Cape were monitored in the positive ionization mode and dansyl-FBAL was in the negative ionization mode. The precursor $[M+H]^+$ or $[M-H]^-$ ions were consistent with the structures of the DNS-Cl derivatives. The fragment ions selected for MRM were derived from the cleavage of DNS-Cl for 5-FU and FBAL, and the drug molecule itself for Cape.

However, there was poor chromatographic separation between the dansyl-5-FU and interference peaks from plastic residues and other endogenous dansyl derivatives from plasma that prevented the use of the C18 stationary phase for analysis of all dansyl-derivatives. To solve this problem, a procedure including LLE with a mixture of acetonitrile and ethyl acetate was introduced. This procedure allowed the extraction of 5-FU and Cape from plasma while FBAL remained in the aqueous phase. After separation and evaporation of the supernatant, FU and Cape were derivatized with DNS-Cl in acetonitrile. This modification eliminated interferences from plastic residues affecting the dansyl-5-FU MRM channel and most of the interference peaks from plasma. A reversed-phase analytical column "Primesep D" (2.1 mm × 50 mm; 5 μm) with embedded basic ion-pairing groups was evaluated and showed good chromatographic separation of dansyl-5-FU from all interference peaks (Licea-Perez et al. 2009).

21.5 Summary

LC-MS-based methods have become routine for quantitative bioanalysis. However, even with the most advanced instrument platform, limited sensitivity or selectivity is constantly encountered when developing a LC-MS assay for some molecules. Undesirable chromatographic retention and severe interference due to the matrix components can also be detrimental to the quantitative studies. Derivatization is a critical procedure to improve method sensitivity or selectivity, despite the complexity in method development and optimization. The current derivatization strategies mainly include detection-oriented, separation-oriented, and stability-oriented derivatization. Overall, derivatization serves as a supplementary means for the normal sample preparation, and can be very useful in the LC-MS analysis of challenging analytes.

References

Bantscheff, M., Lemeer, S., Savitski, M.M., and Kuster, B. (2012). Quantitative mass spectrometry in proteomics: critical review update from 2007 to the present. *Anal. Bioanal. Chem.* 404 (4): 939–965.

Becker, S., Kortz, L., Helmschrodt, C. et al. (2012). LC-MS-based metabolomics in the clinical laboratory. *J. Chromatogr. B Analyt. Technol. Biomed. Life Sci.* 883–884: 68–75.

Bollinger, J.G., Naika, G.S., Sadilek, M., and Gelb, M.H. (2013). LC/ESI-MS/MS detection of FAs by charge reversal derivatization with more than four orders of magnitude improvement in sensitivity. *J. Lipid Res.* 54 (12): 3523–3530.

Bollinger, J.G., Thompson, W., Lai, Y. et al. (2010). Improved sensitivity mass spectrometric detection of eicosanoids by charge reversal derivatization. *Anal. Chem.* 82 (16): 6790–6796.

Borrey, D., Moerman, E., Cockx, A. et al. (2007). Column-switching LC-MS/MS analysis for quantitative determination of testosterone in human serum. *Clin. Chim. Acta* 382 (1–2): 134–137.

Cai, H.L., Zhu, R.H., and Li, H.D. (2010). Determination of dansylated monoamine and amino acid neurotransmitters and their metabolites in human plasma by liquid chromatography-electrospray ionization tandem mass spectrometry. *Anal. Biochem.* 396 (1): 103–111.

Chang, H., Wan, Y., Naile, J. et al. (2010). Simultaneous quantification of multiple classes of phenolic compounds in blood plasma by liquid chromatography-electrospray tandem mass spectrometry. *J. Chromatogr. A* 1217 (4): 506–513.

Chen, M., Liu, K., Zhong, D., and Chen, X. (2012). Trimethylsilyldiazomethane derivatization coupled with solid-phase extraction for the determination of alendronate in human plasma by LC-MS/MS. *Anal. Bioanal. Chem.* 402 (2): 791–798.

Chen, J., Zhang, J., Zhang, W., and Chen, Z. (2014). Sensitive determination of the potential biomarker sarcosine for prostate cancer by LC-MS with *N,N′*-dicyclohexylcarbodiimide derivatization. *J. Sep. Sci.* 37 (1–2): 14–19.

Deng, P., Zhan, Y., Chen, X., and Zhong, D. (2012). Derivatization methods for quantitative bioanalysis by LC-MS/MS. *Bioanalysis* 4 (1): 49–69.

Ding, Z., Kim, S., Dorsam, R.T. et al. (2003). Inactivation of the human P2Y12 receptor by thiol reagents requires interaction with both extracellular cysteine residues, Cys17 and Cys270. *Blood* 101 (10): 3908–3914.

Ding, S., Schoenmakers, I., Jones, K. et al. (2010). Quantitative determination of vitamin D metabolites in plasma using UHPLC-MS/MS. *Anal. Bioanal. Chem.* 398 (2): 779–789.

Faqehi, A.M., Cobice, D.F., Naredo, G. et al. (2016). Derivatization of estrogens enhances specificity and sensitivity of analysis of human plasma and serum by liquid chromatography tandem mass spectrometry. *Talanta* 151: 148–156.

Farid, N.A., McIntosh, M., Garofolo, F. et al. (2007a). Determination of the active and inactive metabolites of prasugrel in human plasma by liquid chromatography/tandem mass spectrometry. *Rapid Commun. Mass Spectrom.* 21 (2): 169–179.

Farid, N.A., Smith, R.L., Gillespie, T.A. et al. (2007b). The disposition of prasugrel, a novel thienopyridine, in humans. *Drug Metab. Dispos.* 35 (7): 1096–1104.

Gao, S., Zhang, Z.P., and Karnes, H.T. (2005). Sensitivity enhancement in liquid chromatography/atmospheric pressure ionization mass spectrometry using derivatization and mobile phase additives. *J. Chromatogr. B Analyt. Technol. Biomed. Life Sci.* 825 (2): 98–110.

Gygi, S.P., Rist, B., Gerber, S.A. et al. (1999). Quantitative analysis of complex protein mixtures using isotope-coded affinity tags. *Nat. Biotechnol.* 17 (10): 994–999.

Han, J., Lin, K., Sequeira, C., and Borchers, C.H. (2015). An isotope-labeled chemical derivatization method for the quantitation of short-chain fatty acids in human feces by liquid chromatography-tandem mass spectrometry. *Anal. Chim. Acta* 854: 86–94.

Hayama, T., Sakaguchi, Y., Yoshida, H. et al. (2012). Binary fluorous alkylation of biogenic primary amines with perfluorinated aldehyde followed by fluorous liquid chromatography-tandem mass spectrometry analysis. *Anal. Chem.* 84 (19): 8407–8414.

Higashi, T., Suzuki, M., Hanai, J. et al. (2011). A specific LC/ESI-MS/MS method for determination of 25-hydroxyvitamin D3 in neonatal dried blood spots containing a potential interfering metabolite, 3-epi-25-hydroxyvitamin D3. *J. Sep. Sci.* 34 (7): 725–732.

Higashi, T., Yamauchi, A., and Shimada, K. (2005). 2-Hydrazino-1-methylpyridine: a highly sensitive derivatization reagent for oxosteroids in liquid chromatography-electrospray ionization-mass spectrometry. *J. Chromatogr. B Analyt. Technol. Biomed. Life Sci.* 825 (2): 214–222.

Hsu, J.L., Huang, S.Y., Chow, N.H., and Chen, S.H. (2003). Stable-isotope dimethyl labeling for quantitative proteomics. *Anal. Chem.* 75 (24): 6843–6852.

Huang, H.J., Chiang, P.H., and Chen, S.H. (2011a). Quantitative analysis of estrogens and estrogen metabolites in endogenous MCF-7 breast cancer cells by liquid chromatography-tandem mass spectrometry.

J. Chromatogr. B Analyt. Technol. Biomed. Life Sci. 879 (20): 1748–1756.

Huang, Y.Q., Ruan, G.D., Liu, J.Q. et al. (2011b). Use of isotope differential derivatization for simultaneous determination of thiols and oxidized thiols by liquid chromatography tandem mass spectrometry. *Anal. Biochem.* 416 (2): 159–166.

Hui, Y., Wong, M., Kim, J.O. et al. (2012). A new derivatization method coupled with LC-MS/MS to enable baseline separation and quantification of dimethylarginines in human plasma from patients to receive on-pump CABG surgery. *Electrophoresis* 33 (12): 1911–1920.

Inoda, H., Nishiyama, T., Yoshikado, T. et al. (2011). Compounds having thiourea moiety as derivatization reagents in liquid chromatography/electrospray ionization-tandem mass spectrometry (LC/ESI-MS/MS): synthesis of derivatization reagents for carboxylic acids. *Biomed. Chromatogr.* 25 (6): 635–640.

Iwasaki, Y., Nakano, Y., Mochizuki, K. et al. (2011). A new strategy for ionization enhancement by derivatization for mass spectrometry. *J. Chromatogr. B Analyt. Technol. Biomed. Life Sci.* 879 (17–18): 1159–1165.

Johanning, J., Heinkele, G., Precht, J.C. et al. (2015). Highly sensitive simultaneous quantification of estrogenic tamoxifen metabolites and steroid hormones by LC-MS/MS. *Anal. Bioanal. Chem.* 407 (24): 7497–7502.

Juang, Y.M., She, T.F., Chen, H.Y., and Lai, C.C. (2013). Comparison of CID versus ETD-based MS/MS fragmentation for the analysis of doubly derivatized steroids. *J. Mass Spectrom.* 48 (12): 1349–1356.

Kalhorn, T.F., Page, S.T., Howald, W.N. et al. (2007). Analysis of testosterone and dihydrotestosterone from biological fluids as the oxime derivatives using high-performance liquid chromatography/tandem mass spectrometry. *Rapid Commun. Mass Spectrom.* 21 (19): 3200–3206.

Karakosta, T.D., Tzanavaras, P.D., and Themelis, D.G. (2013). Determination of glutathione and cysteine in yeasts by hydrophilic interaction liquid chromatography followed by on-line postcolumn derivatization. *J. Sep. Sci.* 36 (12): 1877–1882.

Ke, Y., Gonthier, R., and Labrie, F. (2016). A sensitive and accurate LC-MS/MS assay with the derivatization of 1-amino-4-methylpiperazine applied to serum allopregnanolone, pregnenolone and androsterone in pre- and postmenopausal women. *Steroids* 118: 25–31.

Keski-Rahkonen, P., Desai, R., Jimenez, M. et al. (2015). Measurement of estradiol in human serum by LC-MS/MS using a novel estrogen-specific derivatization reagent. *Anal. Chem.* 87 (14): 7180–7186.

Kushnir, M.M., Rockwood, A.L., Roberts, W.L. et al. (2006). Performance characteristics of a novel tandem mass spectrometry assay for serum testosterone. *Clin. Chem.* 52 (1): 120–128.

Kusmierek, K., Chwatko, G., Glowacki, R. et al. (2011). Ultraviolet derivatization of low-molecular-mass thiols for high performance liquid chromatography and capillary electrophoresis analysis. *J. Chromatogr. B Analyt. Technol. Biomed. Life Sci.* 879 (17–18): 1290–1307.

Kuwabara, T., Takayama, T., Todoroki, K. et al. (2014). Evaluation of a series of prolylamidepyridines as the chiral derivatization reagents for enantioseparation of carboxylic acids by LC-ESI-MS/MS and the application to human saliva. *Anal. Bioanal. Chem.* 406 (11): 2641–2649.

Lamshoft, M., Grobe, N., and Spiteller, M. (2011). Picomolar concentrations of morphine in human urine determined by dansyl derivatization and liquid chromatography-mass spectrometry. *J. Chromatogr. B Analyt. Technol. Biomed. Life Sci.* 879 (13–14): 933–937.

Lavrynenko, O., Nedielkov, R., Moller, H.M., and Shevchenko, A. (2013). Girard derivatization for LC-MS/MS profiling of endogenous ecdysteroids in *Drosophila. J. Lipid Res.* 54 (8): 2265–2272.

Li, S., Jin, Y., Tang, Z. et al. (2015). A novel method of liquid chromatography-tandem mass spectrometry combined with chemical derivatization for the determination of ribonucleosides in urine. *Anal. Chim. Acta* 864: 30–38.

Li, G., Wu, D., Xie, W. et al. (2013). Analysis of amino acids in tobacco by derivatization and dispersive liquid-liquid microextraction based on solidification of floating organic droplet method. *J. Chromatogr. A* 1296: 243–247.

Licea-Perez, H., Wang, S., and Bowen, C. (2009). Development of a sensitive and selective LC-MS/MS method for the determination of alpha-fluoro-beta-alanine, 5-fluorouracil and capecitabine in human plasma. *J. Chromatogr. B Analyt. Technol. Biomed. Life Sci.* 877 (11–12): 1040–1046.

Liem-Nguyen, V., Bouchet, S., and Bjorn, E. (2015). Determination of sub-nanomolar levels of low molecular mass thiols in natural waters by liquid chromatography tandem mass spectrometry after derivatization with p-(hydroxymercuri) benzoate and online preconcentration. *Anal. Chem.* 87 (2): 1089–1096.

Lipkie, T.E., Janasch, A., Cooper, B.R. et al. (2013). Quantification of vitamin D and 25-hydroxyvitamin D in soft tissues by liquid chromatography-tandem mass spectrometry. *J. Chromatogr. B Analyt. Technol. Biomed. Life Sci.* 932: 6–11.

Liu, R., Wang, M., and Ding, L. (2014). A novel liquid chromatography-tandem mass spectrometry method for determination of menadione in human plasma after derivatization with 3-mercaptopropionic acid. *Talanta* 128: 51–57.

Lu, Y., Yao, D., and Chen, C. (2013). 2-Hydrazinoquinoline as a derivatization agent for LC-MS-based metabolomic investigation of diabetic ketoacidosis. *Metabolites* 3 (4): 993–1010.

Lukram, O., Zarapkar, M., Kumar Jha, C. et al. (2012). Electrospray ionization LC-MS/MS validated method for the determination of the active metabolite (R-138727) of prasugrel in human plasma and its application to a bioequivalence study. *Drug Test. Anal.* 4 (2): 158–166.

Maas, A., Maier, C., Michel-Lauter, B. et al. (2017). 1,2-Dimethylimidazole-4-sulfonyl chloride (DMISC), a novel derivatization strategy for the analysis of propofol by LC-ESI-MS/MS. *Anal. Bioanal. Chem.* 409 (6): 1547–1554.

Min, J.Z., Hatanaka, S., Yu, H.F. et al. (2011). Determination of DL-amino acids, derivatized with *R*(-)-4-(3-isothiocyanatopyrrolidin-1-yl)-7-(*N,N*-dimethylaminosulfonyl)-2,1,3-benz oxadiazole, in nail of diabetic patients by UPLC-ESI-TOF-MS. *J. Chromatogr. B Analyt. Technol. Biomed. Life Sci.* 879 (29): 3220–3228.

Muneoka, K., Shirai, Y., Yokoyama, N. et al. (2005). 5-Fluorouracil cardiotoxicity induced by alpha-fluoro-beta-alanine. *Int. J. Clin. Oncol.* 10 (6): 441–443.

Nakanishi, K., Katagi, M., Zaitsu, K. et al. (2012). Simultaneous enantiomeric determination of MDMA and its phase I and phase II metabolites in urine by liquid chromatography-tandem mass spectrometry with chiral derivatization. *Anal. Bioanal. Chem.* 404 (8): 2427–2435.

Niwa, M. (2012). Chemical derivatization as a tool for optimizing MS response in sensitive LC-MS/MS bioanalysis and its role in pharmacokinetic studies. *Bioanalysis* 4 (2): 213–220.

Ogawa, S., Tadokoro, H., Sato, M. et al. (2013). (*S*)-1-(4-Dimethylaminophenylcarbonyl)-3-aminopyrrolidine: a derivatization reagent for enantiomeric separation and sensitive detection of chiral carboxylic acids by LC/ESI-MS/MS. *J. Chromatogr. B Analyt. Technol. Biomed. Life Sci.* 940: 7–14.

Pernica, M., Poloucka, P., Seifertova, M., and Simek, Z. (2015). Determination of alkylphenols in water samples using liquid chromatography-tandem mass spectrometry after pre-column derivatization with dansyl chloride. *J. Chromatogr. A* 1417: 49–56.

Qi, B.L., Liu, P., Wang, Q.Y. et al. (2014). Derivatization for liquid chromatography-mass spectrometry. *TrAC Trends Anal. Chem.* 59: 121–132.

Ranz, A., Korpecka, J., and Lankmayr, E. (2008). Optimized derivatization of acidic herbicides with trimethylsilyldiazomethane for GC analysis. *J. Sep. Sci.* 31 (4): 746–752.

Rao, Y., Xiang, B., Bramanti, E. et al. (2010). Determination of thiols in yeast by HPLC coupled with LTQ-orbitrap mass spectrometry after derivatization with *p*-(hydroxymercuri)benzoate. *J. Agric. Food Chem.* 58 (3): 1462–1468.

Rebane, R., Oldekop, M.L., and Herodes, K. (2012). Comparison of amino acid derivatization reagents for LC-ESI-MS analysis. Introducing a novel phosphazene-based derivatization reagent. *J. Chromatogr. B Analyt. Technol. Biomed. Life Sci.* 904: 99–106.

Rebane, R., Oldekop, M.L., and Herodes, K. (2014). Matrix influence on derivatization and ionization processes during selenoamino acid liquid chromatography electrospray ionization mass spectrometric analysis. *J. Chromatogr. B Analyt. Technol. Biomed. Life Sci.* 955–956: 34–41.

Rehmel, J.L., Eckstein, J.A., Farid, N.A. et al. (2006). Interactions of two major metabolites of prasugrel, a thienopyridine antiplatelet agent, with the cytochromes P450. *Drug Metab. Dispos.* 34 (4): 600–607.

Ross, P.L., Huang, Y.N., Marchese, J.N. et al. (2004). Multiplexed protein quantitation in *Saccharomyces cerevisiae* using amine-reactive isobaric tagging reagents. *Mol. Cell. Proteomics* 3 (12): 1154–1169.

Sakaguchi, Y., Kinumi, T., Yamazaki, T., and Takatsu, A. (2015). A novel amino acid analysis method using derivatization of multiple functional groups followed by liquid chromatography/tandem mass spectrometry. *Analyst* 140 (6): 1965–1973.

Santa, T. (2011). Derivatization reagents in liquid chromatography/electrospray ionization tandem mass spectrometry. *Biomed. Chromatogr.* 25 (1–2): 1–10.

Santa, T. (2013). Derivatization in liquid chromatography for mass spectrometric detection. *Drug Discov. Ther.* 7 (1): 9–17.

Sentellas, S., Morales-Ibanez, O., Zanuy, M., and Alberti, J.J. (2014). GSSG/GSH ratios in cryopreserved rat and human hepatocytes as a biomarker for drug induced oxidative stress. *Toxicol. In Vitro* 28 (5): 1006–1015.

Star-Weinstock, M., Williamson, B.L., Dey, S. et al. (2012). LC-ESI-MS/MS analysis of testosterone at sub-picogram levels using a novel derivatization reagent. *Anal. Chem.* 84 (21): 9310–9317.

Sun, Y., Yao, T., Guo, X. et al. (2016). Simultaneous assessment of endogenous thiol compounds by LC-MS/MS. *J. Chromatogr. B Analyt. Technol. Biomed. Life Sci.* 1029–1030: 213–221.

Tang, Z. and Guengerich, F.P. (2010). Dansylation of unactivated alcohols for improved mass spectral sensitivity and application to analysis of cytochrome P450 oxidation products in tissue extracts. *Anal. Chem.* 82 (18): 7706–7712.

Tanishima, S. and Morio, Y. (2013). A review of minodronic acid hydrate for the treatment of osteoporosis. *Clin. Interv. Aging* 8: 185–189.

Tarcomnicu, I., Gheorghe, M.C., Silvestro, L. et al. (2009). High-throughput HPLC-MS/MS method to determine ibandronate in human plasma for pharmacokinetic applications. *J. Chromatogr. B Analyt. Technol. Biomed. Life Sci.* 877 (27): 3159–3168.

Veldboer, K., Vielhaber, T., Ahrens, H. et al. (2011). Determination of zoledronic acid in human urine and blood plasma using liquid chromatography/electrospray mass spectrometry. *J. Chromatogr. B Analyt. Technol. Biomed. Life Sci.* 879 (22): 2073–2080.

Vichi, S., Cortes-Francisco, N., and Caixach, J. (2013). Determination of volatile thiols in lipid matrix by simultaneous derivatization/extraction and liquid chromatography-high resolution mass spectrometric analysis. Application to virgin olive oil. *J. Chromatogr. A* 1318: 180–188.

Visser, W.F., Verhoeven-Duif, N.M., Ophoff, R. et al. (2011). A sensitive and simple ultra-high-performance-liquid chromatography-tandem mass spectrometry based method for the quantification of D-amino acids in body fluids. *J. Chromatogr. A* 1218 (40): 7130–7136.

Vitku, J., Chlupacova, T., Sosvorova, L. et al. (2015). Development and validation of LC-MS/MS method for quantification of bisphenol A and estrogens in human plasma and seminal fluid. *Talanta* 140: 62–67.

Wang, H.Y., Chu, X., Zhao, Z.X. et al. (2011). Analysis of low molecular weight compounds by MALDI-FTICR-MS. *J. Chromatogr. B Analyt. Technol. Biomed. Life Sci.* 879 (17–18): 1166–1179.

Wickremsinhe, E.R., Tian, Y., Ruterbories, K.J. et al. (2007). Stereoselective metabolism of prasugrel in humans using a novel chiral liquid chromatography-tandem mass spectrometry method. *Drug Metab. Dispos.* 35 (6): 917–921.

Xu, K., Zhang, Y., Tang, B. et al. (2010). Study of highly selective and efficient thiol derivatization using selenium reagents by mass spectrometry. *Anal. Chem.* 82 (16): 6926–6932.

Xu, F., Zou, L., Liu, Y. et al. (2011). Enhancement of the capabilities of liquid chromatography-mass spectrometry with derivatization: general principles and applications. *Mass Spectrom. Rev.* 30 (6): 1143–1172.

Yang, Y., Liu, C., Zhang, Y. et al. (2015). On-cartridge derivatization coupled with solid-phase extraction for the ultra-sensitive determination of minodronic acid in human plasma by LC-MS/MS method. *J. Pharm. Biomed. Anal.* 114: 408–415.

Zhang, W., Li, P., Geng, Q. et al. (2014). Simultaneous determination of glutathione, cysteine, homocysteine, and cysteinylglycine in biological fluids by ion-pairing high-performance liquid chromatography coupled with precolumn derivatization. *J. Agric. Food Chem.* 62 (25): 5845–5852.

Zhou, Y., He, X., Li, H. et al. (2015). Pharmacokinetics and tolerability of minodronic acid tablets in healthy Chinese subjects and food and age effects on the pharmacokinetics. *Clin. Ther.* 37 (4): 869–876.

Zhu, Y., Deng, P., and Zhong, D. (2015). Derivatization methods for LC-MS analysis of endogenous compounds. *Bioanalysis* 7 (19): 2557–2581.

Zhu, L.S., Lapko, V.N., Lee, J.W. et al. (2006). A general approach for the quantitative analysis of bisphosphonates in human serum and urine by high-performance liquid chromatography/tandem mass spectrometry. *Rapid Commun. Mass Spectrom.* 20 (22): 3421–3426.

Ziegler, J. and Abel, S. (2014). Analysis of amino acids by HPLC/electrospray negative ion tandem mass spectrometry using 9-fluorenylmethoxycarbonyl chloride (Fmoc-Cl) derivatization. *Amino Acids* 46 (12): 2799–2808.

22

Sample Preparation for LC-MS Bioanalysis of Lipids

Samuel Hofbauer, Ashkan Salamatipour, Ian A. Blair, and Clementina Mesaros

22.1 Introduction

Lipid research has generated significant interest due to the role these molecules play in cell membranes and other critical biological areas. Lipids are comprised of eight different classes (Figure 22.1) of biologically derived molecules (fatty acyls, glycerolipids, phospholipids, sphingolipids, sterol lipids, prenol lipids, saccharolipids, and polyketides). The lipids' structures are of high diversity, and have a wide range in polarity, but are typically sparingly soluble in water at neutral pH. Lipids are particularly important due to their structural functions in eukaryotic cell membranes. The importance of lipids in biology is due to their roles as mediators in cellular signaling which regulates various physiological, pharmacological, and pathological processes in the body. Lipids can have biological function both in their unmodified form and after modification. Reactive oxygen species (ROS) and various oxygenases can modify both intact lipids or lipase-derived polyunsaturated fatty acids (PUFAs). Numerous disorders, including inborn errors of metabolism, cardiovascular disease, and obesity have been known to involve lipid-related disorders, and increasing knowledge suggests that lipids play a role in other major human diseases, such as cancer and neurodegenerative disorders (Blair 2008; Han 2007; van Meer 2005).

Fatty acyls, glycerolipids, phospholipids, and sphingolipids share similar biochemical properties due to the presence of similar fatty acid (FA) substituents at the *sn*-1 and *sn*-2 positions. FAs of human origin have an even number of carbon atoms, and PUFAs are susceptible to lipid peroxidation through enzymatic and nonenzymatic processes such as the actions of cyclooxygenases (COXs) and lipoxygenases (LOXs). The mechanisms through which these actions take place, known as lipid peroxidation, as well as the resulting products, have attracted interest due to the negative effects of lipid peroxidation on human health and the general important roles of lipids on biological systems such as those played by prostaglandins (PGs) and leukotrienes. Glycolipids and phospholipids have a hydrophilic head and hydrophobic tails composed of long-chain FAs, which facilitates their self-assembly into membranes and lipid droplets, and enables the coordination of membrane events such as signal transduction (van Meer 2005). Many lipids rely on various transporters that can facilitate translocation of individual lipids across the plasma membrane and organelle membranes, with some lipids, such as phosphoinositides, can themselves play a role as mediators of cellular signal transduction pathways (Hannun and Obeid 2008). Steroids are a group of lipids with complex ring structures and they include important functional molecules for the body such as steroid hormones (the gonadal or sex steroids, corticosteroids, glucocorticoids, and mineralocorticoids), vitamin D, and cholesterols. Fat-soluble vitamins such as vitamin D, E, A, and K are all lipids that play critical functions in maintaining body health and therefore of great interest for quantitative analysis.

22.2 Sample Preparation for LC-MS Bioanalysis of Lipids

22.2.1 LC-MS Bioanalysis of Lipids

Due to their diverse and complex structures, and the biological activity that these structures lend themselves to, there is a great necessity for the development of sensitive and specific LC-MS methodologies for lipid analysis. With advances in atmospheric pressure ionization-based techniques such as electrospray ionization (ESI) and atmospheric pressure chemical ionization (APCI), along with matrix-assisted laser desorption (MALDI), a greater range of lipids can be analyzed by MS.

Sample Preparation in LC-MS Bioanalysis, First Edition. Edited by Wenkui Li, Wenying Jian, and Yunlin Fu.

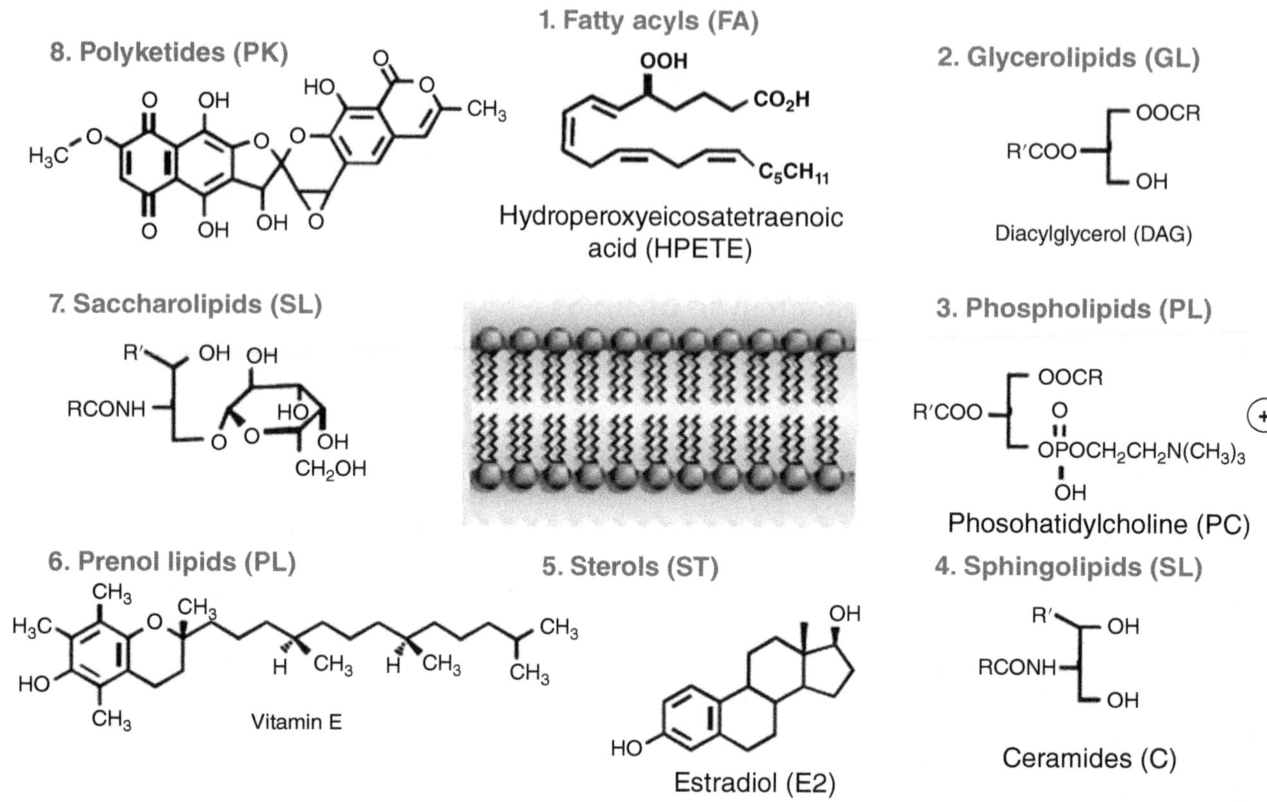

Figure 22.1 Classes of lipids.

In conjunction with collision-induced dissociation (CID), tandem mass spectrometry (MS/MS) and high resolution mass spectrometry (HRMS), LC-MS has revolutionized the possibilities of quantifying and analyzing lipids. The LC-MS methods employed for the analysis of different classes of lipids nowadays include untargeted lipidomics and targeted methods. One of the advantages of using untargeted LC-HRMS methods for lipids analysis is the simultaneous determination of multiple lipids in a single analysis (Bird et al. 2011a, b; Snyder et al. 2013) making the most efficient method for large studies, where several classes of lipids can be quantified. This type of untargeted analysis is able to detect lipids from almost all the classes and their precision can be improved by the addition of internal standards. However, untargeted methods often lack the sensitivity to detect less abundant lipids. When one is interested in less abundant species, targeted methods need to be employed and they usually involve stable isotopically labeled internal standards and in some cases, derivatization steps, in order to enhance sensitivity.

LC separations represent one of the most time-consuming aspects of lipids' bioanalysis. Reducing the LC run time is desirable in order to increase the number of samples that can be run in a day but it is important that isomeric lipids are separated from each other. This is particularly important for lipids because there could be a large number of enantiomers and diastereomers derived from enzymatic or nonenzymatic pathways and they could interfere with the detection of the analyte of interest. The chromatographic run must be balanced against the accuracy that arises from the clear separation of all the isomers and efficiency of the analysis.

ESI makes it possible to analyze medium to highly polar lipids over a wide mass range. The analyte in solution is sprayed through the ESI needle where a high voltage is applied to form charged droplets, which generate gas-phase analyte ions upon evaporation before entering the ion source. Lipids are often analyzed in the positive ESI mode but they can be detected in the negative ESI mode too. When the flow into the ESI source is decreased to 10–100 nl min^{-1}, the technique is called nanospray ionization. A significant increase in detection sensitivity is observed with nanospray because the smaller droplets that are formed in the source ionize more efficiently.

APCI has been used for analysis of relatively nonpolar lipids with molecular weight up to 1500 Da. In APCI, a high-voltage needle is used to produce a corona discharge for ionization of the evaporated solvent and for reaction with the analytes to form molecule ions. To

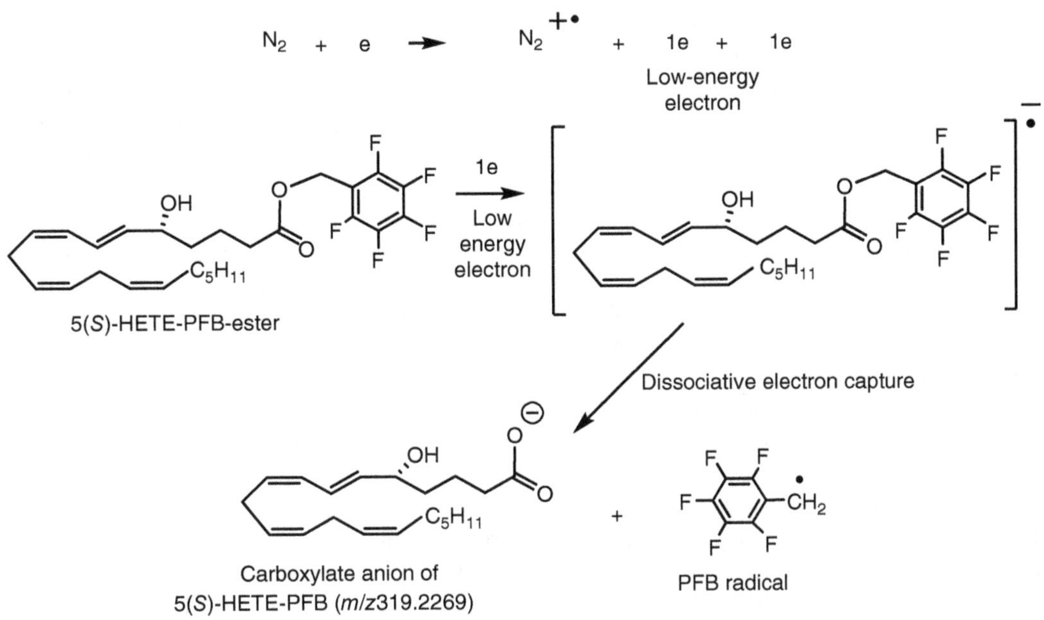

Figure 22.2 ECAPCI/MS/HRMS of 15-HETE-PFB. The low-energy electrons generated from corona discharge interact with the nitrogen sheath gas and generate radical cations. Dissociative electron capture results in a very strong product ion corresponding to 15-HETE anion *m/z* 319.2269. *Source:* From Mazaleuskaya et al. (2018).

increase the sensitivity of lipid detection, a derivatization strategy using electron capturing reagent such as pentafluorobenzyl (PFB) has been developed to facilitate a process known as electro capture APCI (ECAPCI), as illustrated in Figure 22.2 for analysis of 5-hydroxyeicosatetraenoic acid (5-HETE), a metabolite of arachidonic acid. In APCI, the low-energy electrons generated in the APCI source (through interaction of the corona discharge with the nitrogen nebulizing gas) can be captured with a suitable electron-capturing group such as PFB. The initially formed radical anion then dissociates through dissociative electron capture into a carboxylate anion which can be detected in negative mode with high sensitivity (Lee et al. 2003).

22.2.2 Sample Preparation for LC-MS Bioanalysis of Lipids

22.2.2.1 Key Considerations for LC-MS Bioanalysis of Lipids

Many of the difficulties in sample preparation for liquid chromatography–mass spectrometry (LC-MS) come from the particular issues surrounding lipids. To improve the sensitivity for lipid analysis by LC-MS, different cleaning-up protocols for different biological samples are used. A comprehensive review of all those techniques by classes of lipids can be found in the 4th edition of *Lipid Analysis: Isolation, Separation, Identification and Lipidomic Analysis* (Oily Press Lipid Library Series) by W. W. Christie and X. Han (Christie and Han 2010). Sample cleanup is one of the most critical steps in bioanalysis of lipids since interference coming from matrix and other isomers present in the biological fluids is a major challenge in lipid analysis. The availability of the stable isotope (e.g. [^{13}C])-labeled internal standards for a large number of lipid metabolites makes it possible to reliably track extraction recovery, compensate for matrix effects, and identify artifacts (such as oxidation) coming from extraction. It is particularly important to have blank samples with just the internal standards added from the beginning (Jian et al. 2010) that are then carried through the entire workflow, particularly for the determination of the low abundance lipids. Special attention should be paid to the reagents, as well as the glassware, which can increase the background levels that could interfere with the analyte' signals, especially for the low abundant metabolites. In addition, it is advisable to check each internal standard individually for the presence of unlabeled lipids, since cases were encountered where one internal standard was contaminated with a different unlabeled lipid analyte.

22.2.2.2 Common Sample Preparation Methods for Lipids

The most commonly used technique for lipid extraction is liquid–liquid extraction procedure known as the Folch extraction (Folch et al. 1951) and its slight Bligh–Dyer modification (Bligh and Dyer 1959). It employs a tri-solvent extraction, one being water, since all biological relevant matrices are water based. The other two solvents are organic with one polar (methanol most of the times, but can be substituted for a less polar organic alcohol like butanol) and one nonpolar (chloroform most of the time, but can be replaced by dichloro-methane). The extraction involves two steps: The first one is mixing the sample with water, methanol, and chloroform to form a monophasic ternary solvent mixture where the lipids together with other biomolecules are extracted from biological fluids. The second step is adding additional water followed by centrifugation when the initial single phase is separated into two phases: an organic-rich phase containing the total lipids on the bottom and a water-rich phase containing other biomolecules (proteins, sugars, small polar metabolites, etc.) on the top. The principle of this method is the partial miscibility of the chloroform in the water phase and the partial miscibility of water in the organic phase. In this manner all the biomolecules, independent of their volume and "hydrophilicity," are solubilized in either one of the coexisting phases. By adding sufficient amount of water, a biphasic system is then spontaneously formed, leading to the partitioning of lipids into the organic chloroform rich phase (Breil et al. 2017).

Beside the tri-phasic extraction, a simple two-phase extraction can be employed for extraction of lipids, particularly in the case of targeted bioanalysis (Jian et al. 2010, 2013). One could use different water-immiscible organic solvents (ether, ethyl acetate, methyl-*tert*-butyl ether, hexanes, or mixtures of the above solvents) for optimal recovery and minimal interferences from the background for other unwanted endogenous components. One thing to consider is the pH adjustment for the extraction. Most of the lipids will be extracted in the pH range 2–6, but a different pH may be optimal for removing the matrix interference, hence it is recommended that once a solvent is chosen for extraction, the whole pH range will be tested in increments of 1.

In recent years, solid-phase extraction (SPE) has been widely used for high-throughput analysis of lipids. The principle of SPE extraction of lipids is their preferential retention on certain types of stationary phase (typically C18) that is prepacked in a cartridge. By varying the pH and ionic strength of an aqueous mixture, other biomolecules can be washed from the cartridge and the lipids of interest can be then eluted in an appropriate organic solvent. One of the appeals of SPE purifications is the availability of the SPE cartridges in a 96-well format making the methodology ideal for a large number of samples (Gu et al. 2018), while there is also an increased trend of conducting LLE in the 96-well format by using liquid handler (Jian et al. 2010, 2013).

Alternatively, the protein precipitation (PPT) method can be applied for quick and simple removal of protein contents in the samples to enable LC-MS analysis of the lipid contents in the supernatant. The choice of solvent or combination of solvent used for precipitation of the protein should be evaluated for their effectiveness of recovering the lipids. Incorporating of stable isotope-labeled internal standard is highly recommended in order to effectively compensate for matrix effects, which is usually high for PPT-processed samples. One example of using PPT for lipid exaction can be found in the example presented in Section 22.3.1.

22.2.2.3 Derivatization

Chemical derivatization can introduce functional groups into an analyte to improve its chromatographic retention, ionization efficiency, and fragmentation pattern in MS/MS detection (Qi et al. 2014). It is a frequently used strategy for facilitating isomer separation and enhancing detection sensitivity for LC-MS analysis of lipids. The functional groups in lipids such as carboxylic acid, phenol, and alcohol can readily react with derivatization reagents to generate products that exhibit better LC and MS behavior. The ECAPCI utilizing electron capturing reagent PFB for analysis of fatty acid metabolites is one of such examples. The PBF derivatives of arachidonic acid metabolites are more readily retained and therefore separated on chiral HPLC condition and show significantly enhanced ionization, leading to highly sensitivity and selective detection of different enantiomers (Figure 22.3). Another type of lipids that frequently require derivatization for LC-MS analysis are steroids. Steroids exhibit strong biological activities at extremely low concentration. They require not only sensitive but also selective detection because coexistence of their precursors and metabolites in biological matrix. Numerous derivatization strategies have been developed for LC-MS analysis of steroids, which are comprehensively summarized in the literature. (Higashi and Ogawa 2016) A separate chapter (Chapter 21) in this book has also been dedicated for discussion of derivatization. When working with the derivatization method, thorough evaluation should be done to assess the formation of interfering by-products, potential nonquantitative reaction, and stability of the derivatives. In theory, the reaction should be fast, specific, efficient, and generate stable products. One example of using derivatization for analysis of steroid can be found in the typical protocol presented in Section 22.2.3.4.

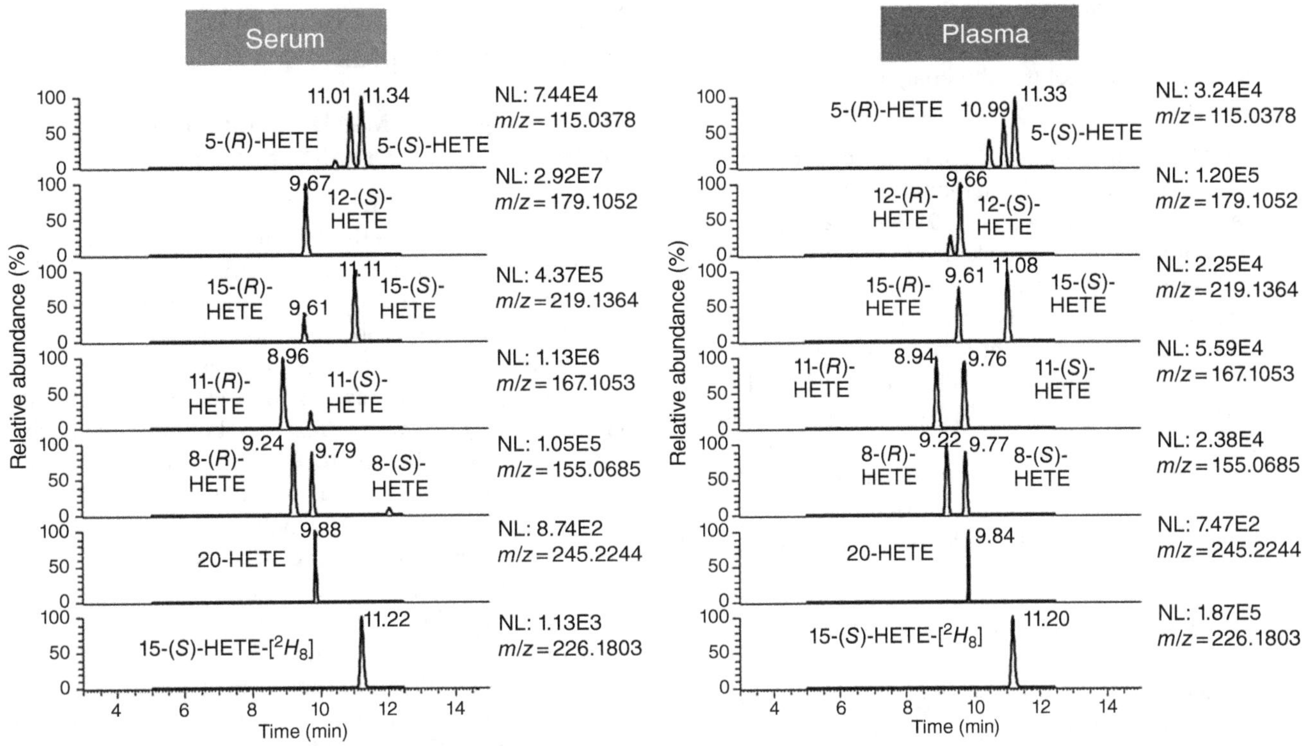

Figure 22.3 Typical LC-ECAPCI/MS/HRMS chromatograms of HETEs as PFB derivatives extracted from 0.2 ml human plasma or serum, spiked with a synthetic heavy isotope analog internal standard (1 ng of 15 [$^2H_8$])-(S)-HETE. *Source:* From Mazaleuskaya et al. (2018).

22.2.3 Typical Protocols of Sample Extraction for Lipid LC-MS Bioanalysis

22.2.3.1 A Typical Protocol of Folch LLE

A typical Foch LLE protocol (for analysis of lipid profile from cultured cells by Ulmer et al. (2015))

- Ice-cold methanol (0.5 ml) and chloroform (1 ml) were added directly to cell pellet.
- The suspension was vortexed occasionally to bring about physical mixing and incubated on ice for 20–40 minutes.
- After the addition of 0.38 ml of water, allowing for a chloroform : methanol : water ratio of 8 : 4 : 3 (v/v/v) according to the original Folch procedure, the suspension was incubated on ice for an additional 10 minutes.
- Samples were centrifuged at 311 × *g* for five minutes at 4 °C.
- The lower-phase (organic) layer was transferred to a new tube.
- The aqueous layer was re-extracted with 2 : 1 chloroform/methanol.
- The organic layers were combined for analysis.
- Samples were then dried under nitrogen and reconstituted in 50 μl of isopropanol (IPA) and transferred to an LC vial with glass insert for LC-MS analysis.

22.2.3.2 A Typical Protocol of LLE

A typical LLE protocol (for the analysis of fatty acid ethanolamides in human plasma by Jian et al. (2010))

- An aliquot (150 μl) of each plasma sample was pipetted into the wells of a 1 ml 96-well polypropylene plate.
- 20 μl of internal standard spiking solution in acetonitrile was added to each well.
- The plate was vortexed at a setting of 1000 rpm for one minute, followed by addition of 500 μl of 90/10 : *n*-hexane/isopropanol (v/v) to each well using a Tomtec liquid handling robot (Tomtec Inc., Hamden, CT).
- The plate was sealed, vortexed on a plate shaker at a setting of 1600 rpm for 10 minutes, and then centrifuged to separate the phases.
- The organic layer (400 μl) was immediately transferred to a new 96-well plate using the Tomtec.
- The samples were evaporated to dryness under a gentle flow of nitrogen at ambient temperature and reconstituted using 100 μl of ethanol.
- The plate was vortexed at a setting of 1000 rpm for one minute and placed into the autosampler set at 10 °C. The injection volume was 10 μl.

22.2.3.3 A Typical Protocol of SPE

A typical SPE protocol (for the analysis of endogenous PUFAs in human serum by Gu et al. (2018))

- HLB SPE cartridges were conditioned successively with ethyl acetate (2 ml), methanol (2×2 ml), and 0.1% acetic acid in water/methanol (95 : 5, v/v, 2 ml).
- 100 μl of human serum spiked with 2 μl of IS working solution in methanol was loaded onto the cartridge.
- The cartridge was washed with 1.5 ml of acetic acid/methanol (95 : 5, v/v) to remove the endogenous impurities.
- The PUFAs were eluted out with 1 ml of methanol and 2 ml of ethyl acetate containing 6 μl of glycerol/methanol (30 : 70, v/v).
- The residue dried by nitrogen was reconstituted in 100 μl of methanol followed by centrifugation at 13 000*g* for seven minutes before injected for LC-MS analysis.

22.2.3.4 A Typical Protocol of LLE with Derivatization

A typical LLE protocol with derivatization (for the analysis of ethinyl estradiol in human plasma by Li et al. (2005))

- Samples were briefly vortex-mixed and aliquots of 0.50 ml of samples were then transferred from the vials into 16×125 mm glass test tubes with screw caps.
- Internal standard working solution in methanol (25.0 μl) was then added.
- 4.0 ml of *n*-butyl chloride were added, and the tubes were capped and vortexed at high speed for three minutes.
- The samples were then centrifuged at 4000 rpm for 10 minutes.
- The aqueous layer was frozen in a dry ice-acetone bath and the organic layer was decanted into pre-labeled 13×100 mm glass tube.
- The organic layer was evaporated to dryness under a stream of nitrogen at 40 °C.
- To the residue, 100 μl of sodium bicarbonate buffer (pH 11) was added, followed by vortexing at high speed for three minutes and then by adding 100 μl of dansyl chloride in acetone (1.00 mg ml^{-1}).
- The tubes were vortexed for one minute and kept in a water bath set at 60 °C for six minutes to facilitate derivatization.
- Tubes were then placed into another water bath at room temperature.
- After vortexing for one minute, the samples were transferred into a 0.7 ml plastic HPLC vial and 30.0 μl was injected onto the LC-MS/MS system for analysis.

Note: The product ion of *m/z* 171 by fragmentation from dansyl derivative greatly enhanced the sensitivity and an LLOQ of 2.5 pg ml^{-1} in plasma was achieved.

22.3 Case Studies of LC-MS Bioanalysis of Lipids

22.3.1 Targeted LC-MS/MS Analysis of Lipid Metabolites Involved in Inflammation

A recent study by Gandhi et al. (2017) successfully fine-tuned an extraction method to monitor the lipid metabolites involved in inflammation. Inflammation, which is the main mechanism leading to an immune response after infection by microbes, is also thought to play a primary role in various diseases such as Alzheimer's disease, Parkinson's disease, and rheumatoid arthritis (Ganesh 2014). Common features behind the mechanisms of various inflammatory conditions are the activation of immune cells such as microglia in the brain and macrophages in the lung, which can induce oxidative stress through production of cytokines and chemokines (Lucas et al. 2006; Minghetti 2005; Tabas and Glass 2013). Additionally, the release of lipid mediators from the COX pathways including PGs, endocannabinoids, leukotrienes, and thromboxanes can also lead to inflammatory reactions (Hall et al. 2009; Kidd and Urban 2001; Ricciotti and FitzGerald 2011). Both clinical and preclinical studies suggest that PGs such as PGD_2, PGE_2, $PGF_{2\alpha}$, and thromboxane B_2 (TXB_2) are strongly associated with the onset and exacerbation of pain, fever, and allergic reactions (Hall et al. 2009; Kawabata 2011; Kunori et al. 2011). While the majority of these compounds cannot remain stable for long in plasma, they are significantly more stable in the brain and spinal cord, thus allowing for accurate quantitation (Eguchi et al. 1988; Fisher and Zipser 1985). Consequently, Gandhi et al. created an LC-MS/MS method for the analysis of these metabolites in order to enable the monitoring and elucidation of inflammatory pathways, and compared their method's effectiveness with that of the more traditional ELISA assay. Although ELISAs allow for analysis of these compounds with minimal sample preparation, they have very high rates of false-positives due to their inability to effectively distinguish between isomeric species such as PGE_2, PGE_1, and PGD_2 (Faupel-Badger et al. 2010).

The *in vitro* experiment was conducted in isolated and cultured microglia cells from 3-day-old Sprague–Dawley rats. For the *in vivo* experiment, the brain and spinal cord from Sprague–Dawley rats were homogenized in 3x (brain) and 4x (spinal cord) w/v aqueous MSD lysis buffer, respectively. Sample preparation for both the *in vivo* and *in vitro* samples involved addition of 10 μl of the deuterated internal standard (ISTD) mixture (containing PGE_2, PGD_2, TXB_2, LTB_4, and 2-AG [2-arachidonoylglycerol]) in acetonitrile to each sample. The samples were then protein precipitated with 300 μl each of ethyl acetate and acetonitrile. This was followed by ~30 seconds of vortexing and centrifugation at 3300 rpm for 20 minutes at 4 °C. The supernatant was collected, dried under nitrogen gas at 40 °C and resuspended in 125 µl of 30% acetonitrile (in water) containing 0.01% formic acid. 20 µl of this final solution was then injected onto the LC column.

For the PPT method, a variety of different solvent combinations including chloroform : methanol : water, hexane : ethyl acetate, acetonitrile : methyl-*tert*-butyl ether, acetonitrile : ethyl acetate, or hexane : isoproply alcohols were tested for obtaining sufficient and reproducible analyte recoveries for this study. The analyte recoveries ranged from 50 to 80% for all analytes in cell culture medium, brain and spinal cord tissues using only 1 : 1 acetonitrile : ethyl acetate mixture. The recoveries were much lower when other solvent mixtures were used (data not shown). Thus, 1 : 1 acetonitrile : ethyl acetate was used as the solvent combination for the final method. IS recoveries were also in the same range (50–80%) as deuterated analogs of analyte standards were spiked in the matrix. It was also found that incorporation of deuterated internal standard for respective lipid was able to effectively correct for matrix effects.

The LC-MS/MS system consisted of a triple quadrupole mass spectrometer in multiple reaction monitoring mode, with the ESI operating in the negative mode except when analyzing 2-AG, for which the positive mode was used. The LC method allowed for the resolution of all six analytes (PGE_2, PGD_2, $PGF_{2\alpha}$, LTB_4, TXB_2, and 2-AG) and their deuterated ISTDs within five minutes. Matrix effects were minimal and the sample carry-over for all analytes was negligible. When comparing the detection of PGE_2 standard using this LC-MS/MS method side-by-side with ELISA, it was seen that the concentration measured by LC-MS/MS is approximately two to three folds lower than that obtained by ELISA assay, suggesting likely increased specificity in PGE_2 detection. The method was successfully used to measure lipid levels in the brain and spinal cord tissue of healthy rats and rats with Chronic Constriction Injury before or after treatment with analgesic compounds. Overall, the method by Gandhi et al. allowed for the sensitive and specific detection of CNS lipid biomarkers of inflammatory-mediated neuropathic pain with a simple sample preparation technique and a run time of less than five minutes.

22.3.2 LC-ECAPCI/MS Analysis of Bioactive Eicosanoid Lipids

The biosynthesis of eicosanoids can occur enzymatically via LOXs, COXs, and cytochrome P450, or nonenzymatically through free-radical reactions. The enzymatic routes are highly enantiospecific and therefore require chiral separation and high-sensitive detection. A particularly sensitive and selective method based on chiral UPLC-MS/HRMS coupled to ECAPCI was used to scan the effect of drugs on the bioactive eicosanoid lipidome mediated by COX-1/2 and microsomal prostaglandin E synthase-1 (mPGES-1) (Mazaleuskaya et al. 2016, 2018).

The LC-MS-based assay was applied to measure the eicosanoid lipidome in stimulated human whole blood. The samples were processed by SPE cleanup and dried under nitrogen gas. The PFB derivatives were prepared by reconstituting the dried SPE extracts using diisopropylethylamine in 100 µl of acetonitrile (1 : 19, v/v) and 50 µl of PFB-Br in acetonitrile (1 : 9, v/v). After a short derivatization for 30 minutes at 60 °C, the solution was evaporated to dryness under a nitrogen stream at room temperature, and redissolved in 100 µl of hexane/ethanol (97 : 3, v/v) for mass spectrometry analysis. It was not necessary to purify the PFB-ester derivatives prior to LC-MS analysis. The resuspension solvent after derivatization was mostly hexanes, which helped to get rid of any leftover salts from the sample cleanup procedure.

LC-ECAPCI/MS analyses of PFB-derivatives of eicosanoid were conducted using normal-phase solvents and chiral columns with high sensitivity. This made it possible to separate complex mixtures of eicosanoid enantiomers and to quantify them using a stable isotope dilution methodology in a single run. A typical chromatogram obtained from the targeted chiral LC-MRM/MS analysis of all the HETEs and ^{2}H-labeled analog 15-(*S*)-HETE using ECAPCI is shown in Figure 22.3. Detection limits for individual eicosanoids were found to be in the fg range. The LC-ECAPCI/MS method made it possible to precisely quantify a range of lipids and unequivocally show product specificity of different enzymes.

22.4 Summary

Overall, there have been significant recent advances in sample preparation techniques for lipid bioanalysis through LC-MS. A major challenge in this field is the loss of sensitivity and specificity due to matrix effects, which has made sample preparation workflow a priority in lipid analysis. Additionally, extensive effort has been made toward increasing the specificity of the extraction for the targeted analyte, either by pulling out the analytes of interest, or by removing as much of the matrix components and other interfering compounds as possible. Here, we have presented a few recent examples of advancements in sample preparation techniques for lipids analysis, although there is still much room for the advancement of workup methodologies and thus the effectiveness of LC-MS as a bioanalytical tool.

There are still many pitfalls that can be encountered in the analysis of lipids particularly when quantifying them from complex biological mixtures. Due to the advancement in areas such as sample extraction techniques, chromatographic separation, and high-resolution mass spectrometry, there will almost certainly be substantial improvement to the existing assays by improving specificity and sensitivity.

References

Bird, S.S., Marur, V.R., Sniatynski, M.J. et al. (2011a). Lipidomics profiling by high-resolution LC-MS and high-energy collisional dissociation fragmentation: focus on characterization of mitochondrial cardiolipins and monolysocardiolipins. *Anal. Chem.* 83: 940–949.

Bird, S.S., Marur, V.R., Sniatynski, M.J. et al. (2011b). Serum lipidomics profiling using LC-MS and high-energy collisional dissociation fragmentation: focus on triglyceride detection and characterization. *Anal. Chem.* 83: 6648–6657.

Blair, I.A. (2008). DNA adducts with lipid peroxidation products. *J. Biol. Chem.* 283: 15545–15549.

Bligh, E.G. and Dyer, W.J. (1959). A rapid method of total lipid extraction and purification. *Can. J. Biochem. Physiol.* 37: 911–917.

Breil, C., Abert, V.M., Zemb, T. et al. (2017). "Bligh and Dyer" and Folch methods for solid-liquid-liquid extraction of lipids from microorganisms. Comprehension of Solvatation mechanisms and towards substitution with alternative solvents. *Int. J. Mol. Sci.* 18 (4): 708.

Christie, W.W. and Han, X. (2010). *Lipid Analysis: Isolation, Separation, Identification and Lipidomic Analysis, Oily Press Lipid Library Series*, 4e, vol. 24. Bridgwater, England: Oily Press.

Eguchi, N., Hayashi, H., Urade, Y. et al. (1988). Central action of prostaglandin E2 and its methyl ester in the induction of hyperthermia after their systemic administration in urethane-anesthetized rats. *J. Pharmacol. Exp. Ther.* 247: 671–679.

Faupel-Badger, J.M., Fuhrman, B.J., Xu, X. et al. (2010). Comparison of liquid chromatography-tandem mass spectrometry, RIA, and ELISA methods for measurement of urinary estrogens. *Cancer Epidemiol. Biomarkers Prev.* 19: 292–300.

Fisher, M. and Zipser, R. (1985). Increased excretion of immunoreactive thromboxane B2 in cerebral ischemia. *Stroke* 16: 10–14.

Folch, J., Ascoli, I., Lees, M. et al. (1951). Preparation of lipide extracts from brain tissue. *J. Biol. Chem.* 191: 833–841.

Gandhi, A.S., Budac, D., Khayrullina, T. et al. (2017). Quantitative analysis of lipids: a higher-throughput LC-MS/MS-based method and its comparison to ELISA. *Future Sci. OA* 3: FSO157.

Ganesh, T. (2014). Prostanoid receptor EP2 as a therapeutic target. *J. Med. Chem.* 57: 4454–4465.

Gu, W.Y., Liu, M.X., Sun, B.Q. et al. (2018). Profiling of polyunsaturated fatty acids in human serum using off-line and on-line solid phase extraction-nano-liquid chromatography-quadrupole-time-of-flight mass spectrometry. *J. Chromatogr. A* 1537: 141–146.

Hall, A., Brown, S.H., Budd, C. et al. (2009). Discovery of GSK345931A: an EP(1) receptor antagonist with efficacy in preclinical models of inflammatory pain. *Bioorg. Med. Chem. Lett.* 19: 497–501.

Han, X. (2007). Neurolipidomics: challenges and developments. *Front. Biosci.* 12: 2601–2615.

Hannun, Y.A. and Obeid, L.M. (2008). Principles of bioactive lipid signalling: lessons from sphingolipids. *Nat. Rev. Mol. Cell Biol.* 9: 139–150.

Higashi, T. and Ogawa, S. (2016). Chemical derivatization for enhancing sensitivity during LC/ESI-MS/MS quantification of steroids in biological samples: a review. *J. Steroid Biochem. Mol. Biol.* 162: 57–69.

Jian, W., Edom, R., Weng, N. et al. (2010). Validation and application of an LC-MS/MS method for quantitation of three fatty acid ethanolamides as biomarkers for fatty acid hydrolase inhibition in human plasma. *J. Chromatogr. B Analyt. Technol. Biomed. Life Sci.* 878: 1687–1699.

Jian, W., Edom, R., Xue, X. et al. (2013). Quantitation of leukotriene B4 in human sputum as a biomarker using UPLC-MS/MS. *J. Chromatogr. B* 932: 59–65.

Kawabata, A. (2011). Prostaglandin E2 and pain – an update. *Biol. Pharm. Bull.* 34: 1170–1173.

Kidd, B.L. and Urban, L.A. (2001). Mechanisms of inflammatory pain. *Br. J. Anaesth.* 87: 3–11.

Kunori, S., Matsumura, S., Okuda-Ashitaka, E. et al. (2011). A novel role of prostaglandin E2 in neuropathic pain: blockade of microglial migration in the spinal cord. *Glia* 59: 208–218.

Lee, S.H., Williams, M.V., DuBois, R.N., and Blair, I.A. (2003). Targeted lipidomics using electron capture atmospheric pressure chemical ionization mass spectrometry. *Rapid Commun. Mass Spectrom.* 17: 2168–2176.

Li, W., Li, Y.H., Li, A.C. et al. (2005). Simultaneous determination of norethindrone and ethinyl estradiol in human plasma by high performance liquid chromatography with tandem mass spectrometry – experiences on developing a highly selective method using derivatization reagent for enhancing sensitivity. *J. Chromatogr. B* 825: 223–232.

Lucas, S.M., Rothwell, N.J., and Gibson, R.M. (2006). The role of inflammation in CNS injury and disease. *Br. J. Pharmacol.* 147 (Suppl 1): S232–S240.

Mazaleuskaya, L.L., Lawson, J.A., Li, X. et al. (2016). A broad-spectrum lipidomics screen of antiinflammatory drug combinations in human blood. *JCI Insight* 1: https://doi.org/10.1172/jci.insight.87031.

Mazaleuskaya, L.L., Salamatipour, A., Sarantopoulou, D. et al. (2018). Analysis of HETEs in human whole blood by chiral UHPLC-ECAPCI/HRMS. *J. Lipid Res.* 59: 564–575.

van Meer, G. (2005). Cellular lipidomics. *EMBO J.* 24: 3159–3165.

Minghetti, L. (2005). Role of inflammation in neurodegenerative diseases. *Curr. Opin. Neurol.* 18: 315–321.

Qi, B.L., Liu, P., Wang, Q.Y. et al. (2014). Derivatization for liquid chromatography-mass spectrometry. *TrAC Trends Anal. Chem.* 59: 121–132.

Ricciotti, E. and FitzGerald, G.A. (2011). Prostaglandins and inflammation. *Arterioscler. Thromb. Vasc. Biol.* 31: 986–1000.

Snyder, N.W., Khezam, M., Mesaros, C.A. et al. (2013). Untargeted metabolomics from biological sources using ultraperformance liquid chromatography-high resolution mass spectrometry (UPLC-HRMS). *J. Vis. Exp.* e50433.

Tabas, I. and Glass, C.K. (2013). Anti-inflammatory therapy in chronic disease: challenges and opportunities. *Science* 339: 166–172.

Ulmer, C.Z., Yost, R.A., Chen, J. et al. (2015). Liquid chromatography-mass spectrometry metabolic and lipidomic sample preparation workflow for suspension-cultured mammalian cells using Jurkat T lymphocyte cells. *J. Proteomics Bioinform.* 8 (6): 126–132.

23

Sample Preparation for LC-MS Bioanalysis of Peptides

Long Yuan

23.1 Introduction

Peptides are short chains of amino acids that are linked together by peptide (amide) bonds. Structurally, peptides are similar to proteins. Size is the basic factor distinguishing peptides from proteins, however, there is no clear-cut view. Typically, peptides consist of fewer than 50 amino acids or have molecular weights <10 kDa. In addition, peptides usually are less complex in structure with a single chain, while proteins can form complex tertiary and quaternary conformations with two or more peptide chains joined together via disulfide bounds.

Peptides have gained increasing interest in biomedical and pharmaceutical research. More than 7000 peptides have been identified from natural sources, and many of them play important roles in regulating biological functions, e.g. acting as hormones, growth factors, neurotransmitters, ion channel ligands, and antibiotics (Fosgerau and Hoffmann 2015). Endogenous peptides have also drawn great interest to be used as biomarkers or diagnostic tools (Diamandis 2004). Peptide therapeutics, often synthetic analogs of endogenous peptides, is an important class of therapeutic agents. In 2012 alone, there were six peptide therapeutics approved as new molecular entities (Kaspar and Reichert 2013). Currently, there are >60 peptide therapeutics approved by the US Food and Drug Administration (FDA) and ~140 peptide drugs in clinical development (Kaspar and Reichert 2013). Quantification of peptides in biological samples plays an essential role in peptide research and development, for example, in providing reliable pharmacokinetics (PK), pharmacodynamics (PD), and toxicological data. Consequently, the demand of sensitive, selective, accurate, and robust bioanalytical methods for peptides are high and increasing.

Conventionally, immunoassays or liquid chromatography (LC) combined with UV, fluorescence, or electrochemical detection were used for the quantification of peptides (John et al. 2004; van den Broek et al. 2008). However, these methods often lack the selectivity or require lengthy sample pretreatment (e.g. derivatization required for fluorescence detection). Liquid chromatography–mass spectrometry (LC-MS) combines chromatographic separation with highly sensitive and selective mass spectrometric detection, which results in superior sensitivity and selectivity compared to the traditional methods. In recent years, LC-MS has become the preferred method for the quantitative bioanalysis of peptides (John et al. 2004; van den Broek et al. 2008). It has been widely applied for the bioanalysis of peptides, including both endogenous peptides (Berna et al. 2006; Oe et al. 2006; Li et al. 2007b; Kumar et al. 2010; Lame et al. 2011; Chappell et al. 2013, 2016; Taylor et al. 2016) and peptide therapeutics (van den Broek et al. 2006, 2007; Li et al. 2007a; Halquist et al. 2012; Xu et al. 2017).

Sample preparation is a critical component for ensuring the selectivity, sensitivity, reproducibility, and robustness of a LC-MS bioanalytical method. Biological matrices (plasma, serum, blood, etc.) contain abundant endogenous components such as proteins, phospholipids, and salts. These components can cause significant matrix effect (ion suppression or enhancement) to the mass spectrometric detection of the analyte (King et al. 2000; Jemal et al. 2010). Furthermore, the abundant proteins and phospholipids may buildup or even precipitate in the column, which will rapidly block the column and deteriorate its performance. Therefore, the biological samples cannot be directly injected into a LC-MS system for analysis and require sample preparation to remove most of (ideally all of) the matrix components. For quantitative bioanalysis of peptides, in addition to matrix effect, sensitivity (or quantification limit) is often another challenge. The concentration of peptide analytes in biological matrices can be very low, especially for some endogenous peptides. An appropriate sample preparation strategy can effectively extract and enrich the target

Sample Preparation in LC-MS Bioanalysis, First Edition. Edited by Wenkui Li, Wenying Jian, and Yunlin Fu.

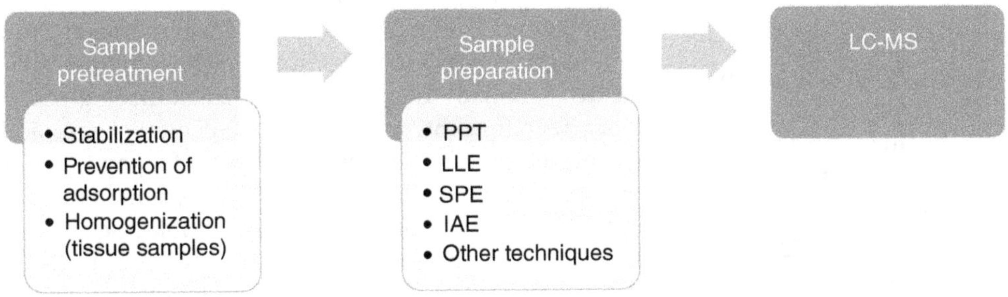

Figure 23.1 General overview of sample preparation (including sample pretreatment) strategies for LC-MS bioanalysis of peptides. PPT, protein precipitation; LLE, liquid–liquid extraction; SPE, solid-phase extraction; IAE, immunoaffinity extraction.

Table 23.1 Comparison of different sample preparation strategies for peptides.

	Purification efficiency	Cost	Pro	Con
PPT	+	+	Simple and fast operation; easy to combine with other methods	Limited sample cleanup
LLE	++	+	Simple operation; relatively efficient purification	Only applicable for hydrophobic peptides
SPE	++	++	Various separation mechanisms to select	Requires more thorough method development
IAE	+++	+++	Highly selective and efficient purification	Requires a suitable reagent; reagent generation may take long time

PPT, protein precipitation; LLE, liquid–liquid extraction; SPE, solid-phase extraction; IAE, immunoaffinity extraction.
+, Low; ++, moderate; +++, high.

peptide, while removing other components from the biological matrices, therefore, significantly reducing the matrix effects and improving the sensitivity, selectivity, and reproducibility of the LC-MS assay.

Sample preparation is often the most challenging part in the development of a LC-MS bioanalytical method. An overview of sample preparation strategies as well as sample pretreatment considerations for LC-MS bioanalysis of peptides is shown in Figure 23.1. A variety of sample preparation techniques have been developed for the extraction of peptides, including protein precipitation (PPT), liquid–liquid extraction (LLE), solid-phase extraction (SPE), and immunoaffinity extraction (IAE). Their pros and cons are briefly summarized in Table 23.1 and will be discussed in detail in this chapter. In addition, peptides are well known to tend to adsorb on vessel surfaces. Peptides may also undergo chemical or enzymatic degradation in biological samples ex vivo. Therefore, appropriate sample pretreatment is often required prior to sample extraction to prevent the adsorption (nonspecific binding) and/or ensure the stability of the peptide analyte during sample collection, processing, and storage. In this chapter, these various sample pretreatment considerations and strategies will also be introduced and discussed. The strategies and methodologies discussed here are also useful for LC-MS bioanalysis of proteins using the "bottom-up" approach, in which surrogate peptide(s) of the proteins of interest are extracted and analyzed.

23.2 Properties of Peptides and Sample Pretreatment

The physicochemical properties (hydrophobicity, polarity, pI, solubility, etc.) of a peptide are determined by its amino acid composition. Amino acids have different properties depending on their specific side chains. There are hydrophobic (nonpolar) amino acids, such as phenylalanine, tyrosine, and tryptophan; and polar and uncharged amino acids like serine, threonine, asparagine, and glutamine. Lysine, histidine, and arginine are basic, while aspartic acid and glutamic acid are acidic. The presence of the N-terminal amine and C-terminal carboxyl groups also makes peptides amphoteric and can form zwitterions in solutions. As a result, depending on the properties of the target peptides, their sample preparation could be quite different and need to be optimized accordingly.

23.2.1 Stability

Stability is a critical parameter for a bioanalytical assay. The stability of an analyte during sample collection, storage, and preparation needs to be evaluated and controlled to ensure the integrity of the analyte, therefore, ensuring the accuracy, reliability, and robustness of the assay. Peptides usually are stable in solid state (e.g. as lyophilized form). In solution, peptides may undergo various types of chemical degradation, such as, oxidation, reduction, deamidation, and hydrolysis (Reubsaet et al. 1998). Moreover, in plasma, serum, blood, or other biological matrices, the proteases and other enzymes often can quickly metabolize peptides and cause their instability. Therefore, it is critical to evaluate and ensure the stability of peptides in solutions and biological matrices during method development.

Chemical degradation of peptides is related to the specific amino acid residue in the peptide sequence. For example, methionine, tryptophan, and cysteine residues are sensitive to oxidation. Peptides containing asparagine-glycine and aspartic acid-glycine are prone to deamidation and isomerization. Caution needs to be taken when these specific residues are present in the peptides. Temperature control, pH control, and addition of antioxidants are commonly used strategies to prevent or minimize the chemical degradation (Li et al. 2011c). As a precaution, exposure to elevated temperature and atmospheric oxygen and freeze–thaw cycles should be minimized.

Various approaches have been applied to inhibit the degradation enzyme(s), therefore, stabilizing the enzymatic degradation of peptides (Ewles and Goodwin 2011). Lowering the temperature (e.g. on ice) is a useful approach, since it can significantly slow down the enzymatic and chemical reactions during sample processing and preparation. Often, acidifying the samples with acid (formic acid, hydrochloric acid, trifluoroacetic acid, etc.) can inhibit proteases and stabilize peptides. Various protease inhibitors or inhibitor cocktails have also been evaluated and applied to reduce or eliminate proteolytic degradation of peptides (Wolf et al. 2001; Belenky et al. 2004; Fanciulli et al. 2007). For a specific peptide, various classes or types of protease inhibitors may have different inhibition activities. Belenky et al. (2004) evaluated the effect of different protease inhibitors, including competitive inhibitors, irreversible inhibitors, and kallikrein-specific PPACK II, on the stabilization of B-type natriuretic peptide (BNP) in human plasma. The stability of BNP was improved by the evaluated inhibitors to various extents. PPACK II was the most effective inhibitor when used singly. Using a mixture of different types of inhibitors (inhibitor cocktail) is an effective approach to maximize the stabilization efficiency, since the cocktail can cover various types of enzymes. For peptides that can be degraded by other enzymes, the corresponding enzyme inhibitors also need to be included in the inhibitor cocktail to stabilize the samples. For instance, glucose-dependent insulintropic peptide (GIP) can be rapidly hydrolyzed by the exopeptidase dipeptidyl peptidase IV (DP IV). For the quantification of GIP in human plasma, in addition to a cocktail of protease inhibitors, DP IV inhibitor valine-pyrrolidide was added to prevent the DP IV-catalyzed N-terminal degradation (Wolf et al. 2001). In another example, bradykinin, a potential breast cancer peptide biomarker, can be rapidly degraded by angiotensin I converting enzyme (ACE) and carboxypeptidase N in human plasma. For the analysis of bradykinin, inhibitors of ACE (captopril) and carboxypeptidase N (MEGETPA) were added to plasma to stabilize the samples (van den Broek et al. 2010a). Sometimes, the enzymatic degradation may be so fast that it is extremely difficult or even impossible to find effective enzyme inhibitors to stabilize the peptide. Direct denaturation of the samples (direct collection of samples into organic solvent) can be applied as an effective strategy to stabilize the samples, since the enzymes will be denatured and inactivated immediately after the collection (Haynes et al. 2011).

23.2.2 Adsorption

Peptides are well known to often have nonspecific binding to vessel surfaces (sample tubes, pipette tips, 96-well plates, autosampler injection syringes, etc.) (van Midwoud et al. 2007; Maes et al. 2014). Adsorption could happen in all steps of a bioanalytical method: from the collection and storage of biological samples, the preparation of standard stock solution (dissolution of the solid standard material) and subsequent solutions, the sample preparation, until the injection of the extracted samples to a LC-MS system. Adsorption in these steps may severely affect the linearity, accuracy, sensitivity, and reproducibility of the bioanalytical assay, therefore, it should be carefully investigated at the start of method development. The degree of adsorption is peptide dependent and is influenced by many factors such as peptide amino acid composition, structure, charge, and hydrophobicity of the peptide. Moreover, the composition of the dissolution solvent and the type of surface material (e.g. glass vs. polypropylene) can also affect peptide adsorption (Grohganz et al. 2004; Vatansever et al. 2010; Goebel-Stengel et al. 2011). Due to the unique characteristics of different peptides, the optimal solution and the vessel material used should always be experimentally determined for each peptide.

Generally, more severe adsorption occurs to peptides in aqueous solutions, especially at low concentration.

Thus, it is a good practice to avoid using low-concentration aqueous peptide solutions. For the preparation of matrix-free peptide solutions (reference standard stock solutions, diluted working solutions, etc.), to prevent adsorption, it is important to evaluate and select the appropriate solvents, additives, and the type of surface material for the peptide analyte. The addition of organic solvents (acetonitrile, methanol, dimethyl sulfoxide [DMSO], etc.) in dissolution solvents can improve the solubility of peptides, therefore, minimizing and preventing the adsorption to the vessel surfaces. Charged molecules usually have better solubility in aqueous solutions, while uncharged molecules have stronger adsorption to hydrophobic surfaces. Thus, adjusting the pH of the solution can also reduce the adsorption by changing the charge state of the peptides.

For peptides in serum or plasma samples, the adsorption is much less significant and usually less a concern, since the abundant endogenous proteins in serum/plasma can effectively block the adsorption sites. For the preparation of peptide calibration standards and QC samples in plasma/serum, it is recommended to avoid the serial-dilution in matrix-free solutions, but to directly spike a high-concentration stock solution into plasma/serum, and subsequently do the serial-dilution in the biological matrix. For biological matrices containing low level of proteins (e.g. urine and cerebrospinal fluid [CSF]), adsorption could be a severe issue for peptide analytes (Lame et al. 2011; Meng et al. 2016). Careful evaluation should be conducted to prevent the adsorption during sample collection, processing, and preparation. The addition of surfactants, e.g. Triton X-100, Tween-20, and CHAPS (Grohganz et al. 2004; Meng et al. 2016) or proteins, e.g. bovine serum albumin and plasma (Schwartz et al. 1997; Lame et al. 2011) are effective approaches to prevent the adsorption.

Severe adsorption could occur to the extracted samples, since there would be much less proteins remaining to block the adsorption sites after sample preparation. In addition, the drying step in sample preparation can cause significant loss of peptides (Pezeshki et al. 2009). Therefore, it still needs to be kept in mind to prevent the peptide adsorption after sample extraction, until submitting for LC-MS analysis. Maintaining a certain percentage of organic solvent in the extracted or reconstituted samples (e.g. 20% or greater acetonitrile or methanol, optimal percentage needs to be experimentally determined) is an effective approach to improve the solubility of the peptide analyte in the extraction solution and prevent the adsorption. The addition of surfactants or protein can be used prior to sample preparation, but is not recommended after sample preparation, since they are not compatible with LC-MS analysis.

23.3 Sample Preparation Strategies

23.3.1 Protein Precipitation

23.3.1.1 Overview of Protein Precipitation

PPT is a simple, fast, and inexpensive sample preparation technique. It utilizes precipitating reagents to remove proteins and thus cleanup the samples (Chang et al. 2007). PPT is commonly used for the extraction of small-molecule analytes, especially in drug discovery. PPT can also be applied to the extraction of peptides, since peptides are small and usually will not be precipitated by the precipitating reagents like the much larger endogenous proteins (albumins, globulins, etc.) in plasma/serum. For a typical PPT method, a precipitating reagent (organic solvents, acids, salts. etc.) is first added to the biological samples to denature and thus precipitate the proteins out of solution. The precipitated proteins are then separated from liquid components by centrifugation, and the liquid supernatant is used for the subsequent processing and analysis. Commonly used precipitating reagents include water-miscible organic solvents (e.g. acetonitrile, methanol, ethanol), acids (e.g. trichloroacetic acid), and salts (e.g. zinc sulphate, ammonium sulphate) (Polson et al. 2003). The acids and salts used for precipitation may interfere with chromatographic separation or cause ion suppression during mass spectrometric detection. Thus, for LC-MS methods, water-miscible organic solvents are the preferred and most commonly used precipitating reagents.

23.3.1.2 PPT in Peptide Bioanalysis

PPT can efficiently remove most of the proteins. For example, at 2 : 1 volumes of acetonitrile to plasma, >90% of proteins were removed from dog, rat, mouse, and human plasma (Polson et al. 2003). However, PPT alone has limited ability in removing phospholipids and salts. Significant amounts of these components will still be present in the sample extracts, which could cause matrix effect to the analyte and severely affect the sensitivity and quality of the assay. Furthermore, the peptide analytes, especially large peptides, may have poor solubility or would coprecipitate with proteins in high-percentage organic solvents, thus, resulting in the loss of the analyte and low recovery (John et al. 2004). In these cases, low solvent volumes (2 : 1 volumes or even lower) can be used to improve the recovery of the peptide (Chang et al. 2005; Hu and Kamberi 2009), but with the cost of reduced efficiency in removing proteins. Therefore, PPT usually is applied to exploratory assays or assays that need less sensitivity. One advantageous application of PPT is for the extraction of unstable peptide analytes, since it can quickly denature the active enzymes in the biological matrices, and therefore, prevent the enzymatic degradation. It is also much faster compared to LLE, SPE, or other techniques, which can minimize the potential degradation during operation (Yuan et al. 2014a). With its simple and fast operation, PPT is easy to be used as a pretreatment method in combination with other sample preparation techniques. For example, PPT has been used as a pretreatment prior to SPE (Chambers et al. 2013, 2014; Miyachi et al. 2017) and LLE (Celli et al. 2004).

23.3.1.3 A Representative PPT Protocol

Here, a representative PPT protocol for the extraction of human neutrophil peptides from human plasma and serum is shown (adapted from van den Broek et al. (2010b)).

- Add 100 μl of standards, QCs, blanks, and unknown samples into each well of a 96-well plate
- Add 10 μl of internal standard (IS) working solution
- Mix well and then let sit for at least five minutes
- Add 200 μl of precipitation solvent (1% TFA [v/v] in acetonitrile)
- Vortex vigorously and then centrifuge at high speed (e.g. 4000 × *g*) for five minutes
- Transfer the supernatant and evaporate it under nitrogen
- Reconstitute the residues in reconstitution solution (400 μl of acetonitrile/water/formic acid [25/75/0.25, v/v/v])

Notes

- The extraction can also be performed using individual tubes. However, it will be labor-intensive and time-consuming. The combination of 96-well plates and a robotic liquid handler can achieve high-throughput and semiautomatic extraction (Yuan and Ji 2012; Zheng et al. 2014). When using the plate format, it is critical to ensure the plate is caped and sealed well during vortexing, therefore, avoiding leakage and cross-contamination between samples.
- The precipitating reagent and the volumes need to be optimized to achieve the satisfactory recovery and matrix effect for the peptide analyte.
- The original method did not specify the equilibration time after adding IS. Some equilibration time (e.g. the five-minutes time added in the protocol) should be given to allow adequate distribution of IS into the matrices. This practice as well as the separate addition of IS prior to adding precipitation solvent would help ensure the IS can maximally mimic and track the analyte during extraction, and therefore improve the assay accuracy and quality.
- Acidification of the precipitation solvent is for preventing coprecipitation of the peptide analytes.
- The supernatant could also be directly submitted to LC-MS analysis or diluted with water (or buffer solution) and then submitted to analysis to avoid the evaporation and reconstitution steps.
- The evaporation and reconstitution steps in this case are beneficial since some interferences that are insoluble in the reconstitution solution could be removed by centrifugation. However, for peptides that have adsorption (nonspecific binding) to the container, the additional evaporation and reconstitution steps may cause the loss of the analytes. In addition, some peptides are difficult to be redissolved in the solution after evaporating to full dryness. For these peptides, it is recommended to avoid the evaporation step. If really needed, the complete dryness should be prevented by carefully controlling the evaporation time or by adding a small volume of DMSO or low concentration of BSA (e.g. 10 μl of 0.1% BSA) prior to evaporation (Zheng et al. 2016).

23.3.2 Liquid–Liquid Extraction

23.3.2.1 Overview of LLE

LLE is a sample extraction method that utilizes the differential distribution of an analyte into a two-phase system (typically an aqueous solution and an immiscible organic solvent or solvent mixture) to selectively extract the analyte from the biological matrices (Chang et al. 2007). It is widely used for the extraction of small-molecule analytes and is often the preferred method for bioanalytical assays the in development stage (Yuan et al. 2016). This is because LLE usually generates much cleaner extracts than PPT with its high efficiency in removing endogenous proteins, lipids, and salts from biological samples, and the cost of LLE is relatively lower compared to SPE. LLE is also easy to be automated to achieve simple and high-throughput extraction with the use of the 96-well plate format and robotic liquid handling systems (Zheng et al. 2014). However, one limitation of LLE is that it usually does not work well for polar or ionic analytes. These compounds often dissolve well in the water phase but have poor solubility and thus low extraction recovery in the organic phase.

23.3.2.2 LLE in Peptide Bioanalysis

Most peptides have poor solubility in organic solvents due to their ionic nature, and therefore, are not suitable to be extracted by LLE. The application of LLE to peptide analysis is limited to non-ionic (hydrophobic) peptides, such as cyclic peptides (Li and Chan 2000; Liu et al. 2000; Celli et al. 2004; Shin et al. 2005; Li et al. 2007a) or C- and N-termini-modified peptides (Garteiz et al. 1998; Niwa et al. 1999; Ochiai et al. 2001). These peptides do not contain the charged terminal groups, therefore, are relatively easy to dissolve and distribute into the organic phase. To achieve the maximum recovery for the analyte while minimizing the matrix effect from the matrix components, the LLE conditions, including organic extraction solvent, extraction buffer, pH, and ratio of sample and extraction solvent volumes, need to be carefully assessed. Polar organic solvents such as *n*-butyl chloride (Garteiz et al. 1998), ethyl acetate (Li and Chan 2000) and methyl-*tert*-butyl ether (Shin et al. 2005) are often used as the extraction solvent for peptides. For the extraction of aqueous biological samples, usually, the analyte is controlled to be at its non-charged status by adjusting the pH of the extraction solution, therefore, the analyte can be maximally extracted to the organic phase, while the endogenous proteins and salts are separated and retained in the aqueous phase.

23.3.2.3 A Representative LLE Protocol

Here, a representative LLE protocol for the extraction of a cyclic depsipeptide from plasma is shown (adapted from Li and Chan (2000)).

- Add 0.5 ml of standards, QCs, blanks, and unknown samples into individual tubes
- Add 0.1 ml of IS working solution
- Add 2.0 ml of extraction buffer (50 mM potassium phthalate buffer, pH 4.0)
- Cap the tubes. Mix well and then let sit for at least one minute
- Add 6 ml of extraction solvent (ethyl acetate)
- Cap the tubes. Mix well, then centrifuge at high speed for five minutes
- Transfer the organic layer to a new tube and evaporate to dryness under nitrogen
- Reconstitute the residues in reconstitution solution

Notes

- The extraction solvent, extraction buffer, pH, and ratio of sample and extraction solvent volumes need to be carefully evaluated to achieve the satisfactory recovery and matrix effect of the peptide analyte.
- The original method used individual tubes for the extraction, which was tedious and time-consuming. High-throughput LLE can be achieved by using 96-well plates and a robotic liquid handler. The most commonly used volume for the 96-well plates is about 1 ml for each tube. So the sample volume needs to be scaled down accordingly for the assay, and the plate cap should be sealed well during extraction to avoid leakage.
- The mixing step after adding the extraction solvent is critical for facilitating the extraction of analytes into the organic layer. For extraction using 96-well plates, it is recommended to shake the plate at high speed for 15 minutes using a linear shaker to ensure adequate mixing.

23.3.2.4 Salting-out Assisted LLE

It is well known that the addition of salts (or concentrated salt solution) can induce the separation of water-miscible organic solvent from water to form a two-phase system (so-called "salting out"). Salting-out assisted liquid–liquid extraction (SALLE) is a LLE technique that utilizes this salting-out effect for the extraction of analyte into water-miscible organic phase (Tang and Weng 2013). For a SALLE method, typically, a concentrated salt (e.g. magnesium sulfate, zinc sulfate, ammonium acetate) solution and a water-miscible organic solvent (e.g. acetonitrile, methanol) are added into aqueous biological samples. The high concentration salts can significantly decrease the solubility of the target analyte in the aqueous solution, and therefore enhance its extraction into the organic phase. The sample is then mixed well to facilitate the extraction, and centrifuged to expedite the separation (salting-out) of the organic phase. The organic layer is then separated and used for LC-MS analysis. Similar to the conventional LLE, SALLE usually generates much cleaner extracts than the PPT technique, since the organic solvents and concentrated salts used can precipitate proteins before the phase separation, thus, efficiently removing most of the endogenous matrix components. One advantage of SALLE is its ability to achieve good extraction for a wide range of compounds, especially for polar compounds that are difficult to be extracted using the conventional LLE technique. SALLE does not require the evaporation and reconstitution steps because its resulting extraction solution can be directly diluted with water or other buffers. For optimization of a SALLE method, the organic solvents and salts used, the pH, and the ratio of aqueous and organic solvent volumes all need to be carefully evaluated.

As discussed in Section 23.3.2.2, most peptides are polar, and therefore, are not suitable to be extracted using the conventional LLE technique. SALLE, with its ability to extract relatively polar compounds, could be a potential extraction method for peptide analytes. So far, there is very limited application of SALLE for the extraction of peptides, and the applications are mainly for lipophilic peptides, for example, cycloporine-A, a cyclic peptide (Rustum 1990; Annesley and Clayton 2004). Rustum (1990) reported the SALLE method for the extraction of cycloporine-A from whole blood. He first used acetonitrile to precipitate the whole blood samples and then salted-out the acetonitrile phase from the supernatant using zinc sulfate and cadmium sulfate, followed by ammonium sulfate. Annesley and Clayton (2004) developed a simpler SALLE method for the extraction of cycloporine-A and other immunosuppressant drugs (tacrolimus, sirolimus, and everolimus) in whole blood. They used water to help lyse the blood cells and combined this step and the salting-out step using zinc sulfate and methanol into one step. The inclusion of water hemolysis helped yield cleaner extracts and more consistent extraction.

23.3.2.5 A Representative SALLE Protocol

Extraction of cycloporine-A, a cyclic peptide, from whole blood (adapted from Annesley and Clayton (2004)).

- Add 50 µl of standards, QCs, blanks, and unknown samples into individual tubes
- Add 250 µl of water, 250 µl of 0.1 M zinc sulfate in water solution, and 500 µl of methanol containing the IS
- Vortex for 30 seconds
- After 5–10 minutes, centrifuge the tubes
- Transfer the supernatant out for analysis

Note: The addition of water is to help lyse the blood cells and dissolve the blood components in the blood samples. For extraction of plasma or serum samples, this may not be needed.

23.3.3 Solid-Phase Extraction

23.3.3.1 Overview of SPE

SPE is a sample preparation technique that utilizes the same principles as chromatographic separation to purify the target analyte(s) in a sample (Hennion 1999). SPE can provide efficient extraction of various types of analytes by using sorbents with different separation mechanisms (e.g. reversed-phase for nonpolar compounds and ion-exchange for polar or ionizable compounds). SPE usually obtains much cleaner extracts than PPT, and has better or similar extraction performance compared to LLE methods, since it can achieve more selective extraction of the analytes by optimizing sorbents or loading/washing/elution conditions toward the unique physicochemical properties of the analytes. Another advantage of SPE is its ability to conveniently enrich or preconcentrate samples with large volumes, which is difficult for LLE or PPT methods using the 96-well plate format. A large volume of sample can pass through the SPE plate, the analyte can be retained and concentrated on the sorbent, and then eluted using a small volume of solution. This approach is very useful for improving assay sensitivity when a large volume of sample is available. However, SPE methods require extensive and sometimes challenging method development to identify the suitable sorbent, as well as the washing and elution conditions. The cost of SPE is also much higher than PPT and LLE methods.

23.3.3.2 SPE in Peptide Bioanalysis

SPE has been widely used for the extraction of peptide analytes from biological matrices. Since the sizes of the endogenous proteins (albumin, immunoglobulins, etc.) are much larger than peptides, they can be easily removed by using sorbents with appropriate pore size (e.g. 50–100 Å), while the peptide analytes being retained on the sorbents. Phospholipids and salts can also be efficiently removed from the samples using appropriate washing and elution conditions. Various types of sorbents including reversed phase (RP, [Ji et al. 2003, 2005; Baralla et al. 2011; Chambers et al. 2013; Meng et al. 2016]), strong cation exchange (SCX, [Zhang et al. 2007; Stoyanov et al. 2011]), mixed-mode cation exchange (MCX, [Heudi et al. 2008; Lame et al. 2011; Yuan et al. 2013]), mixed-mode anion exchange (MAX [Wilson et al. 2010; Halquist et al. 2012; Gong et al. 2015; Miyachi et al. 2017]), mixed-mode weak cation exchange (WCX, [van den Broek et al. 2010a]), and mixed-mode weak anion exchange (WAX, [Bronsema et al. 2015]) have been used for the purification of peptides. The selection of an appropriate SPE sorbent is peptide dependent, and need to be carefully evaluated and optimized based on the physicochemical properties (polarity, basicity, charge, etc.) of the target peptide.

SPE has also been applied for the purification of the tryptic peptides after the digestion of proteins in biological samples (Yang et al. 2007; Heudi et al. 2008; Yuan et al. 2013; Bronsema et al. 2015; Gong et al. 2015). For samples without immunoaffinity pretreatment, since the abundant endogenous proteins in plasma/serum are digested along with the target analyte and generate large amounts of background peptides, the resulting digested samples are highly complex. It is challenging to develop a SPE method that can effectively remove the abundant background peptides and retain the target peptides. Using a SPE separation mechanism that is orthogonal to the chromatographic separation is an effective approach for purification of complicated samples (Yuan et al. 2013). The orthogonal SPE sample preparation provides an additional dimension of separation to the chromatographic separation, and therefore, usually can achieve better sample cleanup than using SPE with the same separation mechanism. For example, MCX SPE, RP SPE, and two-dimensional (2D) SPE (RP SPE followed by MCX SPE) were compared for the cleanup of digested serum samples (Yuan et al. 2013). MCX SPE, with its orthogonal mechanism to the RP LC separation, was found to be more efficient than RP SPE in removing the background peptides from serum tryptic digests. The orthogonal SPE sample preparation efficiently removed most of the background peptides, and therefore, greatly reduced the ion suppression and background noise, and improved the assay sensitivity. To achieve further purification of the peptides of interest, 2D SPE can be applied, since it adds an additional dimension of SPE sample cleanup. Yang et al. (2007) applied 2D SPE (RP SPE followed by SCX SPE) for the purification of the tryptic digests of a therapeutic protein in serum samples. RP SPE was first applied to remove salts and highly hydrophobic constituents, and then SCX SPE to remove the background peptides that had significantly different basicity from the peptides of interest. As a result, they successfully eliminated an interference peak that previously could not be removed by using RP SPE alone, and obtained a much cleaner sample extract. The background noise and ion suppression were also significantly reduced, and the signal-to-noise ratio (*S*/*N*) was highly improved in LC-MS/MS analysis.

23.3.3.3 A Representative SPE Protocol

Extraction of oxyntomodulin, a therapeutic peptide composed of 37 amino acids, from plasma using MAX SPE (adapted from Halquist et al. (2012)).

Oasis MAX microelution SPE plates (2 mg, 30 µm particle size) in 96-well format were used for the extraction.

- Add 100 µl of standards, QCs, blanks, and unknown samples into a 96-well plate
- Add 100 µl of 4% phosphoric acid, and 20 µl of IS working solution, then mix well
- Condition the SPE plate with 200 µl of methanol followed by 200 µl of 4% phosphoric acid
- Load the samples onto the SPE plate
- Wash the plate sequentially with 200 µl of 5% ammonium hydroxide and then 200 µl of 5% acetonitrile in water
- Elute the extracted samples into a new 96-well collection plate using 2 × 25 µl of 5% trifluoroacetic acid (TFA) in acetonitrile/water (75 : 25, v/v)

Notes

- Microelution plates have reduced surface area and less amount of sorbents compared to the regular SPE plates, which helps to reduce the nonspecific binding and sample loss during extraction. In addition, microelution plates require much less elution volume (as low as 25–50 µl). This allows eliminating the evaporation and reconstitution steps, as well as increasing the pre-concentration factor. Microelution plates are particularly suitable for peptides with nonspecific binding issue or samples with limited volume.
- Prior to loading the plasma samples onto the SPE plate, 4% phosphoric acid were added into the samples to disrupt the binding of target peptide to plasma proteins. Sometimes, chaotropes such as guanidine and urea can also be used to disrupt the binding of analytes to proteins.
- Washing and elution solutions need to be carefully optimized to achieve good recovery and matrix effect. In this case, using higher percentage of organic solvent (20% acetonitrile in water) as the washing solution would result in significantly lower recovery. A stronger elution solution (5% TFA) was also used to improve the recovery.
- Vacuum or positive pressure is usually needed to facilitate the conditioning, washing, and elution steps.

23.3.3.4 Online SPE

There is an increasing need of highly sensitive assays for the analysis of peptide, especially peptide biomarkers. Moreover, the volumes or amounts of the samples sometimes are very limited (e.g. microsampling blood or plasma samples, tumor microbiopsy samples), which raises even higher requirements on sample preparation. Off-line SPE sample preparation usually involves multiple steps, such as sample loading, washing, elution, evaporation, and reconstitution. The loss of analyte during these steps will cause decreased recovery and thus reduced sensitivity. Especially, since peptides are well known to often have nonspecific binding to vessel surfaces, the sample loss during these steps could be even severe with increased steps in sample preparation. Online SPE sample preparation is a useful approach in addressing these challenges (Rogeberg et al. 2014). Online SPE eliminates the evaporation and reconstitution steps, and also reduces the sample transfer steps. As a result, it can significantly reduce the potential sample loss during these steps and increase the extraction recovery. Unlike off-line sample preparation in which only part of the extracted sample can be injected for LC-MS analysis, the entire sample loaded into the online SPE system are purified and analyzed. This allows the maximal utilization of the sample, which also helps improve the assay sensitivity, especially for samples with limited amount. Another advantage of online SPE is its ability to be fully automated, which can reduce the labor-intensive manual operations and minimize the risk of human errors.

Column-switching technique using multiport valves is the most commonly used online SPE sample preparation approach (Rogeberg et al. 2014). Typically, the sample is first injected into an extraction column (it can be a SPE column or even an analytical column for better separation). After the washing and the chromatographic separation on the extraction column, endogenous proteins and other matrix components are removed, while the target analyte is retained on the column. The multiport valve is then switched to allow the analyte to be flushed to a second column (analytical column) for LC-MS analysis. In the meantime, the extraction column is washed and conditioned for the next injection. For online SPE, the biological samples can be directly injected into the online system for purification and analysis (Wang et al. 2010). The samples can also be pretreated with PPT or other extraction methods, and then the extract is submitted to the online system to achieve further purification of the sample (Liu et al. 2011; Wang et al. 2012). For the purification of peptides from highly complex samples (e.g. digested biological samples), SPE columns may not provide enough sample cleanup due to their limited separation power. An analytical column can be used as the extraction column to achieve better separation and cleanup (this also transforms this system into a simple 2D-LC system). Orthogonal separation mechanisms are often selected for the first and second column to achieve better separation and therefore achieve enhanced purification (Liu et al. 2009; Xu et al. 2010; Shen et al. 2015). For example, Liu et al. (2009) used RP for the first column and hydrophilic interaction (HILIC) for the second column for the quantitative analysis of 11 therapeutic peptides in plasma. Xu et al. (2010) applied SCX for the first dimension and RP for the second dimension for the sensitive quantification of a PEGylated therapeutic peptide in human plasma. Shen et al. (2015) achieved the orthogonal chromatographic separation by using a high pH (pH 8) mobile-phase RP separation for the first dimension and a low pH (pH 3) RP separation for the second dimension.

23.3.4 Immunoaffinity Extraction

23.3.4.1 Overview of Immunoaffinity Extraction

IAE or immunocapture is a sample preparation technique that utilizes the immunoaffinity binding of a capturing reagent to the analyte of interest. The selective capture of the analyte by the reagent enables the efficient purification and enrichment of the analyte from biological samples. Compared to the above-mentioned PPT, LLE, and SPE methods, IAE has superior selectivity and efficiency, making it highly suitable for assays that require exceptional sensitivity. IAE has become the current method of choice for developing ultrasensitive LC-MS methods for the analysis of peptides and proteins. Usually, the more selective the reagent is, the better the sample purification, and as a result, the cleaner the sample extract generated and the better the sensitivity. Antibodies (monoclonal or polyclonal) against the target analytes are the most popular immunocapture reagents. Alternatively, the receptor or ligand of the analyte can also be used as the reagent. One potential limitation of IAE is its requirement of a suitable capture reagent, which may not be easily available and could take months to generate.

The most commonly used format for IAE is magnet beads (Wolf et al. 2001; Oe et al. 2006; Thomas et al. 2011; Kinumi et al. 2014; Lee et al. 2016; Taylor et al. 2016; Xu et al. 2017), due to their simple and flexible operation, and little requirements on instrumentation. The beads' format is also compatible with 96-well plate format for automation and high-throughput bioanalysis. A typical method consists of three simple steps: bind, wash, and elute. The capture antibodies are immobilized on the surfaces of magnet beads. The beads are then added into the samples (with incubation buffer). After incubation, the target analytes in the samples will be bind to the beads by the capture antibodies, while the other matrix components will be retained in the supernatant and removed. The beads are washed, and then the analytes are eluted from the beads for LC-MS analysis. In addition to the magnet beads format, cartridge or tip-based method (Liu et al. 2015a) and plate-based method (Judák et al. 2018) have also been used for IAE. For these methods, the capture antibodies are immobilized on the surface of the cartridges/tips packing material or on the wells of a plate.

23.3.4.2 IAE in Peptide Bioanalysis

IAE has been widely applied for the extraction and enrichment of peptides from biological samples (Berna et al. 2006; Oe et al. 2006; Li et al. 2007b; Kumar et al. 2010; Chappell et al. 2013, 2016; Kinumi et al. 2014; Lee et al. 2016; Taylor et al. 2016; Xu et al. 2017). Oe et al. (2006) developed a sensitive IAE LC-MS/MS method for the analysis of amyloid beta peptides ($A\beta_{1-40}$ and $A\beta_{1-42}$), biomarkers for Alzheimer's disease, in CSF. By using a monoclonal antibody (mAb) to selectively capture the target peptides, they achieved efficient purification of the analytes and eliminated the interference from the endogenous components (which was previously observed when they used a LLE method). The method was validated with a LLOQ of 200 pg ml^{-1} for $A\beta_{1-42}$ and 200 pg ml^{-1} for $A\beta_{1-40}$. In another example, Xu et al. (2017) needed to develop a highly sensitive method (target LLOQ 0.1 ng ml^{-1}) for the simultaneous analysis of insulin glargine, a therapeutic peptide analog of human insulin, and its two active metabolites in human plasma. They evaluated PPT and SPE methods first, but could only achieve a LLOQ of 1 and 0.5 ng ml^{-1}, respectively, mainly due to the high background and severe interferences from complex matrix components. An IAE method was then developed using an anti-insulin mAb to selectively extract glargine and the two metabolites. The method achieved the desired sensitivity at 0.1 ng ml^{-1} and demonstrated superior specificity against the interference from endogenous insulin, exogenous insulin analogs, and potential preexisting anti-insulin antibodies in patient samples. They also fully automated the IAE method on a robotic liquid handling system to improve the throughput and consistency of the sample preparation. The method was successfully validated based on regulatory requirements and applied to biosimilar clinical studies.

IAE has also been used for the purification and enrichment of peptides from highly complicated samples, for example, digested plasma or serum samples (Anderson et al. 2004; Whiteaker et al. 2007, 2010; Kuhn et al. 2009). IAE, with its high selectivity, can efficiently extract the target peptides (even at low concentration) from the abundant background proteins/peptides and remove other endogenous components, thus, can greatly improve the sensitivity and selectivity of the LC-MS assays. Anderson et al. (2004) developed stable-isotope standards with capture by anti-peptide antibodies (SISCAPA) technique for the quantification of peptides in complex plasma digests. They used anti-peptide antibodies to specifically capture the peptides of interest in digested samples. They were able to achieve an average of 120-fold enrichment of the target peptides, which could contribute to about two orders of magnitude improvement of sensitivity. In another example, Whiteaker et al. (2007) demonstrated that by using IAE with an anti-peptide antibody, they achieved >1000-fold enhancement of the sensitivity for the analysis of an endogenous serum protein, α1-antichymotrypsin, in human serum.

It has been demonstrated that multiplexed immunocapture of multiple peptides can be achieved by using a combination of anti-peptide antibodies targeting individual peptides (Kuhn et al. 2009, Whiteaker et al. 2010). Thus, multiple peptide or protein analytes can be simultaneously purified and analyzed using a single assay with high efficiency and reproducibility. Kuhn et al. (2009) developed a multiplexed SISCAPA method for simultaneous analysis of two small proteins (troponin I and interleukin-33) in plasma. The plasma samples were digested with trypsin and the resulting digests were purified by the multiplexed IAE method using anti-peptide antibodies targeting specific peptides generated from the target proteins. They compared the results generated from the multiplexed assay to those from the separate assays using individual anti-peptide antibody, and observed excellent agreement. The multiplexed assay also matched well ($R = 0.89$) with a commercial immunoassay for the analysis of patient samples. Whiteaker et al. (2010) reported the simultaneous analysis of nine protein biomarkers in plasma by using a multiplexed, magnetic bead-based IAE method. The multiplexed IAE process offered significant sensitivity improvement (100–1000 fold) with sufficient precision (median coefficient of variation, 12.6%) for all the target protein analytes.

Multiplexed immunocapture have also been applied for the quantitative bioanalysis of peptides (Wolf et al. 2004; Thomas et al. 2011; Lee et al. 2016; Taylor et al. 2016). For example, Thomas et al. (2011) reported the simultaneous extraction of insulin-like growth factor (IGF) 1, 2, their degradation products, and synthetic analogues in urine using anti-IGF1 and anti-IGF2 antibodies. Lee et al. (2016) developed a multiplexed IAE assay for the quantification of four proglucagon-derived peptides in plasma and achieved a LLOQ of 0.78 pM for all the peptides. Taylor et al. (2016) used anti-intact insulin B chain and anti-C-peptide antibodies for the extraction of insulin and C-peptide from serum samples. For both peptides, their multiplexed method achieved equivalent sensitivity compared to those of current commercial immunoassays. The measured insulin results using this method matched very well with those from a FDA-approved commercial immunoassay (correlation $r = 0.97$).

Online IAE sample preparation is also applied to the purification and enrichment of peptides (Berna et al. 2006; Nemirovskiy et al. 2007; Li et al. 2007b; Neubert et al. 2010) and proteins (Hoos et al. 2006; Cingöz et al. 2010). Similar to online SPE, online IAE can be performed by using the column-switching technique with an immunoaffinity column as the first column (Dufield and Radabaugh 2012). The capture antibodies are immobilized on the packing particles in the immunoaffinity column for the selective extraction and enrichment of the analytes (Moser and Hage 2010). Online IAE has the same advantages as online SPE sample preparation in improving the assay efficiency and reproducibility, minimizing the adsorption loss of the analyte during transfer, and eliminating the risk of human errors (through automation).

23.3.4.3 A Representative IAE Protocol

Extraction of insulin glargine and its two active metabolites from plasma using IAE (adapted from Xu et al. (2017)).

Preparation of antibody-coated magnetic beads

- Wash the magnetic beads (Dynabeads M280 streptavidin beads) three times with phosphate-buffered saline with 0.05% Tween-20 (PBST)
- Resuspend the beads into the same volume of phosphate-buffered saline (PBS)
- Add biotinylated anti-insulin mAb to the bead suspension to reach the concentration of 10 μg Ab ml^{-1} beads
- Gently rotate at room temperature for one hour
- Wash the Ab-coated beads three times with PBST
- Resuspend the beads in PBS and store at 4 °C

Note: This preparation procedure is for streptavidin beads, and biotinylated antibody is required for the coating of the beads. For other types of beads (protein A, protein G, tosylactivated, etc.), the procedures and required antibodies are different. Follow the manufacturer's instruction on how to prepare the corresponding Ab-coated beads.

IAE sample preparation using the Ab-coated beads

- Add 0.5 ml standards, QCs, blanks, and unknown samples into a 96-well plate
- Add 25 μl of IS working solution, and mix well
- Add 100 μl aliquot of the above-prepared Ab-coated beads into each sample
- Vortex at ambient temperature for about two hours to allow adequate binding of the analytes to the capture antibody
- Pull down the beads using a magnetic block, and transfer and discard the supernatant
- Wash the beads with 300 μl of PBST twice and then with 200 μl PBS twice
- Add 120 μl of elution solution (2% acetic acid/50% methanol in water) to each well and shake vigorously at room temperature for five minutes
- Transfer the supernatant into a new 96-well plate for analysis

Notes

- To minimize the nonspecific binding caused by loss of analytes, it is recommended to use low binding plates and vials for the sample preparation.
- To achieve good and consistent recovery of the peptide analyte, the IAE conditions, including the beads to capture antibody ratio, the amount of beads (antibody) needed, the binding time, as well as the washing and elution conditions need to be optimized.
- Severe nonspecific binding could also occur to the eluted peptide analytes, since after IAE there would be much less proteins remaining to block the adsorption sites. It is a good practice to include a certain percentage of organic solvent in the elution solution (50% methanol in this case, optimal percentage needs to be experimentally determined for other peptides) to prevent the adsorption. Sometimes, proteins (e.g. BSA) or surfactants can also be added in the elution solution to minimize the adsorption.

23.3.5 Other Sample Preparation Techniques for Peptides

23.3.5.1 Immobilized Metal Ion Affinity Chromatography

Transition metal ions (e.g. Ni^{2+}, Zn^{2+}, Cu^{2+}, and Co^{2+}) are well known to be able to chelate with amino acids that have strong electron donor groups, such as histidine, cysteine, and tryptophan (Cheung et al. 2012). Immobilized metal ion affinity chromatography (IMAC) immobilizes the metal ions on column support and utilizes this chelating effect to selectively extract peptides or proteins having these types of amino acids exposed on the surface (Posewitz and Tempst 1999; Kokubu et al. 2005; Zheng et al. 2005; Cheung et al. 2012; Wilffert et al. 2013). IMAC is popularly used for the purification of histidine-tagged fusion proteins, natural metal-binding proteins, as well as peptides or proteins containing metal-binding amino acids (Cheung et al. 2012). For example, Wilffert et al. (2013) developed an IMAC method for the purification and enrichment of recombinant human tumor necrosis factor-related apoptosis-inducing ligand (rhTRAIL), a 60 kDa protein with histidines exposed on the surface, in serum. They achieved selective and efficient extraction of rhTRAIL (72% extraction recovery with 95% of serum proteins removed) because of the strong interaction between histidines and the immobilized Ni^{2+} ions on the IMAC resin. A LLOQ of 20 ng ml^{-1} (~350 pM) was reached and the method was fully validated according to regulatory agencies' guidelines for bioanalysis.

IMAC has also been used for the selective purification and enrichment of phosphopeptides or phosphoproteins based on their affinity for ions such as Fe^{3+}, Ti^{4+}, and Ga^{3+}, especially in proteomics for qualitative work (Posewitz and Tempst 1999; Kokubu et al. 2005; Zheng et al. 2005; Lai et al. 2012; Tape et al. 2014; Ruprecht et al. 2015). For instance, Ruprecht et al. (2015) developed a robust workflow for the enrichment of phosphopeptides using a commercially available Fe-IMAC column. Their approach offered selective, comprehensive, and reproducible enrichment of phosphopeptides out of complex lysates, and scaled linearly over a wide range of sample quantities (100 μg to 5 mg of digest). Using this approach, they were able to identify >7500 unique phosphopeptides with 90% selectivity and good reproducibility (median CV 15%).

23.3.5.2 Molecularly Imprinted Polymers

Molecularly imprinted polymers (MIPs) are synthetic materials that are capable of selectively recognizing specific target molecules (Janiak and Kofinas 2007; Stevenson et al. 2016). MIPs, sometimes referred to as plastic antibodies, mimic antibodies by using affinity recognition sites that can recognize target analytes by a lock and key mechanism, and therefore, achieving selective purification based on affinity (Rossetti et al. 2014). MIPs are synthesized by forming the cross-linking of functional monomers or polymers around a template molecule (usually the target analyte). After removing the template, cavities having the size and shape complementary to the target molecule are formed, and thus can selectively rebind the template molecule. Compared to antibodies, MIPs have their unique advantages, such as easy and fast generation of the materials, cheap price, good stability, and tolerance to harsh conditions (wide range of pH, various solvents, etc.). MIPs have been used as SPE sorbents for the extraction of a wide range of small-molecule analytes in complex matrices. In recent years, MIPs have also drawn great attention and interest for the purification of peptides and proteins (Qader et al. 2014; Rossetti et al. 2014, 2016; Liu et al. 2015b; Chen et al. 2016; Stevenson et al. 2016; Yang et al. 2016, 2017). Rossetti et al. (2014) explored the application of MIPs for the cleanup and enrichment of a proteotypic signature peptide of pro-gastrin releasing peptide (ProGRP, a peptide biomarker) from digested serum samples. MIPs complementary to the ProGRP signature peptide were used as the SPE sorbents for the extraction. They achieved selective extraction of the target peptide while removing other tryptic peptides from digested serum samples, thus, resulting in reduced background and obtaining an estimated limit of detection of 625 pM. They also directly compared an online MIP method with IAE (using mAb-coated magnetic beads) and PPT (using acetonitrile in a ratio of 1 : 1) for the analysis ProGRP in serum (Rossetti et al. 2016). MIP extraction was able to offer very close performance to IAE in terms of matrix effect, sensitivity (limit of detection 11 and 14 pM, respectively), and reproducibility (RSD ≤11% and ≤16%, respectively) for the extraction of 50 μl of sample. PPT, as expected, suffered from high ion suppression, low sensitivity, and poor reproducibility (RSD ≥ 34%). Their results demonstrated the promising potential of MIPs as an alternative to antibodies for the purification of peptides, especially in proteomic workflows.

23.3.5.3 Ultrafiltration

Ultrafiltration is a separation technique that utilizes membrane to separate analytes based on their molecular weights (MW). Depending on the size of the target peptide analytes, membranes with different MW cutoffs (e.g. 10, 30, and 100 kDa) can be selected. Several applications of ultrafiltration have been reported for the purification of peptides from various types of biological matrices, such as urine (Fierens et al. 2000), CSF (Wang et al. 2016), serum (Kobold et al. 2008; Su et al. 2008), and plasma (Goda et al. 2008). Ultrafiltration has also been used in combination with other sample preparation methods. For example, it was applied after PPT to further cleanup the sample extracts (Hatziieremia et al. 2007; Sobhi et al. 2011), and as a pretreatment to preconcentrate the peptide analyte prior to IAE (Judák et al. 2018). One limitation of ultrafiltration is that peptide analytes may suffer from severe adsorption during filtration, causing low recovery and high variability (Heinig et al. 2013). In addition, the filtration step is often time-consuming, especially for serum or plasma samples, which could even take one hour or more (Sobhi et al. 2011; Judák et al. 2018). Thus, ultrafiltration usually is not appropriate for high-throughput bioanalytical assays of peptides.

One unique application of ultrafiltration is for the determination of the unbound (free) fraction of the analyte (Li et al. 2011b). Ultrafiltration can only allow the free analyte molecules to pass through the membrane, and filter out the bound fraction of the analyte. For example, Li et al. (2014) employed ultrafiltration technique to separate free vancomycin, a glycopeptide antibiotic drug, from the protein-bound fraction in human plasma. The free drug was separated from plasma samples using filters with 10-kDa MW cutoff, and the filtrates were injected for 2D-LC analysis. Kratzer et al. (2014) developed an ultrafiltration method for the determination of the free fraction of vancomycin and three beta-lactam drugs (ertapenem, ceftriaxone, and cefazolin). They evaluated various filter types and experimental conditions (MW cutoff, centrifugal force and time, pH and temperature), and found that physiological conditions (37 °C and pH 7.4) and low-to-moderate centrifugal forces generated more reliable results.

23.3.5.4 Derivatization

Derivatization is a commonly used strategy for improving the sensitivity, stability, or chromatographic behavior of the analytes in LC-MS analysis (Iwasaki et al. 2011; Xu et al. 2011; Niwa 2012; Zhu et al. 2015). Derivatization of peptides has been applied for improving protein identification and characterization (Foettinger et al. 2006; Xu et al. 2008; Suh et al. 2010), de novo sequencing of protein (Madsen and Brodbelt 2009), and quantification of proteins (Sandra et al. 2008; Ji et al. 2009; Qiao et al. 2014; Zhou et al. 2014; Yuan et al. 2014b). Derivatization of peptides can enhance the electrospray ionization efficiency of peptides (e.g. incorporating an easily ionizable moiety such as quaternary ammonium into the molecules) and/or improve their fragmentation pattern (e.g. generating a more abundant product ion), therefore, can also be used for increasing the sensitivity for the analysis of the peptides (Mirzaei and Regnier 2006; Wang et al. 2009; Kulevich et al. 2010). In one study, ionization efficiency of peptides was increased roughly 10-fold after derivatization with a quaternary ammonium reagent, and in one extreme case, the ionization efficiency of a peptide was increased 500-fold (Mirzaei and Regnier 2006). Kinumi et al. (2014) developed a highly sensitive method for quantification of human serum C-peptide, a peptide biomarker consisting of 31 amino acids, in serum using IAE and derivatization. The serum samples were purified by IAE using mAb against human C-peptide, and then followed by derivatization with 6-aminoquinolyl-*N*-hydroxysuccinimidyl carbamate. The derivatization of C-peptide significantly increased the sensitivity (more than fivefold). In combination with IAE, they achieved a 23-fold enhancement of the sensitivity compared to using SPE without derivatization. The generated results also correlated well ($r^2 = 0.9994$) with those from an established commercial immunoassay.

23.3.5.5 Digestion

Direct LC-MS analysis of large peptides often suffers from poor sensitivity due to the decreased ionization efficiency, the presence of broad isotope distributions and multiply charged ions, and poor chromatography with their increased size. To improve the assay sensitivity, an alternative approach is to digest the large peptide to form smaller peptide fragments. Then one unique peptide fragment with good specificity and sensitivity will be selected as a surrogate analyte to represent the intact peptide of interest, and be analyzed by LC-MS. This surrogate peptide approach is a well-established strategy for the absolute quantification of proteins (Ewles and Goodwin 2011; Li et al. 2011a; van den Broek et al. 2013). Its feasibility has also been demonstrated for the quantification of large peptides (van den Broek et al. 2007). One limitation of this approach is that since only a small fragment of the target peptide is monitored, it may not truly represent the intact peptide. For example, the degradation or biotransformation products of the target peptide may contain the same fragment, therefore, cannot be differentiated from the intact peptide. This will cause the inaccurate determination of the peptide analyte concentration. Another potential concern is the complexity of the biological sample after digestion. Peptides are relatively easy to be separated from the endogenous proteins in the biological matrices. However, with the digestion approach, the abundant endogenous proteins are also digested along with the target peptide, generating huge amounts of background peptides. Purifying surrogate peptides from all these background peptides becomes much more challenging. Depletion of the abundant endogenous proteins (e.g. using IAE) prior to the digestion is highly recommended for this approach.

This digestion approach relies on the analysis of the surrogate peptide for the quantification of the intact target peptide. Thus, the selection of an appropriate surrogate peptide is critical for ensuring the specificity, sensitivity, and reproducibility of the assay. The selected surrogate peptide needs to be a signature peptide that is unique to the peptide of interest and should not be present in the inactive biotransformation products or in the background peptides from the digested biological matrices. Peptides prone to chemical or enzymatic modification (e.g. peptides containing methionine, tryptophan, or cysteine) should be avoided to be used as the surrogate peptide (Wu et al. 2011).

Enzymatic digestion is the predominantly used digestion method, and trypsin is the most commonly used enzyme due to its good specificity and relatively low cost. Other enzymes (Lys-C, Asp-N, Glu-C, etc.) can also be used, especially when no appropriate surrogate peptide can be identified using tryptic digestion (Switzar et al. 2013). Digestion conditions, including enzyme to protein ratio, incubation buffer, incubation time, and temperature need to be optimized to achieve more complete and reproducible digestion.

23.4 Conclusions

Sample preparation is a critical, and often the most challenging, part in LC-MS bioanalysis of peptides. An appropriate sample preparation strategy can effectively purify and enrich the target peptide, while removing other matrix components from the biological samples. Precautions in operation and appropriate sample pretreatment (as needed) also ensure the stability and minimize the adsorption loss of the peptide analyte during sample preparation. All of these will play an important role in improving the sensitivity, selectivity, and reproducibility of the LC-MS assay for peptide bioanalysis. With the continuous improvement in sample preparation techniques and methodologies, LC-MS analysis of peptide will become more sensitive, specific, and reliable. LC-MS-based assay will increasingly become the method of choice for the quantitative bioanalysis of peptides.

Acknowledgements

The author thanks Drs. Qin C. Ji and Naiyu Zheng for their insightful comments and careful review of this manuscript.

References

Anderson, N.L., Anderson, N.G., Haines, L.R. et al. (2004). Mass spectrometric quantitation of peptides and proteins using stable isotope standards and capture by anti-peptide antibodies (SISCAPA). *J. Proteome Res.* 3 (2): 235–244.

Annesley, T.M. and Clayton, L. (2004). Simple extraction protocol for analysis of immunosuppressant drugs in whole blood. *Clin. Chem.* 50 (10): 1845–1848.

Baralla, E., Nieddu, M., Boatto, G. et al. (2011). Quantitative assay for bradykinin in rat plasma by liquid chromatography coupled to tandem mass spectrometry. *J. Pharm. Biomed. Anal.* 54 (3): 557–561.

Belenky, A., Smith, A., Zhang, B. et al. (2004). The effect of class-specific protease inhibitors on the stabilization of B-type natriuretic peptide in human plasma. *Clin. Chim. Acta* 340 (1): 163–172.

Berna, M., Schmalz, C., Duffin, K. et al. (2006). Online immunoaffinity liquid chromatography/tandem mass spectrometry determination of a type II collagen peptide biomarker in rat urine: investigation of the impact of collision-induced dissociation fluctuation on peptide quantitation. *Anal. Biochem.* 356 (2): 235–243.

Bronsema, K.J., Bischoff, R., Bouche, M.-P. et al. (2015). High-sensitivity quantitation of a Nanobody® in plasma by single-cartridge multidimensional SPE and ultra-performance LC-MS/MS. *Bioanalysis* 7 (1): 53–64.

Celli, N., Mariani, B., Di Carlo, F. et al. (2004). Determination of Aplidin®, a marine-derived anticancer drug, in human plasma, whole blood and urine by liquid chromatography with electrospray ionisation tandem mass spectrometric detection. *J. Pharm. Biomed. Anal.* 34 (3): 619–630.

Chambers, E.E., Fountain, K.J., Smith, N. et al. (2014). Multidimensional LC-MS/MS enables simultaneous quantification of intact human insulin and five recombinant analogs in human plasma. *Anal. Chem.* 86 (1): 694–702.

Chambers, E.E., Lame, M.E., Bardsley, J. et al. (2013). High sensitivity LC-MS/MS method for direct quantification of human parathyroid 1–34 (teriparatide) in human plasma. *J. Chromatogr. B* 938: 96–104.

Chang, M.S., Ji, Q., Zhang, J., and El-Shourbagy, T.A. (2007). Historical review of sample preparation for chromatographic bioanalysis: pros and cons. *Drug Dev. Res.* 68 (3): 107–133.

Chang, D., Kolis, S.J., Linderholm, K.H. et al. (2005). Bioanalytical method development and validation for a large peptide HIV fusion inhibitor (Enfuvirtide, T-20) and its metabolite in human plasma using LC-MS/MS. *J. Pharm. Biomed. Anal.* 38 (3): 487–496.

Chappell, D.L., Lee, A.Y.H., Bernstein, H.S. et al. (2016). Development and validation of an IA-LC/MS method to quantitate active and total B-type natriuretic peptide in human plasma. *Bioanalysis* 8 (22): 2341–2349.

Chappell, D.L., Lee, A.Y.H., Castro-Perez, J. et al. (2013). An ultrasensitive method for the quantitation of active and inactive GLP-1 in human plasma via immunoaffinity LC-MS/MS. *Bioanalysis* 6 (1): 33–42.

Chen, J., Shinde, S., Subedi, P. et al. (2016). Validation of molecularly imprinted polymers for side chain selective phosphopeptide enrichment. *J. Chromatogr. A* 1471: 45–50.

Cheung, R.C.F., Wong, J.H., and Ng, T.B. (2012). Immobilized metal ion affinity chromatography: a review on its applications. *Appl. Microbiol. Biotechnol.* 96 (6): 1411–1420.

Cingöz, A., Hugon-Chapuis, F., and Pichon, V. (2010). Total on-line analysis of a target protein from plasma by immunoextraction, digestion and liquid chromatography–mass spectrometry. *J. Chromatogr. B* 878 (2): 213–221.

Diamandis, E.P. (2004). Mass spectrometry as a diagnostic and a cancer biomarker discovery tool: opportunities and potential limitations. *Mol. Cell. Proteomics* 3 (4): 367–378.

Dufield, D.R. and Radabaugh, M.R. (2012). Online immunoaffinity LC/MS/MS. A general method to increase sensitivity and specificity: how do you do it and what do you need? *Methods* 56 (2): 236–245.

Ewles, M. and Goodwin, L. (2011). Bioanalytical approaches to analyzing peptides and proteins by LC-MS/MS. *Bioanalysis* 3 (12): 1379–1397.

Fanciulli, G., Azara, E., Wood, T.D. et al. (2007). Liquid chromatography–mass spectrometry assay for quantification of Gluten Exorphin B5 in cerebrospinal fluid. *J. Chromatogr. B* 852 (1): 485–490.

Fierens, C., Thienpont, L.M.R., Stöckl, D. et al. (2000). Quantitative analysis of urinary C-peptide by liquid chromatography–tandem mass spectrometry with a stable isotopically labelled internal standard. *J. Chromatogr. A* 896 (1): 275–278.

Foettinger, A., Leitner, A., and Lindner, W. (2006). Derivatisation of arginine residues with malondialdehyde for the analysis of peptides and protein digests by LC-ESI-MS/MS. *J. Mass Spectrom.* 41 (5): 623–632.

Fosgerau, K. and Hoffmann, T. (2015). Peptide therapeutics: current status and future directions. *Drug Discov. Today* 20 (1): 122–128.

Garteiz, D.A., Madden, T., Beck, D.E. et al. (1998). Quantitation of dolastatin-10 using HPLC/electrospray ionization mass spectrometry: application in a phase I clinical trial. *Cancer Chemother. Pharmacol.* 41 (4): 299–306.

Goda, R., Masumoto, H., Okazaki, O., and Sudo, K. (2008). Development of a pretreatment method for amyloid β-protein analysis based on the effect of acetic acid on the dissolution of plasma polypeptides. *Biomed. Chromatogr.* 22 (11): 1279–1287.

Goebel-Stengel, M., Stengel, A., Taché, Y., and Reeve, J.R. (2011). The importance of using the optimal plasticware and glassware in studies involving peptides. *Anal. Biochem.* 414 (1): 38–46.

Gong, C., Zheng, N., Zeng, J. et al. (2015). Post-pellet-digestion precipitation and solid phase extraction: a practical and efficient workflow to extract surrogate peptides for ultra-high performance liquid chromatography – tandem mass spectrometry bioanalysis of a therapeutic antibody in the low ng/mL range. *J. Chromatogr. A* 1424: 27–36.

Grohganz, H., Rischer, M., and Brandl, M. (2004). Adsorption of the decapeptide Cetrorelix depends both on the composition of dissolution medium and the type of solid surface. *Eur. J. Pharm. Sci.* 21 (2–3): 191–196.

Halquist, M.S., Sakagami, M., and Karnes, H.T. (2012). Determination of oxyntomodulin, an anorectic polypeptide, in rat plasma using 2D-LC-MS/MS coupled with ion pair chromatography. *J. Chromatogr. B* 903: 102–111.

Hatziieremia, S., Kostomitsopoulos, N., Balafas, V., and Tamvakopoulos, C. (2007). A liquid chromatographic/tandem mass spectroscopic method for quantification of the cyclic peptide melanotan-II. Plasma and brain tissue concentrations following administration in mice. *Rapid Commun. Mass Spectrom.* 21 (15): 2431–2438.

Haynes, J.J., Jones, H., Gibson, D., and Clark, G.T. (2011). Bioanalytical determination of unstable endogenous small peptides: RFRP3 and its metabolites in rat blood. *Bioanalysis* 3 (7): 763–778.

Heinig, K., Wirz, T., Schick, E., and Guenzi, A. (2013). Bioanalysis of therapeutic peptides: differentiating between total and anti-drug antibody bound drug using liquid chromatography–tandem mass spectrometry quantitation. *J. Chromatogr. A* 1316: 69–77.

Hennion, M.-C. (1999). Solid-phase extraction: method development, sorbents, and coupling with liquid chromatography. *J. Chromatogr. A* 856 (1): 3–54.

Heudi, O., Barteau, S., Zimmer, D. et al. (2008). Towards absolute quantification of therapeutic monoclonal antibody in serum by LC-MS/MS using isotope-labeled antibody standard and protein cleavage isotope dilution mass spectrometry. *Anal. Chem.* 80 (11): 4200–4207.

Hoos, J.S., Sudergat, H., Hoelck, J.-P. et al. (2006). Selective quantitative bioanalysis of proteins in biological fluids by on-line immunoaffinity chromatography–protein digestion–liquid chromatography–mass spectrometry. *J. Chromatogr. B* 830 (2): 262–269.

Hu, X. and Kamberi, M. (2009). Quantification of apolipoprotein A-I mimetic peptide D-4F in rabbit plasma by high-performance liquid chromatography coupled with electrospray ionization tandem mass spectrometry. *J. Chromatogr. B* 877 (20): 1946–1950.

Iwasaki, Y., Nakano, Y., Mochizuki, K. et al. (2011). A new strategy for ionization enhancement by derivatization for mass spectrometry. *J. Chromatogr. B* 879 (17–18): 1159–1165.

Janiak, D.S. and Kofinas, P. (2007). Molecular imprinting of peptides and proteins in aqueous media. *Anal. Bioanal. Chem.* 389 (2): 399–404.

Jemal, M., Ouyang, Z., and Xia, Y.Q. (2010). Systematic LC-MS/MS bioanalytical method development that incorporates plasma phospholipids risk avoidance, usage of incurred sample and well thought-out chromatography. *Biomed. Chromatogr.* 24 (1): 2–19.

Ji, Q.C., Rodila, R., Gage, E.M., and El-Shourbagy, T.A. (2003). A strategy of plasma protein quantitation by selective reaction monitoring of an intact protein. *Anal. Chem.* 75 (24): 7008–7014.

Ji, Q.C., Rodila, R., Morgan, S.J. et al. (2005). Investigation of the immunogenicity of a protein drug using equilibrium dialysis and liquid chromatography tandem mass spectrometry detection. *Anal. Chem.* 77 (17): 5529–5533.

Ji, C., Sadagopan, N., Zhang, Y., and Lepsy, C. (2009). A universal strategy for development of a method for absolute quantification of therapeutic monoclonal antibodies in biological matrices using differential dimethyl labeling coupled with ultra performance liquid chromatography-tandem mass spectrometry. *Anal. Chem.* 81 (22): 9321–9328.

John, H., Walden, M., Schäfer, S. et al. (2004). Analytical procedures for quantification of peptides in pharmaceutical research by liquid chromatography–mass spectrometry. *Anal. Bioanal. Chem.* 378 (4): 883–897.

Judák, P., Van Eenoo, P., and Deventer, K. (2018). Utilizing ELISA-plate based immunopurification and liquid chromatography-tandem mass spectrometry for the urinary detection of short- and long acting human insulin analogues. *J. Pharm. Biomed. Anal.* 153: 76–81.

Kaspar, A.A. and Reichert, J.M. (2013). Future directions for peptide therapeutics development. *Drug Discov. Today* 18 (17): 807–817.

King, R., Bonfiglio, R., Fernandez-Metzler, C. et al. (2000). Mechanistic investigation of ionization suppression in electrospray ionization. *J. Am. Soc. Mass Spectrom.* 11 (11): 942–950.

Kinumi, T., Mizuno, R., and Takatsu, A. (2014). Quantification of serum C-peptide by isotope-dilution liquid chromatography–tandem mass spectrometry: enhanced detection using chemical modification and immunoaffinity purification. *J. Chromatogr. B* 953–954: 138–142.

Kobold, U., Dülffer, T., Dangl, M. et al. (2008). Quantification of hepcidin-25 in human serum by isotope dilution micro-HPLC–tandem mass spectrometry. *Clin. Chem.* 54 (9): 1584–1586.

Kokubu, M., Ishihama, Y., Sato, T. et al. (2005). Specificity of immobilized metal affinity-based IMAC/C18 tip enrichment of phosphopeptides for protein phosphorylation analysis. *Anal. Chem.* 77 (16): 5144–5154.

Kratzer, A., Liebchen, U., Schleibinger, M. et al. (2014). Determination of free vancomycin, ceftriaxone, cefazolin and ertapenem in plasma by ultrafiltration: impact of experimental conditions. *J. Chromatogr. B* 961: 97–102.

Kuhn, E., Addona, T., Keshishian, H. et al. (2009). Developing multiplexed assays for troponin I and interleukin-33 in plasma by peptide immunoaffinity enrichment and targeted mass spectrometry. *Clin. Chem.* 55 (6): 1108–1117.

Kulevich, S.E., Frey, B.L., Kreitinger, G., and Smith, L.M. (2010). Alkylating tryptic peptides to enhance electrospray ionization mass spectrometry analysis. *Anal. Chem.* 82 (24): 10135–10142.

Kumar, V., Barnidge, D.R., Chen, L.S. et al. (2010). Quantification of serum 1-84 parathyroid hormone in patients with hyperparathyroidism by immunocapture in situ digestion liquid chromatography-tandem mass spectrometry. *Clin. Chem.* 56 (2): 306–313.

Lai, A.C.Y., Tsai, C.F., Hsu, C.C. et al. (2012). Complementary Fe^{3+}- and Ti^{4+}-immobilized metal ion affinity chromatography for purification of acidic and basic phosphopeptides. *Rapid Commun. Mass Spectrom.* 26 (18): 2186–2194.

Lame, M.E., Chambers, E.E., and Blatnik, M. (2011). Quantitation of amyloid beta peptides Aβ1–38, Aβ1–40, and Aβ1–42 in human cerebrospinal fluid by ultra-performance liquid chromatography–tandem mass spectrometry. *Anal. Biochem.* 419 (2): 133–139.

Lee, A.Y.H., Chappell, D.L., Bak, M.J. et al. (2016). Multiplexed quantification of proglucagon-derived peptides by immunoaffinity enrichment and tandem mass spectrometry after a meal tolerance test. *Clin. Chem.* 62 (1): 227–235.

Li, Z. and Chan, K.K. (2000). A subnanogram API LC/MS/MS quantitation method for depsipeptide FR901228 and its preclinical pharmacokinetics. *J. Pharm. Biomed. Anal.* 22 (1): 33–44.

Li, F., Fast, D., and Michael, S. (2011a). Absolute quantitation of protein therapeutics in biological matrices by enzymatic digestion and LC-MS. *Bioanalysis* 3 (21): 2459–2480.

Li, W., Lin, H., Smith, H.T., and Tse, F.L.S. (2011b). Developing a robust ultrafiltration-LC-MS/MS method for quantitative analysis of unbound vadimezan (ASA404) in human plasma. *J. Chromatogr. B* 879 (21): 1927–1933.

Li, W., Luo, S., Hayes, M. et al. (2007a). Determination of *N*-methyl-4-isoleucine-cyclosporin (NIM811) in human whole blood by high performance liquid chromatography-tandem mass spectrometry. *Biomed. Chromatogr.* 21 (3): 249–256.

Li, W., Nemirovskiy, O., Fountain, S. et al. (2007b). Clinical validation of an immunoaffinity LC-MS/MS assay for the quantification of a collagen type II neoepitope peptide: a biomarker of matrix metalloproteinase activity and osteoarthritis in human urine. *Anal. Biochem.* 369 (1): 41–53.

Li, X., Wang, F., Xu, B. et al. (2014). Determination of the free and total concentrations of vancomycin by two-dimensional liquid chromatography and its application in elderly patients. *J. Chromatogr. B* 969: 181–189.

Li, W., Zhang, J., and Tse, F.L.S. (2011c). Strategies in quantitative LC-MS/MS analysis of unstable small molecules in biological matrices. *Biomed. Chromatogr.* 25 (1): 258–277.

Liu, G., Ji, Q.C., Jemal, M. et al. (2011). Approach to evaluating dried blood spot sample stability during drying process and discovery of a treated card to maintain analyte stability by rapid on-card pH modification. *Anal. Chem.* 83 (23): 9033–9038.

Liu, A., Kozhich, A., Passmore, D. et al. (2015a). Quantitative bioanalysis of antibody-conjugated payload in monkey plasma using a hybrid immuno-capture LC-MS/MS approach: assay development, validation, and a case study. *J. Chromatogr. B* 1002: 54–62.

Liu, A., Tweed, J., and Wujcik, C.E. (2009). Investigation of an on-line two-dimensional chromatographic approach for peptide analysis in plasma by LC-MS-MS. *J. Chromatogr. B* 877 (20): 1873–1881.

Liu, Z., Welin, M., Bragee, B., and Nyberg, F. (2000). A high-recovery extraction procedure for quantitative analysis of substance P and opioid peptides in human cerebrospinal fluid. *Peptides* 21 (6): 853–860.

Liu, L., Zhong, T., Xu, Q., and Chen, Y. (2015b). Efficient molecular imprinting strategy for quantitative targeted proteomics of human transferrin receptor in depleted human serum. *Anal. Chem.* 87 (21): 10910–10919.

Madsen, J.A. and Brodbelt, J.S. (2009). Simplifying fragmentation patterns of multiply charged peptides by N-terminal derivatization and electron transfer collision activated dissociation. *Anal. Chem.* 81 (9): 3645–3653.

Maes, K., Smolders, I., Michotte, Y., and Van Eeckhaut, A. (2014). Strategies to reduce aspecific adsorption of peptides and proteins in liquid chromatography–mass spectrometry based bioanalyses: an overview. *J. Chromatogr. A* 1358: 1–13.

Meng, M., Wang, L., Liu, S. et al. (2016). Simultaneous quantitation of polymyxin B1, polymyxin B2 and polymyxin B1-1 in human plasma and treated human urine using solid phase extraction and liquid chromatography–tandem mass spectrometry. *J. Chromatogr. B* 1012–1013: 23–36.

Mirzaei, H. and Regnier, F. (2006). Enhancing electrospray ionization efficiency of peptides by derivatization. *Anal. Chem.* 78 (12): 4175–4183.

Miyachi, A., Kobayashi, M., Mieno, E. et al. (2017). Accurate analytical method for human plasma glucagon levels using liquid chromatography-high resolution mass spectrometry: comparison with commercially available immunoassays. *Anal. Bioanal. Chem.* 409 (25): 5911–5918.

Moser, A.C. and Hage, D.S. (2010). Immunoaffinity chromatography: an introduction to applications and recent developments. *Bioanalysis* 2 (4): 769–790.

Nemirovskiy, O.V., Dufield, D.R., Sunyer, T. et al. (2007). Discovery and development of a type II collagen neoepitope (TIINE) biomarker for matrix metalloproteinase activity: from in vitro to in vivo. *Anal. Biochem.* 361 (1): 93–101.

Neubert, H., Gale, J., and Muirhead, D. (2010). Online high-flow peptide immunoaffinity enrichment and nanoflow LC-MS/MS: assay development for total salivary pepsin/pepsinogen. *Clin. Chem.* 56 (9): 1413–1423.

Niwa, M. (2012). Chemical derivatization as a tool for optimizing MS response in sensitive LC-MS/MS bioanalysis and its role in pharmacokinetic studies. *Bioanalysis* 4 (2): 213–220.

Niwa, M., Enomoto, K., and Yamashita, K. (1999). Measurement of the novel decapeptide cetrorelix in human plasma and urine by liquid chromatography-electrospray ionization mass spectrometry. *J. Chromatogr. B Biomed. Sci. Appl.* 729 (1–2): 245–253.

Ochiai, H., Mori, H., Murata, H. et al. (2001). Validation of an analytical method for a potent antitumor agent, TZT-1027, in plasma using liquid chromatography-mass spectrometry. *J. Chromatogr. B Biomed. Sci. Appl.* 762 (2): 155–163.

Oe, T., Ackermann, B.L., Inoue, K. et al. (2006). Quantitative analysis of amyloid β peptides in cerebrospinal fluid of Alzheimer's disease patients by immunoaffinity purification and stable isotope dilution liquid chromatography/negative electrospray ionization tandem mass spectrometry. *Rapid Commun. Mass Spectrom.* 20 (24): 3723–3735.

Pezeshki, A., Vergote, V., Van Dorpe, S. et al. (2009). Adsorption of peptides at the sample drying step: influence of solvent evaporation technique, vial material and solution additive. *J. Pharm. Biomed. Anal.* 49 (3): 607–612.

Polson, C., Sarkar, P., Incledon, B. et al. (2003). Optimization of protein precipitation based upon effectiveness of protein removal and ionization effect in liquid chromatography–tandem mass spectrometry. *J. Chromatogr. B* 785 (2): 263–275.

Posewitz, M.C. and Tempst, P. (1999). Immobilized gallium(III) affinity chromatography of phosphopeptides. *Anal. Chem.* 71 (14): 2883–2892.

Qader, A.A., Urraca, J., Torsetnes, S.B. et al. (2014). Peptide imprinted receptors for the determination of the small cell lung cancer associated biomarker progastrin releasing peptide. *J. Chromatogr. A* 1370: 56–62.

Qiao, X., Qin, X., She, D. et al. (2014). Mass spectrometry-based tag and its application to high efficient peptide analysis – a review. *Talanta* 126: 91–102.

Reubsaet, J.L.E., Beijnen, J.H., Bult, A. et al. (1998). Analytical techniques used to study the degradation of proteins and peptides: chemical instability. *J. Pharm. Biomed. Anal.* 17 (6): 955–978.

Rogeberg, M., Malerod, H., Roberg-Larsen, H. et al. (2014). On-line solid phase extraction–liquid chromatography, with emphasis on modern bioanalysis and miniaturized systems. *J. Pharm. Biomed. Anal.* 87: 120–129.

Rossetti, C., Abdel Qader, A., Halvorsen, T.G. et al. (2014). Antibody-free biomarker determination: exploring molecularly imprinted polymers for pro-gastrin releasing peptide. *Anal. Chem.* 86 (24): 12291–12298.

Rossetti, C., Levernæs, M.C.S., Reubsaet, L., and Halvorsen, T.G. (2016). Evaluation of affinity-based serum clean-up in mass spectrometric analysis: plastic vs monoclonal antibodies. *J. Chromatogr. A* 1471: 19–26.

Ruprecht, B., Koch, H., Medard, G. et al. (2015). Comprehensive and reproducible phosphopeptide enrichment using iron immobilized metal ion affinity chromatography (Fe-IMAC) columns. *Mol. Cell. Proteomics* 14 (1): 205–215.

Rustum, A.M. (1990). Estimation of cyclosporin-A in whole blood by simple and rapid reversed-phase HPLC utilizing a salting-out extraction procedure. *J. Chromatogr. Sci.* 28 (11): 594–598.

Sandra, K., Moshir, M., D'Hondt, F. et al. (2008). Highly efficient peptide separations in proteomics. Part 1. Unidimensional high performance liquid chromatography. *J. Chromatogr. B Analyt. Technol. Biomed. Life Sci.* 866 (1–2): 48–63.

Schwartz, M., Kline, W., and Matuszewski, B. (1997). Determination of a cyclic hexapeptide (L-743 872), a novel pneumocandin antifungal agent in human plasma and urine by high-performance liquid chromatography with fluorescence detection. *Anal. Chim. Acta* 352 (1–3): 299–307.

Shen, Y., Zhang, G., Yang, J. et al. (2015). Online 2D-LC-MS/MS assay to quantify therapeutic protein in human serum in the presence of pre-existing antidrug antibodies. *Anal. Chem.* 87 (16): 8555–8563.

Shin, B.S., Kim, J., Yoon, C.H. et al. (2005). Development of a liquid chromatography/electrospray tandem mass spectrometry assay for the quantification of apicidin, a novel histone deacetylase inhibitor, in rat serum: application to a pharmacokinetic study. *Rapid Commun. Mass Spectrom.* 19 (3): 408–414.

Sobhi, H.R., Vatansever, B., Wortmann, A. et al. (2011). Generic approach for the sensitive absolute quantification of large undigested peptides in plasma using a particular liquid chromatography–mass spectrometry setup. *J. Chromatogr. A* 1218 (47): 8536–8543.

Stevenson, D., El-Sharif, H.F., and Reddy, S.M. (2016). Selective extraction of proteins and other macromolecules from biological samples using molecular imprinted polymers. *Bioanalysis* 8 (21): 2255–2263.

Stoyanov, A.V., Rohlfing, C.L., Connolly, S. et al. (2011). Use of cation exchange chromatography for human C-peptide isotope dilution – mass spectrometric assay. *J. Chromatogr. A* 1218 (51): 9244–9249.

Su, S., Yu, Y., Mo, W. et al. (2008). A development of LC-MS method combining ultrafiltration and lyophilization for determination of r-RGD-Hirudin in human serum. *J. Chromatogr. B* 870 (1): 27–31.

Suh, M.-J., Clark, D.J., Parmer, P.P. et al. (2010). Using chemical derivatization and mass spectrometric analysis to characterize the post-translationally modified *Staphylococcus aureus* surface protein G. *Biochim. Biophys. Acta* 1804 (6): 1394–1404.

Switzar, L., Giera, M., and Niessen, W.M.A. (2013). Protein digestion: an overview of the available techniques and recent developments. *J. Proteome Res.* 12 (3): 1067–1077.

Tang, Y.Q. and Weng, N. (2013). Salting-out assisted liquid–liquid extraction for bioanalysis. *Bioanalysis* 5 (12): 1583–1598.

Tape, C.J., Worboys, J.D., Sinclair, J. et al. (2014). Reproducible automated phosphopeptide enrichment using magnetic TiO_2 and Ti-IMAC. *Anal. Chem.* 86 (20): 10296–10302.

Taylor, S.W., Clarke, N.J., Chen, Z., and McPhaul, M.J. (2016). A high-throughput mass spectrometry assay to simultaneously measure intact insulin and C-peptide. *Clin. Chim. Acta* 455: 202–208.

Thomas, A., Kohler, M., Schanzer, W. et al. (2011). Determination of IGF-1 and IGF-2, their degradation products and synthetic analogues in urine by LC-MS/MS. *Analyst* 136 (5): 1003–1012.

van den Broek, I., Niessen, W.M.A., and van Dongen, W.D. (2013). Bioanalytical LC-MS/MS of protein-based biopharmaceuticals. *J. Chromatogr. B* 929: 161–179.

van den Broek, I., Sparidans, R.W., Huitema, A.D.R. et al. (2006). Development and validation of a quantitative assay for the measurement of two HIV-fusion inhibitors, enfuvirtide and tifuvirtide, and one metabolite of enfuvirtide (M-20) in human plasma by liquid chromatography–tandem mass spectrometry. *J. Chromatogr. B* 837 (1–2): 49–58.

van den Broek, I., Sparidans, R.W., Schellens, J.H.M., and Beijnen, J.H. (2007). Enzymatic digestion as a tool for the LC-MS/MS quantification of large peptides in biological matrices: measurement of chymotryptic fragments from the HIV-1 fusion inhibitor enfuvirtide and its metabolite M-20 in human plasma. *J. Chromatogr. B* 854 (1–2): 245–259.

van den Broek, I., Sparidans, R.W., Schellens, J.H.M., and Beijnen, J.H. (2008). Quantitative bioanalysis of peptides by liquid chromatography coupled to (tandem) mass spectrometry. *J. Chromatogr. B* 872 (1–2): 1–22.

van den Broek, I., Sparidans, R.W., Schellens, J.H.M., and Beijnen, J.H. (2010a). Quantitative assay for six potential breast cancer biomarker peptides in human serum by liquid chromatography coupled to tandem mass spectrometry. *J. Chromatogr. B* 878 (5): 590–602.

van den Broek, I., Sparidans, R.W., Schellens, J.H.M., and Beijnen, J.H. (2010b). Validation of a quantitative assay for human neutrophil peptide-1, -2, and -3 in human plasma and serum by liquid chromatography coupled to tandem mass spectrometry. *J. Chromatogr. B* 878 (15): 1085–1092.

van Midwoud, P.M., Rieux, L., Bischoff, R. et al. (2007). Improvement of recovery and repeatability in liquid chromatography–mass spectrometry analysis of peptides. *J. Proteome Res.* 6 (2): 781–791.

Vatansever, B., Lahrichi, S.L., Thiocone, A. et al. (2010). Comparison between a linear ion trap and a triple quadruple MS in the sensitive detection of large peptides at femtomole amounts on column. *J. Sep. Sci.* 33 (16): 2478–2488.

Wang, J., Cunningham, R., Zetterberg, H. et al. (2016). Label-free quantitative comparison of cerebrospinal fluid glycoproteins and endogenous peptides in subjects with Alzheimer's disease, mild cognitive impairment, and healthy individuals. *Proteomics Clin. Appl.* 10 (12): 1225–1241.

Wang, D., Fang, S., and Wohlhueter, R.M. (2009). N-terminal derivatization of peptides with isothiocyanate analogues promoting Edman-type cleavage and enhancing sensitivity in electrospray ionization tandem mass spectrometry analysis. *Anal. Chem.* 81 (5): 1893–1900.

Wang, Y., Qu, Y., Bellows, C.L. et al. (2012). Simultaneous quantification of davalintide, a novel amylin-mimetic peptide, and its active metabolite in beagle and rat plasma by online SPE and LC-MS/MS. *Bioanalysis* 4 (17): 2141–2152.

Wang, Q.-Q., Xiang, S.-S., Jia, Y.-B. et al. (2010). An improved on-line solid phase extraction coupled HPLC-MS/MS system for quantification of Sifuvirtide in human plasma. *J. Chromatogr. B* 878 (21): 1893–1898.

Whiteaker, J.R., Zhao, L., Anderson, L., and Paulovich, A.G. (2010). An automated and multiplexed method for high throughput peptide immunoaffinity enrichment and multiple reaction monitoring mass spectrometry-based quantification of protein biomarkers. *Mol. Cell. Proteomics* 9 (1): 184–196.

Whiteaker, J.R., Zhao, L., Zhang, H.Y. et al. (2007). Antibody-based enrichment of peptides on magnetic beads for mass-spectrometry-based quantification of serum biomarkers. *Anal. Biochem.* 362 (1): 44–54.

Wilffert, D., Reis, C.R., Hermans, J. et al. (2013). Antibody-free LC-MS/MS quantification of rhTRAIL in human and mouse serum. *Anal. Chem.* 85 (22): 10754–10760.

Wilson, S.F., Li, H., Rose, M.J. et al. (2010). Development and validation of a method for the determination of a therapeutic peptide with affinity for the human B1 receptor in human plasma using HPLC-MS/MS. *J. Chromatogr. B* 878 (9): 749–757.

Wolf, R., Hoffmann, T., Rosche, F., and Demuth, H.-U. (2004). Simultaneous determination of incretin hormones and their truncated forms from human plasma by immunoprecipitation and liquid chromatography–mass spectrometry. *J. Chromatogr. B* 803 (1): 91–99.

Wolf, R., Rosche, F., Hoffmann, T., and Demuth, H.-U. (2001). Immunoprecipitation and liquid chromatographic–mass spectrometric determination of the peptide glucose-dependent insulinotropic polypeptides GIP1-42 and GIP3-42 from human plasma samples: new sensitive method to analyze physiological concentrations of peptide hormones. *J. Chromatogr. A* 926 (1): 21–27.

Wu, S.T., Ouyang, Z., Olah, T.V., and Jemal, M. (2011). A strategy for liquid chromatography/tandem mass spectrometry based quantitation of pegylated protein drugs in plasma using plasma protein precipitation with water-miscible organic solvents and subsequent trypsin digestion to generate surrogate peptides for detection. *Rapid Commun. Mass Spectrom.* 25 (2): 281–290.

Xu, Y., Mehl, J.T., Bakhtiar, R., and Woolf, E.J. (2010). Immunoaffinity purification using anti-PEG antibody followed by two-dimensional liquid chromatography/tandem mass spectrometry for the quantification of a PEGylated therapeutic peptide in human plasma. *Anal. Chem.* 82 (16): 6877–6886.

Xu, Y., Sun, L., Anderson, M. et al. (2017). Insulin glargine and its two active metabolites: a sensitive (16pM) and robust simultaneous hybrid assay coupling immunoaffinity purification with LC-MS/MS to support biosimilar clinical studies. *J. Chromatogr. B* 1063: 50–59.

Xu, Y., Zhang, L., Lu, H., and Yang, P. (2008). Mass spectrometry analysis of phosphopeptides after peptide carboxy group derivatization. *Anal. Chem.* 80 (21): 8324–8328.

Xu, F., Zou, L., Liu, Y. et al. (2011). Enhancement of the capabilities of liquid chromatography–mass spectrometry with derivatization: general principles and applications. *Mass Spectrom. Rev.* 30 (6): 1143–1172.

Yang, F., Deng, D., Dong, X., and Lin, S. (2017). Preparation of an epitope-imprinted polymer with antibody-like selectivity for beta2-microglobulin and application in serum sample analysis with a facile method of on-line solid-phase extraction coupling with high performance liquid chromatography. *J. Chromatogr. A* 1494: 18–26.

Yang, Z., Hayes, M., Fang, X. et al. (2007). LC-MS/MS approach for quantification of therapeutic proteins in plasma using a protein internal standard and 2D-solid-phase extraction cleanup. *Anal. Chem.* 79 (24): 9294–9301.

Yang, K., Li, S., Liu, J. et al. (2016). Multiepitope templates imprinted particles for the simultaneous capture of various target proteins. *Anal. Chem.* 88 (11): 5621–5625.

Yuan, L., Aubry, A.-F., Arnold, M.E., and Ji, Q.C. (2013). Systematic investigation of orthogonal SPE sample preparation for the LC-MS/MS bioanalysis of a monoclonal antibody after pellet digestion. *Bioanalysis* 5 (19): 2379–2391.

Yuan, L., Aubry, A.-F., and Ji, Q.C. (2016). A simple, effective approach for rapid development of high-throughput and reliable LC–MS/MS bioanalytical assays. *Bioanalysis* 8 (17): 1809–1822.

Yuan, L. and Ji, Q.C. (2012). Automation in new frontiers of bioanalysis: a key for quality and efficiency. *Bioanalysis* 4 (23): 2759–2762.

Yuan, L., Jian, W., Zhang, D. et al. (2014a). Application of a stabilizer cocktail of *N*-ethylmaleimide and phenylmethanesulfonyl fluoride to concurrently stabilize the disulfide and ester containing compounds in a plasma LC-MS/MS assay. *J. Pharm. Biomed. Anal.* 88: 552–561.

Yuan, L., Mai, A., Aubry, A.-F. et al. (2014b). Feasibility assessment of a novel selective peptide derivatization strategy for sensitivity enhancement for the liquid chromatography/tandem mass spectrometry bioanalysis of protein therapeutics in serum. *Rapid Commun. Mass Spectrom.* 28 (7): 705–712.

Zhang, T., Watson, D.G., Azike, C. et al. (2007). Determination of vancomycin in serum by liquid chromatography–high resolution full scan mass spectrometry. *J. Chromatogr. B* 857 (2): 352–356.

Zheng, H., Hu, P., Quinn, D.F., and Wang, Y.K. (2005). Phosphotyrosine proteomic study of interferon α signaling pathway using a combination of immunoprecipitation and immobilized metal affinity chromatography. *Mol. Cell. Proteomics* 4 (6): 721–730.

Zheng, N., Jiang, H., and Zeng, J. (2014). Current advances and strategies towards fully automated sample preparation for regulated LC–MS/MS bioanalysis. *Bioanalysis* 6 (18): 2441–2459.

Zheng, N., Zeng, J., Manney, A. et al. (2016). Quantitation of a PEGylated protein in monkey serum by UHPLC-HRMS using a surrogate disulfide-containing peptide: a new approach to bioanalysis and in vivo stability evaluation of disulfide-rich protein therapeutics. *Anal. Chim. Acta* 916: 42–51.

Zhou, Y., Shan, Y., Zhang, L., and Zhang, Y. (2014). Recent advances in stable isotope labeling based techniques for proteome relative quantification. *J. Chromatogr. A* 1365: 1–11.

Zhu, Y., Deng, P., and Zhong, D. (2015). Derivatization methods for LC-MS analysis of endogenous compounds. *Bioanalysis* 7 (19): 2557–2581.

24

Sample Preparation for LC-MS Bioanalysis of Proteins

Nico van de Merbel

24.1 Introduction

The rapid growth in the development of macromolecular drugs by the pharmaceutical industry and the ever-increasing interest in protein biomarkers in the medical and pharmaceutical sciences have resulted in a growing need for analytical methods for the accurate quantitation of proteins in biological samples. Traditionally, proteins are quantified by ligand-binding assays (LBAs), but LC-MS is increasingly being used as an alternative platform. This technique has some well-demonstrated advantages, such as short method development time, high assay specificity, wide assay dynamic range and multiplexing capability, and it can also be used to obtain information that is complementary to that obtained by LBA (Li et al. 2011; van den Broek et al. 2013; Yang et al. 2013). Compared to the quantitation of small molecules, LC-MS bioanalysis of proteins typically is more challenging, from both practical and theoretical point of view. Proteins have a large and complicated three-dimensional structure and often are not single, well-defined molecular species but rather heterogeneous mixtures of structurally related isoforms. This may give rise to a complex and sometimes unpredictable behavior in sample processing and analysis. In addition, it is usually only a part of the protein molecule that responds to an analytical technique and, therefore, different concentration results may be obtained for a given sample via different techniques, depending on their analytical principle and the part of the protein to which they are directed (Bults et al. 2015).

Sample preparation is a crucial aspect of the quantitative bioanalysis of proteins by LC-MS. Protein analytes typically need to be quantified among a multitude of matrix proteins that often have similar physicochemical properties and analytical behaviors and occur at much higher concentrations and that may thus strongly interfere with the analysis. From an analytical perspective, it is therefore essential to reduce the complexity of the biological sample extract prior to LC-MS analysis and there are many different approaches for sample preparation that can be taken (Bischoff et al. 2013; Van de Merbel 2015; Wilffert et al. 2015). From a more conceptual point of view, the choice for a particular sample preparation approach will determine which part of the protein is the basis for quantitation and, therefore, what the concentration result actually represents. It is the aim of this chapter to review and discuss current techniques and potential future trends in sample preparation for quantitative LC-MS bioanalysis of proteins from both practical and theoretical perspectives.

Since proteins are chains of amino acids, just like peptides that are discussed in Chapter 23, it is important to first distinguish the two classes of molecules. There is no clear definition of what a protein exactly is, but very generally it is larger and structurally more complex than a peptide. According to IUPAC (1997) a protein is a naturally occurring or synthetic polypeptide that has a molecular weight greater than about 10 000 Da, but this limit is rather arbitrary. A more practical distinction is related to the way of production. Peptides can be chemically generated using solid-phase synthesis, with chains of over 100 amino acids currently being possible, which may correspond to molecular masses exceeding 10 000 Da. Proteins, on the other hand, need biological systems (cell cultures) for their production because of their structural complexity, even if their mass is below 10 000 Da. Altogether, the molecular mass range between ~5000 and 15 000 Da is a gray zone and compounds within this range may either be regarded as large peptides or as small proteins. In this chapter, examples will mainly be focused on proteins with molecular masses above 10 000.

Sample Preparation in LC-MS Bioanalysis, First Edition. Edited by Wenkui Li, Wenying Jian, and Yunlin Fu.

24.2 Intact Versus Digested Protein Analysis

Just like for small molecules, liquid chromatography coupled to tandem mass spectrometry (LC-MS/MS) in the single reaction monitoring (SRM) mode, often also called the multiple reaction monitoring (MRM) mode, is used for the vast majority of LC-MS methods for protein quantitation, mainly because of its sensitivity and its widespread availability in the pharmaceutical and related industries. A limitation for protein quantitation is that LC-MS/MS is not directly compatible with molecules with a mass above around 5000 Da, because the ions are distributed over many different charge states and also often do not fragment efficiently, which greatly reduces sensitivity. Therefore, protein analytes are typically digested into a mixture of smaller peptides and one or more of these, the signature or surrogate peptides, are then quantified as a measure for the intact protein. This means that the digestion of a protein (see Section 24.3) is at the very heart of most sample preparation approaches for quantitative LC-MS bioanalysis of proteins. Depending on the required sensitivity and selectivity, digestion may be complemented by extraction steps at the level of the intact protein (Sections 24.4 and 24.5), at the level of the surrogate peptide (Section 24.6,) or both (Section 24.7). Figure 24.1 shows a schematic representation of these options.

Quantitation of intact proteins by LC-MS/MS is typically restricted to relatively small molecules with masses below 10 000 Da (Van den Broek and van Dongen 2015). Often, these small proteins hardly fragment at low collision energies, while at higher collision energies, many fragments are generated of low abundance and in both cases sensitivity is limited. As an alternative, the same *m/z* value can be selected for the precursor and product ions, an approach known as pseudo-SRM. Although less attractive in terms of selectivity and sensitivity than normal SRM, it can be successfully applied down to low ng ml^{-1} levels, as was shown for the 7.5-kDa biomarker IGF-1 in plasma (Bredehoft et al. 2008). With regard to sample preparation, small intact proteins typically behave like peptides from digested large proteins and most peptide extraction techniques that are discussed in this chapter can be used to extract these analytes from a biological matrix.

Recently, reports have started to appear on the use of LC coupled to high-resolution mass spectrometry (HRMS) for the quantitative determination of larger intact proteins, as an alternative to LC-MS/MS after digestion. An advantage of this approach is the absence of a digestion step, which is considered as one of the most time-consuming parts of the bioanalytical workflow and needs to be carefully optimized to ensure completeness and reproducibility. In addition, a surrogate peptide represents only a small part of the protein, sometimes as little as a few percent and quantitation of a surrogate peptide inherently means a loss of information that might be essential to understand the *in vivo* fate of a protein.

The selectivity of quantitation by HRMS is based on its high resolving power, which allows for the detection of ions in a very small mass window with a high mass accuracy and thus separates the analyte ions from matrix interferences to a much larger extent than conventional MS/MS. In its simplest form, for all data points across a chromatogram a full-scan spectrum is acquired, which contains multiple ions corresponding to the different charge states of a protein and the responses of one or more of these ions are then combined into an extracted ion chromatogram. This approach works best for proteins with a molecular mass up to around 15 000 Da, such as lysozyme (14.7 kDa) that has been quantified in human plasma down to 0.5 μg ml^{-1} by summing the 10 most intense ions (Ruan et al. 2011). For larger molecules, the

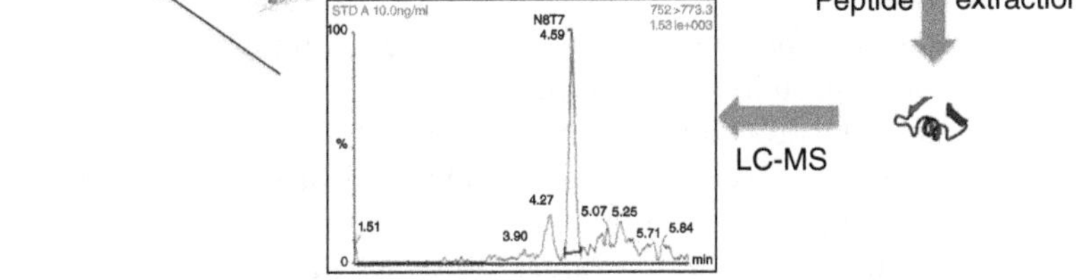

Figure 24.1 Schematic representation of the different components of sample preparation for quantitative LC-MS bioanalysis of proteins.

mass spectra become increasingly complex because of the large number of charge states and the occurrence of isotope distribution patterns across all these charge states. The MS signal is therefore divided over more and more ions and gets too much diluted to be of practical value. This can be potentially overcome by using dedicated software to deconvolute a complete HR mass spectrum and convert all multiply charged ions into the corresponding and much simplified neutral spectrum, which is subsequently used as the basis for quantitation. This approach no longer uses the actual raw data for quantitation and, therefore, the deconvolution process needs careful attention, but it has been shown that proteins as large as antibody fragments (Liu et al. 2011) and even intact monoclonal antibodies (mAbs, Jian et al. 2016) can be reliably quantified in biological samples at the low $\mu g\,ml^{-1}$ level. HRMS of large intact proteins requires a much more thorough sample cleanup than LC-MS/MS after digestion, mainly because the high levels of large endogenous proteins will also produce highly complex mass spectra that will easily interfere with the analyte ions used for quantitation. For that reason, a very selective extraction of the protein analyte, typically using immunoaffinity materials, is essential.

24.3 Enzymatic Digestion

Cleavage of a protein into peptides usually involves enzymatic treatment. The most widely used enzyme for protein digestion is trypsin, which cleaves the peptide chains after a lysine or arginine moiety, except when there is a C-terminal side proline. It is well suited for quantitative purposes because it generates peptides of a size (500–2000 Da) that is very compatible with MS/MS detection, while it is also readily available at a reasonable price. Trypsin digestion is widely used in the field of proteomics and much knowledge has been built up regarding its use for protein analysis by LC-MS. The speed of trypsin digestion depends on a number of experimental parameters (Hustoft et al. 2012). The optimal pH range is 7.5–8.5, but under these slightly alkaline conditions *in vitro* degradation (deamidation) of peptides containing asparagine or glutamine may occur. Lowering the pH to 7.0 will help reduce this effect, albeit at the expense of a slower digestion (Bults et al. 2016). A digestion temperature of 37 °C is commonly used, as trypsin has its highest activity at this temperature. In the proteomic field, digestion is often performed overnight to ensure complete cleavage of all proteins in a sample, but the process can be accelerated by procedures such as heating to 50–60 °C, the use of microwaves, or immobilizing trypsin on a solid support (Capelo et al. 2009). For quantitation of one or a limited number of target proteins, the time required for complete digestion is experimentally determined and often in the range of two to three hours, even at standard conditions as shown in some examples (Bronsema et al. 2015a; Bults et al. 2016). The amount of trypsin that is added to a biological sample has to be sufficient to ensure rapid digestion, but low enough to prevent nonspecific digestion or excessive formation of trypsin auto-digestion products. A ratio of trypsin to protein of 1 : 20 can be used as a rule of thumb (Hustoft et al. 2011), but this may need to be optimized based on the specifics of a method, because the biological sample may contain an excessive amount of endogenous proteins, which compete with the target protein for trypsin digestion. Finally, adding an organic solvent or a detergent during digestion is often found to improve digestion efficiency, probably because it induces or maintains protein denaturation and thus improves the accessibility of the cleavage sites, while it also helps solubilize nonpolar peptides after their formation (Hervey et al. 2007).

For optimal digestion, it is helpful to unfold the protein to optimally expose the cleavage sites to trypsin, especially when the surrogate peptide is located at the inside of the protein. This is usually performed by denaturation of the protein prior to digestion, for example, by the addition of urea, followed by reduction of intramolecular disulfide bridges and subsequent alkylation of the resulting free thiol groups (López-Ferrer et al. 2006; Proc et al. 2010). This approach, however, does add to the complexity and length of the sample preparation step and may not be strictly necessary. If the surrogate peptide does not contain a disulfide-forming cysteine, reduction and alkylation might be omitted and denaturation can be simply achieved by adding an organic solvent to the sample and digesting the resulting protein pellet (Yuan et al. 2012). Next to trypsin, a number of other proteases can be used, but these usually are more expensive and have a lower digestion activity. Lys-C and Glu-C cleave proteins at the C-terminal side of lysine and glutamic acid, respectively. Likewise, Asp-N causes cleavage N-terminally to aspartic acid. These enzymes typically give rise to larger peptides than trypsin does, which may or may not be an advantage for use as surrogate peptides. When necessary, several enzymes can be combined to further cleave the peptides into smaller sizes (Swaney et al. 2010).

A disadvantage of digestion is that it greatly increases the complexity of a biological sample. Plasma, for example, contains around $80\,mg\,ml^{-1}$ of proteins, which are all cleaved into peptides of similar size with more or less comparable physicochemical and analytical properties, because they all consist of combinations of the same 20 amino acids. Without additional cleanup, chromatograms of digested biological samples often contain many background peaks, originating from endogenous peptides that show a response at the same mass transition as

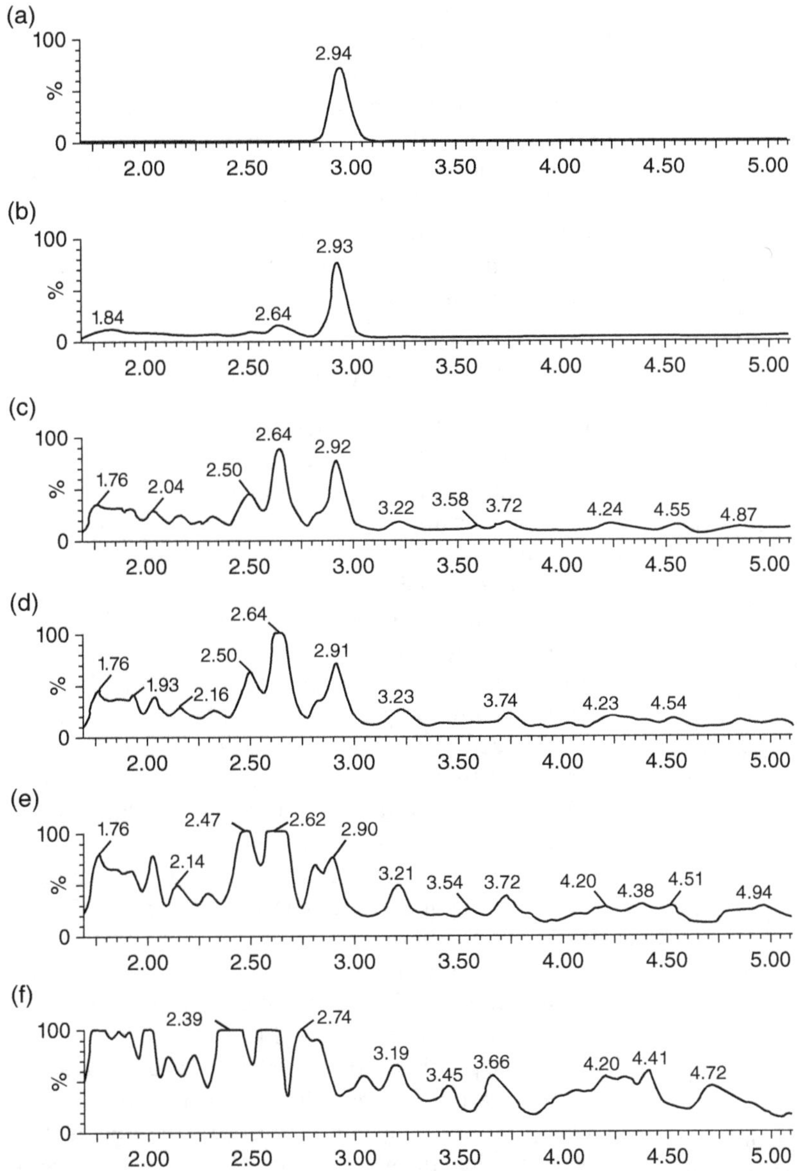

Figure 24.2 LC-MS/MS chromatograms for the surrogate peptide of 2.0 ng ml^{-1} salmon calcitonin (retention time ~2.9 minutes). Increasing amounts (volume percentage) of trypsin-digested plasma were added to a digest of an aqueous solution of salmon calcitonin: 0% (a), 1% (b), 5% (c), 10% (d), 20% (e), and 50% (f). *Source:* Reproduced with permission from Bronsema et al. (2013). Copyright (2013) American Chemical Society.

the surrogate peptide. This effect is illustrated in Figure 24.2 for digested salmon calcitonin in the presence of increasing amounts of digested plasma (Bronsema et al. 2013). Method selectivity and sensitivity are considerably affected by the presence of the endogenous peptides: the achievable lower limit of quantitation (LLOQ) increases 100-fold, from 0.2 ng ml^{-1} (60 pM) in the absence of matrix peptides to 20 ng ml^{-1} (6 nM) in the presence of 50% of digested plasma. The use of digestion without additional sample cleanup can therefore only be used for applications that do not need very high sensitivity. The typical LLOQ for a protein in plasma or serum, when only treated by digestion, is in the high ng ml^{-1} to low μg ml^{-1} range. This may be sufficient for biopharmaceuticals, which are given at a high dose, such as many mAbs. For example, Ouyang et al. (2012) quantified an unspecified mAb in monkey plasma down to 2.6 μg ml^{-1} by precipitating the proteins from a 25-μl sample with methanol, isolating the protein pellet by centrifugation and digesting the pellet using modified trypsin (1.3 mg ml^{-1}) in the presence of 10% methanol at 60 °C for one hour. Similarly, the enzyme α-glucosidase was determined in human plasma with an LLOQ of 0.5 μg ml^{-1} by adding 400 μl of methanol to 20 μl of sample and digesting the resulting protein pellet at pH 8.2 with 0.2 mg ml^{-1} of trypsin in the presence of 0.05% Tween-20 at 37 °C for three hours (Bronsema et al. 2015b).

24.4 Protein Depletion

In plasma and serum, albumin is present at ~45 mg ml^{-1} and it accounts for more than 50% of the total protein concentration. Therefore, its removal is advantageous for sensitive protein quantification. A number of albumin-depletion kits are commercially available, which contain affinity resins based on dyes or anti-albumin antibodies. Typically, a biological sample is loaded onto the resin, where albumin molecules are trapped by the affinity ligands and thereby removed from the sample. Upon digestion, the sample with albumin depleted is expected to be less complex and lower analyte concentrations can usually be determined. Hagman et al. (2008) compared several kits for albumin depletion from serum, but a considerable amount of albumin was generally still present after treatment. Using the most efficient kit, 49% of the total plasma protein content was removed, which allowed for quantification of a therapeutic mAb down to 2 µg ml^{-1} in human and monkey serum. For the removal of multiple endogenous proteins, several affinity materials can be combined. To quantify the rheumatoid-arthritis marker C-reactive protein, three affinity columns were coupled in series to remove albumin, immunoglobulin G (IgG), and haptoglobin from diluted human serum and close to 90% of endogenous proteins could be depleted (Kuhn et al. 2004). Analyte recovery was just 61%, which indicates that a substantial amount of analyte was lost during processing, probably as a result of co-depletion. In addition to the possibly considerable loss of protein analytes, other disadvantages of depletion kits include relatively high cost and often limited sample throughput. For this reason, it has not become very popular for routine analysis.

24.5 Protein Extraction (Before Digestion)

The targeted extraction of a protein analyte from the matrix typically helps increase the sensitivity and selectivity of an analytical method. Still, there are some issues that need to be taken into account. When present in a biological sample, most proteins are bound to other macromolecules to some extent, such as specific carrier proteins or, for biopharmaceuticals, their pharmacological target receptors and/or anti-drug antibodies(ADAs). It very much depends on the applied extraction technique whether only the unbound fraction of the protein analyte, or the total protein analyte (bound and unbound) is extracted from the sample, or a concentration that is in between the two. However, it is often difficult to demonstrate or confirm which fraction of the analyte is actually extracted. Another practical challenge is related to the internal standard. To adequately normalize variability in the protein extraction step, a protein internal standard is required, but this is usually difficult or very costly to obtain (Bronsema et al. 2012).

24.5.1 Differential Protein Precipitation

In some cases, protein analytes can be separated from the majority of endogenous proteins by organic solvent-based precipitation. Upon addition of an organic solvent to a biological sample most of the matrix proteins precipitate and can be removed by centrifugation while small or very hydrophobic protein analytes are recovered in the supernatant. One example is the 8.6-kDa protein interferon-gamma-inducible protein-10, which was quantitatively extracted from serum by the addition of acetonitrile containing 2% formic acid to serum in a 2 : 1 (v/v) ratio (Zhang et al. 2014). PEGylated proteins usually have a better solubility in organic solvents than their non-PEGylated counterparts. PEGylated adnectin, for example, was recovered up to 60% in the supernatant after addition of 0.1% formic acid in isopropanol to plasma, allowing quantitation down to 10 ng ml^{-1} (Dawes et al. 2013). Similarly, a PEGylated protein could be extracted from plasma and several tissue homogenates by protein precipitation, with *n*-butanol giving the best recovery (>80%) (Gong et al. 2014).

Differential precipitation may also be induced by selective denaturation of proteins by pH manipulation, because proteins are generally best precipitated at their isoelectric point (pI) where their solubility is the lowest. The small endogenous protein α-cobratoxin was precipitated from horse plasma by adding ammonium sulfate at pH 8.6, with 78% of matrix proteins remaining in the supernatant (Bailly-Chouriberry et al. 2013). This principle can also be applied the other way around, i.e. by precipitation of matrix proteins and retaining the protein analyte in the supernatant. The immunosuppressive protein drug alefacept was extracted from plasma in this way: by adjusting the pH to 5.1 and increasing the temperature to 45 °C, albumin (pI: 5.3) was removed by precipitation while the analyte (pI: 7.2) was recovered for 25% (Halquist and Karnes 2011).

An elegant way to selectively remove the most abundant plasma protein, albumin, is the addition of a 10-fold volume excess of 2-propanol containing 1% trichloroacetic acid (TCA) to plasma and discarding the resulting supernatant (Liu et al. 2014; see Figure 24.3). Because of

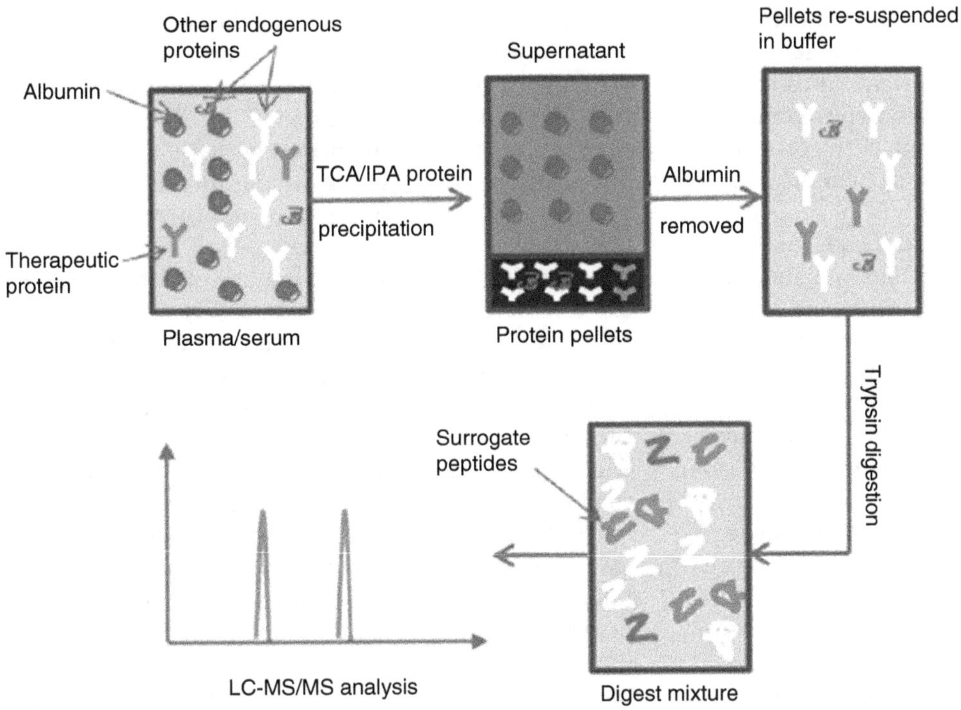

Figure 24.3 Schematic representation of the removal of albumin from plasma by trichloroacetic acid (TCA) in 2-propanol (IPA). *Source:* Reproduced with permission from Liu et al. (2014). Copyright (2014) American Chemical Society.

the good solubility of the albumin–TCA complex in the supernatant, about 50% of the plasma protein content could be removed from the sample, while three protein analytes had a recovery of close to 100% in the protein pellet. This compared favorably to a commercial albumin removal kit, which caused the loss of 50% of the analytes. For one of the proteins, an 80-kDa fusion protein, sensitivity was increased 10-fold (final LLOQ of 50 ng ml^{-1}), compared to a regular pellet digestion after protein precipitation with 75% of methanol.

Altogether, differential protein precipitation is a simple and cheap approach, but it usually does not lead to a high degree of selectivity. It is mainly attractive for proteins with molecular properties that are clearly different from those of the bulk of the background plasma proteins, such as proteins that are small or PEGylated.

24.5.2 Immunocapture

The most rigorous cleanup of a biological sample can probably be achieved by applying immunocapture (IC) techniques (Fung et al. 2014). Typically, magnetic beads or other resins are coated with a protein that has a high binding affinity toward the analyte, such as an antibody against the analyte or the pharmacological target to which a biopharmaceutical is directed. Upon mixing of the biological sample with a suspension of the beads or passing it through a cartridge filled with the resin, the protein analyte selectively binds to the IC material and the other matrix components are subsequently removed by washing steps.

A common approach is to attach the small-molecule biotin to an analyte-directed antibody or pharmacological target and add this biotinylated protein to magnetic beads that are coated with streptavidin. Because of the very high-affinity binding of biotin and streptavidin, incubation of streptavidin-coated beads with the biotinylated protein will result in an IC reagent that is strongly bound to the beads. Kaur et al. (2016) employed this approach for the extraction of a therapeutic mAb from monkey serum. Streptavidin-coated magnetic beads were conjugated with a biotinylated antibody directed to human IgG1 and an aliquot of 10 µl of serum was subsequently mixed with 25 µl of a suspension of the beads. After overnight incubation, the beads were isolated by applying a magnet, the supernatant was discarded and the beads were subsequently washed with a buffered saline solution. The extracted protein analyte was digested on-bead and a signature peptide was then quantified by LC-MS/MS. An IgG1-type mAb, containing ^{13}C-^{15}N-labeled arginine and lysine, was used as the internal standard. The average extraction recovery was 92% for the analyte and 99% for the internal standard and the LLOQ was 0.1 µg ml^{-1}. Alternatively, analyte and biotinylated IC reagent can be preincubated to form a complex before addition to streptavidin-coated beads. To give an example, a proprietary protein drug was determined in human plasma down to 0.5 μg ml^{-1} by incubating 20 μl of sample with 50 μl of a buffer containing a biotinylated antibody against the drug for two hours and then mixing it with 50 μl of a suspension of streptavidin-coated magnetic beads for another hour (Duggan et al. 2016). After four washing steps, the drug was eluted from the beads with acid, a stable-isotope labeled peptide was added as an internal standard, and the samples were reduced, alkylated, and digested overnight. An extraction recovery of 85–95% was obtained. A third approach is to incubate all the components (sample, IC reagent, and beads) together. Wright and Dufield (2014) quantified alpha-1 antitrypsin in serum using such a single-stage IC technique. The LLOQ of their method was 1 ng ml^{-1} from a 50 μl serum sample, which was a 500-fold improvement compared to direct digestion without further cleanup.

IC reagents can also be directly bound to a supporting material. Tosylated beads can bind proteins by a covalent reaction of the tosylate groups on the beads with a protein's amino groups. The mAb Erbitux, for example, was extracted from human serum by its pharmacological target soluble epidermal growth factor receptor (sEGFR) that had been attached to tosylated magnetic beads (Dubois et al. 2008). The reaction took four hours at room temperature and was followed by washing steps to remove non-covalently bound sEGFR and overnight blocking of non-reacted tosylate groups. The serum sample and a murine anti-sEGFR antibody internal standard were then added and incubated with the functionalized beads for two hours. After elution under acidic conditions, the extracted sample was reduced and digested at 40 °C for three hours. The entire procedure resulted in a 10-fold increase in sensitivity compared to digestion only and allowed quantitation of the analyte down to 20 ng ml^{-1}.

Antibody-based reagents can also be bound to a support by the action of protein A/G, which selectively binds to the constant region (Fc) of all subclasses of human IgG antibodies. Berna and Ackermann (2009) used a commercial protein A/G-coated 96-well enzyme-linked immunosorbent assay (ELISA) plate as the support and added an antibody directed against the biomarker N-terminal pro-B-natriuretic peptide (NTproBNP) to form a complex between protein A/G and the antibody (one-hour incubation). The antibody was cross-linked to protein A/G (one-hour incubation) and after blocking of unconjugated protein A/G, 100 μl of rat serum and the internal standard, a stable-isotope labeled form of the analyte, were added and the plate was incubated for two hours, washed three times, and eluted with acid. The eluate was dried, redissolved in digestion buffer containing trypsin and digested in a microwave at 55 °C for 30 minutes. The procedure allowed quantitation with an LLOQ of 100 pg ml^{-1}. They used a similar procedure for the determination of another biomarker, Myl3, down to 0.95 ng ml^{-1}.

A generic IC approach is the use of antibodies that are directed against the constant (Fc) part of human IgG and therefore can capture any mAb of the human IgG family. This is a straightforward way to quantify humanized mAbs in animal samples. One example of this approach is the extraction of three different mAbs and an Fc-fusion protein from mouse serum and tissue homogenates using a commercially available biotinylated anti-human IgG Fc-specific antibody that was attached to streptavidin-coated magnetic beads (Sleczka et al. 2014). Sample aliquots of 25 μl were mixed with the beads in a 96-well plate for up to three hours and after four washing steps, the captured protein analytes were denatured with acetonitrile and digested overnight at 37 °C with tryspin. Accurate quantitation was achieved down to 20 ng ml^{-1}. A limitation of this approach is that it cannot be efficiently applied to biological samples of human origin, as the IC reagent will also capture the entire human IgG fraction, which is present in plasma at concentrations as high as 15 mg ml^{-1}.

ADAs that are formed *in vivo* as a result of the exposure of the body to an exogenous protein drug can also be determined by IC and LC-MS/MS. Chen et al. (2016) added a biotinylated form of a proprietary drug (protein Z) to human plasma samples containing ADAs to form (biotinylated) drug–ADA complexes. Next, streptavidin-coated magnetic beads were added and the drug–ADA complexes were bound to the beads and subsequently isolated using a magnet. After washing, the ADAs were eluted from the beads, digested by trypsin, and assayed by LC-MS/MS. Specific signature peptides were quantified to distinguish several ADA subclasses: IgM, IgE, and the IgG isotypes 1–4.

Sample treatment by IC can result in methods with an impressive sensitivity. The endogenous 9.4-kDa protein parathyroid hormone (PTH), for example, was quantified in human serum down to 40 pg ml^{-1} by treating a 1-ml sample with commercially available polystyrene beads coated with murine anti-PTH antibodies (Kumar et al. 2010). A completely ^{15}N-labeled form of PTH was added to the sample as an internal standard at the very beginning of the sample-handling procedure. After an incubation time of four hours, the beads were washed twice with a saline solution and digested on-bead with trypsin for 30 minutes (Figure 24.4).

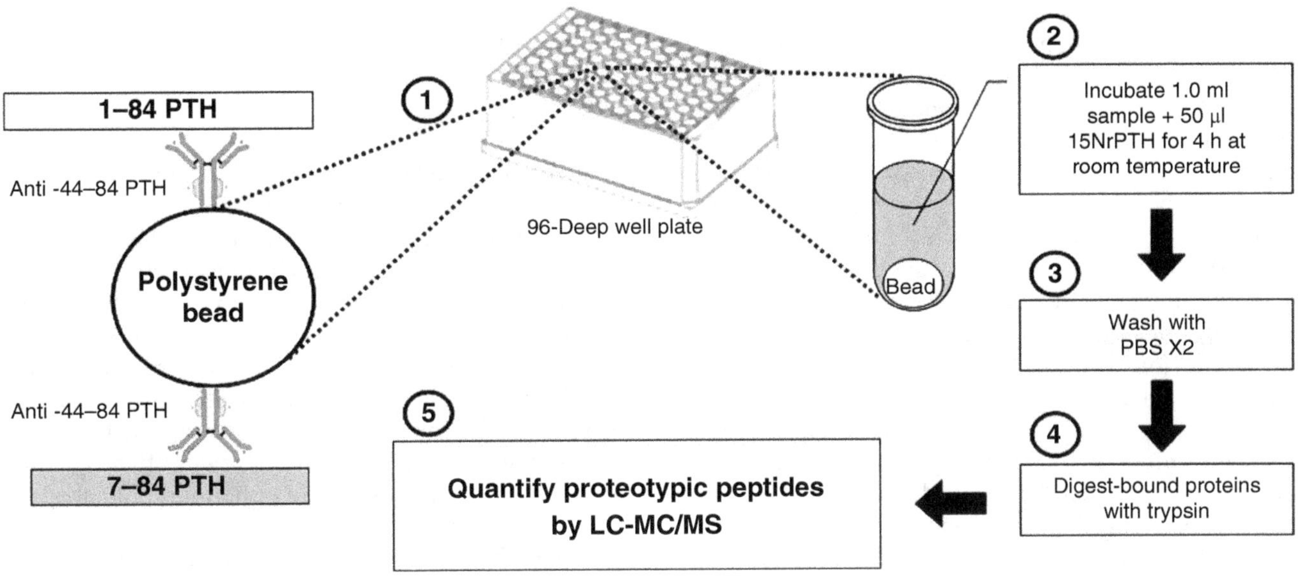

Figure 24.4 Schematic representation of the immunocapture of PTH from human serum. *Source:* From Kumar et al. (2010).

24.5.3 Antibody-free Extraction

Despite the widely reported success of sample cleanup by IC techniques, the approach is not generic because it heavily relies on the availability of specific anti-analyte-directed proteins which may not be readily available and can also suffer from poor lot-to-lot consistency. Other, so-called antibody-free extraction approaches can therefore be an attractive alternative (Wilffert et al. 2015). An interesting option for protein extraction is immobilized metal ion affinity chromatography (IMAC), which is based on the interaction of metal ions like Ni^{2+} with amino acids with strong electron donor groups such as histidine. Proteins containing these amino acids on their surface will be selectively captured by IMAC resins. For example, recombinant human tumor necrosis factor-related apoptosis-inducing ligand (rhTRAIL) was quantified in human and mouse serum down to 20 ng ml^{-1} by recovering more than 70% of the analyte with IMAC, while removing 95% of matrix proteins (Wilffert et al. 2013). Serum samples were loaded onto the IMAC resin in the presence of 15 mM of the competing ligand imidazole and the material was subsequently washed with a buffer containing 15 mM imidazole to reduce the binding of endogenous proteins. Elution of rhTRAIL from the resin was accomplished with 140 mM imidazole. Similarly, a modified form of the endogenous protein endostatin, extended with six histidines at its N-terminus, was extracted from rat plasma with a Ni^{2+}-loaded IMAC resin with an LLOQ of 50 ng ml^{-1} (Li et al. 2012).

Solid-phase extraction (SPE) with ion-exchange materials is another option for protein extraction. In this case, sample cleanup is based on differences in the pI of analyte and background proteins. Proteins with a relatively high pI are positively charged at neutral or slightly alkaline pH and can be trapped on a cation-exchange resin, while many endogenous proteins with a lower pI will be negatively charged and thus not be recovered in the extract. Strong cation-exchange (SCX) SPE was used to extract rhTRAIL (pI: 8.9) from human serum at a pH of 7.5 and with a sodium chloride concentration of 40 mM in both the loading and washing solvents (Wilffert et al. 2016a). Elution was performed with 100 mM sodium chloride and at a pH of 8, which resulted in a protein removal efficiency of 99% and an analyte recovery of 72%. An even higher cleanup efficiency (>99.9% protein removal) was obtained by sequential extraction with SCX SPE and IMAC. It allowed quantification of rhTRAIL down to 0.5 ng ml^{-1}, which shows that antibody-free extraction approaches can result in assay sensitivities comparable to those obtained using IC techniques. The orthogonality of SCX SPE and IMAC was demonstrated by identification of the endogenous plasma proteins removed by the cleanup: β2-glycoprotein (pI: 8.3), for example, was extracted by SCX but not IMAC, while apolipoprotein A-I (pI: 5.6) was retained by IMAC but not SCX. Because of the relatively large size of the analyte (58 kDa), special attention had to be paid to the extraction process: SPE materials with a large pore size (250 Å) were used to allow penetration of rhTRAIL into the pores. In addition, the sample was mixed with the SPE material for 10 minutes to make sure that the protein analyte had sufficient time for binding.

SPE with a reversed-phase material also is a possibility for sample preparation prior to digestion, as was shown for human growth hormone in human serum (Pritchard et al. 2014). Before applying the serum sample to the SPE cartridge, the protein analyte was denatured by the addition of 4 M urea at pH 10 to eliminate three-dimensional structural differences, which would have otherwise led to a difference in extraction behavior between the analyte and its stable-isotope labeled internal standard. The sample was then subjected to two subsequent SPE steps on a 300-Å C18 phase. The first was performed at pH 10 to break the strong interaction between the analyte and its endogenous binding protein GHBP and to ensure a high extraction recovery (79%), which was much lower (<20%) at neutral or acidic pH. With this approach, the total (free plus GHBP-bound) concentration of human growth hormone is thus obtained. The SPE eluent was then acidified with 5% formic acid and applied to a second C18 SPE column. This combination proved to be beneficial in terms of sensitivity and reproducibility, probably because of the removal of proteins that, if not removed, would have caused ion suppression after digestion. The optimized method allowed the quantification of physiologically relevant low-ng ml^{-1} concentrations of growth hormone in serum. Reversed-phase SPE is also potentially attractive for PEGylated proteins, which are relatively hydrophobic. To give an example: a monolithic C18 SPE material was used to extract PEGylated interferon α-2a from human serum down to 2 ng ml^{-1} with a recovery of around 50%, while removing more than 80% of endogenous plasma proteins (Yang et al. 2009).

24.6 Peptide Extraction (After Digestion)

The removal of interfering matrix components after the digestion step has some advantages. If a sample is first subjected to digestion, all proteins will be cleaved to peptides and any protein–protein interaction will therefore be broken. This means that the total concentration of a protein will, in theory, always be obtained. However, it may be more difficult to achieve sufficient selectivity with peptide extraction than with protein extraction, because the peptides in a plasma digest are much more similar to each other than the plasma proteins were before digestion. Still, the optimization of a peptide extraction procedure typically is more straightforward because of the wide availability of commercial materials that are commonly used for small-molecule extractions and the more predictable extraction behavior of smaller peptides compared to that of intact proteins. Finally, peptide extraction does not necessarily require a protein-based internal standard and has been shown to perform well when using a stable-isotope labeled form of the signature peptide (e.g. Bronsema et al. 2013), which is less expensive and easier to obtain.

24.6.1 Immunocapture

IC at the peptide level is comparable to that for intact proteins. In this approach, the immobilized antibodies are raised against the surrogate peptide instead of the intact protein (Becker and Hoofnagle 2012). One of the advantages of IC of peptides is its compatibility with more extreme experimental conditions, as there is no risk of denaturation of the analyte during the sample preparation step. This procedure is particularly beneficial for processing tissue sample for analysis (Neubert et al. 2016). IC of peptides after digestion is very widespread in the field of biomarker analysis. It can be performed using magnetic beads, which are coated with protein G to allow binding of the anti-peptide antibodies. Whiteaker et al. (2007) used this approach for the extraction of the surrogate peptides of α1-antichymotrypsin (AAC) and tumor necrosis factor alpha (TNFα) from a human serum digest. Protein G-coated magnetic beads were incubated for one hour with a solution containing the antibody and after a washing step the antibody was covalently attached to the beads by a cross-linking reaction. Tryptic digests of serum were incubated overnight with a suspension of the functionalized beads, which were subsequently washed and the captured surrogate peptides were eluted in a 5% acetic acid solution. A 1453- and 573-fold increase in the LC-MS/MS response was obtained for the surrogate peptide of AAC and TNFα, respectively, by this approach as compared to that with digestion only (Figure 24.5).

Automation is attractive for the improvement of efficiency and reproducibility of quantitative LC-MS bioanalysis of proteins via digestion followed by IC (Razavi et al. 2016). A typical workflow is a two-hour digestion step with trypsin, addition of stable-isotope labeled peptide internal standard(s), followed by a one-hour incubation with magnetic beads coated with the antibodies, several washing steps, and elution at low pH. By using this approach, five endogenous model proteins with widely varying plasma concentrations, from 10 ng ml^{-1} (mesothelin) to 50 mg ml^{-1} (albumin), were quantified in plasma by extraction with specific anti-peptide mAbs with coefficients of variation (%CV) below 5% in all cases. The high selectivity of the IC steps may even allow analysis without an LC separation step, such as direct quantitation by MALDI-TOF mass spectrometry. Human protein C inhibitor (PCI), a biomarker for prostate cancer, was quantified in human serum. During sample preparation, the surrogate peptide-directed antibodies were added to Protein G-coated magnetic beads.

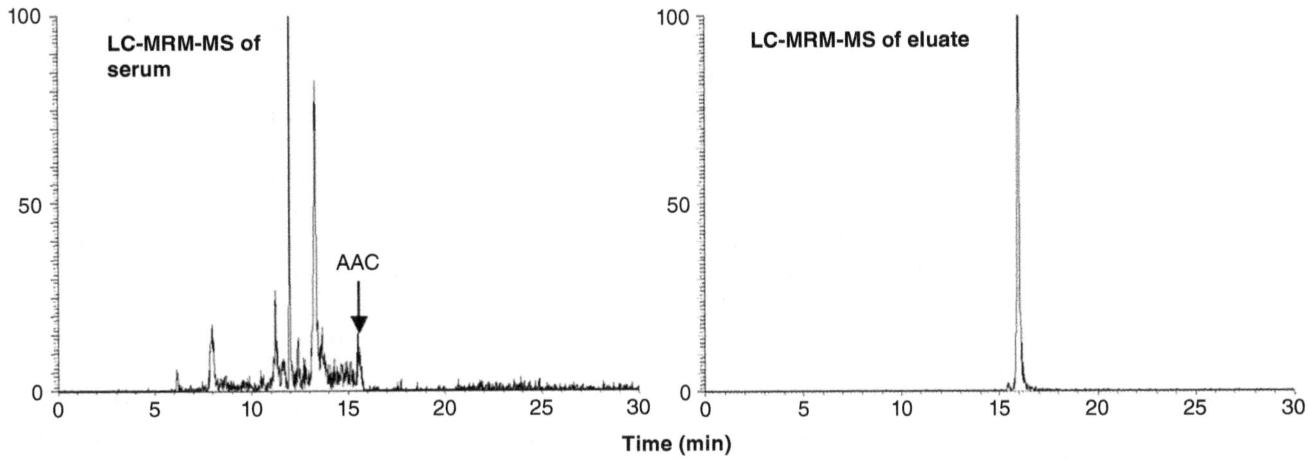

Figure 24.5 LC-MS/MS chromatograms for the signature peptide of α1-antichymotrypsin in a human serum digest without (left panel) or with (right panel) immunocapture of the peptide. *Source:* Reproduced with permission from Whiteaker et al. (2007). Copyright (2007) Elsevier.

This was followed by incubation of 10-μl serum digests with the functionalized beads. After two wash steps, the surrogate peptide was eluted using 13 μl of 0.1% formic, half of which was spotted onto a MALDI plate (Razavi et al. 2013). The assay was linear between 4 and 1000 fmol of the analyte (per 10 μl serum), with a CV typically <10%. The instrument cycle time was <10 seconds per sample.

IC materials can also be online coupled to LC-MS (Dufield and Radabaugh 2012). A number of applications based on this principle have been reported. For example, human neonatal Fc receptor (FcRn) was quantified in several tissues of transgenic mice by homogenization of the tissue samples, followed by pellet digestion with trypsin and online IC LC-MS (Fan and Neubert 2016). Five antibodies, each directed toward an FcRn tryptic peptide, had been covalently cross-linked to the protein-G coated resin packed in a cartridge. The digest was loaded at pH 7 and, after a washing step, eluted with 0.4% trifluoroacetic acid, after which the cartridge was re-equilibrated at pH 7 for the next analysis. The eluted peptides were trapped on a C18 cartridge and, next, chromatographically separated on a C18 analytical column. The total run time was 15 minutes and the assay LLOQ was 0.89 pmol g^{-1} tissue. Similarly, total pepsin/pepsinogen was measured in human saliva down to 0.165 ng ml^{-1} by loading saliva digests onto an IC column packed with an immobilized anti-surrogate peptide antibody, which was online coupled to a C18 trapping cartridge and a C18 nanoLC column (Neubert et al. 2010).

24.6.2 Antibody-Free Extraction

Ion-exchange SPE, using commercially available materials, has not been widely adopted for the extraction of surrogate peptides from the digests of biological samples, but it can be used in cases when this approach offers sufficient selectivity. The digests of protein-rich samples, such as plasma, contain many peptides, which all have carboxylic and amine groups, and the surrogate peptide of interest can only be separated from the endogenous background peptides if its pI value is sufficiently different. This means that the pH and ionic strength of the loading, washing, and elution solvents have to be optimized for a selective extraction. To give an example, weak anion-exchange SPE was used for the extraction of a tryptic surrogate peptide from a proprietary biopharmaceutical protein down to 10 ng ml^{-1} in rabbit and human plasma (Bronsema et al. 2015a). Since the surrogate peptide contained three carboxylic acid groups, it was strongly retained by the positively charged SPE phase at pH 5, while endogenous peptides with less negative charges were not trapped during sample loading or were removed from the SPE material by a washing step with a high ionic strength (300 mM sodium chloride). Elution took place at a high pH by neutralizing the SPE material. The mixed-mode SPE phase also contained reversed-phase groups, but these did not capture the relatively polar surrogate peptide, while some less polar endogenous peptides did remain bound by reversed-phase interactions. Two dimensions of selectivity (ion-exchange and reversed-phase) were thus used to isolate the surrogate peptide from the plasma digest.

Yang et al. (2007) used reversed-phase SPE to extract the surrogate peptides of somatropin and bovine fetuin (the internal standard) from a plasma digest. After loading the digest onto a C18 SPE phase, a washing step with 0.5 ml^{-1} of 10% acetonitrile was applied and the peptides were eluted with 1.5 ml^{-1} of 35% acetonitrile. With this extraction, insufficient selectivity was observed and a mixed-mode cation exchange (MCX) SPE step had to be added for further removal of interfering matrix peptides. This was achieved by loading the entire eluate onto an MCX phase, followed by washing with 0.5 ml^{-1} of 10% ammonia in acetonitrile and elution with 1.5 ml^{-1} of 20% ammonium hydroxide in acetonitrile. The eluate was evaporated to dryness and redissolved in 50 µl of 0.1% formic acid prior to analysis by LC-MS/MS. The extraction recovery of the entire procedure was around 57% and an LLOQ of 1 µg ml^{-1} was obtained.

24.7 Combined Protein and Peptide Extraction

When extraction at the protein or the peptide level alone does not lead to sufficient selectivity and sensitivity, combined protein and peptide extractions can be considered. In principle, these extractions can be combined in many ways and, as long as the separation mechanisms are orthogonal, improved selectivity and sensitivity can be expected as compared to the single-extraction approach.

Even without using IC materials, excellent selectivity and sensitivity can be obtained, as was shown for the biomarker soluble tumor necrosis factor-related apoptosis-inducing ligand (TRAIL) in saliva and sputum (Wilffert et al. 2016b). An aliquot of 100 µl of the sample was extracted at the protein level by IMAC, the eluate was subjected to trypsin digestion, and the tryptic peptides were further purified using cation-exchange SPE under acidic conditions. The surrogate peptide contained four basic amino acids and was therefore positively charged at low pH and trapped on the SPE phase. Endogenous peptides with fewer positive charges were removed by washing with 200 mM sodium chloride and the surrogate peptide was eluted at alkaline pH and quantified using LC-MS/MS. TRAIL concentrations as low as 0.2 ng ml^{-1} could be quantified in saliva and sputum from asthma patients.

An example of the combination of IC at the protein level and antibody-free SPE at the peptide level was reported by Winther et al. (2009). The small cell lung cancer biomarker pro-gastrin releasing peptide (ProGRP) was extracted from human serum by using a microtiter plate coated with an analyte-specific antibody. Sample aliquots of 200 µl were applied to the plate, which was subsequently incubated for one hour and thoroughly washed, six times by a plate washer and twice manually. In-well digestion of the extract was performed overnight at 37 °C. In-house-prepared SPE pipette tips containing C18 material were used for the further cleanup. Up to 125 µl of digest was applied and, after a washing step with 15% acetonitrile, the surrogate peptides were eluted with 50 µl of 45% acetonitrile. This yielded an extraction recovery of 65% and allowed quantitation down to 0.5 ng ml^{-1}.

The most rigorous combination of protein and peptide extraction is IC of the protein analyte followed by digestion and IC of the surrogate peptide. This approach requires two specific antibodies and is certainly not generic, but it may lead to impressive sensitivities. An interesting example is the determination of the biomarker interleukin-21 (IL-21) in human serum and monkey tissues down to 0.78 pg ml^{-1} by combining off-line magnetic bead-based protein extraction using an anti-IL21 antibody and online enrichment of the surrogate peptide using immobilized anti-peptide antibodies (Palandra et al. 2013; see Figure 24.6).

24.8 Summary

Sample preparation is an essential part of any LC-MS-based method for the quantitative determination of proteins in biological samples. Currently, most LC-MS methods employ an enzymatic digestion step to convert the protein analyte into a mixture of peptides and quantify one or more of these as a surrogate for the intact protein. Low-µg ml^{-1} to high-ng ml^{-1} concentrations can typically be measured if a digested sample is directly analyzed with LC-MS/MS. To improve the selectivity and sensitivity, extraction of the protein before digestion and/or extraction of the surrogate peptide after digestion can be applied. Extraction approaches based on IC provide the best selectivity and sensitivity, but require specific analyte-directed reagents such as antibodies and are not generic. Analyte-free extraction methods such as SPE are a universally applicable but less selective alternative. By combining different, carefully optimized extraction techniques, low- to sub-pg ml^{-1} sensitivities can be reached. Digestion of the untreated biological matrix results in obtaining the total concentration of a protein analyte, while protein extraction before digestion might result in the free or total fraction or a concentration in between the two, depending on the extraction principle.

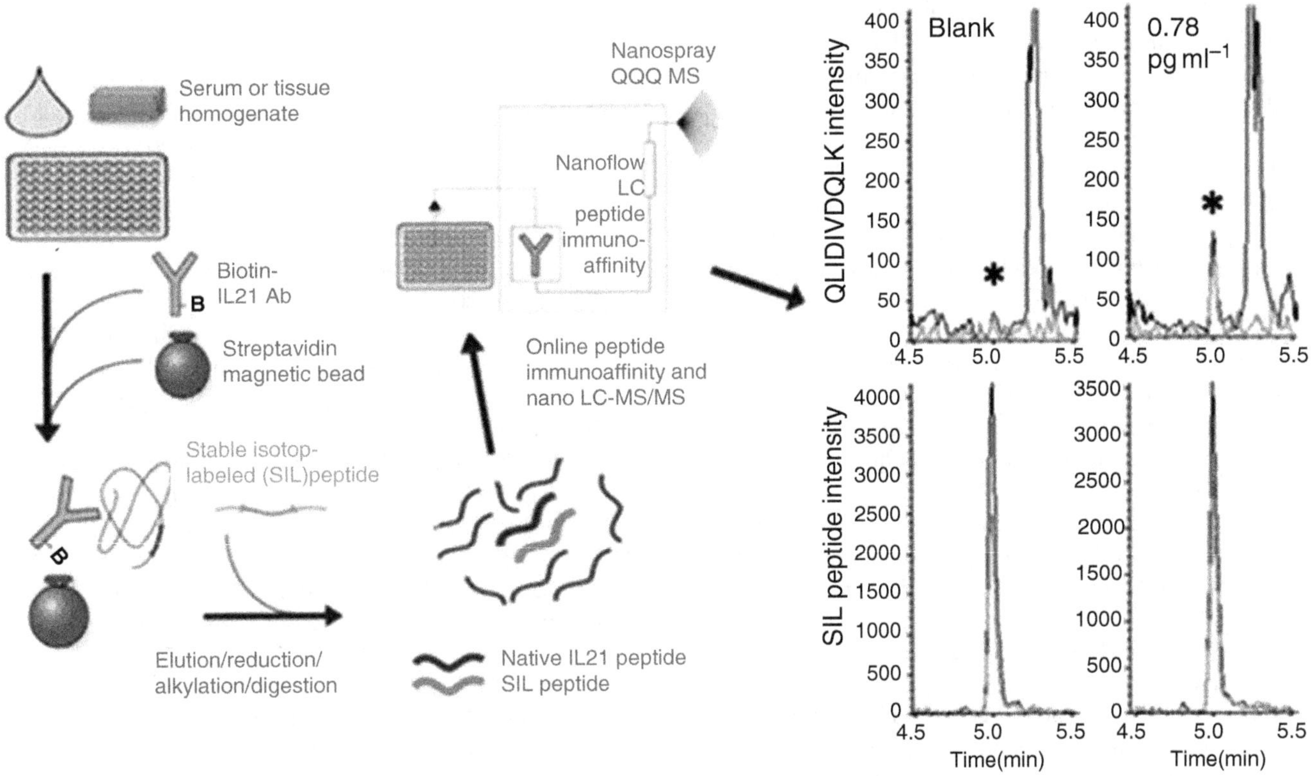

Figure 24.6 Schematic representation of the combination of off-line protein immunocapture and online peptide immunocapture for the quantitation of interleukin-21 (IL-21) plus corresponding chromatograms of blank and LLOQ samples. *Source:* Reproduced with permission from Palandra et al. (2013). Copyright (2013) American Chemical Society.

References

Bailly-Chouriberry, L., Cormant, F., Garcia, P. et al. (2013). Identification of α-cobratoxin in equine plasma by LC–MS/MS for doping control. *Anal. Chem.* 85 (10): 5219–5225.

Becker, J.O. and Hoofnagle, A.N. (2012). Replacing immunoassays with tryptic digestion-peptide immunoaffinity enrichment and LC–MS/MS. *Bioanalysis* 4 (3): 281–290.

Berna, M. and Ackermann, B. (2009). Increased throughput for low-abundance protein biomarker verification by liquid chromatography/tandem mass spectrometry. *Anal. Chem.* 81: 3950–3956.

Bischoff, R., Bronsema, K.J., and van de Merbel, N.C. (2013). Analysis of biopharmaceutical proteins in biological matrices by LC-MS/MS – I. Sample preparation. *Trends Anal. Chem.* 48: 41–51.

Bredehoft, M., Schanzer, W., and Thevis, M. (2008). Quantification of human insulin-like growth factor-1 and qualitative detection of its analogues in plasma using liquid chromatography/electrospray ionisation tandem mass spectrometry. *Rapid Commun. Mass Spectrom.* 22 (4): 477–485.

Bronsema, K.J., Bischoff, R., Bouche, M.P. et al. (2015a). High-sensitivity quantification of a Nanobody in plasma by single-cartridge multi-dimensional SPE and ultra-performance LC-MS/MS. *Bioanalysis* 7 (1): 53–64.

Bronsema, K.J., Bischoff, R., Pijnappel, W.W.M.P. et al. (2015b). Absolute quantification of the total and antidrug antibody-bound concentrations of recombinant human α-glucosidase in human plasma using protein G extraction and LC-MS/MS. *Anal. Chem.* 87 (8): 4394–4401.

Bronsema, K.J., Bischoff, R., and van de Merbel, N.C. (2012). Internal standards in the quantitative determination of protein biopharmaceuticals using liquid chromatography coupled to mass spectrometry. *J. Chromatogr. B* 893–894: 1–14.

Bronsema, K.J., Bischoff, R., and van de Merbel, N.C. (2013). High-sensitivity LC-MS/MS quantification of peptides and proteins in complex biological samples: the impact of enzymatic digestion and internal standard selection on method performance. *Anal. Chem.* 85 (20): 9528–9535.

Bults, P., Bischoff, P., Bakker, H. et al. (2016). LC-MS/MS-based monitoring of in vivo protein biotransformation: quantitative determination of trastuzumab and its deamidation products in human plasma. *Anal. Chem.* 88 (3): 1871–1977.

Bults, P., van de Merbel, N.C., and Bischoff, R. (2015). Quantification of biopharmaceuticals and biomarkers in complex biological matrices: a comparison of liquid chromatography coupled to tandem mass spectrometry and ligand binding assays. *Expert Rev. Proteomics* 12 (4): 355–374.

Capelo, J.L., Carreira, R., Diniz, M. et al. (2009). Overview on modern approaches to speed up protein identification workflows relying on enzymatic cleavage and mass spectrometry-based techniques. *Anal. Chim. Acta* 650 (2): 151–159.

Chen, L.Z., Roos, D., and Philip, E. (2016). Development of immunocapture-LC/MS assay for simultaneous ADA isotyping and semiquantitation. *J. Immunol. Res.* 2016: 7682472.

Dawes, M.L., Gu, H., Wang, J. et al. (2013). Development of a validated liquid chromatography tandem mass spectrometry assay for a PEGylated adnectin in cynomolgus monkey plasma using protein precipitation and trypsin digestion. *J. Chromatogr. B* 934: 1–7.

Dubois, M., Fenaille, F., Clement, G. et al. (2008). Immunopurification and mass spectrometric quantitation of the active form of a chimeric therapeutic antibody in human serum. *Anal. Chem.* 80 (5): 1737–1745.

Dufield, D.R. and Radabaugh, M.R. (2012). Online immunoaffinity LC/MS/MS. a general method to increase sensitivity and specificity: how do you do it and what do you need? *Methods* 56 (2): 236–245.

Duggan, J., Ren, B., Mao, Y. et al. (2016). LC-MS quantification of protein drugs: validating protein LC-MS methods with predigestion immunocapture. *Bioanalysis* 8 (18): 1951–1964.

Fan, Y.Y. and Neubert, H. (2016). Quantitative analysis of human neonatal Fc receptor (FcRn) tissue expression in transgenic mice by online peptide immuno-affinity LC-HRMS. *Anal. Chem.* 88 (8): 4239–4247.

Fung, E.N., Bryan, P., and Kozhich, A. (2014). Techniques for quantitative LC–MS/MS analysis of protein therapeutics: advances in enzyme digestion and immunocapture. *Bioanalysis* 6 (18): 2335–2337.

Gong, J., Gu, X., Achanzar, W.E. et al. (2014). Quantitative analysis of polyethylene glycol (PEG) and PEGylated proteins in animal tissues by LC-MS/MS coupled with in-source CID. *Anal. Chem.* 86 (15): 7642–7649.

Hagman, C., Ricke, D., Ewert, S. et al. (2008). Absolute quantification of monoclonal antibodies in biofluids by liquid chromatography-tandem mass spectrometry. *Anal. Chem.* 80 (4): 1290–1296.

Halquist, M.S. and Karnes, H.T. (2011). Quantification of Alefacept, an immunosuppressive fusion protein in human plasma using a protein analogue internal standard, trypsin cleaved signature peptides and liquid chromatography tandem mass spectrometry. *J. Chromatogr. B* 879 (11–12): 789–798.

Hervey, W.J., Strader, M.B., and Hurst, G.B. (2007). Comparison of digestion protocols for microgram quantities of enriched protein samples. *J. Proteome Res.* 6 (8): 3054–3061.

Hustoft, H.K., Malerod, H., Wilson, S.R. et al. (2012). A critical review of trypsin digestion for LC-MS based proteomics. In: *Integrative Proteomics* (ed. H.C. Leung), 73–92. Rijeka, Croatia: InTech.

Hustoft, H.K., Reubsaet, L., Greibrokk, T. et al. (2011). Critical assessment of accelerating trypsination methods. *J. Pharm. Biomed. Anal.* 56 (5): 1069–1078.

IUPAC (1997). *Compendium of Chemical Terminology*, 2e (the "Gold Book"). Compiled by A. D. McNaught and A. Wilkinson. Oxford: Blackwell Scientific Publications.

Jian, W., Kang, L., Burton, L., and Weng, N. (2016). A workflow for absolute quantitation of large therapeutic proteins in biological samples at intact level using LC-HRMS. *Bioanalysis* 8 (16): 1679–1691.

Kaur, S., Liu, L., Cortes, D.F. et al. (2016). Validation of a biotherapeutic immunoaffinity-LC–MS/MS assay in monkey serum: "plug-and-play" across seven molecules. *Bioanalysis* 8 (15): 1565–1577.

Kuhn, E., Wu, J., Karl, J. et al. (2004). Quantification of C-reactive protein in the serum of patients with rheumatoid arthritis using multiple reaction monitoring mass spectrometry and 13C-labeled peptide standards. *Proteomics* 4 (4): 1175–1186.

Kumar, V., Barnidge, D.R., Chen, L.S. et al. (2010). Quantification of serum 1–84 parathyroid hormone in patients with hyperparathyroidism by immunocapture in situ digestion liquid chromatography–tandem mass spectrometry. *Clin. Chem.* 56 (2): 306–313.

Li, F., Fast, D., and Michael, S. (2011). Absolute quantitation of protein therapeutics in biological matrices by enzymatic digestion and LC–MS. *Bioanalysis* 3 (21): 2459–2480.

Li, Y.C., Liang, Y., Tang, Z.Y. et al. (2012). Quantification of endostar in rat plasma by LC–MS/MS and its application in a pharmacokinetic study. *J. Pharm. Biomed. Anal.* 70: 505–511.

Liu, H., Manuilov, A.V., Chumsae, C. et al. (2011). Quantitation of a recombinant monoclonal antibody in monkey serum by liquid chromatography–mass spectrometry. *Anal. Biochem.* 414 (1): 147–153.

Liu, G., Zhao, Y., Angeles, A. et al. (2014). A novel and cost effective method of removing excess albumin from plasma/serum samples and its impacts on LC–MS/MS bioanalysis of therapeutic proteins. *Anal. Chem.* 86 (16): 8336–8346.

López-Ferrer, D., Cañas, B., Vázquez, J. et al. (2006). Sample treatment for protein identification by mass spectrometry-based techniques. *Trends Anal. Chem.* 25 (10): 996–1005.

Neubert, H., Fan, Y.Y., and Fernandez Ocaña, M. (2016). Quantification of protein biomarkers in tissues: new capabilities with pellet digestion peptide immunoaffinity LC–MS/MS. *Bioanalysis* 8 (15): 1551–1555.

Neubert, H., Gale, J., and Muirhead, D. (2010). Online high-flow peptide immunoaffinity enrichment and nanoflow LC-MS/MS: assay development for total salivary pepsin/pepsinogen. *Clin. Chem.* 56 (9): 1413–1423.

Ouyang, Z., Furlong, M.T., Wu, S. et al. (2012). Pellet digestion: a simple and efficient sample preparation technique for LC–MS/MS quantification of large therapeutic proteins in plasma. *Bioanalysis* 4 (1): 17–28.

Palandra, J., Finelli, A., Zhu, M. et al. (2013). Highly specific and sensitive measurements of human and monkey interleukin 21 using sequential protein and tryptic peptide immunoaffinity LC-MS/MS. *Anal. Chem.* 85 (11): 5522–5529.

Pritchard, C., Groves, K.J., Biesenbruch, S. et al. (2014). Quantification of human growth hormone in serum with a labeled protein as an internal standard: essential considerations. *Anal. Chem.* 86 (13): 6525–6532.

Proc, J.L., Kuzyk, M.A., Hardie, D.B. et al. (2010). A quantitative study of the effects of chaotropic agents, surfactants, and solvents on the digestion efficiency of human plasma proteins by trypsin. *J. Proteome Res.* 9 (10): 5422–5437.

Razavi, M., Anderson, N.L., Pope, M.E. et al. (2016). High precision quantification of human plasma proteins using the automated SISCAPA immuno-MS workflow. *N. Biotechnol.* 33 (5A): 494–502.

Razavi, M., Johnson, L.D.S., Lum, J.J. et al. (2013). Quantification of a proteotypic peptide from protein C inhibitor by liquid chromatography–free SISCAPA-MALDI mass spectrometry: application to identification of recurrence of prostate cancer. *Clin. Chem.* 59 (10): 1514–1522.

Ruan, Q., Ji, Q.C., Arnold, M.E. et al. (2011). Strategy and its implications of protein bioanalysis utilizing high-resolution mass spectrometric detection of intact protein. *Anal. Chem.* 83 (23): 8937–8944.

Sleczka, B.G., Mehl, J.T., Shuster, D.J. et al. (2014). Quantification of human mAbs in mouse tissues using generic affinity enrichment procedures and LC–MS detection. *Bioanalysis* 6 (13): 1795–1811.

Swaney, D.L., Wenger, C.D., and Coon, J.J. (2010). Value of using multiple proteases for large-scale mass spectrometry-based proteomics. *J. Proteome Res.* 9 (3): 1323–1329.

Van de Merbel, N.C. (2015). Advances in LC-MS-MS based quantitation of biopharmaceuticals in biological samples. *LC-GC Eur. Suppl.* 38–44.

Van den Broek, I., Niessen, W.M.A., and van Dongen, W.D. (2013). Bioanalytical LC-MS/MS of protein-based biopharmaceuticals. *J. Chromatogr. B* 929: 161–179.

Van den Broek, I. and van Dongen, W.D. (2015). LC-MS-based quantification of intact proteins: perspectives for clinical and bioanalytical applications. *Bioanalysis* 7 (15): 1943–1958.

Whiteaker, J.R., Zhao, L., Zhang, H.Y. et al. (2007). Antibody-based enrichment of peptides on magnetic beads for mass-spectrometry-based quantification of serum biomarkers. *Anal. Biochem.* 362 (1): 44–54.

Wilffert, D., Asselman, A., Donzelli, R. et al. (2016a). Highly sensitive antibody-free μLC-MS/MS quantification of rhTRAIL in serum. *Bioanalysis* 8 (9): 881–890.

Wilffert, D., Bischoff, R., and van de Merbel, N.C. (2015). Antibody-free workflows for protein quantification by LC-MS. *Bioanalysis* 7 (6): 763–779.

Wilffert, D., Donzelli, R., Asselman, A. et al. (2016b). Quantitative antibody-free LC–MS/MS analysis of sTRAIL in sputum and saliva at the sub-ng/mL level. *J. Chromatogr. B* 1032: 205–210.

Wilffert, D., Reis, C.R., Hermans, J. et al. (2013). Antibody-free LC–MS/MS quantification of rhTRAIL in human and mouse serum. *Anal. Chem.* 85 (22): 10754–10760.

Winther, B., Nordlund, M., Paus, E. et al. (2009). Immuno-capture as ultimate sample cleanup in LC-MS/MS determination of the early stage biomarker ProGRP. *J. Sep. Sci.* 32 (17): 2937–2943.

Wright, K. and Dufield, D. (2014). Minimalistic sample preparation strategies for LC-MS quantification of large molecule biopharmaceuticals: a case study highlighting alpha-1 antitrypsin protein. *Bioanalysis* 6 (13): 1813–1825.

Yang, Z., Hayes, M., Fang, X. et al. (2007). LC-MS/MS approach for quantification of therapeutic proteins in plasma using a protein internal standard and 2D-solid-phase extraction cleanup. *Anal. Chem.* 79 (24): 9294–9301.

Yang, Z., Ke, J., Hayes, M. et al. (2009). A sensitive and high-throughput LC-MS/MS method for the quantification of pegylated-interferon-alpha2a in human serum using monolithic C18 solid phase extraction for enrichment. *J. Chromatogr. B* 877 (18–19): 1737–1742.

Yang, Z., Li, W., Smith, H.T., and Tse, F.L.S. (2013). LC-MS bioanalysis of proteins. In: *Handbook of LC-MS Bioanalysis* (ed. W. Li, J. Zhang and F.L.S. Tse), 601–605. Hoboken, NJ: Wiley.

Yuan, L., Arnold, M.E., Aubry, A.-F., and Ji, Q.C. (2012). Simple and efficient digestion of a monoclonal antibody in serum using pellet digestion: comparison with traditional digestion methods in LC–MS/MS bioanalysis. *Bioanalysis* 4 (24): 2887–2896.

Zhang, H., Xiao, Q., Xin, B. et al. (2014). Development of a highly sensitive liquid chromatography/tandem mass spectrometry method to quantify total and free levels of a target protein, interferon-gamma-inducible protein-10, at picomolar levels in human serum. *Rapid Commun. Mass Spectrom.* 28 (13): 1535–1543.

25

Sample Preparation for LC-MS Bioanalysis of Oligonucleotides

Michael G. Bartlett, Jaeah Kim, Babak Basiri, and Ning Li

25.1 Introduction

Most bioanalytical chemists can tell you that well-developed sample preparation is the key to any successful method. It follows one of the most well-used axioms in the field. You have won or lost the battle before the injection is made. This outlook is especially true for large complex biomolecules like oligonucleotides. Oligonucleotides have many unique properties which both create challenges and also present opportunities during the development of bioanalytical methods. Of these properties three are particularly noteworthy when compared to the other major classes of biomolecules. In proteins each amino acid weighs ~100 Da and with oligosaccharides each carbohydrate unit weights ~180 Da. However, each nucleotide adds ~350 Da to the size of the oligonucleotide. Therefore, even relatively small oligonucleotides are fairly large biomolecules. The phosphate backbone creates a substantial amount of negative charge to oligonucleotides and also dramatically increases their water solubility. Finally, many oligonucleotides have quite complex higher order structure such as forming duplexes with their complimentary strand or forming hairpins.

25.2 Properties of Oligonucleotides and Associated Challenges in LC-MS Bioanalysis

In general, oligonucleotides are polymers with each repeating unit being composed of three basic parts, the base, the sugar, and the backbone (see Figure 25.1) (Belmont et al. 2001; Blackburn 2006; Egli and Saenger 2013; Sinden 2012). The differences in the various classes of oligonucleotides can be attributed to the variations among these three portions of the molecules. The base is a small organic heterocycle which is generally classified as either pyrimidines or purines depending on the ring structure (Blackburn 2006; Egli and Saenger 2013; Sinden 2012). In terms of the biological function of oligonucleotides, the bases control the majority of the most critical functions such as carrying the basic genetic information within organisms (Lodish et al. 2000). Among the physiochemical properties of the bases there are two significant issues that should be understood. The first is the effect of the base guanine on aggregation and secondary structure. A higher percentage of guanines in sequence will promote aggregation and sequences of four consecutive guanines are motifs often associated with secondary structures such as G-quadruplexes (Belmont et al. 2001; Kern and Kerwin 2002; Kutyavin et al. 2002; Sen and Gilbert 1988, 1992; Wu et al. 2007). The second major consideration is the pK_a of the bases. Most LC-MS methods involving oligonucleotides use mobile phases where $pH > 8$ (Dai et al. 2005; Erb et al. 2012; Johnson et al. 2005). This effectively neutralizes the bases with the exception of thymine which has a much higher pK_a than the other natural bases (Verdolino et al. 2008). This creates an interesting dichotomy in the LC-MS of oligonucleotides. Many reports urge greater use of higher pHs to attempt to neutralize thymine; however, the presence of charged thymine appears to inherently improve the response of oligonucleotides when analyzed using electrospray ionization (Mitchell et al. 2009). Therefore, while the pK_a of thymine is likely important for recovery during sample preparation, it appears to be less of an issue once the sample is introduced into an LC-MS system. This distinction is also important because many articles in the literature make use of oligonucleotides which are homopolymers of thymine (Apffel et al. 1997; Huber and Krajete 1999; Schneider and Chait 1993). It is wise to remember that all other base sequences of the same size will have lower LC-MS response. However, in terms of bioanalysis the bases are not the major source of method complexity.

Sample Preparation in LC-MS Bioanalysis, First Edition. Edited by Wenkui Li, Wenying Jian, and Yunlin Fu.

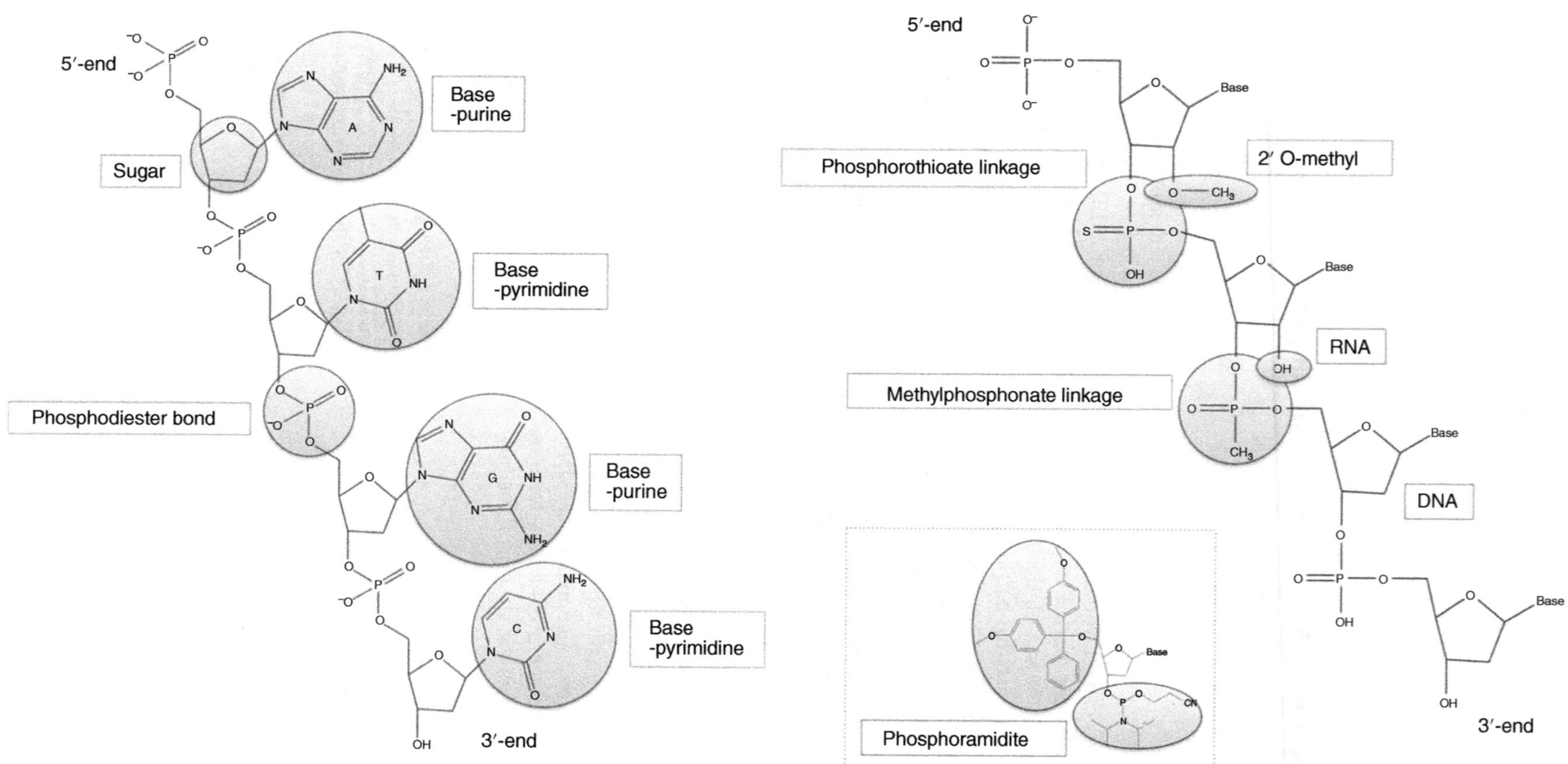

Figure 25.1 The structure of oligonucleotide.

Sugar is one of the most important parts of an oligonucleotide and accounts for a great deal of the diversity among the various types (Belmont et al. 2001; Lescrinier et al. 2003; Rich et al. 1978). The sugar portion of an oligonucleotide monomer is composed of the cyclized five carbon carbohydrate, ribose (Blackburn 2006; Egli and Saenger 2013; Sinden 2012). Among the ribose structure, the 2′-carbon is where the greatest variety of structure can be found. The structure is generally shown in a two-dimensional (2D)Haworth projection form. This often leads to the erroneous impression that this simple five-membered ring is planar. Ribose ring pucker arises because a flat pentagon puts all ring carbon atoms into eclipsed conformations. Puckering relieves this steric strain at the expense of slight deviations from the tetrahedral bond angle. The pucker is described in terms of the ribose ring atom that is displaced furthest out of the plane. The lowest energy conformations are 2′-endo and 3′-endo (Figure 25.2) (Rao and Kollman 1985). Among the major classes of oligonucleotides, B-form DNA adopts the 2′-endo conformation and A-form RNA adopts the 3′-endo conformation (Blackburn 2006). Therefore, the ring pucker is one of the major determinants of the secondary structure of oligonucleotides (Blackburn 2006). It is generally important to consider the impact of any substituent changes to the ribose ring because if they alter the ring pucker there can be dramatic changes in the three-dimensional structure of the oligonucleotide (Dickerson et al. 1982). This can also be important because some secondary structures (such as G-quadruplexes) also incorporate alkali metal atoms. In mass spectrometry, a great deal of effort is spent trying to minimize the amount of alkali metal adduction observed in the mass spectra (Birdsall et al. 2016; Nozaki et al. 2010). It is therefore important to recognize that these adducts may not simply arise from the mobile phase but may be intrinsic features of the oligonucleotide and can often explain the difference observed in the degree of adduction between various oligonucleotides.

The backbone generally consists of a phosphodiester linkage as shown in Figure 25.1 (Belmont et al. 2001; Blackburn 2006; Egli and Saenger 2013; Sinden 2012). This linkage is critical because it is the chemical bond between each monomer unit and creates the oligonucleotide polymer. The linkage is also important because it carries a negative charge and therefore impacts most of the physiochemical properties that are attributed to oligonucleotides (i.e. that they are highly polar and polyanionic). It is also this portion of the molecule that is the most susceptible to chemical and enzymatic degradation. Hydrolysis can occur at the phosphodiester linkage under either acidic or basic conditions (Bahr et al. 2009). Therefore, it is important to store most at neutral pH values. Hydroxylation at the 2′ position greatly increases the susceptibility of the phosphodiester bond to undergo hydrolysis (Blackburn 2006). This difference is principally why RNA is so much less stable than any other type of oligonucleotide. It is also why there is so much effort made to chemically modify the 2′ position of synthetic oligonucleotides (Faria and Ulrich 2008). Enzymatic hydrolysis occurs via two major classes of hydrolase enzymes, DNases and RNases (Blackburn 2006). While it is apparent from the names of the two classes of enzymes which type of oligonucleotides they preferentially target, both enzymes work by increasing the rate of hydrolysis of the phosphodiester linkage. There are several types of chemical modifications of the phosphodiester backbone that can significantly decrease the rate of hydrolysis (De Mesmaeker et al. 1995; Deleavey and Damha 2012; Milligan et al. 1993; Uhlmann and Peyman 1990; Varma 1993). Several involve removing the charge from the phosphodiester linkage such as methylphosphonates (Blackburn 2006; Miller et al. 2000; Miller and Ts'o 1987) or phosphoramidites (Figure 25.2) (Blackburn 2006; Dagle and Weeks 1996; Sauer et al. 2000). These improve stability by limiting the proton transfer to the negatively charged oxygen atom that is the initial event in the hydrolysis reaction. One of the most successful modifications of the phosphodiester linkage is the replacement of the non-bridging oxygen with a sulfur atom creating a phosphorothioate linkage (Figure 25.1) (Blackburn 2006; Eckstein 2000). The phosphorothioate bonds significantly reduce the ability of both endo- and exonucleases to hydrolyze this bond (Sayers et al. 1989).

Although these specialized bonds confer stability against indiscriminate nuclease digestion, the repeated introduction of these bonds can also create limitations to their function as antisense oligonucleotides. Each phosphorothioate bond introduced into the oligo creates a chiral center at each bond which is designated as either a

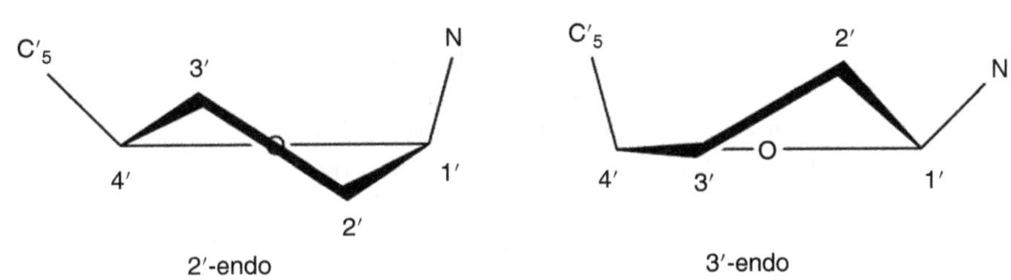

Figure 25.2 2′-Endo and 3′-endo conformations of sugar.

"Sp" or "Rp" conformation. This ultimately leads to multiple isomers of the oligo generated during synthesis. These individual isomers have been shown to have different characteristics and functional properties. It has been suggested that this heterogeneity is responsible for the limited success that oligonucleotides have had as therapeutic agents. This has created a great deal of recent effort aimed at methods to synthesize phosphorothioate oligonucleotides with a single optimized configuration.

OIigonucleotides are quite polar, especially for biomolecules. To take advantage of the polar and highly charged nature of oligonucleotides they are typically analyzed using ion-pair chromatography (Beverly et al. 2005; Dai et al. 2005; Huber et al. 1993; Huber and Krajete 1999; McCarthy et al. 2009; Oberacher et al. 2001). There has been substantial debate in the literature about the optimal ion pairing agent to use for the LC-MS analysis of oligonucleotides. Recently, this has even extended to studies involving the fluorinated alcohol mobile-phase additives (Apffel et al. 1997; Basiri et al. 2017; Beverly et al. 2005; Chen and Bartlett 2012, 2013; Dai et al. 2005; Erb and Oberacher 2014; McGinnis et al. 2013b). It does appear that the composition of the mobile phase can be optimized depending on the type of oligonucleotide under investigation. However, what should be noted is that even when using ion-pair chromatography, relatively low percentages of organic solvent should be used in the mobile phase in order to retain these molecules. Since electrospray sensitivity is significantly influenced by the amount of organic solvent (Dams et al. 2002) this is one of the major limitations to the determination of low concentrations of oligonucleotides by LC-MS. In terms of hydrophobicity, RNAs are the most polar followed by DNA and then therapeutic oligonucleotides. This also generally corresponds to the inherent response of these molecules during LC-MS analysis. Therapeutic oligonucleotides provide the highest response followed by DNA and then RNA.

25.3 Classes of Oligonucleotides

Oligonucleotides can be viewed as belonging to three distinct classes: DNA, RNA, and therapeutic oligonucleotides. Each class has distinct physiochemical properties and often requires different sample handling to ensure the stability of the analyte during analysis.

The most widely studied class of oligonucleotides is DNA. They are mainly made as primers for polymerase chain reactions or for insertion into various vectors. LC-MS has been widely used for the confirmation of synthesis and to determine purity for small DNAs (Nikcevic et al. 2011; Oberacher et al. 2001; Ravikumar et al. 2003). Another significant application for LC-MS of oligonucleotides is the determination of DNA methylations (Hu et al. 2013; Le et al. 2011; Li and Franke 2011; Lin et al. 2016; Quinlivan and Gregory 2008). The DNA bases cytosine and adenine can be methylated (Klose and Bird 2006; Low et al. 2001; Wilson and Murray 1991). Cytosine methylation appears to occur widely and to have a distinct role in the regulation of gene transcription (Bird 1992; Bird and Wolffe 1999; Walsh and Bestor 1999). Cytosine methylation differs widely between species with nearly 14% of cytosines methylated in *Arabidopsis thaliana* but only 0.3% in *Drosophila* (Capuano et al. 2014). Adenine methylation has received considerably less scrutiny since it has primarily been limited to prokaryotes. The recent observation of 6-methyl adenine in mice and its role in transcriptional silencing during stem cell differentiation have shown that this area likely deserves far greater attention (Wu et al. 2016). Other types of DNA alkylation reactions such as those from environmental carcinogens or several types of chemotherapeutics are generally not analyzed as oligonucleotides. Since these reactions are not sequence specific, the exposure to such chemicals is assessed by conducting LC-MS on the nucleosides following enzymatic hydrolysis.

LC-MS has been widely used to study posttranslational modifications in RNA (Cao and Limbach 2015; Douthwaite and Kirpekar 2007; Kowalak et al. 1993; Su et al. 2014; Taniguchi and Hayashi 1998). This approach generally involves using a T1 enzymatic digest to create small oligonucleotides (Douthwaite and Kirpekar 2007; Kowalak et al. 1993; Taniguchi and Hayashi 1998). This allows modifications to be detected using tandem mass spectrometry by deviations from the expected molecular weights and for confirmation of their identity and placement to occur. While almost every major class of RNA has been studied using this approach to some extent, transfer RNA has received the greatest attention (Wetzel and Limbach 2012). Transfer RNAs are typically 75–80 nucleotides in length (Kahl 2015) and fall near the upper end of the size range for LC-MS analysis of oligonucleotides (Apffel et al. 1997). They also contain a substantial number of modifications (between 30 and 50% of the nucleosides) which makes mass spectrometry a particularly well-adapted tool to interrogate the biological significance of these modifications. More recently LC-MS has been used to look at smaller biological RNA molecules such as siRNA and miRNA (Beverly et al. 2005, 2006; McCarthy et al. 2009). With LC-MS methods now rivaling hybridization immunoassays for sensitivity, it is expected that there will be a dramatic increase in the number of applications in this area. This is especially true since several miRNA molecules have been shown to have either methylations or an additional nucleotide on either the 3′ or 5′ end of the oligonucleotide, which were not anticipated or detected when using deep sequencing of other gene arrays (Chen et al. 2005; Cheng et al. 2009). The methylations have been shown to have biological significance but it is unclear if the $n+1$ miRNA species have biological significance such as acting as a reservoir awaiting activation by removal of the additional nucleotide.

Therapeutic oligonucleotides are those that aim to alter a biological event, typically through an antisense mechanism (Crooke 1992; Crooke and Bennett 1996; Dai et al. 2005; Kurreck 2003; Lin et al. 2007; Milligan et al. 1993; Uhlmann and Peyman 1990). They are typically 18–24 nucleotides in length. Their extreme specificity has always made them attractive, although they have met with limited clinical success with only three compounds having reached the market to date. There are a myriad of significant technical challenges associated with therapeutic oligonucleotides with stability and delivery to the site of action ranking among the most significant. However, when viewing them as an analyte for method development they are often easier to handle than their naturally occurring counterparts. Most therapeutic oligonucleotides contain phosphorothioate backbones to improve their stability and have a wide variety of 2′-modifications on the ribose ring (De Mesmaeker et al. 1995; Deleavey and Damha 2012). Early generation therapeutic oligonucleotides contained 2′-modifications at all of the positions (Manoharan 1999; Monia et al. 1993) but more recent ones generally only have the first 3–4 nucleotides at both the 3′ and 5′ ends modified (Hoke et al. 1991). These 2′-modifications tend to decrease RNase H binding of the sense–antisense complex which is an essential part of the antisense interference pathway and therefore explains the shift to only placing these modifications at the ends of the strands (Monia et al. 1993). These modifications do create complications for verifying structure using tandem mass spectrometry since abstraction of the 2′ hydrogen is necessary for fragmentation. Therefore, these molecules tend to not provide much structural information from either end of the molecule. Perhaps this lack of fragmentation could be considered evidence of successful 2′-modification.

25.4 Major Types of Sample Extraction Techniques

There have been many approaches used successfully to isolate oligonucleotides prior to LC-MS analysis. They can generally be divided into six different approaches. These approaches cover the range of general sample preparation strategies but generally have a few particular caveats that are specific to this class of biomolecules.

25.4.1 Liquid–Liquid Extraction

One of the earliest molecular biology methods for isolating oligonucleotides is the classic phenol–chloroform extraction (typically using a 2 : 1 ratio) (Figure 25.3) (Beverly et al. 2006; Griffey et al. 1997; Murphy et al. 2005). This approach is enhanced by neutralizing the analytes at high pH using ammonium hydroxide. If the sample is biologic, a lysis buffer is added to release the oligonucleotides from within cells or extracellular vesicles. Next the phenol–chloroform mixture is added to the sample. This is followed by collection of the aqueous supernatant layer, which is washed with an immiscible organic solvent like methyl *tert*-butyl ether or isopropyl ether. The organic layer is removed and discarded. Prior to this step, biological samples are still often slightly colored due to the presence of other unwanted biomolecules. However, following this washing step the aqueous layer is typically clear. To the remaining aqueous layer ethanol and glycogen are added and mixed. The mixture is store at −80 °C to precipitate the oligonucleotides. In this process, glycogen serves as an inert carrier for the oligonucleotides and its presence generally improves the overall recovery from the alcohol precipitation process. Upon formation of precipitate, the supernatant is removed and the precipitate is air dried. The final sample can then be stored at −80 °C and reconstituted in the mobile phase prior to LC-MS analysis.

The phenol–chloroform extraction has been widely used by biologists for decades. It provides reliable isolation of a wide variety of oligonucleotides and therefore can be considered as a proven approach to almost any type of oligonucleotide. In general, relatively clean extracts are expected with recoveries around 70% for most oligonucleotides in the range of 10–30 bases in length (Chen and Bartlett 2013). A major concern with this approach is that it is difficult to automate and would be impractical for large-scale use. Several of the reagents used in the approach require personal protective equipment in order to safely handle the process. Among these reagents, phenol is caustic; chloroform is a probable human carcinogen and ethers are quite volatile and need to be used in a fume hood to avoid inhalation.

25.4.2 Proteinase K Digestion

Another classic sample preparation method that has been successfully adapted to the LC-MS analysis of oligonucleotides in biological samples involves the use of proteinase K digestion (Figure 25.4) (McGinnis et al. 2013a). One of the most significant features of this approach is the limited number of sample transfer steps that are involved which would facilitate automation of the approach for processing of a large batch of samples.

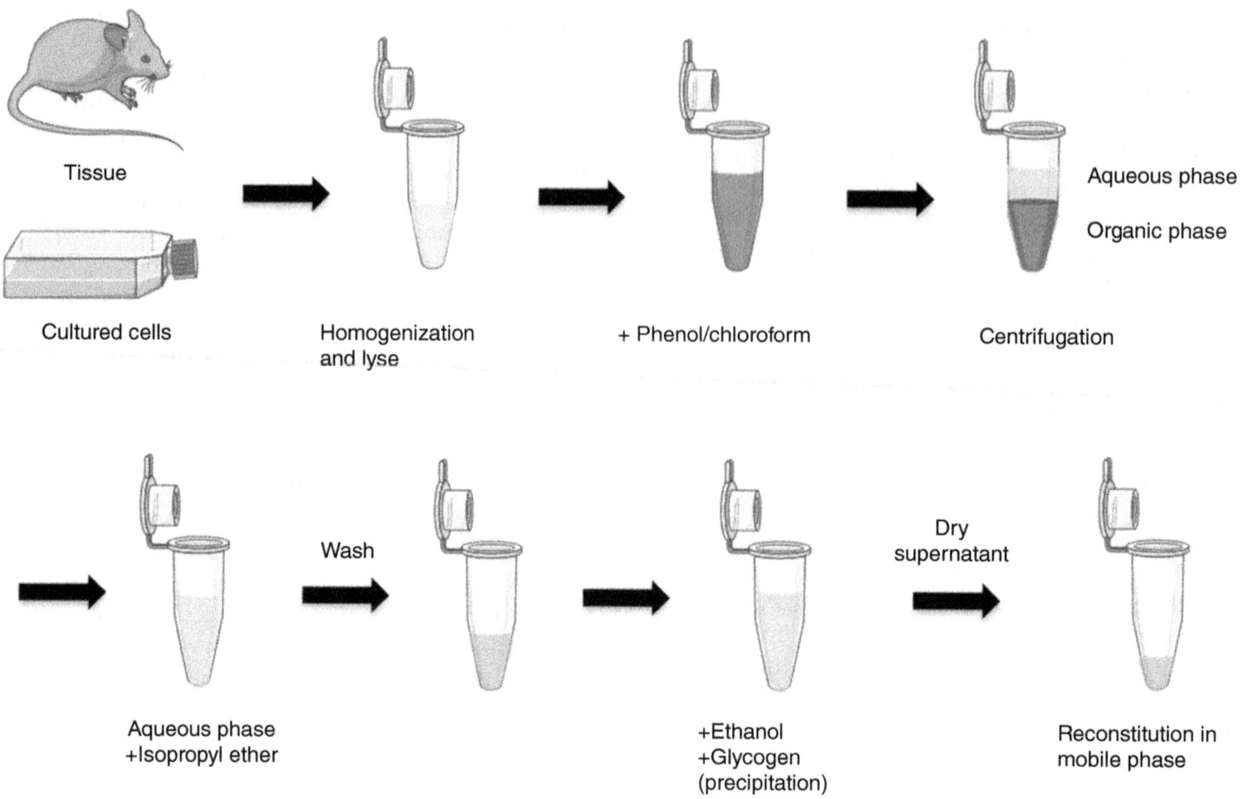

Figure 25.3 Schematic of phenol/chloroform LLE.

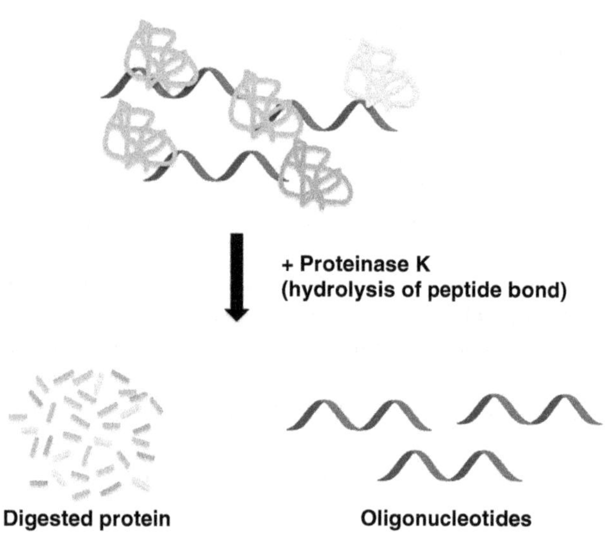

Figure 25.4 Oligonucleotide isolation by proteinase K digestion.

This approach involves the removal of proteins by enzymatically digesting them to amino acids by incubating with proteinase K at physiological temperature. The samples require buffering to maintain the enzymatic activity of proteinase K. Typically, this is accomplished by using tris(hydroxymethyl)aminomethane (Tris) or another equivalent buffer. There are several other additives that are critical to the success of this method and they also need to be added along with the enzyme. Dithiothreitol is needed to reduce disulfide bonds and guanidinium chloride is used as a general chaotropic agent. Together these chemicals disrupt the higher-order structure of proteins and accelerate the digestion process by facilitating access of proteinase K to the peptide bonds of the proteins. To prevent degradation of oligonucleotides during the enzymatic digestion, it is also important to add a metal chelator such as ethylenediaminetetraacetic acid (EDTA) to remove vital metal atom cofactors from nucleases present in the sample. While the literature states that this digestion should take 12 hours or more, recent work by McGinnis et al. (2013a) has shown that it is complete in as little as 3 hours.

Similar to the phenol–chloroform extraction, the proteinase K digestion has been used successfully for many years for the isolation of oligonucleotides. This approach can also be considered quite generic and can be applied to many different types of oligonucleotides including some large ones. The recovery for small oligonucleotides using this approach has been shown to exceed 90% (McGinnis et al. 2013a). The one concern with this approach is that phospholipids will not be separated from the oligonucleotides. However, it is unclear if these phospholipids will adversely impact the robustness of

LC-MS methods. No rigorous validation of a method using this approach has been reported to assess the potential concern due to phospholipids in LC-MS bioanalysis of oligonucleotides.

25.4.3 Solid-phase Extraction

There have been many successful approaches to isolate oligonucleotides involving solid-phase extraction (SPE) (Figure 25.5) (Chen and Bartlett 2012; Dai et al. 2005; Johnson et al. 2005). These SPE approaches generally involve the use of either a mix-mode or a hydrophilic–lipophilic balanced cartridge sorbents. Both approaches attempt to exploit the highly polar and charged nature of oligonucleotides. In general, samples are buffered to around a pH of 5.5 prior to being loaded onto the cartridge to improve retention (Chen and Bartlett 2012). This allows samples to be washed using a 1 : 1 mixture of this buffer along with an organic solvent such as methanol or acetonitrile. Elution is carried out by using another buffer with a pH of ~9.5 that is mixed in a 1 : 1 ratio with a 4 : 1 mixture of acetonitrile and tetrahydrofuran. Following elution, the solvent is evaporated and the residues are reconstituted in either water or mobile phase prior to LC-MS analysis.

SPE has a long history of successful implementation for the determination of oligonucleotides. The recovery using these approaches is generally in the range of 60–80% (Chen and Bartlett 2012). In addition, SPE is easily automatable and can be scaled for high-throughput by using 96- and 384-well plates. One of the general difficulties for SPE is that it is often subject to nonspecific adsorptive losses due to the substantial surface area contained within the cartridges. This can result in difficulty in using it for trace analysis. It should also be noted that SPE is often used in combination with the phenol–chloroform extraction and ethanol precipitation.

25.4.4 Ion-exchange Magnetic Bead Extraction

Magnetic beads have become more widely used for the isolation of molecules during sample preparation (Figure 25.6) (Ye and Beverly 2011). Magnetic beads facilitate both the washing and isolation steps due to their ability to contain analytes during these processes by the transient application of a magnetic field. The first application of magnetic beads to sample preparation of oligonucleotides involved the use of magnetic ion-exchange beads. For these samples, an internal standard (IS) can be easily added since the isolation mechanism would not discriminate against other oligonucleotides. It is also beneficial to dilute the sample by adding a weak acidic solution to ensure that the phosphates are in a highly charged state. The magnetic beads are then added to the diluted sample. The sample can be agitated to ensure proper mixing by repeated pipetting. The beads can be washed using an acidic solution (to ensure the oligonucleotides remain bound) containing a small quantity of a nonionic surfactant such as polyoxyethylenesorbitan (Tween).

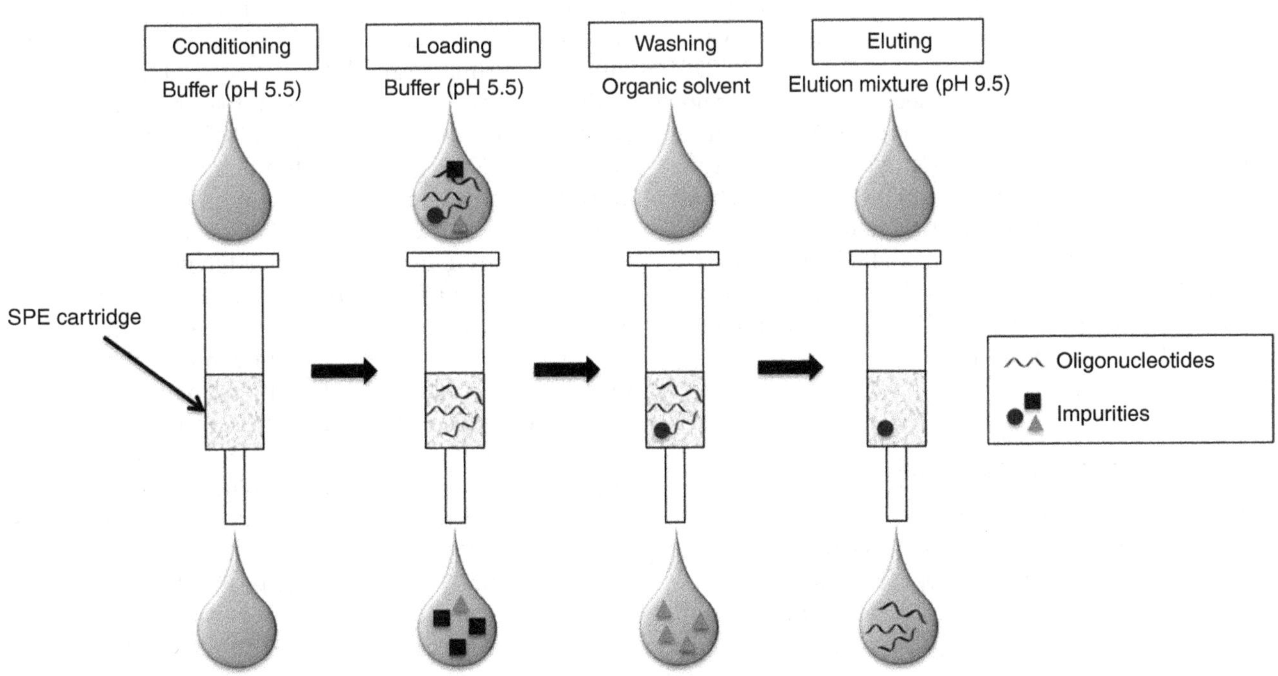

Figure 25.5 Solid-phase extraction.

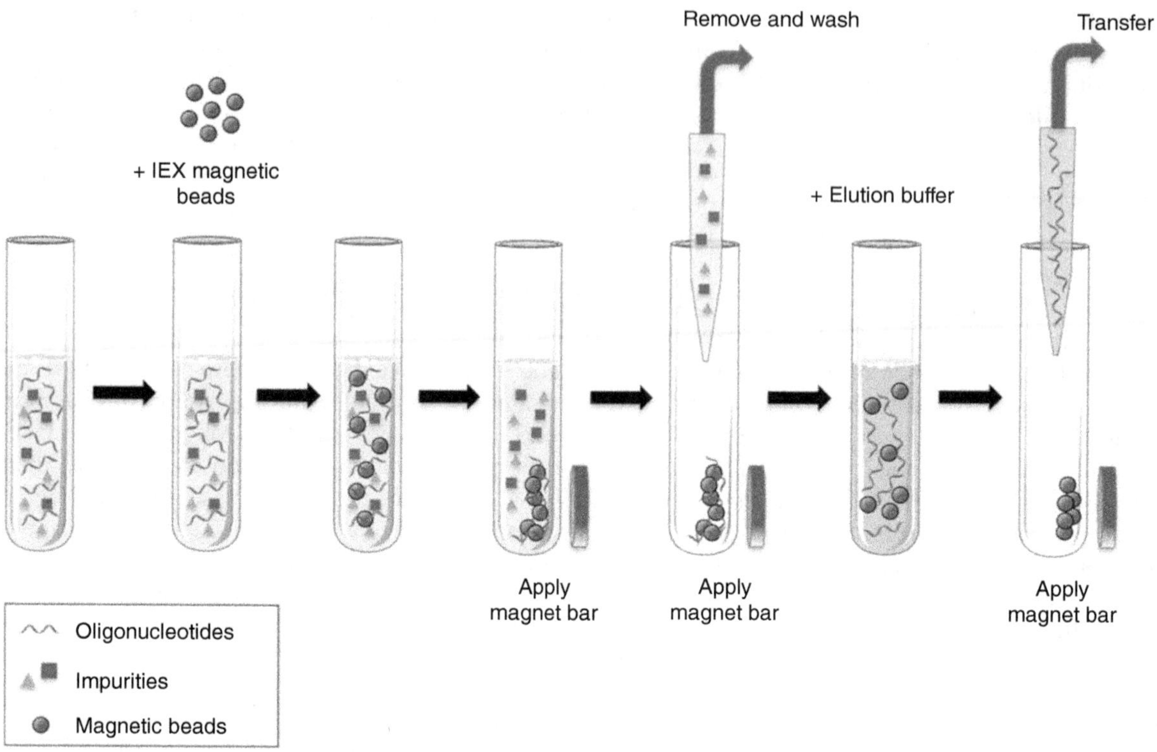

Figure 25.6 Ion-exchange magnetic bead extraction.

Oligonucleotides can then be eluted from the beads using a high-concentration-displacing solution such as ammonium chloride. It is important to remember that these samples will need to be filtered prior to LC-MS analysis to make sure that the magnetic beads are properly removed and will not potentially damage the chromatographic system.

The use of an ion-exchange mechanism with magnetic beads makes this a general approach for a large variety of oligonucleotides. The exception to this approach is for the uncharged oligonucleotides such as methylphosphonates or phosphoramidites. For quantitation, it allows for easy addition of an IS. It also has very few sample transfer steps which decreases losses through nonspecific adsorption. The recoveries using this approach have been shown to range from 80 to 99% (Ye and Beverly 2011).

25.4.5 Immunoaffinity Capture Extraction

As opposed to the general approaches described in the prior sections, there have been two recent studies that involve the creation of sample preparation schemes targeting either a subset of oligonucleotides or even a specific oligonucleotide sequence (Figure 25.7). The first approach involves the creation of an antibody for the Argonaute 2 (Ago2) protein followed by attachment of this antibody to a magnetic bead (Hauptmann et al. 2015; Li et al. 2012; Portnoy et al. 2016; Yang et al. 2014). Attachment of this antibody to the bead is fairly straightforward. Commercially available protein G magnetic beads can be used to capture either the anti-mouse Ago2 (for rat or mouse samples) or the anti-human Ago2 (for human or primate samples) antibodies. The Ago2 antibody can then be used to immunoaffinity-capture the RNA-induced silencing complex (RISC) which would entrain any siRNA or miRNA fragments that would be actively incorporated into this protein complex (Flores-Jasso et al. 2013). Sample preparation begins by placing the Ago2–antibody complex-loaded magnetic beads in a physiological buffer of Tris and sodium chloride. The activated beads can then be incubated with samples overnight. It is beneficial to add a metal chelator like EDTA to the samples to reduce RNase activity and add a lysis buffer to disrupt cells and other vesicles. This facilitates the exposure of the RISC protein complex to the antibody. Following three hours of incubation at 4 °C, the beads are washed four times with a buffer of Tris, EDTA, NaCl, Glycerol, NaF, Nonidet, and DTT. Once captured, the small miRNA and siRNA are released from the beads using acetic acid to denature the protein complex. Samples are then neutralized and buffered with the addition of a typical LC-MS ion-pairing agent such as triethylamine prior to LC-MS analysis.

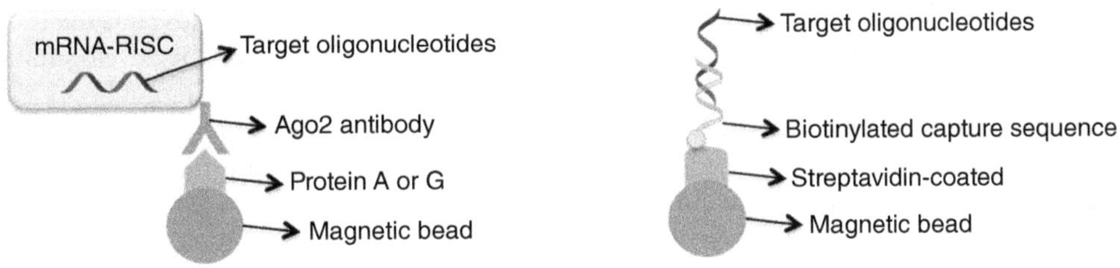

Figure 25.7 Immunoaffinity capture approaches.

The second is an even-more targeted approach involving the addition of a complementary capture sequence. For this approach hydrophilic streptavidin magnetic beads are used (Flores-Jasso et al. 2013; Portnoy et al. 2016). The beads are added to an RNase-free microcentrifuge tube with a washing buffer containing a high concentration of sodium chloride, Tris, and EDTA. A biotinylated complimentary capture oligonucleotide sequence, dissolved in the washing buffer, is added to the beads; vortex-mixed, and then incubated few several minutes to allow formation of the complex. Upon application of the magnetic field, the supernatant is removed. The beads are rinsed in triplicate using the washing buffer and the supernatant is removed after each rinse. A blank sample is added to the washing buffer and heated to 65 °C for several minutes, which is followed by quick chilling in an ice bath for a few minutes to allow endogenous compounds to occupy nonspecific binding sites and complete the preparation of the beads. Once the beads are prepared, the sample is added and vortexed to ensure proper mixing. The mixture is incubated for 20–30 minutes. Finally the magnetic field is applied and the supernatant is removed. The beads were rinsed again using the washing buffer, the magnetic field is applied, and the supernatant is discarded. An elution buffer composed of a low concentration of sodium chloride, Tris, and EDTA at 65 °C is added to the sample and incubated for a few minutes at 65 °C. The eluent containing the targeted oligonucleotide can be removed with the supernatant following application of the magnetic field. It may be beneficial to repeat the elution process and combine the two fractions prior to LC-MS analysis.

In general, both approaches can greatly reduce the complexity of biological samples. This reduces the burden on the chromatographic separation and can lead to shorter run times and greater throughput of the method. However, both approaches have only been used for qualitative studies. The high specificity of these methods makes the incorporation of an IS difficult which would certainly increase the difficulties associated with using these approaches for quantitation.

25.4.6 Preparative Chromatography

Recently, Isobe and co-workers have been using chromatographic fractionation of complex biological samples with great success (Yamauchi et al. 2013). This approach involves the use of a polystyrene-divinylbenzene based chromatography to obtain fractions of oligonucleotides. There have been varying applications of this approach depending on the application (Yamauchi et al. 2016). A macro-porous resin can be used to fractionate samples of oligonucleotide ranging from 20 to nearly 10 000 bases in length.

Monolithic and nonporous resins have been used successfully for isolation of oligonucleotides with <1000 bases in length. This is ideal for the isolation of miRNAs. The fractions can then be further analyzed using nano-LC with a C_{18} trapping column to remove lower molecular-weight compounds from the samples prior to separation using a C_{30}-UG_3 column. The approach has been used successfully to survey miRNAs from cancer cells.

25.5 Key Considerations in Sample Preparation for LC-MS Bioanalysis of Oligonucleotides

There are some particular attributes of oligonucleotides that require attention during the sample preparation process in order to have an optimal outcome. One of the most important attributes is the choice of sample containers. Oligonucleotides are particularly susceptible to nonspecific adsorption. Containers that are designed to resist binding of oligonucleotides are highly recommended. These containers are generally made from materials with low zeta potentials that do not develop surface charges and, therefore, avoid sample loss due to ionic interaction with the analytes. In addition, it is also recommended to minimize the total steps of sample transfer during the sample preparation process.

Nucleases are a ubiquitous class of enzymes that are, in general, responsible for the degradation of oligonucleotides.

Nucleases are secreted from the skin. Simply touching any of the materials used for testing can introduce these enzymes into the system. This can dramatically reduce recovery and also cause a general decrease in method precision and accuracy. Choosing reagents that are nuclease-free such as your sample containers, buffers, and solvents is required.

ISs are critical for absolute quantification of oligonucleotides. Unfortunately, the stable isotope-labeled oligonucleotide is not always readily available. Analogs have been commonly used. For methods involving liquid–liquid extraction or SPE, the analog ISs, in general, do well in tracking the analytes of interest. Furthermore, a higher concentration of ISs might also act as a "sacrificial" oligonucleotide to coat nonspecific binding sites (Chen and Bartlett 2012), leading to increased assay sensitivity. However, for methods using sequence-specific agents (i.e. antibodies or capture strands), the use of an analog IS is challenging as the reagent might not bind to the IS although it might capture the analytes well.

The use of heparin di-sodium, instead of EDTA di-sodium, as anticoagulant has been reported with a severe interference (10 minutes wide or more), which overlaps both the analyte and IS (Korfmacher 2009). The cause might have been due to the polyanionic structure of heparin, which is similar to that of oligonucleotides. Taking this into consideration, polyanionic surfactants such as Triton-X 100 should be avoided during sample preparation. If there is a need, nonionic surfactants such as Tween-80 are recommended.

The sulfur atom in phosphorothioate oligonucleotides is prone to oxidation (Chen and Bartlett 2012). This is particularly true if the sample preparation procedure involves an evaporation step (Chen and Bartlett 2012; Erb et al. 2012; Rudge et al. 2011). The addition of an antioxidant such as Tris(2-carboxylethyl)phosphine (TCEP) is advisable to reduce this phenomena.

The above issues are all significant issues that need to be seriously considered prior to beginning the development of any sample preparation method for oligonucleotides. All of these issues can cause significant deviations in method performance up to and including total method failure. It is encouraging that among all of the various sample preparation options described above, none of them have increased sensitivity to any of the critical issues presented in this section.

25.6 Representative Protocols

A range of sample preparation techniques are available for the LC-MS bioanalysis of oligonucleotides. Two representative protocols are discussed below to facilitate the understanding of the key elements in LC-MS bioanalysis of the challenging molecules.

25.6.1 Quantitation of a Therapeutic Antisense Oligonucleotide from Plasma

Among the many therapeutic oligonucleotides the most studied ones are anti-sense DNAs. The greatest success in this class has come from single-stranded phosphorothioate-modified oligonucleotides in the range of 18–24 mers (Graham et al. 2013; Murray et al. 2012; Wan and Seth 2016). However, oligonucleotides of this type have several unique challenges in LC-MS quantification, including increased hydrophobicity and susceptibility to oxidation. In addition, evaporation and certain storage conditions may promote the conversion of the phosphorothioate backbone back to the normally occurring phosphodiester linkage. Therefore, addition of antioxidants is common in many of these methods. Of course, evaluation should be conducted to assess the possible impact of high concentrations of antioxidant on the method.

Protocol 1: Determination of an antisense DNA from plasma using LC-MS (Chen and Bartlett 2012).

Key supplies: The 24mer phosphorothioate DNA oligonucleotide (5′-TCGTGCTTTTGTTGTTTTCGCGTT-3′), its 23mer 3′$n-1$ metabolite, and IS pdC18 (PAGE purification, purity >90%) were purchased from Integrated DNA Technologies (Coralville, IA). Chemicals such as phenol, 1,1,1,3,3,3-hexafluoro-isopropanol (HFIP), diisopropylethylamine (DIEA), triethylamine (TEA), dimethylbutylamine (DMBA), and glycogen(blue mussel) were purchased from Sigma-Aldrich Inc. (St. Louis, MO). Ammonium hydroxide solution (35%) was purchased from J.T. Baker (Phillipsburg, NJ). HPLC-grade methanol, ethanol, isopropanol, acetonitrile, and chloroform were obtained from Fisher Scientific (Pittsburgh, PA). DNA Lobind centrifuge tubes were purchased from Eppendorf (Hauppauge, NY). UHPLC-MS/MS analysis was performed on a Waters Acquity UHPLC system using a 1.7 μm Waters Acquity BEH C18 100 mm × 1.0 mm column coupled with a Waters Synapt G2 qTOF mass spectrometer (Milford, MA).

Part 1: Sample Preparation

Because of the nonspecific binding of analytes to storage containers, all stock solutions, working solutions, and samples were prepared in DNA Lobind tubes.

1) To 200 μl plasma samples a 100 μl of 5% ammonium hydroxide and a 40 μl of phenol-chloroform (w/v = 2 : 1) were added.
2) The mixture was vortex-mixed for 1 minute and centrifuged at 14 000 × *g* for 10 minutes.
3) The supernatant was transferred to a new tube and washed with 1 ml isopropyl ether.
4) The aqueous phase was then mixed with 2 μl of a 20 mg ml^{-1} glycogen solution and 800 μl of ethanol before storage at −80 °C for an hour.
5) The solution was then centrifuged at 14 000 × *g* for 30 minutes (0 °C). The supernatant was removed and the precipitated DNA was air-dried in a fume hood.
6) A 50 μl of 10 mM DIEA in deionized water was used as the reconstitution solution and a 20 μl was injected into the UHPLC system.

Part 2: LC-MS Analysis

1) Chromatographic separations were performed at a flow rate of 0.15 ml min^{-1} on a 1.7 μm Waters Acquity BEH C18 100 mm × 1.0 mm column.
2) Mobile phase A consisted of 15.7 mM DIEA and 20 mM HFIP in water, and mobile phase B consisted of the same concentration of DIEA and HFIP in water: ethanol (50 : 50).
3) A 20 μl injection of each sample was loaded onto the column and separated using the following gradient conditions (time [minutes], % mobile phase B): (0, 5) (1, 5) (2, 29) (4, 29) (4.01, 5) (5, 5).
4) The column temperature was maintained at 60 °C. The column eluent from 0 to 1.5 minutes was diverted to waste.
5) The weak and strong wash for the autosampler needle are 1000 μl of water and 800 μl 5% methanol, respectively.
6) The mass spectrometry system was operated in the negative-ion MS/MS mode with a one second scan time. The MS/MS transition optimized for 23mer, 24mer, and IS are *m/z* 590 → 319, 667 → 319, and 536 → 304, respectively.
7) The capillary voltage was set at 2.3 kV, the cone voltage was 20 V, the extraction cone voltage was 2 V, the source temperature was 120 °C, the desolvation temperature was 450 °C, cone gas was 6 l h^{-1}, and desolvation gas was 700 l h^{-1}.

25.6.2 Targeted LC-MS of Cellular Messenger RNA

Cellular messenger RNA (mRNA) is modified on the 5′ end by the incorporation of a N^7-methyled guanosine triphosphate cap (m^{7}GpppG) (Lewis and Izaurflde 1997; Mignone et al. 2002). The 5′ cap is an important transcriptional regulatory element (Lewis and Izaurflde 1997; Mignone et al. 2002; Van Der Velden and Thomas 1999). Its role in the translation of mRNA has been widely studied (Van Der Velden and Thomas 1999). An mRNA that does not contain the m^{7}GpppG cap is not recognized by the protein complex that initiates translation and is not exported from the nucleus to the cytoplasm (Lewis and Izaurflde 1997). The cap also protects the 5′ end of mRNA from degradation by nucleases (Van Der Velden and Thomas 1999).

When mRNAs are produced using *in vitro* transcription (IVT) from a DNA template they are required to have the 5′ cap added in order to be translated (Jani and Fuchs 2012). There are a few basic approaches to adding the cap to mRNA produced using IVT (Jani and Fuchs 2012; Martin and Moss 1976; Wei and Moss 1977). However, neither is 100% effective (Beverly et al. 2016; Tsuda and Suzuki 1982). One possibility is that the cap can be incorporated with the reverse orientation (GpppGm7) such that the 7-methylguannine is no longer at the 5′ end (Beverly et al. 2016). The development of anti-reverse cap analogs can be used to ensure proper orientation (Jemielity et al. 2003; Peng et al. 2002; Stepinski et al. 2001) but there can still be undesirable capping products formed during this process (Beverly et al. 2016). The most common of these include the loss of the 7-methyl guanosine due to simple competition with GTP during the addition of the cap (Beverly et al. 2016). This results in a cap structure of pppG (Beverly et al. 2016).

Protocol 2: Determination of mRNA capping efficiency using magnetic bead capture and LC-MS (Beverly et al. 2016).

Key supplies: RNase H, RNase reaction buffer, and RNA 5′-pyrophosphohydrolase with NEbuffer 2 were obtained from New England Biolabs (Ipswich, MA). Streptavidin-coated magnetic beads (Dynabeads MyOne Streptavidin C1) were obtained from Invitrogen (Grand Island, NY). The separation was conducted using a Waters Acquity UPLC system (Milford, MA) interfaced to a Thermo Scientific Q-Exactive orbitrap mass spectrometer (Grand Island, NY). The column was a Waters 2.1 × 50 mm C18 (Milford, MA).

Part 1: Sample Preparation

1) Remove 0.1 M NaCl storage solution from beads by applying a magnetic field or by centrifugation at 15 000 × *g* for five minutes.
2) Add 120 μl of a solution containing the mRNA and the biotinylated probe sequence to the streptavidin beads and incubate for 30 minutes at room temperature. Gently rotate the tube to ensure complete binding of the biotinylated probe sequence.
3) Add 10 μl of RNase H (50 units) and mix via pipetting. Incubate the mixture for three hours at 37 °C.
4) Magnetically condense the beads and discard the supernatant without touching the beads.
5) Add 100 μl of a solution containing 5 mM Tris-HCl, pH 7.5; 0.5 mM EDTA; and 1 M NaCl to wash the beads. Mix by pipetting. Magnetically condense the beads and remove the supernatant.
6) Repeat the washing step three times to remove the RNase H, residual mRNA, and the reaction solution.
7) Wash the beads three times using distilled water to remove salts from the sample.
8) To release the bound RNA from the beads, remove the distilled water and replace it with 100 μl of 75% methanol that has been heated to 80 °C. Heat this mixture for three minutes at 80 °C and then place on a magnet for three minutes.
9) Remove the supernatant which contains the RNA and the biotinylated capture sequence.
10) Evaporation the supernatant down to 10 μl using an evaporative centrifuge at room temperature (~45 minutes).
11) Reconstitute the sample residue in 50 μl of 100 μM EDTA/1% methanol prior to LC-MS analysis.

Part 2: LC-MS Analysis

1) The analysis of the 5′ end containing mRNA fragments is conducted using a mobile phase consisting of A: 200 mM hexafluoroisopropanol, 8.5 mM trimethylamine, pH 7.9 and B: methanol.
2) The column is held at 75 °C at a flow rate of 300 μl min^{-1}.
3) A gradient for elution starts at 5% B for 1 minute followed by a linear increase to 25% over 12 minutes. At 12 minutes there is a sudden increase to 90% B for 1 minute to wash the column followed by a return to 5% B at 13 minutes to equilibrate the column prior to injecting the next sample.
4) The mass spectrometry system is operated in the negative ion mode over a *m/z* range from 600 to 3000 *m/z* with the resolution set to either 70 000 or 140 000.
5) The source and capillary temperatures are set to 400 °C.

25.7 Summary

When developing a LC-MS method for the determination of oligonucleotides, a number of analyte-specific factors need to be considered in selecting the most appropriate method of sample preparation. Summarized below are some of the key considerations.

- Be aware of the possible losses of analytes due to nonspecific adsorption, particularly at low concentration levels. Therefore, it is important to minimize the use of filters and volumetric transfers. Device with a deactivated surface is strongly recommended. An analog oligonucleotide at a relatively high concentration can serve as an IS and also as sacrificial oligonucleotide to minimize the impact due to nonspecific adsorption.
- A combination of liquid–liquid extraction and SPE needs to be considered in order to obtain high recovery of the analyte of interest with minimized ion suppression.
- For the analysis of oligonucleotides containing phosphorothioate linkages, it is important to treat the samples with an antioxidant to prevent the possible conversion of the phosphorothioate back to a phosphodiester.
- Care must be taken when handling RNA due to its susceptibility to degradation from RNases. Therefore, sterile materials and/or solutions free from RNases have to be considered to ensure the robustness of the assay.

References

Apffel, A., Chakel, J.A., Fischer, S. et al. (1997). Analysis of oligonucleotides by HPLC-electrospray ionization mass spectrometry. *Anal. Chem.* 69: 1320–1325.

Bahr, U., Aygün, H., and Karas, M. (2009). Sequencing of single and double stranded RNA oligonucleotides by acid hydrolysis and MALDI mass spectrometry. *Anal. Chem.* 81: 3173–3179.

Basiri, B., Hattum, H.V., van Dongen, W.D. et al. (2017). The role of fluorinated alcohols as mobile phase modifiers for LC-MS analysis of oligonucleotides. *J. Am. Soc. Mass Spectrom.* 28 (1): 190–199.

Belmont, P., Constant, J.-F., and Demeunynck, M. (2001). Nucleic acid conformation diversity: from structure to function and regulation. *Chem. Soc. Rev.* 30: 70–81.

Beverly, M., Dell, A., Parmar, P., and Houghton, L. (2016). Label-free analysis of mRNA capping efficiency using RNase H probes and LC-MS. *Anal. Bioanal. Chem.* 1–10.

Beverly, M., Hartsough, K., and Machemer, L. (2005). Liquid chromatography/electrospray mass spectrometric analysis of metabolites from an inhibitory RNA duplex. *Rapid Commun. Mass Spectrom.* 19: 1675–1682.

Beverly, M., Hartsough, K., Machemer, L. et al. (2006). Liquid chromatography electrospray ionization mass spectrometry analysis of the ocular metabolites from a short interfering RNA duplex. *J. Chromatogr. B* 835: 62–70.

Bird, A. (1992). The essentials of DNA methylation. *Cell* 70: 5–8.

Bird, A.P. and Wolffe, A.P. (1999). Methylation-induced repression – belts, braces, and chromatin. *Cell* 99: 451–454.

Birdsall, R.E., Gilar, M., Shion, H. et al. (2016). Reduction of metal adducts in oligonucleotide mass spectra in ion-pair reversed-phase chromatography/mass spectrometry analysis. *Rapid Commun. Mass Spectrom.* 30: 1667–1679.

Blackburn, G.M. (2006). *Nucleic Acids in Chemistry and Biology*. Cambridge, UK: Royal Society of Chemistry.

Cao, X. and Limbach, P.A. (2015). Enhanced detection of post-transcriptional modifications using a mass-exclusion list strategy for RNA modification mapping by LC-MS/MS. *Anal. Chem.* 87: 8433–8440.

Capuano, F., Mülleder, M., Kok, R. et al. (2014). Cytosine DNA methylation is found in *Drosophila melanogaster* but absent in *Saccharomyces cerevisiae*, *Schizosaccharomyces pombe*, and other yeast species. *Anal. Chem.* 86: 3697–3702.

Chen, B. and Bartlett, M. (2012). A one-step solid phase extraction method for bioanalysis of a phosphorothioate oligonucleotide and its 3′ n-1 metabolite from rat plasma by uHPLC-MS/MS. *AAPS J.* 14: 772–780.

Chen, B. and Bartlett, M.G. (2013). Evaluation of mobile phase composition for enhancing sensitivity of targeted quantification of oligonucleotides using ultra-high performance liquid chromatography and mass spectrometry: application to phosphorothioate deoxyribonucleic acid. *J. Chromatogr. A* 1288: 73–81.

Chen, C., Ridzon, D.A., Broomer, A.J. et al. (2005). Real-time quantification of microRNAs by stem–loop RT–PCR. *Nucleic Acids Res.* 33: e179–e179.

Cheng, A., Li, M., Liang, Y. et al. (2009). Stem-loop RT-PCR quantification of siRNAs in vitro and in vivo. *Oligonucleotides* 19: 203–208.

Crooke, S.T. (1992). Therapeutic applications of oligonucleotides. *Annu. Rev. Pharmacol. Toxicol.* 32: 329–376.

Crooke, S.T. and Bennett, C.F. (1996). Progress in antisense oligonucleotide therapeutics. *Annu. Rev. Pharmacol. Toxicol.* 36: 107–129.

Dagle, J.M. and Weeks, D.L. (1996). Positively charged oligonucleotides overcome potassium-mediated inhibition of triplex DNA formation. *Nucleic Acids Res.* 24: 2143–2149.

Dai, G., Wei, X., Liu, Z. et al. (2005). Characterization and quantification of Bcl-2 antisense G3139 and metabolites in plasma and urine by ion-pair reversed phase HPLC coupled with electrospray ion-trap mass spectrometry. *J. Chromatogr. B* 825: 201–213.

Dams, R., Benijts, T., Günther, W. et al. (2002). Influence of the eluent composition on the ionization efficiency for morphine of pneumatically assisted electrospray, atmospheric-pressure chemical ionization and sonic spray. *Rapid Commun. Mass Spectrom.* 16: 1072–1077.

Deleavey, G.F. and Damha, M.J. (2012). Designing chemically modified oligonucleotides for targeted gene silencing. *Chem. Biol.* 19: 937–954.

De Mesmaeker, A., Altmann, K.-H., Waldner, A., and Wendeborn, S. (1995). Backbone modifications in oligonucleotides and peptide nucleic acid systems. *Curr. Opin. Struct. Biol.* 5: 343–355.

Dickerson, R.E., Drew, H.R., Conner, B.N. et al. (1982). The anatomy of a-, b-, and z-DNA. *Science* 216: 475–485.

Douthwaite, S. and Kirpekar, F. (2007). Identifying modifications in RNA by MALDI mass spectrometry. *Methods Enzymol.* 425: 1–20.

Eckstein, F. (2000). Phosphorothioate oligodeoxynucleotides: what is their origin and what is unique about them? *Antisense Nucleic Acid Drug Dev.* 10: 117–121.

Egli, M. and Saenger, W. (2013). *Principles of Nucleic Acid Structure*. New York: Springer Science & Business Media.

Erb, R., Leithner, K., Bernkop-Schnürch, A., and Oberacher, H. (2012). Phosphorothioate oligonucleotide quantification by μ-liquid chromatography-mass spectrometry. *AAPS J.* 14: 728–737.

Erb, R. and Oberacher, H. (2014). Comparison of mobile-phase systems commonly applied in liquid chromatography-mass spectrometry of nucleic acids. *Electrophoresis* 35: 1226–1235.

Faria, M. and Ulrich, H. (2008). Sugar boost: when ribose modifications improve oligonucleotide performance. *Curr. Opin. Mol. Ther.* 10: 168–175.

Flores-Jasso, C.F., Salomon, W.E., and Zamore, P.D. (2013). Rapid and specific purification of Argonaute-small RNA complexes from crude cell lysates. *RNA* 19: 271–279.

Graham, M.J., Lee, R.G., Bell, T.A. et al. (2013). Antisense oligonucleotide inhibition of apolipoprotein C-III reduces plasma triglycerides in rodents, nonhuman primates, and humans. *Circ. Res.* 112: 1479–1490.

Griffey, R.H., Greig, M.J., Gaus, H.J. et al. (1997). Characterization of oligonucleotide metabolism in vivo via liquid chromatography/electrospray tandem mass spectrometry with a quadrupole ion trap mass spectrometer. *J. Mass Spectrom.* 32: 305–313.

Hauptmann, J., Schraivogel, D., Bruckmann, A. et al. (2015). Biochemical isolation of Argonaute protein complexes by Ago-APP. *Proc. Natl. Acad. Sci.* 112: 11841–11845.

Hoke, G.D., Draper, K., Freier, S.M. et al. (1991). Effects of phosphorothioate capping on antivense oligonucleotide stability, hybridization and antiviral efficacy versus herpes simplex virus infection. *Nucleic Acids Res.* 19: 5743–5748.

Hu, C.-W., Lee, H., Chen, J.-L. et al. (2013). Optimization of global DNA methylation measurement by LC-MS/MS and its application in lung cancer patients. *Anal. Bioanal. Chem.* 405: 8859–8869.

Huber, C.G. and Krajete, A. (1999). Analysis of nucleic acids by capillary ion-pair reversed-phase HPLC coupled to negative-ion electrospray ionization mass spectrometry. *Anal. Chem.* 71: 3730–3739.

Huber, C.G., Oefner, P.J., and Bonn, G.K. (1993). High-resolution liquid chromatography of oligonucleotides on nonporous alkylated styrene-divinylbenzene copolymers. *Anal. Biochem.* 212: 351–358.

Jani, B. and Fuchs, R. (2012). In vitro transcription and capping of Gaussia luciferase mRNA followed by HeLa cell transfection. *J. Vis. Exp.* 61: e3702.

Jemielity, J., Fowler, T., Zuberek, J. et al. (2003). Novel "anti-reverse" cap analogs with superior translational properties. *RNA* 9: 1108–1122.

Johnson, J.L., Guo, W., Zang, J. et al. (2005). Quantification of raf antisense oligonucleotide (rafAON) in biological matrices by LC-MS/MS to support pharmacokinetics of a liposome-entrapped rafAON formulation. *Biomed. Chromatogr.* 19: 272–278.

Kahl, G. (2015). *The Dictionary of Genomics, Transcriptomics and Proteomics*. Hoboken, NJ: Wiley.

Kern, J.T. and Kerwin, S.M. (2002). The aggregation and G-quadruplex DNA selectivity of charged 3, 4, 9, 10-perylenetetracarboxylic acid diimides. *Bioorg. Med. Chem. Lett.* 12: 3395–3398.

Klose, R.J. and Bird, A.P. (2006). Genomic DNA methylation: the mark and its mediators. *Trends Biochem. Sci.* 31: 89–97.

Korfmacher, W.A. (2009). *Using Mass Spectrometry for Drug Metabolism Studies*. Boca Raton, FL: CRC Press.

Kowalak, J.A., Pomerantz, S.C., Crain, P.F., and McCloskey, J.A. (1993). A novel method for the determination of posttranscriptional modification in RNA by mass spectrometry. *Nucleic Acids Res.* 21: 4577–4585.

Kurreck, J. (2003). Antisense technologies. *Eur. J. Biochem.* 270: 1628–1644.

Kutyavin, I.V., Lokhov, S.G., Afonina, I.A. et al. (2002). Reduced aggregation and improved specificity of G-rich oligodeoxyribonucleotides containing pyrazolo [3, 4-d] pyrimidine guanine bases. *Nucleic Acids Res.* 30: 4952–4959.

Le, T., Kim, K.-P., Fan, G., and Faull, K.F. (2011). A sensitive mass spectrometry method for simultaneous quantification of DNA methylation and hydroxymethylation levels in biological samples. *Anal. Biochem.* 412: 203–209.

Lescrinier, E., Froeyen, M., and Herdewijn, P. (2003). Difference in conformational diversity between nucleic acids with a six-membered "sugar" unit and natural "furanose" nucleic acids. *Nucleic Acids Res.* 31: 2975–2989.

Lewis, J.D. and Izaurflde, E. (1997). The role of the cap structure in RNA processing and nuclear export. *Eur. J. Biochem.* 247: 461–469.

Li, X. and Franke, A.A. (2011). High-throughput and cost-effective global DNA methylation assay by liquid chromatography-mass spectrometry. *Anal. Chim. Acta* 703: 58–63.

Li, L., Zhu, D., Huang, L. et al. (2012). Argonaute 2 complexes selectively protect the circulating microRNAs in cell-secreted microvesicles. *PLoS One* 7: e46957.

Lin, Z.J., Li, W., and Dai, G. (2007). Application of LC–MS for quantitative analysis and metabolite identification of therapeutic oligonucleotides. *J. Pharm. Biomed. Anal.* 44: 330–341.

Lin, X.-C., Zhang, T., Liu, L. et al. (2016). Mass spectrometry based ultrasensitive DNA methylation profiling using target fragmentation assay. *Anal. Chem.* 88: 1083–1087.

Lodish, H., Berk, A., Zipursky, S.L. et al. (2000). Structure of nucleic acids. In: *Molecular Cell Biology*, 4e. New York: W. H. Freeman.

Low, D.A., Weyand, N.J., and Mahan, M.J. (2001). Roles of DNA adenine methylation in regulating bacterial gene expression and virulence. *Infect. Immun.* 69: 7197–7204.

Manoharan, M. (1999). 2′-Carbohydrate modifications in antisense oligonucleotide therapy: importance of conformation, configuration and conjugation. *Biochim. Biophys. Acta* 1489: 117–130.

Martin, S.A. and Moss, B. (1976). mRNA guanylyltransferase and mRNA (guanine-7-)-methyltransferase from vaccinia virions. Donor and acceptor substrate specificites. *J. Biol. Chem.* 251: 7313–7321.

McCarthy, S.M., Gilar, M., and Gebler, J. (2009). Reversed-phase ion-pair liquid chromatography analysis and purification of small interfering RNA. *Anal. Biochem.* 390: 181–188.

McGinnis, A.C., Cummings, B.S., and Bartlett, M.G. (2013a). Ion exchange liquid chromatography method for the direct determination of small ribonucleic acids. *Anal. Chim. Acta* 799: 57–67.

McGinnis, A.C., Grubb, E.C., and Bartlett, M.G. (2013b). Systematic optimization of ion-pairing agents and hexafluoroisopropanol for enhanced electrospray ionization mass spectrometry of oligonucleotides. *Rapid Commun. Mass Spectrom.* 27: 2655–2664.

Mignone, F., Gissi, C., Liuni, S., and Pesole, G. (2002). Untranslated regions of mRNAs. *Genome Biol.* 3: 1.

Miller, P.S., Cassidy, R.A., Hamma, T., and Kondo, N.S. (2000). Studies on anti-human immunodeficiency virus oligonucleotides that have alternating methylphosphonate/phosphodiester linkages. *Pharmacol. Ther.* 85: 159–163.

Miller, P. and Ts'o, P. (1987). A new approach to chemotherapy based on molecular biology and nucleic acid chemistry: Matagen (masking tape for gene expression). *Anticancer Drug Des.* 2: 117–128.

Milligan, J.F., Matteucci, M.D., and Martin, J.C. (1993). Current concepts in antisense drug design. *J. Med. Chem.* 36: 1923–1937.

Mitchell, C.R., Bao, Y., Benz, N.J., and Zhang, S. (2009). Comparison of the sensitivity of evaporative universal detectors and LC/MS in the HILIC and the reversed-phase HPLC modes. *J. Chromatogr. B* 877: 4133–4139.

Monia, B.P., Lesnik, E., Gonzalez, C. et al. (1993). Evaluation of 2′-modified oligonucleotides containing 2′-deoxy gaps as antisense inhibitors of gene expression. *J. Biol. Chem.* 268: 14514–14522.

Murphy, A.T., Brown-Augsburger, P., Yu, R.Z. et al. (2005). Development of an ion-pair reverse-phase liquid chromatographic/tandem mass spectrometry method for the determination of an 18-mer phosphorothioate oligonucleotide in mouse liver tissue. *Eur. J. Mass Spectrom.* 11: 209–216.

Murray, S., Ittig, D., Koller, E. et al. (2012). TricycloDNA-modified oligo-2′-deoxyribonucleotides reduce scavenger receptor B1 mRNA in hepatic and extra-hepatic tissues – a comparative study of oligonucleotide length, design and chemistry. *Nucleic Acids Res.* 40: 6135–6143.

Nikcevic, I., Wyrzykiewicz, T.K., and Limbach, P.A. (2011). Detecting low-level synthesis impurities in modified phosphorothioate oligonucleotides using liquid chromatography–high resolution mass spectrometry. *Int. J. Mass Spectrom.* 304: 98–104.

Nozaki, K., Tarui, A., Osaka, I. et al. (2010). Elimination technique for alkali metal ion adducts from an electrospray ionization process using an on-line ion suppressor. *Anal. Sci.* 26: 715–718.

Oberacher, H., Parson, W., Mühlmann, R., and Huber, C.G. (2001). Analysis of polymerase chain reaction products by on-line liquid chromatography-mass spectrometry for genotyping of polymorphic short tandem repeat loci. *Anal. Chem.* 73: 5109–5115.

Peng, Z.-H., Sharma, V., Singleton, S.F., and Gershon, P.D. (2002). Synthesis and application of a chain-terminating dinucleotide mRNA cap analog. *Org. Lett.* 4: 161–164.

Portnoy, V., Lin, S.H.S., Li, K.H. et al. (2016). saRNA-guided Ago2 targets the RITA complex to promoters to stimulate transcription. *Cell Res.* 26: 320–335.

Quinlivan, E.P. and Gregory, J.F. (2008). DNA methylation determination by liquid chromatography–tandem mass spectrometry using novel biosynthetic [U-15N] deoxycytidine and [U-15N] methyldeoxycytidine internal standards. *Nucleic Acids Res.* 36: e119–e119.

Rao, S.N. and Kollman, P. (1985). On the role of uniform and mixed sugar puckers in DNA double-helical structures. *J. Am. Chem. Soc.* 107: 1611–1617.

Ravikumar, V.T., Krishna Kumar, R., Capaldi, D.C., and Cole, D.L. (2003). Synthesis of high quality phosphorothioate oligonucleotides as antisense drugs. Use of I-linker in the elimination of 3′-terminal phosphorothioate monoesters. *Nucleosides Nucleotides Nucleic Acids* 22: 1421–1425.

Rich, A., Wang, A., Quigley, G. (1978). Diversity of sugar pucker in the nucleic acids. Paper presented at the Frontiers of Bioorganic Chemistry and Molecular Biology: Proceedings of the International Symposium on Frontiers of Bioorganic Chemistry and Molecular Biology, Moscow and Tashkent, USSR (25 September–2 October 1978).

Rudge, J., Scott, G., Hail, M., and McGinley, M. (2011). Preparation and LC/MS analysis of oligonucleotide therapeutics from biological matrices. *Chromatography Today* 4 (1): 16–20.

Sauer, S., Lechner, D., Berlin, K. et al. (2000). Full flexibility genotyping of single nucleotide polymorphisms by the GOOD assay. *Nucleic Acids Res.* 28: e100–e100.

Sayers, J.R., Olsen, D.B., and Eckstein, F. (1989). Inhibition of restriction endonuclease hydrolysis by phosphorothioate-containing DNA. *Nucleic Acids Res.* 17: 9495.

Schneider, K. and Chait, B.T. (1993). Matrix-assisted laser desorption mass spectrometry of homopolymer oligodeoxyribonucleotides. Influence of base composition on the mass spectrometric response. *Org. Mass Spectrom.* 28: 1353–1361.

Sen, D. and Gilbert, W. (1988). Formation of parallel four-stranded complexes by guanine-rich motifs in DNA and its implications for meiosis. *Nature* 334: 364–366.

Sen, D. and Gilbert, W. (1992). [10] Guanine quartet structures. *Methods Enzymol.* 211: 191–199.

Sinden, R.R. (2012). *DNA Structure and Function*. St. Louis: Elsevier.

Stepinski, J., Waddell, C., Stolarski, R. et al. (2001). Synthesis and properties of mRNAs containing the novel "anti-reverse" cap analogs 7-methyl (3′-O-methyl) GpppG and 7-methyl (3′-deoxy) GpppG. *RNA* (7): 1486–1495.

Su, D., Chan, C.T., Gu, C. et al. (2014). Quantitative analysis of tRNA modifications by HPLC-coupled mass spectrometry. *Nat. Protoc.* 9: 828.

Taniguchi, H. and Hayashi, N. (1998). A liquid chromatography/electrospray mass spectrometric study on the post-transcriptional modification of tRNA. *Nucleic Acids Res.* 26: 1481–1486.

Tsuda, M. and Suzuki, Y. (1982). Efficient and strand-selective in vitro transcription initiation by purified RNA polymerase II from a unique site of the fibroin gene. *J. Biol. Chem.* 257: 12367–12372.

Uhlmann, E. and Peyman, A. (1990). Antisense oligonucleotides: a new therapeutic principle. *Chem. Rev.* 90: 543–584.

Van Der Velden, A.W. and Thomas, A.A. (1999). The role of the 5′ untranslated region of an mRNA in translation regulation during development. *Int. J. Biochem. Cell Biol.* 31: 87–106.

Varma, R.S. (1993). Synthesis of oligonucleotide analogues with modified backbones. *Synlett* 1993: 621–637.

Verdolino, V., Cammi, R., Munk, B.H., and Schlegel, H.B. (2008). Calculation of pK_a values of nucleobases and the guanine oxidation products guanidinohydantoin and spiroiminodihydantoin using density functional theory and a polarizable continuum model. *J. Phys. Chem. B.* 112: 16860–16873.

Walsh, C.P. and Bestor, T.H. (1999). Cytosine methylation and mammalian development. *Genes Dev.* 13: 26–34.

Wan, W.B. and Seth, P.P. (2016). The medicinal chemistry of therapeutic oligonucleotides. *J. Med. Chem.* 59: 9645–9667.

Wei, C. and Moss, B. (1977). 5′-Terminal capping of RNA by guanylyltransferase from HeLa cell nuclei. *Proc. Natl. Acad. Sci.* 74: 3758–3761.

Wetzel, C. and Limbach, P.A. (2012). Global identification of transfer RNAs by liquid chromatography–mass spectrometry (LC–MS). *J. Proteomics* 75: 3450–3464.

Wilson, G.G. and Murray, N.E. (1991). Restriction and modification systems. *Annu. Rev. Genet.* 25: 585–627.

Wu, Z.-S., Guo, M.-M., Shen, G.-L., and Yu, R.-Q. (2007). G-rich oligonucleotide-functionalized gold nanoparticle aggregation. *Anal. Bioanal. Chem.* 387: 2623–2626.

Wu, T.P., Wang, T., Seetin, M.G. et al. (2016). DNA methylation on N6-adenine in mammalian embryonic stem cells. *Nature* 532 (7599): 329–333.

Yamauchi, Y., Nobe, Y., Izumikawa, K. et al. (2016). A mass spectrometry-based method for direct determination of pseudouridine in RNA. *Nucleic Acids Res.* 44: e59–e59.

Yamauchi, Y., Taoka, M., Nobe, Y. et al. (2013). Denaturing reversed phase liquid chromatographic separation of non-coding ribonucleic acids on macro-porous polystyrene-divinylbenzene resins. *J. Chromatogr. A* 1312: 87–92.

Yang, M., Haase, A.D., Huang, F.-K. et al. (2014). Dephosphorylation of tyrosine 393 in argonaute 2 by protein tyrosine phosphatase 1B regulates gene silencing in oncogenic RAS-induced senescence. *Mol. Cell* 55: 782–790.

Ye, G. and Beverly, M. (2011). The use of strong anion-exchange (SAX) magnetic particles for the extraction of therapeutic siRNA and their analysis by liquid chromatography/mass spectrometry. *Rapid Commun. Mass Spectrom.* 25: 3207–3215.

26

Sample Preparation for LC-MS Bioanalysis of Antibody–Drug Conjugates

Cong Wei and Ragu Ramanathan

26.1 Introduction

Antibody–drug conjugates (ADCs) have emerged as an important class of targeted biological therapeutics. ADCs are composed of monoclonal antibodies (mAbs) and cytotoxic drugs (payloads) linked together through cysteine, lysine, or other engineered residues via a variety of linkers (Grafmuller et al. 2016). Specific delivery of ADC as chemotherapeutic agents to tumor cells can be achieved, therefore, enhancing pharmacological activity at the target site while sparing nontarget tissues from chemotherapeutic damage (Alley et al. 2008; Chari 2008; Jaracz et al. 2005). Due to the complex structure of ADCs, their stability requirements exceed the simple sum of each part of the molecule as the stability of mAb, payload, and linker each has impact on toxicity, efficacy, and potentially anti-drug antibody (ADA)-response of the ADCs (Ross and Wolfe 2016). Furthermore, due to the exquisite potency of their payloads, ADCs must satisfy some demanding criteria with respect to specificity, safety, and stability in order to achieve effective and selective delivery of the cytotoxic agent to the target.

Bioanalysis plays a critical role in ADC drug development for the assessment of safety and efficacy. The unique and complex structural features of ADC require specific bioanalytical strategies that are different from those applied to small-molecule xenobiotics and large-molecule biologics (e.g. peptide or protein therapeutics). Quantification of four major analytes, including unconjugated payload, conjugated payload, ADC (i.e. conjugated mAb), and total mAb, as well as characterization of changes in drug-to-antibody ratio (DAR) distribution in biological matrices, are integral components of ADC bioanalysis. Due to the complexity of ADCs, sample preparation can be challenging and should be carefully considered and executed in the bioanalytical workflows for various moieties of the ADC. This chapter describes different sample preparation techniques and their applications in quantitative analysis of four major analytes mentioned above. In addition, sample preparation for the determination of DAR from biological matrices is discussed, which offers an indication of the ADC stability and possible biotransformation.

26.2 Properties of ADC and Challenges for Sample Preparation

A desirable ADC should remain stable with its three structural components (mAb, linker, and payload) in systematic circulation to achieve the goal of targeted delivery of tumor-fighting payload to the intended site while minimizing adverse effects on nontarget cells (Casi and Neri 2012). However, competing physical, chemical, and biological processes impact on the structural integrity of ADCs in circulation, leading to production of new ADC subspecies which subsequently complicates the bioanalysis (Grafmuller et al. 2016) and pharmacokinetic (PK) assessment (Lin and Tibbitts 2012) during drug development. Normally, the bioanalysis of a therapeutic molecule, either a mAb or a small-molecule drug, is relatively straightforward. In most of the applications, the concentrations of the parent species are determined over time by using the starting material to prepare the calibration standards and QCs. The concentration versus time profile is then used to calculate PK parameters for the therapeutic. In contrast, an ADC therapeutic is typically a complex and dynamically changing mixture *in vivo* due to biotransformation with a different clearance rate associated with a different DAR species, or a combination of these processes (e.g. payload deconjugation and/or other biotransformations) (Kaur et al. 2013; Xu et al. 2013). In this context, even the existing PK language for "therapeutic concentration" versus time becomes ambiguous. Moreover, the calibration standards comprised of the

Sample Preparation in LC-MS Bioanalysis, First Edition. Edited by Wenkui Li, Wenying Jian, and Yunlin Fu.

reference materials may no longer fully represent the analytes present in the study samples *in vivo*, presenting a unique challenge for quantitative assays. Therefore, in ADC bioanalysis, it is important to clearly define the species (e.g. unconjugated payload, conjugated payload, ADC, and/or total mAb) to be measured in the *in vivo* samples as the sample preparation strategy may need to be adjusted from one species to another (Kaur et al. 2013).

Despite improvement in ADC design and synthesis, payload deconjugation still occurs following ADC administration, leading to an increase in unconjugated payload levels and changes in the ADC composition (i.e. the DAR distribution) *in vivo* (Casi and Neri 2012). As a result, efficacy and safety of the ADCs may be impacted. Therefore, the measurement of the conjugated payloads *in vivo* is important for the evaluation of the safety and efficacy of ADCs and for the assessment of drug (payload) exposure at the target site; the determination of change in the DAR distribution and average DAR values are critical for providing the stability profile of the ADCs.

26.2.1 Unconjugated Payload

Payloads incorporated into ADCs are largely cytotoxic small molecules, including maytansinoid (DM1) in trastuzumab-emtansine (anti-HER2) by Genentech/Roche, calicheamicin in inotuzumab-ozogamicin (anti-CD22) by Pfizer, mono-methyl auristatin E (MMAE) in brentuximab-vedotin (anti-CD30) by Seattle Genetics, and pyrrolobenzodiazepine (PBD), etc. The potency of the payload is usually sufficiently high to bring cell killing in the tumor cell (Rao et al. 2015). Tubulin polymerization inhibitors such as auristatins, maytansines (Fanale et al. 2014; Krop et al. 2010), and DNA-damaging agents such as calicheamicin (Li et al. 2014) are clinically validated payloads. Other classes, for example, DNA-alkylating agents which bind to the minor groove of DNA, such as duocarmycins and PBD, have shown preclinical efficacy in mouse models and a few ADCs conjugated with PBD are also currently in clinical trials (Cabanillas 1992; Elgersma et al. 2015).

Unconjugated payload includes the payload deconjugated from the ADCs in the plasma or tumor tissues after administration and those present in trace in the ADC formulation.

The sample extraction methods for unconjugated payload bioanalysis include protein precipitation (Grafmuller et al. 2016; Wei et al. 2016), solid-phase extraction (SPE) (Heudi et al. 2016), liquid–liquid extraction (LLE) (Wei et al. 2012), and supported-liquid extraction (SLE) (Figure 26.1). Protein precipitation method is usually employed for the studies in discovery and early development stage, while SPE, LLE, and SLE are often used for the studies in mid-late stage of drug development. It is worth noting that due to the presence of various proteases in matrix, unconjugated payload levels may be artificially elevated in the sample matrix after collection. Therefore, one consideration is to add protease inhibitor cocktails prior to sample extraction or perform the sample extraction in an ice water bath to prevent or minimize payload deconjugation from ADC in sample preparation.

Sensitivity is one of the challenges for unconjugated payload quantification. Unconjugated payload is often more hydrophobic compared to many small-molecule analytes. For instance, the lower limit of quantification (LLOQ) of an unconjugated payload (auristatin) was achieved down to $0.015\,\text{ng}\,\text{ml}^{-1}$. However, the measured concentrations were still below limit of quantification (BLQ) for the plasma samples obtained at either early time points (five minutes post dose) or late time points (two to three weeks post dose) in an *in vivo* study with Cynomolgus monkeys administered an intravenous bolus dose of $6\,\text{mg}\,\text{kg}^{-1}$ (Grafmuller et al. 2016). The LLOQ of unconjugated payload DM1 in human serum was reported as $0.20\,\text{ng}\,\text{ml}^{-1}$ for clinical samples by a validated bioanalytical assay using online SPE coupled with LC-tandem mass spectrometry (Heudi et al. 2016). However, the unconjugated payload DM1 in clinical patient samples with $0.3\,\text{mg}\,\text{kg}^{-1}$ ADC intravenous administration every three weeks appeared to be non-detectable by such a method (Heudi et al. 2016). Similarly, Wei et al. (2012) described the development of a sensitive and selective LC-MS/MS method for the quantification

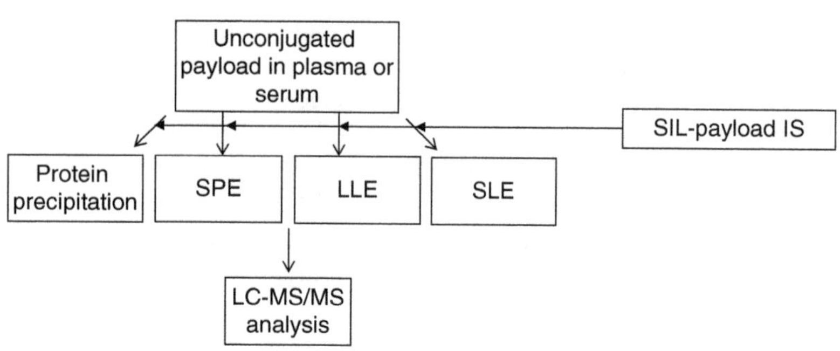

Figure 26.1 Workflow depicting various sample preparation methods for quantitative bioanalysis of unconjugated payload.

of unconjugated DM4 (another maytansinoid derivative with a thiol group, thiol-DM4) in plasma of Cynomolgus monkey and Sprague–Dawley rat, with LLOQ of $0.50\,ng\,ml^{-1}$. However, the unconjugated payload DM4 concentrations were only measured up to 24 hours in a 3-week time-course study in Cynomolgus monkeys (Wei et al. 2012). Birdsall et al. (2016) showed the LLOQ of another unconjugated payload (maleimide-valine-citrulline-dansyl sulfonamide ethyl amine [maleimide-linker-DSEA]) as $0.30\,ng\,ml^{-1}$ for trace-free drug species analysis with online SPE coupled with LC-tandem mass spectrometry. Overall, the reported sensitivity is not sufficient to cover late time-points of a study, and sensitivity remained a challenge when measuring the *in vivo* study samples.

Another challenge for unconjugated payload quantification is that the analytes sometimes need to be stabilized prior to the LC-MS/MS analysis. For example, payload DM1 contains a free thiol moiety, which is likely to dimerize or react with other thiol-containing molecules in serum, samples were pretreated with a reducing agent (tris(2-carboxyethyl) phosphine) (TCEP), and further blocked with *N*-ethylmaleimide (NEM) (Heudi et al. 2016). DM4 as a payload also contains a free thiol moiety, and a LLE procedure in an ice water bath was applied for the sample extraction, which demonstrated sufficient stability for the DM4 payload quantification (Wei et al. 2012). The plasma samples were also treated with chilled methanol before LLE to improve the release of payload from protein binding (Wei et al. 2012).

26.2.2 Conjugated Payload

Conjugated payload is defined as payload conjugated to the antibody. The conjugated payload represents the active ADC drug component and is the preferred analyte when possible in a hybrid immunoaffinity capture-LC-MS assay, which allows accurate quantification of conjugated payload for the ADC having varying DAR (Wang et al. 2016). There is an increasing interest in measuring the conjugated payload in circulation to define the PK of the ADCs (Liu et al. 2015). Moreover, the conjugated payload assay offers direct measurement of the bioactive cytotoxic payload molecules linked to the mAb and thus may provide more accurate assessment of payload exposure at the target site, e.g. in tumor tissues, and possibly exposure vs. efficacy correlation (Kumar et al. 2015).

Payload is conjugated to the antibody by linker which is an important part of ADC. Linkers can be of different types such as hydrazone linkers, disulfide linkers, and cleavable or non-cleavable peptide linkers (Dubowchik et al. 2002; Kaneko et al. 1991; Rao et al. 2015; Senter 2009; Toki et al. 2002). The mode of release of the active payload from these linkers differs markedly *in vivo*. The hydrazone linkers are pH sensitive and are cleaved at lower pH in compartments such as endosome and/or lysosome by releasing the active payload. Disulfide linkers take advantage of the reducing environment of the intracellular milieu to release the active payload (Ducry 2012; Rao et al. 2015). Cleavable peptide linkers are known to be cleaved by lysosomal proteases, whereas the non-cleavable linkers release the active payload by degradation of antibody in lysosomal compartment. Peptide-based linkers (cleavable and non-cleavable) have been found more use recently because they exhibit higher plasma stability and improved activity (Dubowchik et al. 2002; Rao et al. 2015). Overall, linker stability has been improved significantly through the creation of enzymatically cleavable and non-cleavable linkers since the chemically labile hydrazine and certain disulfide linkers were used in the first-generation ADC (Birdsall et al. 2016). Disulfide linker chemistry has also been improved significantly with incorporation of protection groups, for example, interchain disulfide reduction followed by addition of a dithiophenylmaleimide linker containing an alkyne group enables conjugation with payloads containing an azide group (Jackson 2016).

The feature of the linker moiety largely determines the analytical strategy for conjugated payload bioanalysis. For example, depending on whether a linker is enzymatically cleavable or non-cleavable, different sample preparation methods should be applied. If the linker is enzymatically cleavable (e.g. dipeptide-based cleavable linker), a lysosomal enzyme such as cathepsin B is able to specifically cleave the linker to release the payload for LC-MS detection, following anti-idiotype mAb immunoaffinity capture of the ADCs from matrices (Figure 26.2). If the linker is a non-cleavable linker which enzyme is not able to cleave the payload off the linker, then following the anti-idiotype mAb capture of the ADCs from matrices, protease (e.g. trypsin) digestion of the antibody is going to be applied and the peptide-linker-conjugated payload is the surrogate analyte monitored by LC-MS/MS (Figure 26.2). An anti-idiotype mAb capture-based LC-MS/MS assay enables the measurement of the payload conjugated to the mAb. Nevertheless, if using an anti-payload mAb as a capture reagent, one could potentially measure various types of payload molecules in the sample, including unconjugated payload, payload conjugated to the mAb, payload attached to the linker, and any catabolite of the mAb with payload attached (Liu et al. 2015). In addition, Protein A or Protein G can be used as a generic capture reagent, and a Protein A capture LC-MS/MS assay with papain (a cysteine protease) cleavage for conjugated-payload quantification of the valine-citrullin-linked MMAE ADC has recently been reported (Sanderson et al. 2016). Others have reported measuring total

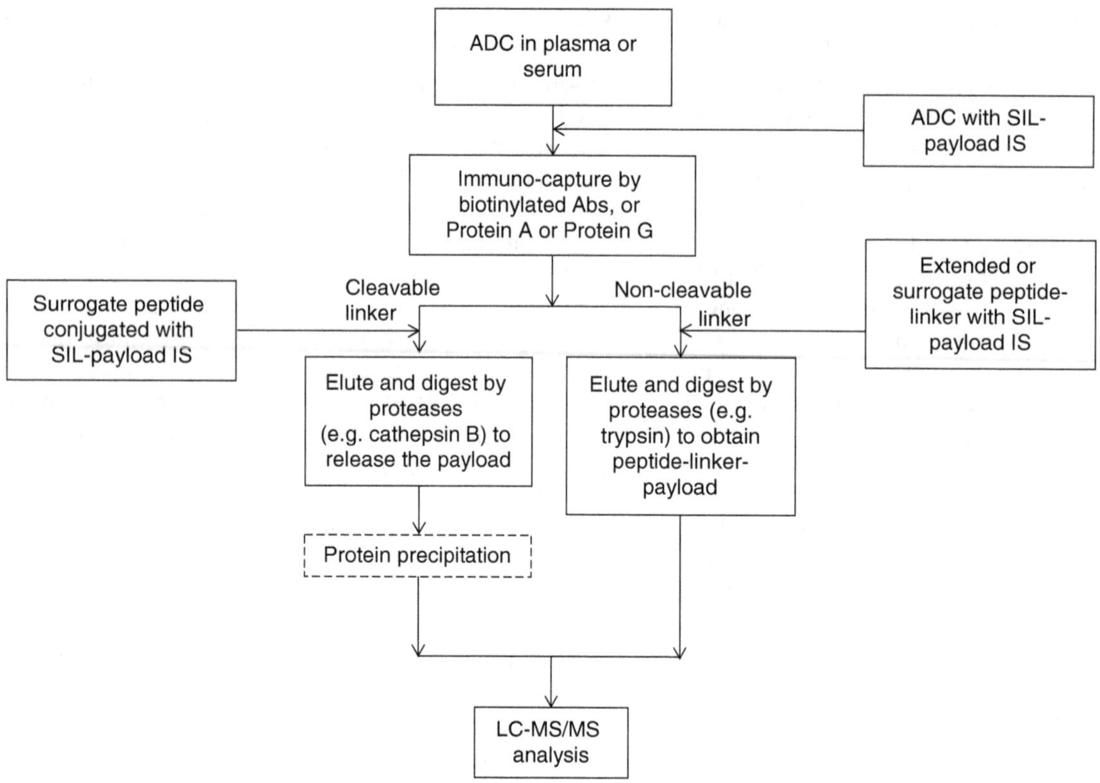

Figure 26.2 Schematic illustration of sample preparation workflows for conjugated payload quantification on ADCs with cleavable linkers and non-cleavable linkers, with SIL-IS options. The step with dash frame is optional.

payload in plasma by chemical cleavage of a labile linker (Buckwalter et al. 2004).

There are several challenges in sample preparation for quantitative analysis of conjugated payload. First, when the linker is not enzymatically cleavable but the calibration reference standard may not readily be available. The reference standard would be the payload conjugated with the peptide that is enzymatically cleaved from the Ab, and the reference standard often needs additional conjugation of the payload to the peptide which would vary depending on the conjugation site and where the protease cleaves in the Ab. In the case of non-cleavable linkers, the conjugated payload approach may not be suitable or be very complicated for lysine- or reduced interchain cysteine-linked ADCs. The conjugated payload assay also needs to include the corresponding internal standard (IS) which may also require additional synthesis. Second, the payload molecule and/or the linker could undergo metabolism without deconjugation from the Ab in circulation, which would result in different analytes to monitor by LC-MS/MS. Furthermore, soluble target in circulation may block the binding sites on the ADC, thus when anti-idiotype mAb is used for immunoaffinity capture in the conjugated payload assay, it may not be able to capture all the ADC, leading to reduced recovery if the assay is intended to measure the total Ab concentrations. In preclinical species, the soluble target may not be cross-reactive to the ADC, or the level of soluble target may be too low relative to the ADC level to have a significant impact on the measurement (Wang et al. 2016). However, the levels of soluble target in the patient populations are projected to be high and variable (Wang et al. 2016).

26.2.3 Conjugated Ab

Conjugated Ab is defined as Ab conjugated with at least one payload. Measurement of conjugated Ab is often employed to obtain both preclinical and clinical PK profiles of ADC. With limited knowledge on ADC performance in a clinical setting, it is still uncertain whether conjugated-Ab or conjugated-payload correlates more closely with efficacy and safety though it has been suggested that conjugated-antibody is the preferred analyte for toxicology correlation while conjugated-payload is the key analyte for efficacy (Kumar et al. 2015; Wang et al. 2016).

While ligand-binding assay (LBA) has been widely employed for conjugated Ab quantification by using anti-payload Ab as the primary Ab, hybrid immunoaffinity

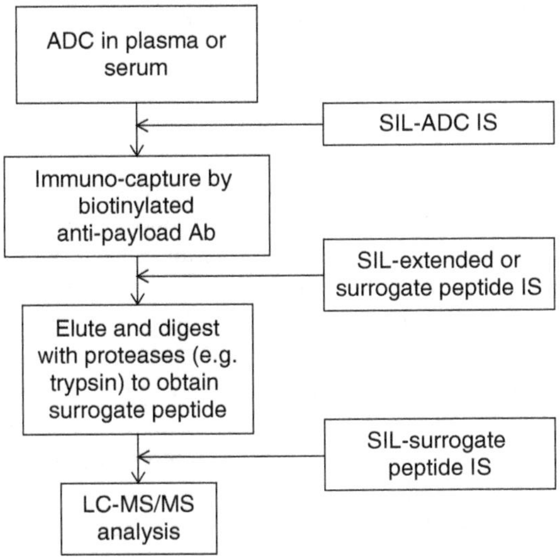

Figure 26.3 Workflow depicting sample preparation methods for conjugated antibody quantification, with SIL-IS options.

capture LC-MS/MS analysis has been playing an increasing role in the bioanalysis of conjugated Ab. For the latter, anti-payload antibody can be used for immunoaffinity capture of the conjugated Ab, and the procedure is then followed by enzymatic (e.g. trypsin) digestion and LC-MS/MS of the surrogate peptide of the Ab (Figure 26.3). A peptide in the Fc region or in the heavy chain complementarity-determining region (CDR) of human IgGs can be used as the surrogate peptide in LC-MS/MS in support of nonclinical studies, while a surrogate peptide in the heavy chain CDR is needed in LC-MS/MS in support of clinical studies. The caveat of this approach is that the anti-payload antibody may preferentially bind to the unconjugated payload in the matrix, thus relatively excess amount of anti-payload antibody is usually needed to ensure the capture of all conjugated Ab.

Another concern for the conjugated Ab assay is that the calibration curve prepared using the ADC drug reference may not fully represent the analytes in the study samples because of the possible changes in *in vivo* DAR over the sampling period of the study. As aforementioned, the ADC *in vivo* is a mixture of ADC species of different DARs, typically 0–8 in the random lysine-conjugated ADCs (Gorovits et al. 2013; Kaur et al. 2013; Stephan et al. 2011). Unfortunately, the individual DAR1–8 ADC reference standards are not readily available and only the drug substance with an average DAR of 3–4 is used as the reference standard. As a result, the distribution of analyte species with different DAR values in the incurred samples is different from that in the standards (Wang et al. 2016). While the conjugated-payload measurement is sensitive to changes in the DAR (i.e. DAR-proportional), the conjugated-Ab assay measures the antibody portion of ADCs and is theoretically DAR independent (Gorovits et al. 2013; Kaur et al. 2013; Stephan et al. 2011; Wang et al. 2016). There was one case that for the random-conjugated ADC of interest, the conjugated-Ab LBA developed using anti-payload capture reagents was DAR-sensitive; in contrast, the hybrid conjugated-Ab LC-MS assays using anti-payload capture in the cartridge format were proven to be DAR-insensitive (Wang et al. 2016), for which the reason was unknown. Nevertheless, the LC-MS/MS approach to measuring the conjugated Ab has not yet been widely practical (when compared with LBA) unless there is a specific need, e.g. to differentiate different types of conjugated Abs, and soluble target-bound conjugated Ab vs. free conjugated Ab.

Another challenge is that the payload or linker in ADC may undergo biotransformation *in vivo* leading to the generation of metabolite-conjugated Ab. For the PK study, both payload-conjugated Ab and metabolite-conjugated Ab would need to be measured (when the metabolite is known to be active), for which the capture antibody that recognizes both payload and metabolite is needed for the immunoaffinity capture. It should be noted that very limited information is available whether the capture antibody could recognize both original payload and its metabolite equivalently, which is similar to the issue of equal/unequal recovery with different DAR species. If metabolite-conjugated Ab is considered as inactive, then only payload-conjugated Ab should be measured, for which a different anti-payload capture antibody that only recognizes payload, not metabolite, should be employed for the payload-conjugated Ab immunoaffinity capture.

26.2.4 Total Ab

Total Ab in ADC bioanalysis includes the antibodies that are conjugated to the payload (conjugated Ab) and also those naked (unconjugated). The Ab part of ADC structure is usually IgG1, IgG2, or IgG4 that shares structural characteristics with most mAbs. The molecular weight of an Ab is ~150 kDa. Total Ab profile is an important aspect of the overall PK assessment of ADC (Gorovits et al. 2013; Stephan et al. 2011). Traditionally, total Ab has been measured by LBA using antigen or anti-idiotype antibody (Stephan et al. 2011). For nonclinical studies, generic capture probes, e.g. anti-human IgGs antibodies, have been used for the total Ab assay in LBA as well. Immunoaffinity capture LC-MS/MS has been increasingly used as an alternative to LBA for protein and antibody quantification (Furlong et al. 2014; Lesur et al. 2010; van den Broek et al. 2013). This alternative total Ab assay approach is by using immunoaffinity capture LC-MS/MS followed by enzymatic digestion and a signature peptide in the heavy-chain

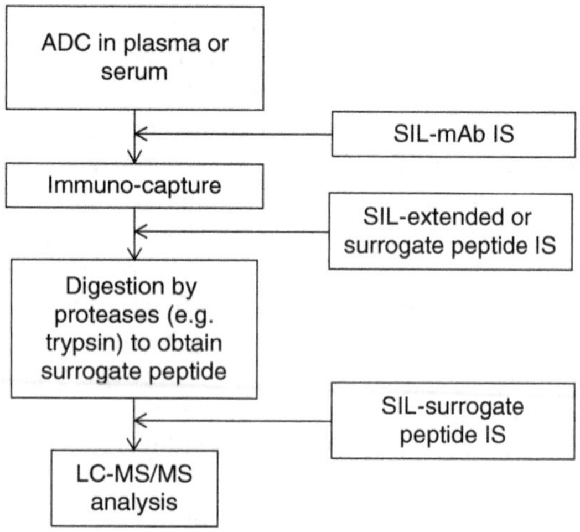

Figure 26.4 Workflow depicting sample preparation methods for total antibody quantification, with SIL-IS options.

CDR is used for quantification (Stephan et al. 2011; Wang et al. 2016) (Figure 26.4). Peptides in the Fc region can also be monitored as signature peptides in nonclinical studies.

For total Ab assay, ideally, a corresponding stable isotope-labeled (SIL) mAb would be used as the IS to track ADC analyte during the entire process of sample preparation, including immunoaffinity capture, trypsin digestion, and LC-MS/MS analysis (Wang et al. 2016). There are many other IS choices that can be used in a quantitative protein or Ab assay. Trade-offs in potential benefits, cost, and availability must be weighed and the ultimate effectiveness of the IS verified during method validation (Jenkins et al. 2015). For nonclinical studies using genetic total Ab assay via LC-MS/MS, the labeled SILu™Mab is usually used as the IS for total Ab measurement of ADCs (Figure 26.4). However, the approach is not applicable to clinical assays due to the reason mentioned above. In the cases where a SIL mAb is not available for the analysis of the ADC of interest, a SIL surrogate peptide with a few extended amino acids can be used as the IS. This IS can be added into the samples prior to trypsin digestion to track the analyte during the digestion process, LC separation, and ionization and detection by MS (Figure 26.4). It should be noted that the digestion efficiency might be different between the SIL-IS and the Ab. The exact SIL peptide may also be used as the IS, which can only track the LC-MS analysis (Figure 26.4). It has been shown that during the immunoaffinity capture where no IS was added, robust performance of the standard curves and quality controls (QCs) was consistently demonstrated (Wang et al. 2016).

For total Ab assay using anti-idiotype capture, the possible interference due to the presence of soluble target, discussed previously in Section 26.2.2, is negligible in preclinical studies but may need to be taken into account in human plasma (Wang et al. 2016).

26.2.5 DAR of ADC

DAR is a unique feature of an ADC molecule. The molecule not only consists of three components (e.g. Ab, linker, and payload) but also represents a heterogeneous mixture of antibodies linked to various numbers of payload drugs, due to multiple lysine residues or cysteine residues derived from interchain disulfide bonds available for conjugation (Xu et al. 2013). In the random lysine-conjugated ADCs, the DAR value is typically 0–8. When conjugation occurs at two site-specific engineered cysteines (also referred to as THIOMAB®) or glutamines, the complexity of the ADC is significantly decreased, resulting in the formation of an ADC with a well-defined DAR of 0–4 (Grafmuller et al. 2016; Xu et al. 2013). The ratio of conjugated-payload and total Ab is defined as the average DAR and its change *in vivo* is considered a good indication of the degree of ADC deconjugation and other biotransformation processes.

Characterization of the DAR distribution identifies changes in the stoichiometric ratio between the payload and the monoclonal antibody. It is important for the ADC to remain stable while in circulation in order to achieve targeted delivery of tumor-fighting payloads and minimize adverse effects on normal cells (Xu et al. 2013). Linker stability is crucial to avoid premature deconjugation of the payload from the ADC prior to reaching the target cells. Even with advanced linker technologies, to a certain extent, drug deconjugation is still expected to occur over time following the ADC administration, causing changes in the ADC composition (i.e. DAR distribution) that may affect not only efficacy and safety but also the bioanalytical strategy as the samples now contain analytes that are different from the starting reference materials. Therefore, characterization of changes in the DAR distribution in plasma/serum is critically needed. In some cases, due to the biotransformation of the payload and if the metabolite is active, both the ratio of conjugated-payload to total-Ab (average DAR, payload) and the ratio of conjugated-(payload + metabolite) to total-Ab (average DAR, payload and metabolite) need to be assessed.

Current DAR assay methods typically measure analyte mixtures that represent a subset of the DAR mixture. The pioneering work by Xu et al. (2007) described the DAR characterization of trastuzumab-MCC–DM1 ADC in plasma by affinity MS. The extracellular domain of the receptor that recognizes the Ab component of the ADC or a mouse mAb to the drug component was immobilized onto a membrane or a bead-based support in a 96-well format. Plasma samples containing a heterogeneous

ADC mixture were incubated with the affinity support, unbound components were removed by washing, and bound ADCs were eluted and analyzed by LC-MS (Figure 26.5a). Individual intact ADCs with specific DARs of interest were characterized and quantified. This method has also been used to quantitate thiomAb anti-MUC16-VC–MMAE by Xu et al. (2011), and it was demonstrated that the stability of the ADC could be effectively assessed by monitoring its DAR distribution in plasma/serum. Another approach of DAR determination by using antihuman Fc Ab affinity capture and reducing the captured intact ADC to light and heavy chains in the sample preparation of preclinical plasma/serum samples was also reported (Grafmuller et al. 2016; Wei et al. 2016) (Figure 26.5b). In another study, Su et al. (2016) used IdeS to cleave ADC to Fab and Fc fragments following the immunoaffinity capture to enhance the sensitivity of DAR measurement in preclinical plasma samples (Figure 26.5b). As an indirect approach of DAR determination, Wang et al. (2016) measured the molar concentrations of conjugated-payload and total-Ab by aforementioned methodology, and the calculated ratio of conjugated-payload to total-Ab was reported as average DAR across the PK time course. Similarly, the molar

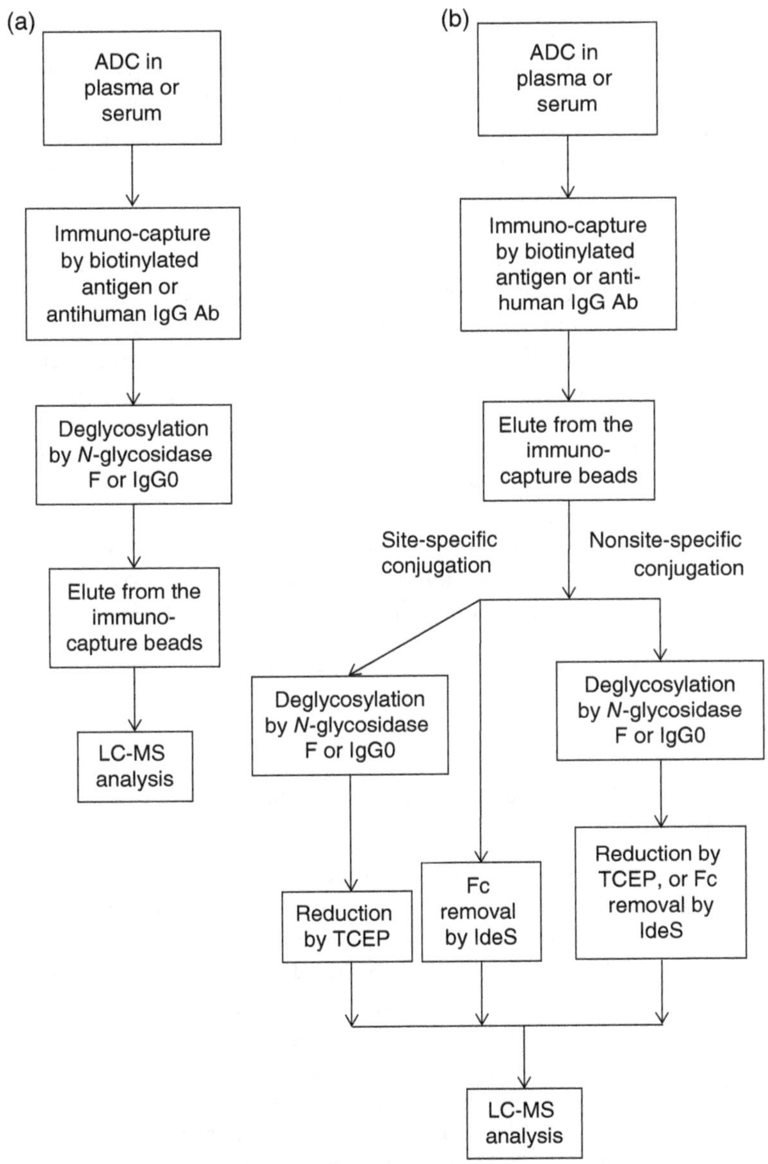

Figure 26.5 Schematic illustration of sample preparation workflows for DAR analysis. (a) Workflow for DAR analysis on intact ADC; (b) workflow for DAR analysis on reduced or fragmented ADC. It is noted that one needs to make sure the linker-payload will not be cleaved off by IdeS digestion, and IdeS will not be suitable in sample preparation for DAR analysis on ADCs with Lys-conjugation. IdeS digestion of removing the Fc fragment can be performed before the elution step followed by the LC-MS analysis.

concentrations of conjugated-(payload + metabolite) and total-Ab were measured, respectively, and the calculated ratio of conjugated-(payload + metabolite) to total-Ab was reported as average DAR (payload + metabolite) across the PK time course (Wang et al. 2016). Relative ratios of individual ADC species were obtained based on their peak areas in the deconvoluted mass spectra. Average DAR is thus estimated with the calculation shown below.

$$\text{Average } DAR = \frac{\pounds\left(\%\text{ peak area} \times \text{No. of conjugated payload}\right)}{100}$$

It is worth noting that in addition to obtaining the avg. DAR values to evaluate the overall ADC stability, critical information of structural changes related to the ADC, linker, and payload can also be generated by this method, which can play a key role in the optimization of the linker–drug combination as well as the selection of lead candidates in the ADC drug discovery.

The current DAR determination methodology with immunoaffinity capture-LC-MS is limited by its relatively low throughput and time-consuming process of the sample preparation. In addition, the relatively low sensitivity of LC-MS on large molecules challenges the DAR determination, leading to that some DAR species are under the limit of detection in matrix samples from late time points in PK studies (e.g. two or three weeks after initial ADC administration).

26.3 Sample Preparation Methods and Common Protocols

26.3.1 Unconjugated Payload

The sample preparation for the analysis of unconjugated payload by LC-MS is often similar to most of the small molecules (Figure 26.1). Depending on the sensitivity requirement of the payload of interest and nature of the sample matric, protein precipitation, LLE, SPE, and others can be employed.

Prior to sample extraction, calibration standards and QC samples can be prepared as follows: Stock solutions of payload standard and its SIL-IS are prepared separately in water : acetonitrile (50:50, v:v) at 0.1 mg ml^{-1} and stored at 4 °C. Calibration standards and QC samples are prepared by serial dilution of a 0.1 mg ml^{-1} stock solution in K3EDTA monkey plasma. Calibration standards are prepared at 0.0150, 0.0300, 0.100, 0.500, 2.00, 8.00, 13.0, and 15.0 ng ml^{-1} and QC samples are prepared at 0.015, 0.045, 0.750, and 12.0 ng ml^{-1}. An IS intermediate solution is prepared at 1000 ng ml^{-1} from the 0.1 mg ml^{-1} stock solution in acetonitrile : water 1:1 (v:v) and an IS working solution is prepared fresh daily in acetonitrile at 0.500 ng ml^{-1}. All matrix samples are thawed and extracted on wet ice.

26.3.1.1 Sample Preparation by Protein Precipitation

The protein precipitation method for sample extraction was reported by Grafmuller et al. (2016) and Wei et al. (2016) for the analysis of auristatin payload (PF-06380101) in the *in vitro* plasma incubated with ADC and in an *in vivo* Cynomolgus monkey study with administered ADC.

1) Vortex all samples.
2) Transfer a 25 μl aliquot of each study sample, calibration standard, QC sample, and blank samples to a 96-well plate on wet ice, and add 25 μl of blank plasma into each sample prior to extraction. Keep all matrix samples on wet ice.
3) Add 200 μl of the SIL payload IS working solution (0.500 ng ml^{-1} in acetonitrile) to the wells of the standards, QCs, control blank, and study samples, or add 200 μl of acetonitrile to double blanks (Note: equilibrate both the IS working solution and acetonitrile at 4 °C prior to add to the 96-well polypropylene plate). Vortex for approximately one minute.
4) Centrifuge at approximately 3000 rpm for approximately five minutes.
5) Transfer 200 μl to a clean plate.
6) Dry down supernatant using heated nitrogen evaporator at 40 °C and reconstitute with 100 μl acetonitrile:water 1:3 (v:v). Vortex for approximately one minute.
7) Inject 20 μl for LC-MS/MS analysis.

26.3.1.2 Sample Preparation by SPE

An online SPE method for the analysis of unconjugated payload DM1 was reported by Heudi et al. (2016). The online SPE system consisted of a Prospekt-2 apparatus (Spark Holland, Emmen, The Netherlands) composed of an autosampler (Endurance), a solvent delivery unit (SDU), and an automatic cartridge exchange (ACE) module. The SPE cartridges used were HySphere C18, HD 7 mm from Spark (Holland, Emmen, The Netherlands). Because the analyte payload DM1 contains a free thiol moiety, likely to dimerize or react with other thiol-containing molecules in serum, samples were pretreated with a reducing agent (e.g. TCEP) and further deactivated with NEM.

Sample Pretreatment

1) Vortex all samples.
2) Transfer a 25 µl aliquot of each study sample, calibration standard, QC sample, and blank samples into a 96-well plate on wet ice. Prior to extraction, the samples are spiked with a 25 µl aliquot of the IS working solution prepared in acetonitrile : water (50 : 50, v : v) with 0.1% formic acid at a concentration of 1.0 ng ml^{-1}.
3) Add 100 µl of reduction buffer (0.125 M TCEP bond-breaker solution in DPBS [Dulbecco's phosphate-buffered saline, Lonza, Walkersville, MD, USA]) to each well.
4) Seal the plate with a film and incubate on a shaker at room temperature for at least 1.5 hours with protection from light.
5) At the end of the incubation, add a 100 µl aliquot of alkylation solution (0.2 M NEM freshly prepared in Dulbecco's phosphate-buffered saline) to each well. Seal the plate with a film and incubate on a shaker at room temperature for another 1.5 hours with protection from light.
6) Add a 250 µl aliquot methanol containing 0.1% formic acid to each well.
7) Briefly shake the plate and add a 200 µl aliquot of water to each well.
8) Seal the plate and centrifuge at 2250 × *g* for 10 minutes at ~4 °C.
9) Inject 200 µl of supernatant into the SPE-LC-MS/MS system.

Sample Extraction by SPE

10) Condition the SPE cartridges with 1000 µl of methanol and subsequently with 1000 µl of deionized water.
11) Successively wash the cartridges with 1500 µl of acetonitrile, 2000 µl of tetrahydrofuran, and 1500 µl of deionized water.
12) Load 200 µl of diluted serum (calibration standards, QCs, and study samples) onto the cartridge using1000 µl deionized water.
13) Wash the cartridges with 1000 µl of acetonitrile/water (20/80, v/v) followed by 1000 µl of methanol/water (5/95, v/v) solution.
14) After completion of the sample preparation cycle, switch the cartridge in-line with the mobile phases to desorb the analytes and transfer them into the LC column.
15) After 30-second of elution period, switch the cartridge back off-line, replace it with a new cartridge, at which the sample preparation cycle starts again with the next sample. The total time of the SPE cycle should be less than the chromatographic run-time.

26.3.1.3 Sample Preparation by LLE

A LLE procedure on an ice water bath for the extraction of unconjugated payload maytansinoid DM4 from monkey and rat plasma sample was reported by Wei et al. (2012).

1) Thaw the frozen plasma samples on an ice water bath prior to use.
2) Transfer 100 µl plasma samples to new 12 mm × 75 mm glass test tubes. Add a 10 µl aliquot of the IS working solution (250 ng ml^{-1} ansamitocin P-3 in acetonitrile : water [1 : 1, v : v]) to standards, QCs, control blank, and incurred dosed samples, or add 10 µl of acetonitrile : water (1 : 1, v : v) to double blank samples.
3) Add a 0.400 ml aliquot of a chilled ammonium formate solution (0.1 M, pH 6.4) to each sample. For rat plasma, add additional 75.0 µl aliquot of chilled ethanol to improve the release of the analyte from protein binding.
4) Add a 1.50 ml aliquot of the extraction solvent (ethyl acetate/methyl *tert*-butyl ether [1 : 1, v : v]) to perform the LLE. Make sure the tubes are capped securely.
5) Invert the tubes several times and vortex-mixed thoroughly for two minutes.
6) After vortexing, centrifuge the samples at 3000 rpm for five minutes at 4 °C to separate the two liquid layers.
7) Transfer the upper organic layer containing DM4 and its IS to new tubes and evaporate the samples to the dryness under nitrogen gas at room temperature.
8) Reconstitute the resulting residue in a 0.100 ml aliquot of ethanol/acetonitrile/water/acetic acid/ammonium formate (700 : 600 : 100 : 0.4 : 1.0; v : v : v : v : w), and transfer the samples into autosampler vials (crimp-capped).
9) Inject a volume of 4.0 µl into the LC-MS/MS system.

26.3.1.4 Sample Preparation by SLE

This following SLE procedure was used to support quantitative analysis of unconjugated payload in both discovery and development stages.

1) Aliquot 25 µl of matrix sample to a 96-well polypropylene plate on wet ice. Keep all samples on wet ice.
2) Add a 75 µl aliquot of water to each sample. Add a 100 µl aliquot of the IS working solution (0.500 ng ml^{-1} in acetonitrile) to standards, QCs, control blank, and incurred samples, or add a 100 µl aliquot of water to double blank samples (Note: equilibrate these solutions at 4 °C prior to being added to the 96-well

polypropylene plate). Centrifuge the plate at approximately 3000 rpm for approximately one minute, which is followed by vortex-mixing for approximately one minute.

3) For SLE, Isolute® SLE+ 200 fixed well plates (Biotage, Charlotte, NC, USA) were used:
   a) Add ~200 µl sample to the assigned well of a SLE plate.
   b) Apply vacuum for 5–15 seconds to initiate sample loading. Samples should be fully loaded in approximately one minute. After samples are completely loaded, wait approximately five minutes.
   c) Elute the SLE plate with 1.0 ml/well of methyl *tert*-butyl ether (MTBE) with gravity for approximately five minutes. Apply vacuum for approximately two minutes. Add a 50 µl aliquot of water to each well of the new plate and mix.
4) Dry down the eluent using heated nitrogen evaporator at 40 °C and reconstitute the residue with 300 µl acetonitrile. Vortex-mix for approximately one minute.
5) Centrifuge the plate at approximately 3000 rpm for approximately five minutes. Transfer 275 µl of the reconstituted sample extract to a clean plate.
6) Dry down supernatant using heated nitrogen evaporator at 40 °C and reconstitute with 100 µl acetonitrile:water 1:3 (v:v). Vortex for approximately one minute.
7) Inject 20 µl for LC-MS/MS analysis.

26.3.1.5 Preparation of Samples with Tissues

For samples in tissues (e.g. tumor tissue), the tissue samples are homogenized prior to the analyte being extracted from the tissue homogenates. Preparation of tumor tissue samples for the analysis of unconjugated payload involves tissue homogenization, dilution, and protein precipitation.

1) Prepare the tissue lysis buffer (ThermoFisher Scientific, Rockford, IL, USA) containing 1% protease inhibitor cocktails (Sigma-Aldrich, St. Louis, MO, USA).
2) Vortex-mix all samples.
3) Add the tissue lysis buffer with the five times volume of tumor weight (1 g equivalent to 1 ml, 6-fold total dilution) to the tumor tissue.
4) Homogenize the tissues in the lysis buffer containing 1% protease inhibitor cocktails by bead beater for three minutes.
5) Vortex-mix the samples, and allow the samples to sit at 4 °C for 20 minutes in order for the lysis buffer to penetrate into the tissue.
6) Centrifuge the samples at 11000 rpm for 10 minutes.
7) Mix a 10 µl of supernatant with 90 µl of blank plasma (10-fold dilution). Transfer 50 µl of the diluted sample to the assigned well of a clean plate.
8) Add a 200 µl aliquot of the IS working solution to standards, QCs, control blank, and incurred samples. (If blank tumor tissues are not available for standard and QC preparation, blank plasma could be used as surrogate matrices.)
9) Vortex-mix the samples and centrifuge at approximately 3000 rpm for approximately five minutes.
10) Transfer 100 µl of the supernatant to a clean plate, and then add 50 µl water to the samples.
11) Inject 20 µl for LC-MS/MS analysis.

26.3.2 Conjugated Payload

26.3.2.1 Sample Preparation Using Anti-idiotype mAb or Anti-human Fc Ab for Immunoaffinity Capture

A conjugated payload assay was fully validated in monkey plasma (Liu et al. 2015), in which a random lysine-conjugated ADC with dipeptide linker was immunoaffinity captured by anti-idiotype mAb, followed by cathepsin B release of the payloads. The ADC contains a cleavable dipeptide linker that can be cleaved specifically by lysosomal enzymes. Cathepsin B is used and the *in vivo* condition is mimicked to release the payload from the ADC. Besides anti-idiotype Ab, anti-human Fc Ab or anti-human IgG Ab can also be used for immunoaffinity capture in preclinical species matrices (Figure 26.2).

1) Biotinylation of capture reagent: A mouse-originated anti-idiotype mAb is used as the specific capture reagent. The capture reagent at a concentration of ~2 mg ml^{-1} in DPBS is biotinylated using the EZ-Link Sulfo-NHS-LC-Biotin according to the manufacturer's instruction (Thermo Scientific, Somerset, NJ, USA). The lyophilized biotin is reconstituted with deionized water immediately prior to mixing with the capture reagent at an optimized molar challenge ratio of 12:1. The mixture was incubated at room temperature for 30 minutes and then purified using Zeba Spin Desalting Column, 7K MWCO (Thermo Scientific). The protein concentration of the purified sample was determined based on its absorbance at 280 nm using NanoDrop 2000 (Thermo Scientific). The biotinylated anti-idiotype capture mAb is aliquoted and stored at −70 °C until use.
2) Immunoaffinity capture of ADC: The immunoaffinity capture of the ADC from biological matrix is conducted on AssayMAP streptavidin (SA-W) cartridges using an

Agilent AssayMAP Bravo platform (Agilent Technologies, Wake-field, MA, USA). The cartridges are equilibrated using 50 µl of DPBS at a flow rate of 25 µl min^{-1}. The biotinylated anti-idiotype capture mAb is diluted to a concentration of 600 µg ml^{-1} using DPBS and an aliquot of 50 µl biotinylated capture reagent was immobilized on the streptavidin cartridge at a flow rate of 5 µl min^{-1}. The ADC calibration standards, QCs, and study samples in monkey plasma are diluted at a ratio of 1:1 with DPBS. 25 µl of the samples are slowly loaded on the cartridges with immobilized capture mAb at a flow rate of 2 µl min^{-1}. The cartridges are washed using 50 µl of washing solution which consisted of 500 mM sodium chloride and 0.1% BSA in DPBS. After multiple washes, the captured analyte is eluted from the cartridges into a collection plate, using 36 µl of 12 mM hydrochloric acid with 0.5% BSA at a flow rate of 5 µl min^{-1}. The cartridges are not reused for assay validation and sample analysis although the performance may be acceptable for the same-day storage and re-usage based on the manufacturer's suggestion. The eluted sample is immediately neutralized with 2 µl of 250 mM ammonium bicarbonate to reach a pH of 5–6, which is favorable for the following enzyme cleavage step. An aliquot of 25 µl sample is subsequently transferred to a new 96-deep-well plate for the following enzyme cleavage and extraction.

3) Cathepsin B cleavage: The cleavage by cathepsin B is conducted in a buffer containing 1.34 units of the enzyme at an optimal pH of 5–6 in the presence of reducing reagent DTT. The detailed steps are as follows. Cathepsin B enzyme from human liver (EMD Millipore, Billerica, MA, USA) is activated in sodium acetate buffer at a pH of 5–6 in the presence of EDTA and DTT. An aliquot of 50 µl cathepsin B stock solution containing approximately 60 units of the enzyme (according to the manufacturer's definition and measurement) is mixed with 400 µl of activation buffer containing 18.75 mM sodium acetate, 4 mM EDTA, and 7.5 mM DTT. To each of the samples obtained from the immunoaffinity capture step, 10 µl of the activated cathepsin B solution containing1.34 units of the enzyme is added. The mixture is incubated in a Thermomixer R model 5355 (Eppendorf, Hamburg, Germany) at 25 °C for three hours with vortexing at a speed of 400 rpm (116 × *g*) which is cycled on and off every one minute. The cleavage reaction is stopped by adding 150 µl of cold acetonitrile containing 1.000 ng ml^{-1} of the SIL-IS of payload. The mixture is vortexed vigorously for six minutes, then centrifuged at ~2200 × *g* for 10 minutes. An aliquot of 75 µl supernatant is transferred and mixed with 75 µl of mobile phase A (5 mM ammonium bicarbonate containing 0.005% ammonium hydroxide in 95:5, v:v, water:acetonitrile) before injecting the sample to the LC-MS/MS system.

26.3.2.2 Sample Preparation Using Protein A or Protein G for Capture

Immunoaffinity capture by Protein A or Protein G cartridge followed by solid-phase enzymatic cleavage of payload using papain has been used in sample preparation for the analysis of conjugated-payload (Sanderson et al. 2016; Wang et al. 2016). The release of the payload by cathepsin B can also be used following the capture by Protein A or Protein G, and the procedure of cathepsin B cleavage is the same as that described in Section 26.3.2.1.

1) Preparation of Protein A resins: Prior to use, Protein A affinity resin is re-equilibrated in 20 resin bed volumes of PBS (pH 7.4) before resuspension in buffer at a slurry ratio of one part affinity resin to three parts buffer (v:v). Slurry (800 µl) is added to a 96-well polypropylene filter microplate and centrifuged at 2000 × *g* for five minutes to remove the aqueous content.
2) Immunoaffinity capture of ADC: Aliquots (200 µl) of human plasma spiked with the ADC IS (SIL payload conjugated to the antibody) are immobilized on Protein A by adding to Protein A affinity resin and placing on a titer-plate shaker for one hour at 4 °C. After washing the samples and prior to enzymatic release of the ADC, papain is reconstituted to a concentration of 2 mg ml^{-1} in papain activation buffer (10 mM EDTA, 20 mM potassium phosphate, 2 mM L-cysteine [pH 7.0]) and incubated for 15 minutes at 37 °C.
3) Enzymatic cleavage by papain: Aliquots (200 µl) of activated papain are added to each Protein A-immobilized ADC sample, equilibrated for five minutes at 37 °C and covered and incubated for an additional four hours at 37 °C with gentle shaking on a titer-plate shaker. Papain-released antibody-conjugated payload is recovered by centrifugation (2000 × *g* for five minutes at 4 °C), and the Protein A affinity resin is washed once with papain activation buffer (200 µl). The combined eluent and wash fractions are acidified with 0.2% formic acid (800 µl) prior to SPE using Oasis MCX plates (Waters). Extraction plates are preconditioned with 80% dichloromethane/20% isopropanol + 1% ammonium hydroxide (400 µl), methanol (400 µl), water (400 µl), and 0.2% formic acid (400 µl). Samples are then applied to the preconditioned extraction plates, which are subsequently washed with 0.2% formic acid (400 µl) and methanol (400 µl). Samples are eluted with 80% dichloromethane/20% isopropanol + 1% ammonium hydroxide (2 × 400 µl), dried under nitrogen at 40 °C and reconstituted in 95% acetonitrile/5% water + 1% formic acid (40 µl) for LC-MS/MS analysis.

26.3.2.3 Sample Preparation for Conjugated Payload Quantification in ADCs with Non-cleavable Linkers

When ADCs contain non-cleavable linkers which cannot be cleaved by enzymatic digestion to release the payload, peptide-linker-payload is then used for the analysis of conjugated payload. Anti-idiotype Ab, anti-Fc Ab, Protein A, or Protein G can be used as the immune-capture reagent for the immune-capture as discussed in the previous sections. Following the immune-capture, enzymatic digestion using proteases (e.g. trypsin) is used to obtain the sample mixture containing the peptide-conjugated payload for LC-MS/MS analysis.

1) Immunoaffinity capture of ADC: the same as in the conjugated-payload assay in Sections 26.3.2.1 and 26.3.2.2.
2) Enzymatic cleavage by proteases (e.g. trypsin): To the aliquots (25 μl) of the samples eluted from the immunoaffinity capture step, 1 M Tris-Cl (~5 μl) was added to adjust pH to about pH 7. Then a 10 μl aliquot of 100 mM ammonium bicarbonate (NH_4HCO_3) and a 25 μl aliquot of 0.2% RapiGest (prepared in 100 mM NH_4HCO_3, Waters, Milford, MA, USA) was added to the samples. This was followed by heating at 80 °C for five minutes. After cooling down, the samples are then added with 10 μl of 100 mM dithiothreitol (DTT) to reduce disulfide bonds, and incubated at 56 °C for 30 minutes. After the reduction by DTT, the samples are cooled to room temperature, and then added with 2.5 μl of 300 mM iodoacetamide (IAM), and incubated in the dark at room temperature for 30 minutes. 5 μl of 0.1 $\mu g\,\mu l^{-1}$ trypsin is then added, and the samples are incubated at 37 °C overnight. After the overnight trypsin digestion, the samples are added with 6 μl 5% trifluoroacetic acid (TFA) to deactivate trypsin and incubated at 37 °C for 30 minutes degrade RapiGest. The samples are then centrifuged at 10,000 rpm at 4 °C for one minute. The supernatants of the centrifuged samples are taken and subjected to the LC-MS/MS analysis.

26.3.3 Conjugated Ab

There is a case reporting a conjugated-Ab assay by immunoaffinity capture LC-MS/MS. Briefly, the samples were prepared with anti-payload antibody for immunoaffinity capture followed by enzymatic (e.g. trypsin) digestion and a surrogate peptide in the heavy-chain CDR was monitored for quantification (Wang et al. 2016) (Figure 26.3). As aforementioned, the anti-payload antibody as the capture reagent may preferentially bind to unconjugated payload rather than the conjugated-Ab in the study samples. When unconjugated payload in the study samples is at a negligible low level, the immunoaffinity capture LC-MS/MS assay using anti-payload Ab could be applied to have a relative true assessment of the conjugated-Ab levels in the samples. An excess amount of the anti-payload antibody with sufficient capacity is needed for the immunoaffinity capture in order to obtain such a true assessment of the conjugated Ab.

26.3.4 Total Ab

Total Ab concentrations have been conventionally determined by LBA. More recently, total Ab measurement by ligand binding LC-MS/MS (or hybrid LBA/LC-MS/MS) has increasingly gained attention (Furlong et al. 2014; Lanshoeft et al. 2016). This platform often utilizes antigen, anti-idiotype antibody, Protein A, or Protein G cartridge for capturing the Abs, which is followed by enzymatic digestion and LC-MS/MS analysis of the surrogate peptide in the heavy chain CDR or Fc region of the Abs. The immunoaffinity capture and enzymatic digestion steps are very similar to that in the conjugated-payload assay and the conjugated Ab assay (Figure 26.4).

1) Immunoaffinity capture of ADC: The immunoaffinity capture of ADC can be performed either by the cartridge format or by magnetic beads format. The immunoaffinity capture procedures using the streptavidin cartridges with specific biotinylated-antigen, biotinylated-anti-idiotype antibody, or Protein A cartridges are the same as in Sections 26.3.2.1 and 26.3.2.2 described for the conjugated-payload assay. The immunoaffinity capture can also be conducted with Dynabeads MyOne streptavidin T1 magnetic beads (Life Technologies, Camarillo, CA, USA) using biotinylated-antigen or biotinylated-anti-idiotype antibody as the procedure described below:
   a) Beads (1 ml of 10 $mg\,ml^{-1}$) are washed three times with 1 ml of PBST solution, and then resuspended with 1 ml of PBST.
   b) Based on the beads' capacity (200 μg of Ab per 1 ml of streptavidin T1 beads), the beads are saturated by adding about 400 μg of Ab for streptavidin T1 beads. The beads and mAb mixture are incubated at room temperature for one hour with shaking.
   c) After incubation, the beads are washed again three times with 1 ml of PBST, and then resuspended with 1 ml of PBST containing 1% BSA (bovine serum albumin). The immobilized beads are stored at 4 °C for future use.
   d) PBST (10 μl) is added to 10 μl of each serum sample in a 96-well plate, and then 50 μl of capture beads is added to each sample.

e) The plate is incubated at room temperature for 30 minutes in a Thermomixer R model 5355 incubator (Eppendorf, Hamburg, Germany) at a vortexing speed of 900 rpm (15 seconds on/off) for the immunoaffinity capture.
f) The plate is first washed three times with 200 µl of PBST, and then one time with 200 µl of 50 mM ammonium bicarbonate using the Janus Mini automated liquid handler from Perkin Elmer (Waltham, MA, USA).

2) Enzymatic cleavage by proteases (e.g. trypsin): To each of the samples obtained from immunoaffinity capture, using either the beads format or cartridge format, 100 µl of 50 mM ammonium bicarbonate containing 10% methanol is added, the samples are then incubated at 90 °C for 30 minutes for thermal denaturation. After the samples are cooled to room temperature, a 25 µl aliquot of Promega trypsin (~50 µg ml^{-1}) is added into each sample for digestion at 60 °C for two hours. The digestion is stopped by adding 10 µl of 10% formic acid in water, and 20 µl of 200 ng ml^{-1} of the SIL surrogate peptide IS working solution in 20% acetonitrile and 80% water is added before LC-MS/MS analysis of conjugated-Ab. The transition of the surrogate peptide is monitored by LC-MS/MS.

26.3.5 DAR of ADC

The strategies by immunoaffinity capture followed by capillary LC-MS to characterize intact ADC DAR distribution in biological matrices have been demonstrated in recent years (Kaur et al. 2013; Xu et al. 2011, 2013). The determination of *in vivo* DAR values can also be conducted on ADC fragments (Grafmuller et al. 2016; Su et al. 2016; Wei et al. 2016), which provides improved sensitivity for *in vivo* DAR analysis (Figure 26.5). Either *N*-glycosidase F or IgG0 can be used for deglycosylation to reduce heterogeneity and simplify the mass spectra. FabRICATOR® (IdeS) can also be used to generate a smaller size F(ab')2 fragment to enhance sensitivity and resolution for better DAR measurement and structural elucidation.

26.3.5.1 Sample Preparation for DAR Analysis on Intact ADC

A method for *in vivo* DAR analysis on intact ADC was reported by Xu et al. (2013):

Briefly, an aliquot (typically 50 µl) of plasma or serum containing the ADC is added to a 96-well plate containing the buffer of 0.01 M HEPES, pH 7.4, 0.15 M NaCl, 3 mM EDTA, and 0.005% P20, followed by addition of streptavidin paramagnetic beads coupled with the biotinylated capture probes; for example, the target antigen or ADA. The plate is then incubated at room temperature for two hours. The ADC analytes captured by the beads are isolated, washed, and subsequently deglycosylated by peptide *N*-glycosidase F (ProZyme, Hayward, CA, USA) overnight at 37 °C. The beads are then isolated and washed with HBS-EP buffer (GE Healthcare, Little Chalfont, UK), water, and 10% acetonitrile in water to ensure that nonspecifically bound proteins are effectively removed. The ADC is eluted by 30% acetonitrile in water with 1% formic acid and injected into LC-MS for analysis.

26.3.5.2 Sample Preparation for DAR Analysis on Reduced ADC

A method for DAR analysis on reduced ADC from *in vitro* and *in vivo* plasma was reported by Grafmuller et al. (2016) and Wei et al. (2016), in which ADC was reduced to light chains and heavy chains for DAR analysis:

A 25 µl aliquot of all matrix samples in a 96-well LoBind plate are deglycosylated with IgG0 (Genovis Inc., Cambridge, MA, USA) at 37 °C for one hour. Following deglycosylation, a capture antibody (biotinylated goat antihuman IgG1 Fcγ fragment specific at 1 mg ml^{-1}) is added and the mixture was heated at 37 °C for one hour followed by gentle shaking at room temperature for a second hour. Dynabead MyOne Streptavidin T1 magnetic beads are added to the samples and incubated at room temperature for one hour with gentle shaking. The sample plate is then washed with 200 µl PBS + 0.05% Tween-20, 200 µl PBS, and HPLC grade water. The bound ADC is eluted from the magnetic beads with 55 µl of 2% FA (v/v). A 50 µl aliquot of each sample is transferred into a new plate and spiked with 5 µl of 200 mM TCEP. After 10 minutes incubation at room temperature, the samples are ready for LC-MS analysis.

26.4 Future Perspective

Significant advancement has been made in the ADC development in the pharmaceutical industry for the past decade. There is no doubt that such an achievement is associated with (i) the introduction of antibodies with strong binding specificity/affinity, (ii) the advancement of linker chemistry that leads to minimal deconjugation of the ADCs in circulation, and (iii) highly potent payloads with different mechanisms of action in tumor cell killing. The complexity of the ADC structure and its dynamically changing composition *in vivo* have presented tremendous challenges in bioanalysis, especially sample preparation, in comparison with other small-molecule drugs or biotherapeutics. It is anticipated that new techniques will continue to evolve to effectively address the challenges. For instance, the new payloads in the next generation of ADCs are expected to be highly potent, thus the expected therapeutic dose will be low. Therefore, efforts will be needed to improve assay sensitivity for the analysis of payloads (unconjugated or conjugated). Other new features including novel antibody modalities, new linkers to enable high DAR conjugation, prodrugs of payloads to reduce systemic toxicity, and coadministered ADCs can all pose new challenges to develop a robust sample preparation procedure for quantitative analysis.

It is expected that more user-friendly automation devices will be available to enhance the sample preparation efficiency and the overall sample analysis throughput. For tissue samples, after homogenization and extraction, the sample may still have debris remaining, which can be re-extracted with fresh lysis buffer and probed to determine if additional analyte remains. If so, repeat extraction may be warranted. It is of note that currently, there is not much advancement in improving throughput of tissue homogenization. Also, improvement on sample preparation efficiency includes on-line digestion to minimize manual treatments for conjugated payload, conjugated Ab, and total Ab. Sometimes one protease (e.g. trypsin) may be sufficient to release the desired sequence and in this case other additional site-specific enzymes such as Asp-N and Glu-C can be applied.

In the foreseeable future, some improvements in manufacturing and purification will be achieved to make standards such as specific DAR species readily available for verification of the assay performance, which could potentially make sample preparation more appropriate. Moreover, measurements of tissue distributions, e.g. characterization or quantification of antibody components or payload moieties, may be conducted more frequently in an effort to better define the mechanism of action and understand correlation with efficacy and toxicity. Overall, with either sample preparation or analyte detection orthogonal to conventional LBAs, the immunoaffinity capture LC-MS or LC-MS/MS approaches will play an increasingly important role in the ADC development. A sensitive and reliable sample preparation process will continue to be fundamental to ensure an accurate evaluation of the fate of ADCs *in vivo*.

Acknowledgements

The authors would like to thank Dr. Keyang Xu for critical review and invaluable comments on the manuscript.

References

Alley, S.C., Benjamin, D.R., Jeffrey, S.C. et al. (2008). Contribution of linker stability to the activities of anticancer immunoconjugates. *Bioconjug. Chem.* 19 (3): 759–765.

Birdsall, R.E., McCarthy, S.M., Janin-Bussat, M.C. et al. (2016). A sensitive multidimensional method for the detection, characterization, and quantification of trace free drug species in antibody-drug conjugate samples using mass spectral detection. *MAbs* 8 (2): 306–317.

Buckwalter, M., Dowell, J.A., Korth-Bradley, J. et al. (2004). Pharmacokinetics of gemtuzumab ozogamicin as a single-agent treatment of pediatric patients with refractory or relapsed acute myeloid leukemia. *J. Clin. Pharmacol.* 44 (8): 873–880.

Cabanillas, F. (1992). Non-Hodgkin's lymphomas: a review of the M.D. Anderson experience. *Semin. Oncol.* 19 (1 Suppl 1): 11–13.

Casi, G. and Neri, D. (2012). Antibody–drug conjugates: basic concepts, examples and future perspectives. *J. Control. Release* 161 (2): 422–428.

Chari, R.V. (2008). Targeted cancer therapy: conferring specificity to cytotoxic drugs. *Acc. Chem. Res.* 41 (1): 98–107.

Dubowchik, G.M., Radia, S., Mastalerz, H. et al. (2002). Doxorubicin immunoconjugates containing bivalent, lysosomally-cleavable dipeptide linkages. *Bioorg. Med. Chem. Lett.* 12 (11): 1529–1532.

Ducry, L. (2012). Challenges in the development and manufacturing of antibody-drug conjugates. *Methods Mol. Biol.* 899: 489–497.

Elgersma, R.C., Coumans, R.G., Huijbregts, T. et al. (2015). Design, synthesis, and evaluation of linker-duocarmycin payloads: toward selection of HER2-targeting antibody-drug conjugate SYD985. *Mol. Pharm.* 12 (6): 1813–1835.

Fanale, M.A., Horwitz, S.M., Forero-Torres, A. et al. (2014). Brentuximab vedotin in the front-line treatment of patients with CD30+ peripheral T-cell lymphomas: results of a phase I study. *J. Clin. Oncol.* 32 (28): 3137–3143.

Furlong, M.T., Titsch, C., Xu, W. et al. (2014). An exploratory universal LC-MS/MS assay for bioanalysis of hinge region-stabilized human IgG4 mAbs in clinical studies. *Bioanalysis* 6 (13): 1747–1758.

Gorovits, B., Alley, S.C., Bilic, S. et al. (2013). Bioanalysis of antibody-drug conjugates: American Association of Pharmaceutical Scientists Antibody-Drug Conjugate Working Group position paper. *Bioanalysis* 5 (9): 997–1006.

Grafmuller, L., Wei, C., Ramanathan, R. et al. (2016). Unconjugated payload quantification and DAR characterization of antibody-drug conjugates using high-resolution MS. *Bioanalysis* 8 (16): 1663–1678.

Heudi, O., Barteau, S., Picard, F., and Kretz, O. (2016). Quantitative analysis of maytansinoid (DM1) in human serum by on-line solid phase extraction coupled with liquid chromatography tandem mass spectrometry – method validation and its application to clinical samples. *J. Pharm. Biomed. Anal.* 120: 322–332.

Jackson, D.Y. (2016). Processes for constructing homogeneous antibody drug conjugates. *Org. Process Res. Dev.* 20 (5): 852–866.

Jaracz, S., Chen, J., Kuznetsova, L.V., and Ojima, I. (2005). Recent advances in tumor-targeting anticancer drug conjugates. *Bioorg. Med. Chem.* 13 (17): 5043–5054.

Jenkins, R., Duggan, J.X., Aubry, A.F. et al. (2015). Recommendations for validation of LC-MS/MS bioanalytical methods for protein biotherapeutics. *AAPS J.* 17 (1): 1–16.

Kaneko, T., Willner, D., Monkovic, I. et al. (1991). New hydrazone derivatives of adriamycin and their immunoconjugates – a correlation between acid stability and cytotoxicity. *Bioconjug. Chem.* 2 (3): 133–141.

Kaur, S., Xu, K., Saad, O.M. et al. (2013). Bioanalytical assay strategies for the development of antibody-drug conjugate biotherapeutics. *Bioanalysis* 5 (2): 201–226.

Krop, I.E., Beeram, M., Modi, S. et al. (2010). Phase I study of trastuzumab-DM1, an HER2 antibody-drug conjugate, given every 3 weeks to patients with HER2-positive metastatic breast cancer. *J. Clin. Oncol.* 28 (16): 2698–2704.

Kumar, S., King, L.E., Clark, T.H., and Gorovits, B. (2015). Antibody-drug conjugates nonclinical support: from early to late nonclinical bioanalysis using ligand-binding assays. *Bioanalysis* 7 (13): 1605–1617.

Lanshoeft, C., Wolf, T., Walles, M. et al. (2016). The flexibility of a generic LC-MS/MS method for the quantitative analysis of therapeutic proteins based on human immunoglobulin G and related constructs in animal studies. *J. Pharm. Biomed. Anal.* 131: 214–222.

Lesur, A., Varesio, E., and Hopfgartner, G. (2010). Accelerated tryptic digestion for the analysis of biopharmaceutical monoclonal antibodies in plasma by liquid chromatography with tandem mass spectrometric detection. *J. Chromatogr. A* 1217 (1): 57–64.

Li, X., Xu, S.N., Qin, D.B. et al. (2014). Effect of adding gemtuzumab ozogamicin to induction chemotherapy for newly diagnosed acute myeloid leukemia: a meta-analysis of prospective randomized phase III trials. *Ann. Oncol.* 25 (2): 455–461.

Lin, K. and Tibbitts, J. (2012). Pharmacokinetic considerations for antibody drug conjugates. *Pharm. Res.* 29 (9): 2354–2366.

Liu, A., Kozhich, A., Passmore, D. et al. (2015). Quantitative bioanalysis of antibody-conjugated payload in monkey plasma using a hybrid immuno-capture LC-MS/MS approach: assay development, validation, and a case study. *J. Chromatogr. B Analyt. Technol. Biomed. Life Sci.* 1002: 54–62.

Rao, C., Rangan, V.S., and Deshpande, S. (2015). Challenges in antibody-drug conjugate discovery: a bioconjugation and analytical perspective. *Bioanalysis* 7 (13): 1561–1564.

Ross, P.L. and Wolfe, J.L. (2016). Physical and chemical stability of antibody drug conjugates: current status. *J. Pharm. Sci.* 105 (2): 391–397.

Sanderson, R.J., Nicholas, N.D., Baker Lee, C. et al. (2016). Antibody-conjugated drug assay for protease-cleavable antibody-drug conjugates. *Bioanalysis* 8 (1): 55–63.

Senter, P.D. (2009). Potent antibody drug conjugates for cancer therapy. *Curr. Opin. Chem. Biol.* 13 (3): 235–244.

Stephan, J.P., Kozak, K.R., and Wong, W.L. (2011). Challenges in developing bioanalytical assays for characterization of antibody-drug conjugates. *Bioanalysis* 3 (6): 677–700.

Su, D., Ng, C., Khosraviani, M. et al. (2016). Custom-designed affinity capture LC-MS F(ab')2 assay for biotransformation assessment of site-specific antibody drug conjugates. *Anal. Chem.* 88 (23): 11340–11346.

Toki, B.E., Cerveny, C.G., Wahl, A.F., and Senter, P.D. (2002). Protease-mediated fragmentation of p-amidobenzyl ethers: a new strategy for the activation of anticancer prodrugs. *J. Org. Chem.* 67 (6): 1866–1872.

van den Broek, I., Niessen, W.M., and van Dongen, W.D. (2013). Bioanalytical LC-MS/MS of protein-based biopharmaceuticals. *J. Chromatogr. B Analyt. Technol. Biomed. Life Sci.* 929: 161–179.

Wang, J., Gu, H., Liu, A. et al. (2016). Antibody-drug conjugate bioanalysis using LB-LC-MS/MS hybrid assays: strategies, methodology and correlation to ligand-binding assays. *Bioanalysis* 8 (13): 1383–1401.

Wei, D., Sullivan, M., Espinosa, O., and Yang, L. (2012). A sensitive LC-MS/MS method for the determination of free maytansinoid DM4 concentrations – method development, validation, and application to the nonclinical studies of antitumor agent DM4 conjugated hu-anti-Cripto MAb B3F6 (B3F6-DM4) in rats and monkeys. *Int. J. Mass Spectrom.* 312: 53–60.

Wei, C., Zhang, G., Clark, T. et al. (2016). Where did the linker-payload go? A quantitative investigation on the destination of the released linker-payload from an antibody-drug conjugate with a maleimide linker in plasma. *Anal. Chem.* 88 (9): 4979–4986.

Xu, K., Liu, L., Dere, R. et al. (2013). Characterization of the drug-to-antibody ratio distribution for antibody-drug conjugates in plasma/serum. *Bioanalysis* 5 (9): 1057–1071.

Xu, K., Liu, L., Saad, O.M. et al. (2011). Characterization of intact antibody-drug conjugates from plasma/serum in vivo by affinity capture capillary liquid chromatography-mass spectrometry. *Anal. Biochem.* 412 (1): 56–66.

Xu K, Saad O, Baudys J. (2007). Bioanalytical strategies for antibody drug conjugate (ADC). Biopharmaceutical development: characterization of trastuzumab–MCC-DM1 in plasma by affinity mass spectrometry. Presented at AAPS National Biotechnology Conference (25 June 2007).

Index

Sample Preparation in LC-MS Bioanalysis, First Edition. Edited by Wenkui Li, Wenying Jian, and Yunlin Fu.

www.ingramcontent.com/pod-product-compliance
Lightning Source LLC
LaVergne TN
LVHW062142200926
848484LV00007B/1844

9781119274292